COMPUTATIONAL ACCELERATOR PHYSICS

The proceedings of this meeting were made possible by the
support of the U.S. Department of Energy,
Division of High Energy Physics.

AIP CONFERENCE PROCEEDINGS 297

COMPUTATIONAL ACCELERATOR PHYSICS

LOS ALAMOS, NM 1993

EDITOR: ROBERT RYNE
UNIVERSITY OF WISCONSIN

American Institute of Physics   New York

Authorization to photocopy items for internal or personal use, beyond the free copying permitted under the 1978 U.S. Copyright Law (see statement below), is granted by the American Institute of Physics for users registered with the Copyright Clearance Center (CCC) Transactional Reporting Service, provided that the base fee of $2.00 per copy is paid directly to CCC, 27 Congress St., Salem, MA 01970. For those organizations that have been granted a photocopy license by CCC, a separate system of payment has been arranged. The fee code for users of the Transactional Reporting Service is: 0094-243X/87 $2.00.

© 1994 American Institute of Physics.

Individual readers of this volume and nonprofit libraries, acting for them, are permitted to make fair use of the material in it, such as copying an article for use in teaching or research. Permission is granted to quote from this volume in scientific work with the customary acknowledgment of the source. To reprint a figure, table, or other excerpt requires the consent of one of the original authors and notification to AIP. Republication or systematic or multiple reproduction of any material in this volume is permitted only under license from AIP. Address inquiries to Series Editor, AIP Conference Proceedings, AIP Press, American Institute of Physics, 500 Sunnyside Boulevard, Woodbury, NY 11797-2999.

L.C. Catalog Card No. 93-074205
ISBN 1-56396-222-5
DOE CONF-9302104

Printed in the United States of America.

CONTENTS

Preface xi
List of Participants xiii
Photographs xvii
Modeling Accelerator Structures and RF Components 1
    K. Ko, C.-K. Ng, and W. B. Herrmannsfeldt
Beam Halo in High-Intensity Beams 9
    T. P. Wangler
Application of MPP to Particle Tracking 19
    G. Bourianoff, B. Cole, and L. Chang
Electromagnetic Modelling of the PEP-II RF Cavity Using
the 3D Finite Element Code MSC/EMAS 27
    R. A. Rimmer and J. J. Burke
Simulation of the CLIC Transfer Structure by Means
of MAFIA 35
    A. Millich
A Boundary-Integral-Method Code for Cavity
RF Mode Analysis 43
    F. P. Adams and M. S. de Jong
ACAO of Cavities 51
    T. Weiland and M. Zhang
Simulating A Singularity-Free Universe Outside
the Problem Boundary in POISSON 58
    K. Halbach and R. D. Schlueter
Calculation of Q-Factors of Lossy Resonators
in the Time Domain Using the Prony-Pisarenko
Approximation 66
    P. Thoma and T. Weiland
Calculating Scattering Parameters of 3-Dimensional
Structures by Broadband Excitation in the Time Domain 74
    H. Wolter and T. Weiland
An Integro-Algebraic Equation for High Frequency
Wake Fields in a Tube with Smoothly Varying Radius 82
    R. L. Warnock
Mode Analysis of Circular Waveguide Structures
Using the Mode Matching Technique 90
    U. van Rienen
Numerical Design and Analysis of a Compact
TE_{10} to TE_{01} Mode Transducer 99
    S. Tantawi, K. Ko, and N. Kroll
Wakefield and Impedance Studies
of a Liner Using MAFIA 107
    W. Chou and T. Barts
ACHRO: A Program to Help Design Achromatic Bends 115
    D. Rusthoi

Calculating Wake Potentials for Ultra Short Bunches 123
    M. Dehler and T. Weiland

Multi-Ion Transport System 131
    B. I. Bondarev, A. P. Durkin, G. T. Nicolaishvili,
    and O. Yu. Shlygin

Beam Line Error Analysis, Position Correction
and Graphic Processing 135
    F. Wang and N. Mao

The Design of a Four Cell Third Order Achromat 143
    W. Wan, E. Goldmann, and M. Berz

Eddy Current Simulations for the SSCL
Low Energy Booster Cavity 150
    Y. Goren and L. Walling

Liner Impedance Calculations Using HFSS 160
    E. Ruiz, L. Walling, Y. Goren, and N. Spayd

The Beam Dynamics Study of Compact Ring
by a New Code 167
    Y. Huang and S. Ohnuma

Three Dimensional δf Simulations of Beams in the SSC 175
    J. Koga, T. Tajima, and S. Machida

Higher-Order Mode (HOM) Damper Design Using HFSS 182
    L. Walling, G. Hulsey, and T. Grimm

Simulation of the Injection Damping and Resonance
Correction Systems for the HEB of the SSC 188
    M. Li, P. Zhang, and S. Machida

A Structured—Modular Approach to Software Design
for High Current Beam Dynamics Simulation 196
    Y. K. Batygin

3D Simulations of an Electrostatic Quadrupole Injector 203
    D. P. Grote, A. Friedman, and S. Yu

Longitudinal Beam Dynamics for Heavy Ion Fusion
Using WARPRZ 211
    D. A. Callahan, A. B. Langdon, A. Friedman,
    and I. Haber

High Power Transit-Time RF Amplifier
for Super Accelerators 219
    M. J. Arman and D. J. Sullivan

Self-Consistent Solution of Multispecies
Charged Particle Optics Problems Using a
Hybrid Fluid/Raytrace/Poisson-Boltzmann Algorithm 227
    J. L. Orthel

Design of Ion Optical Systems by Computer Simulation 235
    B. J. Hall

Numerical Simulations of Input and Output Couplers
for Linear Accelerator Structures.................................. 243
    C.-K. Ng and K. Ko
Fast System Design Using GUIs 251
    M. E. Kress and E. Sohn
GUIs for Scientific Code Usage 261
    N. Dionne
Towards C++ Object Libraries for Accelerator Physics 264
    L. Michelotti
New Features in COSY INFINITY................................... 267
    M. Berz
Zlib and Related Programs
for Beam Dynamics Studies 279
    Y. T. Yan
New Features in the Design Code TLIE 285
    J. van Zeijts
Recent Advances and Applications of
the MAFIA Codes 291
    The MAFIA Collaboration and T. Weiland
Applications of the ARGUS Code in Accelerator Physics 303
    J. J. Petillo, A. Mankofsky, W. A. Krueger, C. Kostas,
    A. A. Mondelli, and A. T. Drobot
Advances/Applications of MAGIC and SOS 313
    G. Warren, L. Ludeking, K. Nguyen, D. Smithe,
    and B. Goplen
Exploiting Periodicity and Other Structural Symmetries
in Field Solvers 323
    E. M. Nelson
Charged Particle Optics Without Detailed Field Maps..................... 333
    D. C. Carey
Overview of WARP, A Particle Code for Heavy Ion Fusion................. 347
    A. Friedman, D. P. Grote, D. A. Callahan, A. B. Langdon,
    and I. Haber
Modelling RF Sources Using 2-D PIC Codes........................... 357
    K. R. Eppley
Numerical Simulation of Relativistic Klystrons 367
    G. M. Fiorentini
LIDOS-Unconventional Helper for Linac Beam Designing 377
    B. I. Bondarev, A. P. Durkin, B. P. Murin,
    G. T. Nikolaishvili, and O. Yu. Shlygin
Numerical Simulations at CEBAF Using PARMELA 385
    H. Liu
New Features and Applications of ABCI............................... 393
    Y. H. Chin

SYNCH-Status and Recent Use at SSCL 403
    A. A. Garren, A. S. Kenney, E. D. Courant,
    A. D. Russell, M. J. Syphers, T. Sen, and T. Barts

Code Transportability and Simple Maintenance Tools 408
    H. Grote

Full-Turn Symplectic Map from a Generator in a
Fourier-Spline Basis 413
    J. S. Berg, R. L. Warnock, R. D. Ruth,
    and É. Forest

Particle Distribution Generator in 4D Phase Space 419
    Y. K. Batygin

Wider Availability of PARMILA and Recent Improvements
to PARMILA 427
    J. L. Merson and L. J. Rybarcyk

ADLIB—A Simple Database Framework for Beamline Codes 435
    C. T. Mottershead

DILUTE: A Code for Studying Beam Evolution
Under RF Noise 442
    H.-J. Shih, J. A. Ellison, and W. E. Schiesser

XORBIT—An X-Windows Accelerator Simulation 450
    K. Evans, Jr.

The Simpsons Program: 6-D Phase Space Tracking
with Acceleration 459
    S. Machida

An Efficient Symplectic Approximation
for Fringe-Field Maps 467
    G. H. Hoffstätter and M. Berz

Simulating Dark Current Effects
in Linear Collider Structures 477
    U. Becker, M. Dehler, and T. Weiland

A High-Order Moment Simulation Model 485
    K. T. Tsang, C. Kostas, and A. Mondelli

A General Purpose Relativistic Beam Dynamics Code 493
    R. True

Graphical User Interface for AMOS and POISSON 500
    T. L. Swatloski

Analysis of Space Charge Calculation in PARMELA
and its Application to the CEBAF FEL Injector Design 508
    H. Liu

Status of the BEDLAM Optics Code 516
    W. P. Lysenko

High-Order Optics with Space Charge:
The TOPKARK Code 524
    D. L. Bruhwiler and M. F. Reusch

PIC Space-Charge Emission with Finite Dt and Dz 532
    D. W. Hewett and Y.-J. Chen
Envelope Model of Beam Transport in ILSE 540
    W. M. Sharp, J. J. Barnard, D. P. Grote, S. M. Lund,
    and S. S. Yu
**Envelope Code for Electrostatically Accelerated Beam
with ESQ Focusing** 549
    L. Soroka and O. A. Anderson
Integration of OLE into the TACL Control System 557
    B. Bowling, D. Douglas, J. Kewisch, P. Kloeppel,
    and G. A. Krafft
**Evolution of Simple Phase Space Distributions Using
the Vlasov Equation** 560
    J. O'Connell, S. Butler, and J. O'Connell
**Solution of Laplace's Equation by the Method
of Moments with Applications
to Charged Particle Transport** 568
    C. K. Allen, S. K. Guharay, and M. Reiser
**The Shell for Particle Accelerator Related Codes
(SPARC): A Unique Graphical User Interface** 576
    G. H. Gillespie
Upgrades to the LBL Lattice Program 584
    J. W. Staples
**BMAP Dipole Magnet Field Analysis
and Orbit Tracking** 590
    S. Humphries, Jr., R. M. Baltrusaitis,
    C. Ekdahl, C. Young, and C. Warn
**TRAK Charged Particle Tracking in Electric
and Magnetic Fields** 597
    S. Humphries, Jr.
Author Index 605

Preface

The 1993 Computational Accelerator Physics Conference (CAP93) was held February 22-26, 1993 at the Pleasanton Hilton Hotel. This conference was the third in a series of related meetings: The first, called the Linear Accelerator and Beam Optics Codes Conference, was held in La Jolla, California, in 1988; the second, called the Conference on Computer Codes and the Linear Accelerator Community, was held in Los Alamos, New Mexico in 1990.

The purpose of CAP93 was "to bring together members of the accelerator community who use and/or develop computer codes for the design and analysis of particle accelerators and beam transport systems." It was co-sponsored by the Los Alamos Accelerator Code Group (LAACG) and the National Energy Research Supercomputer Center (NERSC). There were 162 attendees, predominantly from the United States, but also from France, Germany, Italy, Japan, Switzerland, and the Former Soviet Union. Attendees came from Universities, laboratories and industry.

There were approximately 100 oral and poster papers presented at CAP93. Many presentations discussed the status of new and existing computer codes. In addition, the CAP93 Program Committee had felt that CAP93 should be a "forward looking" conference. For this reason, there were also presentations on massively parallel and distributed processing, object oriented programming, graphical user interfaces, and scientific visualization. Feedback from the attendees indicated that this material was very favorably received, and it is expected that this will continue in future CAP conferences.

Besides the conference program, CAP93 had other interesting activities. Many participants attended an open house at AccSys Technology. Also, the banquet at Wente Brothers winery, starting with a reception in the winery caves, was a memorable event, as was the after dinner speech by Frank Eeckman on "Trends in Computational Neuroscience."

This conference would not have been possible without the efforts of many people. This includes the conference co-chairmen, Robert Ryne and Susarla Murty; the conference editorial staff, Carolyn Beckmann (conference editor) and Anita Rodriguez; members of the organizing committee, including Jose Milovich, Virginia Northam, Arthur Paul and Kirby Fong (who was also the conference photographer); and members of the program committee, Alex Chao, Richard Cooper, John DeFord, Alex Dragt, George Gillespie, William Hermannsfeldt, Leo Michelotti, Alfred Mondelli, David Rice, Robert Ryne and Andrew Sessler.

Special thanks go to Carolyn Beckmann, CAP93 conference editor, who is responsible for these proceedings.

Lastly, I want to thank Dr. David Sutter of the US Department of Energy, Office of High Energy Physics, who for many years has been a supporter of the accelerator physics community, and who provided funds to publish these proceedings.

Robert Ryne

PARTICIPANTS

Adams, Fred—Chalk River Laboratories
Allen, Christopher—University of Maryland
Anderson, Oscar—Lawrence Berkeley Laboratory
Arman, M. J.—Phillips Laboratory
Bailey, Vernon—Pulse Sciences, Inc.
Bane, Karl—Stanford Linear Accelerator Center
Barnard, John—Lawrence Livermore National Laboratory
Barnes, Debra—CRAY Research
Barts, Therese—Superconducting Super Collider Laboratory
Bengstsson, Johan—Lawrence Berkeley Laboratory
Berg, J. S.—Stanford Linear Accelerator Center
Berz, Martin—Michigan State University
Bhandari, Rakesh—Superconducting Super Collider Laboratory
Bharadwaj, Ulnod—Fermi National Accelerator Laboratory
Bondarev, B. I.—Moscow Radiotechnical Institute
Bourianoff, George—Superconducting Super Collider Laboratory
Boyd, John—Lawrence Livermore National Laboratory
Bruhwiler, David—Grumman Corporate Research Center
Burke, John—MacNeal-Schwendler Corporation
Buzbee, B.—Director, Scientific Computing Division
Byrd, John—Lawrence Berkeley Laboratory
Callahan, Debra A.—Lawrence Livermore National Laboratory
Caporaso, George—Lawrence Livermore National Laboratory
Carey, David C.—Fermi National Accelerator Laboratory
Carroll, Lewis—CTI Cyclotron
Celata, Christine—Lawrence Berkeley Laboratory
Chan, Chun-Fai—Lawrence Berkeley Laboratory
Chan, Dominic—Los Alamos National Laboratory
Chao, Alex—Superconducting Super Collider Laboratory
Chen, Tong—Stanford Linear Accelerator Center
Chen, Yu-Jiuan—Lawrence Livermore National Laboratory
Chin, Yong Ho—Lawrence Berkeley Laboratory
Chou, Weiren—Superconducting Super Collider Laboratory
Cole, Benjamin—Superconducting Super Collider Laboratory
Cooper, Richard—Los Alamos National Laboratory
Corlett, John—Daresbury Laboratory
Crandall, Kenneth R.—AccSys Technology, Inc.
Crote, Hans—CERN
DeFord, John F.—Lawrence Livermore National Laboratory
Degenhardt, R.—Michigan State University
Dionne, N.—Raytheon Company
Dostie, Craig—MacNeal-Schwendler Corp.
Doucet, Joseph—Schlumberger
Dove, William—U.S. Department of Energy
Dragt, Alex—University of Maryland
Early, Richard A.—Stanford Linear Accelerator Center

Edighoffer, John—Lawrence Berkeley Laboratory
Eppley, K. R.—Stanford Linear Accelerator Center
Evans, Kenneth Jr.—Argonne National Laboratory
Fawley, William—Lawrence Berkeley Laboratory
Ferguson, Patrick—MDS Company
Fiorentini, G.—Lawrence Berkeley Laboratory
Fong, Kirby—Lawrence Livermore National Laboratory/NERSC
Forest, Etienne—Lawrence Berkeley Laboratory
Friedlander, Fred—Varian Microwave Power
Friedman, Alex—Lawrence Livermore National Laboratory
Fuchi, Kyoko—Michigan University
Garnett, Bob—Los Alamos National Laboratory
Garren, Alper A.—Superconducting Super Collider Laboratory
Geller, Alan—Cray Research
Gillespie, George—G. H. Gillespie Associates, Inc.
Goktepe, Omer—U.S. Department of Energy
Goodrich, C.—University of Maryland
Goren, Yehuda—Superconducting Super Collider Laboratory
Grote, David P.—Lawrence Livermore National Laboratory
Guy, Frank—Superconducting Super Collider Laboratory
Haber, Irving—Naval Research Laboratory
Hahn, Kyoung—Lawrence Berkeley Laboratory
Hall, Bernard—Charles Evans and Associates
Henke, Heino—Technische Universitaet
Hermannsfeldt, William—Stanford Linear Accelerator Center
Hoffstatter, G. H.—Michigan State University
Houck, Tim—Lawrence Livermore National Laboratory
Huang, Yunxiang—Superconducting Super Collider Laboratory
Humphries, Stanley Jr.—Stanley, Field Precision
Iwashita, Yoshihisa—Kyoto University
Jackson, Robert—Naval Research Laboratory
Jones, Roger—Stanford Linear Accelerator Center
Kalkanis, Gregory—Varian Associates
Kalos, Malvin—Cornell University
Katsouleas, T.—University of Southern California
Kenney, Ardith—Superconducting Super Collider Laboratory
Kim, Jin-Soo—Duly Consultants
Kitchens, Tom—U.S. Department of Energy
Knight, Thomas—Stanford Linear Accelerator Center
Ko, Kwok—Stanford Linear Accelerator Center
Koga, James—The University of Texas at Austin
Kress, Michael—College of Staten Island
Kugler, Hartmut—CERN
Lagniel, Jean-Michel—CEA SACLAY
Langdon, Bruce—Lawrence Livermore National Laboratory
Li, Mingyang—Superconducting Super Collider Laboratory
Liewer, Paulette—Jet Propulsion Laboratory

Liu, Hongxiu—CEBAF
Lysenko, Walter—Los Alamos National Laboratory
Machida, Shinji—Superconducting Super Collider Laboratory
Manca, Joseph—Titan Beta
Mankofsky, A.—Science Applications International Corporation
Mao, Naifeng—SSC Laboratory
McCoy, Michel—Lawrence Livermore National Laboratory/NERSC
McCurdy, C. W.—National Energy Research Supercomputer
Merson, Jean—Los Alamos National Laboratory
Michelotti, Leo—Fermi National Accelerator Laboratory
Millich, A.—CERN
Milovich, Jose—Lawrence Livermore National Laboratory
Mondelli, Alfred—SAIC
Mori, Wm.—UCLA
Mottershead, Tom—Los Alamos National Laboratory
Murty, Susarla—Lawrence Livermore National Laboratory
Myers, Timothy—Grumman Corporate Research Center
Nelson, E. M.—Stanford Linear Accelerator Center
Neri, Fillipo—Los Alamos National Laboratory
Ng, Cho-Kuen—Stanford Linear Accelerator Center
O'Connell, James S.—Booz Allen & Hamilton, Inc.
Orthel, John L.—G. H. Gillespie Associates, Inc.
Parazzoli, Claudio—Boeing Defense & Space
Paul, Arthur C.—Lawrence Livermore National Laboratory
Payne, Anthony—Lawrence Livermore National Laboratory
Peters, Gerald—U.S. Department of Energy
Prior, Christopher—Rutherford Appleton Lab.
Putnam, Sidney—Pulse Sciences, Inc.
Raparia, Deepak—Superconducting Super Collider Laboratory
Rimmer, Bob—Lawrence Berkeley Laboratory
Rodenz, Gary—Los Alamos National Laboratory
Ruiz, Everardo D.—Superconducting Super Collider Laboratory
Rusthoi, Dan—Los Alamos National Laboratory
Ryne, Robert—Los Alamos National Laboratory
Sagan, David—Cornell Laboratory
Salop, Arthur—Varian Associates
Schlueter, Ross—Lawrence Berkeley Laboratory
Schoessow, Paul—Argonne National Laboratory
Servranckx, Roger—TRIUMF
Sessler, Andrew—Lawrence Berkeley Laboratory
Shang, Clifford—Lawrence Livermore National Laboratory
Sharp, W. M.—Lawrence Livermore National Laboratory
Shih, Hsiuan-Jeng—Superconducting Super Collider Laboratory
Soroka, Ludmilla—Lawrence Berkeley Laboratory
Staples, John—Lawrence Berkeley Laboratory
Strongin, Boris—Varian Associates
Sutter, David—U.S. Department of Energy

Swatloski, Teresa—Lawrence Livermore National Laboratory
Tantawi, S.—Stanford Linear Accelerator Center
Temkin, Richard—MIT
Thoma, P.—THD
Travish, G.—University of California
True, Richard—Litton Systems
van Rienan, Ursula—THD
van Zeijts, Johannes—CEBAF
Vella, Michael—Lawrence Berkeley Laboratory
Wake, Dan—Pulse Sciences
Walling, Linda—Superconducting Super Collider Laboratory
Wang, Changbiao—Lawrence Berkeley Laboratory
Wang, Fuhua—Superconducting Super Collider Laboratory
Wangler, Thomas P.—Los Alamos National Laboratory
Warnock, Robert L.—Stanford Linear Accelerator Center
Warren, Gary—Mission Research Corporation
Weiland, Thomas—THD
Weishi, Wan—Michigan University
Whealton, John—ORNL
Witherspoon, Sue—CEBAF
Yan, Yiton—Superconducting Super Collider Laboratory
Yu, David—DULY Research

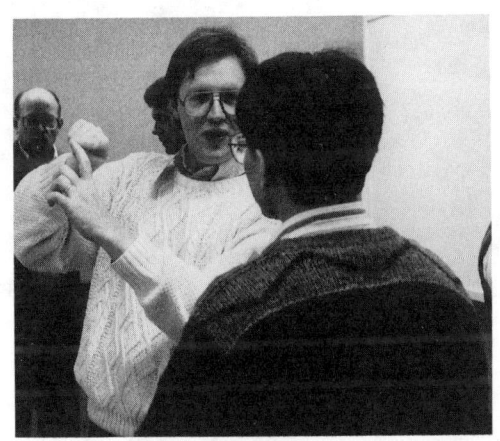

MODELING ACCELERATOR STRUCTURES AND RF COMPONENTS*

K. Ko, C.-K. Ng and W. B. Herrmannsfeldt
Stanford Linear Accelerator Center
Stanford University, Stanford, CA 94309

ABSTRACT

Computer modeling has become an integral part of the design and analysis of accelerator structures and RF components. Sophisticated 3D codes, powerful workstations and timely theory support all contributed to this development. We will describe our modeling experience with these resources and discuss their impact on ongoing work at SLAC. Specific examples from R&D on a future linear collider and a proposed e^+e^- storage ring will be included.

1. INTRODUCTION

There has been a dramatic increase in computer modeling at SLAC due to the R&D on two future accelerators: the Next Linear Collider (NLC) [1] and the Asymmetric B Factory based on PEP (PEP-II) [2]. This is particularly evident in the modeling of accelerator structures and RF components in three dimensions. A test accelerator (NLCTA) [3] is presently under construction and one of its goals is to integrate the new technologies of the X-band RF system (klystron, pulse compression, linac structure) being developed for a TeV scale NLC into a fully engineered sector. For PEP-II, a high power test cavity with a novel higher-order mode (HOM) damping scheme [4] is to be constructed to study its electrical and mechanical performance under operating conditions at high currents. At the moment a substantial modeling effort is devoted to the design and analysis of the many complex cavities and structures involved in these two projects.

Most RF structures in an accelerating system possess geometrical symmetries which allow them to be treated by two-dimensional programs or even by analytical means. There are, however, critical components that are non-symmetrical for which three-dimensional modeling is needed. Previously, the engineering of such components would require repetitive prototyping and cold testing. At SLAC we are making a concerted effort to minimize these time consuming and costly experimental procedures and to move towards nondestructive testing by computer modeling. This alternative approach would not have been feasible without several major developments in modeling resources. They include advances in 3D programs, the speedup in affordable CPU's, and timely theory support. We will discuss each of these resources in the next section.

2. MODELING RESOURCES

(i) 3D Codes

Computer programs to model cylindrically symmetric RF structures have been in routine use for over a decade in accelerator and power tube research. A prime example is SUPERFISH [5] for finding monopole modes in axisymmetric structures. Truly three-dimensional codes for non-axisymmetric structures made their appearances more recently and are just beginning to gain widespread use. A more detailed discussion on 3D RF cavity codes has been presented elsewhere [6]. Here we will only summarize as follows.

Presently available codes are either based on a finite-difference (FD) (Fig. 1a) or finite-element (FE) mesh (Fig. 1b). They can simulate cavity properties in the time and/or frequency domain (td or fd). Several of them also have the particle-in-cell (pic) option to include beam effects self-consistently. Two advances of interest to accelerator design are the quasi-periodic boundary condition for periodic structures and the S-matrix calculation for traveling wave structures. We list below several 3D codes which SLAC has had experience with. Structured on a FD mesh are:

---

* Work supported by Department of Energy, contract DE-AC03-76SF00515.

2 Modeling Accelerator Structures

and on a FE mesh are:
- (a) MAFIA *(CST/Darmstadt)* - td with pic, fd,
- (b) ARGUS *(SAIC)* - td with pic, fd,
- (c) SOS *(MRC)* - td with pic, fd;
- (d) HFSS *(HP/ANSOFT)* - fd,
- (e) EMAS *(MSC)* - td, fd,
- (f) ANTIGONE *(LAL/Orsay)* - td with pic, fd.

The FE codes have an automatic mesh-generator, model geometries better and can interface directly with thermal codes which are also FE based. The FD codes, on the other hand, have enjoyed a much longer lead time in implementation for RF applications. MAFIA [5], in fact, has its origin in the accelerator community; therefore it is not surprising that this set of codes offers the broadest range of capabilities related to accelerator design. ARGUS [5] is under further development for accelerator applications and is available to the community at large through the Los Alamos Accelerator Code Group (LAACG). SOS [5] is similar to MAFIA and ARGUS but is more oriented towards the simulation of microwave devices. Of the FE codes, HFSS [5] is probably the most widely known for S-matrix calculations. EMAS [5] has thus far attracted few users and we are just beginning to take a closer look at ANTIGONE [5]. At present MAFIA and HFSS are our standard programs for modeling cavities and structures although we are always receptive to any code that could provide better results on a specific design.

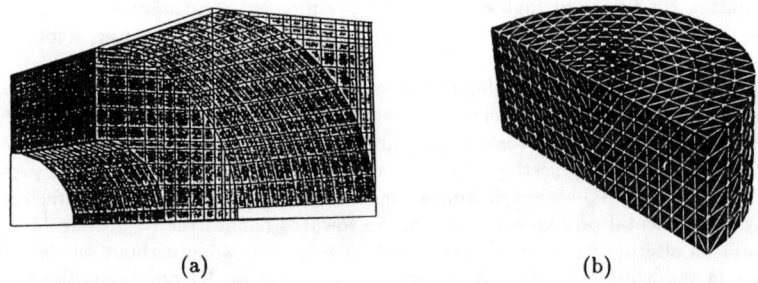

Fig. 1 (a) MAFIA finite-difference mesh and (b) ANTIGONE finite-element mesh.

(ii) Computing Environment

3D modeling used to be considered practical only on supercomputers because of the long run time and large memory it requires. This is no longer the case as the workstation has become competitive in performance. For example, an IBM RISC 6000 model 580 processes at over 40 MFLOPs, allows up to 1024 MB RAM, and provides several GB of disk storage. These specifications come at a small fraction of the cost of a supercomputer. On such a machine with 512MB of RAM, we have modeled one half of the PEP-II RF cavity using MAFIA and calculated forty eigenmodes on a mesh with three-quarter of a million cells in less than 6 hours. This quick turnaround in actual clock time is important as it may translate into shorter design cycle and hence improved efficiency. A mainframe supercomputer is not likely to operate in a dedicated mode on a job to give fast turnaround at a sustained level. In terms of visualization, the workstation has powerful graphics which makes it ideal for 3D modeling since most of the effort involved is in geometry setup of the structure, and post-processing of the results.

Looking at the cost-differential, the effective performance and the open system/graphics advantage, the argument is rather convincing that the workstation could be a viable alternative to the supercomputer. Faced with limited cpu allocation from supercomputers and increasing modeling demand from NLC and PEP-II, we made the choice to go with 3D codes such as MAFIA and HFSS which had support for running on workstations. Today we have built a respectable network of workstations with sufficient computing power and memory to handle reasonably complex structures with relative ease.

(iii) Theory Support

Although great progress has been made with 3D codes there are instances where they are limited in their capabilities. Many cavities in an RF system are coupled to external waveguides for power input or output, such as the input coupler to a linac or the output circuit of a klystron. Recently there has also been strong interest in HOM damping in accelerator cavities using waveguides. Therefore it is of great interest to know the resonant properties like frequency and external Q in such geometries. But this requires a matched load at the end of the waveguide which numerically is not possible since existing codes calculate real eigenfrequencies.

While the waveguide-loading effect on the cavity cannot be evaluated directly, the following theoretical methods have been successfully implemented to find external Q's from closed cavity data (real frequencies) computed with shorting planes as waveguide terminations:

(1) Kroll-Yu/Kroll-Lin methods - find resonant frequency and external Q,
(2) Kroll-Kim-Yu method - calculates S matrix of N-port cavities,
(3) Arcioni method - produces an impedance versus frequency spectrum.

The Kroll-Yu method [7] requires four shorting plane positions in the waveguide to parametrize the resonance which is a complex frequency. Kroll-Lin [8] achieves the same result with two waveguide lengths but needs information about the field amplitudes at the shorting planes as well as the stored energies. For cavities with more than one port, the Kroll-Kim-Yu [9] method derives the S matrix at the frequency points where computer results are obtained. Normally several well chosen waveguide lengths are needed. In contrast, a single length is sufficient for the Arcioni method [10] to produce an impedance spectrum from which the resonant parameters can be deduced. However, modes with frequencies up to 1.5 times the frequency range of interest have to be taken into account for good accuracies. We have applied the Kroll-Yu/Kroll-Lin methods to a wide range of cavities with considerable success. The Kroll-Kim-Yu method is being tried out on some simple components and the results look very encouraging. Arcioni's method has seen limited application so far but an effort is ongoing to explore its potential further. Our experience is that when a direct numerical method is lacking, one can turn to an indirect method that combines theory with existing code capability.

3. MODELING EMPHASIS AND APPROACH

Up to now our 3D efforts on the workstations have focussed primarily on the cold tests of cavities and structures with no beams. The addition of particles vastly increases the complexity of the simulation and quickly pushes the limit of the hardware. We are continuing to work towards a hot test (e.g. klystron interaction), although any meaningful quantitative results will probably have to wait till the next jump in computing power and further code improvements. Meanwhile there is a lot to be learned from modeling cold tests and for most structures, the simulation has proven to be invaluable for design and analysis.

Our cold test modeling emphasizes the following: we require that the geometry be realistic, that is, the mesh reproduces closely the actual dimensions of the structure. We make certain that the simulation is physical; for example, proper boundary conditions are included. And we benchmark the model against established experimental data. These are essential steps to work through before one should consider doing serious design studies with the model.

Numerical cold tests can be performed either in the frequency or time domain. For normal mode analysis of cavities and periodic structures, we consider the time-harmonic solutions at steady-state and find the eigenfrequencies and eigenfunctions of the system. The quasi-periodic boundary condition allows arbitrary phase advance across a single cell in an infinite periodic chain. In waveguide-loaded cavities, theoretical methods described earlier make use of normal mode to produce external Q's. For transmission calculations of traveling structures, we solve for the time response due to a single-frequency externally driven field. Radiating boundary conditions permit power to be input and output through port boundaries at which the S matrix can be determined when steady-state is reached. In certain cases, the transient fields are of interest. Interestingly, HFSS evaluates the S matrix in the frequency domain.

4. THREE-DIMENSIONAL RF STRUCTURES

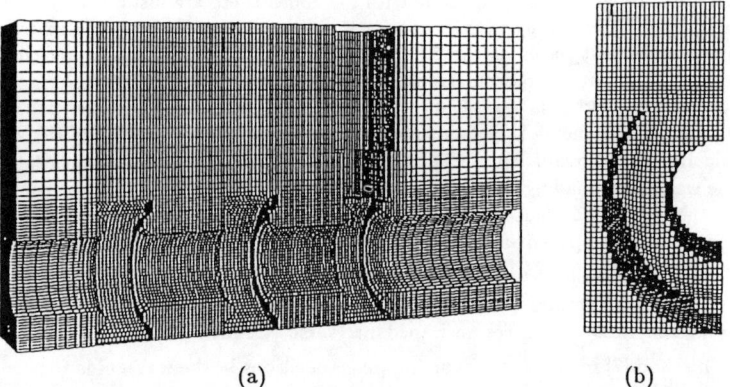

Fig. 2 (a) ARGUS model of klystron output cavity and (b) coupling slot between gaps.

In an RF structure, 3D effects can arise either intrinsically or due to external coupling. We will illustrate this by the 3-gap klystron output cavity shown in Fig. 2a. Even in the absence of the external waveguide, the cavity is intrinsically 3D because of the coupling slots in the common wall between the gaps (Fig. 2b). It is a standing wave cavity and the modes have zero group velocities because the cavity is closed. To extract power, the last gap is coupled to an external waveguide, hence making it a waveguide-loaded cavity. This causes the modes to have non-vanishing group velocities which is necessary for power flow to the waveguide. Strictly speaking the output circuit is no longer a standing wave cavity. The parameters that characterize the circuit are the frequency, the R/Q and the loaded Q or Q_L which is given by

$$\frac{1}{Q_L} = \frac{1}{Q_o} + \frac{1}{Q_e}. \tag{1}$$

where Q_o is due to wall loss and Q_e is due to waveguide loading. In a closed or standing wave cavity Q_L is simply Q_o since Q_e is infinite. In a waveguide-loaded cavity, Q_L can have both contributions if the coupling is weak while for strong coupling, it is dominated by the waveguide loading and Q_L is just Q_e.

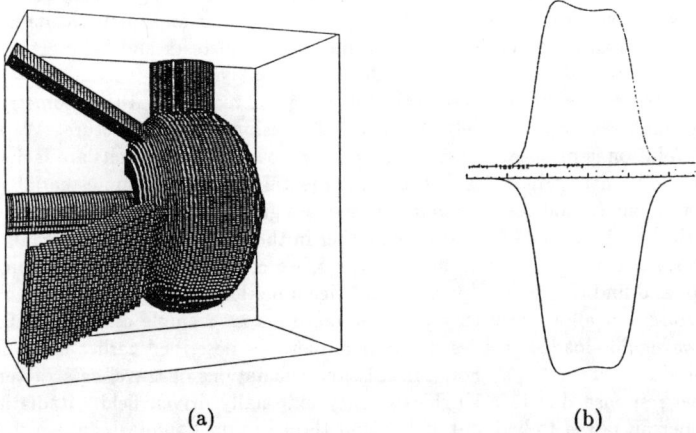

Fig. 3 (a) MAFIA model of PEP-II HOM damped RF cavity and (b) axial longitudinal field from measurement(top) and MAFIA(bottom).

There are two types of modes in waveguide-loaded cavities such as the klystron output cavity of Fig. 2 or the PEP-II RF cavity shown in Fig. 3a. If their frequencies are below the waveguide cutoff then the modes are trapped as standing waves with high Q's. Trapped modes are to be avoided in the klystron output circuit because they can be potentially harmful oscillations. The accelerating mode should ideally be the only trapped mode in the PEP-II cavity and the waveguides are designed not to degrade its shunt impedance appreciably. For trapped modes, standard eigensolutions would suffice provided the waveguides are sufficiently long.

When their frequencies are above the waveguide cutoff, the modes are untrapped and depending on the coupling, their Q's can be moderately to very low. In a klystron output, low Q's are desirable for all the modes but especially for the operating mode to transfer power out efficiently. In the PEP-II cavity, the untrapped modes are the HOM's that are targeted for damping and low Q's are effective in reducing beam instabilities. The theoretical methods described in Section 2 are needed to evaluate the properties of these untrapped modes.

3D RF structures can also be of the traveling wave type. Figure 4a shows a short section of a disk-loaded waveguide with coupler cavities for power input and output. Without the couplers, the linac structure itself is described by the $\omega - \beta$ diagram which can be generated using 2D codes and gives the traveling modes in a periodic structure of infinite extent. The couplers are necessarily 3D for power coupling away from the beam and for this, the 3D time-domain S-matrix calculation we mentioned earlier is required.

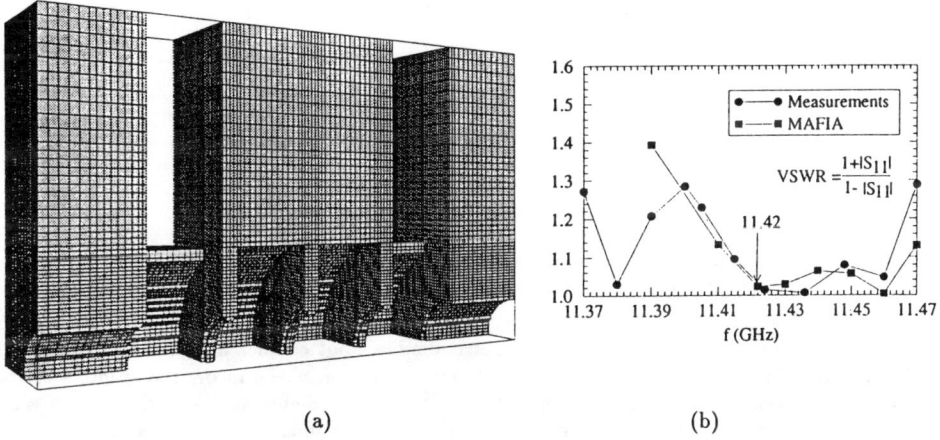

Fig. 4 (a) MAFIA model of NLCTA structure with input/output couplers and (b) VSWR vs. frequency of coupler from MAFIA and measurements.

5. MODELING RESULTS

(i) NLCTA RF System

The NCLTA RF system comprises the klystron, the pulse compression, and the linac all operating at 11.424 GHz. We will show examples of the crucial 3D components from each. For the klystron, the desired peak power (50 MW) and pulse length (1.5 μsec) pose a design challenge on the the output circuit to achieve high efficiency and to avoid gap breakdown. The ongoing approach is to raise the efficiency while lowering the gap voltage through extended interaction in the form of multi-gap structures. Many extended interaction circuits have been tested with varying degree of success. The 3-gap cavity in Fig. 2a was a design for which a cold test comparison was made between experiment and simulation [11]. Table 1 shows good agreement between the two results. The model also indicates uneven gap voltages and large field asymmetries which are undesirable. Consequently, the circuit was not put through hot tests, thus saving valuable time and resources.

6   Modeling Accelerator Structures

	π	$2\pi/3$	2π
Measured	10.07 (47)	10.40 (45)	11.54 (19)
Calculated	10.05 (45)	10.33 (44)	11.53 (20)

Table 1. Cold test results from experiment and modeling - frequency in GHz and Q_L (in parenthesis) of modes in lowest pass band.

To reach the designed accelerating gradient (50MV/m), the klystron output pulse is to be compressed to higher peak power before injection into the linac. During pulse compression, the RF pulse is manipulated and transmitted over substantial lengths at which copper loss can be significant if it were to remain in the TE_{10} mode of the klystron output waveguide. A mode transducer has been designed to mode convert to low-loss TE_{01} mode in overmoded circular guide. Presently in use, the device satisfies the requirements on bandwidth, mode purity and breakdown limit [12]. Figure 5a shows the mode transducer as modeled by HFSS and Fig. 5b shows the input characteristics when compared with experiment. Although exact quantitative agreement was not always possible because the device is sensitive to small dimensional differences, the qualitative results have nevertheless been useful for mode contamination studies.

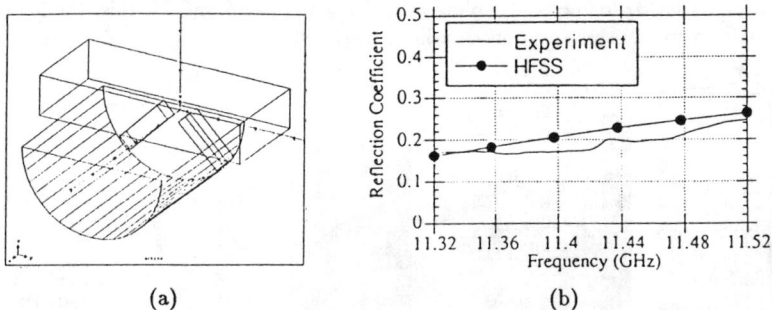

Fig. 5 (a) HFSS model of mode transducer without bifurcation (half geometry) and (b) input characteristics from HFSS and experiment.

The RF power coming out of the pulse compression has to be matched into the linac through coupler cavities. Figure 4a shows a symmetric double-input coupler designed to reduce field asymmetry both in amplitude and phase [13]. This coupler has been incorporated into a 30 cm section and operated up to 100 MV/m without breakdown. Figure 4b shows the transmission data comparing MAFIA with measurement [14]. Similar couplers are being designed for the 75cm and 1.8m accelerator sections and computer modeling is playing an important role in determining the matching and tuning of these cavities.

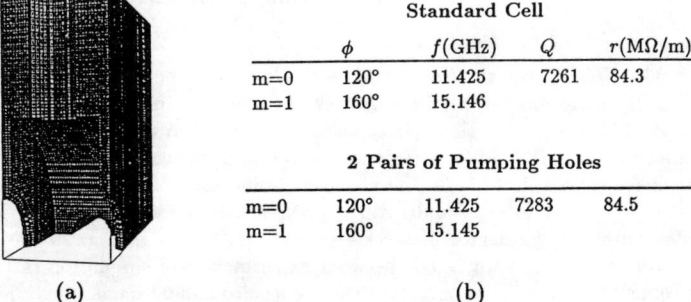

Fig. 6 (a) MAFIA model of pumping cell in detuned structure and (b) tuning results from MAFIA (ϕ is phase advance and r is shunt impedance).

Inside the NLC linac, the high-power RF pulse will accelerate long trains of e^+ or e^- bunches. To suppress multi-bunch beam breakup, two schemes have been proposed to heavily damp the wakefields induced by the bunch trains. One is a HOM damping scheme to couple the wakefields out to external loads through waveguides (as in the PEP-II cavity). Damped structure for the NLC was an area of intense activities [15]. Another scheme which is being adopted for the NLCTA is to detune the cells in a systematic way so as to decohere and hence damp the wakefields. Although the individual cells have varying dimensions, the detuned structure is essentially cylindrically symmetric except for the few cells that also serve as vacuum pumping cells. We have modeled these cells to determine their dimensions so they would retain the original accelerating and dipole mode frequencies in spite of the pumping slots. Figure 7a shows one such cell while Fig. 7b shows the tuning results. In this case, reducing the cell diameter by 3.3 microinches allows us to correct for the frequency shifts in the accelerating and dipole modes due to the slot perturbations.

(ii) PEP-II RF Cavity

A considerable amount of modeling effort has been devoted to the design of the PEP-II HOM damped cavity. The two important features of this cavity are: (a) the waveguide damping scheme to reduce HOM impedances, and (b) the thermal/mechanical design to cope with high power dissipation (150 kW on the cavity walls). A low-power test cavity has been constructed for HOM damping studies. Figure 6a shows the mode spectrum without and with damping waveguides. One sees that essentially all the higher-order modes are damped while the accelerating mode at 476 MHz is practically unchanged. Using the MAFIA model shown in Fig. 3a and the Kroll-Yu method, we calculated a Q_L of 26 for the most dangerous TM_{011} mode as compared with 28 from measurement. The longitudinal fields on axis of the accelerating mode as found from MAFIA and from bead-pull experiment are compared in Fig. 3b which shows good agreement. The field asymmetry is due to the presence of the damping waveguides.

In designing the cooling system for the cavity, power loss data on the cavity wall has to be generated and input into a mechanical code for thermal stress analysis. Figure 6b shows the distribution of the power dissipated. High power loss is seen around the cavity-waveguide junction and more elaborate cooling is planned for this region. The ongoing mechanical design will be tested in the high-power cavity which is scheduled to go to construction fairly soon. In addition to thermal calculations, the MAFIA model is also heavily utilized in the design of the various coupling networks being considered for the cavity.

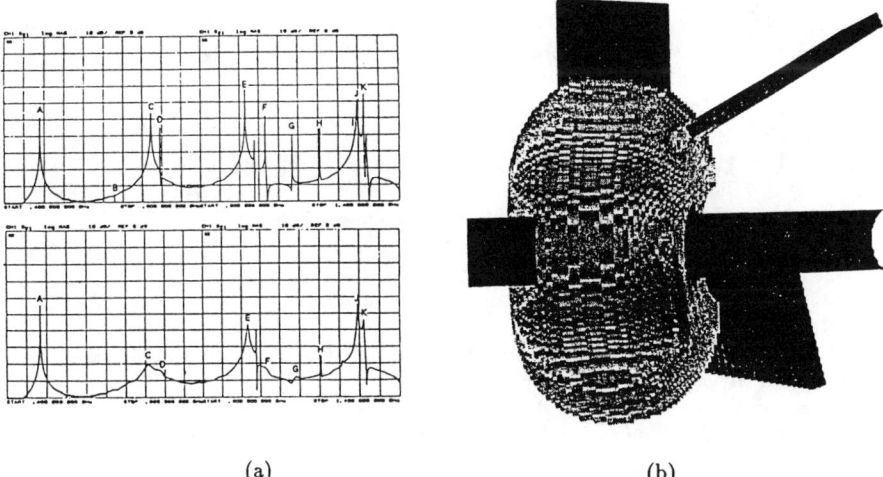

(a)   (b)

Fig. 7 (a) Modes measured from 0.4-1.4 GHz without damping (top) and with damping (bottom) and (b) power loss on cavity wall (lighter shade indicates higher dissipation).

6. CONCLUSION

We have presented an overview of the present status in modeling accelerator structures and RF components. The latest in modeling resources is described and our modeling approach and emphasis explained. The properties of 3D cavities and structures are defined in terms of mode character and external coupling. We have included selected examples from the NLCTA and the PEP-II projects at SLAC to illustrate our principles of application.

ACKNOWLEDGEMENTS

This paper in essence represents the collective efforts of many individuals from the Numerical Modeling group, the Structures group, the Pulse Compression group and the Klystron department and to them we owe our thanks. We also acknowledge the LBL RF group on the PEP-II collaboration. The help from Thomas Weiland and his group on MAFIA 3.1 as well as the effort of Guy Lemeur on ANTIGONE are appreciated.

REFERENCES

[1] R. D. Ruth, 'The Development of the Next Linear Collider at SLAC', *Proc. of Workshop on Physics and Experiments with Linear Colliders*, Saariselka, Finland, September 9-14, 1991; also in SLAC-PUB-5729.

[2] An Asymmetric B Factory, Conceptual Design Report, LBL-PUB-5303, SLAC-372 (1991).

[3] R. D. Ruth et. al., 'The Next Linear Collider Test Accelerator', presented at the Particle Accelerator Conference, Washington, DC, May 17-20, 1993; also in SLAC-PUB-6252.

[4] R. Rimmer et. al., 'An RF cavity for the B Factory', *Proc. of Particle Accelerator Conference*, San Francisco, CA, May 6-9, 1991; also in SLAC-PUB-6129.

[5] H. K. Deavon and K. C. D. Chan, Eds., 'Computer Codes for Particle Accelerator Design and Analysis: A Compendium', Los Alamos Accelerator Code Group, LA-UR-90-1766, 1990.

[6] K. Ko, 'Computer Codes for RF Cavity Design', *Proc. of LINAC 92*, Ottawa, Canada, August 24-28, 1992; also in SLAC-PUB-5888.

[7] N. M. Kroll and D. U. L. Yu, 'Computer Determination of the External Q and Resonant Frequency of Waveguide Loaded Cavities', *Particle Accelerators*, 34, 231 (1990).

[8] N. M. Kroll and X.-T. Lin, 'Computer Determinations of the Properties of Waveguide Loaded Cavities', *Proc. of Linear Accelerator Conference*, Abuquerque, NM, September 10-14, 1990; also in SLAC-PUB-5345.

[9] N. M. Kroll, J.-S. Kim and D. U. L. Yu, 'Computer Determination of the Scattering Matrix Properties of N Port Cavities', *Proc. of LINAC 92*, Ottawa, Canada, August 12-18, 1992; also in SLAC-PUB-5880.

[10] P. Arcioni, 'POPBCI - A Post-Processor for Calculating Beam Coupling Impedances in Heavily Damped Accelerating Cavities', SLAC-PUB-5444, 1991.

[11] K. Ko, T. G. Lee and N. M. Kroll, 'A Three Gap Klystron Output Cavity at X Band', *Proc. of SPIE's Optics, Electro-Optics and Laser Applications in Science and Engineering Symp.*, Los Angeles, CA, January 19-25 1992; also in SLAC-PUB-5760.

[12] S. Tantawi, K. Ko and N. Kroll, 'Numerical Design and Analysis of a Compact TE_{10} to TE_{01} Mode Transducer', these proceedings.

[13] G. A. Loew and R. B. Neal, 'Accelerator Structures', in Linear Accelerators, p39-133, ed. P. M. Lapostolle and A. L. Septier, 1970.

[14] C.-K. Ng and K. Ko, 'Numerical Simulations of Input and Output Couplers for Linear Accelerator Structures', these proceedings.

[15] H. Deruyter et. al., 'Damped and Detuned Accelerator Structures', *Proc. of 1990 Linear Accelerator Conf.*, Albuquerque, NM, September 10-14, 1990; also in SLAC-PUB-5322.

BEAM HALO IN HIGH-INTENSITY BEAMS*

THOMAS P. WANGLER
Accelerator Technology Division, MS-H817, Los Alamos, NM 87545

ABSTRACT

In space-charge dominated beams the nonlinear space-charge forces produce a filamentation pattern, which in projection to the 2-D phase spaces results in a 2-component beam consisting of an inner core and a diffuse outer halo. The beam-halo is of concern for a next generation of cw, high-power proton linacs that could be applied to intense neutron generators for nuclear materials processing. We describe what has been learned about beam halo and the evolution of space-charge dominated beams using numerical simulations of initial laminar beams in uniform linear focusing channels. We present initial results from a study of beam entropy for an intense space-charge dominated beam.

INTRODUCTION

For beams with high average intensity, one may be concerned not with the rms or average phase-space areas, but with the outer part of the distribution, often called the beam halo, which affects particle losses in an accelerator. Relatively small losses in a high-energy accelerator may produce enough radioactivation of the accelerator structure or radiation damage of components to create practical difficulties in maintenance and operation[1]. A major cause of beam-halo growth in low-velocity intense beams is the Coulomb self force. In most accelerator beams this is predominantly a collective force; small-impact-parameter binary collisions usually have little effect on the dynamics. This smoothed or average Coulomb force is called the space-charge force and is described by a repulsive self-electric field and an attractive self-magnetic field. The magnetic term is only important for relativistic beams and its contribution reduces the total space-charge force.

The space-charge force is complicated because the field depends upon the time-varying charge density of the beam, is nonlinear, time dependent, and coupled between the three planes. In the presence of external focusing forces, one observes phenomena that are common in plasma physics, such as plasma oscillations and Debye shielding. The plasma period determines a basic time scale for these phenomena, and the Debye length determines a basic length scale for the particle distribution. The net force, consisting of the external focusing plus the time-dependent space-charge force, may be either attractive or repulsive, and the sign of the net force may even vary across the beam. These conditions can lead to very nonlinear behavior, and one must rely on numerical simulation codes to study the detailed dynamics.

PHASE-SPACE DYNAMICS

Numerical Simulation

We use numerical simulation to look at the multiparticle dynamics to see what changes occur in phase space. We examine the case of a round continuous beam in a uniform linear focusing channel with purely radial focusing. This system represents a smooth approximation for beams in quadrupole focusing channels, and therefore we expect that phenomena observed in the uniform channel will also be observed in the

---

*Work supported by the Los Alamos National Laboratory Institutional Supporting Research under the auspices of the United States Department of Energy.

10 Beam Halo in High-Intensity Beams

quadrupole channel. We use a numerical simulation code[2] for these studies in which the radial space-charge forces are calculated from Gauss's Law. Consequently, we are studying the effect of the collective forces acting on each particle and ignoring the small-impact-parameter binary Coulomb collisions. With radial symmetry this is a 1-D (strictly speaking a single variable) problem. Our computer code has been run with 2000 simulation particles through 56 steps per plasma period, choices that are adequate to represent the main features of the space-charge forces. We have chosen to study the dynamics of an initial rms-mismatched laminar (zero emittance) beam. Laminar beams are idealizations because all real beams have finite emittance. Nevertheless, the laminar beam represents the extreme space-charge limit and allows us to emphasize the effects of the space charge.

In Figs. 1 and 2 we show the distributions of a) the radial or r - r' phase space, b) the projected or x - x' (and y - y') phase space, and c) the x - y beam cross section.

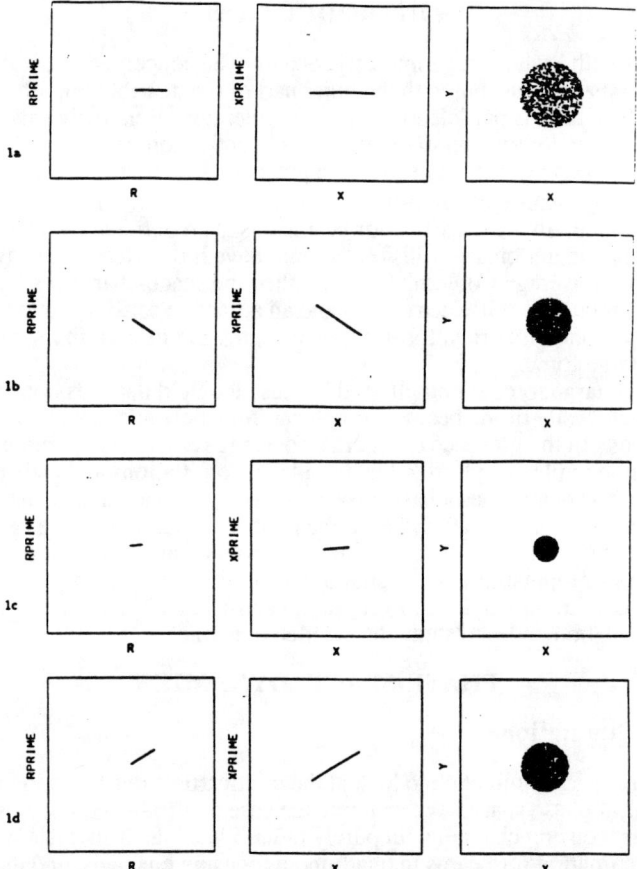

Fig. 1. Radial r - r' phase space, transverse x - x' and y - y' phase space, and cross section x - y from simulation of an initial uniform-density laminar beam in a uniform linear focusing channel for a) t = 0, b) t = 0.25, c) t = 0.50, and d) t = 0.75 in units of beam-plasma periods. The initial rms beam sizes in x and y are 50% larger than the matched size.

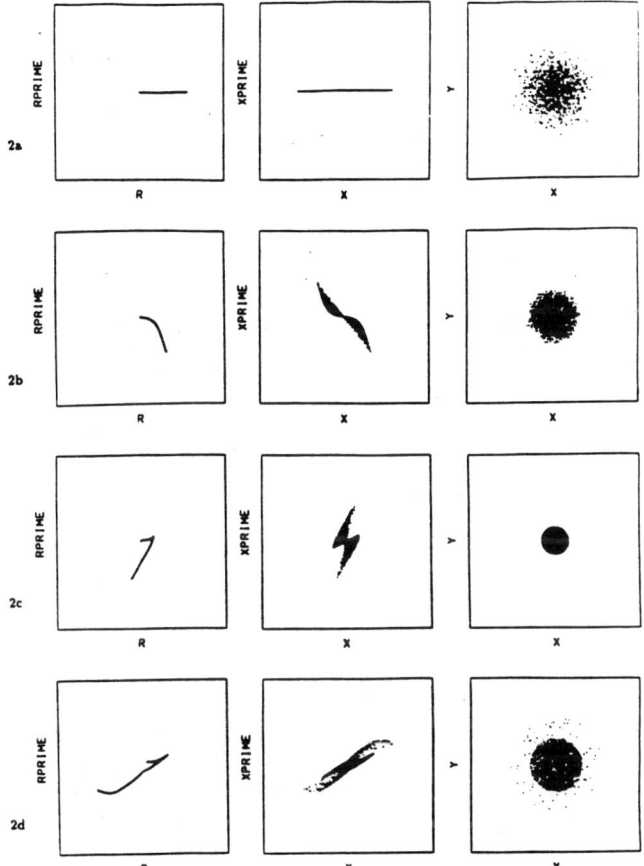

Fig. 2. Radial (r - r') phase space, transverse x - x' and y - y' phase space, and cross section (x - y) from simulation of an initial Gaussian-density laminar beam in a uniform linear focusing channel for a) t = 0, b) t = 0.25, c) t = 0.50, d) t = 0.75, e) t = 1.00, f) t = 1.50, g) t = 2.00, h) t = 3.00, i) t = 4.00, j) t = 5.00, k) t = 10.00, and l) t = 20.00, in units of beam-plasma periods. The initial rms beam sizes in x and y are 50% larger than the matched size.

We show the r - r' phase space because we expect the dynamics to appear simpler in r - r' space when only radial forces act on a laminar beam. We assign an initial positive radius to all particles, but if during the simulation a particle crosses the axis, we change the sign of the radius before plotting a point in r-r' space.

Mismatched Uniform Density Laminar Beam

We begin by studying the dynamics of an initial space-charge dominated uniform-density laminar beam with zero velocity spread, which is rms mismatched so that the initial rms beam size is larger than the matched value by a factor of 1.5. Figures 1a through 1d show the beam characteristics for four different times, 0, 0.25, 0.50, and 0.75, measured in beam-plasma periods. The beam-plasma period for a uniform beam

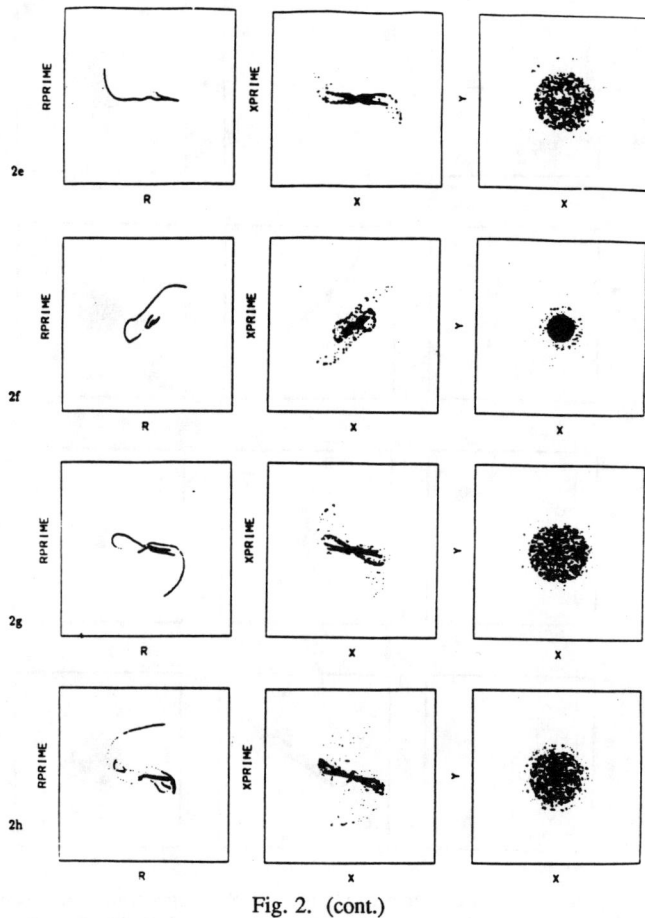

Fig. 2. (cont.)

of density n_0 is defined in the usual way; thus $T_p = 2\pi/\omega_p$, and $\omega_p^2 = q^2 n_o / \varepsilon_o m$ is the beam-plasma frequency. The phase-space plots show density (plasma) oscillations that are excited by the unbalanced external focusing and internal space-charge forces. The total force alternates at the beam-plasma frequency between focusing and defocusing. The charge distribution always remains uniform so that only linear forces act on the beam, and the emittance remains zero. In the absence of space-charge forces particles would experience only the external fields and would cross the axis as they execute betatron oscillations. For a space-charge dominated beam, the particles do not cross the axis, but each particle oscillates about an equilibrium radius.

Gaussian Density Laminar Beam

Next we examine the dynamics of an initial Gaussian-density laminar beam with zero initial velocity spread, which is rms mismatched by the same factor 1.5. Figures 2a through 2l show a sequence of plots for different times in units of the plasma period (defined for the equivalent uniform beam with the same rms size). For this case, the external force is linear, but the space-charge force is nonlinear. Several new features are present. Most of the small amplitude trajectories undergo plasma oscillations (they

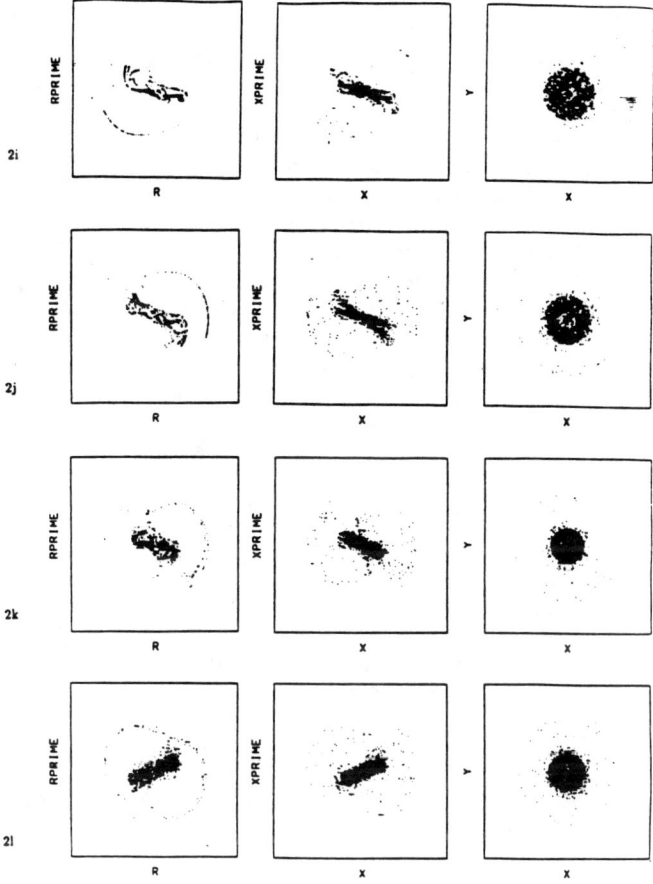

Fig. 2. (cont.)

do not cross the axis) and form an inner core. The large amplitude trajectories correspond to betatron oscillations (they cross the axis) and develop into an outer halo. In r - r' space the halo evolves like a ring-shaped filament. In x - x' space the ring appears as a low-density disk. These differences are the result of the fact that any arbitrary point in r - r' space projects to a straight line in x - x' space that passes through the origin and ranges between (-r,-r') and (r,r'). Although effective emittance growth has often been identified with a process of filamentation, we see that the filament in this problem is observed in the r - r' phase space. In the usual x - x' projected phase space the outer part of the filament becomes a diffuse disk-like halo. This will be discussed in more detail in the following section.

Even within a few plasma periods the nonlinear space-charge force produces a random-looking distribution of points within the core. This randomization or thermalization is the result of a process in which the inner part of the filament in r - r' space is stretched and folded many times. The stretching and folding is associated with variations of the magnitude and sign of the space-charge force. The halo produced after several plasma periods is a common feature of the nonlinear space-charge force. We find that the outer filaments seen in r - r' space contain mostly the particles with large initial amplitudes but also contain a few particles with small initial amplitudes that were launched during the initial stages of randomization of the core. For our example, the halo is a distinctive structure in r - r' space even after 20 plasma periods; unlike the core, the halo is not yet thermalized.

At present there is no established criterion for defining the halo. For the present example of an rms-mismatched Gaussian laminar beam, we find that an ellipse with the same Courant-Snyder parameters as the rms ellipse and with an emittance five times larger than the rms ellipse in r-r' space, appears to enclose most of the core and exclude most of the halo. If we define the core particles to be all those contained within this ellipse, and define halo particles as those outside, we find that after about 10 plasma periods, 6% of the particles are contained within the halo. For this example the core and the halo contribute about equally to the final rms emittance. Furthermore, the rms emittance of the core grows to its final value in about one-quarter of a plasma period, like that of an rms-matched beam[3]. Most of the growth of the halo occurs over about 10 plasma periods[4]. More study is needed to determine how these results vary with the amount of mismatch and to determine what happens when using more realistic beams with nonzero initial emittance. One concern for this problem with pure radial dynamics is that because all particles that cross the axis must pass through a common point at the origin (x=y=0), there may be singular density fluctuations at the origin that may produce unrealistic forces. Previous work on emittance growth for this 2-D mismatched beam is given in Refs. 4 and 5, and for 1-D sheet beams in Ref. 6.

2.4 Filamentation and Beam Halo

The phase-space plots in Fig. 2 show a complex filamentation pattern in r-r' space, where the halo forms an outer ring. The presence of filamentation is a well known effect of nonlinear forces in 1-D, and in this 2-D problem it is observed clearly in r-r' phase-space. However, in the projected phase-spaces x-x', and y-y', the halo does not appear as a ring, but forms the diffuse structures, observed in Fig. 2. This is explained from the fact that each point in r-r' phase space projects into a straight line in x-x' or y-y' phase space, as is shown in Fig. 3.

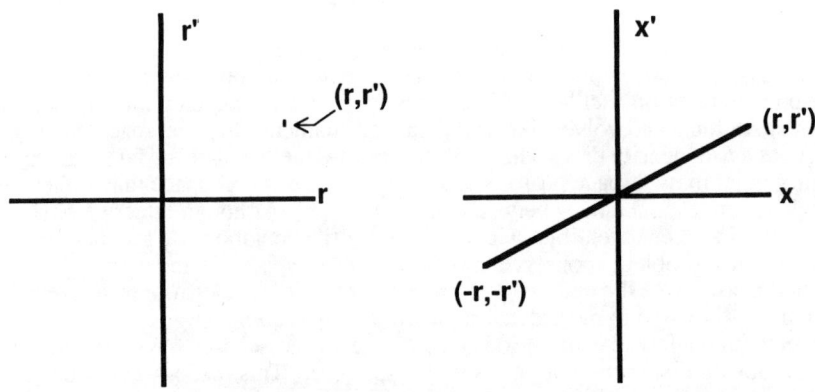

Fig. 3. Projection of a point in r-r' phase space into x-x' phase space as a straight line.

To understand this we consider a particle with polar coordinates (r,q), which moves only radially with divergence r'= dr/dz. In Cartesian coordinates we have x= rcosq, y=rsinq, x'=r'cosq, and y'=r'sinq, and we obtain x'/x = y'/y = r'/r. Therefore, x' = x r'/r and y' = y r'/r, and for fixed r and r' the relationship between x' and x is expressed by a straight line that passes through the origin. The values of x and x' for fixed r and r' depend on the angle q. The extreme values are $x = \pm r$, and $x' = \pm r'$. This results in points or straight lines in the r-r' plane transforming into straight lines in the x-x' plane, while curved lines in the r-r' plane transform into a fanlike butterfly or bow-tie pattern in the x-x' plane. In general, an arbitrary filament in the r-r' plane appears as a more diffuse distribution of points in the x-x' plane. A more complete treatment, including beams with nonzero angular momentum is given in Ref. 7.

BEAM ENTROPY

We are interested in identifying some general principles, which allow us to understand why a charged-particle beam distribution evolves as it does, and towards which steady-state distribution the beam is evolving. We are naturally led to reexamine the concept of beam entropy, which was discussed by Lawson, Lapostolle, and Gluckstern[8] in 1973. In this paper the authors concluded that the beam entropy is a measure of disorder, which is related to emittance, the more conventional measure of disorder in beams. The authors concluded that beam entropy depends on 1) rms emittance, 2) the distribution function, and 3) the phase-space cell size chosen to define the entropy. The dependence of entropy on the distribution function implies the entropy depends not only on the second moments of the distribution, but also on the higher moments. The phase-space cell size of interest may be determined by the experimental resolution. We will see later that the entropy concept may indeed be a useful one if one distinguishes between the microscopic characterization of the distribution and a macroscopic or coarse-grained one. Of course, this distinction has also been important for emittance.

To proceed further we review the entropy concept. First, given an arbitrary phase-space distribution, we divide phase-space into cells of equal volume. One distinguishes between microscopic and macroscopic states. The microscopic state is defined by specifying the cell where each particle is located. The macroscopic state is defined by specifying only the total population of each cell, regardless of which particles are present. The measured properties of the beam, such as the particle distribution and the rms emittance clearly depend only on the macroscopic state. Following Boltzmann's approach, we define the disorder of the macroscopic state as the number of ways of permuting the individual particles between the cells, while maintaining the same macroscopic state. Thus the disorder is the number of distinct microscopic states that result in the same macroscopic state. That this represents a measure of disorder follows from the assumption that more ordered macroscopic states will have fewer microscopic ways in which they can be constructed. Given a system of N particles to be distributed among m cells, and with n_i particles in the ith cell, the disorder W is given by

$$W = \frac{N!}{\prod_{i=1}^{m} n_i!}.$$

The entropy is defined using Boltzmann's constant k, as

$$S = k\log W = k\left[\log N! - \sum_i \log n_i!\right].$$

For example, the most ordered macroscopic state has all particles in one cell, and the above definitions result in S=0. As the particles become distributed in more cells, the disorder and entropy both increase.

It is interesting to consider the limiting case of very large numbers of particles and cells with infinitesimal volume. For very large numbers of particles per cell, it follows from Stirling's formula that

$\log n! \cong n\log n - n$. Then we obtain $\log W \cong N\log N - \sum_i n_i \log n_i$. If the phase-space density is f, and the cell size dV becomes infinitesimal, the number in the ith cell may be expressed as $n_i = fdV$. Changing the summation to an integral, we write

$\log W \cong N\log N - N\log dV + \int dVf \log f$. For fixed N and dV, the first two terms are constant, and if only changes in S with respect to time are of interest, we need only be concerned with the third term. The rate of change of S with respect to time is

$$\frac{dS}{dt} = k\frac{d}{dt}\log W = k\left[\int dV \frac{df}{dt}(1+\log f)\right].$$

For a typical particle-accelerator beam, which satisfies the Vlasov-Poisson equation, df/dt = 0, which implies that dS/dt = 0. Therefore, on a microscopic or fine-grained scale, the entropy is constant. This is not surprising, because from Liouville's theorem the microscopic phase-space density along the trajectory of any particle is constant. However, for entropy defined using phase-space cells with a finite volume, the entropy is not necessarily constant. This situation is not unlike that of emittance, which may change on a coarse-grained scale, even though Liouville's theorem guarantees a constant phase-space volume on a microscopic scale.

Intuitively, a laminar beam is a highly ordered state with zero emittance. Although, particle collisions would produce disorder on a microscopic scale, the space-charge forces, which satisfy the Vlasov-Poisson equation cannot increase the microscopic entropy, as we have seen. However, one may ask whether the nonlinear space-charge forces increase disorder and entropy on a coarse-grained scale. To answer this question, we have carried out a numerical simulation study for an initial Gaussian, rms matched, space-charge-dominated or laminar 2-D continuous beam, in a uniform linear focusing channel with radial symmetry. We divided the phase-space distribution into 100 by 200 equal-volume cells. The initial beam had zero divergence, a Gaussian distribution in real space, truncated at 3s, and populated 74 phase-space cells. Using 2000 particles, at each time step the entropy S = k logW was computed. Fig. 4 shows the computed entropy plotted versus the distance along the channel, expressed in plasma wavelengths. We see oscillations at the plasma period that damp out by about 15 periods. The time-averaged entropy increases most rapidly during the first 15 plasma periods, and continues to increase afterwards very slowly. This example shows that the coarse-grained entropy, averaged over time, does increase. For comparison, the rms emittance (not shown) increases during the first quarter plasma period[3], after which it remains essentially constant. When the same simulation is done for a mismatched, laminar beam, we observe oscillations in the coarse-grained entropy that are associated with the oscillating beam radius. For a uniform-density, laminar beam the time averaged, coarse-grained entropy remains constant.

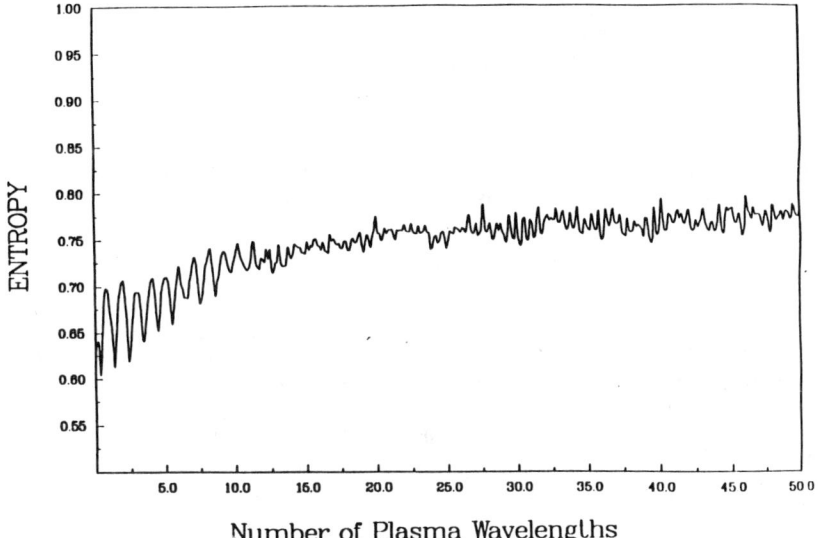

Fig. 4. Beam entropy versus distance measured in plasma wavelengths for an initially rms-matched Gaussian density laminar beam.

A law of coarse grained entropy increase might be used to describe two previously unexplained characteristics of space-charge dominated beams. First, it is observed in numerical simulations that such beams do not appear to evolve toward highly ordered equilibrium states such as the well known K-V distribution, but rather towards more disordered states with a core velocity distribution similar to Maxwellian. Recently for example, the maximum-entropy hypothesis was used to describe the characteristics of the final distribution for a high-intensity expanding beam in free space[9]. Secondly, a tendency for beams with more than one degree of freedom to equipartition has been discussed, especially in high-intensity linacs[10]. Thus, it may be of interest to explore the idea of coarse-grained entropy increase as a means for a conceptual understanding of these important observations.[11]

CONCLUSIONS

Even after more than 20 years we find that there are many questions about space-charge effects that have not been resolved. An important general question concerns the nature of the state of the beam after a few tens to a few hundred plasma periods, a time scale of practical interest for many linear accelerators and transport systems. Is the beam or at least the core of the beam in some approximate equilibrium state? Why are certain equilibrium distributions more representative of real beams than others? Is a coarse-grained Maxwell-Boltzmann distribution a good description? If equipartitioning is a characteristic of the beam, why is it so? One interesting and plausible hypothesis is that enough nonlinearity is provided by the space-charge forces for the beam to approach a state of maximum (coarse-grained) entropy. This may explain why beams in numerical simulations do not evolve toward highly ordered equilibrium states like the K-V distribution, and why we observe a tendency for beams to equipartition their kinetic energies. It is clear that a lot of work still remains before we have a complete understanding of this interesting and important area of charged-particle-beam physics. A more general discussion of space-charge dominated beams, including a more complete set of references on the subject can be found in Ref. 10.

ACKNOWLEDGMENTS

I acknowledge the help of A. Cucchetti on the numerical calculations. I also wish to acknowledge discussions with John Lawson, J. O'Connell, and W. Lysenko on the subject of beam entropy.

REFERENCES

1. T. P. Wangler, G. P. Lawrence, T. S. Bhatia, J. H. Billen, K. C. D. Chan, R. W. Garnett, F. W. Guy, D. Liska, S. Nath, G. H. Neuschaefer, and M. Shubaly, "Linear Accelerator for Production of Tritium: Physics Design Challenges," in "Proceedings of the 1990 Linear Accelerator Conference," Los Alamos National Laboratory report LA-12004-C, September 1990, pp. 548-552.
2. K. R. Crandall, R. S. Mills, and T. P. Wangler, "Simulation of Continuous Beams Having Azimuthal Symmetry to Check the Relation Between Emittance Growth and Nonlinear Energy," Los Alamos National Laboratory Group AT-1 memorandum AT-1:85-216, June 12, 1985.
3. T. P. Wangler, K. R. Crandall, R. S. Mills, and M. Reiser, "Relationship Between Field Energy and RMS Emittance in Intense Particle Beams," in IEEE Trans. Nucl. Sci. **32**, 2196 (1985).
4. A. Cucchetti, M. Reiser, and T. P. Wangler, "Simulation Studies of Emittance Growth in RMS Mismatched Beams," Proceedings of 1991 Particle Accelerator Conference," San Francisco, California, May 6-9, 1991, IEEE Conf. Record 91CH3038-7, p 251.
5. M. Reiser, "Emittance Growth in Mismatched Charged Particle Beams", Proceedings of 1991 Particle Accelerator Conference," San Francisco, California, May 6-9, 1991, IEEE Conf. Record 91CH3038-7, p 2497.
6. O. A. Anderson and L. Soroka, "Emittance Growth in Intense Mismatched Beams", 1987 Particle Accelerator Conference, Washington, D. C., March 16-19, 1987, IEEE Catalog No. 87CH2387-9, p 1043.
7. C. F. Chan, W. S. Cooper, J. W. Kwan, and W. F. Steele, "Dynamics of Skew Beams and the Projectional Emittance,", Nuclear Instruments and Methods in Physics Research **A306,**112(1991).
8. J. D. Lawson, P. M. Lapostolle, and R. L. Gluckstern, "Emittance, Entropy, and Information," Particle Accelerators **5,**61(1973).
9. J. S. O'Connell, "Limiting Density Distribution for Charged Particle Beams in Free Space," Proceedings of 1991 Particle Accelerator Conference," San Francisco, California, May 6-9, 1991, IEEE Conf. Record 91CH3038-7, p 404.
10. T. P. Wangler, "Emittance Growth from Space-Charge Forces," High-Brightness Beams for Advanced Accelerator Applications, AIP Conf. Proc. **253,**21(1991).
11. Courtland L. Bohn, "Transverse Phase-Space Dynamics of Mismatched Changed-Particle Beams," Phys. Rev. Lett. **70**, 932 (1993).

Application of MPP to Particle Tracking

George Bourianoff, Ben Cole, Long Chang
SSC Laboratory, Dallas, Texas 75237

Abstract

The SSC requires massive simulation to support the design, commissioning and operation of the accelerator complex. To this end, the laboratory has made a significant commitment to MPP for this application. A 64 node IPSC/860 was acquired in January of 1991 and has been used extensively in tracking studies of various accelerators of the SSC injector chain. This talk will detail the accomplishments to date and lessons learned.

The most basic observation one can make about tracking on a parallel computer is that for a thin element kick code in the absence of space charge, the problem has a natural granularity that makes it "embarrassingly parallel". One simply distributes the particles over available nodes and tracks. No internode communication is required except for a small amount of diagnostic information that is generated as the run progresses. Hence, the parallel efficiency approaches 100 percent and the problem is scalable to a large number of processors.

This seemingly trivial observation leads immediately to two important conclusions regarding the hardware configuration used to do the tracking. The number of computational nodes shouldn't exceed the number of particles tracked and the overall performance of the calculation will be dominated by single node performance.

The situation becomes less clear as more internode communication is added. The performance of the MPP system on runs where beam emittance is monitored or beam instrumentation is simulated are progressively influenced by message passing overhead. In general, one must be aware that it is sometimes better to abandon the natural granularity and compromise network performance in the interests of optimizing individual node performance.

The addition of space charge forces to the tracking code requires a PIC calculation to be done concurrently with the thin element tracking. A procedure for dynamically sorting particles on to nodes that optimizes machine performance will be described.

The application of a MPP to serve as the engine of real time simulator will be discussed. Such factors as predictability of network collisions and the interrupt response time of the individual node required to write out data become important.

An interactive visualization system designed to display the results from the space charge calculation will be described. It has great flexibility in choice of viewpoint, reference frame and data density.

Strategy of Parallelization

The basic concept of Parallelization is to identify threads of computation which may be done independently, i.e where the calculations within one thread are independent of the data produced in other threads. If each thread is then assigned to a given node, internode communication is minimized and parallel efficiency is maximized. If one is trying to adapt an existing serial code, the procedure is to examine the existing loops within the code and identify the one loop that is most independent. This loop is then chosen to be the one that is distributed across the available computational nodes. In doing this on a distributed memory machine, one must also consider the memory size necessary to support the independent thread and make sure it can be accommodated within the memory available on each node. This procedure allows the code to be ported with minimum restructuring of the basic computational kernels which are typically deeply nested inside of multiple loops.

If this general procedure is applied to thin element kick codes, it is immediately evident that 3 loops control execution of the overall problem. They are loops over particles, turns and elements. The basic concept of Parallelization is to distribute one of these three loops over the computational nodes to achieve the gain associated with parallel processing. It is immediately obvious that the turn loop cannot be distributed since it represents time and must be done sequentially. Hence, either the particle loop or element loop must be distributed across processors. If the element loop is distributed across nodes, all the particle information must be communicated to an adjacent node each time a particle crosses a kick element resulting in a relatively large amount of I/O compared to the arithmetic operations. If the particle loop is distributed across nodes, the result is much more efficient since one particle can advance forward in time indefinitely with no external information *if the lattice information can fit in the memory of one node*.

When the tracking code is parallelized in this way, the application becomes embarrassingly parallel which parallel efficiencies of 98 percent on 32 nodes. This choice of parallelization strategy however does have consequences as far as the single node performance of the tracking code as discussed in the next section.

The performance obtained at the current time on the tracking module is 525 MFOPS on all 64 nodes. This is equivalent to the performance obtained on 2.3 dedicated, single headed CRAY YMPs.

Single Node Performance Optimization

The initial per-node performance of Hypertrack on the individual i860 processors proved to be quite disappointing Even after careful tuning and attempts at cache management, performance of the computational kernel was never better than ~2 MFLOPS double precision, to be compared with a theoretical peak of ~40 MFLOPS. As a part of the SSC purchase of the Hypercube, Intel had allocated some effort on the part of their staff for work on SSC related software development. It was decided that this effort should be put towards improving the performance of this tracking kernel, as this was the basis for the majority of work done by SSC researchers. It was acknowledged that this would require coding in the i860 assembly language, which would compromise code flexibility and make upgrading the kernel difficult. The key issue lay in the fact that the codes used for benchmarking the iPSC/860 were "array-oriented" in that they involved numerous matrix multiplications and dot products. Because the elements of an array are arranged contiguously in memory and such operations use the elements of an array in order, the compiler was able to schedule the off-chip memory accesses. By comparison, the TEAPOT algorithm for a single particle is scalar (i.e. calculates a relatively large number of individual variables), and requires a larger number of memory accesses per calculation (-or- cache reuse is lower). This, coupled with the long latency required for the machine to load data into its cache, explained the slow performance. It was hoped that by utilizing the machine's dual-instruction mode and interleaving the load and arithmetic instructions, better performance could be achieved.

In parallel with this effort, work was underway to utilize the code in its (then current) condition to track numbers of particles larger than the number of processors available (~500 particles on 32 processors, etc.), for the purposes of dynamic aperture calculations and emittance propagation (Ref. 1.) Since at the time each node processed a single particle at a time through all turns before going on to the next available particle, emittance information could only be obtained via post-processing. Also, because each processor could only have one task assigned to it, one processor was required to manage the others, making it unavailable for useful computing. These issues drove the development of a modified version of the code that would track vectors of particles, and this code was named Vectrack for this reason.

In late 1991 Intel personnel had examined the algorithm and started to work, but had made little progress. The technical personnel involved were not wildly hopeful about the final outcome. At that time, it was realized that the Vectrack algorithm was a much more promising candidate. The vector nature of the code made natural the use of pipelined loads and reuse of lattice information already in cache. The SSC personnel involved in the Vectrack effort promised to get a working version of that code ready by early 1992 and to suggest to management that this be used as a starting point. During the spring of 1992 the Vectrack code was brought on-line and put through code checking. After several iterations of tuning by SSC personnel, it obtained performance of between 4 and 5 MFLOPS, a definite improvement. By late March the code was ready for Intel to examine for potential performance improvement, and by midsummer the performance had risen to 8.4 MFLOPS (best case), where the project was declared completed.

The main lessons learned during this effort were:

Vector routines did not always improve performance. (Sometimes, allowing the compiler to pipeline the individual instructions was superior.)

The process of breaking up large loops to allow more efficient pipelining could easily be carried too far, resulting in poorer performance. Determining where this point of diminishing returns lay was very difficult to do at compile time.

The basic algorithm was difficult to optimize, even in the array form. Predominately, this was due to the small cache sizes (both instruction and data) for the i860 chip.

The compiler did a much better job of pipelining if the size of the loop was determined at compile time. (A major inconvenience.)

The compiler was much poorer at handling structure references relative to the handling of the component elements than had been expected.

After the completion of the performance upgrade, an effort was made to relax the restraints imposed by the need to prespecify the size of loops. It is very cumbersome to determine the number of particles per node at compile time; and in the case of dynamic aperture calculations, impossible, as this number changes during the course of the run. Further, allowing the code to operate on uninitialized data ("Nan's") sends the code into the exception handler, which results in exceedingly poor performance as all the processing is being done by an error handler. As a result, the "working version" of the code was modified to handle the array of particles in strides of 4, 8, 16, 32, or 64. The code processes a number of array elements equal to an integer number of strides, the "extras" being filled in with 0. The code itself chooses the stride, minimizing the number of filler entries required. Fig.1 shows the performance of the code after this modification has been made.

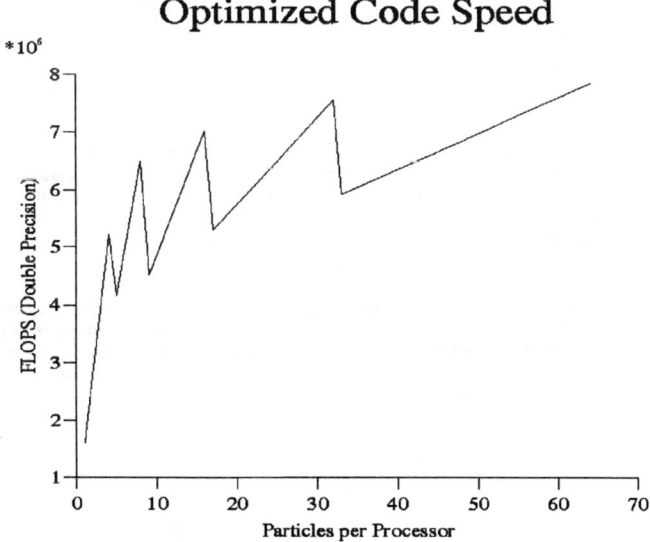

Figure 1 Effect of Matching Nodes to Internal Stride

Space Charge Forces

The inclusion of space charge radically alters the structure of the computational task because much more communication between the nodes is necessary to calculate the electromagnetic fields which act self consistently on each particle. The problem is formulated as a standard Particle In Cell (PIC) model where the fields are calculated on a grid that moves with the centroid of the particle bunch. The procedure is to calculate a charge and current density on the grid by allocating the contribution of each particle to the eight nearest neighboring grid points. The electromagnetic fields are calculated by solving a finite difference approximation of Maxwell's Equations. The value of the fields *at the particle position* is then obtained by interpolating the field quantity from the eight nearest grid points.

The added communication load results from the fact that the grid on which the fields are calculated is decomposed over available nodes. Thus, the memory of each node contains the field information for a certain set of grid points and the particle information for a certain set of particles. When the problem is initialized, the collocated particles and grid points are assigned to the same processor unit. As the problem runs, the particle position gets randomized with respect to the grid points and more internode communication is required.

A detailed description of the PIC module including all equations is contained in Ref.2 and will not be repeated here. For purposes of the following discussion, it is sufficient to know that the code has two major subroutines. Base3D does the element by element tracking calculation and basic particle push. Kick3D does the PIC calculation.

To take advantage of scalable parallel computers, it is necessary to understand the characteristics of the problem so that the program can be implemented correctly and efficiently (with performance scalable to the number of computing nodes available). The characteristics of particle tracking with space charge effects are summarized below.

1. Particles are tracked element-by-element in 6-D phase space. They can be tracked independently in Base3D.

2. Particles are lost when they collide with the wall of the machine.

3. The 3-D PIC method requires a large number of field quantities defined on a 3-D mesh. The memory of a single node is too small to accommodate this, so domain decomposition is required.

4. Communication between a particle and its eight nearest grid points is required to allocate charge and to interpolate the fields.

5. A large data set is required for accurate modeling of the physics, and for the visualization of tracking results. This requires a restart capability, primarily to permit recovery from hardware failure, since the runs may last several days..

For reasons that will be explained in the next section, the main task of parallel implementation is to ensure that each computing node has the same number of particles in order to achieve loadbalancing. This is very important, since tracking the non-interacting particles in the magnetic lattice occupies most of the computational time. The second primary task is to decompose the 3-D grids in Kick3D. The constraints are 1) the space charge code should fit into an 8-Mbyte node; 2) each node should have the same amount of work load in computing the loops over particles for (allocating charge to grids and retrieving field information from grids) as well as the loops over 3-D grids (for computing electromagnetic force); and 3) communication among nodes should be minimized. Below we consider partition schemes for particles in the tracking phase and 3-D grids in space charge.

So that each computing node has the same work load, particles are assigned equally into computing nodes by the block or cyclic method.* Since particles are frequently lost during tracing when they run into a wall, a load imbalance situation will develop. That is, some nodes might have many more particles to track than the others. The cyclic method is usually a better approach to deal with a load-imbalance situation. However, such an approach is not adequate when space charge is introduced. Figure 2(b) demonstrates a case using a cyclic approach, which will produce busy communications between all nodes since it violates the data locality principle. In practice, we found that a block decomposition is a proper way to deal with our problem as long as particles are not lost dramatically.

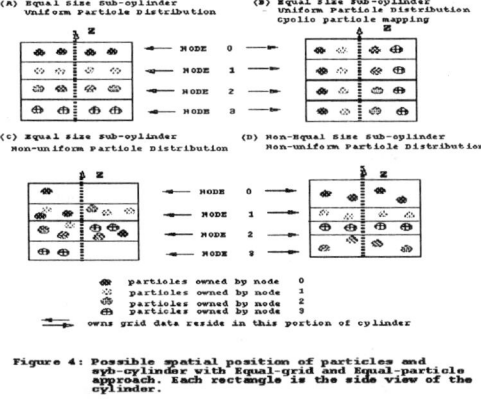

Figure 4: Possible spatial position of particles and sub-cylinder with Equal-grid and Equal-particle approach. Each rectangle is the side view of the cylinder.

Figure 2 Load Decomposition Schemes

There are several ways to map the 3-D grids into computing nodes. It depends on if the Hypercube is connected as a ring, a 2-D mesh, or a 3-D Hypercube. For programming simplicity, we will use block mapping from 3-D grids into a 1-D array of computing nodes. The cylinder is partitioned in the longitudinal direction. The partition method can be based on the equal number of grids (Equal-Grid) strategy or equal number of particles (Equal-Particle) in a sub-cylinder strategy.

In the Equal-Grid approach, each node gets a nearly equal size of sub-cylinder (or the same number of mesh points). If particle distribution is uniform in the longitudinal direction, then each sub-cylinder will contain the same amount of particles. Therefore, there is little or no communication between nodes. Figure 2(a) shows the best-case situation, in which all the particles and their grid neighbors belong to the same node; therefore, little or no communication is needed. In practice, the distribution of particles tends to be nonuniform during simulation. Figure 4(c) shows a nonuniform particle distribution case in which not only is communication necessary, but some nodes have to update grid information for the other nodes as well. As a result, loadbalancing among nodes is uneven.

In the Equal-Particle approach, the cylinder is partitioned into sub-cylinders, each of which contains a nearly equal number of particles. When the particle distribution is nonuniform, each node will have a unequal sub-cylinder (see Figure 2(d). This strategy gains a performance advantage by keeping communication minimal at the expense of uneven grids in each node's domain. To achieve high parallel efficiency, an effective mechanism is necessary to maintain the "Equal-Particle" structures and to minimize the load-imbalance effect of uneven grids.

Both approaches have their advantages and drawbacks. (See Table 1 for a comparison.) In general, there is a trade-off between speed and memory. Since a memory upgrade for iPSC/80 is relatively expensive, it would be very desirable to combine both approaches to compromise parallel efficiency and memory to achieve high parallel efficiency within the limit of node memory. Since we want to keep the code size as small as possible to fit into 8 Mbytes of available memory, we have chosen the simplest strategy to implement particle tracking with space charge code. That is particles are partitioned using the block method, and 3-D grids are decomposed using block mapping with the "Equal-Grid" approach in the z direction. This approach can be implemented more quickly than alternative methods. We made no assumption about spatial relationships between particles and their surrounding grids. Particles can move anywhere (e.g., across several domains (sub-grids) between calculations. This approach is very general and could be implemented with moderate effort should a parallel compiler, which can effectively solve irregular communication within a parallel loop, become available in the future.

Table 1:

	Equal-Grid	Equal-Particle
Load Balance (Loops over Particles)	No (Non-Uniform)	Yes
Load-Balance (Loops over Grids)	Yes	No (Non-Uniform)
Communication Overhead	Large (Non-Uniform)	Small
Memory (Grid) Size Scalability	Yes	No (Non-Uniform)
Programming Effort	Easy	Difficult

In the following, we discuss briefly the techniques used to solve irregular communications in space charge. We consider the case where a particle needs to allocate charge to its eight nearest neighbor grid points (referred to as allocating process). The inverse process of interpolating field information to the particle location (referred to as interpolating process) can be treated similarly. The particles and 3-D grids are partitioned based on the strategies described above. A particle and its nearest neighbors might not belong to the same node. That is, z_i (ipart) and density (ipart) do not necessarily belong to the same node.

The strategy that we use is similar to the approach proposed by Saltz in Ref. 3. The idea is based on block I/O transfer to minimize the communication between nodes. That is, all information that a node needs to communicate with other nodes is accumulated into a buffer. A global communication table that describes how a pair of nodes should communicate with one another is computed first. Each node then sends out self-descriptive information to the other nodes. The information received by a node includes the position and fractional density for each grid that should be updated by this node. An advantage of this approach is that the global communication table needs to be computed only once in Kick3D () at each time step. Therefore, it results in a reduction of communication time that would be very difficult to achieve even using future automatic parallel compilers, since such an automatic parallelizer will not be able to plan ahead and collect operations as a human programmer can. This strategy provides a reliable and effective communication mechanism for FORTRAN implementation.

A particle can move from one grid to another grid between space charge calculations and is in fact unlikely to keep the perfect spatial position seen in Figures 2(a) and 2(d) all the time. It is probable that a situation like in Figure 4(c) will happen during a long run. One way to keep the particles and their associated grid points in the same memory is to sort the particles in the z direction and to re-map into computing nodes. The best sorting algorithm requires order of $(n \log n)/p$ operations ($O(n)/p$ if bucket sort is used), where p is the number of nodes. When n is large, the overhead will be exceptionally high. Since particles change their relative position and surrounding grids in the longitudinal direction gradually, it is necessary to sort these particles only occasionally (about every 50 turns).

Another approach is to have sub-cylinder guards for each node. Here, sorting is reduced to sub-cylinder guard communication between two neighbor nodes. Here, sorting is reduced to sub-cylinder guard communication between two neighbor nodes. This approach usually assumes that particles can move only from a subcylinder to the next neighbor sub-cylinder between space charge calculations. Such a constraint is imposed by numerical stability considerations for any explicit time advance algorithm. Therefore communication is performed only between neighboring nodes. A combination of the above tune-up strategies with our current scheme makes it possible to provide improved performance with only a little extra memory expense.

Interactive Simulation

Because of its raw processing power, the parallel processor is being used as the engine in an interactive simulator of the Low Energy Booster (LEB). This simulator is being developed for two purposes. The first is to form a platform on which high level correction code can be developed and from which an operator can control the simulator with the same look and feel he or she will experience in the control room. The second purpose is to test the data handling characteristics of the EPICS control system.

The architecture of the simulator is shown in figure 3. The computational model running on the Hypercube is shown in the lower right hand corner adjacent to a box showing LEB instrumentation. The simulation results produced on the Hypercube are transferred over an Ethernet connection to a rack mounted SPARC card which formats the data into the same form produced by the A to D converter which actually receives the data from the LEB instrumentation. A software driver for the SPARC card then reads or writes the data into the EPICS control system where it is handled in the same manner that data produced by the actual accelerator instrumentation.

There are several aspects of driving a "real time" simulation with a parallel processor that are worthy of attention. The most obvious of these is simply the rate at which data is produced in the simulator relative to the rate at which data is produced by the real experiment. In the present case, using the Hypercube as the simulator engine, data is produced 3000 times faster in the experiment than in the simulator. However, this is completely adequate for the first of the two purposes described above since it is possible to decompose the overall control problem into a slow time scale which requires human intervention to close the loop and a fast time scale which requires automated procedures to close the loop. The simulator produces results which require human interaction while the fast loop closure is inherent in the computational model.

The second purpose for simulator development mentioned above requires that data be generated in the Hypercube, stored in a disk connected to the SPARC board a read back in a burst mode. This will test the data handling conflicts, latencies and throughput which is vital information for control loop design. This aspect will not be further discussed here.

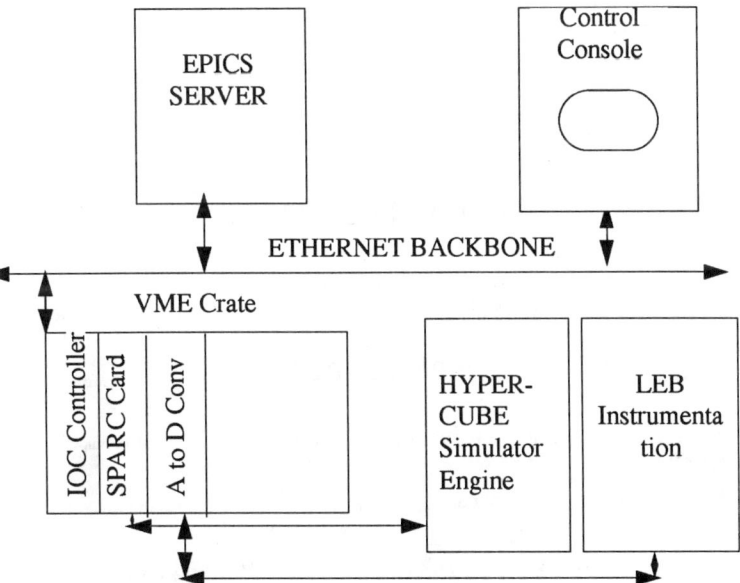

Figure 3 Schematic Diagram of LEB Simulator

Another aspect of real time operation is the ability to communicate to the Hypercube in an asynchronous manner. Since UNIX does not in general support interrupt driven operation, the async communication is handled in a synchronous way by periodically checking for a semaphore that communication needs to take place. This is true at both the Hypercube end and the console workstation end since both of these systems run UNIX. The EPICS software that connects the Hypercube to the control console handles interrupt driven data transfers and data polling at predetermined intervals.

The other issue relevant to "real time" operation is that the nodes must be synchronized after each turn. The computational model has been "instrumented" so that the output from each of the 180 BPMs in the lattice are simulated. It is therefore necessary that the bunch of particles circulate together so that the BPM reading of the bunch can be sent back together. This means that the simulation cannot proceed faster than its slowest node and makes node balancing very important.

The issues that relate to calculating and transferring instrumental data have already been addressed with the present code. The initial implementation of the BPM readout code had logic that would track a given particle on its assigned node from one BPM to the next, stopping at each BPM and sending back to a master node the location of that particle at that time step at that particular BPM. This slowed the processing down greatly (a factor of 4) due to basic I/O overheads and the interruption of the processor stack as described in second section. The logic which is currently implemented accumulates all the BPM output at a given node for a full turn and then sends one large block to the master node. This improves the performance significantly.

Conclusions

1 Distributed memory parallel processors are very well suited to particle tracking. The Intel Hypercube is being successfully used for a significant fraction of the simulations at the SSC.

2. Never believe a MFLOP rating until you try your own code on the machine.

3. Cache size and access latency (number of machine cycles for a load from memory) can be a real bottleneck.

4. A good compiler and a faster chip are of comparable worth. (I.e. The compiler changes made a difference of 4x, comparable with the increases in processing power from one generation of chips to the next.)

5. Language does matter. When first beginning to design the code, the choice of language naturally arose.and C was chosen as being nicer and more readable. However, the C compiler on this machine proved to be unable to optimize strucurebased algorithms (perhaps due to their greater generality) and Intel in general, continues to pay more attention to its Fortran compilers than it does to C.

Acknowledgments

The authors would like to acknowledge the support of the Defense Advanced Research Projects Agency. Some of the support was a joint DARPA/DOE program in the area of high performance computing.

References

1. B. Cole, G. Bourianoff, F. Pilat, and R Talman, Particle Orbit Tracking on a Parallel Computer, *Proceedings of the 1991 IEEE Particle Accelerator Conference*, San Francisco, Ca.

2. L. C. Chang, G. Bourianoff, B. Cole, S. Machida, A Parallel Implementation of Particle Tracking with Space Charge Effects on an Intel IPSC/860, SSCL Note 581, Aug. 1992.

3. J. Saltz st. al., "Run Time Scheduling and Execution of Loops on Message Passing Machines," Journal of Parallel and Distributed Computing, Vol. 8, No. 4, April 1990, Academic Press, pp. 303-312.

ELECTROMAGNETIC MODELLING OF THE PEP-II RF CAVITY USING THE 3D FINITE ELEMENT CODE MSC/EMAS

R. A. Rimmer*
Lawrence Berkeley Laboratory, Berkeley CA94720

J. J. Burke[†]
MacNeal Schwendler Corporation, Electromagnetic Applications Dept.,
4300 W. Brown Deer Road, Suite 300, Milwaukee WI53223-2465

ABSTRACT

The paper describes the modelling of the RF cavity proposed for the PEP-II *B* Factory using the MSC/EMAS 3D finite element electromagnetic analysis software. The mode spectrum of the cavity below the beam-pipe cutoff frequency is explored and the coupling of parasitic higher order modes (HOMs) to external damping loads is estimated. Results are compared to experimental measurements on a test model and to calculations using finite difference codes.

INTRODUCTION

The PEP-II *B* factory is a proposed asymmetric (unequal beam energies), electron-positron collider [1], planned for the PEP tunnel at the Stanford Linear Accelerator Center, SLAC. The design is a collaboration between SLAC, Lawrence Berkeley Lab., Lawrence Livermore National Lab., and several other contributors. To study the *B* meson decay physics requires very high luminosity (collision rate) and reliability for the accelerator. To achieve the desired luminosity requires very high beam currents (1.55 A for the high-energy ring and 2.25 A for the low-energy ring), while high reliability necessitates thorough engineering of all the accelerator components and systems. The high beam currents provide many challenges in the design of the RF systems for the two rings. The number of RF cavities must be kept as low as possible to minimize the impedances of the rings and keep the beam instability growth rates to a manageable level. This means the 476 MHz RF cavities and windows must run reliably at very high power levels (up to 150 kW cavity dissipation, 1 MV gap voltage and 500 kW window transmission) [2], requiring detailed electromagnetic, thermal and stress analyses of these components.

* This work was supported by the Director, Office of Energy Research, Office of High Energy and Nuclear Physics, High Energy Physics Division of the U.S. Department of Energy, under contract number DE-AC03-76SF00098 and [†] MacNeal-Schwendler Corporation.

2D ANALYSIS OF THE RF CAVITY

The cavities are basically axisymmetric structures so the preliminary design was done in 2D using the URMEL code [3]. At this stage the transit–time corrected shunt impedance of the fundamental mode was maximized and the frequencies and impedances of all of the higher–order modes (HOMs) below the beam–pipe cut off frequency were found, see table I and II. These are compared with results from the axisymmetric model in MSC/EMAS [4]. The calculations showed that the worst longitudinal (monopole) HOM was the TM_{011}–like mode at 770 MHz and that the worst deflecting (dipole) mode was the TE_{121}–like mode at 1065 MHz. These modes were targeted using a broad–band HOM damping scheme employing waveguides on the cavity wall positioned so as to couple to the strongest HOMs while not missing any of the others. Figure 1 shows the field of the TM_{011} mode inside the cavity and all the positions where the HOMs have zero magnetic field. These positions would be unsuitable for the location of the damping waveguides, which must couple to the magnetic fields of all the modes. Figure 2 shows the position at 30° up the curved section of the wall chosen for the waveguides. This avoids any HOM zeros and still couples strongly to the TM_{011} mode. The cut–off frequency of the damping waveguides is chosen to be far enough above the fundamental mode so as to strongly attenuate its fields. Three damping waveguides are used, equally spaced around the azimuth, to capture all orientations of the dipole modes while preserving symmetry in the fundamental mode. The damping structures break the axial symmetry of the cavity and fully 3D analysis is required to model this geometry.

Table I monopole HOMs in unloaded cavity

	URMEL			MSC/EMAS 2D	
	FREQ (MHz)	URMEL Q's	Rs (MΩ)	FREQ (MHz)	Q
TM010	489.57	46306	5.036	491.70	50643
TM011	769.78	39625	1.782	776.11	43111
TM020	1015.38	41383	0.000	1019.75	27976
	1291.02	90188	0.692	1295.27	55870
TM021	1295.61	40326	0.265	1303.78	26442
	1585.46	42724	0.216		
	1711.62	85135	0.404		
	1821.89	107874	0.006		
	1890.98	44492	0.075		
	2103.39	66780	0.235		
	2161.89	84386	0.002		
	2252.16	55944	0.068		

Table II dipole HOMs in unloaded cavity

	URMEL		
	FREQ (MHz)	Q	R/k(r)² (MΩ/m)
TE111	679.57	47520	0.001
TM110	795.46	61076	15.531
TE121	1064.81	50048	30.794
	1133.16	49771	0.287
	1208.21	87745	0.573
	1313.21	50189	8.090
	1429.01	38150	3.280
	1541.02	102408	2.809
	1586.23	76118	5.171
	1674.16	36130	6.512
	1704.41	52856	0.181
	1761.93	92516	0.355

3D ANALYSIS BY FINITE–DIFFERENCE CODES

HOM power coupled out into the damping ports is absorbed in broad–band terminations at the end of the waveguides. Ideally these loads would be included in the numerical simulation of the cavity so the loaded Q's of the HOMs could be calculated directly. This facility was not available in the MAFIA code [5] used to perform the initial 3D cavity analysis or the ARGUS code [6] that was subsequently used to continue the study. In these codes a perturbation method must be employed to find the coupling to the waveguides. In this case the Kroll–YU method [7] was employed, which is a development of the Slater perturbation method. In this technique Real eigenvalue solutions are found using MAFIA or ARGUS with several different lengths of waveguide terminated in simple (short–circuit or open–circuit) boundary conditions. Using the frequency variation with waveguide length, the external Q of the model can be derived and hence the loaded Q. Because of the limited number of modes that could be calculated using MAFIA and ARGUS only the first few HOMs could be modelled in this way. The results are shown in tables III and IV. The HOM Q's calculated in this way looked to fulfill the *B* factory requirements so a low–power test cavity (LPTC) was built to confirm the predictions by measuring the damping on these and the other troublesome HOMs.

The low-power test cavity, see Figure 3, was made by electro-forming copper onto an aluminum mandril. Two circular ports were included as alternative locations for the 476 MHz drive loop coupler. When not in use these were blanked off with plugs shaped to maintain the curved inside surface of the cavity. Small probes were inserted through the beam pipes to excite the cavity and the response was measured using a Hewlett Packard 8510C network analyzer. Results from the cold tests showed very strong damping of the worst HOMs, more than three orders of magnitude in the case of the TM_{011} mode, and that the scheme was generally very successful [8]. The measured Q's with low-power models of the waveguide loads agreed well with the Kroll-Yu predictions, which assume perfect terminations. The slightly higher measured Q's could be due to reflections from the load material in the damping waveguides. The results were sufficiently encouraging that a high-power test cavity (HPTC) is being designed using the same basic layout.

2D and 3D ANALYSES BY MSC/EMAS FINITE-ELEMENT CODE

An analysis of an axisymmetric model without damping waveguides, using MSC/EMAS was performed. The model consisted of approximately 5,174 grids. See Table 1 for frequency and Q results.

An analysis of the three dimensional cavity without damping waveguides, using MSC/EMAS was also performed. The model consisted of approximately 5,000 grids. The dominant mode, the TM_{010} like mode, of the cavity was found. The resonant frequency calculated by MSC/EMAS was 489 MHz. Figure 4 shows a contour plot of the z directed electric field in the cavity, using MSC/EMAS. The axisymmetric and three dimensional models employed both real and complex eigenvalue solutions.

CONCLUSIONS

As contrasted with other methods, use of the complex eigenvalue solution in MSC/EMAS afforded a more direct approach to extracting the Q, rather than using a perturbational technique. A logical next step would be to analyze a 180 degree, three dimensional model including the damping waveguides. A complex eignevalue solution of the more comprehensive model is within the capabilities of the MSC/EMAS program and would, additionally, provide current density results around the irises.

Fig.1. TM$_{011}$ mode B field + HOM zeros.

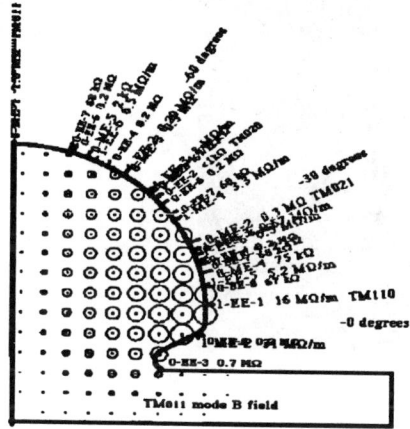

Fig.2. Location of damping waveguide.

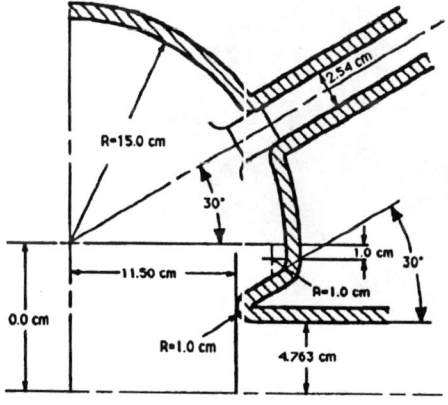

32 Electromagnetic Modelling of the PEP-II RF Cavity

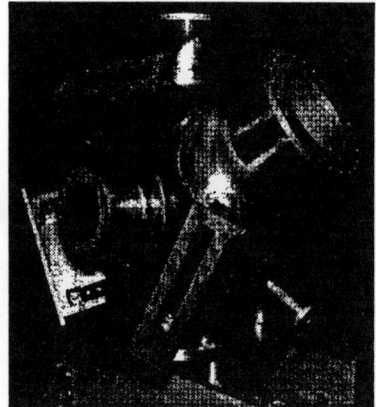

Fig. 3 LPTC showing damping ports to which the HOM waveguides attach. The large circular ports are for the RF drive coupler.

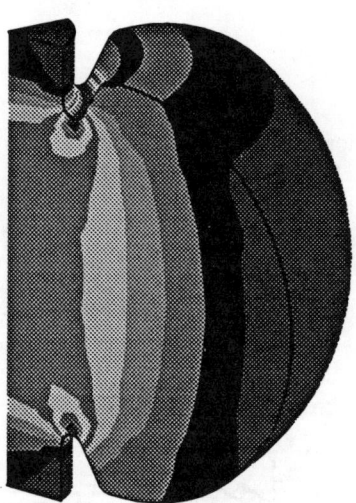

Fig. 4 MSC/EMAS contour plot of the z–directed E field, TM_{010} mode, for the 3D 1/6th model.

Table III monopole HOMs with three 1" x 10" damping waveguides

	ARGUS/Kroll-Yu		MEASUREMENT	
	FL (MHz)	QL	Meas F (MHz)	MEAS Q's
TM010	481	41922	484	*31926
TM011	746	19.5	758	28
TM020	1008	193	1016	246
			invisible after damping	
TM021			1296	907
			1588	178
			invisible after damping	
			1821	295
			invisible after damping	
			2109	233
			2168	201
			2253	500

* fundamental-mode Q for test model not expected to be representative of final cavity

Table IV dipole HOMs with three 1" x 10" damping waveguides

	ARGUS/Kroll-Yu		MEASUREMENT	
	FREQ (MHz)	Q	Meas F (MHz)	MEAS Q's
TE111	671.6	52	invisible after damping	
TM110	791.8	86	invisible after damping	
TE121			invisible after damping	
			1141	112
			1203	1588
			1311	498
			1435	3955
			1554	59
			1588	178
			1674	2134
			1704	444
			1757	7129

REFERENCES

[1] "An Asymmetric B–Factory based on PEP," Conceptual Design Report, LBL PUB–5303, SLAC 372, February 1991.

[2] R. Rimmer et. al., "An RF Cavity for the B–Factory", Proc. PAC, San Francisco, May 6th–9th, 1991, pp819–21.

[3] U. Laustroer, U. van Rienen, T. Weiland, "URMEL and URMEL–T Userguide," DESY M–87–03, Feb 1987.

[4] MSC/EMAS Users' Manual, B.E. MacNeal ed., Los Angeles: The MacNeal–Schwendler Corporation, 1991.

[5] "Reports at the 1986 Stanford Linac Conference., Stanford, USA, June 2–6 1986," DESY M–86–07, June 86

[6] A. Mondelli, et.al. "Application of the ARGUS Code to Accelerator Design Calculations," Proc. 1989 PAC, Chicago IL, March 20th–23rd, 1989.

[7] N. Kroll, D. Yu, "Computer Determination of the External Q and Resonant Frequency of Waveguide Loaded Cavities," SLAC–PUB–5171.

[8] R. Rimmer, "RF Cavity Development for the PEP–II B Factory", Proc. Int. Workshop on B–Factories, BFWS92, KEK, Japan, Nov. 17–20, 1992.

Simulation of the CLIC Transfer Structure by Means of MAFIA.

Antonio Millich
CERN
1211 Geneva 23, Switzerland

ABSTRACT

The function of the CTS is to extract 30 GHz power from the drive beam and to make it available for the acceleration of the main beam. The simulation of a six cells section of the CTS using the MAFIA set of codes has provided the designers of the structure with a set of RF parameters at 30 GHz. The frequency domain analysis has allowed the plotting of the dispersion curves for the first few pass bands, whereas the time domain analysis has provided results on the shape and magnitude of the longitudinal and transverse wake fields and of the loss factors.

1. INTRODUCTION

The CLIC Transfer Structure serves the purpose of extracting 30 GHz power from the drive beam in the two-beam accelerator scheme of CLIC [1]. This note describes the results obtained by simulation using the code MAFIA [2], on a section of 12 cells of the 30 GHz CTS. The frequency domain analysis has provided the designers with useful information on the RF properties of the structure, while the results of the time domain computations of the beam induced wake fields have allowed testing the drive beam stability by means of tracking programs.

2. CTS GEOMETRY AND FUNCTION

The CTS essentially consists of a smooth cylindrical beam chamber of 12 mm diameter, which is coupled by means of diametrically opposite slots to two periodically loaded rectangular (8 x 4 mm) waveguides (fig. 1). The periodicity of the waveguide 'teeth' is such that the phase velocity of the $2\pi/3$ mode at 30 GHz is equal to the speed of light in vacuum.

The ultra relativistic drive beam creates in the waveguides a field that propagates in phase with the exciting bunch, so that constructive transfer of energy to the waveguide mode is possible all along the structure. The drive beam is made up of a train of bunches spaced by one wavelength of the 30 GHz mode, which is 10 mm, so that each bunch contributes to the coherent excitation of the mode, the energy of which increases until the last bunch has left the structure. At that moment the waveguides are filled with energy which propagates at the group velocity of about one third the speed of light and which is transferred to the main linac disc-loaded structure.

The length of the CTS is such that the duration of the energy discharge pulse fills the gap between two successive bunch trains spaced 2.84 nsec or one 352 MHz period. Taking into account the bunch train

36 Simulation of the CLIC Transfer Structure

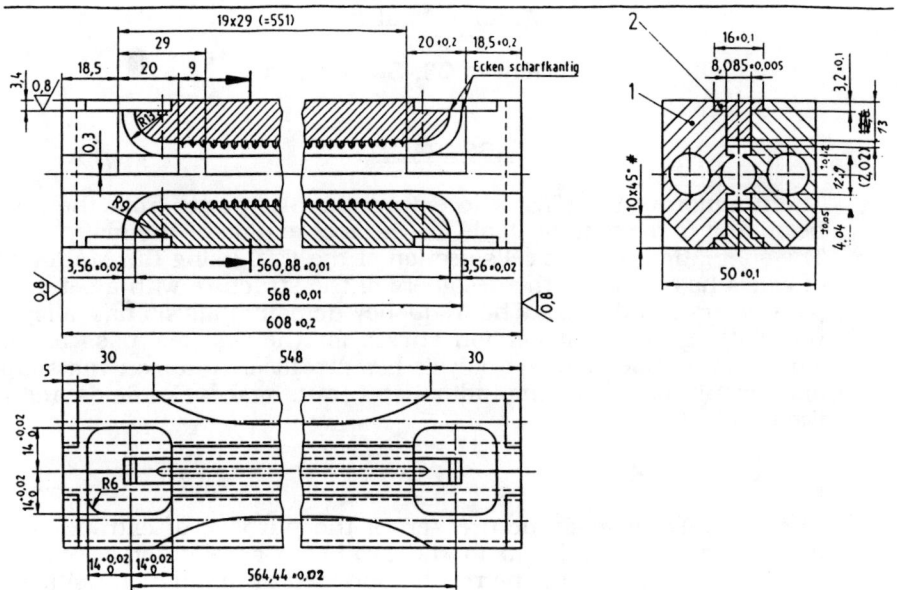

Fig. 1. CLIC Transfer Structure Geometry
Courtesy of L. Thorndahl and G. Carron

transit time and its time span, one CTS is about 0.50 m long and presents 144 rounded teeth in the waveguides. Four bunch trains are therefore necessary to provide a pulse of duration longer than 11.1 nsec, which is the main linac disc-loaded waveguide filling time. The energy stored in the CTS waveguides by one train of bunches must supply a power level of 80 MW during 2.84 nsec. Approximate CTS dimensions and wake fields magnitudes were initially obtained with model work based on the wire method of beam simulation. [3]

3. MESH GENERATION AND FREQUENCY DOMAIN ANALYSIS

By means of the MAFIA module M310 the CTS geometry was simulated as shown in fig. 2. Thanks to symmetry, only one fourth of the structure needs to be retained in the simulation. The memory space available in the workstation being limited, only six cells were used in the frequency domain computations with 64000 mesh points, whereas 12 cells with 128000 mesh points were used in the time domain computation of wake fields. The average resolution in the three dimensions is 0.3 mm. The beam trajectory is the z axis which coincides with the axis of the cylindrical chamber.

By means of modules R310 and E310 the resonant modes of the CTS were computed. The solutions found varying the boundary conditions of the z end planes present a phase shift per cell from 0 to π in steps of $\pi/6$. They are shown in Table 1, while the dispersion characteristic is plotted in fig. 3.

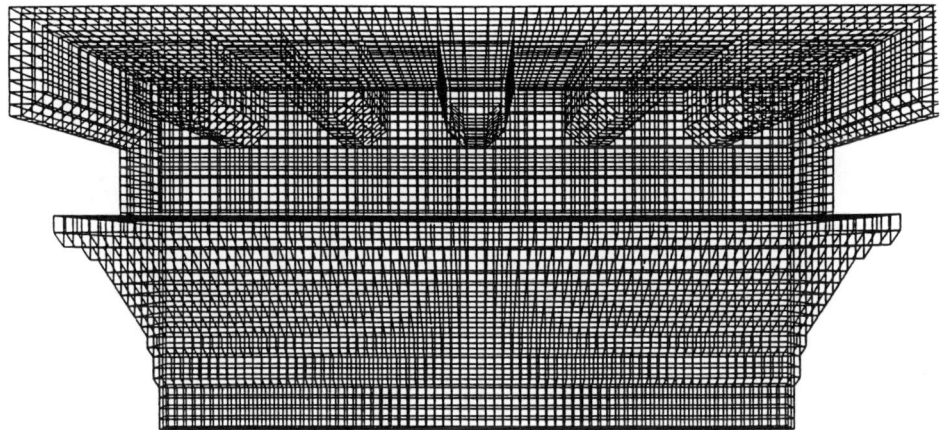

Fig. 2. Mesh of a six cells section of CTS (only one quarter shown).

Table 1. Normal modes found by MAFIA for the first CTS pass band.

Mode number	Frequency (GHz)	Phase shift/Cell
1	19.729	0
2	20.759	$\pi/6$
3	23.430	$\pi/3$
4	26.858	$\pi/2$
5	30.000	$2\pi/3$
6	32.027	$5\pi/6$
7	32.690	π

The dimensions of the CTS were chosen such that the intersection of the line representing a phase velocity equal to the speed of light with the dispersion curve occurs at the frequency of the $2\pi/3$ mode, which is 30.00 GHz. The group velocity plot, computed from the dispersion curve by means of numerical derivation, is also shown in fig. 3. The group velocity of the $2\pi/3$ mode is 32% of the speed of light.

By means of the post processor P310, the Q of the structure was computed together with the shunt impedance per unit length R' and the r'=R'/Q parameter. Table 2 gives the numerical values including those of the longitudinal loss factor per unit length and per structure for the $2\pi/3$ mode.

Table 2. CTS RF parameters

Synchronous mode frequency	=	30.00	GHZ
Q factor	= 3808		
Shunt impedance R' (true ohms)	=	12.6	KΩ/m
r' = R'/Q	=	3.30	Ω/m
Loss factor k'	=	0.156	V/pCm

38 Simulation of the CLIC Transfer Structure

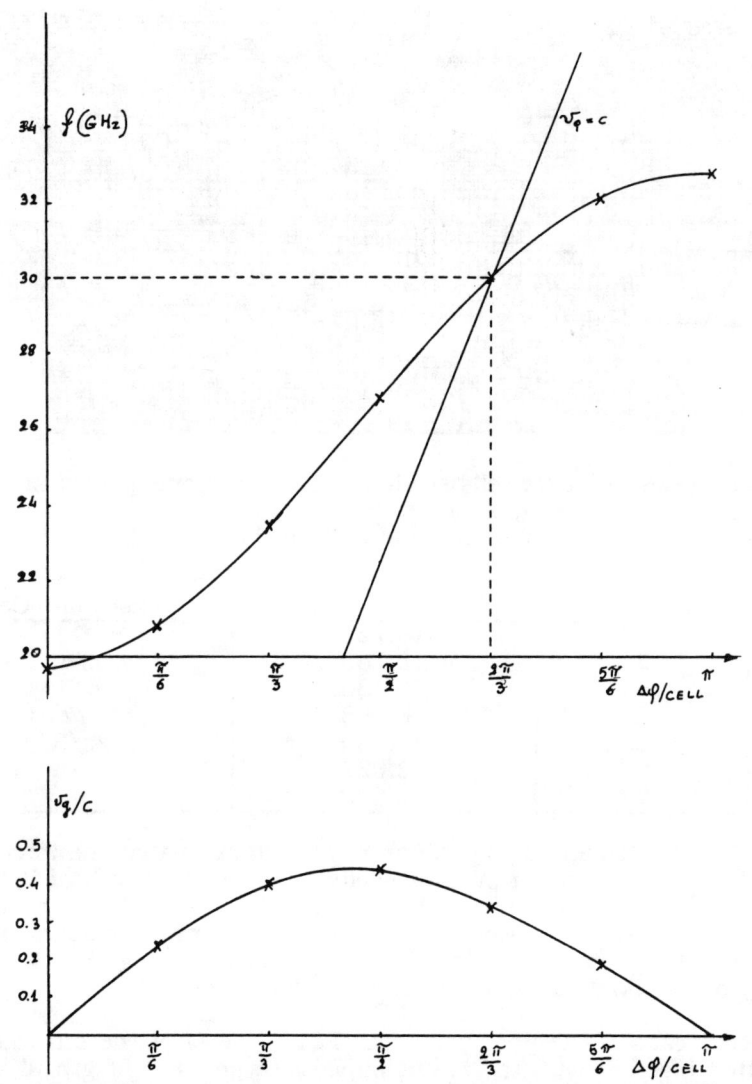

Fig. 3. Dispersion characteristic of the CTS and group velocity plot.

4. WAKE FIELDS COMPUTATION IN TIME DOMAIN

4.1 Longitudinal wake fields

For this analysis the 12 cells geometry was used in module T3310. The boundary conditions were chosen as perfect magnetic conductors for the $x = 0$ and $y = 0$ symmetry planes, while for the z end planes the waveguide condition was imposed. For the computation of the longitudinal wake field, a bunch of $\sigma_z = 1$ mm and charge normalised to 1 pC was placed in the centre of the beam chamber. The longitudinal wake is shown in fig. 4. It is a damped sinusoid with zero crossings distant exactly one 30 GHz wavelength or 10 mm. The centre of the bunch which generates the wake is slightly to the left of the first negative peak, so that the bunch 'sees' the decelerating voltage it induces in the 12 cells structure. Each successive bunch arrives at one subsequent negative peak and will experience the decelerating effect of its own wake plus the sum of those of the preceding bunches. The last bunch in the train will of course experience the largest decelerating field, so that a quasi-linear energy depletion occurs within the bunch train, which corresponds to the quasi-linear build up of energy in the waveguide mode.

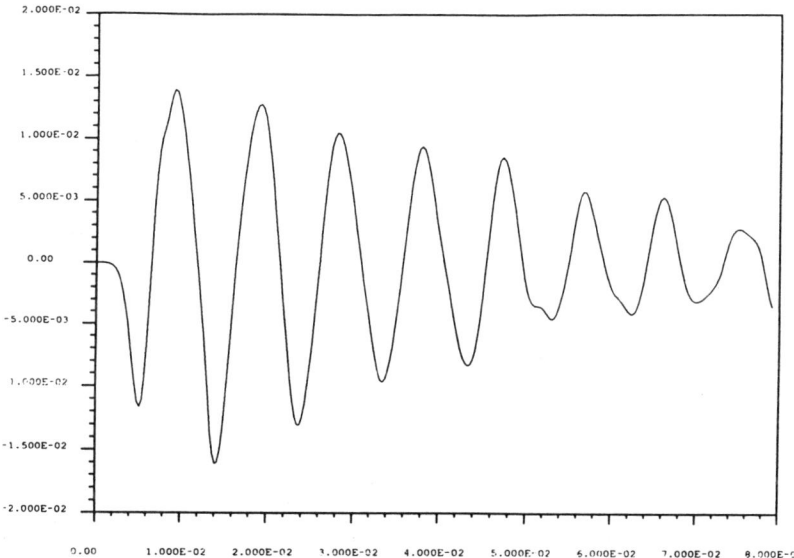

Fig. 4. Longitudinal wake field of a bunch of $\sigma = 1$ mm, charge 1pC, on the axis and traversing 12 CTS cells.

The peak field seen by the second bunch in the train is found to be $1.5 \, 10^{-2}$ V/pC for the 12 cells length, or 0.375 V/pCm. A bunch of 160 nC generates a peak wake field of 60 KV/m at the second bunch position. The last bunch in the train, say the 11th, experiences ten times this peak field plus its own wake, that is about 630 KV/m, so that on average the bunch train sees a decelerating voltage of 330 KV/m with a loss of 0.58 J/m, which is the energy that gets transferred to the waveguide modes.

4.2 Transverse wake fields

The knowledge of the amplitude and shape of these wake fields is of paramount importance for the studies of the transverse drive beam stability by means of tracking programs. [4]

Setting the y = 0 symmetry plane as perfectly conducting, only the deflecting modes are excited by the bunch placed one mm off centre in the y direction. The resulting wake field, shown in fig. 5, is also a damped sinusoid. The bunch exciting the wake experiences its own deflecting action which reaches a maximum at its tail. Since the transverse wake is offset by $\pi/2$ with respect to the longitudinal one, the subsequent bunches arrive at the nodes of the wake field and therefore no cumulative deflecting effect occurs for particles near the bunch centre. The peak transverse wake field is found to be 2.9×10^{-3} V/pC for 1 mm beam displacement and 12 cells length, which corresponds to 32 V/pCm for one CTS structure.

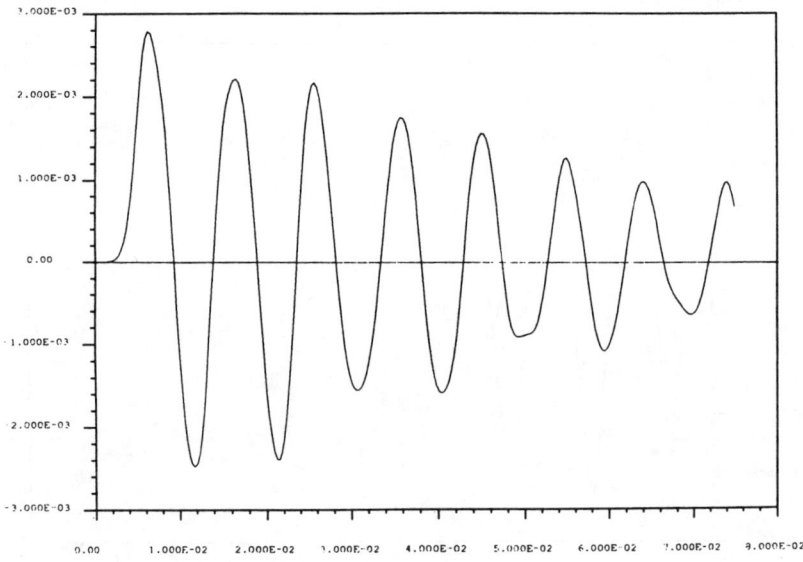

Fig. 5. Transverse wake field of a bunch 1 mm off axis vertically, with σ = 1 mm, charge 1 pC, traversing 12 CTS cells.

5. HIGHER ORDER MODES

The present version of the MAFIA program does not allow to explore the resonant modes of a structure in a user defined frequency interval, but it finds all the modes starting with the lowest one. This feature sets a limitation on the number of higher bands one can explore given a fixed amount of computer memory available. Using the six cells geometry of the rectangular waveguide without the cylindrical beam chamber, it was possible to find some forty modes in the frequency band from 18 GHz to 72 GHz. Fig 6 shows the resulting dispersion diagram. The intersections of the straight line representing a phase velocity equal to the speed of light with the dispersion curves indicates the synchronous modes that may be harmful to the beam. The relative importance of the shunt impedance of these modes with respect to the fundamental one, the synchronous $2\pi/3$ at 30 GHz, has been computed and the results have shown that only the TE30 mode has an appreciable effect, of the order of 5%.

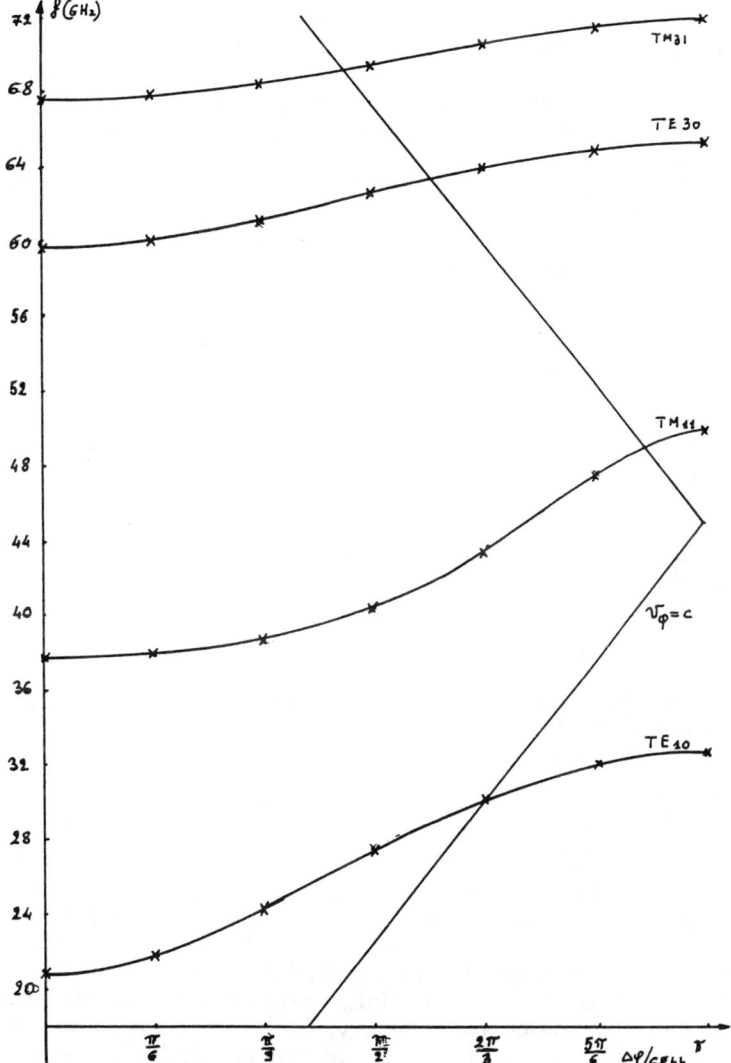

Fig. 6. Pass bands of the periodically loaded CTS waveguide in the range 18 GHz to 72 GHz.

6. MECHANICAL TOLERANCES

MAFIA has proved to be a very helpful tool in the determination of the sensitivity of the CTS RF properties to the mechanical parameters of the structure [5]. To this end small variations have been applied to the four main parameters of the waveguides, namely their width and height as well as to the teeth height and length in the z direction, and the resulting variation in the $2\pi/3$ mode frequency computed by means of the E310 module. Care has been taken in order that the small variation of one dimension, of the order of 0.2 mm, would not modify the other mechanical parameters thus giving independent results. As an example the sensitivity of the frequency to the length of the teeth in z has been computed while keeping the periodicity of the teeth constant at 3.332 mm. Table 3 shows the results of these computations. One interesting point is that, contrary to our expectations, the height of the waveguide teeth is a crucial parameter while the distance of the same teeth from the coupling slot is a less sensitive parameter. This fact becomes clear when considering the field lines distribution of the $2\pi/3$ mode, which are crowded in the spacing between successive teeth and rarefy towards the coupling slot.

Table 3. Sensitivity of the $2\pi/3$ mode frequency to the mechanical parameters : Wh, Ww, Th, Tl, respectively height and width of the waveguide and height and length of teeth.

Wh = 4.0 mm	$\Delta f/\Delta$Wh	= 0.44 MHz/μm
Ww = 8.0 mm	$\Delta f/\Delta$Ww	= 1.67 MHz/μm
Th = 2.0 mm	$\Delta f/\Delta$Th	= 4.30 MHz/μm
Tl = 1.666 mm	$\Delta f/\Delta$Tl	= 2.45 MHz/μm

7. ACKNOWLEDGEMENT

L. Thorndahl and G. Carron have supplied the information of the CTS mechanical design. T. Weiland and his team have helped me understand MAFIA. W. Schnell and G. Guignard have supported this study.

REFERENCES

1. W. Schnell, The CLIC Study of an Electron-Positron Collider, Contribution to LC92 ECFA Workshop on e^+e^- Linear Colliders, Garmish Partenkirchen, August 1992. CERN SL /92-51 and CLIC Note 184
2. MAFIA User Handbook, The MAFIA Collaboration
3. L. Thorndahl, G. Carron, Impedance and Loss Factor Measurements on a CLIC Transfer Structure (CTS), Proceedings of the Third European Particle Accelerator Conference, EPAC 92, Berlin, 24-28 mars 1992, pp 913,915.
4. G. Guignard, Beam Dynamics Investigations for the CLIC Drive Beam, CERN SL/92-22(AP) and CLIC Note 157.
5. L. Thorndahl, G. Carron, RF Hardware Development Work for the CLIC Drive Beam, Contribution to the 1993 Particle Accelerator Conference, Washington D.C. (to be published)

A Boundary-Integral-Method Code for Cavity RF Mode Analysis

F.P. Adams and M.S. de Jong
Accelerator Physics Branch, AECL Research
Chalk River Laboratories, Chalk River, Ontario, K0J 1J0, Canada

Abstract

A 3-dimensional boundary-integral-method code has been developed for rf cavity analysis. The code is based on the magnetic field integral equation (MFIE):

$$\frac{1}{2}\cdot\vec{H}(\vec{r}) = -\int_S \hat{n}'\times\vec{H}(\vec{r}')\times\vec{\nabla}_{r'}\left(\frac{e^{ik|\vec{r}-\vec{r}'|}}{4\pi|\vec{r}-\vec{r}'|}\right) dA'$$

The rf envelope of a cavity is input as a set of directed surface patches, and the magnetic field is taken to be piece-wise constant on the surface patches. A frequency-dependent, linear matrix equation is generated from the MFIE, where the domain of integration is the rf envelope described by the surface patches. A variant of forward iteration is used to find the largest complex eigenvalue of the matrix at a specific frequency. This eigenvalue is 1 when the frequency coincides with a cavity rf resonance, and has positive or negative complex phase when slightly above or below resonance, respectively. The eigenvector describes the magnetic field distribution of the corresponding rf resonant mode. The code can take advantage of symmetries such as reflection and rotation to reduce the size of the matrix equation. The use of surface patch elements allows the modeling of complex cavity geometries, and simplifies transfer of data to finite-element heat-transfer and stress-analysis codes. The cavity analysis code includes a solver and a graphics display utility, both of which run on microcomputers.

Introduction

The magnetic field integral equation (MFIE) is commonly used for the analysis of antennas of arbitrary shape.[1,2,3] The MFIE relates the magnetic field $\vec{H}(\vec{r})$ to an integral over a surface S enclosing the point $\vec{r}$ at which the magnetic field is being evaluated:

$$\vec{H}(\vec{r}) = -\int_S \hat{n}'\times\vec{H}(\vec{r}')\times\vec{\nabla}'\psi(\vec{r}'-\vec{r})\, dA' \quad (1)$$

where $\hat{n}'$ is the inward-facing unit vector normal to the surface S at $\vec{r}'$, and where

$$\psi(\vec{r}'-\vec{r}) = \frac{e^{ik|\vec{r}'-\vec{r}|}}{4\pi|\vec{r}'-\vec{r}|}, \quad (2)$$

k being the free-space wave number of the rf.

The use of finite element models based on the MFIE for the analysis of rf resonances in cavities of arbitrary shape has previously been suggested.[2] Since the MFIE is based on surface quantities only, it follows that a boundary element method code may be developed on the basis of equation (1). Such a code has been developed at the Chalk River Laboratories. The code is a boundary-element code in which a frequency-dependent, linear matrix equation is generated from the MFIE.

BOUNDARY INTEGRAL METHOD

Note that equation (1) has only three variables, the three components of the magnetic field on the surface. If the surface is defined to be coincident with a conducting surface, such as the metal wall of an accelerating cavity, then the component of the magnetic field normal to the surface must be zero. The number of variables in the problem is then reduced to the two perpendicular components of $\vec{H}(\vec{r}')$ that are tangential to the surface S.

The MFIE as expressed in equation (1) has a simple form that suggests an iterative solution. The suitability of the MFIE for boundary element analysis of cavities was studied using MathCAD software[4] on a microcomputer. This software can be used to perform numerical evaluation of integral equations such as appear in equation (1). The test cases were simple spherical and right-cylindrical cavities. $\vec{H}(\vec{r}')$ was assumed to be azimuthally constant and azimuthally directed. The profile of the cavity was described by N joined segments, over which the magnetic field was assumed to be piece-wise constant. The description of the magnetic field distribution was therefore reduced to a set of N non-zero, complex values forming an array H. The azimuthal part of the surface integral was performed numerically, and the integral over the profile of the cavity was replaced by a summation over the N discrete segments:

$$H_j = \hat{\phi}_j \cdot \sum_{i=1}^{N} \frac{a_i}{2\pi R_0} \left(\int_0^{2\pi} \hat{n}_i \times (H_i \hat{\phi}) \times \vec{\nabla}_i \psi(\vec{r}_i - \vec{r}_j) \, d\phi \right) = A(H) \qquad (3)$$

where a_i is the area of the surface described by segment i of the cavity profile and R_0 is the radius of the cavity.

The integration and summation were performed N times to obtain the magnetic field at each of the N points around the cavity profile. Solution of the equation was achieved by performing this calculation iteratively, beginning with an array H^0 of N pseudo-random complex numbers. Near the TM101 resonance of the cavity the array H converged to a solution after a few iterations. After convergence the elements of H were found to all have the same complex phase, and these numbers were found to match the analytically-derived magnetic field distribution of the mode to within ±1%. The effect of further iterations on H was to multiply each of the elements of H by a complex factor

$$A(H) \approx e^{i\alpha} H. \qquad (4)$$

The quantity α was found to be linearly frequency-dependent, and for each of the lowest few resonances there was a frequency f_0 near the resonant frequency for which

$$\alpha(f) \approx \frac{(f-f_0)}{\Phi}. \qquad (5)$$

The resonances of the cavity were thus indicated by zeroes of α. Comparison with analytic solutions for a 20 cm diameter spherical cavity showed that the calculations of the frequencies

Table I   RF Modes in a spherical cavity 20 cm in diameter

Resonant Mode	Frequency (MHz)		
	Boundary Integral Calculations	SUPERFISH[5] (50 × 50 mesh)	Analytically Derived Frequency[6]
TM101	1309.13	1309.14	1309.12
TM102	2918.52	2919.02	2918.52
TM103	4445.28	4446.42	4445.28
TM104	5957.50	5963.50	5957.47
TM301	2372.99	2373.17	–
TM501	3406.81	3407.57	–
TM302	4161.42	4163.35	–

of TMm0p resonant modes were quite accurate for $N = 10$, as indicated in Table I. The quantity Φ was 287 MHz (5.0 MHz/degree) for the TM101 mode, and was slightly greater for the other modes.

The boundary integral method developed under MathCAD was then modified to model azimuthally symmetric cavities of arbitrary (r, z) profile. The profile was again constructed using N discrete segments, and the integration over azimuth in equation (3) was replaced by a summation over M discrete points:

$$H_j = \hat{\phi}_j \cdot \sum_{i=1}^{N} \frac{a_i}{N} \left(\sum_{k=1}^{M} \left(\hat{n}_i \times (H_i \hat{\phi}) \times \vec{\nabla}_i \psi (\vec{r}_i - \vec{r}_j) \right) \right) = A(H) \qquad (6)$$

where a_i is the area of the surface generated by rotating segment i around the axis of symmetry of the cavity. This was reduced to the simple matrix equation

$$H_j = \sum_{i=1}^{N} A_{i,j} H_i, \qquad (7)$$

where

$$A_{i,j} = \frac{a_i}{N} \hat{\phi}_j \cdot \left(\sum_{k=1}^{M} \left(\hat{n}_i \times \hat{\phi} \times \vec{\nabla}_i \psi (\vec{r}_i - \vec{r}_j) \right) \right). \qquad (8)$$

Thus, the boundary integral problem becomes a matrix eigenvalue problem. This model was used to study TM0np resonances in a simple right-cylindrical cavity with a height of 10 cm and a diameter of 20 cm. Convergence of equation (7) was reached after a few tens of iterations for most modes.

The resonant frequencies of the lower-frequency TM0np modes were predicted using iterative calculations based on equations (7) and (8) and finding the zeroes of α. The accuracy of mode frequencies calculated this way was better than ±0.5% for $M = 30$ and

$N = 30$, as shown in Table II. The magnetic field distribution described by H was studied for the TM010 mode, and was found to agree with analytic predictions of the magnetic field distribution for right-cylindrical cavities, except at the corners where the flat top and bottom of the cavity meet the cylindrical side wall.

Table II   RF modes in a cylindrical cavity 10 cm high and 20 cm in diameter

Resonant Mode	Frequency (MHz)		
	Boundary Integral Calculations (Equation (8))	Analytically Derived Frequency[7]	Boundary Integral Code (103 patches)
TM010	1148.47	1147.43	1152.23
TM011	1896.23	1887.72	1889.36
TM020	2637.59	2633.82	2641.41
TM021	3036.98	3030.49	3037.78
TM022	3976.02	3990.56	3996.37

BOUNDARY ELEMENT CODE

A boundary element code was developed from a generalization of equations (7) and (8) for cavities of arbitrary shape. This code reads in a file that contains a description of the cavity geometry in the form of a set of directed quadrilateral surface patches. The quadrilateral surface patches are described by ordered sets of node numbers in a list, and the coordinates in space of each node are specified in a list. This cavity description format was selected to ease the conversion of mesh and field data between the rf code and MARC, a finite-element code that is used at Chalk River for the thermal and mechanical analysis of rf cavities.[8]

The rf magnetic field is assumed to be constant over each surface patch, and to be tangent to the patch. A basis is defined for each surface patch i, where $\hat{n}_i^1$ and $\hat{n}_i^2$ are orthogonal unit vectors parallel to the patch and $\hat{n}$ is a unit vector normal to the patch. Since the four nodes defining a surface patch are not constrained to lie in a plane, the unit vector $\hat{n}$ is defined to be parallel to the cross product of vectors joining opposite corners of the patch. The other two unit vectors, $\hat{n}_i^1$ and $\hat{n}_i^2$, are then defined to be orthogonal to each other and to $\hat{n}$. Equation (7) then becomes

$$H_j^1 = \sum_{i=1}^{N}\left(A_{i,j}^{11} H_i^1 + A_{i,j}^{21} H_i^2\right) \text{ and } H_j^2 = \sum_{i=1}^{N}\left(A_{i,j}^{12} H_i^1 + A_{i,j}^{22} H_i^2\right). \quad (9)$$

where

$$\vec{H} = H^1 \hat{n}^1 + H^2 \hat{n}^2. \tag{10}$$

Note that the problem has two degrees of freedom per surface patch. The boundary integral method is used to calculate the terms in the four matrices

$$A_{i,j}^{11} = \hat{n}_j^1 \cdot \sum_{i=1}^{N} \left(\int_{S_i} \hat{n}_i \times \hat{n}_i^1 \times \vec{\nabla}_i \psi \left(\vec{r}_i - \vec{r}_j \right) dA \right), \quad A_{i,j}^{12} = \hat{n}_j^1 \cdot \sum_{i=1}^{N} \left(\int_{S_i} \hat{n}_i \times \hat{n}_i^2 \times \vec{\nabla}_i \psi \left(\vec{r}_i - \vec{r}_j \right) dA \right),$$

$$A_{i,j}^{21} = \hat{n}_j^2 \cdot \sum_{i=1}^{N} \left(\int_{S_i} \hat{n}_i \times \hat{n}_i^1 \times \vec{\nabla}_i \psi \left(\vec{r}_i - \vec{r}_j \right) dA \right) \text{ and } A_{i,j}^{22} = \hat{n}_j^2 \cdot \sum_{i=1}^{N} \left(\int_{S_i} \hat{n}_i \times \hat{n}_i^2 \times \vec{\nabla}_i \psi \left(\vec{r}_i - \vec{r}_j \right) dA \right). \tag{11}$$

Note the similarity to equation (8), except that the integral is performed over each surface patch S_i. Gaussian integration is used in the code. Matrices A^{11}, A^{21}, A^{12} and A^{22} may be combined into a single 2N-by-2N matrix

$$K = \begin{bmatrix} A^{11} & A^{21} \\ A^{12} & A^{22} \end{bmatrix} \tag{12}$$

and the two vectors H^1 and H^2 may be combined into a single vector of length 2N

$$h = \begin{bmatrix} H^1 \\ H^2 \end{bmatrix} \tag{13}$$

to provide a new equation of the same the form as equation (7),

$$h_j = \sum_{i=1}^{2N} K_{i,j} h_i, \tag{14}$$

so that solving for a magnetic field distribution at a given frequency is a matrix eigenvalue problem.

The code developed is based on the solution of equation (7). A variant of forward iteration is used in the code to find the eigenvector of the matrix with the largest complex eigenvalue.[9] When the frequency is near the frequency of a resonant mode of the cavity, the eigenvector describes the magnetic field distribution of the rf resonant mode. The resonant frequencies of the cavity modes may be found by repeating the solution of the eigenvalue problem at various frequencies and searching for the frequency at which the complex part of the eigenvalue becomes zero.

For a cavity problem that is described by a set of N surface patches, the vector h will contain 2N complex elements and the matrix K will contain $4N^2$ complex elements. Since the boundary integral method does not involve taking differences between similar values, as is the case with methods based on differential equations, the code stores all elements in single (32-bit) precision. For a cavity problem described by 300 surface patches, only 2.9 megabytes of memory is required to store the matrix K. For 1000 surface patches, 32 megabytes of memory will be required to store the matrix. If insufficient memory is available for the matrix, the code can store the matrix in a temporary file on a hard disk.

This temporary file will be 290 megabytes for a problem described by 3000 surface patches, which is not practical on most microcomputers.

The time required to construct the matrix K and the time required to perform each iteration on h are also proportional to N^2. Some simple timing tests have been performed on microcomputers. Scaling from our results, approximate times for execution on a microcomputer using a 33 MHz Intel 80486 microprocessor are given in Table III. Matrix K is stored on the hard disk in all cases.

Table III Execution times on a 33-MHz, Intel 80486-based microcomputer

Problem Size (patches)	Code Execution Time	
	Build Matrix K	Per Iteration on h
300	1.5 minutes	8 seconds
1000	20 minutes	1.5 minutes
3000	4 hours	20 minutes

Two classes of rf mode are possible in a cavity with reflective symmetry across some plane: those whose magnetic-field distribution is symmetric under reflection across that plane and those whose magnetic-field distribution is anti-symmetric under such reflection. The vector h thus contains redundant information for such problems. One can reduce the size of the vector h by a factor of two by removing the redundant surface patches and using the appropriate symmetry in generating the matrix K. Note that this also reduces the size of the matrix K by a factor of four for each instance of reflection. In the case of a simple right-cylindrical cavity, the problem has reflective symmetry across three mutually orthogonal planes. Using this symmetry reduces the number of surface patches by a factor of eight, and reduces the size of the matrix K by a factor of 64. Note that the time required to calculate the matrix K is reduced by approximately a factor of eight, since the symmetry of the problem must be expanded and the effect of each of the reflected surface patches must be considered. The summation in equation (11) would thus be eight times as large.

As part of the cavity problem description, the code developed reads in three parameters that specify the reflective symmetry of the problem. The user is then prompted to decide whether the rf mode to be calculated should be symmetric or anti-symmetric under reflection across the plane of symmetry. After the mode has been calculated, the user-supplied symmetry information is stored in a computer file along with the magnetic field distribution. The results of some typical mode frequency calculations made with the code are shown in Table II above, where the model has been reduced to 103 patches by the use of eight-fold reflective symmetry. Although results are not shown, the code is equally effective at calculating TMmnp and TEmnp modes of all symmetries.

It is also possible to make use of rotational symmetry to reduce the size of a problem. Thus, the code also reads in two parameters that specify the axis and order of the rotational symmetry of the problem. An azimuthally-symmetric cylindrical cavity could be described by a narrow wedge of surface patches that is specified to be rotated around the z-axis 50 times. A more complicated cavity might have three-fold rotational symmetry about the z-axis and reflective symmetry across the x-y plane. The number of surface patches required to describe the problem could thus be reduced six-fold by the use of symmetry, and the size of the matrix K could be reduced 36-fold.

By comparison with 3-dimensional finite-element techniques based on differential equations, the boundary integral method can greatly reduce the number of variables in large problems. Consider a cylindrical cavity with a height of 10 cm and a diameter of 20 cm, to be modeled with a resolution of one centimetre. The cavity surface area is roughly 1260 cm^2, and the volume is roughly 6300 cm^3. Using surface patches and the boundary integral method, the cavity can be modeled using roughly 1260 patches, and each patch will have two degrees of freedom (H^1 and H^2) for a total of 2520 problem variables. Alternatively, 6300 volume elements would be required, with typically six degrees of freedom per volume element (E_x, E_y, E_z, H_x, H_y, and H_z) for a total of 37 800 problem variables. Although this fifteen-fold reduction in the number of variables is offset somewhat by the fact that the matrix K is generally complex, dense, non-symmetric and non-Hermitian, the iterative methods developed for use in the code are quite efficient in finding solutions to equation (14).

After sufficiently many iterations have been performed on the problem and the magnetic field data are stored to a file, a second program is used to display the data graphically. This program is also useful for viewing the cavity problem when constructing a cavity model. The code provides screen display only. Printed output is obtained by capturing the screen image to a computer file, which may then be printed. Excellent results may be achieved through the use of high-resolution (1024 × 768 × 256-colour) screen graphics and thermal-wax-transfer colour printers. The resolution of the output is therefore limited to the maximum screen resolution that can be displayed.

Post-processing of magnetic field data can be used to find quantities of interest. Surface heating in rf cavities, for example, is proportional to the square of the rf magnetic field tangent to the surface. Results from the code have been used as input to thermal-mechanical stress calculations performed at Chalk River. The electric field may also be calculated by

$$\vec{E}(\vec{r}) = \vec{\nabla} \times \vec{H}(\vec{r}) \tag{15}$$

Rather than take differences of the local magnetic field, using equation (1) one can substitute for $\vec{H}(\vec{r})$ to obtain

$$\vec{E}(\vec{r}) \propto \vec{\nabla} \times \int_S \hat{n}' \times \vec{H}(\vec{r}') \times \vec{\nabla}\psi(\vec{r}'-\vec{r}) \, dA', \tag{16}$$

or substitute using equation (14) to obtain

$$\vec{E}_j(\vec{r}_j) = \left(\hat{n}_j \cdot \sum_{i=1}^{N} \left(\int_{S_i} \vec{\nabla}_j \times \hat{n}_i \times \left(H_i^1 \hat{n}_i^1 + H_i^2 \hat{n}_i^2 \right) \times \vec{\nabla}_i \psi(\vec{r}_i - \vec{r}_j) \, dA \right) \right) \hat{n}_j. \tag{17}$$

Note that equation (17) is less general than equation (16), which would allow for the calculation of the electric field at any arbitrary point inside the cavity. Equation (17) has been implemented in the code, allowing calculation of electric fields for solved modes. The magnetic field data for the mode are retrieved by the code from the file in which they have been stored, and a new file is generated which contains the electric field data. This second file can also be read by the graphical display package, allowing the user to view the electric fields. The ability to calculate electric and magnetic fields at arbitrary points inside the cavity will be added later, as part of the ongoing development of the code.

Conclusions

A 3-dimensional boundary-integral-method code has been developed for rf cavity analysis. Magnetic and electric surface field distributions are also calculated by the code. The code allows the modeling of three-dimensional cavities of arbitrary geometry, and can be used to calculate cavity mode frequencies with good accuracy. The use of symmetry properties in the code allows complicated cavities to be modelled using microcomputers.

References

1. J.A. Stratton, Electromagnetic Theory (McGraw-Hill Book Company, 1941), p. 464.
2. D.S. Jones, Methods in Electromagnetic Wave Propagation (Clarendon Press, 1979), p. 463.
3. I. Wolff, The Waveguide Model for the Analysis of Microstrip Discontinuities, in Numerical Techniques for Microwave and Millimeter-Wave Passive Structures, edited by T. Itoh (John Wiley and Sons, 1989), p. 91.
4. MathSoft Inc., Cambridge, Massachusetts, MathCAD Version 3.1, 1992.
5. K. Halbach and R.F. Holsinger, SUPERFISH, A Computer Program for the Evaluation of RF Cavities with a Cylindrical Symmetry, Particle Accelerators 7, 213 (1976).
6. S. Ramo, J.R. Whinnery and T. Van Duzer, Fields and Waves in Communications Electronics (John Wiley and Sons, 1965), p.556.
7. S. Ramo, J.R. Whinnery and T. Van Duzer, Fields and Waves in Communications Electronics (John Wiley and Sons, 1965), p.548.
8. MARC Analysis Research Corporation, Palo Alto, California, MARC.
9. J.H. Wilkinson, The Algebraic Eigenvalue Problem (Clarendon Press, 1965), p. 571.

ACAO OF CAVITIES

T. Weiland M. Zhang
Technische Hochschule Darmstadt, Fachbereich 18, FG TEMF,
Schloßgartenstr. 8, 6100 Darmstadt, Germany

ABSTRACT

An example for automatic Computer-Aided Optimizations (aCAO) is presented. The aCAO is performed with a newly-born MAFIA optimization driver (OO), which is operated within the MAFIA software environment[1]. The whole optimization procedure with OO is a fully hardware-independent and automatic operation.

In the example, the PETRA cavity, now used in the PETRA storage ring, DESY, Germany, is optimized with respect to a goal function of $max\{R_s/\,|k_t|\}$. With a view to demonstrating the capability and practicality of the presented optimzation driver–OO, we concentrate ourselves mainly on the procedure itself, not on the very essence of how to build a pratical PETRA cavity.

The obtained results are briefly discussed at the end of the paper.

INTRODUCTION

The designing of cavities is very important for the performance of an entire accelerator. And hardware fabrications of them are, for most cases, quite expensive and time-consuming.

Recently, since the rapid development of computers in both hardwares and softwares is achieved, CAD techniques are stepping up to another higher level, yielding the new branch of automatic Computer-Aided Optimizations (aCAO).

What is presented here is a *machine-independent* automatic Computer-Aided Optimization procedure, which is performed within the MAFIA environment. The optimization procedure is driven and managed via command files in which optimization strategies are organized and written in the MAFIA command language. Therefore the files are transportable to any computer where MAFIA is properly installed. MAFIA can be operated either interactively or non-interactively.

MAFIA is a general purpose software package which solves the Maxwell's equations by means of the FIT[2] algorithm. For more about MAFIA software's capabilities see[3,4].

OO–The MAFIA Optimization Driver

The newly born MAFIA module, which is named OO for optimization, is actually a "big executable" of all that MAFIA possesses as dedicated modules. The following "equation" shows what the MAFIA family is made up of, which is the up-to-date status as foreseen for version 3.2.

MAFIA= M(Mesh-generator)
+ E(Eigenvalue solver)
+ S(Static solver)
+ W(Wirbel=eddycurrent solver)
+ T2/T3(2d/3d Time-domain solver)
+ TS2/TS3($2\frac{1}{2}$d/3d PIC solver)
+ P(Postprocessor)
+ OO(Optimization driver)

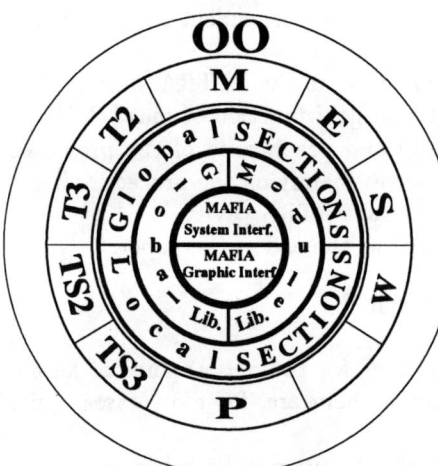

Figure 1: *This figure shows the MAFIA layers. The innermost core consists of two MAFIA underlying libraries: the MAFIA system library and the MAFIA graphic library, where all hardware-dependent calls or operations are handled to ensure other higher-level libraries from being hardware-dependent.*

The first nine modules (solvers/processors) are dedicated ones, since they each accomplish a certain definite task such as generating supporting meshes (M), solving transient time-domain responses in 3d (T3) and so on. While the optimization driver (OO) does things like organizing, managing and synchronizing the whole MAFIA kernel operations and communications between all the dedicated members. The constitution of the optimization driver is shown below:

OO=M+E+S+W+T2+T3+TS2+TS3+P

The hierachical struture of the optimization driver is depicted in Fig 1.

The following example of optimization is performed with the above-mentioned OO, which is driven by a command file in which all things like field calculation, looping, decision making and overall optimizing strategies are written in the MAFIA command language.

OPTIMIZATION OF PETRA CAVITIES

Here we take the optimization of PETRA cavities as an example. In Fig 2, the "PETRA" shows a quarter of the profile of the PETRA cavities now used in DESY, Germany.

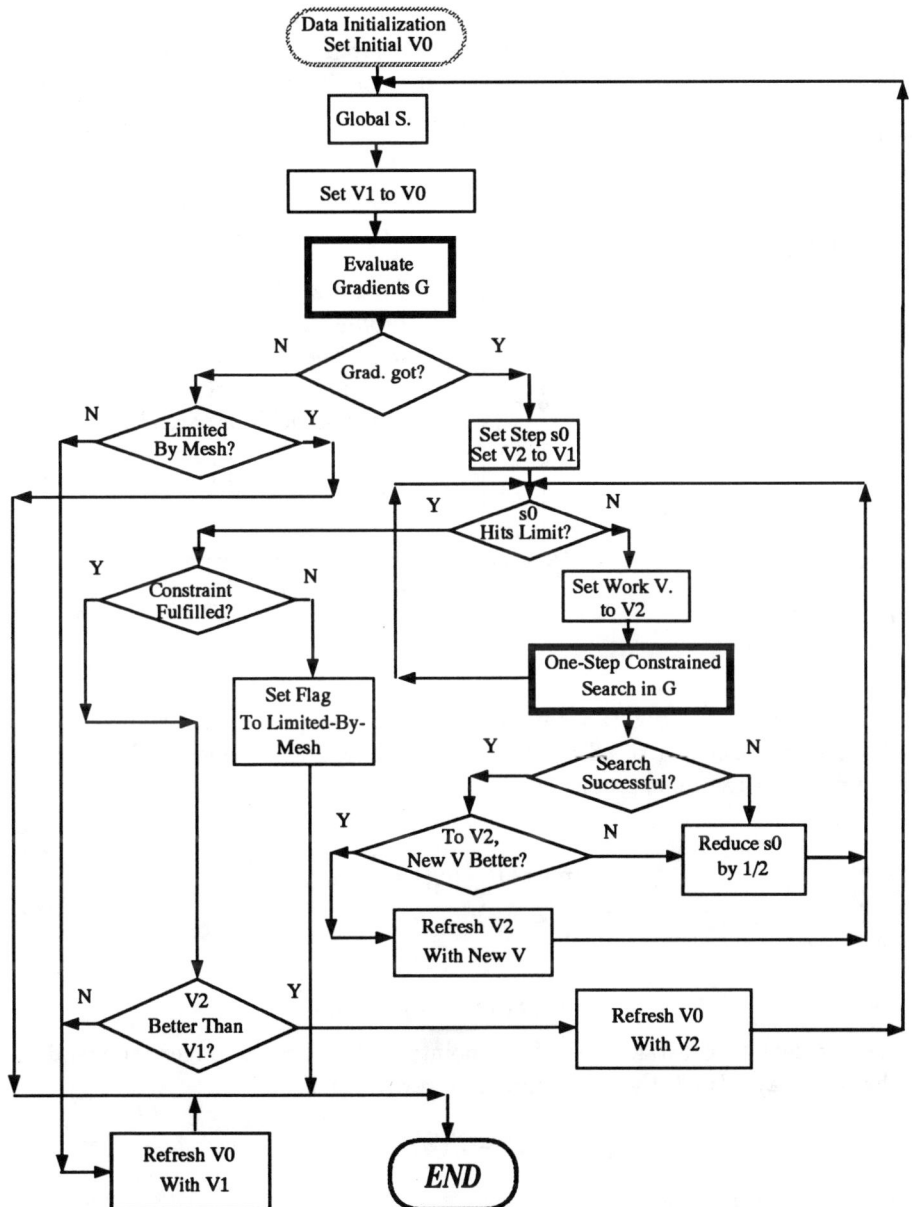

Figure 3: *Flow-diagram for the Optimization of the PETRA Cavity.*

For such standing wave accelerating cavities, two important parameters are R_s (shunt impedance) and k_t (transverse kick). From them we formulate a downright goal function: $max\{R_s/|k_t|\}$, under the condition that the resonance frequency (f) of the accelerating mode (TM010) be fixed to that of the used PETRA's (f_0).

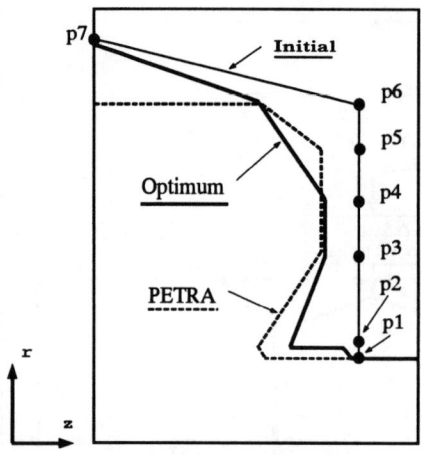

Figure 2: *Quarter of the profile of the Petra cavity. The bottom side of the enclosing rectangle is the ideal orbit of particles accelerated and the real cavity is formed by rotating the profile around this axes; The right side is the symmetry plane.*

Points p1 to p7 (Fig 2) are used to optimize the profile with respect to the goal function. Points p1 to p6 can be changed along z-axis while p7 along r-axis. So the problem can be expressed as follows:

$$g := max\{R_s/|k_t|\} = u(\mathbf{Z}, r_{p_7}) \qquad (1)$$
$$f = v(\mathbf{Z}, r_{p_7}) \qquad (2)$$

where

$$\mathbf{Z} := (z_{p_1}, z_{p_2}, z_{p_3}, z_{p_4}, z_{p_5}, z_{p_6})^T \qquad (3)$$

Since f is kept to constant, one of the points p1 to p7 is no longer independent of the rest. Say it's p7, then Eq (2) can be rewritten as

$$r_{p_7} = V(\mathbf{Z}). \qquad (4)$$

And therefore Eq (1) becomes

$$g = u(\mathbf{Z}, V(\mathbf{Z})) = U(\mathbf{Z}). \qquad (5)$$

Eq (5) means that as long as Eq (4) is fulfilled, the optimization becomes a six-dimensional problem. To fulfill Eq (4) we use p7 to draw the frequency f to the given one, i.e. to f_0. Naturally, the selection of such a point is really quite arbitrary. One may also select two or even more points to do so instead of only one. But our selection here is justified by the fact that changing p7 in r brings the maximal impact on the frequency while incurs little distortions to the field distribution near the z-axis and it is the most economical selection with respect to the cpu consumption.

We use the gradient method (constrained and non-constrained) to carry out local extreme searches and random-walk procedures to determine possible global best(s).

"Constrained gradients" := $\partial U(\mathbf{Z})/\partial \mathbf{Z}$;
"Non-constrained gradients" := $\partial u(\mathbf{Z}, r_{p_7})/\partial \mathbf{Z}$;
"Random walking procedure" := after a local extreme is reached, there are two schemes of the random walking: 1) generate random search steps and do one-dimensional searches in the current $\mathbf{Z}$ and 2) generate random $\mathbf{Z}$s, and check whether there is any better $\mathbf{Z}$ than the current one. If true, then start another round of the local extreme searches, otherwise (by a given limited number of trials), stop the whole procedure and return the current local extreme $\mathbf{Z}$ as the optimum sought.

Since the goal has a very strong dependency upon f, the non-constrained scheme cannot effectively provide local gradient information, hence leading to zigzag or even false search paths. So it's comparatively time-consuming and sometimes even misses local extremes. Because of this we use the "constrained" scheme, which is a pure gradient method as Eq (5) is concerned. It takes longer time to evaluate gradients than the non-constrained one. But taking the optimization procedure as a whole, it's more economical instead.

The flow-diagram of the optimization procedure is given in Fig 3.

RESULTS

MAFIA can be used to solve a large varity of problems: static, time-harmonic, transient, eddy currents, particle-in-cells and many others. For each of these aspects MAFIA has one dedicated module: S (static), E (time-harmonic), T2/T3 (2d/3d transient), ...

For the problem here only E and T2 are used. We use E to evaluate R_s and T2 for k_t. Adding M (the mesh-generator in MAFIA), we form the very basic element of the driving command file as below:

$$\mathbf{newa} = \begin{cases} \text{do} & \text{M} & \text{(generate supporting meshes)} \\ \text{do} & \text{E} & \text{(evaluate } R_s) \\ \text{do} & \text{T2} & \text{(evaluate } k_t) \\ \text{return} & \ldots & g = R_s/|k_t| \end{cases}$$

These four major steps are grouped into a MAFIA macro "newa" (say). Anytime "newa" is invoked the computer automatically does them all. The MAFIA *macro* functions in almost the same way as the *subroutine* in FORTRAN.

Initial Structure : PILLBOX
Calculated Structure : 1/2 Generatrix of Cavity
Coordinate System Used : rz of Cylindrical C.S.
Meshes : r * z = 83 * 88 = 7304 (points)
CPU Time Used : 7 Days on SUN SPARC 1
Total Number of Structure Modifications : 370

Structure \ Item	*Initial*	*Optimum*
R_S (MΩ)	3.45329	4.07297
$\|k_t\|$ (V/C)	$3.22895*10^{10}$	$2.99309*10^{10}$
$R_S/\|k_t\|$ (Ω*C/V)	$1.06948*10^4$	$1.36079*10^4$
$f_{(TM-EM-0-1)}$ (MHz)	507.379872	507.631424
Relative Improvement on R_S	=	17.94%
Relative Improvement on $\|k_t\|$	=	7.30%
Relative Improvement on $R_S/\|k_t\|$	=	27.24%
Frequency Deviation	<	5 ‰

Structure \ Item	*PETRA*	*Optimum*
R_S (MΩ)	3.82604	4.07297
$\|k_t\|$ (V/C)	$3.30542*10^{10}$	$2.99309*10^{10}$
$R_S/\|k_t\|$ (Ω*C/V)	$1.15750*10^4$	$1.36079*10^4$
$f_{(TM-EM-0-1)}$ (MHz)	507.807168	507.631424
Relative Improvement on R_S	=	6.45%
Relative Improvement on $\|k_t\|$	=	9.45%
Relative Improvement on $R_S/\|k_t\|$	=	17.56%
Frequency Deviation	<	3.5 ‰

Figure 4: *Results of the Optimization*

The drawback of the generality of our approach, namely the possibility to generate problem specific optimization procedures, is that debugging may become a major task! Therefore we implemented a DEBUG option into OO that works similarly to debuggers known from compilers.

The results we obtain for the optimization of the PETRA cavity are listed in Fig 4. The "optimum" (Fig 2) shows the final structure.

From the final optimal profile of the cavity, one may argue that it looks a little bit strange. There exists a sharp angle at the nose and a blunt transition at the symmetry plane. The reason for this is the limited number of the optimizing points. And what is more, we restricted, in the optimization, that the adjustment of the points could only be done in one direction, points 1 to 6 in z and point 7 in r. If we select more points along the profile and allow the modification of their locations in full degrees of freedom, then we could, for sure, obtain much more realistic strutures. So this example is a proof of principle demonstrating the capabilities and the structure of MAFIA-OO.

CONCLUSION

The presented procedure serves as a good example for the new aCAO option in MAFIA. For most practical design problems, such an aCAO procedure is really an indispensible tool. The MAFIA environment provides a very wide spectrum of capabilities, from static to particle-in-cell. It means that OO, the MAFIA optimization driver, can be used to optimize much broader scopes of practical problems in designing electromagnetic devices than what is presented here. As a tool for aCAO, OO together with its supporting MAFIA environment will find wide applications in designing electromagnetic devices in the near future. And it has actually already been used to design automatically many different types of cavities needed for the DESY/THD S-Band collider[5].

REFERENCES

1. T. Weiland and M. Zhang *A Tool for Automatic Computer-Aided Optimization with MAFIA*, Proceedings of the 5th International IGTE Symposium, Graz, Austria, Sept. 1992, pp255-260
2. T. Weiland *A Discretisation Method for The Solution of Maxwell's Equations for Six Component Fields*, AEU, 31, (1977), p116
3. M. Bartsch et al. *MAFIA in Practice: the Capabilities of the MAFIA CAD System*, International Conference on Electromagnetic Field Problems and Applications, Oct. 1992, Hangzhou, China
4. The MAFIA Collaboration. *Recent Advances and Applications of the MAFIA Codes*, this conference
5. T. Weiland (for the S-Band Collider Group), *Status of the DESY/THD 500 GeV S-Band Linear Collider Study*, talk at the LC92, Garmisch Partenkirchen, to be published

SIMULATING A SINGULARITY-FREE UNIVERSE OUTSIDE THE PROBLEM BOUNDARY IN *POISSON* [1]

K. Halbach and R.D. Schlueter
Lawrence Berkeley Laboratory, Berkeley, CA 94720

ABSTRACT

An exact analytical solution developed from the Dirichlet problem exterior to a circle is employed in the magnetostatics code POISSON[1] to provide a boundary condition option which simulates a singularity-free universe external to the problem domain. Problems with domains of large unequal extents in perpendicular directions are treated by first conformally mapping the exterior of an ellipse onto the exterior of the unit circle. Problems exhibiting symmetry in one or two planes are modeled using a semi-circle or quarter-circle, respectively, in conjunction with the singularity-free rest-of-universe boundary condition.

INTRODUCTION

Currently POISSON allows for a fixed potential or a zero normal derivative of the potential on problem boundaries. The analysis of many magnetostatics problems would benefit from the ability to simulate a singularity-free universe external to the problem domain. Here, the analytical development and numerical adaptation for POISSON of a free-floating potential boundary condition that simulates a singularity-free rest-of-the-universe is described. An approximate technique addressing this general problem has been already been developed[2] whereby the harmonic description of the potential function on an arc just interior to the problem domain's boundary is used to approximate and update the boundary potentials in a relaxation procedure. Herein, a solution is developed from the Dirichlet problem exterior to a circle which is analytical and exact (except for the discrete nature of the mesh); no approximations are employed. Problems with domains of large unequal extents in perpendicular directions may also be handled by first conformally mapping the exterior of an ellipse onto the exterior of a circle; the analytical development and implementation in POISSON are treated below. Problems exhibiting symmetry in one or two planes may be modeled using a semi-circle or quarter-circle, respectively, in conjunction with the singularity-free rest-of-universe boundary condition.

The complete analytical and numerical procedure of the next two sections can be summarized as follows, starting with the simulation of a singularity-free universe outside a circle, to be followed by a general description of the process for outer boundaries other than circles: POISSON is instructed to make a mesh with meshpoints on a circular boundary and with a set of meshpoints just outside that circle, connected to the points on the circle by regular point-to-point meshlines. Subject to the condition that one has only vacuum outside that circle, we derive below an exact, and fortunately, linear relationship between the vector potentials

---

[1] This work was supported by the Director, Office of Energy Research, Office of Basic Energy Sciences, Material Sciences Division, of the U.S. Department of Energy under Contract No. DEAC03-76SF00098.

on the circle and both the scalar and vector potentials at any point outside that circle. The matrix that allows the calculation of the vector potential for the points just outside the circle from the value of the vector potential of the points located on the circle is then computed. The program then goes through a modified iteration procedure to satisfy the Poisson equation everywhere: first, using current vector potential values of the meshpoints outside the circle, it calculates 'new' values of the vector potential for all points on and inside the circle; then, using the above mentioned matrix, it calculates 'new' values of the vector potential for the meshpoints outside the circle from those on the circle. The program continues this two-step cycle until the solution has converged.

If an elliptical boundary can better accommodate the geometry under investigation, the procedure is identical to that of the case of a circular boundary, with one additional step: After the meshpoints on and just outside the ellipse have been generated, a coordinate transformation is used that maps the ellipse and the points just outside the ellipse conformally (without any singular point on or outside the ellipse) onto a circle and points just outside the circle, respectively. The matrix relating the vector potential at the points on the ellipse to the points just outside the ellipse is then calculated as above. If one wants to use yet a different boundary, an appropriate conformal map to a circle and points outside the circle must be used. So far, the ellipse is the only non-circular boundary that has been implemented, but the program is set up to use any user-provided conformal map.

THE DIRICHLET PROBLEM OUTSIDE A CIRCLE

In this section an expression is derived for the complex magnetic potential $F(z)$ outside a circle from the given vector potential A (or alternatively scalar potential V) on the circle when there are no external singularities. This problem is analogous to that of determining the complex potential at locations in a singularity-free region inside a circle from the values of the vector potential on the bounding circle, found in many complex analysis texts.

Outside a circle the complex potential can be expressed by a Laurent series expansion as follows, where $z \equiv x + iy$ and the a_n are complex constants:

$$F(z) = A + iV = A_\infty + iV_\infty + \sum_{n=1}^{\infty} \frac{a_n}{z^n} \qquad (1)$$

The real part of Eq. (1) gives for the vector potential on a circle of radius r_0:

$$A(r_0, \theta) = A_\infty + \sum_{n=1}^{\infty} \left\{ \frac{Re(a_n)\cos(-n\theta)}{r_0^n} - \frac{Im(a_n)\sin(-n\theta)}{r_0^n} \right\} =$$

$$A(r_0, \theta) = A_\infty + \frac{1}{2}\sum_{n=1}^{\infty} \left\{ \frac{Re(a_n)(e^{-in\theta} + e^{in\theta})}{r_0^n} + i\frac{Im(a_n)(e^{-in\theta} - e^{in\theta})}{r_0^n} \right\} \qquad (2)$$

The vector potential at infinity A_∞ and the complex constants a_n can be determined using the orthogonality properties of $e^{im\theta}$ whereby multiplying both sides of Eq. (2) by $r_0^m e^{im\theta}$ ($\equiv z_0^m$), where m is a nonnegative integer, and integrating over θ from 0 to 2π gives for $m = 0$ and for m a positive integer, respectively:

60   Simulating A Singularity-Free Universe

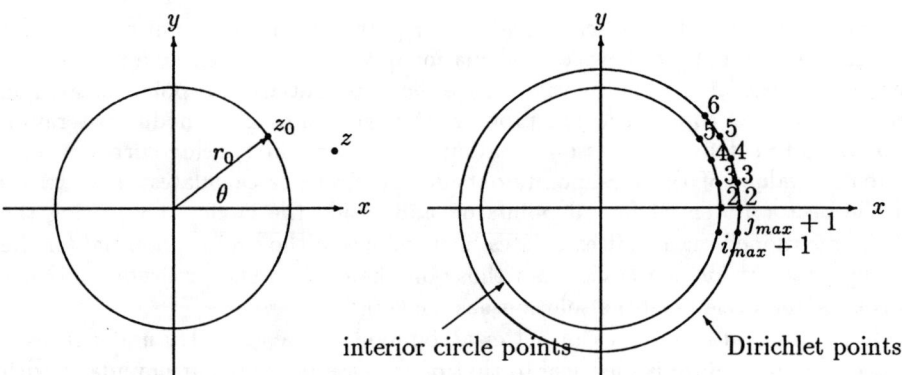

Figure 1. Geometry for Dirichlet problem outside a circle

Figure 2. Mesh discretization and indices

$$A_\infty = \frac{1}{2\pi}\int_0^{2\pi} A(r_0,\theta)d\theta \quad \text{and} \quad a_m = \frac{1}{\pi}\int_0^{2\pi} A(r_0,\theta)r_0^m e^{im\theta}d\theta \qquad (3)$$

Substituting the expressions derived for A_∞ and a_n into Eq. (1) we have for $F(z)$ outside the circle:

$$F(z) = \frac{1}{2\pi}\int_0^{2\pi} A(r_0,\theta)d\theta + iV_\infty + \frac{1}{\pi}\sum_{n=1}^\infty \int_0^{2\pi} A(r_0,\theta)\frac{z_0^n}{z^n}d\theta, \qquad |z|>|z_0| \quad (4)$$

V_∞ can be set to zero without loss of generality. Using the closed form expression for the infinite geometrical series $\sum_1^\infty (z_0/z)^n$ results in the following expression for the complex potential $F(z)$ outside a circle in terms of the known vector potential on the circle:

$$F(z) = A_\infty + \frac{1}{\pi}\int_0^{2\pi} A(r_0,\theta)\frac{z_0}{z-z_0}d\theta = A_\infty + \frac{1}{\pi}\int_0^{2\pi} A(r_0,\theta)\frac{r_0 e^{i\theta}}{re^{i\phi}-r_0 e^{i\theta}}d\theta \quad (5)$$

where $z \equiv re^{i\phi}$ is a point outside the circle, $z_0 \equiv r_0 e^{i\theta}$ is a point on the circle and $A_\infty = \frac{1}{2\pi}\int_0^{2\pi} A(r_0,\theta)d\theta$.

NUMERICAL DISCRETIZATION OF ANALYTICAL SOLUTION TO DIRICHLET PROBLEM OUTSIDE A CIRCLE

Here we derive a discretized expression relating the known vector potentials A_{int_i}, $2 \le i \le i_{max}+1$ of a finite set of points $z_{int_i} \equiv r_0 e^{i\theta_i}$ on a[n interior] circle to the vector potentials A_{dir_j}, $2 \le j \le j_{max}+1$ of a finite set of [Dirichlet] points $z_{dir_j} \equiv r_j e^{i\phi_j}$ exterior to that circle. This discretized expression takes the form of a i_{max} by j_{max} matrix M which is independent of the values of the vector potentials (i.e. need be determined only once): $\vec{A}_{dir} = [M]\vec{A}_{int}$.

We interpolate the vector potential $A(r_0, \theta)$ between known vector potentials on the interior circle as follows, where $\theta_i \leq \theta \leq \theta_{i+1}$: [2] [3] :

$$A(r_0, \theta) = A_{int_i} \left[\frac{\sin(\theta - \theta_{i+1})}{\sin(\theta_i - \theta_{i+1})} \right] + A_{int_{i+1}} \left[\frac{\sin(\theta - \theta_i)}{\sin(\theta_{i+1} - \theta_i)} \right] \quad (6)$$

Substituting the interpolated expression for $A(r_0, \theta)$ into Eq. (5), expanding the trigonometric functions in exponential form, and integrating yields:

$$\frac{1}{2\pi} \int_0^{2\pi} A(r_0, \theta) d\theta = \frac{1}{2\pi} \sum_{i=2}^{i_{max}+1} A_{int_i} \left\{ \tan\left(\frac{\Delta \theta_i}{2}\right) + \tan\left(\frac{\Delta \theta_{i-1}}{2}\right) \right\} \quad \text{and} \quad (7)$$

$$\frac{1}{\pi} \int_0^{2\pi} A(r_0, \theta) \frac{r_0 e^{i\theta}}{r e^{i\phi} - r_0 e^{i\theta}} d\theta \equiv f(\theta_{i-1}, \theta_i, \theta_{i+1}, z_{dir_j}) = \frac{1}{2\pi} \sum_{i=2}^{i_{max}+1} A_{int_i}$$

$$\left\{ \left(\frac{1}{\sin(-\Delta \theta_i)} \right) \left(1 - e^{-i\Delta \theta_i} - i(-\Delta \theta_i) B_{i+1,j}^{-1} + \ln[D_{i,j}]^{[B_{i+1,j}^{-1} - B_{i+1,j}^{-1}]} \right) + \right. \quad (8)$$

$$\left. \left(\frac{-1}{\sin(\Delta \theta_{i-1})} \right) \left(1 - e^{i\Delta \theta_{i-1}} - i\Delta \theta_{i-1} B_{i-1,j}^{-1} + \ln[D_{i-1,j}]^{[B_{i-1,j}^{-1} - B_{i-1,j}]} \right) \right\}$$

where $\Delta \theta_i \equiv \theta_{i+1} - \theta_i$, $2 \leq i \leq i_{max} + 1$, $2 \leq j \leq j_{max} + 1$, and

$$D_{i,j} \equiv \frac{z_{dir_j} - z_{int_{i+1}}}{z_{dir_j} - z_{int_i}}, \qquad B_{i,j} \equiv \frac{z_{dir_j}}{z_{int_i}}, \qquad z_{dir_j} \equiv r_j e^{i\phi_j}, \qquad z_{int_i} \equiv r_0 e^{i\theta_i},$$

The vector potential A_{dir_j} and the matrix M become, respectively:

$$A_{dir_j} = \frac{1}{2\pi} Re \left\{ \sum_{i=2}^{i_{max}+1} A_{int_i} \left[\tan\left(\frac{\Delta \theta_i}{2}\right) + \tan\left(\frac{\Delta \theta_{i-1}}{2}\right) + f(\theta_{i-1}, \theta_i, \theta_{i+1}, z_{dir_j}) \right] \right\} \quad (9)$$

$$M(i,j) = \frac{1}{2\pi} Re \left\{ \tan\left(\frac{\Delta \theta_i}{2}\right) + \tan\left(\frac{\Delta \theta_{i-1}}{2}\right) + f(\theta_{i-1}, \theta_i, \theta_{i+1}, z_{dir_j}) \right\} \quad (10)$$

POISSON has been modified[3] to allow for free-floating 'Dirichlet boundary points' to simulate a singularity-free universe outside problem boundaries. Given an initial assumed set of potentials on a circular Dirichlet outer boundary, the

---

[2] A simpler interpolation expression $A(r_0, \theta) = A_i + \frac{(A_{i+1} - A_i)(\theta - \theta_i)}{\theta_{i+1} - \theta_i}$ is not employed because the linear term in θ, when substituted into Eq. (5) cannot be integrated in closed form. An alternative interpolation expression $A(r_0, \theta) = A_i + \frac{(A_{i+1} - A_i)(\sin(\theta - \theta_i))}{\sin(\theta_{i+1} - \theta_i)}$ when substituted into Eq. (5) can be integrated in closed form. The interpolation expression given in Eq. (7) also allows closed form integration of Eq. (5) and is symmetric with respect to (A_{int_i}, θ_i) and $(A_{int_{i+1}}, \theta_{i+1})$. Finally note that the three interpolation expressions discussed are equivalent in the limit $\Delta \theta_i \equiv \theta_{i+1} - \theta_i \longrightarrow 0$. Errors introduced into the solution are of exactly the same order as those already present in POISSON where the potential is assumed linear between nodes.

[3] Another attractive interpolation technique is to recast Eq. (5) in the form $F(z) = A_\infty + \frac{1}{i\pi} \oint_0^{2\pi} A(z_0) \frac{dz_0}{z - z_0}$ with the integration path on the interior circle. Approximating this path by that of a polygon passing through all node points on the interior circle and interpolating $A(z)$ along each line segment by $A(z) = A_i + \frac{(A_{i+1} - A_i)(z - z_i)}{z_{i+1} - z_i}$, where z lies on the segment with endpoints z_i and z_{i+1}, gives closed expressions for all integrals.

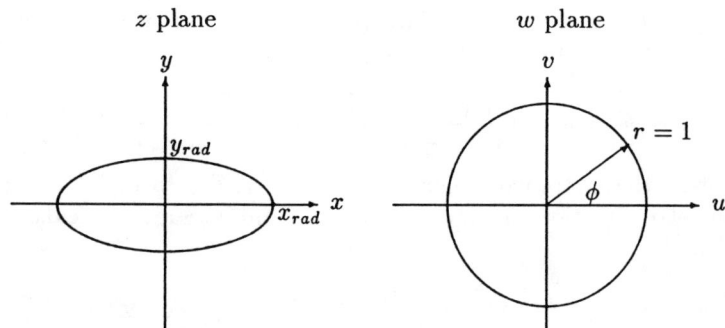

Figure 3. Ellipse in z plane mapped to unit circle in w plane

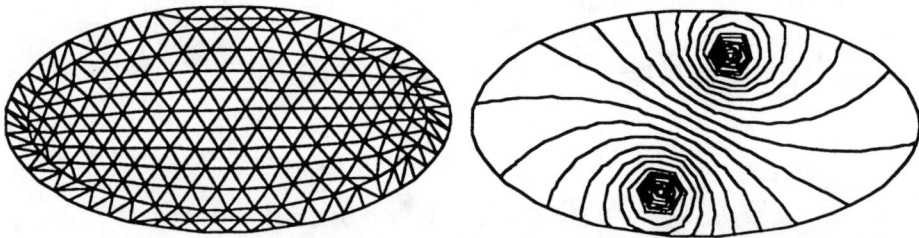

Figure 4. Sample problem mesh for an elliptical domain and corresponding field line plot (a filament pair, $I_1 = -I_2$ with a singularity-free external universe)

Poisson equation is [indirectly] solved in the usual manner as given in [1], yielding vector potential A values at all interior points, including the 'interior circle points'. Then the vector potential solution on the set of 'interior circle points' is used along with the matrix of Eq. (10) to reset the potentials on the outer 'Dirichlet boundary points'. The procedure is iterated until both the relaxed solution over the entire problem domain for the given current set of Dirichlet potentials converges and the iterated Dirichlet potentials themselves converge.

PROBLEM DOMAINS OF HIGH ASPECT RATIOS

An ellipse in the $z(x,y)$ plane can be represented by $z = x_{rad} \cos\phi + iy_{rad} \sin\phi$, $0 \leq \phi \leq 2\pi$. (ϕ is not to be confused with the angular location θ of a point on the ellipse with respect to the x-axis.) Introducing a complex w plane through the polar representation $w = \rho e^{i\phi}$, the ellipse and its exterior in the z plane can be conformally mapped onto the unit circle and its exterior in the w plane with $z = x_{rad}\left(\frac{w+w^{-1}}{2}\right) + y_{rad}\left(\frac{w-w^{-1}}{2}\right)$. All singular points of this map (given by $dz/dw = 0$ or $dw/dz = 0$) are clearly located inside the unit circle in the w plane: $dz/dw = 0 \implies 0 = x_{rad}(1 - \frac{1}{w^2}) + y_{rad}(1 + \frac{1}{w^2}) \implies w^2 = \frac{x_{rad}-y_{rad}}{x_{rad}+y_{rad}}$. Thus the exterior of an ellipse in z is indeed mapped to the exterior of the unit circle in w.

POISSON has been modified[3] to allow for boundaries of the shape of an ellipse.

This feature is especially useful for simulating a singularity-free universe outside problem boundaries when the problem domain itself is longer in one direction than in a perpendicular direction. In such cases an ellipse just interior to the problem's exterior boundary is first mapped onto a unit circle in the w plane and the exterior boundary itself is also mapped to the w plane (not necessarily onto a circle). Since the map is conformal, the vector potentials of the interior circle in w are identical to those of the interior ellipse in z and likewise for the exterior boundaries in w and in z. The matrix M of Eq. (10) is calculated and the iterative procedure described previously for solving the Poisson equation, with the free-floating Dirichlet potentials on the exterior boundary simulating the singularity-free external universe, is executed.

Figure 4 illustrates the meshing used to solve a problem of the type described above along with a field line plot for a current filament pair in a small domain utilizing the singularity-free rest-of-universe boundary condition.

SPECIAL FEATURES

Special case symmetries for the singularity-free external universe boundary condition have been incorporated[3] into POISSON; quarter-circles or half-circles may be employed to reduce mesh area size and enhance the accuracy of field harmonic component calculations.

For a problem with the Dirichlet midplane symmetry $A(x,-y) = -A(x,y)$, a half-circle may be employed (see Figure 5-a,b). Since current sources below the midplane are identical in magnitude but opposite in sign to those above the midplane and the rest of the universe is singularity-free, no restriction is imposed on net current in the half-circle domain with this symmetry type (e.g. Figure 5-b).

Again, for a problem with the Neumann midplane symmetry $A(x,-y) = A(x,y)$, a half-circle may be employed (see Figure 6). Since current sources below the midplane are identical in both magnitude and sign to those above the midplane and the rest of the universe is singularity-free, the net current must be zero in the half-circle problem domain with this symmetry type.

For a problem exhibiting symmetry along both the x- and y-axes, a quarter-circle may be employed. For H-magnet symmetry as shown in Figure 7, i.e. $A(-x,y) = -A(x,y)$ and $A(x,-y) = A(x,y)$, no restriction is imposed on the net current in the quarter-circle problem domain. For an elliptical quadrupole symmetry with Neumann boundary conditions along both the x- and y-axes, i.e. $A(-x,y) = A(x,y)$ and $A(x,-y) = A(x,y)$, the net current must be zero in the quarter-circle problem domain. For the quarter-plane symmetry with Dirichlet boundary conditions along both the x- and y-axes, i.e. $A(-x,y) = -A(x,y)$ and $A(x,-y) = -A(x,y)$, no restriction is imposed on the net current in the quarter-circle problem domain.

NONTRIVIAL *POISSON* TEST CHECK CASES

Here we check the accuracy of the new singularity-free rest-of-universe boundary condition in POISSON with two test cases: (a) current filaments in free-space

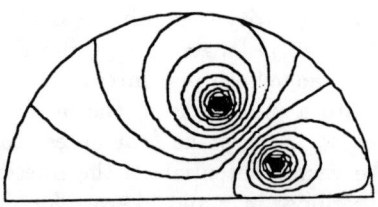
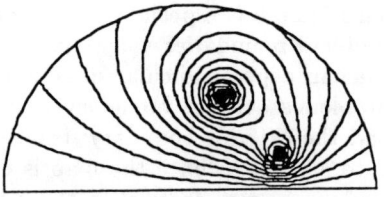

Figure 5. Field line plots for midplane symmetry with $A(x,-y) = -A(x,y)$ (no restriction on $\sum I$ in half-circle), with a singularity-free external universe

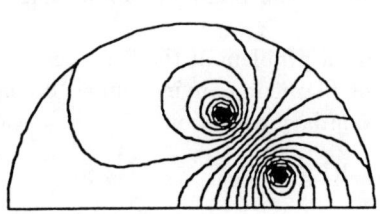
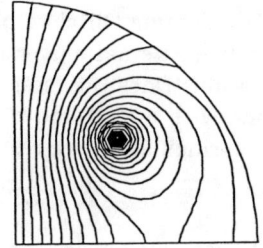

Figure 6. Field line plot for midplane symmetry with $A(x,-y) = A(x,y)$ (two current filaments, $I_1 = -I_2$)

Figure 7. Field line plot for H-magnet symmetry, $A(-x,y) = -A(x,y)$, $A(x,-y) = A(x,y)$ (no $\sum I$ restriction)

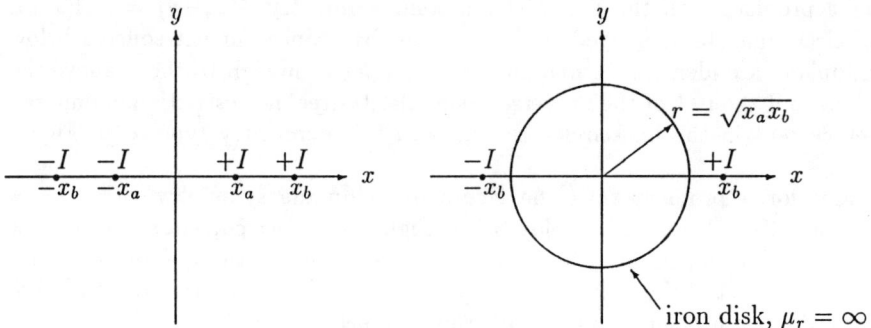

Figure 8a,b. POISSON test case geometries, (a) iron-free and (b) iron present

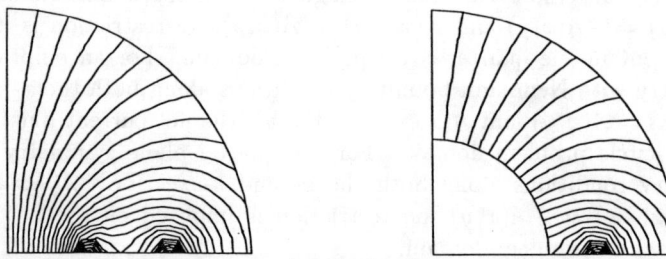

Figure 9a,b. Field line plots using new singularity-free rest-of-universe boundary condition for the geometries of Figure 8a,b.

and (b) current filaments in the presence of infinitely permeable iron, which though computationally difficult, has an exact closed form solution.

(a) **Current filaments in free-space.** The complex potential for the current filament distribution shown in Figure 8a is $F(z) \equiv A + iV = \sum_{i=1}^{4} \frac{-\mu_0 I_i}{2\pi} \ln(z - z_i)$ where $z_1 = (-x_b, 0) = -z_4, z_2 = (-x_a, 0) = -z_3$ and $I_1 = I_2 = -1, I_3 = I_4 = +1$.

A POISSON run for the geometry of Figure 8a, using the quarter-circle H-magnet symmetry of the previous section and the singularity-free rest-of-universe boundary condition with the singularities input in the first quadrant gives the field line plot of Figure 9a. Examination of the corresponding POISSON output file shows that the computationally obtained vector potential A agrees everywhere with the closed form solution given by the real part of $F(z)$ given above.

(b) **Current filaments in the presence of infinitely permeable iron.** For the current filament distribution shown in Figure 8a, we have when the scalar potential $V = 0$: $Im\left\{\ln[\frac{(z-x_a)(z-x_b)}{z}] - \ln[\frac{(z+x_a)(z+x_b)}{z}]\right\} = 0$. Now defining the scalar $\rho \equiv \sqrt{x_a x_b}$, we have $\frac{(z \pm x_a)(z \pm x_b)}{z} = z + \frac{x_a x_b}{z} \pm (x_a + x_b) = \rho\left(\frac{z}{\rho} + \frac{\rho}{z} \pm \frac{x_a + x_b}{\rho}\right)$ which, along the circle defined by $z = \rho e^{i\phi}$, reduces to $2\rho \cos\phi \pm (x_a + x_b)$. The log of this expression, being pure real, implies that along the circle $x^2 + y^2 = x_a x_b$ we have $V = 0$[4].

Thus for the current filament distribution in Fig. 8a (or 9a), the circle defined by $r = \sqrt{x_a x_b}$ is a constant scalar potential surface; outside this radius the field is identical to that of Fig. 8b (or 9b) where an infinitely permeable iron disk of radius $r = \sqrt{x_a x_b}$ centered at $z = 0$ likewise maintains the scalar potential at this radius constant. The location and magnitude of the outermost current filaments in Figure 8b are identical to those of Figure 8a.

A POISSON run for the geometry of Figure 8b (using the quarter-circle H-magnet symmetry of the previous section and the new singularity-free rest-of-universe boundary condition) with $\mu_r = \infty$ in the iron and with the singularity input in the first quadrant gives the field line plot shown in Figure 9b, which is identical to that of Figure 9a outside the radius $r = \sqrt{x_a x_b}$. Examination of the corresponding POISSON output file shows that the computationally obtained vector potential A agrees at all locations $r \geq \sqrt{x_a x_b}$ with the closed form solution given by the real part of $F(z)$ given in section (a) above.

LITERATURE CITED

1. *Reference Manual for the Poisson/Superfish Group of Codes*, LA-UR-87-126, Los Alamos National Laboratory, Los Alamos, NM, 1987.

2. Caspi, S., et.al., *Incorporation of Boundary Condition into the Program POISSON*, Lawrence Berkeley Laboratory Report LBL-19172, Aug. 1985.

3. Schlueter, R.D., POISSON version _DSA123:[ROSS.POIS.MODS]POISSON, Lawrence Berkeley Laboratory, 1990.

---

[4] Or, for $V = 0$, $Im\{\ln[(z - x_a)(z - x_b)] - \ln[(z + x_a)(z + x_b)]\} \equiv Im\{\ln[r_p e^{i\theta_p}] - \ln[r_q e^{i\theta_q}]\} = 0$, where $r_p = \sqrt{[(x - x_a)(x - x_b) - y^2]^2 + [(2x - x_a - x_b)y]^2}$, $r_q = \sqrt{[(x + x_a)(x + x_b) - y^2]^2 + [(2x + x_a + x_b)y]^2}$, $\theta_p = \arctan\frac{(2x - x_a - x_b)y}{(x - x_a)(x - x_b) - y^2}$, and $\theta_q = \arctan\frac{(2x + x_a + x_b)y}{(x + x_a)(x + x_b) - y^2}$. Solving, we have for $V = 0$: $\theta_p = \theta_q$ from which $y = 0$ or $x^2 + y^2 = x_a x_b$, as above.

CALCULATION OF Q-FACTORS OF LOSSY RESONATORS IN THE TIME DOMAIN USING THE PRONY-PISARENKO APPROXIMATION

P. Thoma, T. Weiland
Technische Hochschule Darmstadt, Fachbereich 18, FG TEMF,
Schloßgartenstr. 8, 6100 Darmstadt, Germany

ABSTRACT

Resonator losses are caused either by lossy parts of the structure or by radiation of electromagnetic field energy through an open part of the resonator boundary (e.g. a coupled waveguide).

One way to calculate the Q-factors of the several resonator modes is to excite a short pulse in the time domain and sample the electromagnetic field in the resonator. Using the fast Fourier transform, the Q-factors can be determined by the width of the resonance peaks. The application of this method may be difficult and the results may be less accurate when the peaks are close to one another.

A new way to calculate the resonance frequencies and the Q-factors without using the Fourier Transform is to use the Prony Pisarenko approximation. In most cases, this method requires much less samplings (and CPU time) than the application of the fast Fourier Transform, even when the peaks are close to one another.

We give a short introduction to this new method and show some examples of damped resonant structures now being studied in the context of linear collider design studies.

INTRODUCTION

In this paper we present a new method to determine the Q-factors of damped resonant structures. The algorithm used here is based on the Prony-Pisarenko-approximation[1,2,3]. After a short explanation of the algorithm, three examples are presented to show the application of this method, in the following called PPA, to low, medium and heavily damped structures and the agreement with the results of a fast Fourier Transform (FFT) is demonstrated.

The first example is a resonator box with lossy walls and Q-factors of about 1000. This somehow hypothetical structure has been studied to examine the accuracy of the PPA in calculating large Q-factors.

The next example shows a Bridge Looped Gap Resonator[4] (BLGR) with Q-factors of approximately 100. This resonator has been designed for ESR-spectroscopy to examine paramagnetic compounds.

As an example for the application of the PPA in the context of collider design studies we show an accelerator cavity with higher order mode couplers. The function of the couplers is to heavily damp the higher modes. The Q-factors in this structure are about 30.

A DESCRIPTION FOR THE RESONATOR FIELD

The algorithm for calculating the Q-factors is based on the analytical description of the electromagnetic field in the resonator.

A resonator can be damped either by lossy parts of the structure or by radiation of field energy through an open part of the resonator boundary. The electromagnetic field inside such a resonator can be written as a superposition of a countable number of different eigenmodes and an additional part $\vec{R}$ which can not be written in this form*. For the calculation of the Q-factors usually only the first part of the field consisting of the eigenmodes is of interest.

The time dependence of each eigenmode is a harmonic oszillation which is damped exponentially by the losses. According to this, e.g. the electric field in the resonator can be written as

$$\vec{E}(x,y,z,t) = \sum_{i=1}^{\infty} E_i \cdot \vec{f_i}(x,y,z) \cdot e^{-\alpha_i t} \cdot cos(\omega_i t + \Phi_i) + \vec{R}(x,y,z,t) \qquad (1)$$

where E_i represent the mode amplitudes, $\vec{f_i}$ the eigenmodes of the resonator and $\vec{R}$ the remaining part.

For the calculation of the Q-factors it is sufficient only to record the time signal of one field component in the resonator. To select this field component it has to be considered that not all modes can be observed in each field component. If the field distribution is essentially known, this fact can be used to minimize the contribution of uninteresting modes to the recorded time signal. Corresponding to eq.(1) the recorded time signal of the e_ν component ($\nu \in \{x,y,z\}$) can be described as

$$e_\nu(t) = \sum_{i=1}^{\infty} D_i \cdot e^{-\alpha_i t} \cdot cos(\omega_i t + \Phi_i) + r(t). \qquad (2)$$

Above we mentioned that for the calculation of the Q-factors only the sum of eigenmodes is significant. It can also be shown that the Q-factor of the i-th mode is given by the following equation**

$$Q_i = \omega_i \cdot \frac{\overline{W_i}}{\overline{P_i}} \approx \frac{\omega_i}{2\alpha_i}. \qquad (3)$$

Therefore we can calculate the Q-factors by the resonance frequencies ω_i and the damping constants α_i. The problem of determining this parameters can be solved using the following approximation for the time signal $s(t)$:

$$s(t) = \sum_{i=1}^{\infty} D_i \cdot e^{-\alpha_i t} \cdot cos(\omega_i t + \Phi_i). \qquad (4)$$

*This means that the additional part can not be separated into a countable number of eigenmodes.

**$\overline{W_i}$ is the mean value of the field energy of the i-th eigenmode during one period, $\overline{P_i}$ is the mean value of the power losses during one period.

The main part of the approximation error is caused by neglecting $r(t)$ (see eq.(2)). In the next sections we give an introduction to the Prony Pisarenko approximation which yields good results in this application.

THE APPROXIMATION METHOD

In a first step we show the way to determine the resonance frequencies and the damping constants of the modes. Here we will assume that the number of modes in the time function is known. Later we will show how to obtain this number.

If eq.(4) is written in a complex form and the number of modes is reduced to a finite number M, we get:

$$s(t) = \sum_{i=1}^{2M} \frac{D_i}{2} \cdot e^{-\alpha_i t} \cdot e^{j(\omega_i t + \Phi_i)} = \sum_{i=1}^{2M} \frac{D_i}{2} \cdot e^{j\Phi_i} \cdot e^{(-\alpha_i + j\omega_i)t} \qquad (5)$$

with: $\quad \omega_k = -\omega_{k-M} \quad, \quad \alpha_k = \alpha_{k-M} \qquad (k = M+1, \ldots, 2M).$

Using the samplings $s_n := s(n \cdot \tau)$ of the time function $s(t)$, we can write

$$s_n = \sum_{i=1}^{2M} C_i \cdot z_i^n \quad , \quad z_i = e^{(-\alpha_i + j\omega_i)\tau}, \quad C_i = \frac{D_i}{2} e^{j\Phi_i}. \qquad (6)$$

The transformation of this equation into Z domain yields

$$S(z) = \sum_{i=1}^{2M} \frac{C_i}{1 - \frac{z_i}{z}} = \ldots = \frac{\sum_{i=0}^{2M-1} b_i \cdot z^{-i}}{1 + \sum_{i=1}^{2M} a_i \cdot z^{-i}}. \qquad (7)$$

According to eq.(7), the complex roots of the polynomial $A(z) := 1 + \sum_{i=1}^{2M} a_i \, z^{-i}$ are equal to the z_i. The coefficients a_i are unknown yet. Now we multiply eq.(7) with the denominator polynomial and transform the resulting expression back into time domain:

$$s_n + \sum_{i=1}^{2M} a_i \cdot s_{n-i} = 0 \quad \text{with} \quad n = 2M, \ldots, N \qquad (8)$$

We obtain the a_i by solving this $N - 2M$ equations using a least square approximation (usually the number of samplings N is larger than $4M$) which leads us to the following linear system of equations:

$$\begin{pmatrix} r_{1,1} & r_{2,1} & \cdots & r_{2M,1} \\ r_{1,2} & r_{2,2} & \cdots & r_{2M,2} \\ \vdots & \vdots & & \vdots \\ r_{1,2M} & r_{2,2M} & \cdots & r_{2M,2M} \end{pmatrix} \cdot \begin{pmatrix} a_1 \\ a_2 \\ \vdots \\ a_{2M} \end{pmatrix} = \begin{pmatrix} -r_{0,1} \\ -r_{0,2} \\ \vdots \\ -r_{0,2M} \end{pmatrix} \qquad (9)$$

The $r_{i,j}$ are given by $r_{i,j} = \sum_{n=2M}^{N} 2 \cdot s_{n-j} s_{n-i}.$

The sampling frequency $f_{sam} = 1/\tau$ has to be at least two times the maximum frequency of the signal to avoid alialising in the spectrum. According to this the maximum frequency of the signal has to be limited using a digital low pass filter. We choose a butterworth filter of an adjustable order[5].

So far we showed how to determine the resonance frequencies and the damping constants. In the following we explain the way to get the mode amplitudes and phases. Therefore we write eq.(7) as follows:

$$\sum_{k=1}^{2M} C_k \prod_{\substack{i=1\\i\neq k}}^{2M} \left(1 - \frac{z_i}{z}\right) = \sum_{i=0}^{2M-1} b_i z^{-i}. \tag{10}$$

By replacing z by z_m we get:

$$C_m \cdot \prod_{\substack{i=1\\i\neq m}}^{2M} \left(1 - \frac{z_i}{z_m}\right) = \sum_{i=0}^{2M-1} b_i \cdot z_m^{-i}. \tag{11}$$

Using this equation we can determine the C_m if we know the b_i. So we have to find a way to calculate the b_i. Therefore eq.(7) can be written as:

$$S(z) = \sum_{i=0}^{2M-1} b_i \cdot H(z) \cdot z^{-i} \quad \text{with} \quad H(z) = \frac{1}{1 + \sum_{i=1}^{2M} a_i z^{-i}}. \tag{12}$$

Here the a_i are already known. Because of this we can transform $H(z)$ into time domain and get the pulse response h_n of this filter. The transformation of eq.(12) into time domain yields:

$$s_n = \sum_{i=0}^{2M-1} b_i \cdot h_{n-i}. \tag{13}$$

Using this equation, the b_i can now be calculated by a least square approximation:

$$e := \sum_{n=2M+1}^{N} \left[s_n - \sum_{i=0}^{2M-1} b_i \cdot h_{n-i}\right]^2 \to \text{Min.} \tag{14}$$

This approximation can easily be transformed into a linear system of equations. Solving this we get the coefficients b_i which yield the C_m by eq.(11). Finally the mode amplitudes and phases can be calculated by $C_m = D_m/2 \cdot e^{j\Phi_m}$.

In the previous steps we have always assumed that the number of modes in the time function is known but actually this number has to be determined, too.

We choose a very simple way to do this but it yields good results anyway. First we start the calculation with $M = 1$, determine the mode parameters and finally calculate the approximation error. Then we increase the number M and do the calculation again until a given maximum value for M is reached. Finally, we choose the value of M which yields the smallest approximation error.

When calculating the z_i, unphysically z_i with $|z_i| > 1$ (that corresponds to negative damping constants) may occur. This is caused either by the neglected part of the time function or by values of M larger than the actual one. We obtained the best results by disregarding those z_i and performing the calculation of the amplitudes and phases without those modes.

In the following sections we want to give some examples for the application of this method with Q-factors varying from large ones (about 1000) down to low ones (about 30).

DETERMINING THE Q FACTORS OF A LOW DAMPED RESONATOR

Our first example is a simple cubic resonator box with lossy walls and vacuum inside as shown in Fig.1.

Figure 1: *Geometry of the resonator box. The box dimensions are 0.1m x 0.05m x 0.02m, the conductivity of the walls is $5 \cdot 10^5$ S/m. Due to symmetries only the eighth part of the structure was calculated.*

The electromagnetic field inside the resonator has been calculated using the time domain solver that is a part of the program family MAFIA. The stimulation has been done by an initialfield. Due to symmetries only the eighth part of the structure needs to be considered. During the calculation we recorded one component of the electric field in the middle of the resonator. The result is shown in Fig.2.

We stimulated the first three eigenmodes but according to the symmetries only the first and the third mode contribute to the recorded time function. Because of the quite large Q-factors the decay of the harmonic oscillations is very slow, so within the calculated time range it is not possible to use the FFT to obtain the resonator spectrum with sufficient accuracy.

We analyzed the time function with the PPA-method. The resulting spectrum is shown in Fig.3. The calculated Q-factors are in a good agreement with the results obtained by the usual pertubation method using surface currents derived from tangential magnetic fields (see Tab.1).

Mode	PPA	Pertubation method	relative error
1	954	951	0.3 %
3	1385	1364	1.5 %

Table 1: *Comparison of the Q-factors calculated by the PPA-method with the results of a pertubation method.*

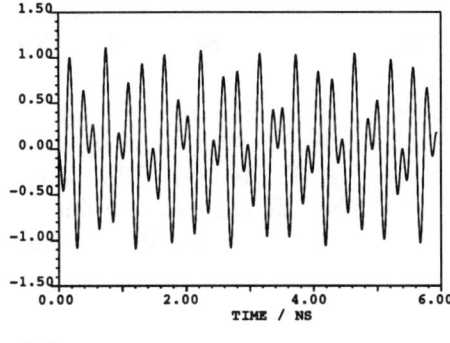

Figure 2: *The time function of one electric field component in the middle of the resonator.*

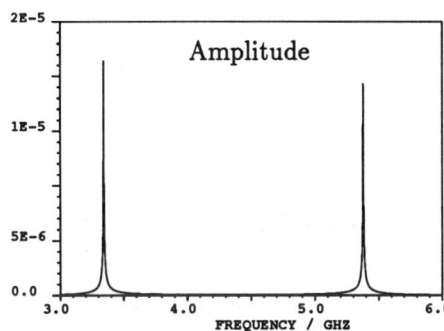
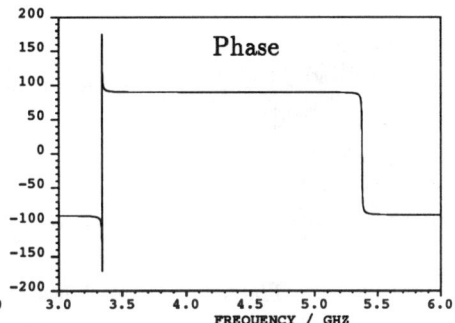

Figure 3: *The spectrum of the recorded time function which has been calculated by the PPA-method.*

DETERMINING THE Q FACTORS OF THE BLGR

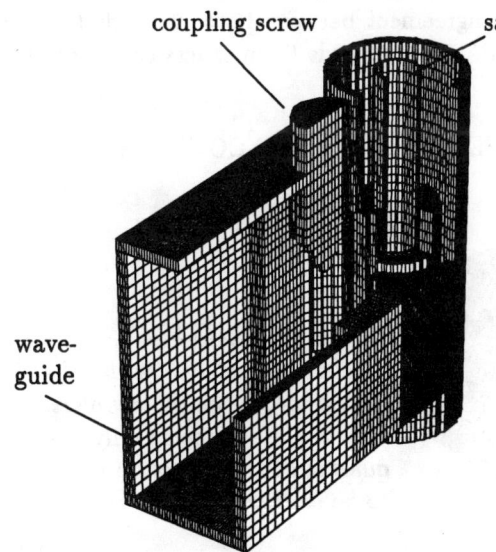

Figure 4: *Geometry of the BLGR resonator. Mw fields are excited by a waveguide mode which propagates into the resonator housing. After a short transient the magnetic field in the resonator is nearly homogeneous. The Q-factor can be adjusted by the coupling screw.*

The second example is a bridged loop-gap resonator which has been developed as a tool for ESR-spectroscopy. The geometry is shown in Fig.4. The resonator field has been calculated in the time domain[4].

The recorded time function of one magnetic field component inside the resonator is shown in Fig.5. We calculated the spectrum of this time signal by the FFT and also by the PPA-method (see also Fig. 5). Here the PPA-method needed three times less samplings than the FFT. The difference between this two spectrums is caused by the transient which is not considered by the PPA-method.

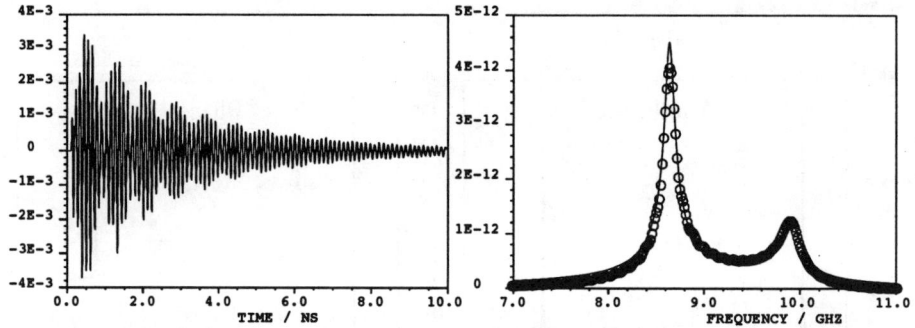

Figure 5: *The left picture shows the recorded time function of a magnetic field component inside the resonator, the right picture shows the spectrum of this time signal. The solid line represents the PPA spectrum, the circles mark the FFT samplings of the spectrum.*

The Q-factor for the main resonance at 8.6 Ghz has been calculated using the FFT by the width of the resonance peak (for this 20 ns have to be calculated) and also by the PPA-method. The agreement between both methods (FFT:86, PPA:86.2) is very good but the PPA in fact needs 6 times less samplings than the FFT.

A CAVITY WITH HIGHER ORDER MODE COUPLERS

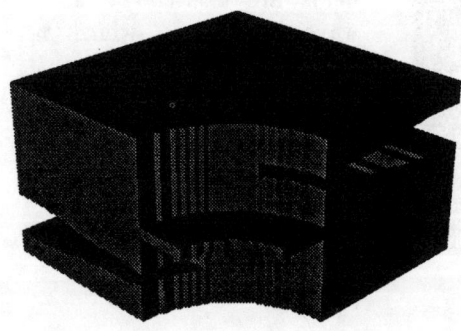

Figure 6: *Geometry of the accelerator cavity with the higher order mode couplers. Due to the symmetry only one quarter of the structure needs to be considered.*

Our last example is an accelerator cavity with higher order mode couplers. We calculated the Q-factor of a higher cavity resonance at 3.47 Ghz by the width of the resonance peak in the FFT spectrum and also by the PPA-method (see Fig. 7). The resulting Q-factors and resonance frequencies are nearly identical (FFT: 34.7, PPA: 35.4) but the PPA method needs 5 times less samplings than the FFT and additional to this the PPA-method yields the Q-factors and resonance frequencies of all resonances from one single computation, even in the case where the peaks are so close to one another that the Q-factors can not be derived from the width of the resonance peaks.

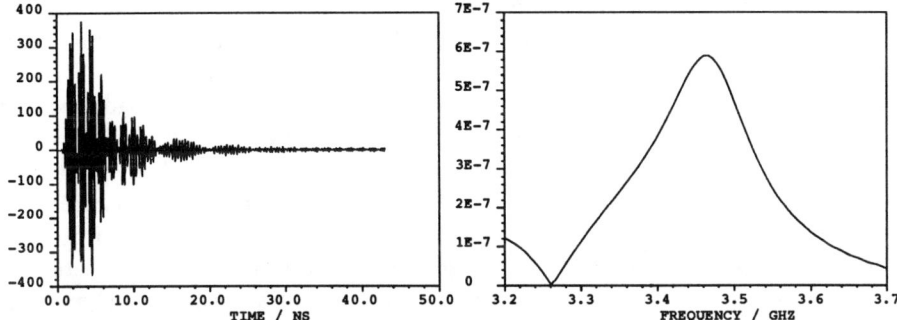

Figure 7: *The left picture shows the recorded time function of the electric field in the middle of the iris, the right picture shows the resonance peak at 3.47 Ghz in the FFT spectrum of the time signal.*

CONCLUSION

In this paper we presented a new method to calculate the resonance frequencies and Q-factors of lossy resonators. For most applications this method needs much less CPU-time than the calculation using the FFT.

The accuracy of the PPA-method has been demonstrated in three examples for a wide range of Q-factors. The PPA method yields good results even when the resonance peaks are close to one another.

REFERENCES

1. J. de L'Ecole de Polytechnique, *Essai expérimental et analytique sur la toi de la dilatabilité des fluides élastiques*, Vol. 1, 1795, p.24-80, R.Prony
2. Geophysics J.R. Astr. Soc., *The retrieval of harmonics from a covariance function*, Vol. 33, 1973, p.347-366, V.F.Pisarenko
3. International Journal of Numerical Modelling, *Electronic Networks Devices & Fields*, Vol.3, 1990, p.269-285, Dubhard, Pompei, Roux, Papiernik
4. 5th International IGTE Symposium, *Numerical field calculation in electrical engineering*, Proceedings, 1992, p.60-65, Wagner, Weiland
5. Digitale Filter, Richard W.Hamming, published by VCH-Informationstechnologie, 1990

CALCULATING SCATTERING PARAMETERS OF 3-DIMENSIONAL STRUCTURES BY BROADBAND EXCITATION IN THE TIME DOMAIN

H. Wolter, T. Weiland

Technische Hochschule Darmstadt, Fachbereich 18, FG TEMF,
Schloßgartenstr. 8, 6100 Darmstadt, Germany

ABSTRACT

Scattering parameters are important characteristics of rf–devices such as couplers. Usually the calculation of these parameters is done in the frequency domain separately for each single frequency point.

The way described here is to excite a broadband pulse in the time domain and perform a fast Fourier transform on the sampled time domain data. With this method it is possible to obtain the scattering parameters for a broad frequency range with one single time domain calculation. The frequency resolution is given by the number of time steps (and step size) and usually is much higher than that of the frequency domain calculation. Thus the resulting spectrum contains more information and even very sharp peaks can be found, which is nearly impossible with frequency domain methods. In most cases the calculation is much faster also.

We will give a short introduction to this method and show some examples of high power devices.

INTRODUCTION

In the time domain the electromagnetic fields inside a waveguide structure can be calculated straightforward, independent of the excitation. At the waveguide boundaries (the 'open ends'), however, the boundary conditions for dispersive waveguides can only be fulfiled for one frequency at a time which means that a broadband signal can not be treated correctly. For the calculation of scattering parameters the excitation of the waveguide structure by a broadband pulse would nevertheless be most advantageous because this yields the results for the whole interesting frequency range in one single calculation. To achieve this the boundary conditions have to be fulfiled for a broad frequency range. This can be done by expanding the boundary field into the 2d-eigenmodes of the waveguide at the boundary. The coefficients of this expansion are the amplitudes of the incoming and reflected waves for these modes. To simplify the calculation the propagation of these waves is approximated by a low order digital filter.

THE METHOD

In the frequency domain the transverse electromagnetic fields in a waveguide are described by a superposition of incoming and reflected waves:

$$\vec{E}(x,y,z,\omega) = \sum_\nu \frac{\vec{E}_\nu(x,y,j\omega)}{f_\nu(j\omega)} \left(\underline{a}_\nu(j\omega) e^{-\gamma_\nu(j\omega)z} + \underline{b}_\nu(j\omega) e^{\gamma_\nu(j\omega)z} \right) \quad (1)$$

with the real power normalization factor

$$f_\nu(j\omega) = \sqrt{\int\int |\vec{E}_\nu(x,y,j\omega) \times \vec{H}_\nu^*(x,y,j\omega)|dA} \qquad (2)$$

and $\vec{E}_\nu(x,y,j\omega)$: Solution of the 2d-eigenvalue problem with the propagation constant $\gamma_\nu(j\omega)$; $\underline{a}_\nu(j\omega), \underline{b}_\nu(j\omega)$: amplitudes of incoming and reflected waves. In the following we will use the weighted wave parameters $\tilde{\underline{a}}_\nu(j\omega) = \underline{a}_\nu(j\omega)/f_\nu(j\omega)$, $\tilde{\underline{b}}_\nu(j\omega) = \underline{b}_\nu(j\omega)/f_\nu(j\omega)$.

Inside the waveguide we can directly calculate the time-dependent electromagnetic field[1,2], but to fulfill the boundary-conditions we have to expand the field at the boundary into the solutions of the 2d-eigenvalue problem, which in homogeneously filled waveguides can be found frequency independent and orthogonal. (1) yields for the field in the layer $z = 0$:

$$\vec{E}(x,y,0,\omega) = \sum_\nu \vec{E}_\nu(x,y)\left(\tilde{\underline{a}}_\nu(j\omega) + \tilde{\underline{b}}_\nu(j\omega)\right)$$
$$=: \sum_\nu \vec{E}_\nu(x,y)\underline{A}_\nu(j\omega) \qquad (3)$$

with $\tilde{\underline{a}}_\nu(j\omega)$: stimulation, $\tilde{\underline{b}}_\nu(j\omega)$: amplitude of reflected wave. $\tilde{\underline{a}}_\nu(j\omega)$ is known, but $\tilde{\underline{b}}_\nu(j\omega)$ has to be calculated. This can be done from the electromagnetic field in the second layer $z = \delta z$:

$$\vec{E}(x,y,\delta z,\omega) = \sum_\nu \vec{E}_\nu(x,y)\left(\tilde{\underline{a}}_\nu(j\omega)\underline{P}_\nu(j\omega) + \tilde{\underline{b}}_\nu(j\omega)\underline{P}_\nu(j\omega)^{-1}\right)$$
$$=: \sum_\nu \vec{E}_\nu(x,y)\underline{B}_\nu(j\omega) \qquad (4)$$

with the propagation filter

$$\underline{P}_\nu(j\omega) = e^{-\gamma_\nu(j\omega)\delta z}. \qquad (5)$$

The propagation constant $\gamma_\nu(j\omega)$ follows from the dispersion equation:

$$\gamma_\nu^2(j\omega) = \left(\frac{\omega_{c\nu}}{c}\right)^2 - \left(\frac{\omega_\nu}{c}\right)^2. \qquad (6)$$

On a discrete grid this equation is changed to:

$$\left(\frac{\sinh(\gamma_\nu \delta z/2)}{\delta z}\right)^2 = \left(\frac{\sinh(\omega_{c\nu}\delta t/2)}{c\delta t}\right)^2 - \left(\frac{\sinh(\omega_\nu \delta t/2)}{c\delta t}\right)^2 \qquad (7)$$

These considerations also apply for the time domain. When transforming (1) into the time domain we get:

$$\vec{E}(x,y,z,t) = \sum_\nu \vec{E}_\nu(x,y)\left(\tilde{a}_\nu(t) * P_\nu(t,z) + \tilde{b}_\nu(t) * P_\nu(t,-z)\right) \qquad (8)$$

with the propagation filter

$$P_\nu(t,z) = \frac{1}{2\pi} \int_{-\infty}^{\infty} e^{-\gamma_\nu(j\omega)z} e^{j\omega t} d\omega \ . \tag{9}$$

From the field $\vec{E}(x,y,\delta z,t)$, which is known, we can calculate the mode amplitudes $B_\nu(t)$ and therefore derive $\tilde{b}_\nu(t)$ from

$$B_\nu(t) = \tilde{a}_\nu(t) * P_\nu(t,\delta z) + \tilde{b}_\nu(t) * P_\nu(t,-\delta z) \tag{10}$$
$$\Rightarrow \tilde{b}_\nu(t) = P_\nu(t,\delta z) * (B_\nu(t) - \tilde{a}_\nu(t) * P_\nu(t,\delta z)) \ . \tag{11}$$

(3) and (11) yields for the mode amplitude at the boundary $z = 0$:

$$A_\nu(t) = \tilde{a}_\nu(t) + P_\nu(t,\delta z) * (B_\nu(t) - \tilde{a}_\nu(t) * P_\nu(t,\delta z)) \ . \tag{12}$$

To simplify the algorithm and save cpu-time we approximate the convolution with P_ν in (12) by a recursion[3]

$$\sum_{\kappa=0}^{N} c_{\nu\kappa} y(n-\kappa) = \sum_{\kappa=0}^{N} c_{\nu(N-\kappa)} x(n-\kappa) \ . \tag{13}$$

The coefficients $c_{\nu\kappa}$ can be chosen in a way that the filter

$$P'_\nu(j\omega) = \frac{\sum_{\mu=0}^{N} c_{\nu\mu} e^{j\mu\omega\delta t}}{\sum_{\mu=0}^{N} c_{\nu(N-\mu)} e^{j\mu\omega\delta t}} \tag{14}$$

is a good approximation to the real propagation filter (5) in the desired frequency range.

The following chart shows the signal flow in the system *filter – waveguide* and illustrates equations (11) and (12):

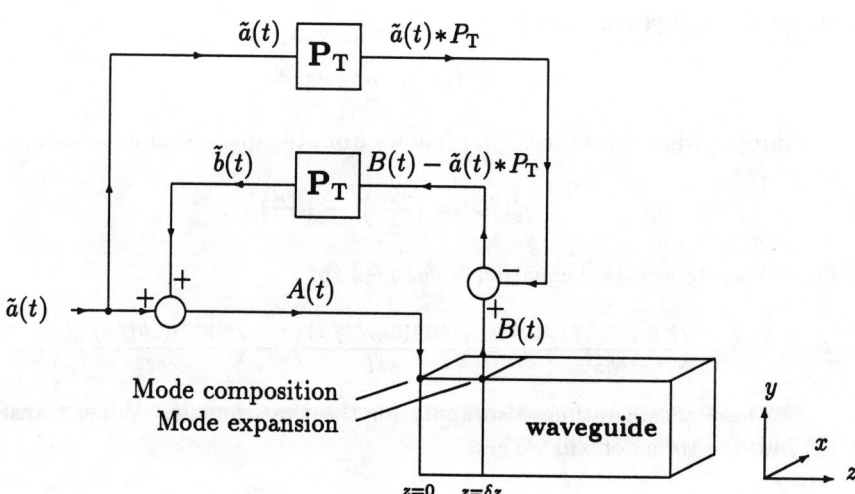

Figure 1: Signal flow in the system *waveguide – filter*

To supress the non-propagating modes the filter P_T consists of the digital filter P' and of a piece of a 1-dimensional transmission line, which is an exact model of the 3-dimsional waveguide for a particular mode and damps the modes below their cutoff-frequency:

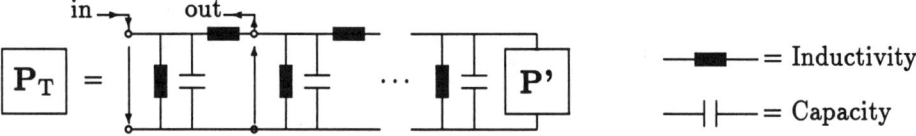

Figure 4: Filter model plus transmission line

EXAMPLES

The first example is a S-band 3db T-splitter designed for very high power. Fig. 1 shows a half of the symmetrical structure. The power coupled in at Port 1 is to be split equally to the Ports 2 and 3 (opposite of Port 2) with an as low as possible reflection around the S-band frequency of 2.9973251 GHz. Figs. 2 and 3 show the exciting pulse at Port 1 in the time domain and its frequency spectrum the range of 2.5 – 4 GHz. The field at the boundary was expanded into the first two (four) 2d-eigenmodes at Port 1 (Port 2), and recursive filters of the order 4 were used to approximate the boundary conditions. Because of the symmetry the problem could be treated as quasi-2d and was discretized into 18876 mesh points. To achieve a high resolution of frequency more than 200000 time steps were performed resulting in a rather high frequency resolution of $\Delta f = \frac{1}{N\Delta t} = 0.006$ GHz. The calculation took about 90 min cpu-time on a HP9000/720 workstation. The cpu-time for the fast Fourier transform of the sampled time data is negligble. Fig. 4 shows the calculated reflexion of the T-splitter in the frequency range of 2.5 – 4 GHz. The sharp peak at the S-band frequency can clearly be seen.

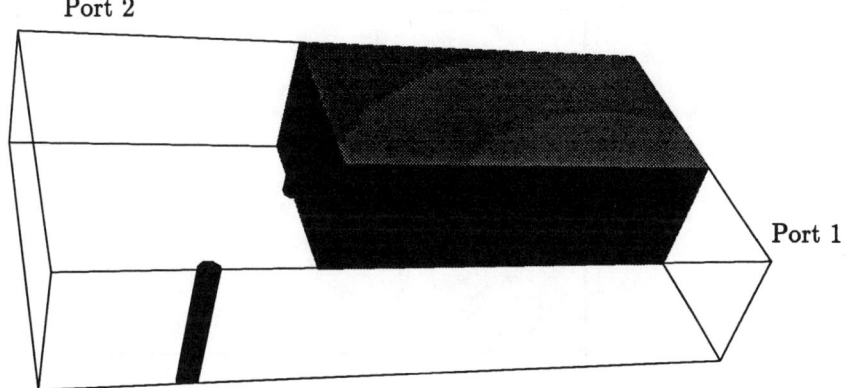

Figure 1: Geometry of the T-splitter (half of the structure)

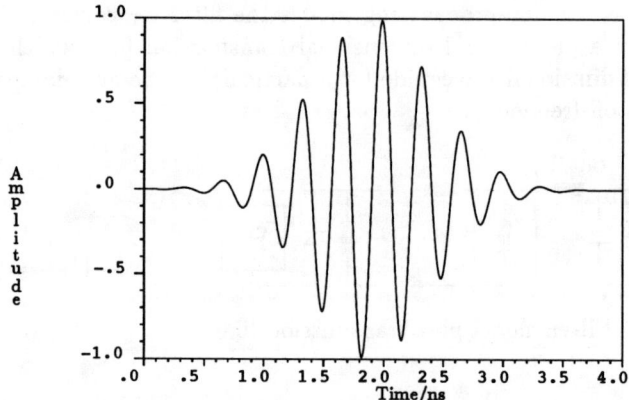

Figure 2: Amplitude over time of the excitation at Port 1

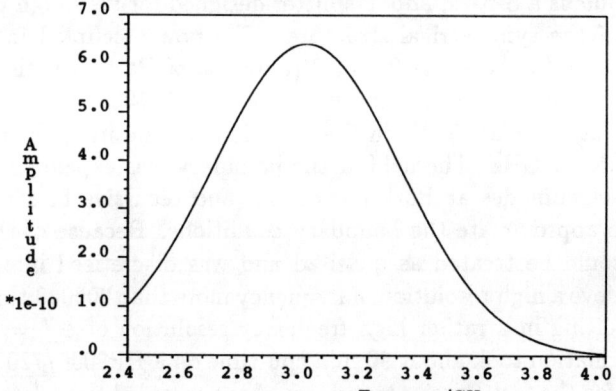

Figure 3: Frequency spectrum of the excitation at Port 1

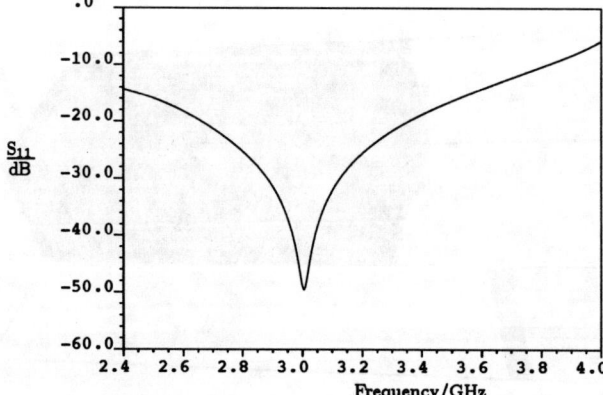

Figure 4: Reflexion S11

The next example is a high power coupler for an S-band linear collider[4]. Fig. 5 shows a quarter of the symmetrical structure. The beampipe runs along the z-axis and is bounded by the waveguide ports Port 1 and Port 3, respectively. The structure is fed symmetrically. This means the power coming from the klystron is split equally to two rectangular waveguides which are coupled to the structure opposite of each other. The upper half of one of those two waveguides is shown in the picture (Port 2). The decoupling of the power is done analogously. Again only the upper part of one decoupling port is shown (Port 4).

This example was discretized into 280000 mesh points. The boundary field at the beampipe ports was expanded into the fundamental mode, at the coupling/decoupling ports the first 7 eigenmodes were taken. The order of the digital filters used was between 3 (lower modes at Ports 2 and 4) and 6 (beampipe and higher modes at Ports 2 and 4). The cpu-time was 23 hours on a HP9000/720 workstation. After about 54000 time steps the signal already was sufficiently decayed so that the FFT could be performed with $2^{18}=262144$ steps which yielded a resolution of $\Delta f = 0.002$ GHz. The reflexion (in: Port 2 ; out: Port 2) is shown in Fig. 6, the transmission (in: Port 2; out: Port 4) in Fig. 7 for the frequency range of 2.8–3.2 GHz.

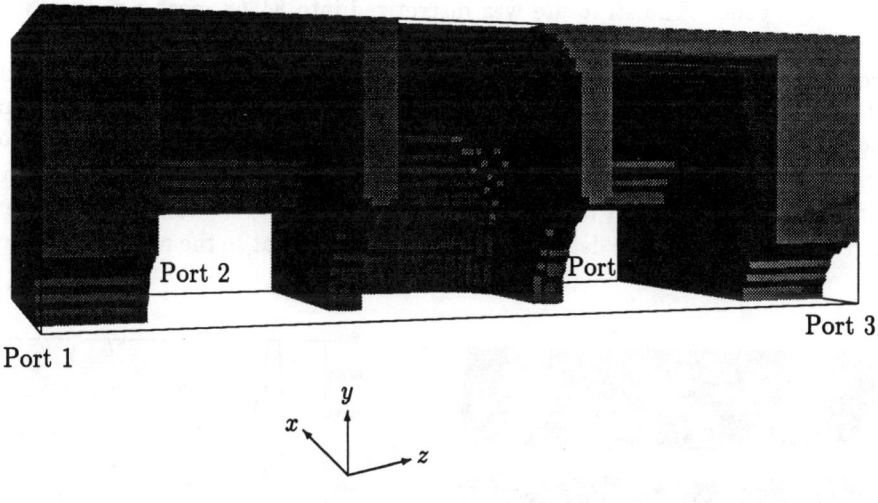

Figure 5: Geometry of the coupler (quarter of the symmetrical structure)

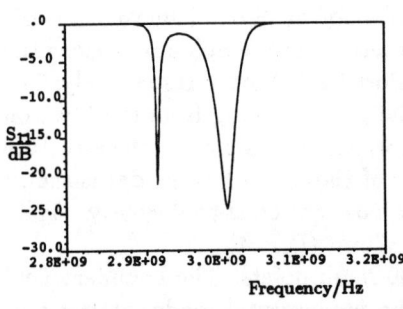

Figure 6: Reflexion (in: Port 2; out: Port 2)

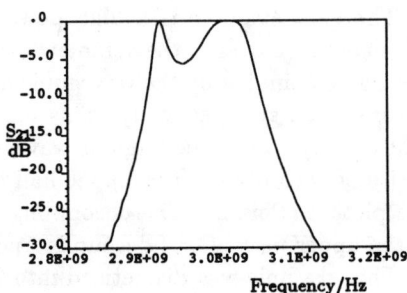

Figure 7: Transmission (in: Port 2; out: Port 4)

The last example is a band-stop filter operating in the X-band. The filter is a rectangular waveguide with 3 resonators on two sides. The geometry (half of the symmetrical structure) is shown in Fig. 8. For accurate results it was sufficient to take the first four 2d-eigenmodes for the expansion of the field at the boundary. The recursive filter was only of the order 3, but already the results were very good. The structure was discretized into 84868 mesh points. 132000 time steps were performed, the cpu-time was 8 hours on a HP9000/720 workstation. The frequency resolution is $\Delta f = 0.002$ GHz. Figure 9 shows the reflexion of the band-stop filter in the frequency range of 9-15 GHz, Fig. 10 the transmission in the same range. Figure 11 for comparison shows the results of the mode matching technique[5] and the results of our simulation in the region of the first peak in the transmission curve (9-11 GHz). The correspondence of these results is remarkably good. Our data are also in good agreement to the measurement by[6].

Figure 8: Geometry of the dual band filter (half of the symmetrical structure)

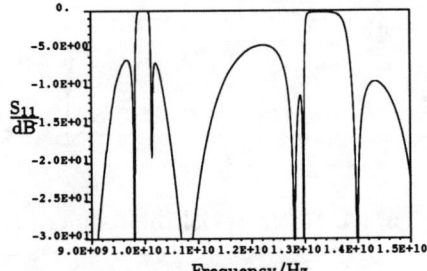

Figure 9: Reflexion S11

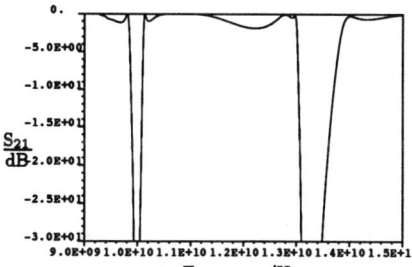

Figure 10: Transmission S21

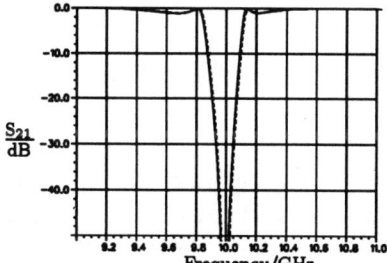

Figure 11: Comparison between orthogonal expansion (dashed line) and the simulation (solid line) for the transmission in the range 9 – 11 GHz

CONCLUSION

We presented a method to calculate scattering parameters in the time domain by a broadband excitation. By performing a large enough number of time steps a very high frequency resolution can be achieved. This method is (for the same resolution) much faster than frequency domain methods.

The method already has been tested in the design of rf-filters where our results always were in very good agreement to measurements and/or the results of the mode matching technique. The accelarator components shown in the examples are principally the same as rf-filter devices so we are quite confident our results are correct even though we could not compare them to measurements.

REFERENCES

1. T. Weiland: Solving Maxwell's Equations by Means of the MAFIA CAD-System. Talk presented at the Beijing International Symposium on Electromagnetic Fields in Electrical Engineering BISEF'88, Beijing, China, October 19-21, 1988
2. M. Bartsch et. al.: Solution of Maxwell's Equations.
   Elsevier, Computer Physics Communications (to be published)
3. M. Dohlus: Ein Beitrag zur numerischen Berechnung elektromagnetischer Felder im Zeitbereich. Dissertation an der TH Darmstadt, Juni 1992
4. T. Weiland (for the S-Band Collider Group),"Status of the DESY-THD 500 GeV S-Band Linear Collider Study", Talk at the LC92 in Garmisch Patenkirchen, to be published
5. B. B. Geib: private communications
6. L. Young, G. L. Matthaei, E. M. T. Jones: Microwave Band-Stop Filters With Narrow Stop Bands. IRE Transactions on Microwave Theory and Techniques, November 1962, pp 416-427

AN INTEGRO-ALGEBRAIC EQUATION FOR HIGH FREQUENCY WAKE FIELDS IN A TUBE WITH SMOOTHLY VARYING RADIUS*

Robert L. Warnock
Stanford Linear Accelerator Center, Stanford University
Stanford, California 94309

ABSTRACT

A technique to find the longitudinal wake field at frequencies above or below the tube cut-off is described. The round tube is infinite in length, and has an arbitrary, smooth variation of radius over a finite interval. A system of integro-algebraic equations enforces the boundary conditions on the wall and the outgoing wave condition at infinity. The first step in an iterative solution of the system, valid for variations of tube radius with small derivative, yields a convenient formula for the impedance as a double integral. At low frequencies the formula gives Yokoya's result plus corrections that can be large. For high frequencies in the case of several wall undulations it gives a sequence of finite-Q resonances. To avoid the limitations of the iterative method, a numerical solution of the system is carried out.

DERIVATION OF THE EQUATION

Other papers of this conference demonstrate the impressive utility of computer codes to determine electromagnetic fields in the presence of conducting walls of general form. Nevertheless, codes designed for arbitrary problems are not likely to answer all questions of interest. One instance is the wake field of a short or irregular bunch, with high-frequency components requiring an extremely fine mesh. The method described here may be a useful complement to general purpose programs. It will handle some problems that would normally require an overly fine mesh, it gives analytic formulas valid for a wide class of mild wall perturbations, and allows tests of accuracy merely by checking of boundary conditions. The method will be described for the case of the longitudinal wake field in a tube with circular cross section and infinite conductivity. It can be extended to treat transverse fields, wall resistance, and pipes of rectangular cross section.

We take cylindrical coordinates (r, ϕ, z) and suppose that the tube radius is given as $R(z) = b - \epsilon s(z)$, where the function $s(z)$ is zero for $|z| > g$, and not necessarily even in z. We assume that s has a continuous first derivative s', normalized so that $\max |s'| = 1$; thus $s'(\pm g) = 0$. With this normalization, ϵ measures the effective strength of the wall perturbation; a perturbative method may succeed if ϵ is small compared to 1.

We work in the frequency domain, with the time dependence $\exp(-i\omega t)$. Attention is restricted to positive values of ω, which suffice to express the wake field, thanks to the reflection property of the impedance, $Z(\omega) = Z(-\omega)^*$. The source is assumed to be axisymmetric, a rigid bunch with total charge q and charge density $\rho(r, \phi, z, t) = (q/2\pi)\lambda(z - \beta ct) f(r)$ where $\int \lambda(z) dz = 1$, $\int f(r) r dr = 1$. It follows that the only non-zero fields are (E_z, E_r, H_ϕ), all independent of ϕ. All fields may be expressed in terms of E_z, which can be written as

$$E_z(r, z, \omega) = \int_\Gamma dk e^{ikz} a(k, \omega) \frac{I_o(\chi r)}{I_o(\chi b)} + e_z(r, z, \omega) \quad , \tag{1}$$

where $\chi^2 = k^2 - (\omega/c)^2$, and I_o is the modified Bessel function of the first kind. The Fourier transform $\hat{e}_z$ of the source term e_z is any particular solution of the inhomogeneous radial wave

---

* Work supported by the Department of Energy, contract DE-AC03-76SF00515.

equation for E_z, regular at $r = 0$. A convenient choice is

$$\hat{e}_z(r,k,\omega) = (i\chi^2 c/2\pi k) Z_o q \hat{\lambda}(\omega/c)\delta(\omega-kv)\left[I_o(\chi r)\int_b^r udu K_o(\chi u) - K_o(\chi r)\int_0^r udu I_o(\chi u)\right]f(u) \quad , \tag{2}$$

where K_o is the modified Bessel function of the second kind, and $\hat{\lambda}$ is the Fourier transform of λ. We work in m.k.s. units, with $Z_o = 120\pi\ \Omega$. The Fourier amplitude of the radial field is

$$E_r(r,k,\omega) = -\frac{ik}{\chi^2}\frac{\partial E_z(r,k,\omega)}{\partial r} \quad . \tag{3}$$

The function $I_o(\chi b)$ has simple zeros in the k-plane at the points $k = \pm k_s, s = 1, 2, \cdots$, where $k_s = ((\omega b/c)^2 - j_{os}^2)^{1/2}$ is defined to be positive for $|\omega b/c| > j_{os}$ and positive imaginary for $|\omega b/c| < j_{os}$; the j_{os} are the positive zeros of the Bessel function J_o. We define s_m to be the number of real k_s. At any frequency above the lowest cutoff ($\omega b/c = j_{o1}$), there are $2s_m$ zeros on the real axis. To dodge the corresponding real poles of the integrand in (1), the contour Γ is indented slightly so as to go above the poles at $k = -k_s$ and below those at $k = k_s$. This choice enforces the outgoing wave boundary condition. We have assumed that $a(k,\omega)$ is analytic in k; our construction of solutions will in fact yield an entire function of k.

The boundary condition on the wall is that $\mathbf{E} = (E_r, E_z)$ be perpendicular to the tangent vector (dR, dz), or

$$E_z(R(z), z, \omega) + R'(z)E_r(R(z), z, \omega) = 0 \quad . \tag{4}$$

This condition leads to an equation for $a(k,\omega)$ through the following steps: (i) write E_z as in (1), and the corresponding expression for E_r constructed from (3); (ii) take the Fourier transform of (4) with respect to z; (iii) subtract $I_o(\chi b)$ from $I_o(\chi R(z))$ in the integrand, and notice that the compensating addition gives $\delta(k-l)$. The result is

$$a(l,\omega) = \int_\Gamma dk M(l,k,\omega) a(k,\omega) + \hat{S}(l,\omega) \quad , \tag{5}$$

where

$$M(l,k,\omega) = \frac{1}{2\pi}\int_{-g}^{g} dz \frac{e^{i(k-l)z}}{I_o(\chi b)}\left[I_o(\chi b) - I_o(\chi R(z)) + \frac{ik}{\chi}R'(z)I_o'(\chi R(z))\right] \quad , \tag{6}$$

and $\hat{S}$ is the Fourier transform of

$$S(z,\omega) = -\int_{-\infty}^{\infty} dk e^{ikz}\left[e_z(R(z), z, \omega) - \frac{ik}{\chi^2}R'(z)\frac{\partial e_z}{\partial r}(R(z), z, \omega)\right] \quad . \tag{7}$$

Henceforth we treat only the relativistic limit. In that limit the source term simplifies: (2) and (7) yield

$$S(z,\omega) = \frac{\epsilon Z_o s'(z)}{2\pi R(z)}\hat{\lambda}(\omega/c) e^{i\omega z/c} \quad . \tag{8}$$

Below cutoff, (5) is an integral equation [1] for $a(k,\omega)$. Above cutoff it is an integro-algebraic equation, since the values $a(k,\omega)$ at the poles on the real axis constitute a discrete set of unknowns to be determined along with the continuous, nonpolar part. These values determine the amplitudes of outgoing waves.

By reversing the order of integrals we see that any solution of (5) may be written in the form

$$a(k,\omega) = \frac{1}{2\pi} \int_{-g}^{g} e^{-ikz} \Phi(z,\omega) dz \quad . \tag{9}$$

Since the integral is finite, $a(k,\omega)$ is an entire function of k, as promised.

FORMULAS FOR THE IMPEDANCE, FROM THE LOWEST ITERATE

An integration by parts on the first two terms of (6) puts the kernel in the form

$$M(l,k,\omega) = \frac{\epsilon}{2\pi} \frac{kl - (\omega/c)^2}{\chi(k-l)} \int_{-g}^{g} dz e^{i(k-l)z} R'(z) \frac{I'_o(\chi R(z))}{I_o(\chi b)} \quad . \tag{10}$$

This shows that the kernel is formally $\mathcal{O}(\epsilon)$, and therefore suggests that the equation (5) might be solved by iteration when ϵ is small. The first approximation is obtained by putting $a = \hat{S}$ under the integral in (5). Since the impedance is proportional to $a(\omega/c,\omega)$, and $\hat{S}(\omega/c,\omega) = 0$, the lowest order impedance is $\mathcal{O}(\epsilon^2)$.

To evaluate the approximated integral of (5) at the synchronous point $k = \omega/c$ we express $\hat{S}$ in terms of its Fourier transform and reverse integration order to obtain

$$a(\omega/c,\omega) \approx \frac{1}{2\pi} \int_{-g}^{g} dz e^{-i\omega z/c} R'(z) \int_{-g}^{g} dz' S(z',\omega) K(z,z',\omega) \quad , \tag{11}$$

where

$$K(z,z',\omega) = \frac{1}{2\pi i} \frac{\omega}{c} \int_{\Gamma} dk e^{ik(z-z')} \frac{I'_o(\chi R(z))}{\chi I_o(\chi b)} \quad . \tag{12}$$

The integral (12) converges exponentially if $R < b$, but diverges for $R > b$. The divergence is an unwanted limitation since we wish to allow arbitrary R. By performing a contour distortion one can continue the integral analytically from $R < b$ to $R > b$, and incidentally gain other benefits. Taking $R < b$ we let the contour become an infinite semi-circle in the upper (lower) half-plane for $z - z'$ positive (negative). The result is

$$K(z,z',\omega) = \frac{\omega}{c} \sum_{s=1}^{\infty} \frac{J_1(j_{os}R(z)/b)}{k_s(\omega) b J_1(j_{os})} e^{ik_s(\omega)|z-z'|} \quad . \tag{13}$$

For $z \neq z'$ the sums converge exponentially, regardless of the value of R. At $z = z'$ and $R = b$ the sum diverges, but if the integral on z' is performed first there will be an extra inverse power of k_s and quadratic convergence, uniform in R.

The formula (11) now involves powers of ϵ higher than the second through the presence of $R(z)$ in the denominator of (8) and in (13). To pick out just the ϵ^2 part we put $R(z) = b$ in both locations. Invoking the usual definition of the impedance in terms of the wake potential, we find $Z(\omega) = -2\pi a(\omega/c,\omega)/(q\hat{\lambda}(\omega/c))$. Then from (8), (11), and (13) we have the impedance to lowest order in ϵ as

$$Z(\omega) = \frac{\omega Z_o \epsilon^2}{2\pi c b^2} \sum_{s=1}^{\infty} \frac{1}{k_s(\omega)} \int_{-g}^{g} dz \int_{-g}^{g} du s'(z) s'(u) e^{ik_s(\omega)|z-u| - i\omega(z-u)/c} \quad . \tag{14}$$

Below cutoff the k_s are all positive imaginary, and the impedance is reactive as required; (the integral is real, since the integrand goes into its complex conjugate on $z \to u$).

A closer look shows that the formula (14) is actually invalid for the frequency ω in a small neighborhood of each traveling wave cutoff, where $\omega b/c = j_{os}$. The kernel K has an inverse square-root singularity at such points, owing to the factor $1/k_s$ in (13). It is therefore not small near such frequencies, and the iterative method fails. The same singularity appears in the field expansion (1), from the residue of the pole as it strikes the real axis. It is cancelled by a corresponding zero of $a(k,\omega)$ at $k = \pm k_s$, so that the amplitude of the newly appearing outgoing wave is finite. We have verified that this mechanism operates in the numerical calculation of the following section, but it is a "nonperturbative" effect that cannot take place in a lowest order calculation. In plotting results from (14), we delete small neighborhoods of the bad points, and let the plotting program interpolate nearby values to fill in the gaps. This is justified by the smooth behavior of Z found in the numerical solutions.

As a first application of (14) we take the frequency below cutoff and perform a partial integration of the u-integral. For this we assume that s'' is at least piecewise continuous. The integrated term yields an s-sum that can be evaluated (namely $\sum_s j_{os}^{-2} = 1/4$), and the corresponding contribution to the impedance is the result obtained previously by Yokoya [2] through a much more involved method:

$$Z(\omega) = -i\frac{\omega}{4\pi c}Z_o\epsilon^2 \int_{-g}^{g} dz[s'(z)]^2 \quad . \tag{15}$$

The term remaining after partial integration can be large, however, and is nonlinear in ω. One expects some correction in any case, since the fields in general must depend on b/g, while there is no such dependence in (15). As an example we take $s(z) = d(1+\cos(\pi z/g))/2$, $|z| < g$, for which $\epsilon = \pi|d|/(2g)$. In the small ω limit a full evaluation of (14) gives in this case

$$Z(\omega) = -i\frac{\omega g}{c}Z_o\frac{\epsilon^2}{4\pi}\left[1 + \left(1 - 2\frac{I_1(\pi b/g)}{(\pi b/g)I_0(\pi b/g)}\right) + 4\pi^2\left(\frac{b}{g}\right)^3 \sum_{s=1}^{\infty}\frac{1 - e^{-2j_{os}g/b}}{j_{os}(j_{os}^2 + (\pi b/g)^2)^2}\right] \quad . \tag{16}$$

The first term in the square bracket is the Yokoya term, the second term vanishes as $(b/g)^2$ and the third as $(b/g)^3$, when $g \to \infty$. The correction to the Yokoya term (mainly from the second term) is about 50% at $2g/b = 2$, and approaches 100% as $g \to 0$. Figure 1 shows the sum of the second and third terms as a function of $2g/b$.

As an example for arbitrary frequency, we generalize the previous example to allow many oscillations of the radius, taking $s(z) = d(1+\cos(\pi pz/g))/2$, where p is an odd integer, thus $\epsilon = \pi p|d|/(2g)$. An exact evaluation of (14) for this model yields

$$Z(\omega) = \frac{\omega b}{c}Z_o\frac{\epsilon^2}{2\pi}\left[-i\frac{g}{2b}[h(\omega) + h(-\omega)] + \left(\frac{\pi pb}{g}\right)^2 \sum_{s=1}^{\infty}\frac{1}{bk_s(\omega)}[f(k_s(\omega),\omega) + f(k_s(\omega),-\omega)]\right] \quad , \tag{17}$$

where

$$h(\omega) = \frac{I_1(\xi)}{\xi I_0(\xi)} \quad , \quad \xi = \left[\left(\frac{\pi pb}{g}\right)^2 + 2\frac{\omega b}{c}\left(\frac{\pi pb}{g}\right)\right]^{1/2} \quad , \quad f(k,\omega) = \frac{1 - \exp[2i(k-\omega/c)g]}{b^4[(k-\omega/c)^2 - (\pi p/g)^2]^2} \quad . \tag{18}$$

The first term is imaginary at all ω, and so is the sum for $s > s_m$. The sum for $s \le s_m$ is complex, and of course present only above cutoff. Its real part is nonnegative as it should be, corresponding to energy lost by the bunch to outgoing waves. The term $h(-\omega)$ has poles, but they are cancelled by corresponding poles in the sum on s.

Figure 2 shows a plot of formula (17) for $p = 1$, $b = 1$cm, $g = 6$cm, $d = 0.3$cm, thus $\epsilon = 0.078$. When the number of wall undulations is increased, a sequence of resonances of decreasing Q appears. Figures 3 and 4 show results for $p = 5$, $b = 1$cm, $g = 6$cm, $d = 0.06$cm, thus the same $\epsilon = 0.078$. Since the impedance simply scales with d^2, it retains its complicated form for arbitrarily small d, while decreasing in magnitude quadratically.

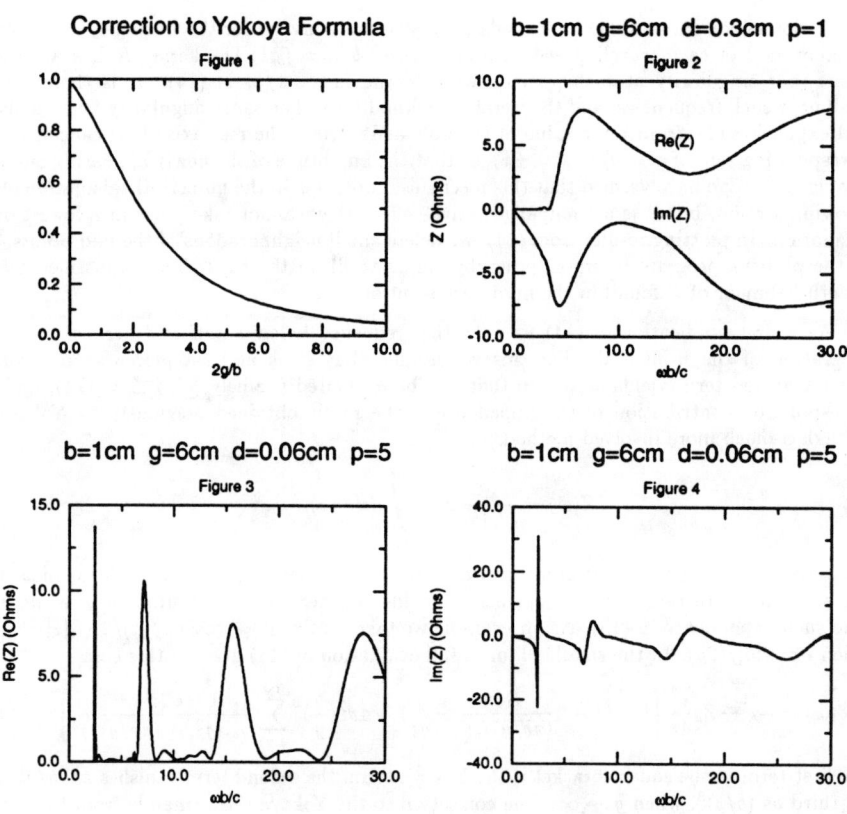

NUMERICAL SOLUTION

For a numerical solution we formulate the integral equation in a different way. In (1) we translate the contour a *finite* distance α into the upper (lower) half-plane for $z > 0$ ($z < 0$), respectively. Thus we have two separate representations of E_z for the two signs of z, as follows:

$$E_z(r,z,\omega) - e_z(r,z,\omega) = e^{-\alpha|z|} \int_{-\infty}^{\infty} dv \left[\frac{a(k,\omega)I_o(\chi r)}{I_o(\chi b)} \right]_{k=v \pm i\alpha} e^{ivz}$$
$$\pm 2\pi i \sum_{s=1}^{n} \frac{j_{os} J_o(j_{os}r/b)}{bJ_1(j_{os})} \left[\frac{a(k,\omega)e^{ikz}}{kb} \right]_{k=\pm k_s(\omega)} \quad (19)$$

The second term represents outgoing waves moving to the right (left), plus any evanescent waves with $0 < |\text{Im } k_s| < \alpha$. The integral decays exponentially as $|z| \to \infty$. Each traveling or evanescent wave in the second term of (19) automatically satisfies the boundary condition $E_z(b,z,\omega) = 0$ in the region $|z| > g$ where the tube has constant radius b. Moreover, in some region $|z| > h > g$ beyond the wall perturbation the decaying integral will be negligible if α is sufficiently large. Thus to satisfy the wall boundary condition for the entire infinite tube we need only take steps to impose it on the finite region $|z| < h$. That can be done numerically by taking a discrete Fourier transform of the boundary condition over this region.

Let us first derive the exact continuum equations, and later discretize. We substitute the composite representation (19) of E_z and the corresponding form for E_r in the boundary condition (4), subtract and add $I_o(\chi b)$ as we did before, then move the contour of the added term (which entails no poles) back to the real axis. Then take the Fourier transform on z, and get a free-standing $a(l,\omega)$ on the left hand side. Finally continue the result analytically to $l = u \pm i\alpha$ and $l = \pm k_t$ to obtain the following system of integro-algebraic equations:

$$a(u \pm i\alpha) = \int_{-\infty}^{\infty} dv \left[K_+(u \pm i\alpha, v + i\alpha)a(v + i\alpha) + K_-(u \pm i\alpha, v - i\alpha)a(v - i\alpha) \right]$$

$$+ \sum_{s=1}^{n} [L_+(u \pm i\alpha, k_s)a(k_s) + L_-(u \pm i\alpha, -k_s)a(-k_s)]$$

$$+ \hat{S}(u \pm i\alpha) \quad , \qquad u \in [-\infty, \infty] \quad ,$$

$$a(\pm k_t) = \int_{-\infty}^{\infty} dv \left[K_+(\pm k_t, v + i\alpha)a(v + i\alpha) + K_-(\pm k_t, v - i\alpha)a(v - i\alpha) \right]$$

(20)

$$+ \sum_{s=1}^{n} [L_+(\pm k_t, k_s)a(k_s) + L_-(\pm k_t, -k_s)a(-k_s)]$$

$$+ \hat{S}(\pm k_t) \quad , \qquad t = 1, 2, \cdots, n \quad .$$

We have suppressed all reference to ω, and have defined the kernel functions as

$$K_\pm(l,k) = \frac{\pm 1}{2\pi} \int_0^{\pm g} dz e^{i(k-l)z} \frac{1}{I_o(\chi b)} [I_o(\chi b) - I_o(\chi R(z)) + \frac{ik}{\chi} R'(z) I'_o(\chi R(z))] \quad ,$$

(21)

$$L_\pm(l,k_s) = \pm i \int_0^{\pm g} dz e^{i(k_s - l)z} \frac{j_{os}}{k_s b^2 J_1(j_{os})} [-J_o(j_{os} R(z)/b) + \frac{ik_s b}{j_{os}} R'(z) J_1(j_{os} R(z)/b)] \quad .$$

In the numerical realization we first redefine the integration variable v by a translation so that its origin corresponds to the point $k = \omega/c$, since the source term is small outside a neighborhood of that point. We then write each v-integral of (19) as a discrete Fourier transform on a finite interval $[-V, V]$. The inverse discrete transform to enforce the boundary condition is defined for $z \in [-h, h]$. The choice of the parameters h, V, α requires a little experimentation, but is not very critical. The contour displacement α must be chosen so that the contour is not too close to an imaginary k_s. In examples studied to date, values of αg in the range 3 to 10 have been satisfactory. All experimentation with solution control parameters is guided by checking satisfaction of the wall boundary condition. Since the fields we construct are automatically solutions of Maxwell's equations, and satisfy the outgoing wave condition, this is a definitive test of accuracy of the solution.

Since we do numerical Fourier analysis of functions with a factor $s'(z)$, the method works best if s'' is continuous. Accordingly, we give results for a model with such continuity, $s(z) = (d/4)(1 + \cos(\pi p z/g))^2$, $|z| < g$, with p odd.

Figure 5 shows the impedance computed for $p = 1$, $b = 1$cm, $g = 6$cm, $d = 0.3$cm, up to a high enough frequency (134 GHz) to compute the wake field of a 1.2mm Gaussian bunch. Note the similarity to Figure 2, recalling that $s(z)$ for Figure 2 is the square root of the present $s(z)$. The computation used 192 complex Fourier amplitudes to represent the two discretized integrals. The computation time was 4 to 5.5 seconds per frequency on the IBM 3090; the larger times occur when there are several outgoing waves. Figure 6 shows the corresponding wake voltage at a distance x from the center of the bunch, in units of bunch length $\sigma = 1.2$mm. Figure 7 shows a test of accuracy for this solution, namely a plot of $\eta = |E_t(R(z), z, \omega)/E_n(R(z), z, \omega)|$ at $\omega b/c = 4.8$ where E_t and E_n are the tangential and normal components of **E** on the wall. The boundary condition is well satisfied at this frequency, with $\eta < 10^{-6}$; the accuracy degrades somewhat at higher frequencies but overall $\eta < 10^{-4}$. Figure 8 derives from the field on axis; it is a plot of $\text{Re}[E_z(r = 0, z, \omega)\exp(i\omega z/c)]$ at $\omega b/c = 14.4$. The oscillations at positive z correspond to the superposition of several outgoing waves with incommensurate wavelengths. The bunch and these waves move to the right; left-going waves are present but very weak, and become strong only in certain narrow frequency bands. Figure 9 shows part of the solution of the equations (20), the continuum component $\text{Re}[a(k, \omega)/I_o(\chi b)]$ for $k = v + i\alpha$ on the upper contour, at $\omega b/c = 14.4$. Figure 10 shows an axial field plot like that of Figure 8 but for $p = 5$ (with d reduced by a factor of 5 to give the same ϵ), at $\omega b/c = 4.8$.

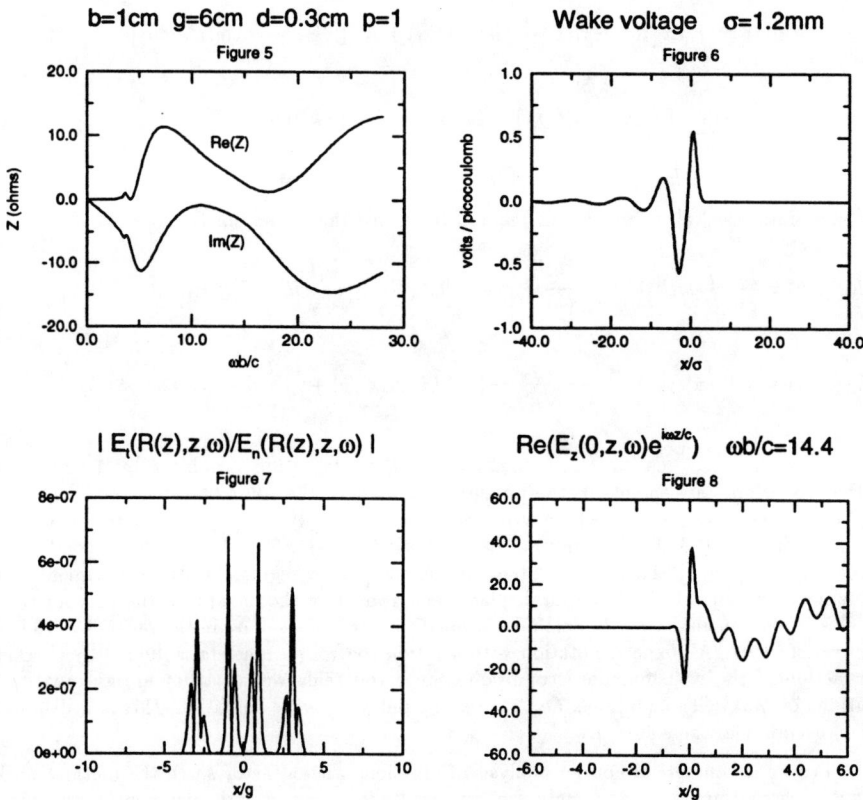

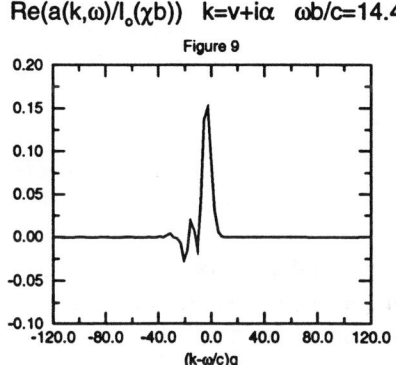
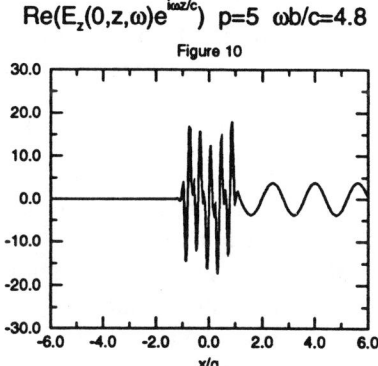

CONCLUSION

We have shown that formula (14) predicts a rich pattern of high frequency effects. A detailed study of this formula, which is easy to evaluate for many different forms of the wall perturbation, should lead to useful insights regarding parameter dependence of wake fields. We have shown that the numerical treatment of the integro-algebraic system can produce very accurate solutions of the boundary value problem. Comparison and parallel further development of the analytic and numerical approaches should be rewarding.

I wish to thank Robert Gluckstern, Karl Bane, and J. Scott Berg for much good advice and technical help.

REFERENCES

1. An equation with the same kernel was derived by S. S. Kurennoy and S. V. Purtov, Particle Accelerators **36** 223 (1992). Their form of the source term is incorrect, however, as is their treatment of frequencies above cutoff.

2. K. Yokoya, CERN SL/90-88 (AP) (1990).

MODE ANALYSIS OF CIRCULAR WAVEGUIDE STRUCTURES USING THE MODE MATCHING TECHNIQUE

U. van Rienen
Technische Hochschule Darmstadt, Fachbereich 18, FG TEMF,
Schloßgartenstr. 8, 6100 Darmstadt, Germany

ABSTRACT

In the future High Energy Physics will need very long linear colliders for e^+e^--collisions in the TeV range to verify the laws of elementary particles with sufficiently high luminosity. Most designs currently under discussion use multibunch operation. Therefore the study of higher order modes excited by previous bunches in the train becomes even more important for the optimal design of the accelerator components. Usually the electromagnetic properties of these components are calculated by computer codes that discretize Maxwell's equations. For long tapered disc-loaded waveguides however, these methods would need the solution of extremely large algebraic eigenvalue problems with many clustered eigenvalues. This is numerically difficult.

In this paper the electromagnetic waves in tapered circular multi-cell structures are calculated by a modal field matching method. The structure is subdivided into subregions of constant circular cross section. In each subregion the electromagnetic fields are expanded in orthogonal series over discrete modes. The field solutions are obtained by field matching at the transverse interfaces between the subregions. The scattering matrix formulation is used to calculate the amplitudes of the field expansion. This formulation assures numerical stability. Numerical results are obtained by appropriate restriction to a finite number of modes. Some convergence investigations are presented as well as comparisons with a grid-oriented numerical method and an equivalent circuit model. The field distribution of deflecting dipole-modes is analyzed for a 180-cell tapered disc-loaded S-band-structure.

INTRODUCTION

A very important aspect in the design of accelerating components is their influence on the beam. The beam can be deflected by the (so-called) higher order modes. Mainly modes of the dipole passband are of great importance. This subject can be studied by the Wake Potential which is calculated from the transverse loss factors of the dipole fields. Besides the quantitative Wake Potential also qualitative knowledge about the field distribution of the modes inside the accelerating structure is important in order to design higher order mode couplers.

In tapered multicell structures, as they are often used in linear accelerators, the dipole fields are expected to be captured in only a small part of the structure. In order to damp the modes with the highest loss factors it is important to calculate the fields in the *whole* structure.

While the ideas about the qualitative behaviour of the higher order modes in tapered multicell structures could be approved in using established grid-oriented computer codes like MAFIA[1] or URMEL(-T) [2,3] for short models with strong tapering, it is necessary to use non-grid-oriented methods to calculate the fields in the complete structure with realistic dimensions. One possible method are equivalent circuits models[4] but the influence of the chosen model on the results is not neglectable. Another method, often used for periodic structures, is the mode matching technique. This method presents a very capable tool for long tapered waveguide structures.

SOME CHARACTERISTICS OF TAPERED WAVEGUIDES

A tapered waveguide structure is matched for the accelerating $2\pi/3$-mode. All other modes have in general an aperiodic field distribution. This fact is of special interest for the higher order modes. The fields of these parasitic modes do not extend over the whole structure but are trapped in relatively few cells. Consequently their maximal loss-factor is smaller than in a corresponding periodic structure. However the sum of all loss-factors in one passband approximately is equal for both cases. Thus the aperiodic structure has many modes with substantial loss-factor while the periodic one has only few modes that significantly interact with the beam. This general behaviour is displayed in Fig. 1.

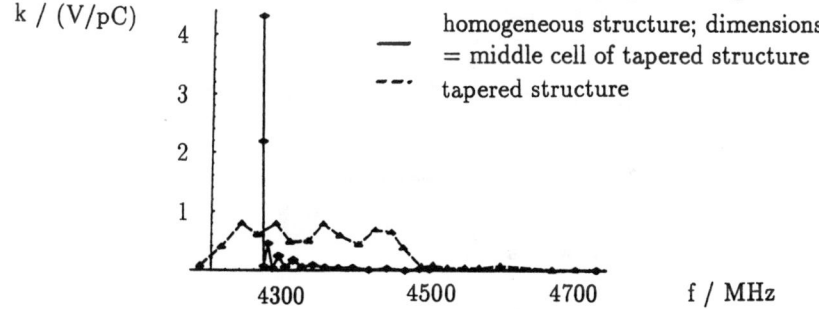

Figure 1: *Loss-factor calculated by URMEL–T for a tapered structure with 18 cells and for a periodic structure with 18 times the middle cell of the previous one.*

At SLAC, which uses a 3m long S-Band structure of 84 cells, it was found[5] that the most severe dipole modes are "π-like" modes which only extend over a few cells and are trapped in the first part of the structure. The question was whether this behaviour would also be found for an S-Band structure of 180 cells with smaller cell-to-cell deviations in the geometry than the SLAC structure.

This subject is very important in order to design features for damping of these modes. The so-called Higher Order Mode Couplers are outgoing waveguides above the cutoff-frequency of the accelerating mode in order to couple out the parasitic dipole modes. If placed near the beginning of the structure they couple out modes with fields nearby. Thus modes trapped near the middle of the structure are not damped by couplers located near the end.

THE MODE MATCHING METHOD

Only a short overview over the most important features concerning the Modal Field Matching are given here since it is a well known method in electrodynamics[6]; for more details see e.g. Steinigke [7].

The Modal Field Matching Technique can be applied for structures which split up into subregions where analytic solutions of Maxwell's Equations can be given as expansion in a series over orthogonal discrete modes. The field solutions for the whole structure are then obtained by field matching at the interfaces between the subregions. The placement of these interfaces is problem-dependant. The tapered multicell structures as used for acceleration of elementary particles have in general to be simplified to apply some Mode Matching Technique. For the structure studied below all rounded corners were approximated by sharp rectangular corners. Now the tapered structure can be subdivided by *transverse* interfaces into subregions of constant circular cross section. The transverse electric field in each subregion is given by some Bessel-Fourier-Series as[1] :

$$\mathbf{E}_t(\rho,\varphi,z) = \sum_{n=1}^{\infty} U_n(z)\,\mathbf{e}_n(\rho,\varphi). \tag{1}$$

Regard a step in a cylindrical waveguide as shown in Fig. 2. The coefficients $U_n(z)$

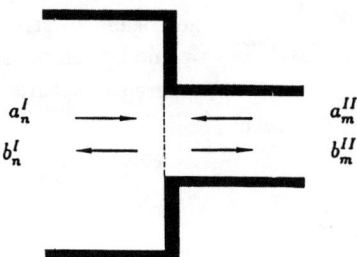

Figure 2: *Step in a cylindrical waveguide.*

and $I_n(z)$ in the Fourier-series, the so-called voltage and current amplitudes, correspond to a superposition of waves in positive and negative z-direction. Exploiting the continuity condition for the tangential fields $\vec{E}_t^{I,II}$ and $\vec{H}_t^{I,II}$ at common cross-sectional areas and the orthonormality of the eigenfunctions finally leads to the connection of U and I by the coupling matrices: $\mathbf{U}^I = \underline{\mathbf{C}}\,\mathbf{U}^{II}$; $\mathbf{I}^{II} = \underline{\mathbf{D}}\,\mathbf{I}^I$ with the current and voltage amplitudes written in vector form[2]. While this formulation is well suited for periodic structures it is more appropriate to use the scattering matrix formulation for aperiodic structures in order to avoid numerical instability. The wave amplitudes a and b of the incident and reflected resp. transmitted waves

---
[1] Similar expressions hold for the other electromagnetic field components.
[2] Note that these systems of equations are of infinite dimension.

are connected in a linear system by the scattering matrix:

$$\begin{pmatrix} \mathbf{b}^I \\ \mathbf{b}^{II} \end{pmatrix} = \begin{pmatrix} \underline{S}_{I,I} & \underline{S}_{I,II} \\ \underline{S}_{II,I} & \underline{S}_{II,II} \end{pmatrix} \begin{pmatrix} \mathbf{a}^I \\ \mathbf{a}^{II} \end{pmatrix}, \qquad (2)$$

This formulation is obtained from the coupling matrices by some transformations. The scattering matrix depends on the geometry and the frequency. In a structure with multiple changes in diameter the scattering matrix is constructed for each subregion. These matrices can be concatenated to the scattering matrix for the total structure or parts of it: $\underline{S} = \underline{S}_1 \odot \underline{S}_2 \odot \cdots \odot \underline{S}_n$. The outgoing amplitudes are calculated from the scattering matrix for the total structure. The inner amplitudes are calculated from the scattering matrices for the parts left and right of their location. The electromagnetic fields are calculated by the (truncated) Bessel-Fourier-Series from the inner amplitudes.

VALIDATION OF THE MODE MATCHING CODE

Even though the mode matching technique is used for many applications these formulations are different from the one presented here for higher order mode analysis. Only recently Heifets and Kheifets[8] calculated *monopole* modes by mode matching technique. The program ORTHO calculates *monopole, dipole and higher order modes*. In the following the results of ORTHO are compared with other methods of calculating the electromagnetic fields of higher order modes and the corresponding loss-factor[9]. Steinigke[7] presents further validations of the underlying formulation of mode matching. For more details see van Rienen[10].

Convergence investigations

The mode matching technique yields by virtue of its methods the exact solutions of Maxwell's equations. In order to use it for numerical field calculations the infinite Fourier-Bessel series (1) has to be truncated after a finite number of summands (i.e. modes). The well-known geometrical convergence criterium, defined for periodic structures, that the mode ratio should equal the ratio of the sections' radii was proven for the convergence of the accelerating mode in a tapered 9-cell structure.

The truncation of the Fourier expansion implies that the field strength will show discontinuities at the junction of adjacent sections. From system theory it is well known that overshooting occurs if a function with discontinuities is represented by a Fourier series. This overshooting *cannot* be eliminated by taking more modes into account. But the problem can be cured by a filter to weight either the eigenfunctions or the coupling matrix. Here a cosine filter for the coupling matrix was used.

The computing time needed by ORTHO is essentially determined by the matrix operations. Since the matrix dimensions depend quadratically on the number of modes which are included in the expansion this quadratic dependence is reflected in the functional dependence of computing time on number of modes.

Comparison with equivalent circuit model

In the equivalent circuit model each cell of the waveguide structure is represented by a resonant circuit. The complete structure is modelled by a chain of these resonant circuits. The modes of the circuit are computed by MAFIA. The coupling between the cells is performed by capacitive and inductive circuit elements[11]. The currents are obtained by solution of an eigenvalue problem. Many other models of equivalent circuits are possible e.g. Bane[12].

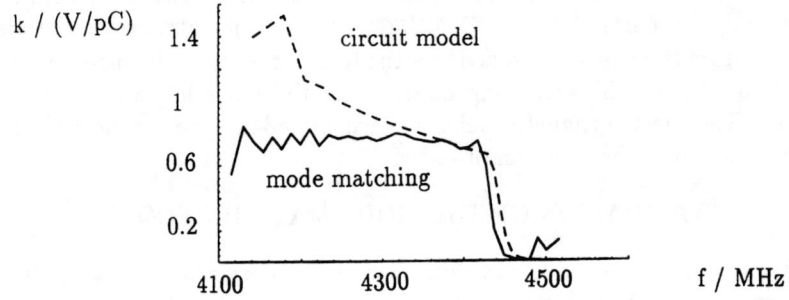

Figure 3: *Loss-factor as a function of frequency for the 36 lowest dipole modes in a tapered 36-cell-structure. Results of ORTHO and the equivalent circuit model are shown.*

A tapered 36-cell structure and a 6m long S-Band structure have been investigated. Fig. 3 displays the loss-factor as a function of frequency, determined by ORTHO or the equivalent circuit model, respectively. Fig. 4 and Fig. 5 show mode patterns for the S-Band structure described in the next section. The loss-factor is displayed in Fig. 6.

The disadvantage of the equivalent circuit model is the very low number of dipole bands whose effects are taken into account while it must be supposed that also higher band modes have an important effect [3]. The model with few dipole bands assumes only a small coupling between the cells. Also the assumption that the structure differs only slightly from a periodic one from cell to cell restricts its application.

APPLICATION TO S-BAND STRUCTURE

Some calculations will be presented for the constant gradient structure of the 500 GeV Linear Collider Study[13]. The characteristics for the accelerating structure are given in Tab. 1. A somewhat simplified structure with 30 sections made from 6 equal cells and some coupling irises between these parts is used for the computer simulation [4]. This modified geometry for the S-Band accelerator structure with $f_{2\pi/3} = 2997.364$ MHz and attenuation $\tau = 0.560$ Neper was studied.

---

[3]Going to more and more bands reduces the speed of the method. Note that the mode matching is equivalent to a circuit model with infinitely many bands.

[4]This technique is well known from existing 5-6 m long structures in injector linacs.

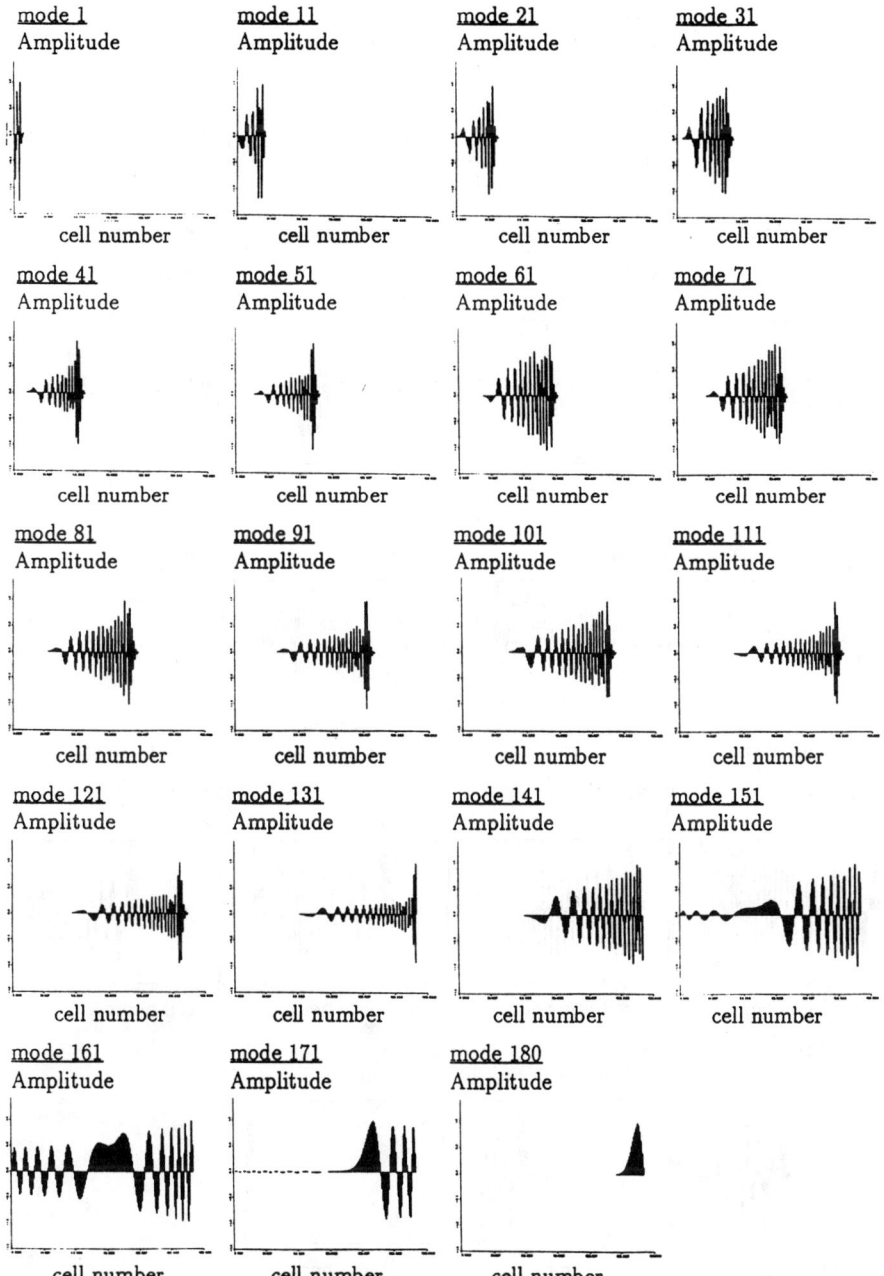

Figure 4: *Some mode patterns of dipole modes for the mode matching code. The plots show the normalized voltage amplitudes. The abcissas give the cell numbers $n = 1, ..., 180$.*

96 Mode Analysis of Circular Waveguide Structures

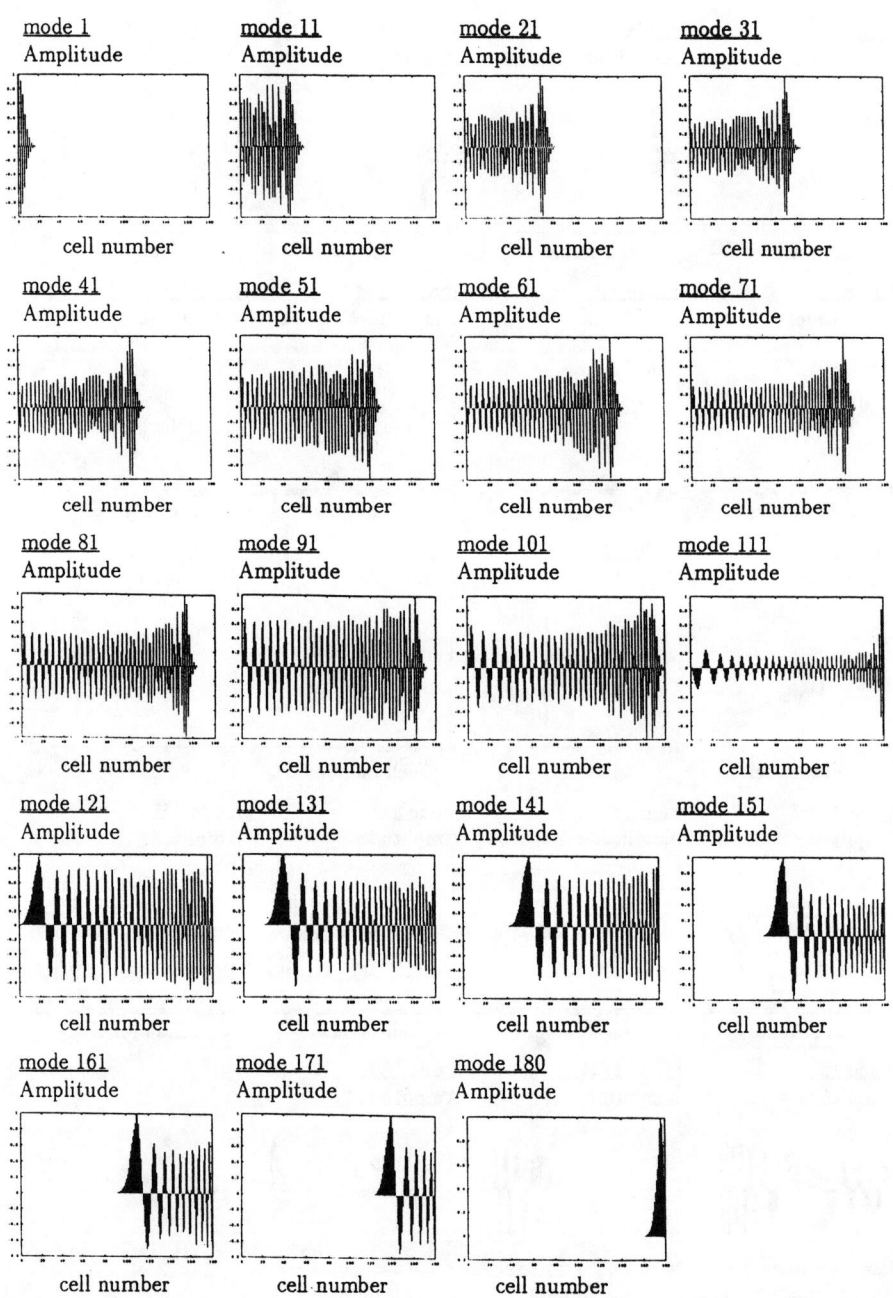

Figure 5: *Some mode patterns of dipole modes for the equivalent circuit model. The plots show the normalized voltage amplitudes. The abcissas give the cell numbers $n = 1, ..., 180$.*

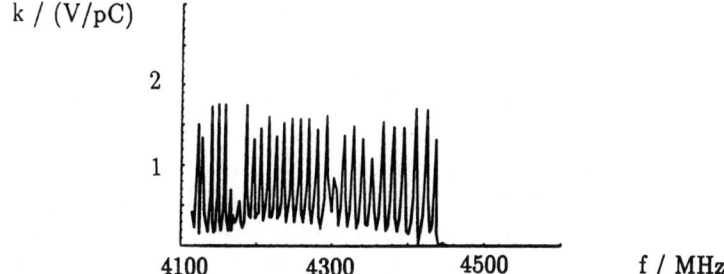

Figure 6: *Loss-factor k as a function of frequency calculated by the mode matching technique.*

accelerating frequency	f = 2997 MHz
accelerating mode	$2\pi/3$-Mode
wave length	$\lambda = 0.1$m
total length	L=6m (180 cells)
mean shunt impedance	$r = 53.6 M\Omega/m$
attenuation factor	$\tau = 0.57$ Neper
Q-value of $2\pi/3$-mode	$Q \approx 13000$
group velocities	$v_g = 4.1...1.3\% \cdot c_0$

Table 1: *Table of characteristic parameters of the accelerating structure*

Fig. 4 in the previous section shows the mode patterns for every tenth dipole mode in the S-band structure. Their characteristics are described here briefly:

- The first ones are π-like modes. The fields are trapped in the first 10% - 20% of the cells. They each have a high loss-factor as depicted in Fig. 6. [5]
- The next modes are trapped in *inner* cells. Their loss-factors are still very high because they all show a π-like field pattern in the rightmost 5-10 of the full cells.
- Then follow some modes which are mainly trapped near the end of the structure and have a very small loss-factor.
- The last modes are either trapped near the end of the structure or they are mixed modes, i.e. they have a portion of a mode of the first passband at the end of the structure and another portion of a mode of the second passband in the first part of the structure.

This result was unexpected as so far no inner trapped modes were assumed. It influences the design plans for damping or tuning strategies in order to treat the deflecting higher order modes.

---

[5]The loss-factors have been calculated at the smallest iris-radius.

CONCLUSION

The Mode Matching Technique has been applicated to calculate *monopole and dipole modes* in a *tapered* circular multicell structure. Fields and quality factors for some dipole modes have been presented.

Comparisons have been performed for the slightly modified geometry of a constant-gradient S-Band structure with 180 cells. The interesting but unexpected result was that 2/3 of the first 180 dipole modes have a high loss-factor. and that not only those modes, which are trapped in the first part of the structure, are among the "severe" modes but also modes which are trapped in the middle part. These results will probably influence the design plans for damping strategies for the S-Band structure. The results of the equivalent circuit model differ slightly from those results obtained by the mode matching code. This can be explained by the fact that higher band modes are not taken into account and that only a small coupling between the cells is assumed.

One important conclusion of the work presented here is that it is very important to do higher order mode analysis for the complete structures rather than to study only very short models because the electromagnetic behaviour of both can differ considerably.

REFERENCES

[1] T. Weiland, Particle Accelerators 17 (1985), 227-242
[2] T. Weiland, NIM 216 (1983), 329-348
[3] U. van Rienen, T. Weiland, Particle Accelerators 20 (1986/87), 239-267
[4] M. Drevlak, *Studienarbeit, FB18 TEMF, TH Darmstadt*, Mai 1991
[5] R.B. Neal, Stanford Univ., W.A. Benjamin, Inc., New York, 1968
[6] H.-G. Unger, Dr. A. Hüthig Verlag 1981
[7] K. Steinigke, Fortschritt-Berichte R. 21, Nr. 108, VDI Verlag 1992
[8] S.A. Heifets, S.A. Kheifets, SLAC-PUB-5907, Sept. 1992
[9] P.B. Wilson, AIP Vol.127 (1983), p. 875-928
[10] U. van Rienen, submitted for publication to Particle Accelerators
[11] T. Weiland, B. Zotter, Particle Accelerators 11 (1981), p. 143
[12] K.F. Bane, R.L. Gluckstern, SLAC-PUB-5783, March 1992
[13] T. Weiland et al., Proc. of the Workshop on Physics and Experiments with Linear Colliders, Saariselkä, Finnland, Sept. 1991

NUMERICAL DESIGN AND ANALYSIS OF A COMPACT TE_{10} to TE_{01} MODE TRANSDUCER*

S. Tantawi, K. Ko, and N. Kroll
Stanford Linear Accelerator Center
Stanford University, Stanford, CA 94309

ABSTRACT

A high-power low-loss mode transducer design has been proposed to adapt the output of the X-Band klystron, WR90 rectangular waveguide, to the input of the pulse compression system, SLED II, which utilizes overmoded circular waveguides operating in the low-loss TE_{01} mode. This device is much more compact than the conventional Marie' type mode converters. The device splits the incoming klystron output into two separate rectangular guides that are then fed into a circular guide through a four-slot arrangement. We will use both MAFIA and HFSS to calculate the transmission properties of the three-dimensional structure. We will also determine the extent of mode contamination and compare the numerical results with experiments.

1. INTRODUCTION

Various types of transducers have been developed for converting the rectangular TE10 mode into the low-loss circular TE_{01} mode. Most of these transducers are based on a design proposed by Marie'[1]. In general, they consist of three sections with adiabatic tapering from one cross-section to the other. The first is a transition from TE_{10} rectangular to TE_{20} rectangular, the second from TE_{20} to a cross-shaped guide, and the third is from that shape to a TE_{01} circular guide. Because the transitions need to be smooth, the transducer length is usually long, leading to unnecessary losses. The excessive length also leads to undesirable closely spaced weak resonances which result from mismatches at the transitions.

The Marie' type mode transducers just described are mostly suited to applications in communication systems where broad bandwidth is required. For accelerator systems, such as the Next Linear Collider Test Accelerator (NLCTA), the bandwidth requirement is much less. The RF components utilized for the NLCTA pulse compression (SLED II) are desired to be compact and low-loss. Based on these requirements, a more compact design with adequate bandwidth was proposed [2] and a sketch of it is shown in Fig. 1. It consists of a bifurcation which splits an incoming rectangular guide into two identical guides. The latter are side-coupled to a circular guide through a four-slot (flower-petal) arrangement in the common wall before terminating in a short at the ends. Such a device is intrinsically three-dimensional and the flower-petal geometry is not amenable to theoretical analysis. As a result, the design approach has primarily been experimental.

In recent years, however, 3-D computer modelling has shown to be both feasible and practical for the design and analysis of devices of complex geometries. In the present paper, we will report our experience with the 3-D electromagnetics codes, HFSS and MAFIA, in simulating the new transducer mentioned above. The motivations behind the numerical effort are to gain insight into the operation of the device, to confirm experimental observations and to assess the accuracies of HFSS and MAFIA in modelling this type of RF components. The paper is organized as follows. In Section 2 we explain the underlying principle for the design while in Section 3 we describe the experimental procedure used to evaluate the device. The numerical models are discussed in section 4 and the comparison between the measurements and simulation results is presented in Section 5. We conclude with a summary on the usefulness of the numerical effort in section 6.

2. PRINCIPLE OF OPERATION

---

* Work supported by Department of Energy, contract DE-AC03-76SF00515.

Fig 1. shows a schematic diagram of the mode transducer. The basic operation of the device can be understood by the following considerations. First, the structure has a reflective symmetry about the Y-Z plane which bisects it. Because the incoming mode in the WR90 guide is TE_{10}, this symmetry plane can be taken to be an electric boundary. This has the consequence that in the circular section TM_{0n} modes are not excited and only one of the two polarizations of the azimuthally varying modes can be excited. Now if one ignores the coupling slots and terminates the WR90 guides at the half-wavelength point from the X-Z plane, then about that plane, there is an additional reflective symmetry for the fields in the region where the WR90 and the circular guides overlap, also equivalent to an electric boundary. To the exent that neglect of the perturbation is vaild, modes with odd azimuthal variation are not excited in the circular section. In the limit in which the slots are narrow and radial with respect to the circular section, the magnetic fields in the slots are radial and provide a drive in the circular waveguide with azimuthal symmetry of the form 4n. If the circular guide radius is such that it cuts off modes with m=4 and above, then TE_{01} is the only propagation mode excited as desired.

A mode transducer operating as above in such a way that the coupling slots are a small perturbation would have excellent mode purity but very poor power conversion. That is to say, most of the power would be reflected. An essential feature of the device is to transfer all the incident power to the circular guide in the TE_{01} mode. While it would be possible to accomplish this objective by means of a strongly resonant matching section, such a solution is unsatisfactory because it would be excessively narrow band and lead to excessive field strengths in the matching section. The alternative is to open up the slots to increase the coupling. The limiting conditions are then significantly compromised, which may be expected to lead to some mode impurity. The design program was directed towards minimizing this effect while at the same time optimizing the power transfer. This was accomplished by the shaping and sizing of the coupling slots and slightly adjusting the distance of the shorting plane. The final dimensions shown in Fig. 1 are arrived at after much empirical experimentation [3].

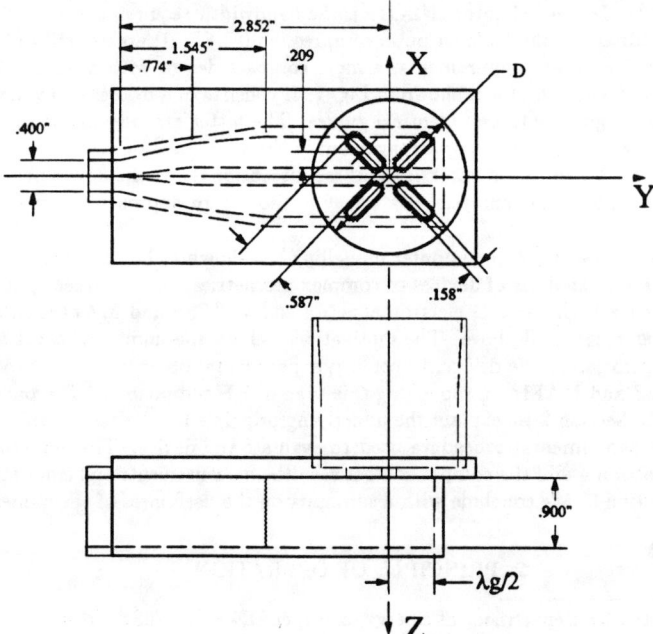

Fig. 1 TE_{10} to TE_{01} Mode Transducer.

3. EXPERIMENTAL PROCEDURES

Fig. 2 shows the experimental setup used for measuring the transducer properties. The transducer circular port is terminated with a slowly tapered horn. The horn acts as a matched load for the transducer. This method of matching the transducer proved to be considerably better than matching it with a tapered lossy cone. The inner surface of the horn has a special curve to reduce as much as possible mode conversion. A sample of H_z (the magnetic field in the Z direction in the circular guide) at the wall is measured through a small circular hole that opens to a rectangular waveguide with its large dimension parallel to the axis of the circular guide. The hole size is chosen such that the coupling between the TE_{01} mode in the circular guide and the TE_{10} in the rectangular guide is small enough so that it does not perturb the fields in the circular guide and at the same time it is large enough to accommodate the dynamic range of the network analyzer and it's test sets (HP 8510C). A reasonable number for that is -45 dB. The wall thickness between the rectangular sampling guide and the circular guide is .03 inches, the smallest size possible for machining the hole without distorting the inner wall of the circular guide. The hole size for the previously mentioned coupling was calculated using a modified Bethe formula [4] to get a approximately 0.1 inch diameter hole. Mode conversion due to this hole for all modes can also be calculated as shown in [5] and is negligible for most modes.

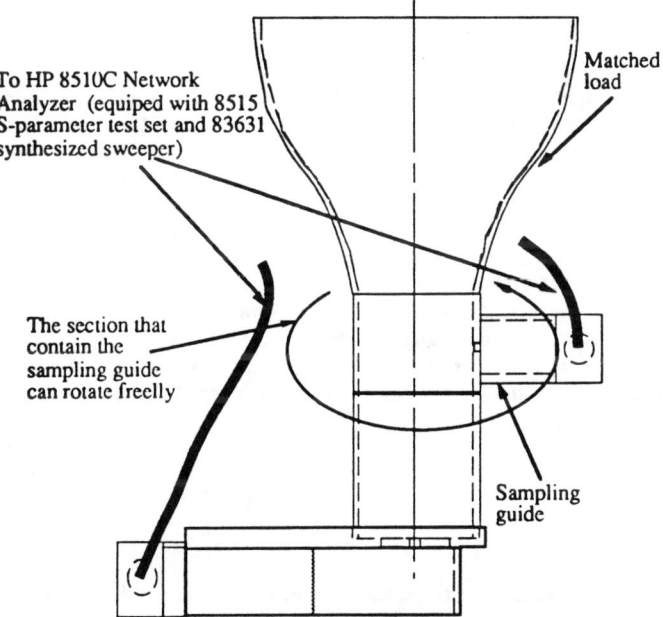

Fig. 2 Experimental Setup for Measuring the Transducer Properties.

The section that contains the sampling rectangular guide is free to rotate around the axis of the circular guide. H_z is sampled at 16 different points around the circular guide. At each point the transmission coefficient, as a function of frequency, between the input of the transducer and the output of the sampling guide is dumped on a disk through GPIB link between the network analyzer and a PC. Both phase and magnitude of the *complex* transmission coefficient (Ti ; 0 i 15) is stored at each point i. The azimuthal harmonics (Hzn -7 n 7) of Hz can be calculated from

$$H_z^n = \sum_{i=0}^{15} T_i \, e^{-j\frac{i n \pi}{8}}$$

Eq. (1) represents a circularly polarized component of the total field H_z. The power contained in a mode that has azimuthal variation m can be calculated from

$$\frac{Power\ in\ TE_{n1}}{Power\ in\ TE_{01}} = \left(\frac{x_0^{'}}{x_n^{'}}\right)^2 \left(1-\left(\frac{n}{x_n^{'}}\right)^2\right) \frac{\sqrt{1-\left(\frac{f_c^n}{f}\right)^2}}{\sqrt{1-\left(\frac{f_c^0}{f}\right)^2}} \frac{|H_z^{+n}|^2 + |H_z^{-n}|^2}{|H_z^0|^2},$$

where x_m is the first zero of the derivative of Bessel function of first kind and mth order, f_c^m is the cutoff frequency of TE_{m1} mode, and f is the operating frequency.

This method of measurements is suitable only for TE modes. In the case of TM modes there is no H_z. If we change the polarization of the sampling guide to sample the azimuthal magnetic field there will be no reference to compare with since TE_{01} has no azimuthal magnetic field also. However, the presence of TM modes has no effect on the measurements. The method also breaks down for the TE_{m1} if the circular guide diameter is large enough to allow TE_{m2} to propagate.

4. NUMERICAL MODELS

Both HFSS and MAFIA are 3D electromagnetic codes capable of calculating S parameters of many port devices. There are several differences between the two codes. HFSS is a finite-element code which evaluates in the frequency domain while MAFIA is based on a finite-difference scheme and solves in the time domain. Fig. 3 shows half the transducer geometry we constructed with MAFIA. It includes both the bifurcation and the flower-petal cut along the Y-Z symmetry plane that we referred to in Section 1. In practice, we simulate the two parts separately. Figs. 4a and 4b show the HFSS model for the bifurcation and the flower-petal respectively. We point out that the coupling slots as modeled by both codes have sharp edges whereas all the edges are rounded in the actual design. This approximation is mainly for numerical expediency and it turns out to have an non-negligible effect on the results.

With either codes, we calculate the input characteristics for each part separately. From the individual scattering matrices we can construct the input characteristics for the total device as follows:

$$S_{11}^t = S_{11}^b + \frac{S_{11}^s \left(S_{12}^b\right)^2}{1 - S_{22}^b S_{11}^s}$$

The superscript t stands for the transducer as a whole, b for the bifurcation, and s for the slot section respectively. In Eq. (3), all the matrix elements are complex quantities in which the relative phases have been retained.

Fig. 3 MAFIA Model for One Half of the Mode Transducer.

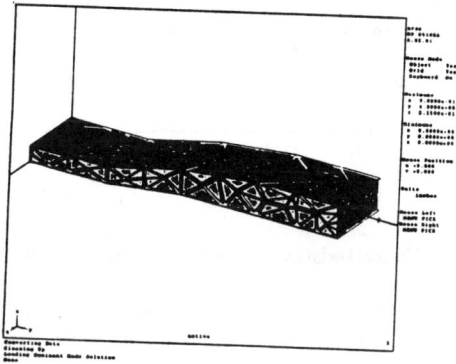

Fig. 4a HFSS Model for One Half of the Bifurcation.

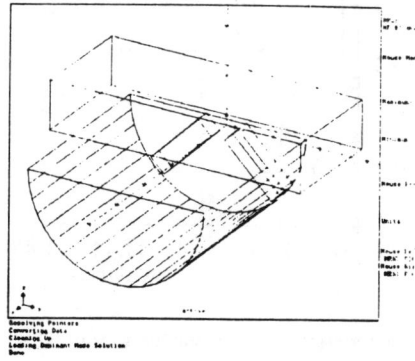

Fig. 4b HFSS Model for One Half of the Flower-petal.

5. COMPARISON BETWEEN SIMULATIONS AND EXPERIMENTS

The operating frequency of the NLCTA is 11.424 GHz (X-Band). The input characteristics of the bifurcation around this frequency as calculated with HFSS and MAFIA are shown in Fig. 5. Both simulations indicate that the section is well-matched. They also agree reasonably well with each other.

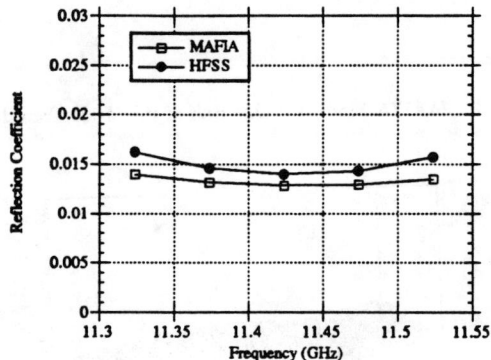

Fig. 5 Input Characteristics of the Bifurcation from HFSS and MAFIA.

In SLED II, the diameter of the circular waveguide chosen to manipulate the power in the TE_{01} mode is 1.75" and the TE_{41} mode will propagate in this guide size. Fig. 6a shows the input characteristics of the total device at this diameter as found from HFSS and from measurements. The agreement is very good.

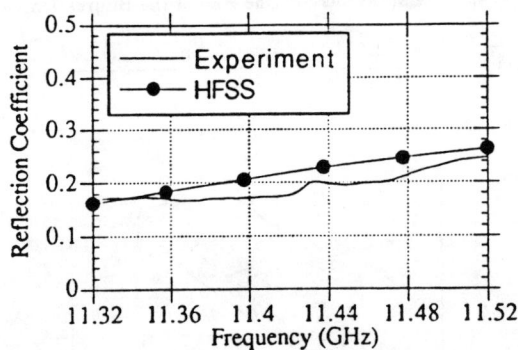

Fig. 6a Input Characteristics of the Transducer from HFSS and Experiment (1.75").

The agreement is less satisfactory when one considers the mode contamination by the TE_{41} mode. As Fig. 6b indicates, measurements show a peak around the cutoff frequency of the TE_{41}. This is not observed in the HFSS results possibly due to the large frequency spacings we have taken in the simulations.

To reduce the TE_{41} contamination, one may start the circular guide with a smaller diameter, 1,5" and then slowly taper up to 1.75" after a distance over which the unwanted modes have decayed sufficiently. For this diameter guide, the reflection coefficient and mode contamination

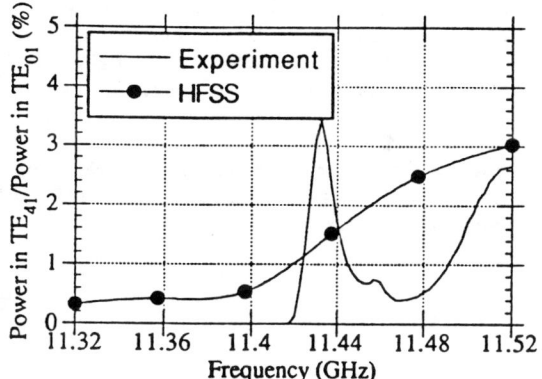

Fig. 6b Mode Impurity in the Transducer from HFSS and Experiment (1.75").

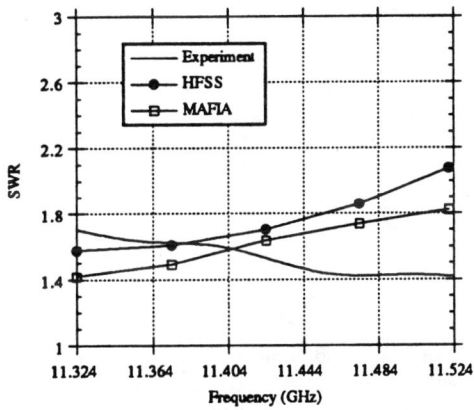

Fig. 7a Input Characteristics of the Transducer from HFSS, MAFIA and Experiment (1.50").

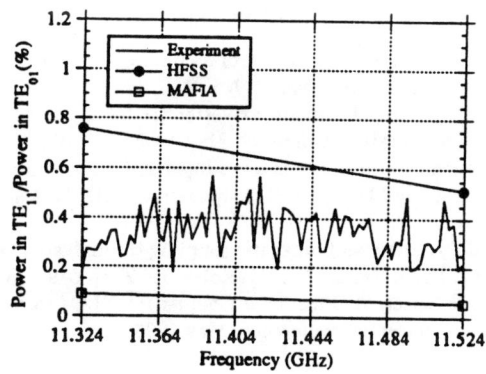

Fig. 7b Mode Impurity in the Transducer from HFSS, MAFIA and Experiment (1.50").

are shown respectively in Figs. 7a and 7b. Again the two codes agree with each other but not as well with the measurements. Regarding mode contamination, both HFSS and MAFIA find that the impurities are mainly due to the odd modes, namely TE_{11} and TM_{11}. Fig. 7b shows how the TE_{11} mode contamination compares between simulations and experiments. We attribute a probable cause for the observed differences to the inaccurate description of the flower-petal by either the HFSS or the MAFIA meshes.

It is worth mentioning that the device can easily be matched with a post in the rectangular guide. The result of such operation is shown in Fig. 8. The SWR is ¡ 1.05 at 11.424 GHz, and the bandwidth is wide enough for the NLCTA applications.

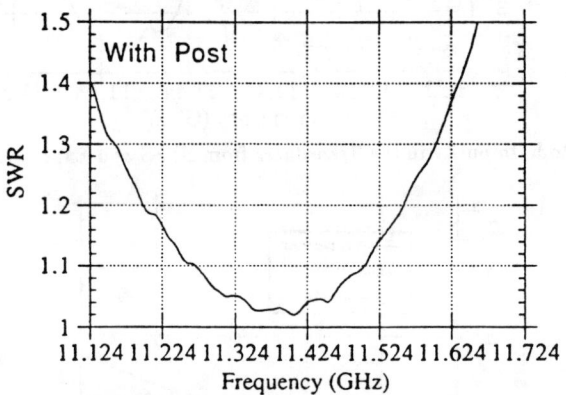

Fig. 8   Input Characteristics of the Transducer from Experiment with an Inductive Post (1.50").

6. CONCLUSION

We tested both experimentally and numerically a compact type of mode transducers. Both simulations and experiments agreed qualitatively. Both showed a very small amount of mode contamination mainly in odd modes. Exact agreement was not possible because of the sensitivity of the device to small variations in geometry, which are difficult to model precisely in the simulations.

REFERENCES

1. S. S. Saad et. al. "Analysis and design of a circular TE01 mode transducer," Microwaves, optics and acoustics, Jan. 1977, vol. 1, No. 2.
2. H. Hoag et. al. "Flower-petal mode converter for NLC," to be published in the 'digest of the 1993 Particle accelerator conference, May 1993.
3. The configuration shown in Fig. 1 is a modified and scaled up version of a 34.5 GHz mode transducer manufactured by the TRG corporation. Z.D. Farkas initiated the SLAC design program.
4. N. A. McDonald, "Electric and magnetic coupling through small apertures in Shielded walls of any thickness," IEEE Trans. MTT October 1972, pp 689-695.
5. S Tantawi, "Mode Selective Directional Coupler for NLC," to be published in the digest of the 1993 Particle accelerator conference, May 1993.

WAKEFIELD AND IMPEDANCE STUDIES OF A LINER USING MAFIA

W. Chou and T. Barts

SSC Laboratory,[†] Dallas, TX 75237, USA

ABSTRACT

The liner is a perforated beam tube which is coaxial with an outer bore tube. The 3D code MAFIA version 3.1 is used to study the wakefields, impedances and resonances of this structure. The short range wakes and low frequency (below the cutoff) impedances are in agreement with the theoretical model. The long range wakes and high frequency resonances are associated with the distribution of the holes (or slots). The dependence of the impedance on the size, shape and pattern of the holes (or slots) is studied. The impact of the liner impedance on the SSC impedance budget is discussed.

I. INTRODUCTION

One novel feature of the SSC Collider is that the synchrotron radiation from the 20 TeV proton beam becomes an important issue. The radiation creates a significant heat load to the cryogenic system. It also causes a large amount of gas load from the photon-induced desorption process. The latter may result in a poor vacuum in the beam tube and an eventual short luminosity life time. One possible solution to the problem is to install a perforated liner inside the bore tube.[1] The principal function of the liner is to decouple the synchrotron radiation from the bore tube. The photo-desorbed gas from the inner surface of the liner would be pumped out by the bore tube (which is at 4 K) and absorbed by the cryo sorber that is located in the co-axial region between the liner and the bore tube.

The introduction of the liner brings up a list of issues that need to be studied. Among them, one is the additional rf impedance. There have been some theoretical studies on the hole impedance in the low frequency region.[2] This paper reports the simulation studies using 3D code MAFIA (version 3.1).[3] It is found that, at the frequencies below the cutoff, MAFIA results are in agreement with the theoretical model. At high frequencies, impedance resonance peaks and long term wakes are observed in the simulations. The size, shape and distribution of the holes (or slots) on the liner surface have significant impact on the liner impedance.

The MAFIA code runs on an IBM RS 6000/560 workstation. The statistics of the code performance is also included.

II. LOW FREQUENCY REGION

A. ANALYTICAL MODEL

For some structures, such as a pipe attached to a small pillbox or a pipe with small holes on its surface (i.e., the perforated liner), the longitudinal and transverse impedances can be approximated by a pure inductance L at low frequencies (below the cutoff):

[†] Operated by the Universities Research Association, Inc., for the U.S. Department of Energy under Contract No. DE-AC35-89ER40486.

Wakefield and Impedance Studies

$$Z_\|(\omega) = i\omega L \qquad (1)$$

$$Z_\perp(\omega) = i\frac{2c}{b^2} L \qquad (2)$$

in which b is the radius of the pipe, c the velocity of light. The corresponding wakefields $W_\|$ and $W_\perp$ are, respectively, the derivative of the δ-function and the δ-function.[4]

$$W_\|(\tau) = \frac{1}{2\pi}\int_{-\infty}^{\infty} d\omega\, e^{-i\omega\tau}\, Z_\|(\omega) = -L\,\delta'(\tau) \qquad (3)$$

$$W_\perp(\tau) = \frac{-i}{2\pi}\int_{-\infty}^{\infty} d\omega\, e^{-i\omega\tau}\, Z_\perp(\omega) = \frac{2c}{b^2} L\,\delta(\tau) \qquad (4)$$

Let $\tau = z/c$, one obtains

$$W_\|(z) = -c^2\, L\,\delta'(z) \qquad (5)$$

$$W_\perp(z) = \frac{2c^2}{b^2} L\,\delta(z) \qquad (6)$$

For a Gaussian bunch with rms length σ, the line charge density is

$$\lambda(z) = \frac{1}{\sqrt{2\pi}\,\sigma} e^{-\frac{z^2}{2\sigma^2}} \qquad (7)$$

The longitudinal and transverse wake potentials generated by this Gaussian bunch are, respectively,

$$w_\|(z) = \int_{-\infty}^{\infty} dz'\, \lambda(z-z')\, W_\|(z') = \frac{c^2 L}{\sqrt{2\pi}\,\sigma^3}\, z\, e^{-\frac{z^2}{2\sigma^2}} \qquad (8)$$

$$w_\perp(z) = \int_{-\infty}^{\infty} dz'\, \lambda(z-z')\, W_\perp(z') = \frac{2c^2 L}{\sqrt{2\pi}\,\sigma\, b^2}\, e^{-\frac{z^2}{2\sigma^2}} \qquad (9)$$

They are shown in Figs. 1a and 1b.

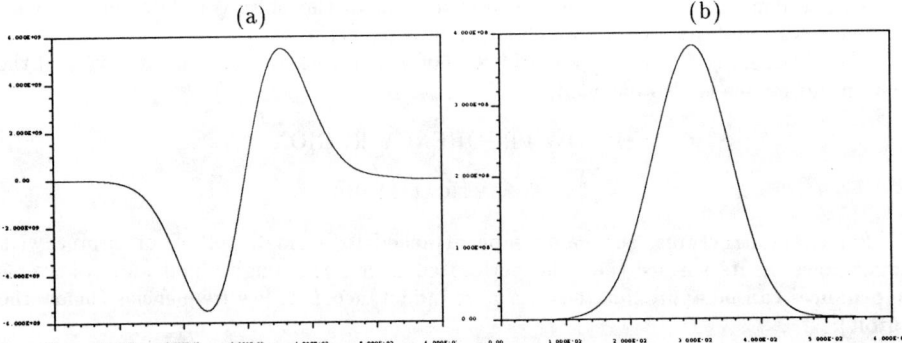

Figure 1: The wake potentials of a purely inductive impedance for a Gaussian bunch: (a) longitudinal, (b) transverse.

The magnitudes and locations of the peaks of the wake potentials are as follows:

$$w_\parallel^{max(min)} = \pm \frac{c^2 L}{\sqrt{2\pi}\,\sigma^2} e^{-1/2}, \quad \text{at } z = \pm\sigma \tag{10}$$

$$w_\perp^{max} = \frac{2c^2 L}{\sqrt{2\pi}\,\sigma b^2}, \quad \text{at } z = 0 \tag{11}$$

Therefore, if the inductance L is known, then Eqs. (8)-(11) give the wake potentials for specified beam pipe radius b and rms bunch length σ.

B. MAFIA RESULTS

The inductance of a small hole with diameter d has been worked out.[2]

$$L = \frac{Z_0}{48\pi^2 c}\frac{d^3}{b^2}, \tag{12}$$

in which $Z_0 = 377\,\Omega$. Therefore, the peaks of the wake potentials of the liner are given by (all dimensions in meters)

$$w_\parallel^{max} = \frac{Z_0 c}{48\pi^{5/2}\sqrt{2e}}\frac{d^3}{\sigma^2 b^2} = 5.77 \times 10^{-2} \times \frac{d^3}{\sigma^2 b^2} \quad V/nC \text{ (per hole)} \tag{13}$$

$$w_\perp^{max} = \frac{Z_0 c}{24\pi^{5/2}\sqrt{2}}\frac{d^3}{\sigma b^4} = 1.90 \times 10^{-4} \times \frac{d^3}{\sigma b^4} \quad V/nC/mm \text{ (per hole)} \tag{14}$$

Equations (13) and (14) can be directly compared with the MAFIA results as shown in Figs. 2a and 2b.

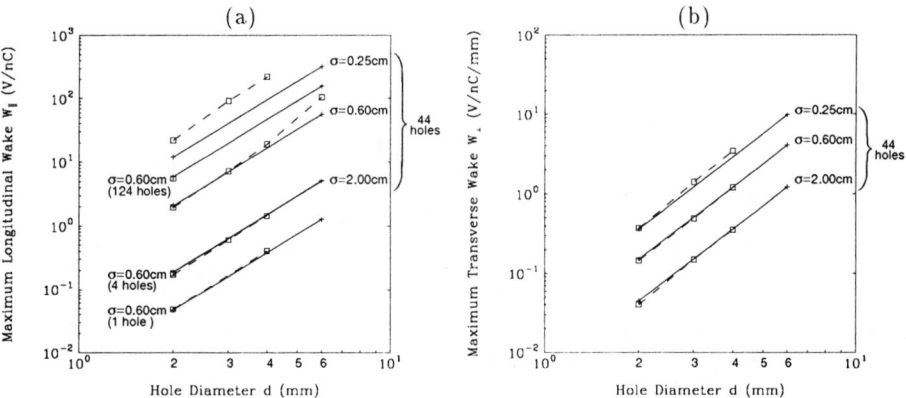

Figure 2: The peak values of the wake potentials of a liner. The solid lines are computed using Eqs. (13)-(14). The squares and dashed lines are the MAFIA results. (a) $w_\parallel^{max}$, (b) $w_\perp^{max}$.

When the hole size is small and bunch length is large, the theory and simulations agree with each other. This implies that the low frequency approximation is valid. However, when the hole becomes larger ($d \geq 4$ mm) or the bunch becomes shorter ($\sigma = 0.25$ and 0.6 cm), the simulation results appear to be larger than what the theory would predict. This probably indicates the breakdown of the low frequency assumption.

The hole shape in the simulations is a square rather than a circle. It would thus give an inductance larger than that of a circular one as predicted by Eq. (12). On the other hand, Eq. (12) is derived from a zero-thickness liner. The finite thickness (1 mm) used in the simulations would lead to a smaller inductance.[5] It is interesting to see from Figs. 2a and 2b that these two effects seem to cancel each other and result in a good agreement between Eqs. (13)-(14) and the MAFIA results.

III. HIGH FREQUENCY REGION

One interesting observation in the simulations is the long term wakes for large holes or short bunches. These long term wakes are potentially dangerous because they may cause coupling among successive bunches, which could be a source of multiple bunch instability and/or emittance growth. In order to understand these wakes, an FFT is carried out to convert the wakes to the impedance in the frequency domain.

A. BENCHMARK TEST

Because of the lack of an appropriate analytic model of the perforated liner in the high frequency region, the MAFIA results are at first compared with the known impedance spectra of a small pillbox. For this simple structure, the impedances obtained from several different methods (field matching,[6] TBCI[7] and transmission line[7]) agree with each other and, therefore, can be used for a benchmark test.

Figs. 3a and 3b are the longitudinal impedance of a small pillbox obtained from the FFT of the wakes computed by MAFIA.

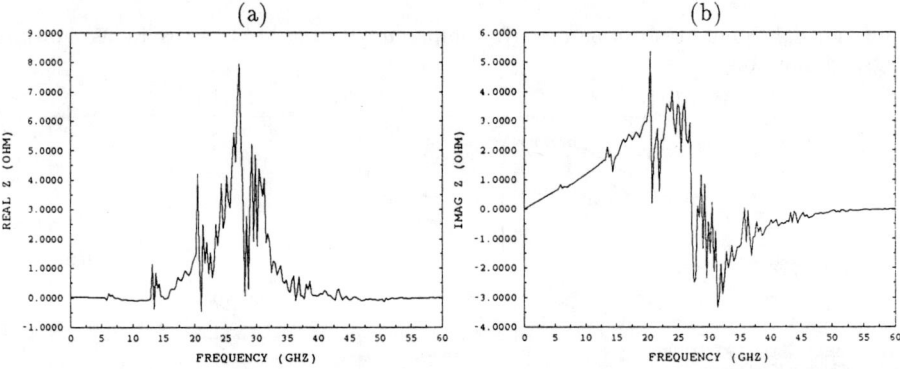

Figure 3: The longitudinal impedance of a small pillbox obtained from MAFIA and FFT: (a) the real part, (b) the imaginary part.

Below the cutoff (which is about 5.75 GHz for a beam pipe radius of 2 cm used in the simulation), the impedance is purely inductive:

$$Z_{\|}(\omega) = i\omega \frac{\mu_0}{2\pi} \frac{4g\epsilon}{b} , \qquad (15)$$

in which $2g$ and 2ϵ are the width and depth of the pillbox, respectively, $\mu_0 = 4\pi \times 10^{-7}$ H/m. The difference between Eq. (15) and the MAFIA results in Fig. 3b is only about 3%. Above the cutoff, the impedance spectra also agree reasonably well with that in Refs. 6 and 7.

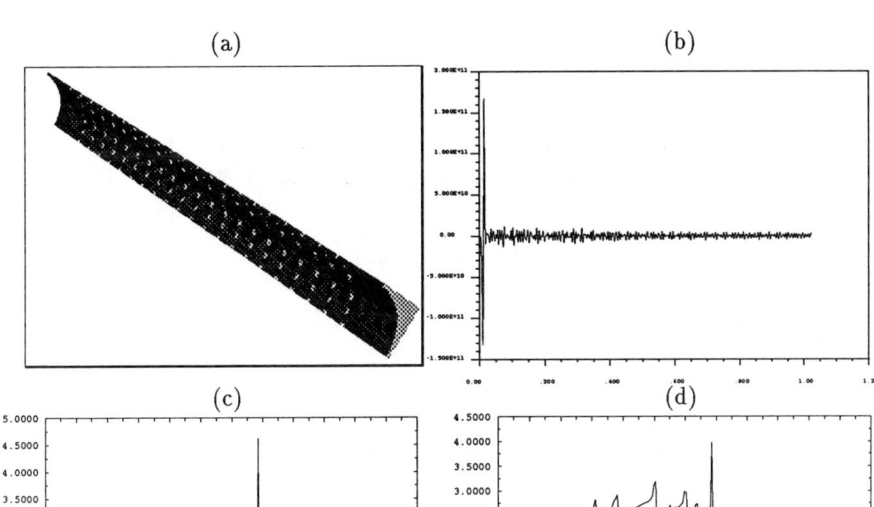

Figure 4: (a) The liner with 420 holes of size 2mm×2mm, periodically distributed. (b) The longitudinal wake of this structure. (c) The real longitudinal impedance. (d) The imaginary longitudinal impedance.

B. Periodic Distributions of Holes on a Liner

The simulations use 20 holes along the azimuthal direction of the liner, which is called a column, and 21 columns along the axial direction. The total number of holes is 420. Each hole is a 2 mm × 2 mm square. The columns are uniformly distributed with 1 cm spacing between each other.

Fig. 4a shows the structure, and 4b the longitudinal wake. The long term wake indicates the existence of resonance impedance, which is clearly seen in Figs. 4c and 4d. The liner radius is 1.65 cm, which gives the longitudinal cutoff at about 7 GHz. Below the cutoff, the impedance is purely inductive. The analytical value is obtained from Eqs. (1) and (12):

$$Z_{\parallel}(\omega) = i\omega \frac{Z_0}{48\pi^2 c} \frac{d^3}{b^2} . \tag{16}$$

The difference between the value given in Fig. 4d and that given by Eq. (16) is again only a few percent, as in the case of the pillbox. The resonance peaks at high frequency is associated with the periodicity of the holes. The first peak occurs near 15 GHz, which corresponds to a half wavelength of 1 cm, the periodic spacing between two columns. The transverse impedance is shown in Figs. 5a and b.

The transverse cutoff is about 11 GHz. Below it, the impedance is also purely

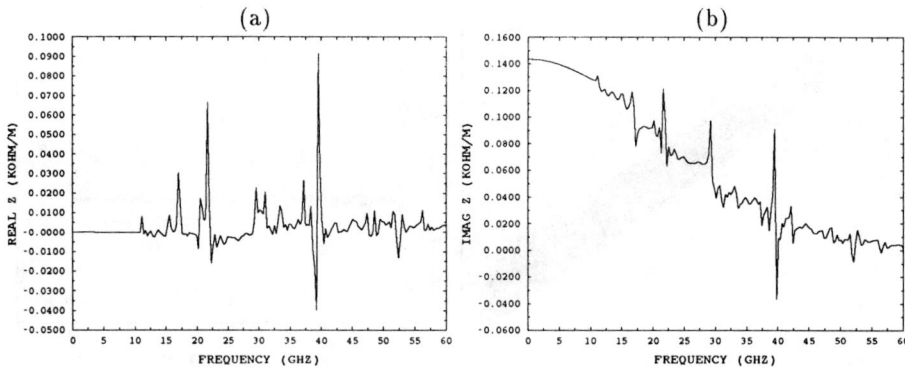

Figure 5: The transverse impedance of the structure in Fig. 4(a): (a) the real part, (b) the imaginary part.

inductive. The analytical value from Eqs. (2) and (12) is

$$Z_\perp(\omega) = i \frac{Z_0}{24\pi^2} \frac{d^3}{b^4} \ . \tag{17}$$

This value agrees with that given in Fig. 5b at zero frequency within a few percent. But the fall off of $Im\ Z_\perp$ from dc to the cutoff is not predicted by the theoretical model. Similar to the longitudinal case, the long term wakes and the resonance impedance at high frequencies are observed.

C. Random distribution of Holes on a Liner

One effective way to reduce the long term wakes and resonance impedance is to destroy the periodicity of the hole distribution. For this purpose, the spacing between two neighboring holes in the axial direction is randomized. Figs. 6a-d show the structure and the resulting longitudinal wakes and impedance. Compared with Figs. 4a-d, the short term wakes and low frequency impedance remain about the same (as they should be due to the additivity), whereas the long term wakes and resonance impedance at high frequencies are greatly decreased. However, by using the same technique, the reduction in the transverse direction is less dramatic. This needs to be understood.

D. Slots on a Liner

When the holes are replaced by the slots that have the same area and have the major axis parallel to the pipe axis, the low frequency impedance are reduced, whereas the long term wakes (i.e., the high frequency resonances) are enhanced because it becomes easier to resonate. This is shown in Figs. 7a and 7b. Therefore, the trade off should be studied carefully. The short slots with rounded edges seem to be a good compromise.

IV. DISCUSSION

Based on the good agreement of the theory with the simulations as well as with some measurements,[8] it is believed that the low frequency part of the impedance of a perforated liner can be estimated accurately. These impedances are plotted in Figs. 8a and 8b for three possible liner IDs. In these calculations, the area coverage of the

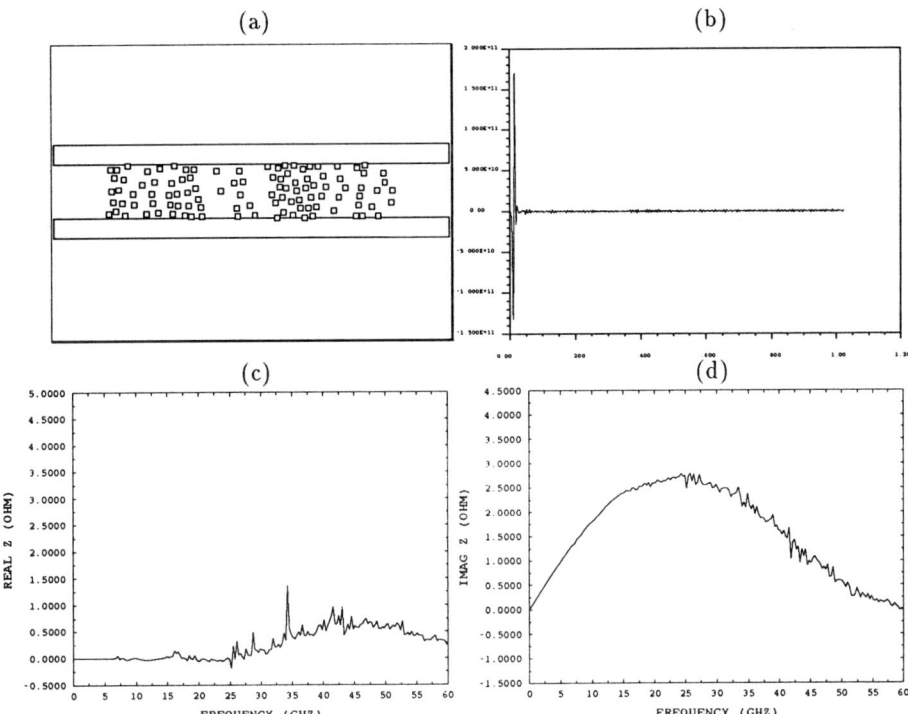

Figure 6: (a) The liner with 420 holes of the size 2mm×2mm, randomly distributed. (b) The longitudinal wake of this structure. (c) The real longitudinal impedance. (d) The imaginary longitudinal impedance.

holes are kept the same for different hole sizes, namely, 1000 holes per meter for $d=$ 2 mm. This is about 3%, 4% and 5% coverage for ID = 33 mm, 24.3 mm and 20.3 mm, respectively.

As a comparison, the impedance budget of the Collider is also plotted. It is seen that, when liner ID is 24.3 mm (as designed for the Accelerator System String Test ASST II) and hole diameter 2 mm, the longitudinal impedance of the liner would be about 34% of the present budget, and transverse impedance about 110%. This represents an significant increment to the budget. Because of a relatively large safety margin reserved in the present impedance budget (which is 6 times greater than the instability threshold impedance), this increment should not jeopardize the beam dynamics at the baseline design current of 72 mA. Rather, its main impact would be to limit the potential of beam current upgrade.

The size, shape and distribution of the holes or slots on the liner surface have significant impact on the impedance. The rounded short slots with random pattern may present a best choice to minimize the impedance while meeting the vacuum pumping requirements.

The code MAFIA version 3.1 runs on an IBM 560 workstation, with 128 MB RAM. It is a virtual memory machine. The maximum number of mesh points in the simulations can reach about 1.8 million. When this number is exceeded, the cpu usage would be

114 Wakefield and Impedance Studies

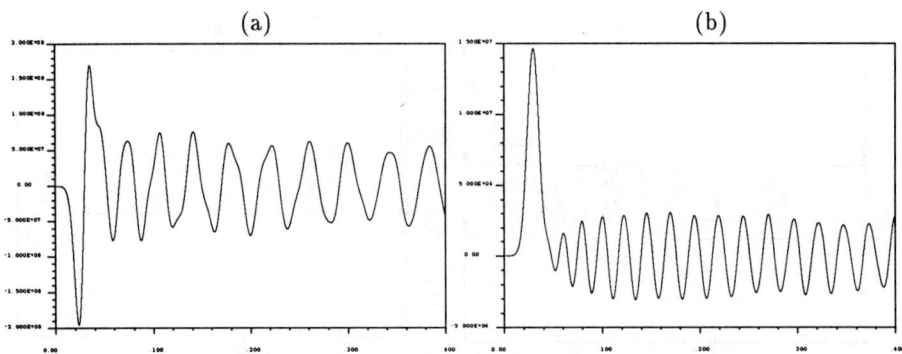

Figure 7: The wake potentials of a liner with four long slots of the size 2mm×22mm: (a) longitudinal, (b) transverse.

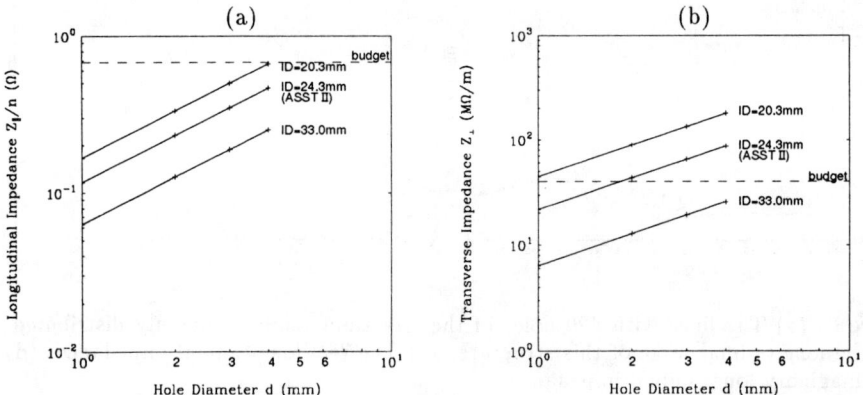

Figure 8: The total impedance of the liner in the SSC Collider for three possible liner IDs: (a) longitudinal, (b) transverse. The dashed lines are the impedance budget.

greatly degraded. A typical job calculating about 1 million mesh points and 1000 wake points takes about 4 hours cpu time. The same job would take about 7 hours cpu time on the NERSC Cray.

REFERENCES

1. H. T. Edwards, SSC Laboratory Report SSCL-N-771 (1991).
2. S. Kurennoy, CERN Report SL/91-29; R. L. Gluckstern, CERN Report SL/92-05.
3. MAFIA is a commercial code for general purpose electromagnetic simulations.
4. W. Chou and T. Barts, SSCL internal technical note PMTN-0057C (1992).
5. R. L. Gluckstern and J. A. Diamond, IEEE MTT, v. 39, p. 274 (1991).
6. H. Henke, CERN Report LEP-RF/85-41 (1985).
7. W. Chou, Proc. 4th Advanced ICFA Beam Dynamics Workshop, KEK Report 90-21, p. 161 (1991).
8. L. Walling, private communication.

ACHRO: A PROGRAM TO HELP DESIGN ACHROMATIC BENDS[*]

D. Rusthoi
Los Alamos National Laboratory, Los Alamos, N M 87545

ABSTRACT

ACHRO is a very simple 2000-line FORTRAN code that provides help for the designer of the achromatic bend. Given a beam momentum, the program calculates the required drift lengths and dipole parameters which it will apply to any one of several different types of achromats. The types of achromats that the code helps to design include the Enge Dual-270°, the Brown 2-Dipole, the Leboutet 3-Dipole, and the Enge 4-Dipole, as well as the periodic systems which can be designed to any order in symmetric, nonsymmetric and stair-step varieties. Given the dimensions into which a bend must fit, ACHRO will calculate the geometrical parameters in an Z-X plane for a single or multiple achromat, and for achromatic "S-bend" configurations where possible. ACHRO makes it very easy to optimize a bend with respect to drift lengths and magnet parameters by allowing the user to change parameter values and see the resulting calculation. Used in conjunction with a beam-transport code, the designer can consider various types of achromatic bends in the same beamline layout in order to compare important bend characteristics such as dispersion at nonachromatic points, isochronicity, sensitivity to errors, geometric or chromatic aberrations, aperture requirements, space for diagnostics, etc., all of which are largely a function of the geometry and the type of achromat selected.

INTRODUCTION

The most laborious task in designing a bend, achromatic or not, is often the bend geometry, and ACHRO was written to do the "dog work" of such calculations. For some bends, especially the periodic types, the geometry can get rather complicated; and to change one value, such as a drift length or a dipole parameter, means that a whole new set of calculations of bend geometry is again required. The code helps the user optimize bend geometry by facilitating change in parameter values while maintaining desired distances or other chosen parameters. For the most part, ACHRO must be used in conjunction with a beam-transport code such as TRACE[1] to fit dipole edge-angles or quadrupole fields where required. For second- or third-order achromats, codes such as TRANSPORT[2] or MARYLIE[3] which have higher order elements must be used.

ACHROMAT DESIGN CAPABILITY

The types of achromats that ACHRO can help design are listed in Table 1. With the given required dimensions, the program calculates all the geometrical parameters in an Z-X plane for a single or multiple achromat,

---

[*] Work supported under the auspices of the U.S.Department of Energy and supported by the U.S.Department of Defense, Army Strategic Defense Command, and the Los Alamos National Laboratory.

and for an "S-bend" configuration where possible. Most of the first-order bends[4] listed in Table I require no additional elements beside the dipoles to be completely achromatic to first order. The periodic systems[5], on the other hand, require quadrupoles to be achromatic to first order, and can be made achromatic to second order with the addition of sextupoles and to third order with the addition of sextupoles and octupoles.

Table I: Types of Achromats Available

First-Order Systems	Periodic Systems
Enge Dual-270°	Stair-Step[6]
Brown 2-Dipole	Symmetric, 1-Dipole/Cell
Leboutet 3-Dipole	Symmetric, 2-Dipoles/Cell
Enge 4-Dipole	Nonsymmetric, 1-Dipole/Cell

A USER-FRIENDLY CODE

ACHRO is extremely user-friendly, prompting for information as needed, and providing a little sketch of the selected achromat. When the code is first executed, the user sees the following display:

```
        *****ACHROMATIC BEND DESIGN HELP*****
        Types of Achromats available are listed below.
     Calculations are done for Single/Multiple Achromats,
         and where possible an S-BEND Configuration.

        1.  ENGE DUAL 270  (1st order)
        2.  2-DIPOLE       (1st order)
        3.  3-DIPOLE       (1st order)
        4.  4-DIPOLE       (1st order)
        5.  STAIR-STEP
        6.  PERIODIC       (Symmetric)
        7.  PERIODIC       (Nonsymmetric)

   Enter Selection [1-7]   (<CR> to Exit): 6
```

Once a choice is entered, the sketch of the achromat is printed to the terminal (the symmetric periodic system with 1-dipole/cell is used as the example):

```
        <---1 cell--->
       "a"    D    "a"       "a"    D    "a"       "a"    D    "a"
       |-----DDD------|  +   |-----DDD------|  +   |-----DDD------|  + ...
             DDDDD                 DDDDD                 DDDDD
              #1                    #2                    #3
```

The user is prompted to enter either a design momentum or the particle mass and kinetic energy. A single entry signifies beam momentum while two entries indicate kinetic energy and mass.

```
       Enter KE and MASS [in MeV], or MOMENTUM [in MeV/c]: 94  .511
```

The desired number of cells, achromats and the total achromatic bend angle is entered, followed by a choice for one of three possible dipole entries (only one entry is required):

```
Choose one of these Entries:
     Dipole Field   (B)
     Dipole Radius  (R)
     Dipole Length  (L)
Enter ["B", "R" or "L"]: B
Enter Dipole Field [Kg]: 2
```

Finally, the desired achromat offset dimension is entered.

```
Enter Req'd Achromat Offset Dimension [m]: 4.3
```

THE OUTPUT

In this example of a periodic bend, no further information was necessary to produce a design. ACHRO displays the resulting calculation as follows:

```
***** Number of Achromats         =           1
***** Total Number of Cells       =           4
***** Total Number of Dipoles     =           4
***** Field of Dipoles      [Kg]=       2.0000000
***** Path Length of Dipoles [m]=       0.2200861
***** Chord Length of Dipoles[m]=       0.2199074
***** Radius of Dipoles RHO  [m]=       1.5762508
***** Angle of Dipoles     [deg]=       8.0000000
***** Angle of Dipoles     [rad]=       0.1396263
***** Total Angle of Bend  [deg]=      32.0000000
***** Beam Momentum      [MeV/c]=      94.5096186
***** Beam Rigidity, BRHO [Kg-m]=       3.1525015
***** Beam Kinetic Energy  [MeV]=      94.0000000
***** Beam Particle Mass   [MeV]=       0.5110000
***** BETA (v/c)                 =       0.9999854
***** GAMMA                      =     184.9530333

<SINGLE ACHROMAT  > Z-Coordinate:    14.9958819 m
<ENDPOINT LOCATION> X-Coordinate:     4.3000000 m
*** Cell Path Length (incl bend):     3.9572794 m
*** Total Drift/Cell (excl bend):     3.7371933 m
*** Drift "a" (half total drift):     1.8685966 m
*** Total Achromatic Path Length:    15.8291175 m

Add Quadrupoles in each Cell to complete ACHROMAT

< MULTI-ACHROMAT>   Z-Coordinate:    29.9917638 m
<S-Bend ENDPOINT>   X-Coordinate:     8.6000000 m
```

All the entered information is "echoed back" in addition to the achromat's dimensions (coordinates, angles, drift lengths), dipole data (path length, chord length, field, radius), and beam parameters (K.E., mass, momentum, rigidity, beta, gamma). The final endpoint coordinates of the achromat are given assuming the bend path traverses in a positive direction in the positive Z-X plane from the origin (0,0). Achromats that require completion with dipole edge-angles or quadrupole focusing are provided with an additional note such as that shown above.

If an S-Bend configuration is possible, the user may be prompted for an additional offset dimension (the height of the "S" or the offset in the bend plane), and as before, the endpoint coordinates are given along with any additional drift distance required between achromats.

If the selected design is periodic (as is this example), the user can request a printout of the centerline coordinates after each cell and dipole:

```
********* CENTERLINE COORDINATES **********
    LOCATION       Z-COORDINATE    X-COORDINATE
    (at exit)        (meters)        (meters)
    Dipole  1        2.0879683       0.0153400
    Cell    1        3.9383799       0.2753984
    Dipole  2        6.0038934       0.5811781
    Cell    2        7.8001037       1.0962331
    Dipole  3        9.8029595       1.6865009
    Cell    3       11.5100074       2.4465276
    Dipole  4       13.4112221       3.3097946
    Cell    4       14.9958819       4.3000000
```

All the above calculations are written to an output file.

MODIFYING DESIGNS

ACHRO allows the user to modify any or all entries in a very user-friendly fashion. All change menus use character data input, so "no change" translates to a simple "carriage return".

```
Enter New Values if desired
                 OLD VALUE      NEW VALUE
                 ----------     ----------
Achromat Angle:   32.00000
  Dipole Field:    2.00000          3
Kinetic Energy:   94.00000
 Particle Mass:    0.51100
No. of Achros:           1
  No. of Cells:          4
  Achro OFFSET:    4.30000
```

If, for example, the dipole field strengths are increased, the dipole physical lengths will decrease, causing the drift lengths to increase in order to maintain the same overall offset. Following the change menu, the code once again displays the calculation. This continues until no further changes are requested, after which the output file is written. The program continues, allowing the user to choose another achromat which is designed using the same beam momentum.

BEYOND DRIFTS AND BENDS

Completing the achromat is done mostly by using a beam-transport code such as TRACE or TRANSPORT as previously mentioned. Using the notation found in TRANSPORT[2], it is only necessary to have the R_{16} and R_{26} terms of the achromat's transfer matrix equal to zero to define a first-order achromat. In our example (the periodic achromat), focusing quadrupoles in both the X and Y planes are required in each cell, and the focusing strength depends on the phase advance per cell which is dependent on the number of cells (for a 4-cell achromat, a phase advance of 90° is needed per cell). Thus, the geometry and dipole parameters calculated in ACHRO are entered into one of the above-mentioned codes, remembering that the quadrupole lengths must be subtracted from the drift lengths calculated by ACHRO. In this case, fitting is done on the quadrupoles, setting the R_{16} and R_{26} terms at the end of the achromat to zero. For second- and third-order achromats (which will not be discussed here), higher-order codes must be used with fitting being done using the higher-order elements.

An additional plus for periodic achromats is that they will produce a unity transfer matrix in the transverse planes. This means that the beam exiting the achromat looks identical to the beam entering the achromat in transverse phase space. This unity matrix is not possible to achieve with the non-periodic types, although there are some things that can be done to produce zeros in certain off-diagonal matrix elements. Each of the achromats is different, and this is discussed below.

COMPARING ACHROMATS

If more than one achromat is designed using ACHRO and a beam-transport code is used to obtain the R-transfer matrices, various characteristics can be compared which may be important to the overall design of the beamline. For example, if the resolving power of a system is important, the maximum values of the R_{16} terms inside the achromat should be compared, since they represent the magnitude of dispersion (referred to as the "Dispersion Index") and are directly proportional to the resolving power. If isochronicity is an important consideration, the R_{56} term can be used as a direct measure of isochronicity. If achromat sensitivity is an issue, one can observe the change (from zero) in the matrix terms R_{16} and R_{26} at the end of the achromat with, say, a 1% change in the field of any magnet element. A 1% change was the value used to compute the sensitivities given in Table II below which shows a comparison of these and other characteristics for several achromats all designed to fit into the same geometry. Note that the "Max Beam Radius" data is normalized to our periodic example, as the same beam was used in each bend design.

TABLE II: Characteristics of Achromats Compared

Achromat	Dispersion Index (m*)	Isochronicity (m*)	Sensitivity		Max Beam Radius		Hardware Req'd (diple/quad)
			R_{16} (m*)	R_{26} (rad*)	X_{max}	Y_{max}	
Dual 270	4.969	-28.386	0.4552	0.0492	7.31	0.72	2/-
2-D	1.497	-0.011	0.0006	0.0027	2.36	0.94	2/1
3-D	1.462	0.704	0.0070	0.0048	2.33	0.40	3/-
4-D	2.372	2.490	0.0511	0.0062	3.98	0.41	4/-
Stair Step	5.461	5.768	0.2171	0.0652	13.29	0.25	4/12
Periodic	0.679	-0.099	0.0063	0.0156	1.00	1.00	4/8

*Units are per fraction momentum spread

As can be seen, the periodic achromat has the smallest Dispersion Index while the Stair-Step has the largest. The 2-Dipole achromat is the most isochronous and the least sensitive to field changes, while the Dual-270° and the Stair-Step appear to be the most sensitive and least isochronous. For aperture requirements, the periodic achromat has the definite advantage. Comparisons such as these can be quite useful in finding both the right achromat and one that is optimized for the specific application. A more detailed discussion of achromats and the comparison of their characteristics, including beam quality considerations, is referenced.[7]

NOTES ON SPECIFIC ACHROMATS

Enge Dual-270°: Unlike other achromats, the Dual-270° has a fixed dipole angle (270°). It is a 0°-bend, similar to an S-bend configuration, useful for a beam offset. The only parameter beside beam momentum needed for a design is the offset distance. It will produce a negative unity R-matrix except for nonzero R_{12} and R_{34} terms. Edge-angle fitting may be required on the inside edges of the dipoles, although the angle should be close to 32.4°. It has a high dispersion index, is most nonisochronous, and has the highest sensitivity factor relative to the other achromats. It requires a large aperture.

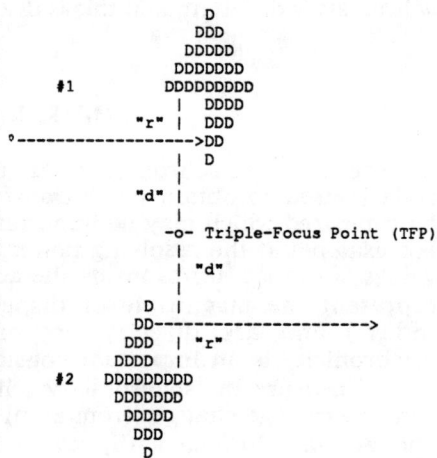

Brown 2-Dipole: The only two parameters needed for the geometry design in addition to particle momentum is the desired angle and a dipole parameter (field, length or radius). A bend-plane focusing quadrupole must be added at the symmetry point (or quadrupoles symmetrically placed) to complete the achromat to first order. If the achromatic distance "d" is observed, a negative unity R-matrix in the bend plane will be achieved except for a nonzero R_{12} term. The 2-Dipole achromat has a moderate dispersion index, is most isochronous and has the least sensitivity to errors.

```
              "d"            "d"
    D                                       D
   DDD-----------------|-----------------DDD
  DDDDD                                   DDDDD
    #1                                      #2
```

Leboutet 3-Dipole: The 3-Dipole achromat is complete with the three dipoles. All that is needed for a design (beside the particle momentum) is the achromat angle and a dipole parameter. Additionally, the geometry results in a negative unity matrix in the bend plane. Qualitatively, the 3-Dipole is similar to the 2-Dipole, but with a shorter achromatic distance and a smaller aperture requirement in the nonbend plane.

```
         "d/2"   D        "d"    DDDDD    "d"     D    "d/2"
        |--------DDD--------------DDD--------------DDD--------|
                DDDDD              D              DDDDD

                 #1                #2               #3
```

Enge 4-Dipole: To complete this achromat, fitting must be done on the entrance and exit edge angles of dipoles #1 and #4. The entrance edge angle should equal the exit edge angle. In addition, the entrance edge angle of dipole #2 and the exit edge angle of dipole #3 should equal the dipole angle. This achromat has great flexibility with respect to its geometry, but has high dispersion, low isochronicity (large R_{56} term), and wins no prizes with respect to aperture requirements. Beside the standard input as required for the 2- and 3-Dipole achromats, a drift length and an offset dimension must be specified. A negative unity R-matrix in the bend plane will be achieved except for a nonzero R_{12} term; however this term can be set to zero if fitting is also allowed on the drift spaces (note that this will change the length and offset of the achromat).

```
                "d1"           "d2"           "d1"
          D            DDDDD          DDDDD           D
         DDD----------DDD------------DDD-----------DDD
        DDDDD           D              D            DDDDD

         #1             #2             #3            #4
```

Stair-Step[6]: The Stair-Step is similar to the Dual-270° in that a single achromat is equivalent to an S-Bend (0°-bend). An offset dimension is a required entry along with the standard beam momentum, achromat angle and dipole parameter. Two geometries are available: one periodic and one nonperiodic. To complete the achromat, two quadrupoles per cell (focusing in both planes) must be added in the cell geometry, while at least two quadrupoles will be needed for achromaticity in the nonperiodic geometry. The Stair-Step has the highest Dispersion Index, has very low isochronicity, is highly sensitive to errors, and requires a very large aperture, the largest relative to the others.

```
              cell
           <---length-->
               D            DDDDD           D            DDDDD
         |-----DDD-----|-----DDD-----|-----DDD-----|-----DDD-----|
              DDDDD           D            DDDDD           D

               #1             #2             #3             #4
```

Periodic: ACHRO will design periodic systems with either one- or two-dipoles per cell. Required input includes the number of cells and achromats in addition to the standard entries of beam momentum, achromat angle and a dipole parameter. A offset dimension must also be entered. If one-dipole per cell is chosen, the dipole is centered in the cell if symmetric, and if nonsymmetric, the user must specify the dipole offset. If the choice is two-dipoles per cell, the user must specify an additional distance within the cell, such as the distance separating the two dipoles. Periodic systems characteristically have low dispersion indices, low sensitivities to errors, and low aperture requirements. As previously mentioned, a unity matrix is obtained in both transverse planes.

```
      <---1 cell--->
      "a"   D   "b"         "a"   D   "b"         "a"   D   "b"
      |----DDD-------|  +   |----DDD-------|  +   |----DDD-------|  + ...
           DDDDD                DDDDD                DDDDD
           #1                    #2                    #3


      <----------1 cell---------->
      "a"   D   "b"      D   "a"        "a"   D   "b"     D   "a"
      |-----DDD-----------DDD-----|  +  |-----DDD----------DDD-----|  + ...
            DDDDD         DDDDD               DDDDD        DDDDD
            #1             #2                  #3           #4
```

CODE AVAILABILITY

ACHRO is available through the Los Alamos Accelerator Code Group, Mail Stop H829, Los Alamos National Laboratory, Los Alamos, New Mexico, 87545, or through E-Mail contact: LAACG@lanl.gov.

---

[1] K.R. Crandall, D.P. Rusthoi, *TRACE-3D Documentation*, Los Alamos National Laboratory report LA-UR-90-4146 (December 1990),

[2] K.L. Brown, K.C. Carey, Ch. Iselin, and R. Rothaker, SLAC 91 (1973 rev), NAL91 and CERN 80-04.

[3] A.J. Dragt, et. al., *MARYLIE 3.0, A Program for Charged-Particle Beam Transport Based on Lie Algebraic Methods*. Univ. of Maryland Technical Report, Mar. (1991).

[4] These first-order achromats are described in *Focusing of Charged Particles, Vol. II*, edited by A. Septier, Academic Press, NY, 1967, pp.203ff.

[5] K.L. Brown and R.V. Servanckx, "First- and Second-Order Charged Particle Optics," in *Physics of High Energy Particle Accelerators*, AIP Conference Proceedings No. 127, M. Month, P.F. Dahl, and M. Dienes Editors, (1985).

[6] Roger Kennedy, Boeing Aircraft Co., private communication.

[7] D.P. Rusthoi, E.A. Wadlinger, "First- Second- and Third-Order Achromatic Bend Systems for Free-Electron Laser Applications", 1991 IEEE Particle Accelerator Conf., Vol 1, p.607, (1991).

CALCULATING WAKE POTENTIALS FOR ULTRA SHORT BUNCHES

M. Dehler, T. Weiland
Technische Hochschule Darmstadt, Fachbereich 18, FG TEMF,
Schloßgartenstr. 8, 6100 Darmstadt, Germany

ABSTRACT

We describe a new algorithm minimizing the storage requirements for wake calculations for ultra short bunches. By the use of a moving window technique, the storage request is no more a function of the overall size of the structure, only depending on the bunch length and the transversal size. Using this technique the wake potentials of a three cell SLAC type cavity have been calculated for bunch lengths as short as $\sigma = 50 \mu m$. With these results the asymptotic behaviour of the longitudinal loss factors for short bunches are examined.

INTRODUCTION

The general trend in accelerator design is to use ultra short bunches with a length below 1 mm as driven charges. In order to calculate the resulting wake potentials using Finite Difference or Finite Integration programs, one needs grids with a small spatial resolution and therefore a high number of grid cells, typically being in the range of 10^7 or more.

This leads to demands of CPU time and storage, which can exceed even the capabilities of modern mainframes. In many cases only the wake potentials within the bunch are of interest, so that field computations can be restricted to a part of the structure. This is already exploited by the program TBCI by the use of a calculation window moving with the bunch through the structure[1], reducing the CPU time consumption considerably. Still problematic is the demand of core space being a linear function of the grid size and because of that, for $2\frac{1}{2}$D programs a quadratic function of the grid resolution.

METHOD

The core space is normally divided up into space for the grid information, the system matrix and the component vectors of the electric and magnetic fields having lengths of a multiple of the grid size. Using a calculation window the system matrix and the component vectors can be restricted to contain only informations about the region inside the window. As the window is moving through the structure the matrix has to be continuously updated with the information on new parts of the structure entering the calculation domain.

The problem still is the storage of the grid information on the material filling inside the cells, which increases linear with the number of grid cells. For most structures this is highly redundant. E. g. when calculating bunches with

$\sigma = 100\mu m$, the grid step has to be smaller than $20\mu m$, whereas the structure can easily be described by coarser mesh step sizes.

These considerations lead to the following setup. The grid, created by the **MAFIA** mesh generator M, is taken only as a kind of information base about the geometric properties of the structure and because of that, is only required to be fine enough to give a good representation of the material boundaries.

The time domain calculation of the electromagnetic fields can be restricted in longitudinal direction to the region, where the bunch is moving and the integration of wake potentials is performed. Only in this region a high resolution grid is needed. Since the integration of the wake fields is done continuously with the time domain iteration, only the field information inside the bunch region has to be available.

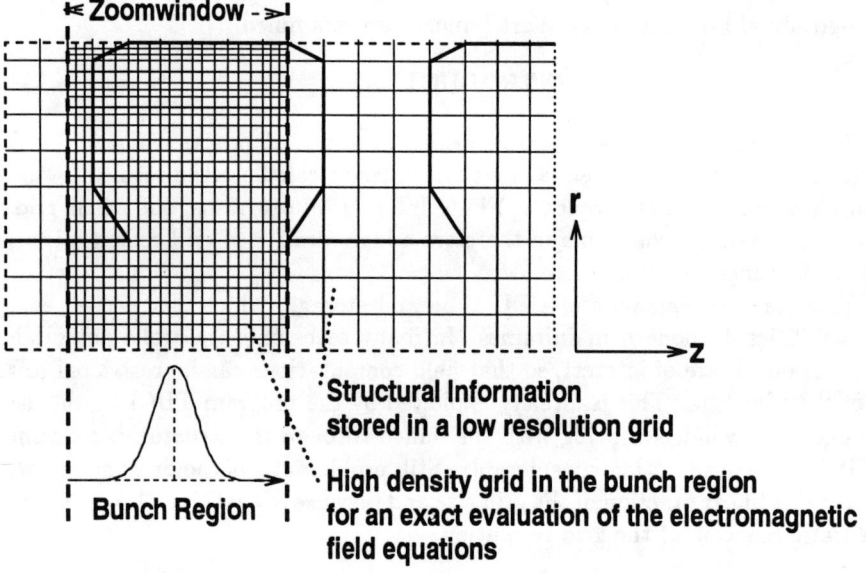

Fig. 1. Adaptive Grid Setup using a moving Zoom Window

The resulting algorithm is indicated in Figure 1. During the time integration loop the grid inside the bunch region is zoomed up by a user defined factor, that is, all grid steps are subdivided into smaller ones, thus creating the necessary resolution for the calculation. A zoom factor of 10 means, that every cell in the original grid describes a rectangular area containing 100 cells of the calculation grid having all the same material fillings.

With this option the storage demand of the original grid is typically negligable compared to that needed for the calculation. It is no more a function of the overall size of the structure, but determined by the length of the calculation region times the cross section.

RESULTS

With this program the wake potentials of a three cell cavity comprised of the average SLAC cell[2] shown in Figure 2 were examined.

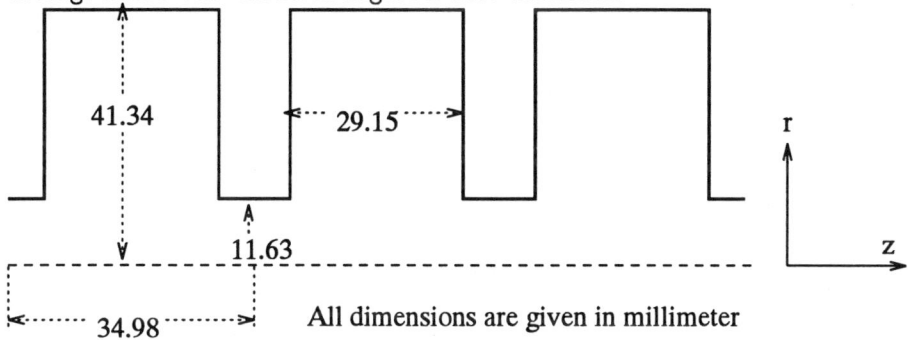

Fig. 2. Three Cell Cavity

We investigated the monopole and dipole wakes for bunches from $\sigma = 1mm$ down to $\sigma = 50\mu m$. For a grid resolution of five grid steps per σ the needed core size and the CPU time for the calculation is given in Table I. In the course of the convergence checks up to ten steps per σ were calculated, yielding an effective grid size of 0.2 billion cells. The core space demand for this size would have been 11.5 GBytes without grid zooming. The actual run needed only 40 MBytes.

Table I. Core size demand and execution times of the wake potential calculation on a SUN SPARC 2. All Values are for a grid resolution of 5 steps per σ.

σ	Core Size/MByte	CPU Time/minutes	Effective Grid Size
Monopole Calculations			
$200\mu m$	4.6	53	$2.8 \cdot 10^6$
$100\mu m$	7.3	273	$11.3 \cdot 10^6$
$50\mu m$	15.6	1023	$45.4 \cdot 10^6$
Dipole Calculations			
$200\mu m$	7.5	156	$2.8 \cdot 10^6$
$100\mu m$	13.9	816	$11.3 \cdot 10^6$
$50\mu m$	25.4	3340	$45.4 \cdot 10^6$

In Figures 3 and 4 the longitudinal and transversal wake potential for a bunch length of $\sigma = 50\mu m$ are shown. In these pictures the influence of large grids with 10^7 and more unknowns can be seen. Normally for simple structures a mesh to bunch length ratio of five grid steps per σ is sufficient to produce accurate results. As the number of grid steps are increasing (for this application 11000 steps in z-drection), even a small numerical dispersion can have visible effects on the result.

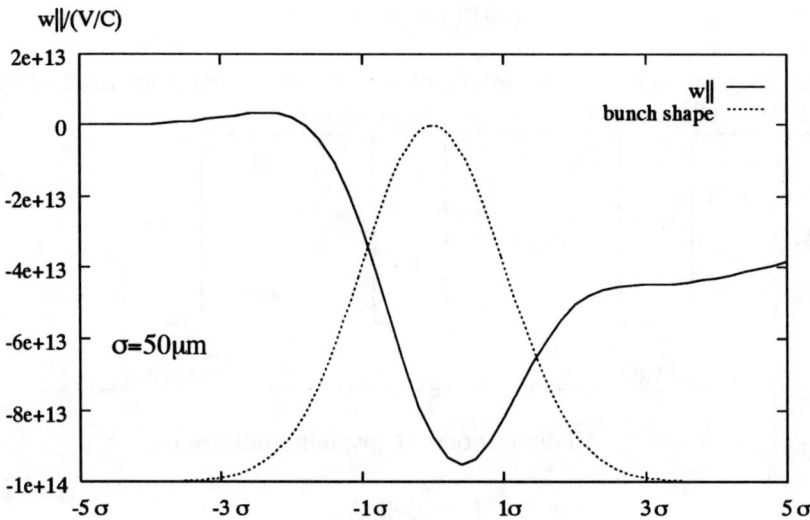

Fig. 3. Longitudinal wake potential for $\sigma = 50\mu m$, dipole calculation. The grid resolution is 5 steps per σ, the wake potential has still not converged yielding (wrong) positive values.

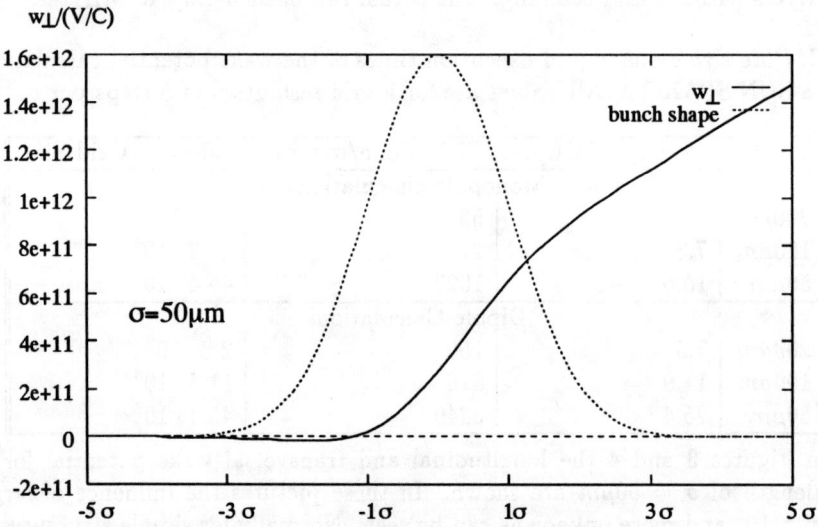

Fig. 4. Transverse wake potential for $\sigma = 50\mu m$, dipole calculation. The grid resolution is 5 steps per σ.

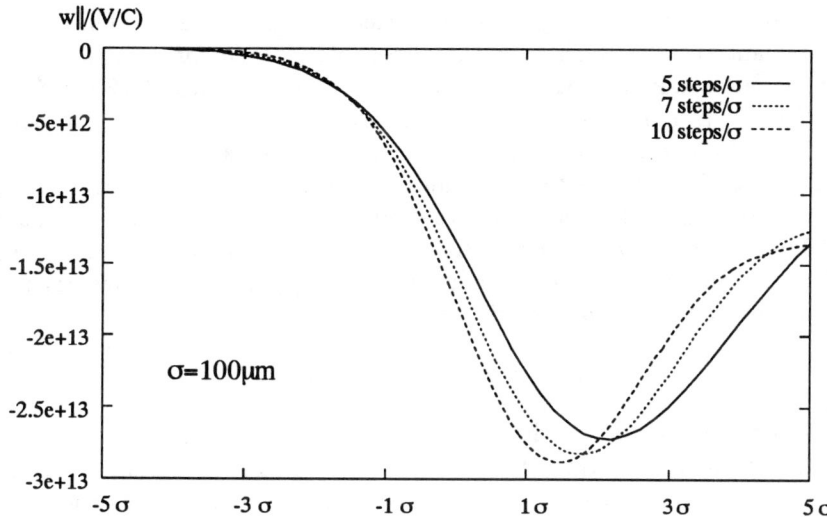

Fig. 5. Monopole Wake Potentials calculated for $\sigma = 100\mu m$ using different grid resolutions.

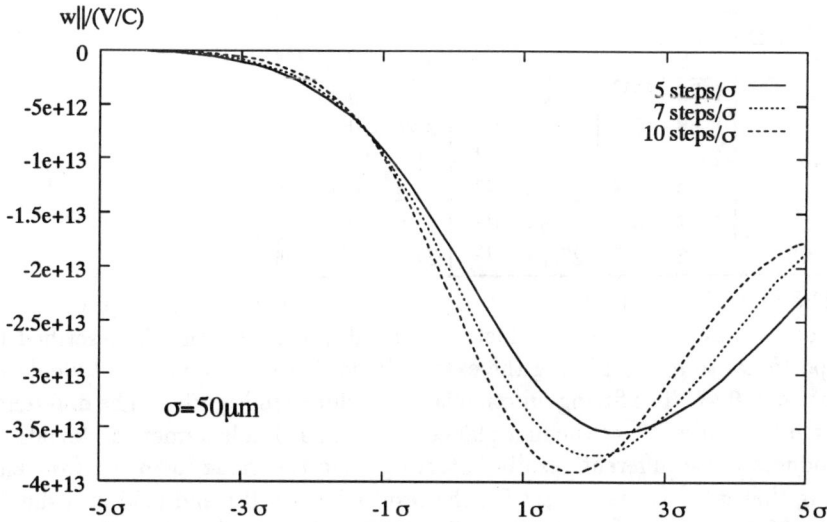

Fig. 6. Monopole Wake Potentials calculated for $\sigma = 50\mu m$ using different grid resolutions.

This dispersion effect shows up also in Figure 5, where the longitudinal wake potential is shown for different grid resolutions. For low resolutions the high frequency, fast varying part of the wake is shifted backwards corresponding to a slower group velocity of the high frequency electromagnetic waves. This is even more dominant in Figure 6 showing the corresponding wake for $\sigma = 50\mu m$. The extremum of the wake potential varies for $\sigma = 100\mu m$ between $2.72 \cdot 10^{13}$ and $2.88 \cdot 10^{13}$ with different grid resolutions, that is about 5%, whereas for $\sigma = 50\mu m$ a variation of 10% between $3.54 \cdot 10^{13}$ and $3.91 \cdot 10^{13}$ occurs.

For the derived loss factors, this means also a degradation of convergence with decreasing bunch length. Since the loss factors for bunch length shorter than $200\mu m$ did not converge sufficiently fast for resolutions up to 10 steps/σ, the following extrapolation scheme was adopted. The numerical error was assumed to be a quadratic function of the grid step h:

$$k_{\|}(h) = k_{0\|} + a \cdot h + b \cdot h^2 \qquad (1)$$

Calculating the loss factor for three different grid sizes gives the function values $k_{\|}(h_1)$, $k_{\|}(h_2)$ and $k_{\|}(h_3)$, which can be used to solve for the asymptotic value $k_{0\|}$. The loss factors for 5 steps/σ, 10 steps/σ as well as the extrapolated results are given in Table II.

Table II. $k_{\|}$ as a function of σ with different grid resolutions and extrapolated. All values are in V/C.

σ/mm	$k_{\|}/(V/C)$		
	5 steps/σ	10 steps/σ	extrapolated
1	$6.73 \cdot 10^{12}$	$6.76 \cdot 10^{12}$	$6.75 \cdot 10^{12}$
.5	$8.88 \cdot 10^{12}$	$9.18 \cdot 10^{12}$	$9.26 \cdot 10^{12}$
.3	$11.1 \cdot 10^{12}$	$11.4 \cdot 10^{12}$	$11.8 \cdot 10^{12}$
.2	$12.9 \cdot 10^{12}$	$13.8 \cdot 10^{12}$	$14.1 \cdot 10^{12}$
.1	$16.7 \cdot 10^{12}$	$18.5 \cdot 10^{12}$	$20.0 \cdot 10^{12}$
.05	$20.6 \cdot 10^{12}$	$23.7 \cdot 10^{12}$	$27.4 \cdot 10^{12}$

The plot of the loss factor versus bunch length is shown in Figure 7. The asymptotic behaviour of the loss factor for small bunch lengths[3] is described to be proportional to $\sigma^{-0.5}$. Fitting the extrapolated loss factors with $\sigma^{-\alpha}$ yields an exponent $\alpha = 0.47$. The fitting function is also included in Figure 7. The difference between $\sigma^{-0.5}$ and $\sigma^{-0.47}$ at the first glance looks like a simple numerical deviation, but produces quite different results. Starting from the value for $\sigma = 1mm$ and assuming that $\sigma^{-0.5}$ is the correct fit, the results for 0.2, 0.1 and 0.05 mm should be $15.1 \cdot 10^{12}$, $21.3 \cdot 10^{12}$ and $30.2 \cdot 10^{12}$ V/C. Whether this deviation is of a physical nature, will be determined by calculations with still higher resolutions of 20 steps/σ and more, which are currently under way.

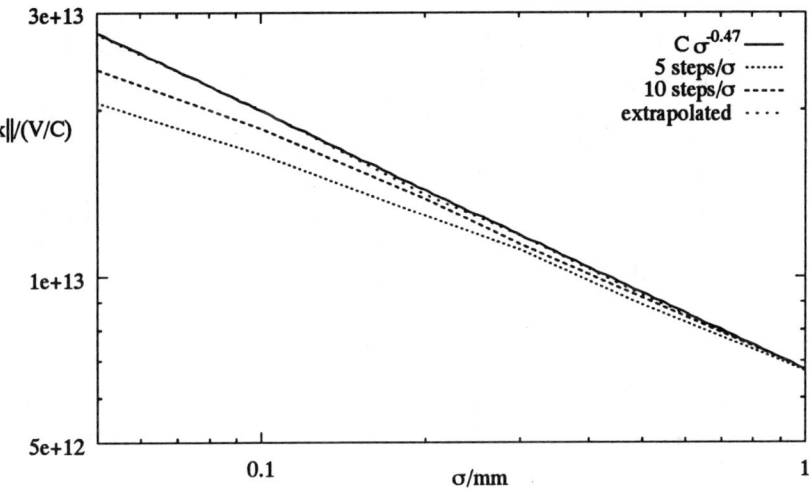

Fig. 7. $k_\parallel$ versus σ calculated with different grid resolutions. Included is also the extrapolated loss factor curve with the fitting curve $k_\parallel \sim \sigma^{-0.47}$

In Table III the calculated loss factors for the dipole case are listed. As all calculations were done with grid resolutions of five steps/σ, the values for small bunch lengths probably still contain significant errors. Because of that an investigation of the asymptotic behaviour is not possible.

Table III. $k_\parallel$ and $k_\perp$ as a function of σ (dipole calculation) calculated with a grid resolution of 5 steps per σ.

σ/mm	$k_\parallel/(V/C)$	$k_\perp/(V/C)$
1.	$12.1 \cdot 10^{12}$	$1.45 \cdot 10^{12}$
0.5	$18.1 \cdot 10^{12}$	$1.07 \cdot 10^{12}$
0.3	$23.8 \cdot 10^{12}$	$0.83 \cdot 10^{12}$
0.2	$29.9 \cdot 10^{12}$	$0.69 \cdot 10^{12}$
0.1	$44.1 \cdot 10^{12}$	$0.51 \cdot 10^{12}$
0.05	$67.2 \cdot 10^{12}$	$0.38 \cdot 10^{12}$

CONCLUSION

We demonstrated a new storage algorithm for calculating the wake potentials of ultra short bunches. By a condensed description of the grid as well as only a partial storage of system matrix and state variables, effective grid sizes exceeding 10^8 cells can be calculated on common desk top workstations.

In the context of the DESY/THD S-Band linear collider study[4] the wake forces inside a three cell structure were investigated. As the results show, the grid dispersion develops with increasing grid size a larger effect on the numerical error. Thus the grid resolution has to increase superlinear with the bunch length. By an extrapolation scheme applied to the loss factors calculated with different resolutions, an asymptotic dependency of the longitudinal loss factor with $k_{\|} \sim \sigma^{-0.47}$ for the monopole case was obtained. Whether this deviation to the traditionally assumed behaviour $k_{\|} \sim \sigma^{-0.5}$ is caused by numerical effects, cannot be answered fully at the moment and will be investigated further.

REFERENCES

1. K. Bane and T. Weiland, "Wake Force Calculation in the Time Domain for long Structures", Proceedings of the 12th International Conference on High-Energy Accelerators, pp. 314 (1983)

2. K. Bane and T. Weiland, "Verification of the SLC Wake Potentials", SLAC/AP-1, Stanford University (1983)

3. K. Bane and M. Sands, "Wakefields of Very Short Bunches in an Accelerating Cavity", Proceedings of the Workshop on Impedances beyond cut-off, preprint SLAC-PUB-4441, Stanford University (1987)

4. T.Weiland (for the S-Band Collider Group), "Status of the DESY-THD 500 GeV S-Band Linear Collider Study", Talk at the LC92 in Garmisch Partenkirchen, to be published

MULTI-ION TRANSPORT SYSTEM

B.I.Bondarev, A.P.Durkin, G.T.Nicolaishvili, O.Yu.Shlygin
Moscow Radiotechnical Institute
Warshavskoe shosse, 132, 113519, Moscow, Russia

Abstract

LIDOS.LBT codes description is reported. The codes permit to calculate transporting of ion with different charge and mass numbers in channel containing quadrupole and solenoidal lenses for ion focusing and bending magnets for its separation. The input/output parameters matching is achieved by optimization procedure. LIDOS.LBT is the part of codes package LIDOS. Some calculation results is discussed.

The heavy ion source consists of several isotopes with different charges and rest masses. In many cases it is necessary to select the main ions and clear the injecting beam from other isotopes before injects to the linac.

Below it will be described how this problem may be solved by the help of LEBT (Low Beam Energy Transport) codes [1].

LEBT codes generate the focusing channel for transporting the beam with preset input and output parameters. LEBT solves the matching and separating problems specified for injection beam. In relation to these problem LEBT has two parts: *Matching/Optimization* and *Separator*.

The *Matching/Optimization* tool provides the designer with the information about mono ion beam transporting and matching, the *Separator* tool - about separating the ions with given charge from multi ion beam.

The first tool was described in the report [2].

The general problem is formulated as follows.

There are multi ion beam with given input parameters (ion types number, charges and masses, ion types currents, total beam emittance, main ion input energy). The focusing channel includes the quadrupole and solenoidal lenses, bending magnets and the slits which function is side ions edging. It is required to calculate channel parameters in order to match in the best way the main ion beam with the channel output parameters.

The ion motion has been described by the equations in which the magnet fields is constant inside channel elements and equal zero outside of them. The input phase ellipsoid has been packed by ions with different masses and charges. The main ions have the greatest particle proportion. The Poisson equation with the boundary relation

$$U \Big|_{x^2 + y^2 = R^2} = 0$$

is used for space charge forces calculation at each step of integration (where U is the space charge potential, R - the aperture radius).

The input beam is shown as an ellipsoid matrix in the (x, x', y, y') phase space

$$\mathbb{E} = \begin{bmatrix} E_{xo} & 0 \\ 0 & E_{yo} \end{bmatrix}$$

where E_{xo}, E_{yo} are the ellipses on the (x,x') and (y,y') planes consequently. The E_{xo} and E_{yo} definitions are based on the beam envelope values r_{xo}, r_{yo} and their tilts r'_{xo}, r'_{yo} at the channel input. In the other case the Twiss parameters α, β have been used. The matrices E_{xo}, E_{yo} can be written in the form as follows (the xo, yo indexes are dropped):

$$E = \begin{bmatrix} \dfrac{1}{r^2} + \dfrac{r'^2}{\varepsilon^2} & \dfrac{-r \cdot r'}{\varepsilon^2} \\ \dfrac{-r \cdot r'}{\varepsilon^2} & \dfrac{r^2}{\varepsilon^2} \end{bmatrix} = \begin{bmatrix} \dfrac{1 + \alpha^2}{\beta \cdot \varepsilon} & \dfrac{-\alpha}{\varepsilon} \\ \dfrac{-\alpha}{\varepsilon} & \dfrac{\beta}{\varepsilon} \end{bmatrix}$$

where ε is the beam emittance.

There are following relations between r, r' and α, β

$$r = \sqrt{\varepsilon \cdot \beta} \qquad r' = \sqrt{\varepsilon/\beta}$$

The slit positions and sizes are added to input data. The slit position are located by the order number and the distance from previous element. The x, y parameters present limitations on the slit size in x,y-direction. If $x = y$ the slit reduced into round window.

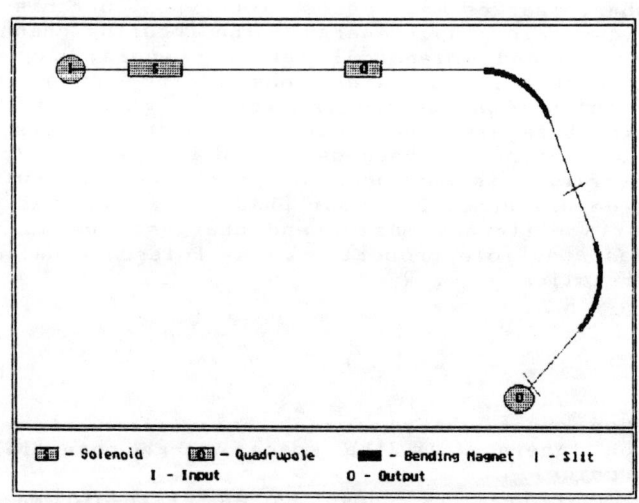

Fig.1. Schematic sketch of the channel

After input data setting the designer sees a schematic sketch of the channel (see Fig.1). The slits are marked by yellow lines. The command *RESTART* is used for the input data correction, the command *CONTINUE* - for beam simulation in the choosing channel version.

During calculation process the beam information has been visualized (see Fig.2).

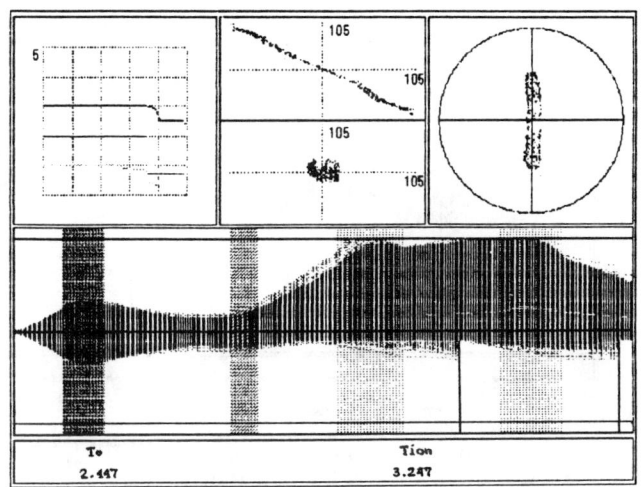

Fig.2. Separator process

The upper half of screen is divided into three sections. In the left side the fractional variation of ion currents are indicated, in the middle part of the halfscreen - the charge distribution and in the right side - the beam phase portraits projection on the (x,y)-plane are indicated.

In the lower half of screen the x-projection (up from axis) and y-projection (down from axis) of ion trajectories are indicated as a function of z on the backgrounds of the channel elements and the aperture. The white colour corresponds to the main ions, which are indicated on the background of the side ion trajectories. Several steps before the beam pass through slit the one is displayed in the upper right window and the "cleaning" process is visualized.

The main ion current value as well as the total current value are displayed at the lowest row. The main ion phase portraits in the (x,x') and (y,y') planes are displayed concurrent with preset output ellipses (red colour).

Provision is not made for the automatically optimization of the slit positions because a solution of Poisson equation is time consuming. That is why the designer to look for the optimum slit position using visual information.

The question *"Would You Like to Represent the Results?"* is asked and in positive case the output phase portraits will be approximated by ellipses. It will be desirable to preset the minimum ion proportion inside of ellipses and the maximum growth of the effective emittance. During the data processing the parameters of ellipse with maximum ion inside are define for growing emittance values. The process is over when the given ion proportion has been exceeded and the ion relative growth is more than the emittance one by factor 1.5. Then the net result will be passed in *Matching/Optimization* tool in order to optimize the channel part placed after the slit (see Fig.3).

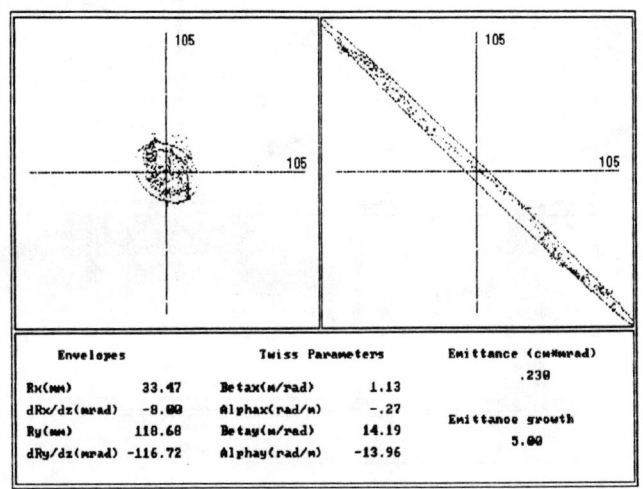

Fig.3. Calculation results representation

The following algorithm may be recommended for a focusing channel optimization when the separator is included.

At the first stage the *Matching/Optimization* tool is used for the preliminary optimization of all channel parameters with the beam can be taken as mono ion and its current as the total ion current. The beam dimension is large when it pass the separator and it is the reason why such calculation gives the satisfactory approximation for fields of focusing lenses placed before separator. Passing this field values into *Separator* tool the designer can define the optimum slit positions. At the next stage the net results (main ion current, emittance and phase ellipse parameters) will be passed in *Matching/Optimization* tool for final optimization of the channel part placed after the slit.

REFERENCES

1. B.I.Bondarev, A.P.Durkin, B.P.Murin, G.T.Nicolaishvili, O.Yu.Shlygin "LIDOS - unconventional helper for linac beam designing",this Conference
2. B.I.Bondarev et al 1992 Linear Accelerators Conference Proceedings v.2, 734(1992)

BEAM LINE ERROR ANALYSIS, POSITION CORRECTION AND GRAPHIC PROCESSING*

Fuhua Wang, Naifeng Mao

Superconducting Super Collider Laboratory
2550 Beckleymeade Avenue, Dallas, TX 75237

ABSTRACT

A beam transport line error analysis and beam position correction code called "EAC" has been developed associated with a graphics and data post processing package for TRANSPORT. Based on the linear optics design using TRANSPORT or other general optics codes, EAC independently analyzes effects of magnet misalignments, systematic and statistical errors of magnetic fields as well as the effects of the initial beam positions, on the central trajectory and upon the transverse beam emittance dilution. EAC also provides an efficient way to develop beam line trajectory correcting schemes.

The post processing package generates various types of graphics such as the beam line geometrical layout, plots of the Twiss parameters, beam envelopes etc. It also generates an EAC input file, thus connecting EAC with general optics codes. EAC and the post processing package are small size codes, that are easy to access and use. They have become useful tools for the design of transport lines at SSCL.

INTRODUCTION

Nowadays we rely on popular general optics codes such as TRANSPORT to design beam line optics. However it is always a desire of beam line designers to have handy tools to look at various aspects of the design e.g. the sensitivity to errors, basic trajectory correction schemes etc. These "tool" codes should be easy to access and to modify to accommodate various real situations. Our work has been done in two steps. First, a TRANSPORT post processing package was developed, which enables us to graphically examine the optical design of the line in various aspects, such as creating plots of Twiss functions, beam envelopes, dispersion vectors and geometrical layout. These are extremely useful in the iteration process of lattice design. The next was the development of the "EAC" code[1] for error effect analysis and graphically interactive beam position correction scheme design. An "EAC" input file is automatically generated during the normal TRANSPORT post processing. Standard drawing software using "TOPDRAWER" is then applied. Users who are familiar with FORTRAN, need little effort to use these graphic packages or for

---
*Work supported by the U.S. Department of Energy

making their own favorite drawings by editing or imitating these codes. We confined ourself mainly to linear optics and limited the scope of work so as to keep the whole program package small.

TRANSPORT POST PROCESSING PACKAGE

This package processes TRANSPORT output in either accelerator notation or traditional notation, which contains all necessary data for each physical element along the beam line. It generates three standard data files for further processing. One includes details of element parameters, such as element names, lengths, apertures, fields, gradients etc., the second one contains various beam parameters, such as β, α, η, ψ functions, and the third one is a comprehensive file as the direct input of the following application codes. Fig.1 is a flow chart which depicts the main functions of this package.

A graphical code, BAEFSYN, along with a title file, BAETITLE.DAT, creates some TOPDRAWER files for plotting β, η, ψ functions, and beam envelopes along the beam line. The phase advance function ψ (ψ_x or ψ_y) are plotted with ψ-$n\pi/2$ as ordinate. As an example, a plot for η and β functions of the SSCL LEB to MEB transfer line is shown in Fig. 2. The GPXA and GPYA codes along with two title files GPXATITLE.DAT and GPYATITLE.DAT, create two TOPDRAWER data files GPXA.TOP and GPYA.TOP for beam line two dimensional geometric layout plotting in both plane view and elevation view. An elevation view of the SSCL LEB to MEB transfer line and dump line layout is shown in Fig. 3. Meanwhile, a three dimensional coordinate data file COD.DAT is created by code CO3-SYN. Coordinates at both ends for each element are given in this file. Code FSCCM creates a file FOR002.DAT which will be the input data file for "EAC".

This post processing package is written in FORTRAN. The total length of the above major processing and application codes is about 1600 lines.

ERROR ANALYSIS AND POSITION CORRECTION CODE "EAC"

The main functions of "EAC" are: Linear beam optics calculations, error effect analysis on the beam central trajectory, the beam phase space ellipse distortion and the beam position correction scheme. A flow chart of EAC and it's associated application codes are shown in Fig.4.

(A) Linear beam optics: this is a useful tool for input data verification. Both graphical and digital outputs of optics are provided.

(B) Beam trajectory calculation: EAC simulates the single particle trajectory in the presence of various errors. For magnet misalignment effects, the mathematical base of TRANSPORT[2] is adopted. Six dimensional magnet displacements are taken into account. Other input errors are the beam initial position errors and bending magnet field errors. All errors can be input directly or

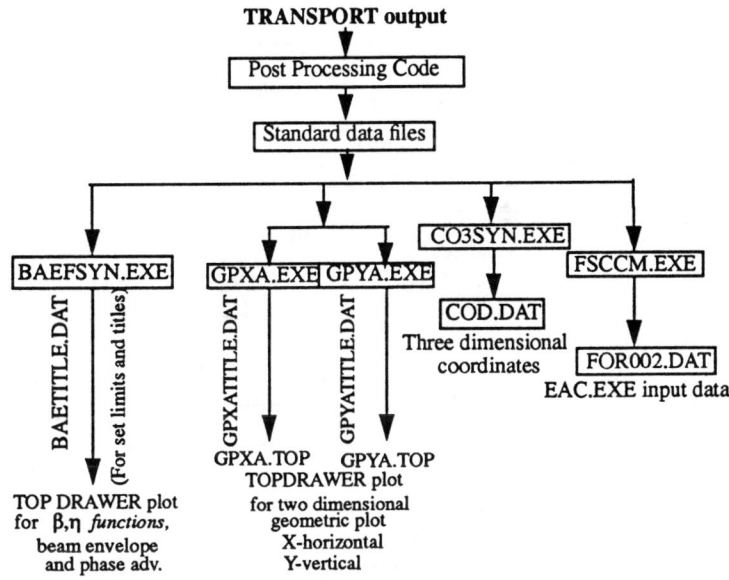

Fig.1 TRANSPORT Post Processing Package

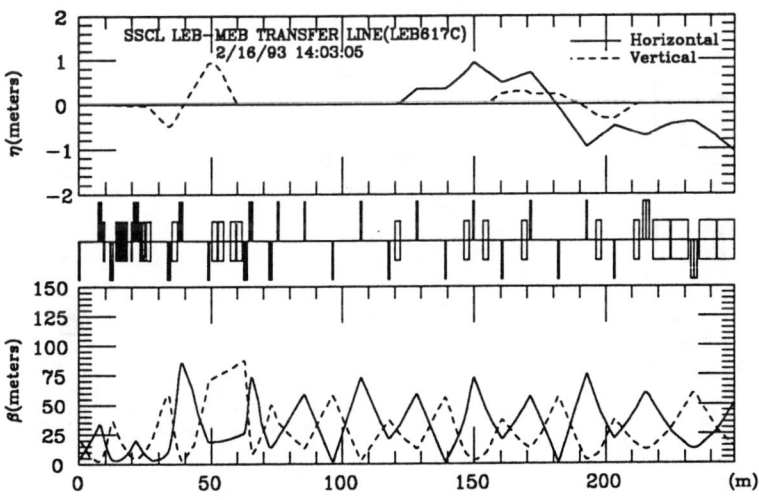

Fig.2 β and η functions

138 Beam Line Error Analysis

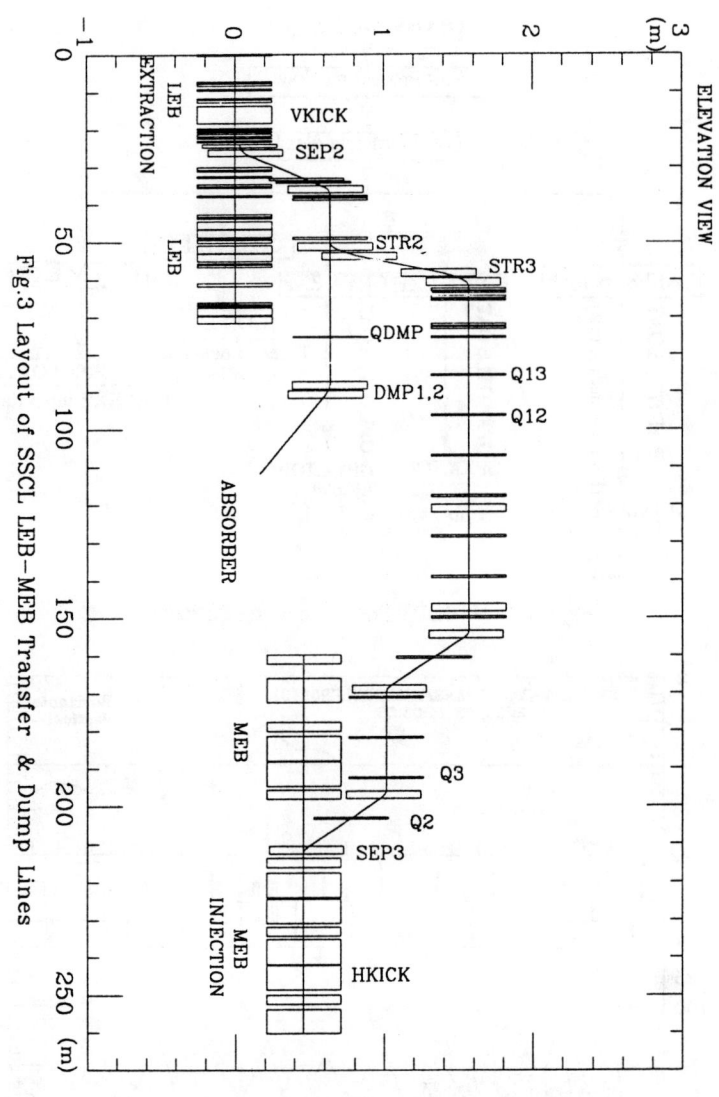

Fig.3 Layout of SSCL LEB-MEB Transfer & Dump Lines

generated by the code(random normal distribution). The magnet misalignment and other errors are uncertain in the design stage, multi-trial simulations on beam lines with random error distributions are applied to determine the tolerances on these errors. The program can assign all these errors from normal distributions with different standard deviation rms values. In this case, rms values for different types of errors are required for input.

Magnet misalignments, field errors and initial beam position errors are responsible for beam centroid displacements from the reference trajectory along the beam line and for the closed orbit mismatching at injection.

Two criteria for tolerance setting may be appropriate: beam line aperture limits and closed orbit mismatching limit. Closed orbit mismatch will cause beam transverse emittance dilution to follow in the accelerator. Closed orbit mismatch caused by systematic errors can be reduced dramatically by a trajectory correction scheme. While orbit mismatch caused by time related statistical errors can't be eliminated by a slow correction process, they should be restrained by regulation tolerance set to transfer line magnet power supplies.

Magnet field variation caused by current ripple and other time related current variations may be coherent if they are powered in series. An option is provided to group magnets in the input data, so that magnets in one group will have the same field variation in such analysis.

The output of a single trial includes the beam positions (a data file and a plot, Fig.5) and error information(RANDOME.dat). The output results from a multi trial run are: (1) Distribution of maximum displacement along the line (Fig.6-1), (2)Equivalent displacement X_{eq} at the concerned point. X_{eq} is a quantity used in determining possible phase area filamentation to follow in the accelerator[3].

Tabulated results are given in file Mplot.dat. One can track back to any of the individual cases with graphics of central trajectory and detailed digital results by selecting corresponding error seed number (see Fig.5).

(C) Beam position correction

There are two correcting methods available in this code[2]: one(corrector) to one(monitor) mode and two(correctors) to two(monitors) mode (orthogonal control). The first way is easy for operating, and may cost less. It is used for rough correction which is sufficient in most part of line. The second method is used in cases where strict position error control are needed.

The error analyzing calculation can be performed with or without a correction scheme. When working with a correction scheme, information of beam position monitors and correction dipoles is needed for input. Monitors and correctors can be defined to work for one transverse direction or for both; correctors can be independent dipoles or auxiliary windings on magnets depending

140 Beam Line Error Analysis

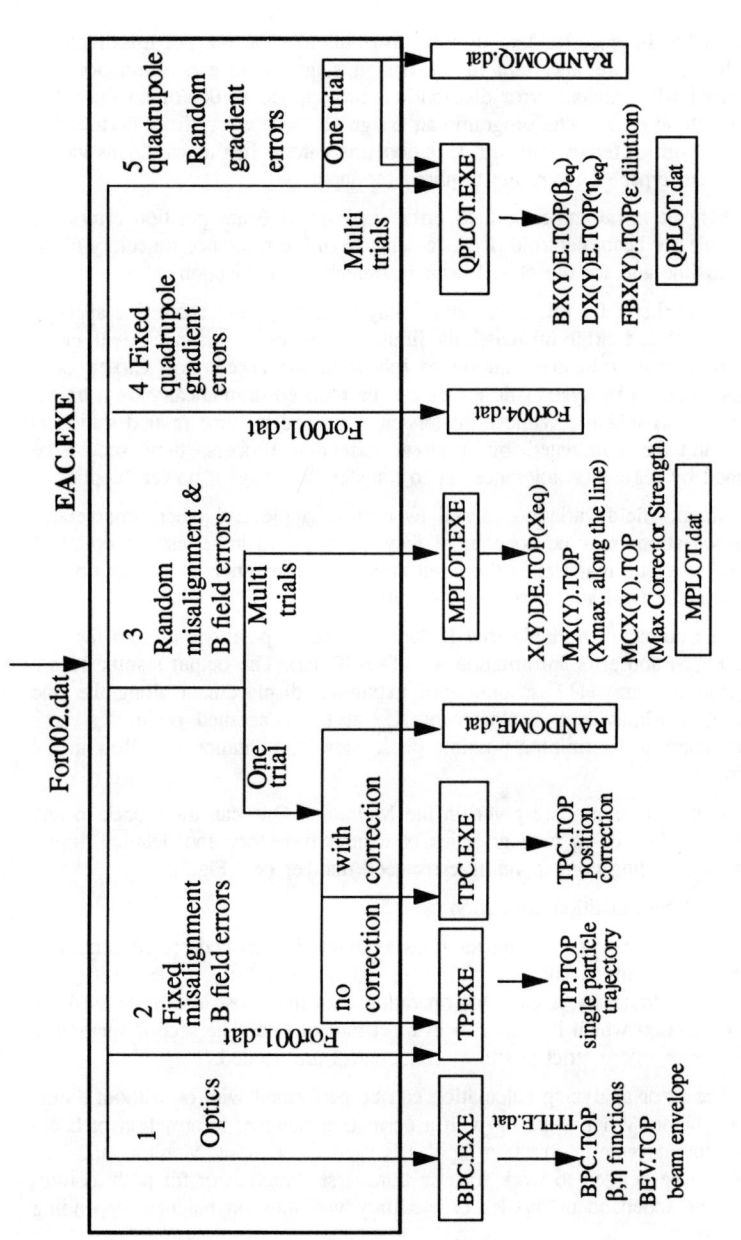

Fig. 4 The EAC Program Package

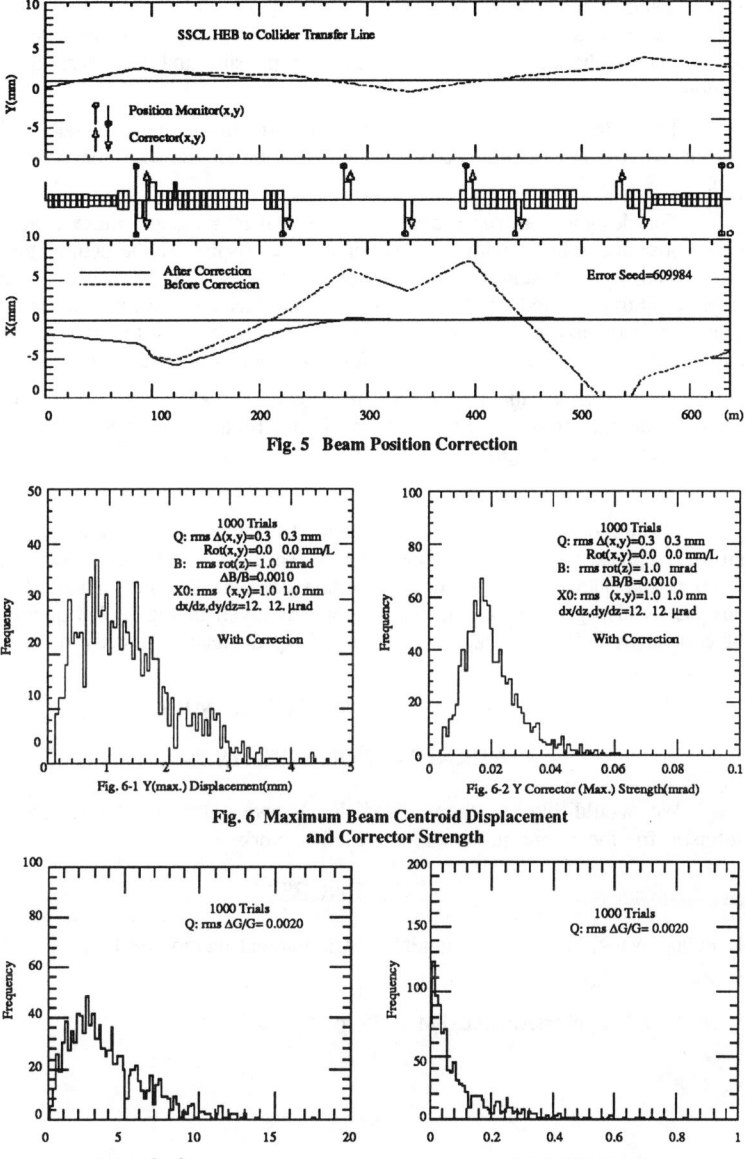

Fig. 5 Beam Position Correction

Fig. 6 Maximum Beam Centroid Displacement and Corrector Strength

Fig.7 Sensitivity to Gradient Errors

on the need in real situations. There is more physics than programing for the arrangement of a correction scheme, however graphically interactive iteration greatly helps the designer to understand the physics and take care of real requirements.

The output of a multi-trials run with a correction scheme includes information on corrector strengths and other information similar to a non correction run.

The designer can run a case of single trial to start, and make a first fit of a correcting scheme with the help of graphic display of the central trajectory. Then by examining the results of multi-trials run, he can track and improve individual undesired cases. The design goals are to meet position error limit requirements and to simplify the correcting scheme to the extent that it can be, so as to reduce cost by using fewer correctors and monitors.

Fig.5 is an example of central trajectories before and after correction. Fig.6 is the statistical distribution of maximum beam displacement along the line and corrector strength needed for the above correcting scheme.

(D)Quadrupole gradient errors

Quadrupole gradient errors will cause distortions of the beam emittance ellipse in phase space. The filamentation of the ellipse will lead to emittance dilution and should be reduced to a certain amount. The flow chart of this analysis is also given in Fig.4. One example is given in Fig.7. The definition of equivalent β function β_{eq} is again like Xeq, a quantity directly related to emittance dilution[3].

"EAC" code has about 2000 standard FORTRAN lines.

ACKNOWLEDGEMENTS

We would like to thank Prof. K. Brown, Drs. J. McGill and D. Johnson for their continuous support of this work.

REFERENCES

1. Fuhua Wang, "EAC user's guide", SSCL internal memo, 1991.
2. Karl L. Brown et al., CERN 80-04, p.229-243, 1980.
3. Michael J. Syphers, FNAL-TM-1456, p.119, 1987.

THE DESIGN OF A FOUR CELL THIRD ORDER ACHROMAT

Weishi Wan, Eyal Goldmann and Martin Berz
Department of Physics and Astronomy, and
National Superconducting Cyclotron Laboratory
Michigan State University
East Lansing, MI 48824

ABSTRACT

A four-cell third order achromatic system with one bending magnet per cell is presented. Instead of repetition of cells, which is used in achromat design based on normal form theory, we utilize cells obtained from the original one through mirror imaging about the x-y plane, which corresponds to a reversion, through mirror imaging about the x-z plane, which corresponds to a switching of the bending direction, and through both of the above processes, which corresponds to a 180° rotation in the x-z plane. The nonlinear part of the map was represented by a Lie exponent, and analytical conditions on this pseudo-Hamiltonian could be found using stochastic methods of theorem proving in COSY INFINITY. A family of systems was determined requiring the least number of conditions. One of them was chosen as an example to design a realistic third order achromat using COSY. The required multipoles are rather weak and the remaining higher order aberrations rather benign.

INTRODUCTION

The search for achromats up to a certain order has always been an interesting problem of optics. While it is not very difficult to design a first order achromat, the correction of all second order aberrations already represents a major effort[1]. In the past few years, various third order achromatic systems consisting of only repetitive identical cells containing five to several hundred bending magnets have been found using normal form theory[2,3,4]. Recently a new theory requiring only four cells with one bending magnet each has been developed to find third order achromats and in principle any higher order ones[5,6]. Two ideas form the core. The first is to introduce mirror symmetry into the consideration. This gives four kinds of cells that can be used in the design rather than the only one in the traditional achromat design based on normal form theory. It means one can make up systems with higher symmetry. The four categories of cells are the original cell (Forward or F), the reverse cell (R) which is the mirror image of the forward cell about the x-y plane (corresponding to a reversion of the order of the elements), the switch cell (S) which is the mirror image of the forward cell about the x-z plane (corresponding to a change of the bending direction) and the combined cell (C) which switches the bending direction and reverses the order of elements as well. The second idea is to find analytical conditions for the Lie exponent in the map

144 The Design of a Four Cell Third Order Achromat

representation $\mathcal{M} = M \circ \exp(: f :)$. Table 1 shows those terms in the Lie exponent which have to be killed to obtain a third order achromat. Every forward cell has a rotation map of phase advance $\theta_{x,y} = n * 90 deg$ ($n = 0, 1, 2, 3$) as its linear map. The conditions are shown in Table 1 and the optimal systems are listed in Table 2 and Table 3.

$$\begin{array}{llllll} x^3 & xa^2 & & & & \\ xy^2 & xb^2 & ayb & & & \\ x^2\delta & x\delta^2 & a^2\delta & y^2\delta & b^2\delta & \end{array}$$

$$\begin{array}{llllll} x^4 & x^2a^2 & a^4 & & & \\ x^2y^2 & xayb & x^2b^2 & a^2y^2 & a^2b^2 & \\ y^4 & y^2b^2 & b^4 & & & \\ x^2\delta^2 & a^2\delta^2 & & & & \\ y^2\delta^2 & b^2\delta^2 & & & & \end{array}$$

Table 1: Second and third order terms that have to be removed from f to obtain a third order achromat

θ_x	$\theta_y = 0$	$\theta_y = 90$	$\theta_y = 180$	$\theta_y = 270$
0	F C S R	F C S R	F C S R	F C S R
90	F C F C	F C F C	F C F C	F C F C
180	F C S R	F C S R	F C S R	F C S R
270	F C F C	F C F C	F C F C	F C F C

Table 2: Achromatic systems with $(x|\delta) = 0$ after the first cell

θ_x	$\theta_y = 0$	$\theta_y = 90$	$\theta_y = 180$	$\theta_y = 270$
0	F R S C	F R S C	F R S C	F R S C
90	F R F R	F R F R	F R F R	F R F R
180	F R S C	F R S C	F R S C	F R S C
270	F R F R	F R F R	F R F R	F R F R

Table 3: Achromatic systems with $(a|\delta) = 0$ after the first cell

The design of a realistic third order achromat relies heavily on COSY INFINITY, using its ability to extract Lie coefficients from a given map and define the objective functions without any restrictions.

In the following sections we will present a design of a four cell third order achromat. Section 2 and 3 present the details of the system and various considerations of our design process. Section 4 concludes this paper with discussions of some potential applications of third order achromats.

THE FIRST ORDER LAYOUT

Our design is aimed at a system for 200MeV protons. The first order forward cell contains a 120 degree inhomogeneous bending magnet and four quadrupoles (Fig. 1). The whole cell is symmetric around the midpoint, which entails $(x|x) = (a|a)$ and $(y|y) = (b|b)$. So we need to fit only five instead of seven conditions in order to obtain a map of the form shown in Table 4. During the process of fitting, the drifts between quadrupoles and dipole are fixed and the variables are the field index of the dipole, the two field strengths of the quadrupoles and the drifts before and between the quads. Table 5 shows the parameters chosen as our starting condition for higher order optimization. The long drifts are spaces where higher order elements will be placed.

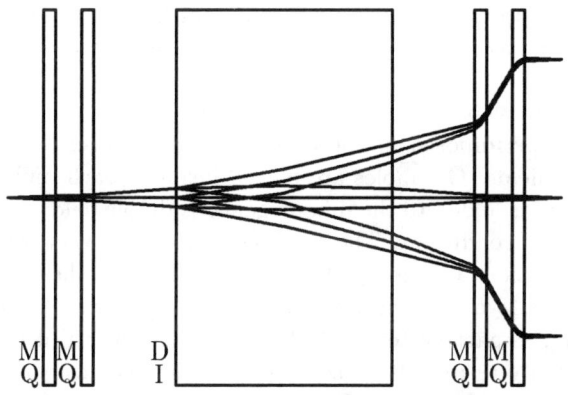

Figure 1: First order forward cell

```
-1.000000        0.0000000E+00  0.0000000E+00 0.0000000E+00  0.2463307E-15 100000
 0.0000000E+00  -1.000000       0.0000000E+00 0.0000000E+00 -7.366987      010000
 0.0000000E+00   0.0000000E+00  0.1387779E-15 1.000000       0.0000000E+00 001000
 0.0000000E+00   0.0000000E+00 -1.000000      0.1526557E-15  0.0000000E+00 000100
 0.0000000E+00   0.0000000E+00  0.0000000E+00 0.0000000E+00  1.000000      000010
 7.366987       -0.2775558E-15  0.0000000E+00 0.0000000E+00  1.733828      000001
```

Table 4: The first order map of the forward cell

THE SECOND AND THIRD ORDER ACHROMAT

To make this system a second order achromat, ten sextupoles were inserted symmetrically with respect to the dipole (Fig. 2). The sextupole strengths provide us the knobs to meet the ten conditions on the Lie coefficients. The required values are rather weak (Table 7), which indicates that the first order layout gives

Element	Field(T)/Length(M)
Drift 1	0.87135375134374986E+00
Quadrupole 1	0.18649849349888584E+00
Drift 2	0.62461177847598495E+00
Quadrupole 2	-0.26717665950046338E+00
Drift 3	2.0
Dipole	0.86(T)
Field index(n_1)	0.39845531733645339E+00

Table 5: The field values and drift lengths of the first order layout

weak higher order aberrations, and the newly introduced sextupoles will not produce strong third order ones either. The map of the whole four cell system was computed. As the theory predicted, the map is free of all second order aberrations.

The last step is to correct the third order aberrations. All the third order elements were superimposed in the existing ones because this tends to require weaker octupole fields. Octupoles are placed inside of each multipole and an octupole component is added to the inhomogeneous dipole field. So there are fifteen variables for fifteen conditions. The results do show that very weak octupoles can meet all the conditions (Table 7). Figure 3 and 4 show the beam envelope and the lab coordinate layout of the whole system. Considering the area it occupies, this is a rather compact system. Table 6 presents the third order map, which shows that the only terms left nonzero are the dependences of time of flight on energy spread up to third order. It makes this system an effective time-of-flight spectrograph.

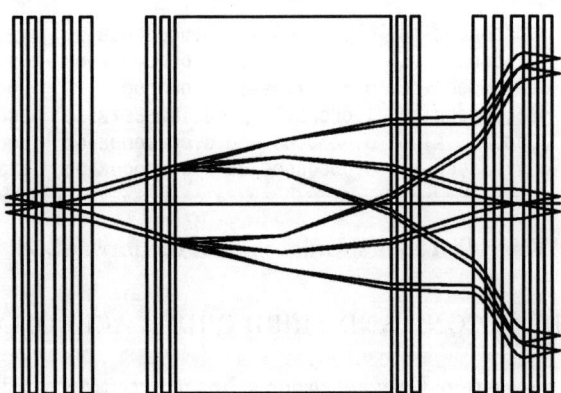

Figure 2: Second order layout of the forward cell. The short elements are the sextupoles.

Using the emittance 1 $mm\ mrad$ and the momentum spread 1%, Figure 5 shows the eighth order beam around the final focal point. The sum of the aberrations at the focal point is about 10 μm horizontally and 3 μm vertically, which is quite small.

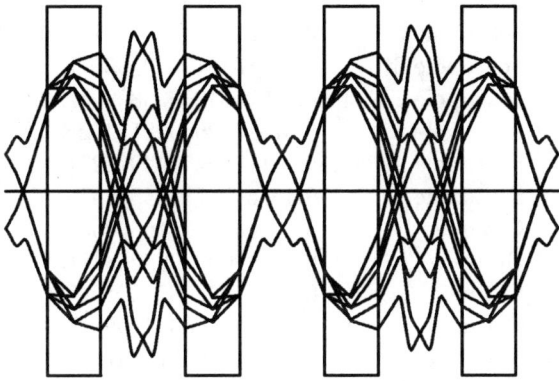

Figure 3: Third order beam envelope with the bending magnets shown

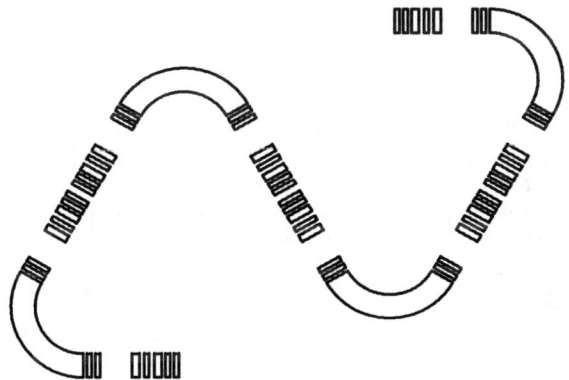

Figure 4: Third order lab layout

```
1.000000        0.0000000E+00 0.0000000E+00 0.0000000E+00 0.0000000E+00 100000
0.0000000E+00 1.000000        0.0000000E+00 0.0000000E+00 0.0000000E+00 010000
0.0000000E+00 0.0000000E+00 1.000000        0.0000000E+00 0.0000000E+00 001000
0.0000000E+00 0.0000000E+00 0.0000000E+00 1.000000        0.0000000E+00 000100
0.0000000E+00 0.0000000E+00 0.0000000E+00 0.0000000E+00 1.000000        000010
0.0000000E+00 0.0000000E+00 0.0000000E+00 0.0000000E+00  6.935312       000001
0.0000000E+00 0.0000000E+00 0.0000000E+00 0.0000000E+00 -21.18904       000002
0.0000000E+00 0.0000000E+00 0.0000000E+00 0.0000000E+00  59.36542       000003
---------------------------------------------------------------------------
```

Table 6: COSY output of third order map of the four cell system (zero means smaller than 1E-11)

The Design of a Four Cell Third Order Achromat

Element	Field(T)
Sextupole 1	-0.63505652509185364E-03
Sextupole 2	0.35855897961217687E-02
Sextupole 3	-0.62171599619533670E-02
Sextupole 4	-0.12117537396329949E-02
Sextupole 5	0.76131423836075176E-02
Sextupole 6	0.38180361792016587E-02
Sextupole 7	0.36827542792682285E-02
Sextupole 8	-0.67593326899855462E-02
Sextupole 9	-0.16119602564297427E-02
Sextupole 10	0.46867037811480167E-02
Octupole 1	-0.84143837081650066E-02
Octupole 2	0.13780092708826341E-01
Octupole 3	-0.44923597587168201E-02
Octupole 4	0.22595586662730582E-02
Octupole 5	-0.63290700557810710E-03
Octupole 6	0.10675946992417028E-02
Octupole 7	-0.99287178779070047E-03
Octupole 8	-0.74085891728212973E-03
Octupole 9	0.76980796352936568E-03
Octupole 10	-0.76725569809791160E-03
Octupole 11	0.40023261493907494E-02
Octupole 12	-0.11080689372217117E-01
Octupole 13	0.35920255962756237E-01
Octupole 14	-0.22219298065280411E-01
Field index(n_3)	-0.11289348840171541E+01

Table 7: The sextupole and octupole strengths at the pole tip

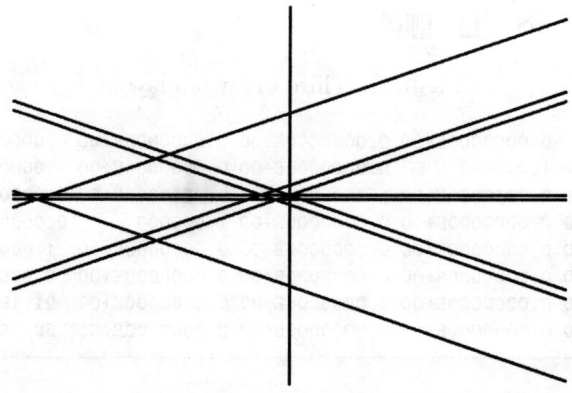

Figure 5: Remaining aberrations up to order eight (scale: $30\mu \times 20\mu$)

CONCLUSION

Based on the theory described in Ref. 6 and used in the environment of COSY INFINITY, a four cell third order achromat was designed. It requires only one dipole and fourteen superimposed multipoles per cell and the residual aberrations are small. Other four cell achromats can be designed using the same method. Such third order achromatic systems can be used as either single pass or circular multi-pass time-of-flight spectrometers, and as bending arms of any circular machine, an interesting topic which deserves more study.

ACKNOWLEDGEMENT

We would like to acknowledge the support of the U.S. National Science Foundation under Grant No. PHY 89-13815, and the Alfred P. Sloan Foundation. Also, we would like to thank G. Hoffstätter and R. Degenhardt for several useful discussions.

REFERENCES

1. R.V.Servrancks and K.L.Brown, Nucl. Instr. Methods A258, 525 (1987).
2. A. J. Dragt, Nucl. Instr. Methods A258, 339 (1987).
3. F. Neri, in Proc. Workshop on High Order Effects. M. Berz and J. McIntyre (Eds.), Technical Report MSUCL-767 (1991).
4. F. Neri, private communication.
5. M. Berz, COSY INFINITY Reference Manual Version 6, Technical Report MSUCL-869 (1993).
6. W. Wan, E. Goldmann and M. Berz, Proc. Workshop on Nonlin. Effects in Accel. Phys., Berlin (1992). M. Berz, S. Martin and K. Ziegler (Eds.), IOP Publishing, Bristol.

EDDY CURRENT SIMULATIONS FOR THE SSCL LOW ENERGY BOOSTER CAVITY*

Y. Goren and L. Walling
Superconducting Super Collider Laboratory,[†] Dallas, TX 75237

ABSTRACT

Eddy currents are developed in the tuner of the Superconducting Super Collider Low Energy Booster (LEB) cavity during the LEB frequency sweep. The two main difficulties created by the eddy currents are excessive tuner-surface heating, and more important, a reduction in the time response of the tuner. We present a detailed analysis of the eddy currents for various tuner designs. The analysis has been done using 2D and 3D time-domain finite element codes (PE2D by Vector-Field and EMAS by MSC). Non-linear analysis was performed utilizing B-H curves. The codes have been benchmarked analytically and by using measured data for different slotted pillbox structures.

INTRODUCTION

The Superconducting Super Collider (SSC) Low Energy Booster (LEB) cavity is designed for frequency sweep of 47.5–59.8 MHz in approximately 20 ms. The frequency sweep is achieved by varying a biased magnetic field perpendicular to the rf magnetic field inside the ferrite-filled cavity tuner. Figure 1 is a 3D view of the LEB tuner. One of the most important aspects of the LEB cavity design is control of the eddy currents developed in the tuner during this frequency sweep. The two main difficulties created by the eddy currents are excessive tuner-surface heating, and more important, a reduction in the response time of the tuner to a triggered control signal. The eddy currents created on the tuner metallic surface can be reduced in two ways: by slotting the surface, which increases their path length, or by equivalently increasing the material electric resistivity. The second approach of using a closed-shell tuner with high resistive alloy is more mechanically suited to the LEB cavity if the ferrites are liquid-cooled. This structure has the electrical disadvantage of reducing the frequency response bandwidth to about 150 Hz for the available resistive alloys. A slotted tuner, on the other hand, is mechanically more complex but has a frequency response bandwidth in excess of 2000 Hz, which is better for the cavity control loop.

---

* To be presented at the Computational Accelerator Physics Conference February, 1993.

[†] Operated by the Universities Research Association, Inc., for the U.S. Department of Energy under Contract No. DE-AC35-89ER40486.

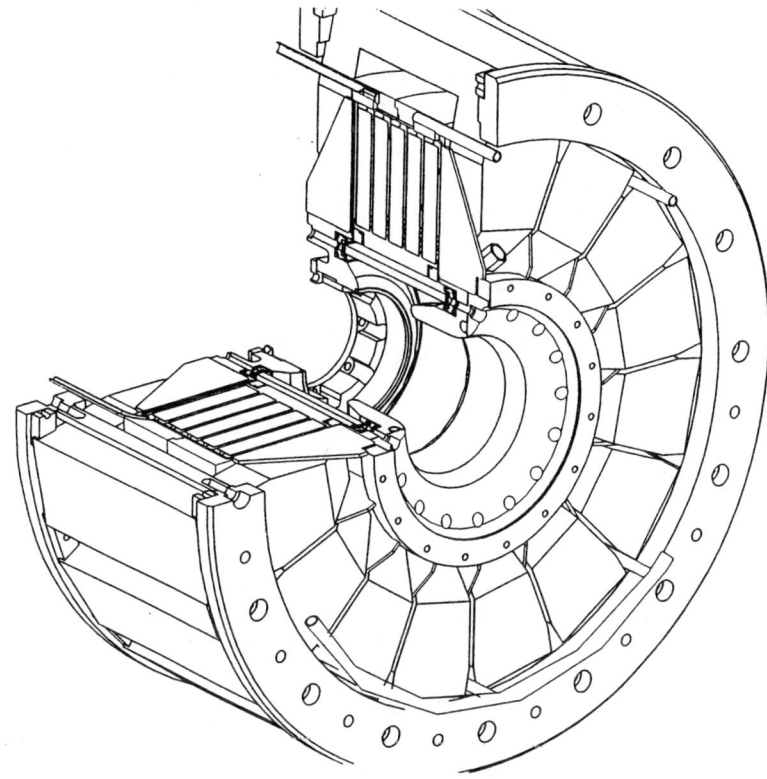

Fig. 1. LEB Tuner Design.

This paper is divided into four sections. The first presents the results of an analytical treatment of the eddy currents developed in an infinitely long metallic cylindrical shell. We show that contrary to the widely held belief that the penetration of magnetic field into metals can be described in terms of a single parameter, the skin depth—in our case the relevant parameter—is the square of the skin depth divided by the shell radius. The second section presents a numerical analysis of a closed-shell tuner design. The analysis has been done using PE2D, which is a time-domain, finite-element 2D code. Utilizing this code we design a tuner made of a Ti–6Al–4V alloy that has a high electrical resistivity as well as very good mechanical strength. This alloy yields a substantial reduction in eddy currents and two orders of magnitude increase in frequency bandwidth compared with a copper tuner. The third section is a numerical analysis of slotted tuner design. The analysis has been done using a 3D time-domain, finite-element code (EMAS by MSC). It shows, contrary to another widely held belief, that eddy current problems can be analyzed by the quasi-stationary approximation—that the rate of penetration of the magnetic field into the tuner through the slots depends on the displacement currents across the slots. The final section is a short summary.

EDDY CURRENTS IN A METALLIC SHELL

First we review the results of long thin metallic shell inside a long solenoid.[1,2] The geometry of this setup is described in Figure 2. The axial magnetic field in region I is given by

$$B_z(t) = \mu \Delta_c \delta_s \exp(-\delta_s t) \int \exp(\delta_s \tau) J(\tau) d\tau \qquad (1)$$

and the eddy current in the metallic shell is

$$J_{eddy} = -J \Delta_c/\Delta_s + \Delta_c/\Delta_s \delta_s \exp(-\delta_s t) \int \exp(\delta_s \tau) J(\tau) d\tau \qquad (2)$$

The results in eqs. (1) and (2) have been obtained assuming a spatially constant current density drive J. The parameter δ_s is defined by

$$\delta_s = 2 / (\mu \sigma_0 \Delta_s R_1) \qquad (3)$$

where σ_0 is the shell conductivity. The parameter δ_s in eq. (3) measures the inverse of the magnetic diffusion time through the metallic shell. A tuner design with low eddy currents and fast magnetic time response will be characterized by the inequality $\delta_s T_{ch} \gg 1$, where T_{ch} is the characteristic time scale of the drive current J. The maximum eddy current obtained for this time scale is

$$J^{max}_{eddy} = J_0 \mu \sigma_0 \Delta_c R_1 /(2 T_{ch}) \qquad (4)$$

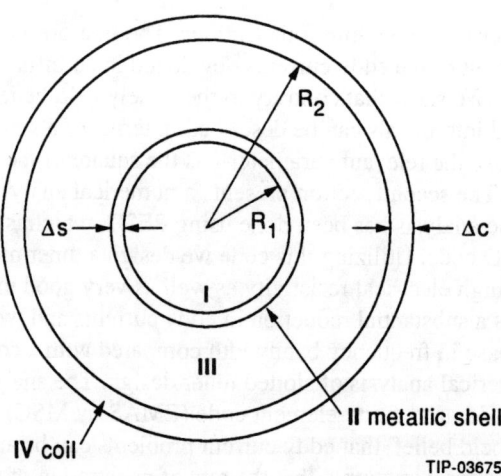

Fig. 2. Infinitely Long Metallic Shell in Solenoidal Magnetic Field.

For equal heating rates of the solenoid and the metallic shell we estimate the shell conductivity $\sigma_0 \cong 5 \times 10^5$ s/m, which is two orders of magnitude lower than copper. The Ti-6Al-4V alloy with conductivity of 5.8×10^5 s/m makes it a good candidate for a closed-shell LEB tuner.

The cavity rf frequency program is achieved by biasing the ferrite. The relationship between this magnetic field and the current drive determines the cavity response to a control signal. It is common to quantify the response in the frequency domain by its 3-dB bandwidth. Fourier decomposing eq. (1) we obtain

$$B_z(\omega) = \mu \Delta_c J(\omega) / (1 + i \Delta_s /(\lambda_s^2 / R_1)) \qquad (5)$$

where $\lambda_s = (2 / \omega \mu \sigma_0)^{1/2}$ is the standard skin depth definition. It can be seen from eq. (5) that the parameter which determines the magnetic penetration through the shell is given by the square of the skin depth divided by the shell radius. The 3-dB frequency bandwidth is given by

$$\Delta_f = 1 / (\pi \mu \sigma_0 \Delta_s R_1) \qquad (6)$$

For a 5-mm-thick titanium alloy shell, this translates into a bandwidth of 292 Hz, about two orders of magnitude greater than for a copper shell.

NUMERICAL ANALYSIS OF A CLOSED-SHELL TUNER

Using the analytical results above as a guide, we numerically simulated the closed-shell LEB tuner. We used the 2D time-domain, finite-element code PE2D by Vector Field. This code is also capable of handling materials with a non-linear B-H curve. The simulated tuner geometry is shown in Figure 3, where the scale is in centimeters. The simulation was done on an old version of the tuner. Higher voltage requirements required us to add one more ferrite ring. The drive current required to follow the frequency program is described in Figure 4. The current reaches approximately 17,000 At at 50 ms, then drops in 30 ms to approximately 4000 At to the end of the cycle. This curve has been determined in two stages. At first the code was run in the steady state mode with various currents to establish a relation between the drive current and rf permeability given by the frequency program. In the second stage the transient analysis with non-linear materials is done. The maximum eddy current is developed at the top of the tuner, where its two half shells are joined (see Figure 3). Figure 4a describes the magnitude of the eddy current at this point. The deviation from the smooth curve is of numerical origin and relates to the way the code handles the derivative of the drive current. The maximum eddy current obtained is 60 A/cm^2 at 17 ms from the begining of the cycle. In comparison the maximum eddy current for a copper tuner is about 2700 A/cm^2. The thermal power, averaged on a cycle, developed at this point is about 0.18 W/cm^3, which can be handled without much difficulty by the tuner internal coolant.

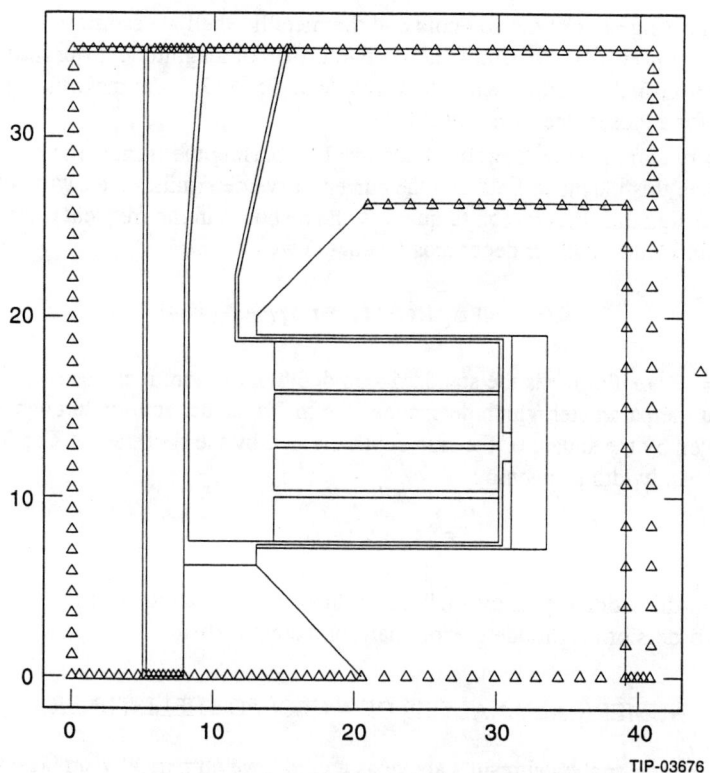

Fig. 3. Tuner Geometry for PE2D.

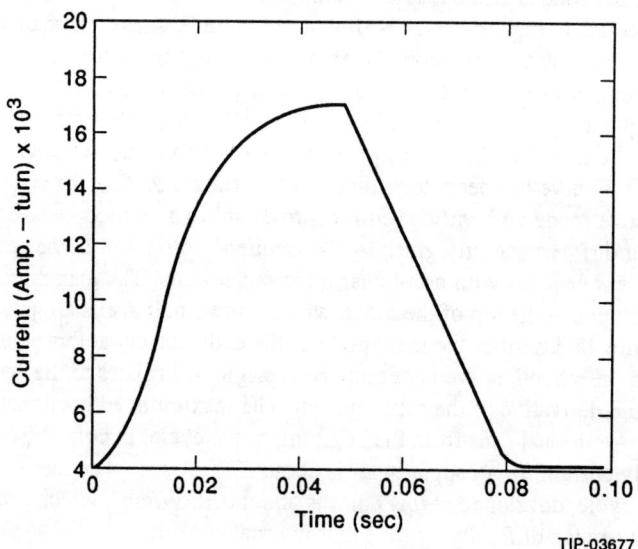

Fig. 4. Tuner Biased Current.

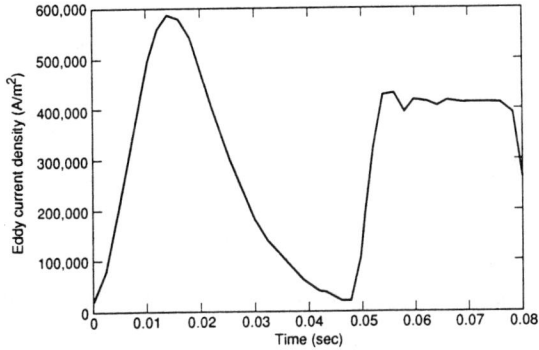

Fig. 4a. Eddy Current in Titanium Tuner.

The analysis of the tuner frequency response is done with the 2D frequency-domain option of EMAS. EMAS is a 3D time-and-frequency domain, finite-element, electromagnetic code. Like PE2D it is able to handle materials with non-linear B-H curve. We find that the frequency response is varied across the tuner cross section with the minimum bandwidth at the bottom of ferrites. The magnetic field vs. frequency at this location is shown in Figure 5. It can be seen from the figure that the 3-dB bandwidth is about 140 Hz, considerably lower than the 292 Hz expected from an infinitely long metallic shell. The discrepancy corresponds to the relatively slow magnetic penetration through the side walls of the tuner. To confirm the above results we benchmarked the code by using measured data of the magnetic field at various locations inside a closed stainless steel can. Figures 6 and 7 are the frequency response of the amplitude and phase of the magnetic field at the center of the can. The experiment and simulation are within the experimental error of about 0.3 dB and 2.0 deg. The narrow frequency response bandwidth led us to abandon the closed-shell tuner and to design a slotted tuner instead.

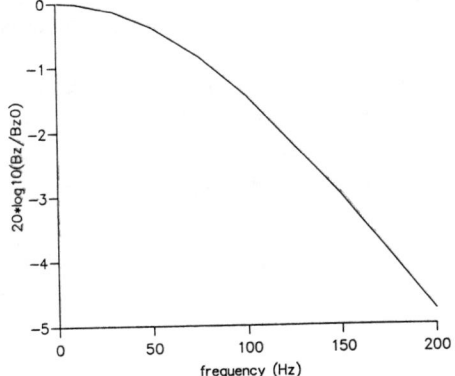

Fig. 5. Frequency Response of Titanium Closed Shell Tuner.

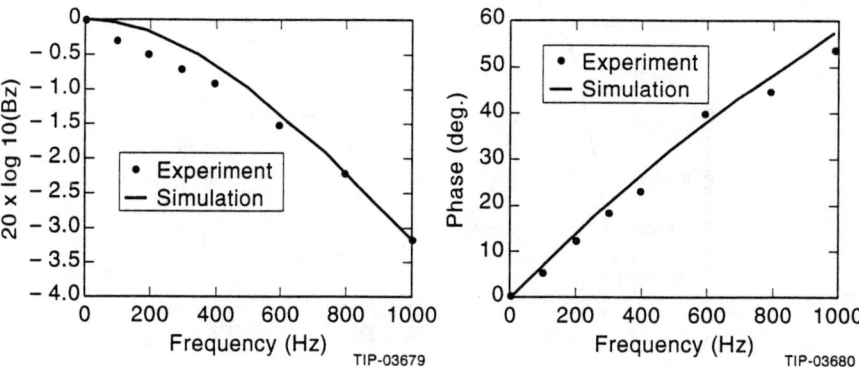

Fig. 6. Magnetic Response of Close Can Amplitude.

Fig. 7. Magnetic Response of Close Can Phase.

NUMERICAL DESIGN OF A SLOTTED-SHELL TUNER

The analysis of the closed-shell tuner in the last section was performed using the quasi-static approximation, which neglects the displacement current in Maxwell's equations. Using this approximation for the 3D problem of a slotted tuner yielded a false solution in which the slots had a very small effect on the rate of magnetic penetration into the tuner. The need for the displacement currents is illustrated in Figure 8. As the eddy currents approach the slot discontinuity, they charge its surface. The charges create electric fields, which oppose the small internal field in the metal and enforce the currents to change direction and bypass the slot.[3] To confirm this assumption we compared the numerical simulation with measured results for a stainless steel can with various numbers of slots. Figures 9 and 10 describes the magnetic frequency response of a can with 8 slots. The simulation results are within the experimental errors of 0.1 dB and 1 deg. Encouraged by these benchmark results, we designed a slotted LEB tuner. The tuner is made of 3-mm stainless steel with 16 5-mm radial slots. The slots are filled with G10 compound (dielectric material) to contain the coolant. The magnetic frequency response of this tuner is shown in Figures 11 and 12 with a 3-dB bandwidth, which exceeds 2000 Hz. Figure 13 describes the eddy current flow across the tuner surface around a slot for a 250-Hz frequency drive. Notice the small elongated elements that must be defined around the slots for the code to converge properly.

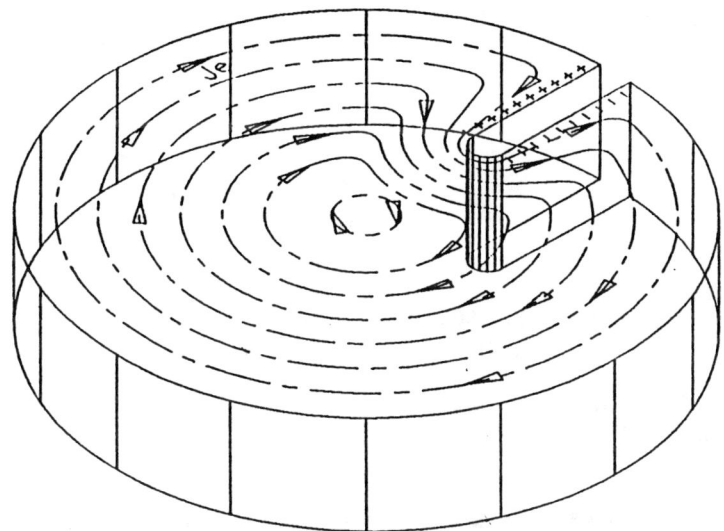

Fig. 8. Eddy Currents Around a Slot.

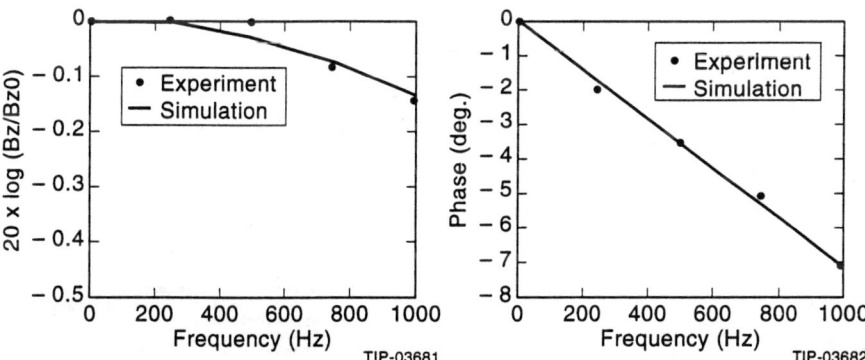

Fig. 9. Magnetic Response of Slotted Can (8 Slots) Magnitude.

Fig. 10. Magnetic Response of Slotted Can (8 Slots) Phase.

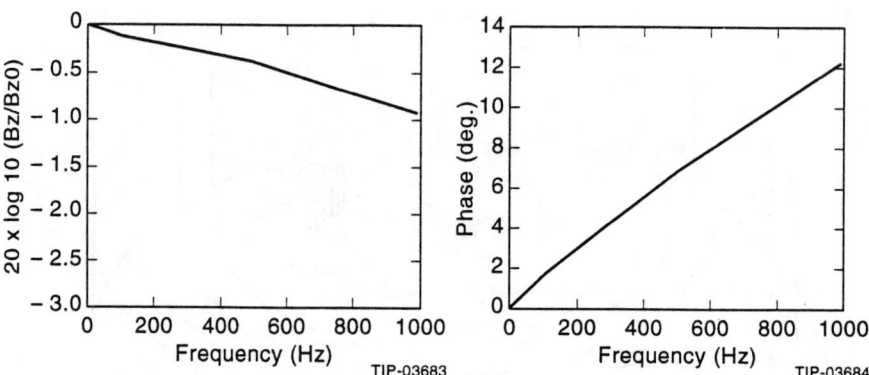

Fig. 11. Magnetic Response of LEB Tuner (16 Slots) Magnitude.

Fig. 12. Magnetic Response of LEB Tuner (16 Slots) Phase.

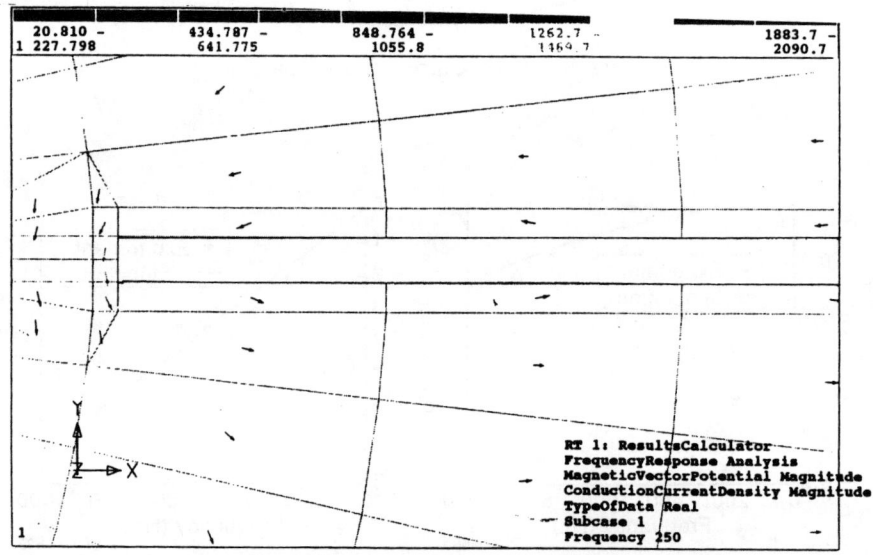

Fig. 13. Eddy Current Around a Slot in the LEB Tuner.

CONCLUSION

We present numerical simulations of two LEB tuner designs. The surface heating due to the eddy currents is controlled in both schemes. The magnetic field frequency response of the closed shell tuner is only marginally acceptable. This leads to the slotted tuner design, which is mechanically more complex, but has the wide bandwidth required for the control system.

REFERENCES

1. S. Fahy, C. Kittel, S. G. Louie, Am. J. Phys. 56, 989 (1988).
2. Y. Goren, B. Campbell, SSC Laboratory, SSCL-505, (1991).
3. Y. Goren, L. Walling, Eddy Current Analysis, RF Workshop, Dallas, June 1992.

LINER IMPEDANCE CALCULATIONS USING HFSS*

E. Ruiz, L. Walling, Y. Goren, N. Spayd
Superconducting Super Collider Laboratory
MS-4010, 2550 Beckleymeade Ave. Dallas, TX 75237

ABSTRACT

We have calculated the leakage and coupling impedance for a holed beam pipe liner for the SSC collider ring using the HP High-Frequency Structure Simulator (HFSS). By using this code we were able to study various hole and slot configurations. We compare simulation results with analytical calculations and measurements.

INTRODUCTION

To avoid reduction of beam lifetime due to photodesorption and to shield the SSC collider ring bore tube from synchrotron radiation, it is proposed to include a holed liner within the bore tube which permits required vacuum while screening the bore tube from synchrotron radiation. Each hole presents a discontinuity to the beam image currents, resulting in beam coupling impedance. Also, due to electromagnetic coupling through the holes in the liner, power will circulate around the accelerator between the liner and bore tube in synchronism with the beam, growing in strength until steady-state amplitude is reached. This steady-state TEM wave would leak back into the beam pipe, thereby possibly presenting an unacceptably high longitudinal impedance. The contribution of pumping holes to the coupling impedance is estimated analytically and numerically and measurement results are presented.

BENCHMARKING HFSS

In order to evaluate the accuracy of using HFSS to calculate impedances of very small obstacles in a beamline, we calculated both low-frequency impedances by simulating the wire measurement technique[1] and high-frequency impedances by wireless excitation of the TM_{01} mode[2] of a small pillbox cavity for which the longitudinal impedances have been calculated by Henke[3] and verified by 2-D numerical simulations by Chou[4]. The beam pipe radius was 20mm, the cavity length 1mm, and its height was 2mm. Very good agreement was found between theory and the HFSS results, except for the case of the low-frequency measurements using a large center conductor (wire radius was 6.3mm, Z_c=50 Ω). The radius of the small center conductor was 1.5mm. The results of these benchmarking simulations are shown in Figures 1 and 2. The TM_{01} simulations are done by modeling a thin slice of the geometry, applying a magnetic boundary condition, and solving for as many axially-symmetric modes at each port as can propagate at the frequency under investigation. The contribution to the impedance of evanescent waves is negligibly small. The impedance is calculated as for coaxial measurements, with

$$Z_c = \frac{377}{4\pi}\left[1 - \left(\frac{f_c}{f}\right)^2\right]^{\frac{1}{2}},$$

and the pertinent transmission parameter is the TM_{01} signal at port 2 due to TM_{01} excitation at port 1.

---

*This work supported by the U.S. Department of Energy under contract No. DE-AC02-89-ER40486.

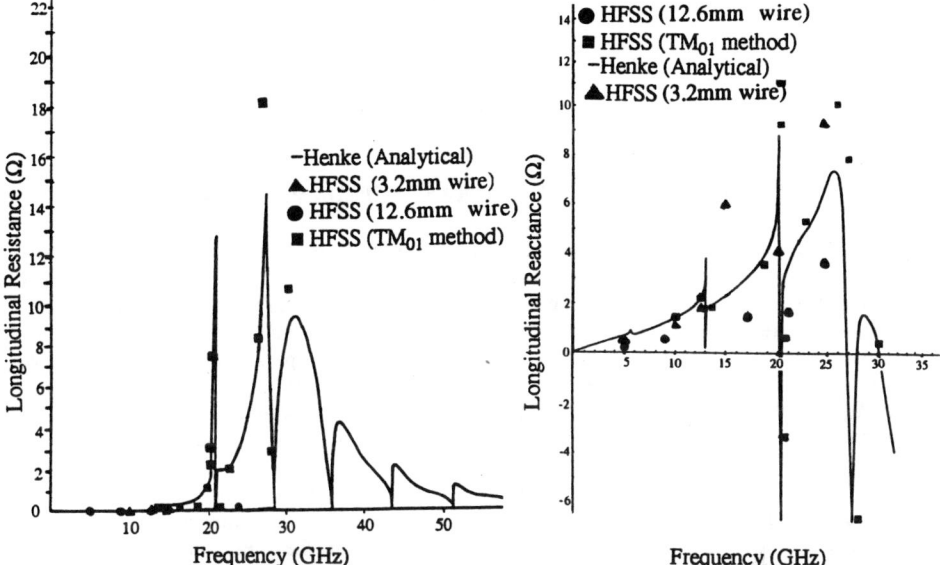

Fig. 1. Results of HFSS Benchmark of Pillbox Cavity - Real[$Z_{//}$]

Fig. 2. Results of HFSS Benchmark of Pillbox Cavity - Imaginary[$Z_{//}$]

ANALYTICAL ESTIMATION OF THE COUPLING IMPEDANCE

Theoretical calculation of the low-frequency coupling impedance of small apertures in a metallic wall have been studied extensively[5-8] but until recently coupling impedance due to their coherent behavior had not been calculated for a beam tube liner. Typically, the pumping holes or slots in an infinitesimally thin conducting boundary excite electric and magnetic dipoles which scatter energy back into the beampipe and couple some energy through the aperture into the coaxial region formed between the liner and bore tube.

Gluckstern[9] has analyzed the impedance due to holes in a liner which includes the effect of wall thickness and the outer coaxial region:

$$Z_{||} = j\frac{Z_o kP}{8\pi b^2}\left[\psi_{in} - \chi_{in} - \frac{(\psi_{out} - \chi_{out})^2}{\psi_{in} - \chi_{in} - j8\pi b^2 (\frac{\alpha}{k}) L ln(\frac{a}{b})}\right] \quad (1)$$

where

$$\frac{\alpha ln}{k}(\frac{a}{b}) = \frac{1}{4}(\frac{\delta_a}{a} + \frac{\delta_b}{b}),$$

Z_o is 377 ohms, $k=2\pi f/c$, P is the number of holes, b is the liner radius, a is the bore tube radius, and δ_a and δ_b are the skin depth for the bore tube and outer surface of the liner, respectively.

For impedance calculations, the electric and magnetic polarizabilities for the inside and outside of the liner, ψ_{out}, χ_{out}, ψ_{in}, and χ_{in}, respectively, were determined using Gluckstern's calculations[10]. The contribution to the impedance from the power circulating in the outer coax is shown by Gluckstern to be negligible for reasonably small holes (4mm diameter or smaller).

To calculate leakage we use Caspers' calculation for transmission coefficient. Caspers calculates $G(\omega)$, the ratio between the current in the outer coax to the beam current. To transform this quantity to our triaxial S-parameter measurements,

$$G_S(\omega) = \sqrt{\frac{Z_c}{Z_i}} G(\omega) \qquad (2)$$

where Z_c and Z_i are the line impedances of the coaxial lines formed by the beam tube and liner, and the liner and center conductor, respectively.

NUMERICAL SIMULATION OF COUPLING IMPEDANCE OF LINER HOLES

First, we tried simulating the triaxial measurement for low-frequency longitudinal impedance. The simulations were for a beam pipe with radius 16.5mm and thickness 1mm. We found that the outer coax contributed insignificantly to the impedance, so to speed up the simulations the outer coax was eliminated. The resulting geometry used in the simulation is shown in Figure 3. Since holes this size represent a very small impedance, it was necessary to detect very small phase shifts due to the holes. We would converge the code until the change in the phase was much smaller than the phase shift ($\theta-\theta_o$), where θ_o was calculated using c specified to about 5 decimal places. However, we learned that there was an error that could not be removed by further adaptive passes when merely subtracting the length of the line. This phase shift is due to small errors in the mapping of the 3-D tetrahedrons to the 2-D port surfaces. We found that we could attain another order of magnitude in phase accuracy from HFSS by modeling the holes and beam tube as separate elements, meshing and running the problem with both specified as vacuum, then, without changing the mesh, re-specifying the holes as metal and re-running over the frequency range. This was used as a "reference pipe" to normalize the data, which removed the port mapping errors as well as the baseline transmission.

Many variations of holes were simulated. These included varying the number of holes axially from 2 to 20, varying the number of holes longitudinally from 2 to 20, and varying the thickness of the holes. These simulations confirmed that for low frequencies, the longitudinal impedance of the holes add, and the impedance is reduced by the wall thickness, as theory predicts. Simulations were also performed with small and large center conductors and compared to measurement results for the same parameters. Measurements were performed using the wire technique[1]. We found very good agreement between HFSS and wire measurements (wire measurements were made of pipes with 1010 holes with the full triaxial geometry). Low-frequency results for 2mm and 3mm diameter holes are shown in Figure 4. The results are presented as impedance per hole, and show measured and simulated data with a 3mm diameter center conductor. As seen in Figure 4, we found better agreement with theory for smaller holes.

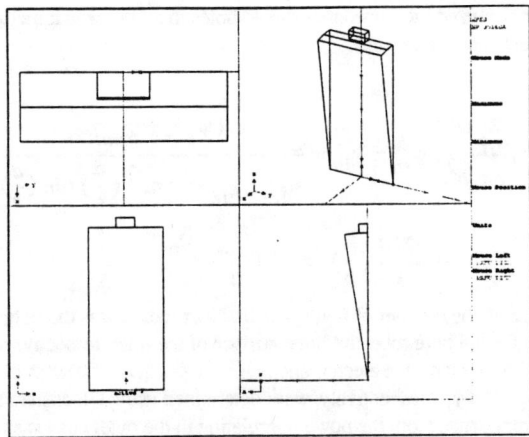

Fig. 3. HFSS Geometry for TM_{01} Simulation
2mm Square Hole Geometry

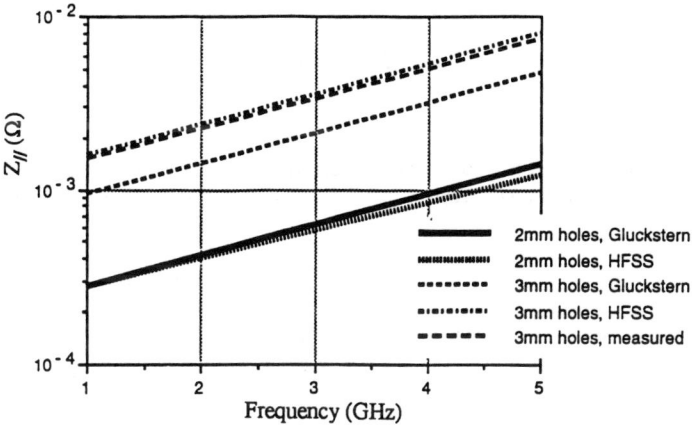

Fig. 4. Low-Frequency Longitudinal Reactance of
2mm and 3mm Holes in a 1mm-thick Liner

Extensive simulations were performed for comparison with MAFIA simulations done by Barts and Chou[11]. We modeled 2mm square holes to replicate the MAFIA geometries. The HFSS simulations showed that the wire method was good to frequencies well above the TM_{01} cutoff frequency. Discrepancies between the MAFIA and HFSS results for high frequency may be due to the effects of interactions between holes longitudinally, as the MAFIA runs were done for many holes (about 22) longitudinally, whereas the HFSS runs were done for only 1 hole longitudinally (20 axially by modeling 9° and employing magnetic mirror symmetry). Barts and Chou have modeled both equal- and randomly-spaced holes. We chose to compare our results to the data from the randomly spaced simulations because the random case would not show large resonances due to longitudinally periodic holes. We plan to run HFSS with the same number of holes longitudinally to better compare to MAFIA simulations. The MAFIA comparison results are shown in figures 5 and 6.

Short slots are being considered as an alternative that would reduce the low-frequency beam impedance. Simulations were done to study the influence of the shape of the slot. Table 1 shows results of simulations using slots 2mm wide and 6mm long compared to 2mm-diameter round holes at 5.0 GHz. Although a large savings in low-frequency impedance could be achieved by using slots, the high-frequency behavior would have to be studied carefully.

Transverse impedance for 3mm diameter holes was also simulated and compared to calculations by Kurennoy[6,7] for zero-thickness walls. We find results about 0.5 times Kurennoy's predictions, which is consistent when the wall thickness is taken into account. These results are shown in Table 2. One hole is simulated longitudinally, and the number of holes axially is varied from 2 to 16. Measurements of the transverse impedance are in progress.

Finally, to estimate the contribution to the longitudinal impedance introduced by the TEM wave travelling in the region outside the liner and 'feeding back' on to the beam, a two-step calculation was performed. A triaxial model, representing 20 holes axially and 22 longitudinally, was driven at the inner coax on one side. The other side has enough blank pipe forming the outer coax such that the maximum radial electric field, E_o, due to forward-leaked power could be determined. Thus, the ratio E_o/I was determined where I is the current on the center conductor within the liner. Then, the center conductor is removed from the model, the outer coax is driven, and the ratio of E_z, the maximum longitudinal electric field on axis, to the radial electric field in the outer coax is determined. This ratio is then multiplied by the scaling ratio l_{att}/l to yield the coupling impedance, where l is the length of the liner segment with holes in it.

Mathematically this becomes

$$\frac{\partial Z_\infty}{\partial l} = \frac{E_{z,\infty}}{I} \approx \frac{E_0}{I}\frac{E_z}{E_0}\frac{l_{att}}{l}. \qquad (3)$$

Numerical simulations were performed for holes of diameter 2mm, 3mm, and 4mm diameter holes. This inquiry confirmed that this contribution was much smaller than the inductive impedance for reasonably-sized holes.

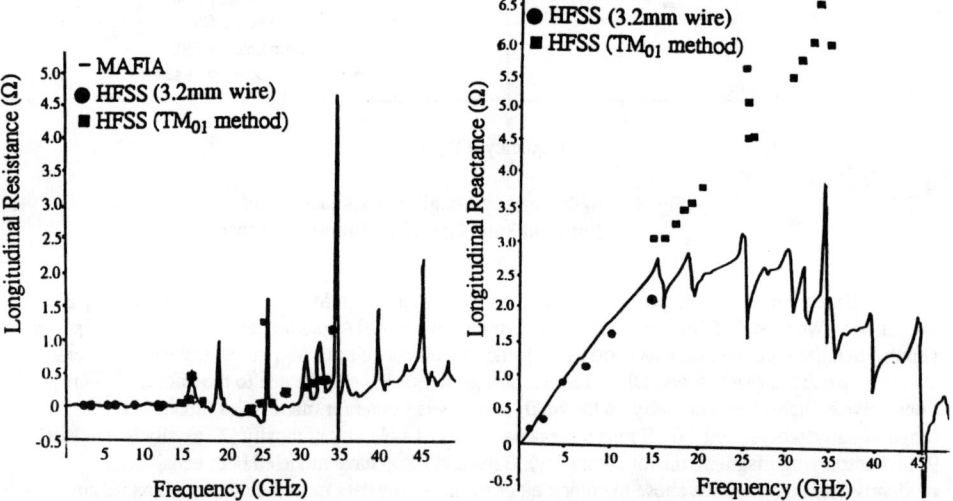

Fig. 5. Results of HFSS Comparison to MAFIA For 2x2mm Square Holes - Real[$Z_{//}$]

Fig. 6. Results of HFSS Comparison to MAFIA For 2x2mm Square Holes - Imaginary[$Z_{//}$]

Table 1: Transverse Impedance of 3mm Diameter Round Holes

Number of Holes Axially	HFSS Transverse Impedance (Ω/m)	Kurennoy Transverse Impedance (Ω/m)
2	1.21	2.62
8	2.43	5.24
16	4.91	10.5

Table 2: Longitudinal Impedance of Various Slot Geometries (2mm x 6mm slots)

Slot Shape	HFSS Longitudinal Impedance per Slot (Ω)	HFSS Longitudinal Impedance/Area (Ω/mm^2)
Hole	0.0012	3.8×10^{-4}
Square Slot	0.0023	1.9×10^{-4}
Rounded Slot	0.0020	1.7×10^{-4}

SIMULATION AND MEASUREMENT OF LINER LEAKAGE

Measurements of leakage were made using a triaxial impedance measurement technique[12, 13]. The test liner is 2 meters long, with 1010 holes in the center 1 meter, and 0.5 meters of solid pipe on both ends. The outer bore tube has a radius of 23mm, while the liner has an inner radius of 16mm and is one mm thick. The center conductor diameter was 12.7mm for the leakage measurements. In the measurement setup the center conductor is placed on axis within the liner and the transmission through the pumping holes, S_{41}, is measured. A transmission measurement is also made through the coaxial regions formed between the liner and bore tube (S_{43}), and between the center conductor and liner (S_{21}), to correct the effects of hardware mismatches and multiple reflections in the triax on the S-parameters. S_{41} is time-gated, and divided by the square root of gated S_{43} and S_{21}. Liners with holes of diameter 1mm, 2mm, 3mm, and 4mm were measured.

For the simulations, 22 holes longitudinally were modeled with nine degrees axially and magnetic symmetry (equivalent to 20 holes axially) was applied. Results of leakage measurements and simulation are shown in Figure 7.

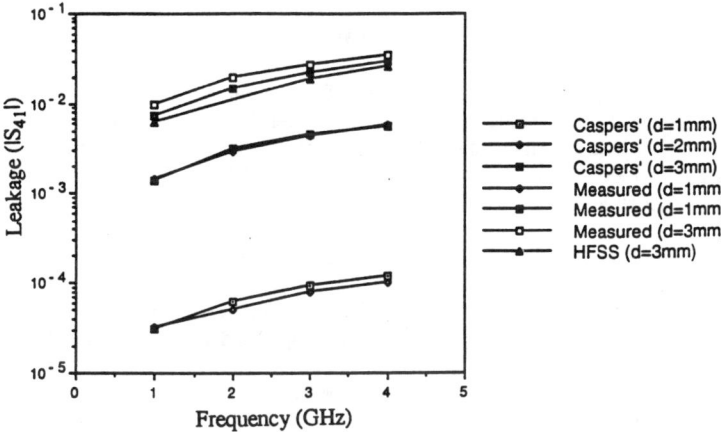

Fig. 7. Leakage Through A Liner with 1010 Holes of Various Diameters

DISCUSSION OF RESULTS

Benchmark

Review of the impedances calculated from HFSS versus Henke's analytical calculation of the longitudinal impedance of the pillbox reveal several subtleties. First, HFSS shows that use of the wire technique with a large diameter center conductor to measure the impedance of a liner with 1.6mm radius is very inaccurate for this type of structure. Calculations using HFSS show that the impedance of the pillbox using the 12.7mm line is roughly 50% of Henke's analytical estimate while the 3.2mm center conductor consistently provided more accurate results. Interesting too is the fact that the smaller diameter center conductor yielded accurate results up to roughly 13 GHz, far above the cutoff of the liner. Even near resonance, HFSS showed very good agreement with the analytical coupling impedance. The HFSS simulation of the TM_{01} (wireless) method for calculating impedance proved to be very accurate above cutoff. This seems to stem in part from the fact that HFSS provides for the launch of a single, higher-order mode into the structure and extraction of S-parameters for that single mode, while allowing energy to be converted into other modes and propagate out of the ports. Coupling this feature with a properly-seeded, fine mesh yields accurate S-parameters and coupling impedances.

Liner

The simulations and the measurements both yielded low-frequency impedances which are 50% higher than Gluckstern's thickness corrected estimate for the 3mm holes, however, for 2mm holes there was good agreement. Further, the simulations confirm the presence of some very high-frequency, high-Q resonances, as predicted by MAFIA. There are high-frequency discrepancies between the codes, which need to be investigated further by adding more holes longitudinally to the HFSS simulation. Also, we intend to investigate high-frequency simulation of transverse impedance using TM_{11} mode excitation. Interesting is the fact that when coupling impedance was calculated by field integration over the surface of the hole, an error of a factor of 2 or more appeared. So, while the S-matrix tended to converge fairly quickly, the fields did not.

CONCLUSION

The finite element electromagnetic solver HFSS has been used to calculate the coupling impedance introduced by the presence of a liner in the collider ring of the SSC. Using this code we were able to study various hole and slot configurations. Good agreement has been demonstrated between HFSS, laboratory measurements, and theory which reveals that HFSS is useful in calculating impedances in accelerator structures, even above cutoff.

ACKNOWLEDGMENT

The authors would like to thank Thomas Scholz for his work in making the majority of the leakage measurements during his two month visit to the Superconducting Super Collider Laboratory.

REFERENCES

[1] L. S. Walling, D.E. McMurry, D.V. Neuffer, and H.A. Thiessen, Nucl. Instr. and Meth. A281 (1989) 433
[2] G. Lambertson, A.F. Jacob, R.A. Rimmer, and F. Voelker, "Techniques for Beam Impedance Measurements Above Cutoff." Proc. of European Particle Accel. Conf., Nice, France, 1990, and LBL-28190
[3] Henke, H., Point Charge Passing a Resonator with Beam Tubes, European Organization for Nuclear Research, Geneva, Switzerland, November, 1985.
[4] Chou, W., Impedance Scaling and Synchrotron Radiation Intercept, SSCL-33, SSCL, November, 1990.
[5] M. Sands, "Energy Loss from Small Holes in the Vacuum Chamber", SLAC/PEP-253, Sept., 1977.
[6] S. Kurennoy, "On the coupling impedance of a hole or slot", CERN SL/91-29 (AP) (1991).
[7] S. Kurennoy, "Beam Coupling Impedance of Holes In Vacuum-Chamber Walls", Institute of High Energy Physics, Protvino. Report 92-84, 1992.
[8] R. L. Gluckstern, "Coupling impedance of a single hole in a thick wall beam pipe", CERN SL/92-05 (AP) (1992)
[9] R. L. Gluckstern, "Coupling impedance of many holes in a liner within a beam pipe", CERN SL/92-05 (AP) (1992).
[10] R. L. Gluckstern and J. A. Diamond, "Penetration of Fields Through a Circular Hole in a Wall of Finite Thickness", IEEE Transactions on Microwave Theory and Techniques, Vol. 39, No.2, February, 1991; p. 274
[11] W. Chou and T. Barts, "Wakefield and Impedance Studies of a Liner Structure Using MAFIA", to be presented at this conference.
[12] F. Caspers, E. Jensen, and F. Ruggiero, "Impedance measurements for the pumping holes in the LHC liner", CERN, 1992.
[13] F. Caspers, "Triaxial line technique", in Proc. of the Coupling Impedance Measurement Workshop, Argonne National Lab APS, August 12-13, 1991

THE BEAM DYNAMICS STUDY OF COMPACT RING BY A NEW CODE

Y. Huang
SSCL, 2550 Beckleymeade Ave., Dallas, TX 75237-3997

S. Ohnuma
Physics Dept., University of Houston, TX 77204-5506

ABSTRACT

A new computer code has been developed for the design of compact synchrotrons. Based on a numerical integration of the exact equations of motion, the beam dynamics study was carried out for a four-cell ring of the Texas Accelerator Center and for a race-track ring of Brookhaven National Laboratory(SXLS), with either computed or measured magnetic field. Significant discrepancies of machine parameters obtained by this code and the standard codes of synchrotrons reveal some unique features of the beam dynamics in small synchrotrons after including the effects of fringe magnetic field. Interactive procedures between the beam dynamics predictions on the one hand and the magnet design on the other are then explored so that cost-effective design can be achieved with the minimum effort.

INTRODUCTION

In early 1977, while working at Brookhaven National Laboratory, Ohnuma noticed that tunes of NSLS calculated with code SYNCH were significantly different from those obtained with code PATRICIA[1]. This problem surfaced again later when people at Fermilab discovered discrepancies in their accumulator ring tunes. At BNL, similar problem was also found when the chromaticities of SXLS were calculated by several standard codes[2]. One source of the ambiguities is the different treatments of the edge field and the combined function field in dipoles. It is concluded that in order to calculate tunes, chromaticities and the dynamic aperture in a small ring with large magnetic bending angles, it is necessary to integrate the exact equations of motion with the realistic magnetic field.

The design objective of a compact ring should be the minimization of machine size and complexity while maintaining the optical quality of the beam. In such machines, edge fields often comprise 20% to 60% of the total machine length. The bending angle of each magnet ranges from 45° to 180° instead of at most a few degrees in high energy synchrotrons. In this case, the fringe multipole field will undoubtedly play a nontrivial role in determining basic machine parameters. Therefore, the standard treatment for simulating particle motion in synchrotrons, which uses the isomagnetic approximation together with thin lens kicks(both linear and nonlinear), no longer accurately models the closed orbit of the machine.

INTERACTIVE MODE IN BEAM DYNAMICS DESIGN

© 1994 American Institute of Physics

An interactive procedures between the beam dynamics predictions and the magnet design has been implemented in our beam dynamics study for the TAC compact ring. First, a new computer code, COMSYN, was developed. This code uses an improved 4th-order Runge-Kutta method to integrate the exact equations of motion. The magnetic field needed is either measured directly or computed by a three-dimensional computer code. Then, based on the orbit tracking calculations, all important lattice parameters(closed orbit, betatron amplitude functions, dispersions, natural chromaticity and dynamic aperture) can be obtained. It is obvious that all these parameters will be different when the effects of the fringe field are included.

The magnetic field of the dipole on the median plane can be expressed in the form

$$B(r,\theta) \approx b_0(\theta) + b_1(\theta)x + b_2(\theta)x^2/2 + b_3(\theta)x^3/6, \qquad (1)$$

where θ is the azimuthal angle and x is the radial deviation from the closed orbit, $x \equiv r - r_0$. The azimuthal dependence of the coefficients $b_i(\theta)$ is found from the fitting by means of a cubic spline. In this way, the field on the median plane and its derivatives can be expressed analytically.

The effort is then made to find the correlations between the lattice property and the multipole components. If some parameters are different from the desired value, one cure may be obtained through Table 1 which shows the influence of multipole field on machine parameters together with the possible cures.

Table 1 Multipole influence and the cures

multipole order	affected parameter	cures
dipole	closed orbit	change field strength or coil length
quadrupole	betatron tunes	adjust gradient in quads
sextupole	natural chromaticity	correct sextupole system
higher multipole	dynamic aperture	change pole shape or add shunt

The relations illustrated in Table 1 are not always so direct depending on many aspects of the coil and iron geometries. The table should therefore be regarded as the initial guidance for the iterative procedures. It is hoped that one should be able to get more insights as the interation proceeds.

THE INTEGRATION CODE–COMSYN

For a long-term tracking calculation, a canonical integration technique is desirable in order to preserve the canonical character of equations of motion. But this is presently limited to the motion where the magnetic field can be described by only one component of the vector potential. Otherwise it will be difficult to obtain an explicit symplectic map[3].

An improved 4th-order Runge-Kutta method is used in COMSYN to integrate the exact equations of motion in a rectangular coordinate system. With time as an independent variable, the equations of motion have the following form:

$$\ddot{x} = \frac{e}{m}(E_x + B_z\dot{y} - B_y\dot{z}) \qquad (2)$$

$$\ddot{y} = \frac{e}{m}(E_y + B_x\dot{z} - B_z\dot{x}) \qquad (3)$$

$$\ddot{z} = \frac{e}{m}(E_z + B_y\dot{x} - B_x\dot{y}). \qquad (4)$$

Because of the simple form of these equations, the CPU time is significantly reduced in comparison with those employing curvilinear coordinates. For 600 integration steps in each superperiod of SXLS, it takes approximately 10 minutes of CRAY time to track a particle over 1000 turns with the realistic 3-D field. The choice of time as an independent integration variable also allows us to avoid the difficulty associated with the infinite slope in the cartesian coordinate system when one integrates over the entire 180° bend, as noted by Moser[4].

Like most of the explicit high order integration procedures, the Runge-Kutta method used here is not symplectic, but the magnitude of the truncation error has been selected so as not to affect the results. The computational accuracy of this code was tested by tracking the reference particle in SXLS for over 10^4 turns (about $\frac{1}{7}$ of the damping time). The deviation of the particle from the ideal orbit is less than 10^{-8} meter in radial distance and 10^{-12} rad in slope, respectively.

CALCULATION AND HARMONIC ANALYSIS OF MAGNETIC FIELD

A computer code for three dimensional field calculation, MAGNUS[5], is used to compute the field distribution in the fringe area of the TAC dipole. The output is stored as a data set $B(r, \theta, z)$ in the computer system for a later use. The field components at the particle position that are needed for the beam tracking are then obtained from the three-dimensional spline interpolation. As an illustration, two different versions of the dipole end winding are studied. B1 has a flat coil arrangement outside the dipole hard edge while B2 has a saddle shape. Fig. 1 shows their mesh generated by computer and the corresponding field profiles along the reference orbits.

It can be seen from Fig. 1 that the field drops below zero outside the hard edge of B1. This is caused by the coil crossing the reference orbit at lower level. In comparison, the field of B2 is above zero along the orbit due to the bending-up of its coils. It is expected that this reversal of the field in B1 would introduce a strong multipole field in the area. This has been confirmed by the harmonic analysis of the field profiles as shown in Fig. 2.

CALCULATION OF LATTICE FUNCTIONS

An iteration is performed on the momentum of the reference particle to find the equilibrium orbit of the real lattice. Four particles are then selected with small deviations in horizontal and vertical coordinates, x, x', y, y' with respect to the reference particle. Based on tracking results over one superperiod, a four-dimensional transfer matrix for the transverse oscillation can be determined numerically. By comparing its elements with those of the parameterized matrix described by Courant and Snyder[6], both β and α functions can be obtained. The values of β, α and γ at any position can then be evaluated by means of a simple rule of transformation. The tunes ν_x, ν_y are derived from the matrix trace. For those particles with large amplitudes, the fast Fourier transform (FFT) is used to analyse their tune spectrum. The dispersion function, η, is calculated through the definition $\eta = x_p/(\delta p/p)|_{\frac{\delta p}{p} \to 0}$ where x_p is the distance between the reference orbit with momentum p and equilibrium orbit of the off-momentum(δp) particle. The natural chromaticity, ξ, is computed from $\xi = \delta\nu/(\delta p/p)|_{\frac{\delta p}{p} \to 0}$, where $\delta\nu$ is the difference of tune between the oscillation around the reference orbit and the off-momentum orbit. The synchrotron radiation integrals I_1 through I_5 and the other beam parameters are evaluated by the procedure described by Helm[7].

Table 2 Machine parameters computed by DIMAD and COMSYN

	DIMAD	COMSYN
horizontal tune ν_x	3.2295	3.229507
vertical tune ν_y	1.1604575	1.1604575
horizontal β at the sextupole center	0.377	0.37681
vertical β at the sextupole center	22.739	22.738759
dispersion η at the sextupole center	0.464	0.464
horizontal natural chromaticity	-3.1	-3.1
vertical natural chromaticity	-8.3	-8.3
natural beam emittance(meter-rad)	2.03×10^{-7}	2.035×10^{-7}
natural energy spread $\frac{\delta E}{E}$	7.25×10^{-4}	7.2503×10^{-4}
damping time(ms) $\tau_x, \tau_y, \tau_\epsilon$	2.49,2.56,1.29	2.51,2.57,1.30
CPU time with SUN (s)	15-20	20-22

Table 2 gives a comparison of the main parameters computed by COMSYN and by DIMAD using isomagnetic field approximation. It can be seen that all the values are in good agreement. With COMSYN, the deviation of the matrix determinant from unity is less than 10^{-8} for horizontal motion and 10^{-12} for vertical motion, demonstrating that the deviation of this code from the exact simplecticity is negligible.

BEAM TRACKING IN TAC AND SXLS WITH REAL FIELD

Trackings with the computed edge field have shown some interesting and quite different results in comparison with those from the conventional treatment of synchrotrons.

First, lattice functions are found to be different from their conceptual design values for TAC due to the effect of the fringe field. For example, the horizontal tune is shifted down by 2.3% and the vertical tune changes by 10.9%, from 1.160 to 1.03. This shows that the vertical tune with the edge field is very close to an integer resonance. To eliminate this problem, the strength of quadrupoles has been readjusted so that the tune is restored to the design value.

Secondly, the dynamic aperture is changed significantly by the fringe field. The dynamic aperture of the TAC machine is defined as a region where particles survive over one thousand turns. Fig. 3 shows the dynamic aperture for the two different edge fields B1 and B2. The rectangle represents the 10 standard deviation area which is needed for the acceptable quantum lifetime of the beam. It can be seen that the second version of the end winding, B2, provides a much bigger dynamic aperture due to its relatively weak multipole, as described above.

Table 3 Machine parameters computed from TOSCA field and fitting field

	TOSCA	FIT	error
energy(MeV)	698.7801394	698.780467	3.3×10^{-4}
horizontal tune ν_x	1.41502944	1.41503188	2.4×10^{-6}
vertical tune ν_y	0.40616711	0.4058189	3.5×10^{-4}
horizontal β_0	2.47880619	2.478453	3.5×10^{-4}
horizontal β_f	0.43621742	0.43612726	9.0×10^{-5}
vertical β_0	1.57095103	1.57287098	1.8×10^{-3}
vertical β_f	7.3114391	7.31409022	2.6×10^{-3}
dispersion η_0	1.4125244	1.412322	2.0×10^{-4}

A 3-D code TOSCA is used to calculate the magnetic field of SXLS along the particle orbit, including the magnetic fringe area. TOSCA is better for this case than MAGNUS since the magnet is free of iron. Table 3 lists some of the lattice parameters obtained in the simulation using the matrix method. Subscript 0 in this table denotes the sextupole center and subscript f the dipole center where β and η functions are evaluated. One serious difficulty in the calculation of the dynamic aperture has been the extensive CPU time required by TOSCA. To make such a study possible, a spline procedure has been employed to create a fitting formula using the TOSCA-generated mesh field. The accuracy of the fitting procedure was checked and it is shown in Table 3. As a result, the CPU time has been reduced from 12 minutes per turn to 8 seconds per turn.

With the fitted magnetic field, the dynamic aperture of SXLS has been calculated as shown in Fig. 4 where the large rectangle represents the beam pipe. It should be noted that the size of the dynamic aperture obtained by code COMSYN is only one third of that calculated by KRACKPOT[8], an integration code which uses two-dimensional isomagnetic field together with kick approximation for higher order terms. If the fringe field is ignored and the same combined function field is used inside the dipole, our calculation yields an aperture quite similar to the one obtained by using KRACKPOT. The significant difference in aperture size should therefore be attributed to effects of the fringe field as well as the longitudinal field component which are ignored in KRACKPOT.

A further study has been carried out by tracking three particles (P1, P2 and P3 shown in Fig. 4) with different initial amplitudes over one thousand turns. FFT is used to find their tunes and the result is shown in Fig. 5. In the tune diagram, $P0$ is the original design operating point, but the real operating point resulting from the TOSCA-generated field is at $P1$. Points $P2$ and $P3$ represent the operating points for particles 2 and 3 in Fig. 4, respectively. It can be seen that, as the amplitude of particle increases, its tunes shift to smaller values and eventually fall within the area of dense resonance lines. This is probably one reason for the narrow dynamic aperture of SXLS. One cure of this is to move the design operating point in the opposite direction to the shift so as to provide a larger space for particles of large amplitudes. An example is given in Fig. 4 when the operating point is moved to $\nu_h = 1.28$ and $\nu_v = 0.48$, the aperture is increased by as much as 30%. Fig. 6 shows the phase space of the horizontal and vertical motions. A large smear can be seen in the vertical space.

ACKNOWLEDGEMENTS

We should like to thank Chuck Swenson and Steven Kramer, Roy Blumberg and Jim Murphy of Brookhaven National Laboratory for many useful discussions.

REFERENCES

1. S. Ohnuma, Fermilab internal report #313 (1983).
2. J.B. Murphy, et al, Design Book of SXLS, BNL Informal Document(1991).
3. Ronald D. Ruth, IEEE Trans, Nucl. Sci., NS-30, No. 4, 1983, pp. 2669-2671.
4. H.O. Moser, A.J. Dragt, Nucl. Instru. Methods, (1987).
5. S. Pissanetsky, Ed. Elsevier Science Public. B. V.(North-Holland).(1986)pp. 121-132.
6. E.D. Courant and H.S. Snyder, Ann. Phys. 3, (1958)pp. 1-48.
7. R.H. Helm, M.J. Lee, P.L. Morton and M. Sands, IEEE Trans. Nucl. Sci. 20(1970)pp. 900-901.
8. M.F. Reusch, E. Forest and J.B. Murphy, *Proceedings of IEEE Prticle Accelerator Conf. in San Francisco*(1991).

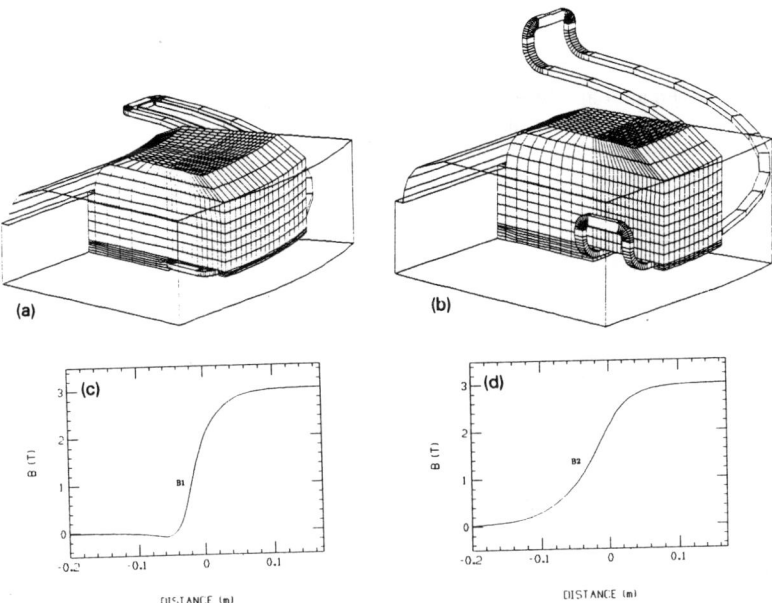

Fig. 1. Mesh generated by computer and the edge field profile of two different versions of end windings B_1(a),(c) and B_2(b),(d) for TAC compact ring.

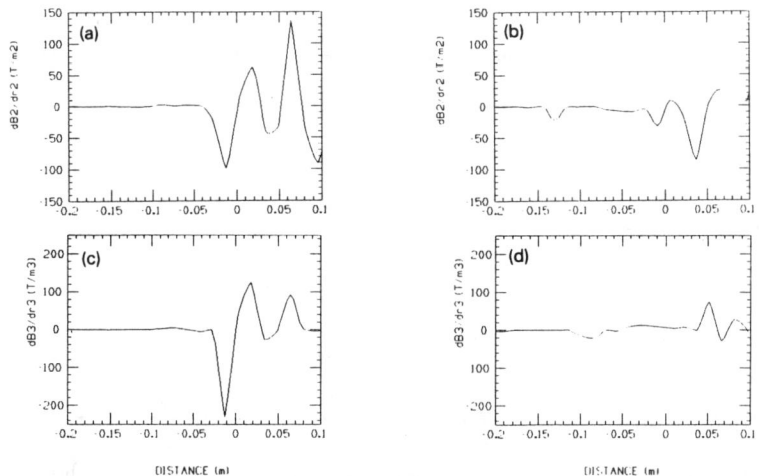

Fig. 2. Sextupole and octupole components of the dipole B_1(a),(b) and B_2(c),(d) along the reference orbit of TAC compact ring.

Beam Dynamics Study of Compact Ring

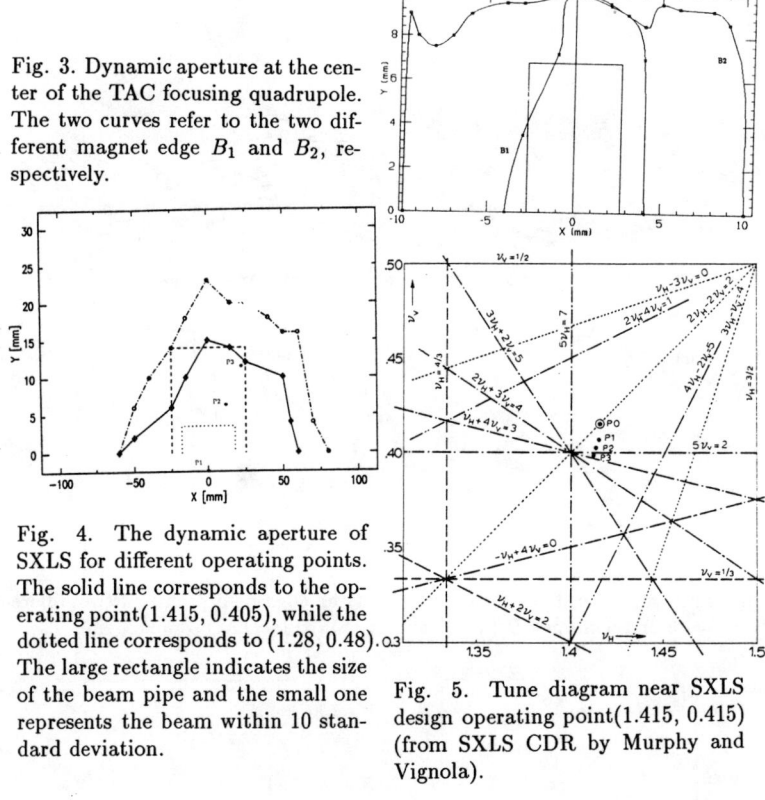

Fig. 3. Dynamic aperture at the center of the TAC focusing quadrupole. The two curves refer to the two different magnet edge B_1 and B_2, respectively.

Fig. 4. The dynamic aperture of SXLS for different operating points. The solid line corresponds to the operating point(1.415, 0.405), while the dotted line corresponds to (1.28, 0.48). The large rectangle indicates the size of the beam pipe and the small one represents the beam within 10 standard deviation.

Fig. 5. Tune diagram near SXLS design operating point(1.415, 0.415) (from SXLS CDR by Murphy and Vignola).

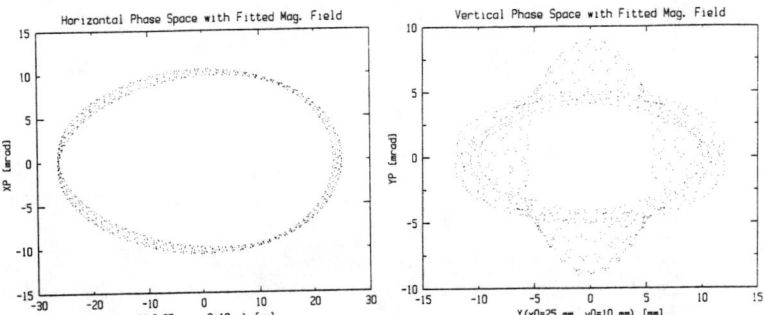

Fig. 6. Phase space of SXLS for the horizontal motion(left) and the vertical motion(right).

THREE DIMENSIONAL δf SIMULATIONS OF BEAMS IN THE SSC

J. Koga and T. Tajima
Institute for Fusion Studies, The University of Texas at Austin Austin, Texas 78712-1060

S. Machida
SSC Laboratory, 2550 Beckleymeade Avenue Dallas, Texas 75237

ABSTRACT

A three dimensional δf strong-strong algorithm has been developed to apply to the study of such effects as space charge and beam-beam interaction phenomena in the Superconducting Super Collider (SSC). The algorithm is obtained from the merging of the particle tracking code Simpsons used for 3 dimensional space charge effects and a δf code. The δf method is used to follow the evolution of the non-gaussian part of the beam distribution. The advantages of this method are twofold. First, the Simpsons code utilizes a realistic accelerator model including synchrotron oscillations and energy ramping in 6 dimensional phase space with electromagnetic fields of the beams calculated using a realistic 3 dimensional field solver. Second, the beams are evolving in the fully self-consistent strong-strong sense with finite particle fluctuation noise is ·greatly reduced as opposed to the weak-strong models where one beam is fixed.

INTRODUCTION

In this paper we present a method for realistically simulating three dimensional beams in the Superconducting Super Collider (SSC). This method merges the ideas of tracking particles (Simpsons code) with the δf algorithm. There are two advantages of this approach. First, the Simpsons code utilizes a realistic accelerator model including synchrotron oscillations and energy ramping in 6 dimensional phase space with electromagnetic fields of the beams calculated using a realistic 3 dimensional field solver. Second, the δf method allows the beams to evolve in the fully self-consistent strong-strong sense with a large reduction of finite particle fluctuation noise.

Macroparticle or Particle-in-Cell (PIC) codes typically use macroparticles to represent the entire distribution of particles. In the beam-beam interaction for the SSC, the beams consist of 10^{10} particles each. Simulating this many particles with the PIC technique is computationally prohibitive. With the conventional PIC code 10^{10} particles are represented by only $10^3 - 10^4$ simulation particles allowing simulation of the beam-beam interaction in a reasonable computation time. However, the fluctuation level of various quantities such as the beam density ρ in the code is much higher than that of the real beam. The fluctuation level $\delta\rho$ goes as approximately:

$$\frac{\delta\rho}{\rho} = N^{-1/2}, \qquad (1)$$

where N is the number of particles. Therefore, the fluctuation level of the PIC code is about 10^3 times higher than that of the real beam. To facilitate the study of subtle effects, a one dimensional δf code has been developed[1]. The one dimensional δf code was used to study single particle diffusion due to the beam-beam interaction in a fully self-consistent way[1]. The particle fluctuation noise was several orders of magnitude below that of standard PIC code[1].

In the following sections we present details on the δf algorithm in one dimension, the Simpsons code, and the full implementation in three dimensions.

δf METHOD ONE DIMENSION

The δf method follows only the fluctuating part of the distribution instead of the entire distribution. In PIC codes a finite number of particles is used to represent the Vlasov equation or Klimontovich equation. In the particular case of the beam-beam interaction[2]:

$$\frac{\partial f}{\partial s} + x'\frac{\partial f}{\partial x} - (K(s)x - F_{bb}(x,s))\frac{\partial f}{\partial x'} = 0, \qquad (2)$$

where $K(s)x$ is the usual magnetic guiding force of a single element and $F_{bb}(x,s)$ is the beam-beam force. The distribution function f is represented by a finite number of particles by:

$$f(x, x', s) = \sum_{i=1}^{N} \delta(x - x_i(s))\delta(x' - x'_i(s)), \qquad (3)$$

where the sum is over the number of simulation particles used. In the δf method only the perturbative part of the distribution is followed. The total distribution function $f(x, x', s)$ is decomposed into

$$f(x, x', s) = f_0(x, x', s) + \delta f(x, x', s), \qquad (4)$$

where $f_0(x, x', s)$ is the steady or slowly varying part of the distribution and $\delta f(x, x', s)$ is the perturbative part. The key to this method is finding a distribution $f_0(x, x', s)$ which is close to the total distribution $f(x, x', s)$. The perturbative part $\delta f(x, x', s)$ then causes only small changes to the distribution, and thus represents only the fluctuation levels. In the particular case of the beam-beam interaction an analytic solution to an equation close to the original Vlasov equation can be found. For a linearized beam-beam force the Vlasov equation can be written in the form:

$$\frac{\partial f_0}{\partial s} + x'\frac{\partial f_0}{\partial x} - (K(s) - F_{bb0}(s))x\frac{\partial f_0}{\partial x'} = 0, \qquad (5)$$

where $F_{bb0}(s)$ is the linear portion of the beam-beam force $F_{bb}(x,s)$. The solution is a Gaussian. Subtracting the linearized equation from the total Vlasov equation–

$$\frac{\partial \delta f}{\partial s} + x'\frac{\partial \delta f}{\partial x} - (K(s)x - F_{bb0}(x,s))\frac{\partial \delta f}{\partial x'} = -(F_{bb}(x,s) - F_{bb0}(s)x)\frac{\partial f_0}{\partial x'}, \quad (6)$$

$F_{bb0}(x,s)$ is the kick from a Gaussian beam.

The perturbative part of the distribution can be represented by a finite number of particles (characteristics):

$$\delta f(x, x', s) = \sum_{i=1}^{N} w_i \delta(x - x_i(s))\delta(x' - x'_i(s)), \quad (7)$$

Substituting this into the equation for δf advance, we obtain:

$$\frac{dw_i}{ds} = -\frac{1}{n}[(F_{bb}(x,s) - F_{bb0}(s)x)\frac{\partial f_0}{\partial x'}]_i \quad (8)$$

where

$$n = \frac{N}{(\Delta x \Delta x')} \quad (9)$$

This density n is calculated on the assumption that the particles are distributed uniformly in phase space. In the δf algorithm x_i, x'_i, and w_i are advanced. The advance of the extra term w_i increases only slightly the number of operations over the PIC method.

THE CODE SIMPSONS

The 6D phase space coordinates; x, x', y, y', z, z' are advanced by the thin lens code Simpsons[3]. The code Simpsons uses *time* as the independent variable so that time dependent parameters such as bending magnet field, rf voltage, power supply ripple and other noise are incorporated in the same way as real machines. The substitution of the thick lens lattice to the thin lens one, the tuning of the betatron phase and the chromaticity, and the introduction of multipole errors and mis-alignment are done by the TEAPOT program[4]. In addition to the lattice file created by TEAPOT, the bending magnet field $B\rho(t)$ and the rf voltage $V(t)$, at least, are read as an external table to define a machine cycle parameters when acceleration as well as synchrotron oscillations is simulated.

Within one time interval, a particle will either remain in drift space or receive one or more momentum kicks due to magnets and/or rf cavities. If there are no lattice elements in the time interval, the positions are updated and the momentum remains constant. If there are magnets in between, first the positions are updated to the magnet location, then the transverse momenta is changed by the thin lens kick. If there is an rf cavity in between, longitudinal and total momentum are changed. We keep track the rf phase by integrating the rf frequency with respect to time so that a particle gets proper energy gain according to the exact time when it passes through a cavity.

THREE DIMENSIONAL FIELD SOLVER[5]

The boundary condition is fixed for circular beam pipe with the perfect conductivity $\sigma = \infty$ at radius $r = b$. Under the Lorentz gauge $\frac{\partial \phi}{\partial t} + c\vec{\nabla} \cdot \vec{A} = 0$, the Maxwell equations are

$$\left(\frac{1}{r}\frac{\partial}{\partial r}(r\frac{\partial}{\partial r}) + \frac{1}{r^2}\frac{\partial^2}{\partial \theta^2}\right)\phi = -4\pi n(r,\theta,z,t), \tag{10}$$

$$\left(\frac{1}{r}\frac{\partial}{\partial r}(r\frac{\partial}{\partial r}) + \frac{1}{r^2}\frac{\partial^2}{\partial \theta^2}\right)\vec{A} = -\frac{4\pi}{c}\vec{J}(r,\theta,z,t), \tag{11}$$

where $n(\vec{x},t)$ and $\vec{J}(\vec{x},t)$ are the charge density and the current, respectively, and the scalar potential ϕ and the vector potential $\vec{A}$ are used. We set the scalar potential $\phi \equiv 0$ at $r = b$ and invoke the ordering

$$\frac{1}{c^2}\frac{\partial^2}{\partial t^2} \sim \frac{\partial^2}{\partial z^2} \ll \frac{1}{r}\frac{\partial}{\partial r}(r\frac{\partial}{\partial r}) + \frac{1}{r^2}\frac{\partial^2}{\partial \theta^2}. \tag{12}$$

This can be justified by the typical dimension of the proton beam.

We further simplify the current density,

$$\vec{J} = J_z \vec{e}_z = \bar{v}_z n \vec{e}_z, \tag{13}$$

where $\bar{v}_z$ is the average velocity of the beam. From the Lorentz force equation $\vec{F} = q(\vec{E} + \frac{\vec{v}}{c} \times \vec{B})$, the electromagnetic force is $\vec{F} = -\frac{q}{\gamma^2}\vec{\nabla}\phi$ and we only need to solve the scalar potential Eq. (10). The charge density and scalar potential are Fourier transformed in θ

$$n(r,\theta,z,t) = \sum_m n_m(r,z,t)\exp(im\theta), \tag{14}$$

$$\phi(r,\theta,z,t) = \sum_m \phi_m(r,z,t)\exp(im\theta), \tag{15}$$

with the inverse transforms

$$n_m(r,z,t) = \frac{1}{2\pi}\int_0^{2\pi} n(r,\theta,z,t)\exp(-im\theta)d\theta, \tag{16}$$

$$\phi_m(r,z,t) = \frac{1}{2\pi}\int_0^{2\pi} \phi(r,\theta,z,t)\exp(-im\theta)d\theta. \tag{17}$$

For each m, the equation assumes the form

$$\frac{1}{r}\frac{\partial}{\partial r}(r\frac{\partial \phi_m}{\partial r}) - \frac{m^2}{r^2}\phi_m = -4\pi n_m. \tag{18}$$

The solution for the equation for $m \geq 0$ is

$$\phi_m = W_m(r) - \left(\frac{r}{b}\right)^m W_m(b), \tag{19}$$

where $W_m(r)$ is

$$W_{m=0}(r) = -4\pi \int_0^r n_0(r,z,t) r' \ln\frac{r}{r'} dr', \tag{20}$$

$$W_{m\geq 1} = -\frac{2\pi r^m}{m} \int_0^r n_m(r,z,t) r'^{(1-m)} dr'$$
$$- \frac{2\pi r^{-m}}{m} \int_0^r n_m(r,z,t) r'^{(1+m)} dr'. \tag{21}$$

The grids are set in the cylindrical coordinate system such that the whole charge distribution is enclosed.

δf ALGORITHM THREE DIMENSIONS

The δf algorithm can be applied to higher dimensions. The application to a storage ring with many elements is similar to the one dimensional beam-beam interaction method. The Vlasov equation is of the form:

$$\frac{\partial f}{\partial s} + \vec{x}' \cdot \frac{\partial f}{\partial \vec{X}} + \vec{A} \cdot \frac{\partial f}{\partial \vec{X}'} = 0 \tag{22}$$

where

$$\vec{x} = \begin{pmatrix} x \\ y \\ z \end{pmatrix} \quad \vec{x}' = \begin{pmatrix} x' \\ y' \\ -\eta z' \end{pmatrix} \tag{23}$$

$$\frac{\partial}{\partial \vec{X}} = \begin{pmatrix} \frac{\partial}{\partial x} \\ \frac{\partial}{\partial y} \\ \frac{\partial}{\partial z} \end{pmatrix} \quad \frac{\partial}{\partial \vec{X}'} = \begin{pmatrix} \frac{\partial}{\partial x'} \\ \frac{\partial}{\partial y'} \\ \frac{\partial}{\partial z'} \end{pmatrix} \tag{24}$$

where $z \equiv \frac{\Delta\phi_{rf}}{2\pi}\lambda_{rf}$ and $z' \equiv \frac{\Delta(pc)}{pc}$ and

$$\vec{A} = \overleftrightarrow{Q}(s) \cdot \vec{x} + \vec{S}(\vec{x}, s) + \vec{RF}(s) + \vec{F}_{sc}(\vec{x}, s) + \vec{F}_{bb}(\vec{x}, s). \tag{25}$$

On the right hand side of Eq. (25), $\overleftrightarrow{Q}(s)$ and $\vec{S}(\vec{x}, s)$ represent the quadrupole and other higher order magnets, respectively, in the storage ring. For example:

$$\overleftrightarrow{Q}(s) = q(s) \begin{pmatrix} 1 & 0 & 0 \\ 0 & -1 & 0 \\ 0 & 0 & 0 \end{pmatrix} \tag{26}$$

where $q(s)$ represents the strengths of the quadrupoles at position s. The quantity $\vec{RF}(s)$ in Eq. (25) represents the contribution from the rf cavity:

$$\vec{RF}(s) = \begin{pmatrix} 0 \\ 0 \\ \frac{f_{rf} \cdot eV}{h\beta^2 c(pc)}[\sin(\phi_s + \Delta\phi_{rf}) - \sin(\phi_s)] \end{pmatrix}. \tag{27}$$

The elements $\vec{F}_{sc}(\vec{x},s)$ and $\vec{F}_{bb}(\vec{x},s)$ in Eq. (25) represent the contributions from space charge and beam-beam forces, respectively.

The steady state equation again takes only the linear terms:

$$\frac{\partial f_0}{\partial s} + \vec{x}' \cdot \frac{\partial f_0}{\partial \vec{X}} + (\overleftrightarrow{A}_0(s) \cdot \vec{x}) \cdot \frac{\partial f_0}{\partial \vec{X}'} = 0, \tag{28}$$

where $\overleftrightarrow{A}_0(s)$ is the linear portion of $\overleftrightarrow{A}$:

$$\overleftrightarrow{A}_0(s) = \overleftrightarrow{Q}(s) + \overleftrightarrow{RF}_0(s) \tag{29}$$

where

$$\overleftrightarrow{RF}_0(s) = \begin{pmatrix} 0 & 0 & 0 \\ 0 & 0 & 0 \\ 0 & 0 & \frac{f_{rf} \cdot eV \cos(\phi_s)}{h\beta^2 c(pc)} \frac{2\pi}{\lambda_{rf}} \end{pmatrix} \tag{30}$$

So $\overleftrightarrow{A}_0(s)$ represents the contributions from the quadrupole magnet elements in the machine and the linear part of the rf force. The solution of the steady state equation is a Gaussian of the form:

$$f_0(\vec{x}, \vec{x}') = N f_{x0}(x, x') f_{y0}(y, y') f_{z0}(z, z') \tag{31}$$

$$f_{x0}(x, x') = \frac{(\gamma \beta)}{2\epsilon_x} \exp(-\frac{\pi A_x^2 (\gamma \beta)}{2\epsilon_x}) \tag{32}$$

$$f_{y0}(y, y') = \frac{(\gamma \beta)}{2\epsilon_y} \exp(-\frac{\pi A_y^2 (\gamma \beta)}{2\epsilon_y}) \tag{33}$$

where $\epsilon_{x,y}$ is the transverse normalized emittance, β and α are the Lorentz factor,

$$A_x^2 = \gamma_x(s)x^2 + 2\alpha_x(s)xx' + \beta_x(s)x'^2, \tag{34}$$
$$A_y^2 = \gamma_y(s)x^2 + 2\alpha_y(s)xx' + \beta_y(s)x'^2, \tag{35}$$

the $\alpha_{x,y}$ and $\beta_{x,y}$ are the Courant-Synder parameters, and:

$$f_{z0}(z, z') = \frac{\beta c \eta}{2\pi \sigma_z^2(s) \omega_s} \exp(-\frac{1}{2\sigma_z^2(s)}(z^2(s) + (\frac{\beta c \eta}{\omega_s})^2 z'^2)) \tag{36}$$

where $\sigma_z = \sigma_{\Delta \phi_{rf}} \lambda_{rf}/2\pi$, and ω_s is the synchrotron frequency. Note that the steady state solution incorporates the linear forces and ignores coupling between x, y and z motion.

The perturbation equation can be written in the form:

$$\frac{\partial \delta f}{\partial s} + \vec{x}' \cdot \frac{\partial \delta f}{\partial \vec{X}} + \vec{A} \cdot \frac{\partial \delta f}{\partial \vec{X}'} = -(\vec{A} - \overleftrightarrow{A}_0(s) \cdot \vec{x}) \cdot \frac{\partial f_0}{\partial \vec{X}'}, \tag{37}$$

So the perturbation equation takes into account the nonlinear force terms and the coupling terms. The weight advance in the finite particle representation is of the form:

$$\frac{dw_i}{ds} = -\frac{1}{n}[((\vec{A}-\vec{A}_0(s))\cdot\vec{x})\cdot\frac{\partial f_0}{\partial \vec{X'}}]_i \tag{38}$$

where

$$n = \frac{N}{(\Delta\vec{x}\Delta\vec{x'})}. \tag{39}$$

The variation of $\frac{\partial f_0}{\partial \vec{X'}}$ with s can be computed from β_x, σ_x, β_y, and σ_y. The betatron oscillation lengths $\beta(s)$ and the beam sigmas $\sigma(s)$ can be determined from the lattice function of the particular machine being studied where

$$\frac{\beta_x(s)}{\beta_{x0}} = \frac{\sigma_x^2(s)}{\sigma_{x0}^2} \tag{40}$$

$$\frac{\beta_y(s)}{\beta_{y0}} = \frac{\sigma_y^2(s)}{\sigma_{y0}^2}, \tag{41}$$

where β_{x0}, σ_{x0}^2, β_{y0}, and σ_{y0}^2 are at some initial point in the collider.

SUMMARY

In this paper we have presented an algorithm for realistic modeling of beams in a hadron collider such as the SSC. The code with this algorithm is expected to be a general study tool for space-charge effects of the Low Energy Booster (LEB) to beam-beam effects of the Collider ring. It will have mainly two parts. In the first part, the 3 dimensional electromagnetic fields are calculated as the sum of the steady part which is solved analytically, and the pertubative part which is computed by the field solver using macroparticles. The code Simpsons is the second part which advances 6 dimensional phase space coordinates to the next location for field calculations. The implementation of the algorithm with the full three dimensional δf method is undertaken.

REFERENCES

1. J. Koga, dissertation, University of Texas at Austin (1992).
2. A. W. Chao, SSCL-346, SSC Laboratory (1991).
3. S. Machida, *this conference*.
4. L. Schachinger and R. Talman, *Part. Accel.* **22**, 35 (1987).
5. The original formalism and approximation in this section was derived by M.L.Sloan., unpublished memo, SSC Laboratory (1991).

HIGHER-ORDER MODE (HOM) DAMPER DESIGN USING HFSS*

L. Walling, G. Hulsey, and T. Grimm
Superconducting Super Collider Laboratory,[†] Dallas, TX 75237

ABSTRACT

This paper reports results of using High-Frequency Structure Simulator (HFSS) to design a Smythe-type broadband longitudinal HOM damper for the SSCL low energy booster (LEB) ferrite-tuned cavity. The code is also used to check the effect of arranging resistive loads in an azimuthally assymetric pattern to damp transverse modes.

INTRODUCTION

The LEB cavity[1] is a ferrite-tuned $\lambda/4$ coaxial cavity that tunes from 47.5 to 59.8 MHz in about 20 ms and requires 127 kV on the gap. R. Baartman of TRIUMF did a coupled bunch mode beam instability analysis[2] for the LEB. According to his analysis, the allowable longitudinal shunt impedance of the cavity HOMs as a function of frequency is shown in Figure 1. Also shown is the calculated achieved shunt impedance of the damper.

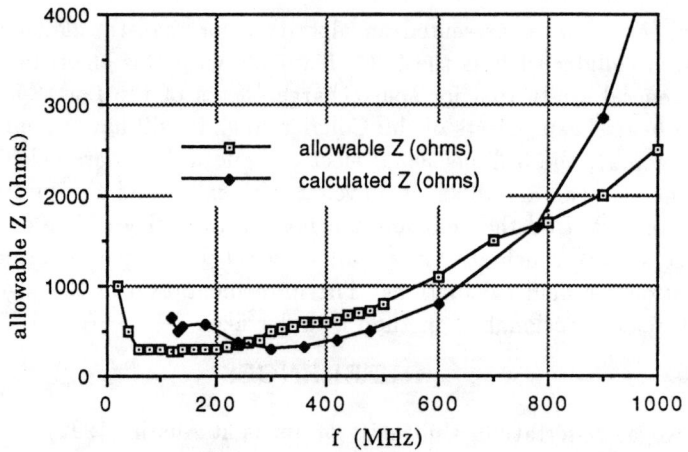

Fig. 1.   Narrow Band Impedance in the SSC LEB.

---

* To be presented at the American Institute of Physics Conference February 22, 1993.

† Operated by the Universities Research Association, Inc., for the U.S. Department of Energy under Contract No. DE-AC35-89ER40486.

For fixed-frequency machines, HOM dampers can be designed to address individual modes. Since the LEB has a large frequency swing, it is more convenient to build a broadband damper. A Smythe-type[3] type broadband damper has been designed using two-element high-pass filters between the damping cavity and four discrete water-cooled loads. A cross-section of the gap end of the cavity with HOM damper is shown in Figure 2. The filter/loads are distributed around the circumference of the damping cavity in such a pattern as to damp transverse as well as longitudinal modes (Figure 3).

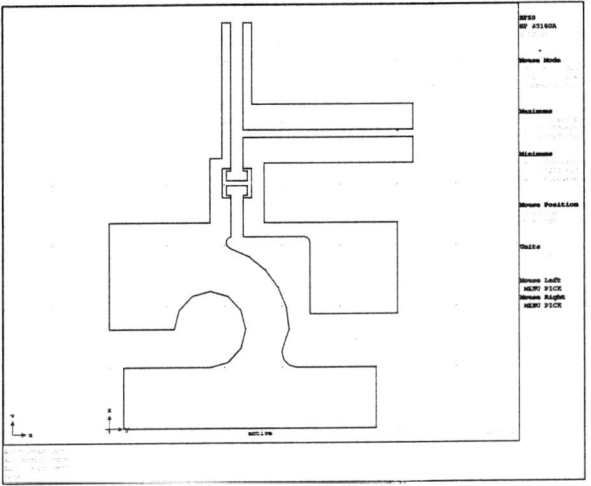

Fig. 2. Cross-section of Final LEB HOM Damper.

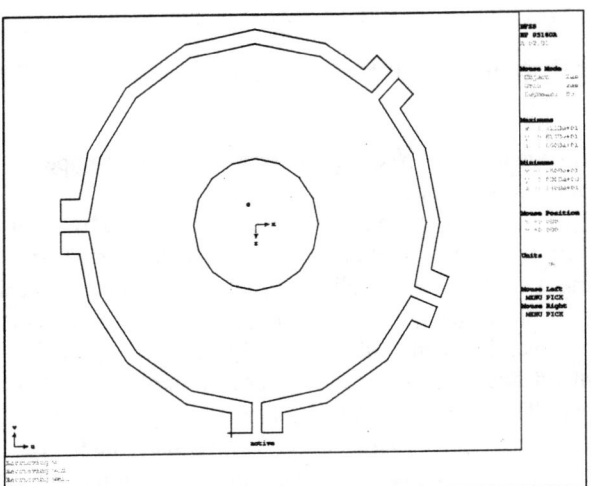

Fig. 3. Axial Arrangement of Filters and Loads on HOM Damper.

DESIGN TECHNIQUE

Smythe showed that his damper could be modeled as a circuit element that is in shunt with the gap of the cavity. The circuit representing the cavity and damper is shown in Figure 4. Then r_{sh}, the shunting resistance of the damper, is

$$r_{sh} = 1/\mathrm{Re}1/z_d \tag{1}$$

The damper impedance, z_d, can be approximated by solving the circuit.

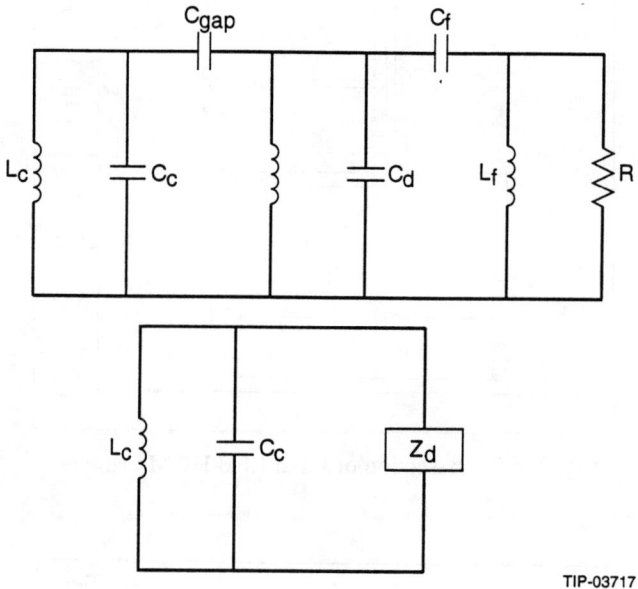

Fig. 4. Circuit Diagram for LEB HOM Damper with High-pass Filters Between Smythe Cavity and Loads.

Originally, we intended to design an simple Smythe damper. However, the $3\lambda/4$ mode occurs at a frequency much lower than three times the fundamental due to the heavy capacitive loading of the cavity. This is aggravated by the large coupling to the damping cavity which is required to damp the low-frequency modes so strongly. After determining the amount of coupling that would be required, we determined that it would be impossible to find a combination of r_d, l_d and c_d that would yield the required damping over the band while leaving the accelerating mode with a large enough shunt impedance to not overdrive the power amplifier (and burn up the 5-kW damping resistors). This could be determined easily using the lumped circuit model. Therefore, we decided to try adding a high-pass filter between the damper and the load. The lumped circuit model for the damper plus filter could be used to estimate the required coupling capacitance to the damping cavity and roughly estimate the circuit parameters. Because the required coupling capacitance turned out to be very large (about 12 pF), the current paths from the gap to the point at which the filter/load is shunted across the Smythe capacitor became long, which reduces the damping of the HOMs. This results in a damping cavity that is not well–represented by any type of circuit approach.

HFSS, with its s-parameter format and finite-element mapping of the rounded surfaces, was the ideal tool for such a design. The main problem was to find an approach that would be fast and efficient, since modeling the entire damper including filter results in a large problem that takes hours to run. Therefore, the design was done in three stages which combined the use of numerical results from HFSS and lumped element analysis in the first two stages to speed up the design process.

Step 1. Pseudo-2d HFSS analysis (Figure 5). We modeled an angular wedge of 2 slices of about 5-10 degrees each (we have found that one thin slice of elements does not yield an accurate solution.) The load is represented by a thin disk of resistive material which represents the four loads in parallel. This is a one-port problem with the port located at a cross section of the cavity just far enough away from the gap such that the fields are approximately transverse. The reflection coefficient is calculated by HFSS and written out to a file. This file was read using a small FORTRAN code which transformed S11 from the port to the gap, then calculated r_{sh}. This was useful to develop a cavity that had approximately the desired r_{sh} over the band, although the fundamental mode shunt impedance was too low.

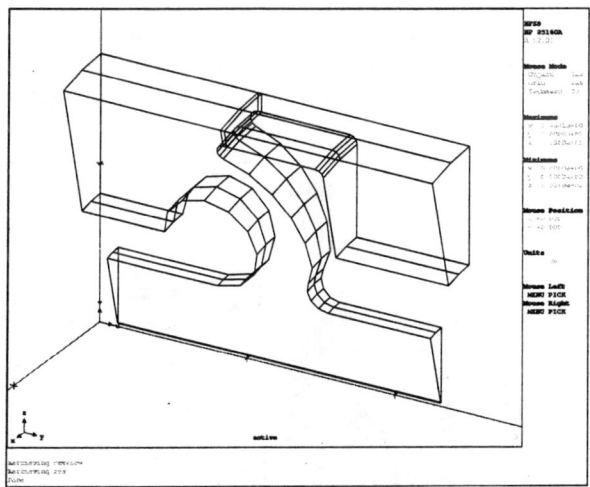

Fig. 5. HFSS Model Using a Disk of Resistive Material to Represent the Loads.

Step 2. Next, another HFSS model (Figure 6) was built which modeled 1/8 of the axial geometry and which included a port at the position the filter/load would be placed. The full 2-port solution was obtained over the frequency range, the s-parameters stored in a file, then read into a FORTRAN code which cascaded the high-pass filter/load parameters onto port 2, then calculated r_{sh} from S11 with the new termination. The damper cavity dimensions were modeled in HFSS, then the filter quickly optimized for that geometry using the FORTRAN program. By varying cavity dimensions systematically, the entire circuit was quickly optimized.

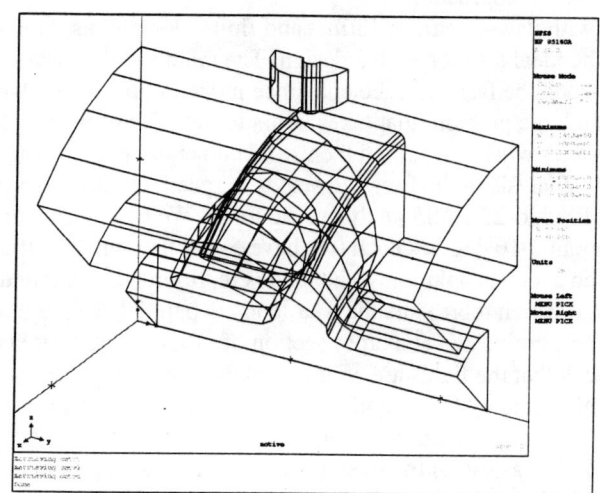

Fig. 6. HFSS Model Using 1/8 Axial Symmetry and Load Represented by Coaxial Port.

Step 3. The actual filter was added to the HFSS model (Figure 7) to do the final checking of the design for damping, field maximums and heating. The shunt impedances of the HOMs as calculated by this HFSS model are shown in Figure 1.

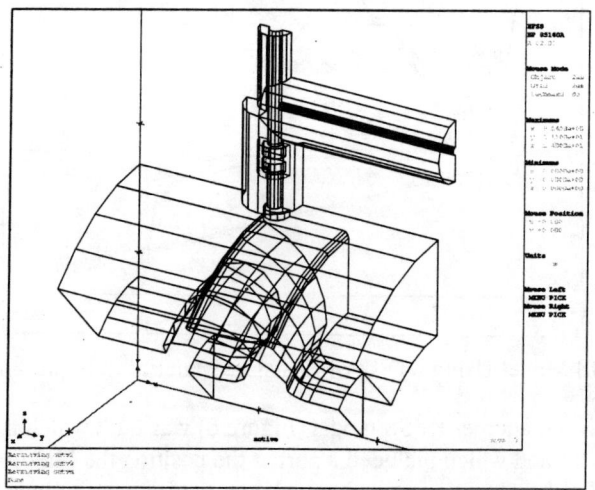

Fig. 7. HFSS Model of Entire HOM Damper.

In parallel to this effort, a hardware model of the filter using high-voltage ceramic capacitors and 5-kW water-cooled loads was built and tested at both low and high power.

TRANSVERSE MODE DAMPING

We then tested the possibility of damping transverse modes by arranging the loads axially as shown in Figure 3. A simplified model of the entire cavity with a simple Smythe damper was built using HFSS. The loads were arranged as suggested by Grimm and the cavity was excited by an off-center wire placed about 1 cm to the right and to the top of the cavity axis, looking down the axis from the gap to the tuner. This method of excitation would excite all longitudinal and transverse modes of the cavity. All materials in the cavity were artifically specified with a loss tangent that would ensure that as the cavity was swept from 300 MHz to 1 GHz at 20 MHz intervals so that any resonances would be identified. Then when resonances were identified, they were identified as to longitudinal or transverse by looking at the fields near resonance. Finally, the loss tangents were reduced to zero and the frequencies were swept around the resonances. A rough analysis determined that the impedances of all transverse modes were below 100 kΩ/m. A similar analysis has not been done with the actual damper with filters.

CONCLUSIONS

HFSS has proven to be an ideal tool for designing HOM dampers because of its s-parameter format, matched port, lossy material capabilities, and finite-element modeling technique which allows close approximation of rounded surfaces. It is very easy to model problems and make changes in models. Measurements of the cavity with damper will be performed in early spring of 1993 to verify the design.

REFERENCES

1. C.C. Friedrichs, L. Walling, B.M. Campbell, "Design of an Accelerating Cavity for the Superconducting Supercollider Low-Energy Booster," IEEE PAC, 1020 (1991).

2. R. Baartman, "Allowed Narrow-Band Impedance in the SSC LEB," TRIUMF Design Note (1992).

3. W.R. Smythe, "Proton Synchrotron RF Cavity Mode Damper Tests," IEEE PAC, 643 (1991).

SIMULATION OF THE INJECTION DAMPING AND RESONANCE CORRECTION SYSTEMS FOR THE HEB OF THE SSC[*]

M. Li, P. Zhang and S. Machida
Superconducting Super Collider Laboratory,[†] Dallas, TX 75237

ABSTRACT

An injection damping and resonance correction system for the High Energy Booster (HEB) of the Superconducting Super Collider (SSC) was investigated by means of multiparticle tracking. For an injection damping study, the code Simpsons is modified to utilize two Beam Position Monitors (BPM) and two dampers. The particles of 200 Gev/c, numbered 1024 or more, with Gaussian distribution in 6-D phase space are injected into the HEB with certain injection offsets. The whole bunch of particles is then kicked in proportion to the BPM signals with some upper limit. Tracking these particles up to several hundred turns while the damping system is acting shows the turn-by-turn emittance growth, which is caused by the tune spread due to nonlinearity of the lattice and residual chromaticity with synchrotron oscillations. For a resonance correction study, the operating tune is scanned as a function of time so that a bunch goes through a resonance. The performance of the resonance correction system is demonstrated. We optimize the system parameters which satisfy the emittance budget of the HEB, taking into account the realistic hardware requirement.

INTRODUCTION

Injection dampers are needed in the High Energy Booster (HEB) of the Superconducting Super Collider (SSC) to remove a large coherent betatron oscillation, which has resulted from unavoidable injection offsets, rapidly to prevent excessive emittance growth.

In concept, a damper system consists of a Beam Position Monitor (BPM) whose position signal is amplified to provide a transverse electric field across a damper, a pair of deflection plates, located about $N + 0.25$ (or 0.75) betatron period downstream of the BPM. (N is an integer.) On successive passes, the oscillation amplitude is reduced until the noise limit of the system is reached.

In practice, the tunes must be changeable in a certain scale, and the damper might not be able to be put at the expected position to get the exact fractional phase advance corresponding to 0.25 (or 0.75) betatron period. So, two BPMs separated by 0.25 betatron period and two dampers are used for one transverse direction.

---

[*] To be presented at the Computational Accelerator Physics Conference February, 1993.

[†] Operated by the Universities Research Association, Inc., for the U.S. Department of Energy under Contract No. DE-AC35-89ER40486.

Because of errors in magnets, betatron resonances of several orders are excited. The way to correct these resonances is well known and it can be done by calculating the bandwidth of a resonance. In real machines, however, one needs to adjust the corrector iteratively by looking at beam behavior because not all the error information, especially concerning random errors, is obtainable. The purpose of the simulation study is to look at the emittance growth and beam loss using macro-particles in the lattice with and without the resonance correction system.

SIMULATION METHOD

We modified the code Simpsons[1] to incorporate the damper system for injection orbit errors and to evaluate performance of the resonance correction system. The code Simpsons is a fully 6-D multiparticle tracking program with acceleration. (In the following study, the particle momentum is fixed to the injection value.) It uses time as the independent variable instead of longitudinal position which is commonly taken by the other programs. One of the advantages of this is that we can change machine parameters such as tune and chromaticity as a function of time just like real machines, which makes it a lot easier to simulate resonance crossing as shown later. The table of quadrupole strength at several times, for example, is read into the code and interpolated at the time when a tracking particle passes the element. Usually, we track only one bunch (although there are 2160 bunches in the HEB) and assume the remainder of the bunches behave identically.

For the damper system simulation, the transverse charge center of macro-particles should be determined every turn. The position of one or two beam position monitors (BPMs) is specified in the lattice beforehand. Because time is the independent variable, all the macro-particles are not necessarily passing BPMs in the same time interval. There is a counter at each BPM which counts the number of particles passing through and a recorder which accumulates the transverse position of particles. When the counter counts the total number of macro-particles, a displacement of the transverse charge center is calculated as the average of all the positions.

Once the beam position is determined at the BPMs, one or two dampers, which locate downstream of the BPMs, are ready to give horizontal and vertical dipole kicks when a particle is passing there. The strength of dipole kicks is determined linearly depending on the beam position at the BPMs, but with a certain maximum strength which should not be exceeded. The sign and amplitude of the linear coefficient is chosen according to the phase advance between the BPMs and dampers.

To check the performance of the proposed resonance correction system, we observed the rms emittance and beam loss due to the resonance crossing. The bandwidth of individual resonance is estimated first, and the strength of correctors is calculated so that the bandwidth is reduced. The simulation of both lattice with and without correctors is then performed. The strength of quadrupoles which are used for tune adjustment is linearly scanned near resonances as a function of time.

INJECTION DAMPING SYSTEM

In general, the centroid oscillation amplitude reduction per turn ΔX is proportional to, $\sqrt{\beta_b \beta_k}\,\theta$ where β_b and β_k are the beta-function at the BPM and the corresponding damper, θ is the kick strength of the damper, when the centroid oscillation itself is larger than a certain amount. Figure 1 shows the beta-functions at one straight section of the HEB. Four BPMs, two horizontal and two vertical, are set into two cells of a dispersion suppressor in such a way that each BPM is right beside either a focusing or defocusing quadrupole to assure that each one has been located at the position with a local peak beta-function. The beta-function at each BPM and damper, and the fractional betatron period corresponding to the phase advance between a BPM and damper pair are shown in the Table I.

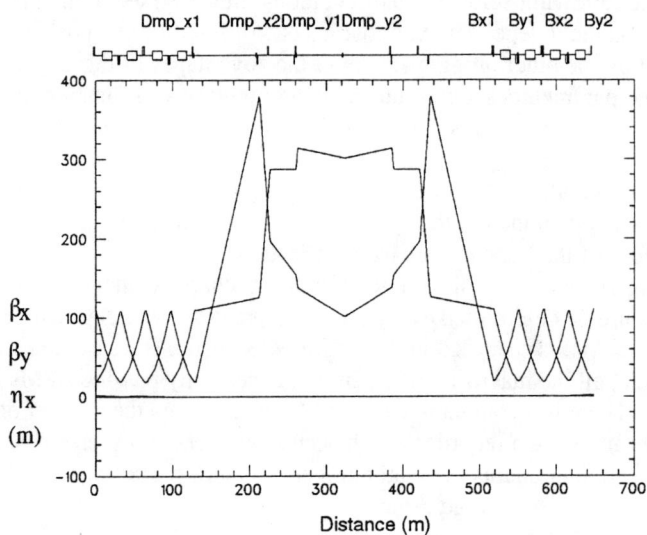

Fig. 1. The Injection Damping Systems in the HEB of the SSC.

Table I Beta–functions and phase advances/2π.

	β_x (m)	β_x (m)	Δv_x to BPM_x1	Δv_x to BPM_x2	Δv_y to BPM_y2	Δv_y to BPM_y2
BPM_x1	120.3					
BPM_x2	95.8					
BPM_y1		102.0				
BPM_y2		110.2				
Damper_x1	98.2		0.05	0.78		
Damper_x2	297.8		0.23	0.96		
Damper_y1		348.4			0.81	0.57
Damper_y1		101.3			0.95	0.71

The maximum voltage across the damper is set up from 1 kV to 3 kV with the assumption of that each damper is 1-meter long and has a 50-millimeter gap between two deflection plates. The initial beam offsets are determined as 2 mm in horizontal and 1.5 mm in vertical, which are the maximum possible values from the transform line coming to the HEB. The normalized emittance of the injection beam is 0.8 mm–mrad in each transverse phase space.

In order to simulate an injected beam, 1024 particles were assigned with a Gaussian distribution in 6-D phase space and then tracked through the ring. The centroid position and the emittance of these particles were calculated after each turn.

Without a damping system, the phase space occupied by a bunch of particles will dilute, which can be seen in Figures 2 and 3. As a result of this, the emittance will increase and then be saturated as the beam dilutes completely, which is shown in Figure 4. In Figure 5 the centroid of the beam oscillates for thousand turns and then gradually comes back to the closed orbit due to the dilution.

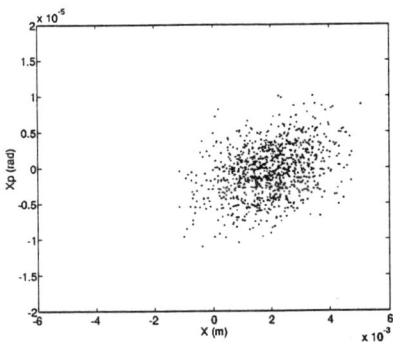

Fig. 2. Horizontal Phase Space at 0 Turn.

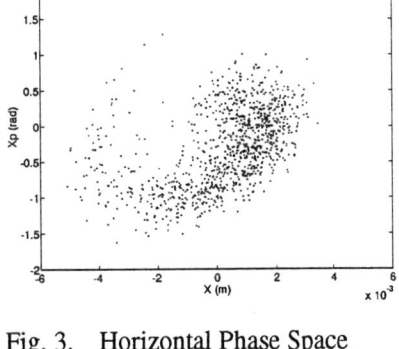

Fig. 3. Horizontal Phase Space at 2000 Turns.

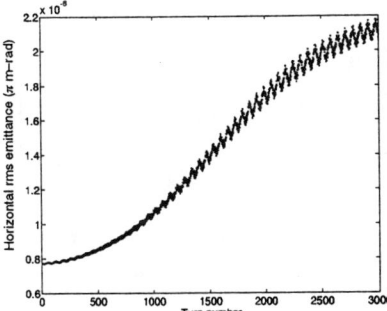

Fig. 4. Emittance Growth Due To Injection Offset.

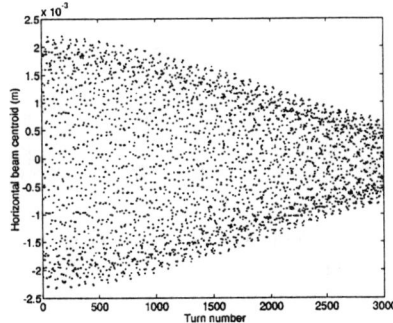

Fig. 5. Beam Centroid Moving Due To Dilution.

With the damping systems applied we have options to turn on one set (one BPM and one damper) only or both sets, or we may set the voltage across the damper to a different value. Figure 6 shows the emittance growths with different cases. The emittance slightly increases as only one set is used with 1-kV voltage. When the voltage is increased to 3 kV, or two sets are turned on with 1-kV voltage at each damper, the emittance keeps its average unchanged but with a small fluctuation. In Figures 7-10 one can see how the beam centroid is damped by differently operated damping systems.

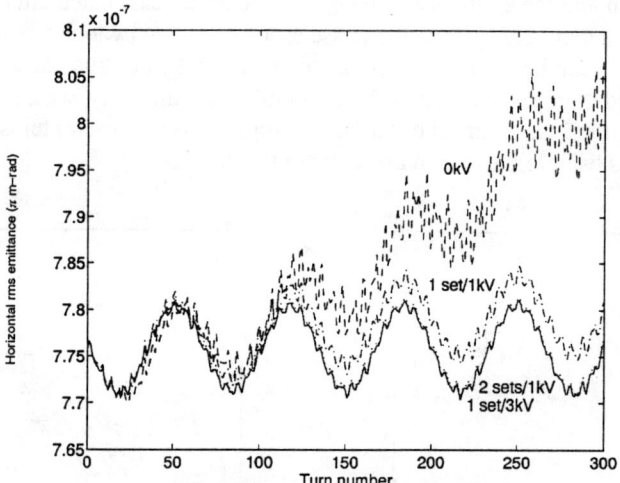

Fig. 6. Emittance Growth With Various Dampers.

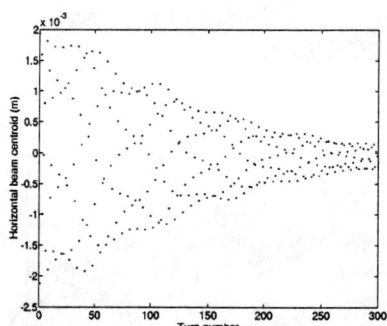

Fig. 7. One Set of Dampers With 1-kV Voltage.

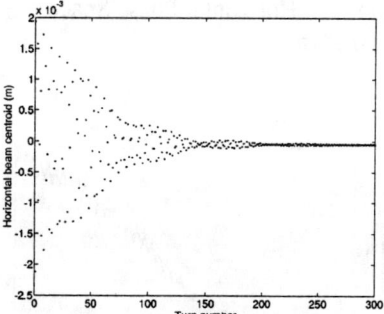

Fig. 8. Two Sets of Dampers With 1 kV on Each.

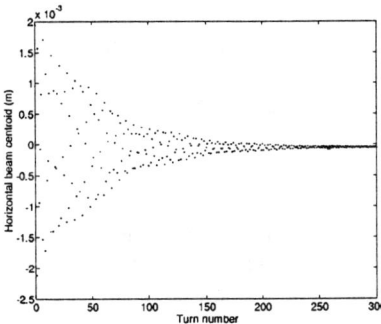
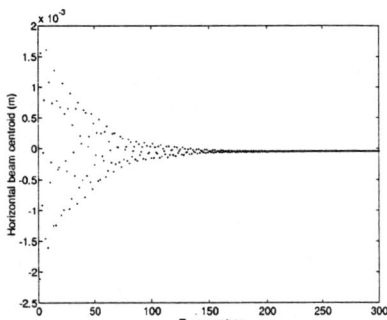

Fig. 9. One Set of Dampers With 3-kV Voltage.

Fig. 10. Two Sets of Dampers With 3 kV On Each.

The fluctuation of the emittance shown in Figure 6 is partly due to the lack of the particles being tracked. By increasing the number of the particles tracked from 1024 to 2048 and 4096 individually, the amplitude of the fluctuation is reduced, which can be seen in Figure 11. The residual oscillation is due to the difference resonance, which cannot be removed.

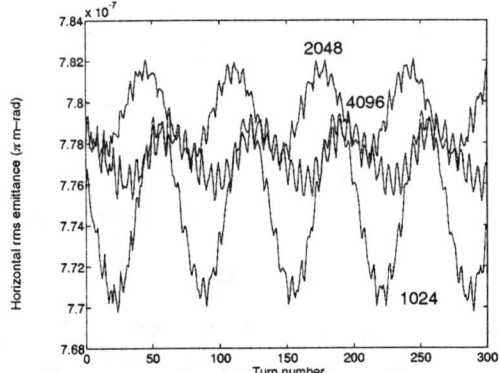

Fig. 11. Emittance Fluctuation With Various Particles.

RESONANCE CORRECTION

The working point of the HEB is (39.42, 38.41) in tune space, which is close to half integer resonance lines. Therefore, a half integer corrector is designed to correct the strong half integer resonance lines.

The corrector consists of eight quadrupole magnets, four of them for resonance line $2\nu_x = 79$ and other four for resonance line $2\nu_y = 77$, respectively. Each of the four magnets is assigned into two groups powered by two adjustable power supplies and those groups of magnets form two orthogonal vectors in phase space to generate harmonics. All of the two magnet sets in each group are wired in series and placed in position with 90° phase advance. Such correctors are able to generate any desired amplitude and phase for harmonics without tune shifts.

The Simpsons, 6-D dynamic simulation code, is used to test the correction scheme. The Simpsons is set up in a tune scan mode to examine the effects of half integer crossing.

In the simulation, the working point is linearly moving from (39.424, 38.414) to (39.543, 38.533) in 10 msec. At about 6 msec, the working point crosses the half integer; 1024 particles are used to calculate emittance. Emittance growth is occurring at half integer crossing while correct circuits are turned off (Figures 12, 13). A smooth half integer crossing is also shown in Figures 12 and 13 for a well-corrected HEB.

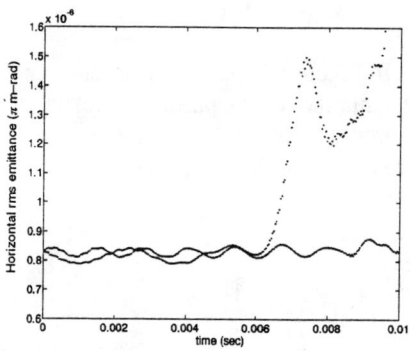

Fig. 12. Horizontal Emittance With and Without Correction.

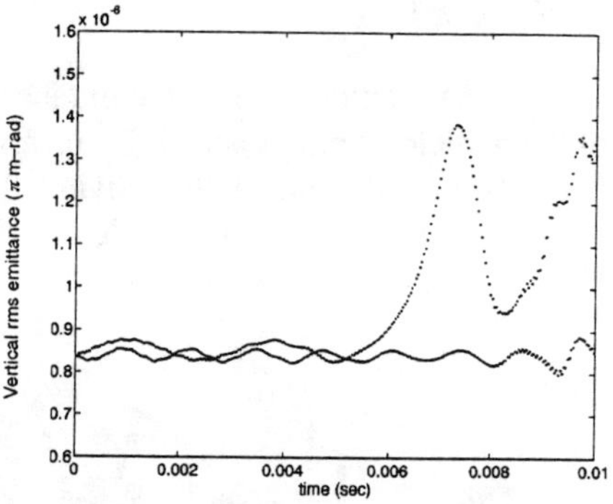

Fig. 13. Vertical Emittance With and Without Correction.

SUMMARY

To verify two design issues in the HEB of the SSC; injection damper and resonance correction systems, we modified the multiparticle tracking code Simpsons and examined the beam emittance growth under several conditions. The location of damper and BPM are chosen with practical constraints, not assuming any ideal situation. One damper and one BPM system turns out to be effective but required voltage for the damper is probably on the margin of the technical feasibility. Two dampers and two BPMs in a system do not require such high voltage but the improvement in terms of beam performance is not significant. The resonance correction for two half–integer resonance, $2v_x = 79$ and $2v_y = 79$, works well with four correctors for each.

ACKNOWLEDGEMENT

The authors would like to express their thanks to Samuel Penner for the useful discussion about injection damping system.

REFERENCE

1. S. Machida, *Computational Accelerator Physics Conference 1993.*

A STRUCTURED - MODULAR APPROACH TO SOFTWARE DESIGN FOR HIGH CURRENT BEAM DYNAMICS SIMULATION

Y.K.Batygin
Electrophysics Department
Moscow Engineering Physics Institute
115409, Moscow, Russia

ABSTRACT

Structured programming is a useful technique for a high quality program design. The method is based on breaking a general problem into independent subtasks which are then combined to achieve the necessary versions of the structure. A modular program BEAMPATH was developed for 2D and 3D particle-in-cell simulation of beam dynamics in a structure containing RF gaps, radio-frequency quadrupoles (RFQ), multipole lenses, waveguides, bending magnets and solenoids.

INTRODUCTION

Particle-in-cell simulation is a well-developed technique for beam dynamics simulation in linear accelerators. Up to now many programs have been developed to investigate the wide range of problems connected with linacs. A new generation of particle accelerators require constant improving of existing codes.

The structure programming technique[1-3] is a useful instrument to provide a flexible software. The method includes a hierarchical program design using program independent modules and a flexible combination of modules to provide a most effective version of the structure for every case of simulation. The structured-modular approach allows to resolve the contradiction between general goals of the project and the efficiency of computing of a separate problem.

NUMERICAL ALGORITHMS DESIGN

The structured -programming technique was originally developed for creating a large program projects containing more than 10^5 lines of source code, but the basic ideas of the method can be applied to a smaller projects as well.

The program design usually starts with control structures layout denoting the main steps of the program execution. The structured approach of program design is based on the following principles:
- "top-down" development of the program
- structured coding
- modular programming.

The "top-down" principle means that the program is created using hierarchical levels. The first stage is to define general goals of the program and to divide the future program into separate modules corresponding to logical completed subproblems. In its turn the separate modules can be divided by subtasks, etc. The typical problem is to find the most effective way to share the subtasks between modules.

In structured coding only a few control structures are used: IF THEN ELSE structure, DO iteration, DO WHILE iteration, DO UNTIL iteration. More complicated structures can consist only of permitted structures (see fig. 1).

Modular programming is a process of dividing the future program into logical completed subprograms. Module is a generalization of the SUBROUTINE concept (in FORTRAN) including the following features:
- independence form calling program
- the possibility to be used as a removable part of any program.

The usefulness of the module can be estimated as N/N_0 where N is a number of problems using this module and N_0 is a general number of problems which can be solved by the software.

One of the advantage of structured approach is a combination of programming and debugging of the program in the same program unit. Testing starts with the programming of the main core of the program while all modules are substituted by imitation statements. New testing data are derived by adding every new module. The most important top level of the program is tested many times from the very begining of the project.

APPLICATION TO PIC CODE

The particle-in-cell code usually consists of several standard steps: generation of the initial distribution of particles in multidimensional phase space, calculation of RF and focusing field, acting on the particles, calculation of the self field of the beam, integration of the equations of motion, output results treatment. Let us see the general solution of the problem as a sequence[4] :

$$B_1 - B_2 - B_3 - \ldots\ldots - B_k \qquad (1)$$

Suppose every standard step of simulation is supported by M standard modules. The result is a modular library:

$$\begin{matrix} B_{11} & B_{12} & B_{13} & \cdots\cdots & B_{1K} \\ \\ B_{21} & B_{22} & B_{23} & \cdots\cdots & B_{2K} \\ \\ \cdots & \cdots & \cdots & \cdots & \cdots \\ \\ B_{M1} & B_{M2} & B_{M3} & \cdots\cdots & B_{MK} \end{matrix} \qquad (2)$$

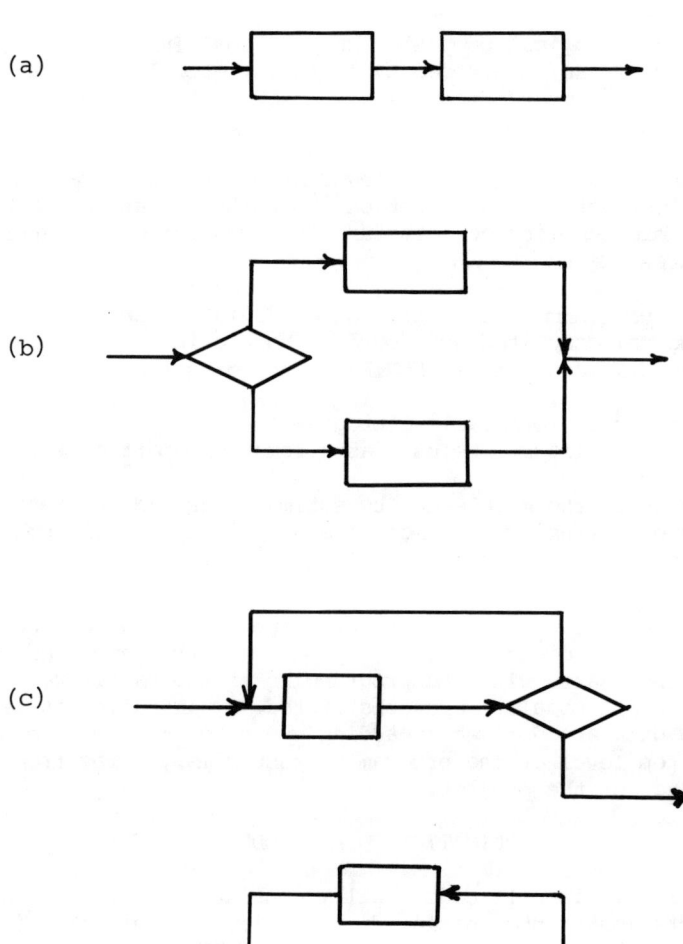

Fig. 1. Control structures used in structured-programming technique:
a) sequence of statements
b) IF THEN ELSE structure
c) DO UNTIL iteration
d) DO WHILE iteration.

where B_{ij} is the i-th module for the j-th standard step of simulation. From eq.(2) it follows that the total number of modules in the library is

$$P = M K \qquad (3)$$

and the number of versions of a system is

$$H = M^K \qquad (4)$$

It illustrates the property of modular structure: the possibility to construct a large number of structure versions using relatively small number of basic modules.

For systematic investigation of beam dynamics in linacs and transport systems the structured modular program library BEAMPATH[4] was developed. The program is used for particle-in-cell simulation of axial-symmetric, quadrupole-symmetric, ribbon and z-uniform beams in a channel containing the following elements: RF gaps, radio-frequency quadrupoles (RFQ), multipole lenses (quadrupoles, sextupoles, octupoles, etc.), waveguides, bending magnets, solenoids, user defined elements. The problem is self-consistent with respect to space charge of the beam. More details about the numerical technique and input/output data of the program are available from the reference manual.

PROGRAM ORGANIZATION

BEAMPATH is a set of basic computational FORTRAN subroutines connected with the global program. Every simulation step in the program can be substituted or modified without affecting the rest of the program. To provide such flexibility of the structure the following standard rules for subroutines are adopted.

1. Every program unit is aimed to solving the complete physical or mathematical problem.

2. Every subroutine can be used as a "black box" i.e. without detailed analysis of a structure of the program and at the same time one can easy understand the program.

3. Every subroutine starts with the short description of the problem containing the name, purpose, usage, input and output parameters, error messages, names of required subroutines and functions. List of parameters contains the error index which is typically assigned as 0 for normal exit and $\neq 0$ for program failures or wrong initial data.

4. All subroutines are free of input/output statements. Subroutines do not contain fixed maximum dimensions for the data arrays. The COMMON statement is not used in the library otherwise

it will result in strong connection between independent parts of the program.

5. Some techniques are used to improve readability of the program: using comments to every completed group of statements, the appropriate choice of the variable names, using empty lines, brackets and indentation in formulas.

NUMERICAL EXAMPLE

The particle distribution of an accelerated beam is usually approximated by Gaussian distribution. In many applications the uniform irradiation of large targets is required. The useful nonlinear optics method of increasing the uniformity of beam distribution was considered by Johnson[5], Meads[6], Sherrill et al.[7], Jason et al.[8] The method is based on nonlinear transverse velocity modulation of particles which force the peripheral particles to move faster to the axis then the inner beam particles. The particle distribution transformation can be described by equation[9]:

$$\rho(x) = \rho_0 \exp(-\frac{x_0^2}{2a^2}) (1 + c_2 + c_3 x_0 + c_4 x_0^2 + \ldots c_n x_0^{n-2})^{-1} \quad (5)$$

where the coefficients c_i, $i = 2, 3, \ldots n$ are connected with the parameters of multipole lenses with $2i$ poles. In fig. 2 the projections of beam distribution into real space (x-y) in the transport channel containing quadrupoles, 8 - poles and 12 - poles magnetic lenses are presented. It is clear that it is possible to improve the uniformity of most of beam particles by using the appropriate channel parameters.

CONCLUSIONS

The application of the structured-modular programming technique to software design for a particle-in-cell simulation of the beams in linacs is described. The method gives the possibility to produce complete, efficient and flexible programs. A modular program BEAMPATH was developed for 2D and 3D simulation of beam dynamics in linear accelerators and transport channels. The feature of the program is the possibility to extend the scope of the problems adding the new versions of standard step of simulation.

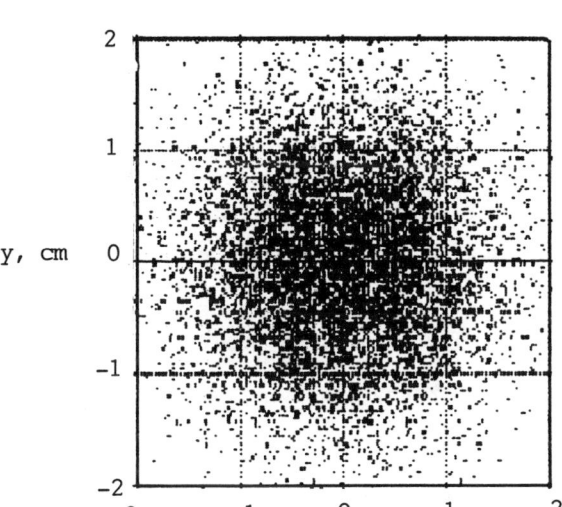

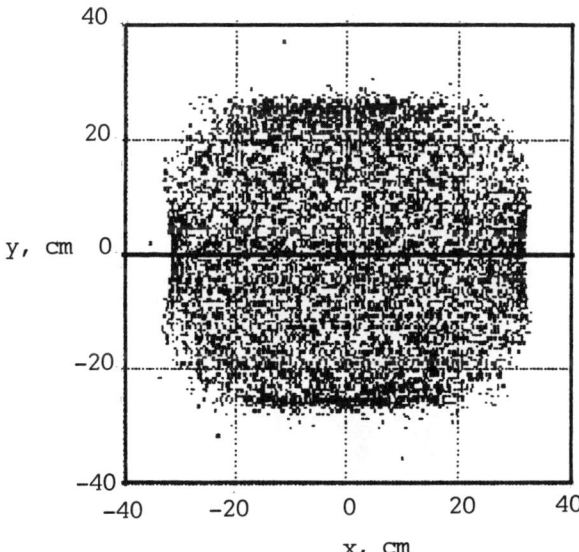

Fig. 2. Projections of computer simulation using code BEAMPATH into real space (x-y) for an initial (upper) and final (lower) beam distribution in a nonlinear optics channel.

REFERENCES

1. E.Yordon, Techniques of Program Structure and Design, Englewood Cliffs, N.J., Prentice-Hall, 1975.
2. D.Tassel, Program Style, Design, Efficiency, Debugging and Testing, Englewood Cliffs, N.J., Prentice-Hall, 1978.
3. W.P.Lysenko, Proc. 1984 Linear Accelerator Conference, Seeheim/Darmstadt, p.327.
4. Y.Batygin, Proc. 3rd Europ. Part. Accel. Conference, Berlin, 1992, p.822.
5. C.H.Johnson, Nucl. Instr. Meth., 127, 1975, p.163.
6. Ph.F.Meads, Jr., IEEE Trans. Nucl. Science, NS-30, No.4, 1983, p.2838.
7. B.Sherrill, J.Bailey, E.Kashy and C.Leakeas, Nucl. Instr. Meth., B40/41, 1989, p.1004.
8. A.Jason, B.Blind, E.Svaton, Proc. 1988 Linear Accelerator Conference, CEBAF - Report - 89-001, p.192.
9. Y.Batygin, Beam Intensity Redistribution in a Nonlinear Optics Channel, presented at the 12th Int. Conf. on the Application of Accecerators in Research & Industry, Texas, 1992; to be published in Nucl. Instr. and Meth.

3D SIMULATIONS OF AN ELECTROSTATIC QUADRUPOLE INJECTOR*

David P. Grote
University of California at Davis
Lawrence Livermore National Laboratory, Livermore CA 94550
Alex Friedman
Lawrence Livermore National Laboratory, Livermore CA 94550
Simon Yu
Lawrence Berkeley Laboratory, University of California
Berkeley California 94720

ABSTRACT

Analysis of the dynamics of a space charge dominated beam in a lattice of electrostatic focusing structures requires a full three-dimensional code that includes self-consistent space charge fields and the fields from the complex conductor shapes. The existing WARP3d[1-3] code, a particle simulation code which has been developed for HIF applications, contains machinery for handling particles in three-dimensional fields. A successive overrelaxation field solver with subgrid-scale placement of boundaries for rounded surfaces and four-fold symmetry has been added to the code. The electrostatic quadrupole (ESQ) injector for the ILSE[4] accelerator facility being planned at Lawrence Berkelay Laboratory is shown as an application. The issue of concern is possible emittance degradation because the focusing voltages are a significant fraction of the particles' energy and because there are significant nonlinear fields arising from the shapes of the quadrupole structures.

INTRODUCTION

WARP3d is a three-dimensional particle-in-cell code designed for heavy-ion fusion (HIF) applications. For many of these problems, it is necessary to model the transport of space charge dominated, heavy-ion beams over long distances. For many simulations of this type, the external focusing elements are modeled as having uniform, sharp edged fields.

Simulations of runs over shorter distances, in particular simulations of the ESQ injector, revealed the need for more accurate, self-consistent, focusing elements which include image, fringe, and other non-linear fields. That has been accomplished by placing the conductors that make up the focusing elements within the grid and including them in the field solution. Simple, widely spaced conductors were first implemented; they were dealt with via a capacity matrix method.[5,6]

More general conductors, such as the closely spaced, "inter-digital" quadrupole focusing conductors of the ESQ, are dealt with via a point successive overrelaxation (SOR) method.[7] This was done because, for the more complex

---

*This work was performed under the auspices of the U.S. D.O.E. by Lawrence Livermore National Laboratory under contract W-7405-ENG-48.

structures the capacity matrix becomes too big for solution in a reasonable amount of time. The matrix becomes bigger when the structures take up more space inside the field grid and require more points to be described. Also, the elements cannot be treated individually so all must be included in one very large matrix. Currently, the input to the code is set up to specifically handle the "inter-digital" structure. The internal representation however, can handle any conductor shape.

Use of four-fold symmetry of the potential ϕ puts the centerline of the beam along the edge of the grid. The normal derivative is set to zero on the symmetry boundaries. The particles still fill the whole system, but they all contribute their charge to, and get the fields from, the one simulated quadrant.

SOR METHOD AS IMPLEMENTED IN WARP3D

In three dimensions, Poison's equation is approximated by a seven point finite difference scheme. After rearrangement and inclusion of the relaxation parameter, the equation that is iterated for each grid point is:

$$\phi_{ijk} \Leftarrow \omega \Delta^2 (\rho_{ijk}/\epsilon_0 + \frac{\phi_{i+1jk}+\phi_{i-1jk}}{\Delta x^2} + \frac{\phi_{ij+1k}+\phi_{ij-1k}}{\Delta y^2} + \frac{\phi_{ijk+1}+\phi_{ijk-1}}{\Delta z^2}) + (1-\omega)\phi_{ijk}. \quad (1)$$

where the subscripts are grid cell indices, ω is the relaxation parameter,

$$\frac{1}{\Delta^2} = \frac{1}{\Delta x^2} + \frac{1}{\Delta y^2} + \frac{1}{\Delta z^2}, \quad (2)$$

and Δx, Δy, and Δz are the grid cell sizes in the various directions. The ordering that is used is even-odd, a straightforward three-dimensional generalization of the familiar red-black scheme.

In the code, for efficiency, this equation is applied to the entire potential array excepting the first and last axial planes. This is done by treating the 3d array as a 1d array and applying the equation to every other point. However, this equation is not correct at the boundaries, thus, before this equation is applied, the potential on the grid boundaries, and the conductor boundaries if needed, is saved. The potential on points internal to the conductors is enforced by setting them to the desired value before each iteration.

SUBGRID-SCALE PLACEMENT OF BOUNDARIES

Subgrid-scale placement of boundaries is handled by explicitly including the boundary location in the finite difference form of Poison's equation. In 1-D, the finite difference form of Poison's equation, $\nabla^2 \phi = -\rho/\epsilon_0$, is

$$\frac{\phi_{i+1} - 2\phi_i + \phi_{i-1}}{\Delta x^2} = -\frac{\rho_i}{\epsilon_0}. \quad (3)$$

If the edge of the conductor surface is between the grid points i and $i+1$, the potential at the edge would be interpolated as

$$\phi_{\text{edge}} = (1-\delta)\phi_i + \delta\phi_{i+1}. \quad (4)$$

Here, δ is the distance between the grid point i and the surface, divided by the grid cell size. The value of ϕ_{edge} is the known voltage on the conductor, and the value ϕ_{i+1} is a free parameter since it is inside the conductor. The previous equation can be rearranged to give a value to ϕ_{i+1}.

$$\phi_{i+1} = \frac{\phi_{\text{edge}} - (1-\delta)\phi_i}{\delta} \quad (5)$$

This is then put back into Poison's equation,
$$\frac{(\phi_{\text{edge}} - (1-\delta)\phi_i)/\delta - 2\phi_i + \phi_{i-1}}{\Delta x^2} = -\frac{\rho_i}{\epsilon_0}. \tag{6}$$
Note that in the limit $\delta \to 0$, ϕ_i approaches ϕ_{edge}. This is then rearranged to bring all of the terms in ϕ_i to the left hand side, and the SOR relaxation parameter, ω, is included to give the final equation:
$$\phi_i = \left(\frac{\phi_{i-1} + \phi_{\text{edge}}/\delta + \Delta x^2 \rho_i/\epsilon_0}{2 + \frac{1-\delta}{\delta}}\right)\omega + \phi_i(1-\omega). \tag{7}$$
This equation is used for the point ϕ_i during iteration. The original equation is still used everywhere else.

The idea extends easily to two and three dimensions. The three-dimensional version has been implemented in WARP3d. The seven point difference scheme is used and any points that are inside the conductor are handled as above. For the allowed classes of conductors, there are eight cases, with up to three, non-opposite, nearby points inside the conductor. The case when one point is in the conductor is handled as above but with the differences in the other two directions included. For example, if the point that is in the conductor is ϕ_{i+1jk}, the equation for ϕ_{ijk} would look like

$$\phi_{ijk} = \frac{\omega \Delta^2 (\rho_{ijk}/\epsilon_0 + \frac{\phi_{\text{edge}}/\delta_x + \phi_{i-1jk}}{\Delta x^2} + \frac{\phi_{ij+1k} + \phi_{ij-1k}}{\Delta y^2} + \frac{\phi_{ijk+1} + \phi_{ijk-1}}{\Delta z^2})}{1 + \frac{(1-\delta_x)\Delta^2}{\delta_x \Delta x^2}} \tag{8}$$
$$+ (1-\omega)\phi_{ijk}.$$

Here, δ_x is the unitless distance in the x direction between the grid point ijk and the surface.

When there are two points inside the conductor, the interpolations are done for each of the two points and then put back into the finite difference equation. If the points ϕ_{i+1jk} and ϕ_{ij+1k} are inside the conductor, for example, the equation for ϕ_{ijk} would be

$$\phi_{ijk} = \frac{\omega \Delta^2 (\rho_{ijk}/\epsilon_0 + \frac{\phi_{\text{edge}}/\delta_x + \phi_{i-1jk}}{\Delta x^2} + \frac{\phi_{\text{edge}}/\delta_x + \phi_{ij-1k}}{\Delta y^2} + \frac{\phi_{ijk+1} + \phi_{ijk-1}}{\Delta z^2})}{1 + \frac{(1-\delta_x)\Delta^2}{\delta_x \Delta x^2} + \frac{(1-\delta_y)\Delta^2}{\delta_y \Delta y^2}} \tag{9}$$
$$+ (1-\omega)\phi_{ijk}.$$

Here, δ_x and δ_y are the respective unitless distances in the x and y directions between the grid point ijk and the surface. The extension to three points within the conductor is similar.

In the code, to ease bookkeeping, there are two sets of arrays for the points near the conductor surfaces: one set for the points that have $i+j+k$ even, and one for the points that have $i+j+k$ odd. The values that are saved in the arrays are the location of the point, the magnitude of the distances from the surface in the x, y, and z directions, the sign of the direction to the point that is inside the conductor, and the previous value of the potential at the point.

In the simulations of the ESQ, without the subgrid-scale placement of boundaries, small grid cell sizes are needed to resolve the surfaces of the conductors. Without resolving the conductors, the focusing fields and therefore the beam envelope were inaccurate. The emittance growth, which is very sensitive to the shape of the envelope, was then also incorrect. Subgrid-scale placement of boundaries alleviated this problem since the focusing fields are correct, almost independent of the grid cell size. The restriction on the grid cell size is then greatly relaxed.

INJECTOR DESCRIPTION

The injector must produce a 2 megavolt, 0.8 Amp beam with low normalized transverse emittance, near 0.5 π-mm-mrad. The source emits into a diode region, or gap, with a potential drop; the beam then enters the first quadrupole, then passes through several more quadrupoles with common end plates, and exits the injector. These quadrupoles have an "inter-digital" structure; two rods placed along the axial direction are attached to the left plate and two to the right plate, with the sets of rods overlapping.

With the high currents needed from the injector, large focusing fields are needed to confine the beam. Since the particles have low energy in the injector, the focusing potentials are a large fraction of the particle energy. This results in what is called the "energy effect". It is significant here because, at low energy, the motion is not paraxial. Note that this affect is present in all electrostatic quadrupoles, but the focusing fields are generally much smaller then the particle energy so the resulting distortion is very small.

The focusing structure produces fields that are not purely quadrupolar. A variety of multipoles arise of the form $\phi_{nm} = V_{nm}(r/R)^n cos(m\theta)$, where R is the quadrupole aperture and $\phi = \sum_{n,m} \phi_{nm}$. The quadrupole field is the ϕ_{22} term and the octupole term is ϕ_{44}.

Several injector designs have been studied. Because of breakdown, the voltages that can be applied to the conductors are limited. The maximum voltage that can reasonably be applied across the diode is 1 Megavolt, but 0.5 MV would be better. Parameters are given in table 1, along with those of a scaled experiment that is being carried out at LBL, which also has been simulated.

DESCRIPTION OF COMPUTER RUNS

The beam is injected into the diode with constant current from a planer injection surface which is assumed to be an equipotential and is followed to the end of the injector. The runs are quasi time dependent; for efficiency, the time interval between field solutions is longer than the time step by which the particles are advanced. The transient behavior of the particles is lost and only steady-state is obtained. Generally, steady-state is reached shortly after one transit time across the injector.

The size of the field grid cells is limited by the accuracy of the self-fields. With the subgrid-scale placement of the boundaries, the fields from the conductors are accurate with fairly large grid cell sizes. The field grid must be large enough to include enough of the quadrupole conductors for good representation of the focusing fields. To meet these requirements, on the order of 50 grid cells are needed in each tranverse direction; the beam occupies about a third and at least half of the rod is inside the grid. The number of grid cells in the axial direction, between two hundred and six hundred, depends on the number of quadrupoles being simulated.

The injected particle distribution is a semi-Gaussian with circular transverse profile. Typically, the total number of particles used ranges from 70,000 to 300,000, with several hundred injected at each time step. The transit time of typical runs is between 300 and 1000 time steps, depending on the time step

	ILSE injector	Scaled experiment
Beam current (Amps)	0.8	0.008
Inital beam radius (cm)	6.0	10.0
Quad voltages[a] (kV)	220, 260 ,260	10.3, 12.2, 12.2
	255, 255, 250	11.97
Quad length[b] of first quad (cm)	31.0	7.15
of others (cm)	47.0	11.155
Rod length of first quad (cm)	25.5	5.89, 6.195[c]
of others (cm)	39.5	9.7
Apertures[d] (cm)	13, 12, 11	3.25, 3.0, 2.75
	10, 9, 8	2.5
Plate width (cm)	1.	0.6

[a]The ILSE injector has either four, five, or six quads depending on the diode voltage. The scaled injector has four quads.
[b]Distance between center of plates.
[c]The rods of the first quad are not the same length.
[d]Distance from beam center to the rod. Rod radii are 8/7 of the aperture.

Table 1: Parameters for ILSE injector and scaled injector experiment.

size. Another 100 to 200 steps are done to ensure convergence. The number of cycles, or field solves, is five to ten. The shortest runs with fewer time steps and particles used just under 3 minutes of CPU time on the NERSC Cray C-90 machine. The longer runs done for checks of numerical errors use 20 to 30 minutes.

ILSE INJECTOR

The first simulations that were done were of the ILSE injector with a 0.5 MV diode. The initial run showed a very large emittance growth. Figure 1 shows the normalized emittance along with the beam profiles and potential drop on axis. Note that the sharp rise in emittance in the y direction follows shortly after the widest part of the y envelope, and similarly in the x direction.

This emittance growth is not true emittance growth, but rather results from the twisting or "essing" of phase space, which can be undone. The $y-y'$ phase space at several points along z are shown in figure 2. The shape of the phase space, the essing, indicates action of fourth order (in potential) fields, ϕ_{4m}. The same runs as above were done, but cancelling out the various fourth order fields by applying their negative (calculated via a least squares fit). Cancelling the pure octupole field, ϕ_{44}, and the ϕ_{40} field had little effect on the emittance growth. Cancelling the ϕ_{42} field however, had a dramatic effect on the emittance, which was cut in half.

Both V_{40} and V_{44} change sign over one quadrupole element. Coupled with the fact that the energy changes little inside a quadrupole, their effect on the beam nearly vanishes with integration over one quadrupole. V_{42}, on the other hand, does not change sign over one quadrupole element. With integration over one quadrupole, its effect builds up. However, its sign does change from quadrupole to quadrupole, partially cancelling itself, leading to the reduction in emittance. The cancellation is not complete since the envelope and the

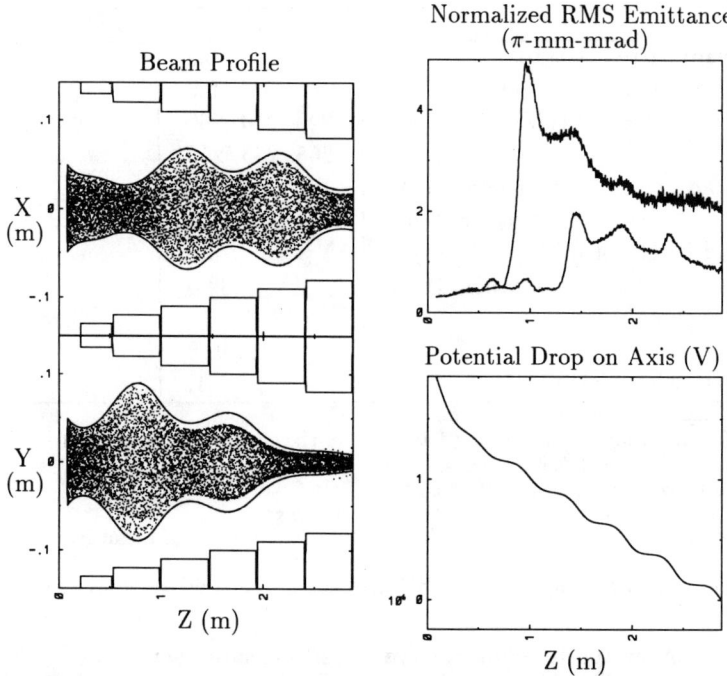

Figure 1: ILSE injector results. The top curve in the emittance corresponds to the y profile. Note that the peak in emittance follows shortly after the profile has reached its widest. The curves surrounding the beam are the RMS envelopes. The gap between the envelope and the particles is due to beam hollowing. The rectangles are the quadrupoles.

energy of the beam change.

The remaining emittance growth is caused by the energy effect. That effect was analyzed by expanding to fourth order in potential the equation of motion of a particle in the fields. The resulting expression for the perturbation in v_x is

$$\frac{dv_{x1}}{dz} = \frac{v_{z0}}{2} \left\{ \left[\frac{1}{8} \left(\frac{V_{22}}{T_0/e} \right)^2 + \left(\frac{V_{44}}{T_0/e} \right) \right] \left[-\frac{\partial}{\partial x} \left(\frac{r}{R} \right)^4 \cos 4\theta \right] + \right.$$
$$\left[\frac{1}{2} \left(\frac{V_{22}}{T_0/e} \right) \left(\frac{V_{20}}{T_0/e} \right) + \left(\frac{V_{42}}{T_0/e} \right) \right] \left[-\frac{\partial}{\partial x} \left(\frac{r}{R} \right)^4 \cos 2\theta \right] + \quad (10)$$
$$\left. \left[\frac{1}{8} \left(\frac{V_{22}}{T_0/e} \right)^2 + \frac{1}{4} \left(\frac{V_{20}}{T_0/e} \right)^2 + \left(\frac{V_{40}}{T_0/e} \right) \right] \left[-\frac{\partial}{\partial x} \left(\frac{r}{R} \right)^4 \right] \right\}$$

The terms in V_{2m} arise from the interaction of the terms $\frac{\partial \phi}{\partial x}$ and $\frac{1}{v_z}$ which represents the energy effect. The terms in V_{4m} arise from $\frac{\partial \phi}{\partial x}$, the quadrupole structure.

The energy effect can be cancelled like the fourth order fields by applying

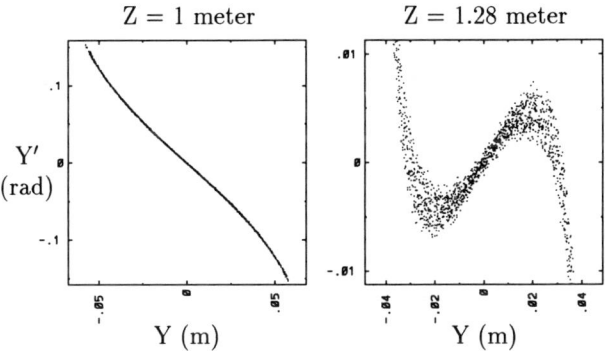

Figure 2: ILSE injector phase space at various locations along the injector.

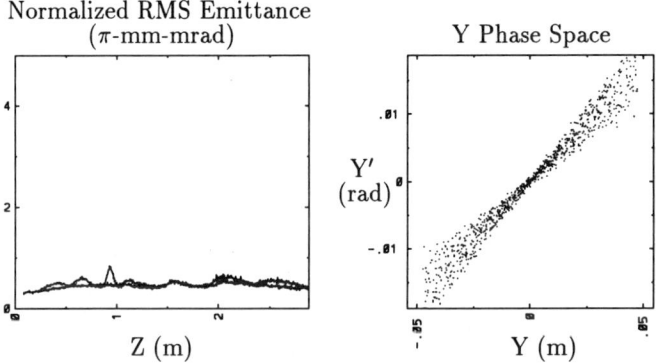

Figure 3: Injector results with both corrections. The emittance becomes flat and the essing of the phase space goes away. The "bow tie" shape in the phase space results from using particles from a finite range of z.

the appropiate fields to the beam. The correcting fields should, on integration over one quadrupole, cancel the energy effect. Over one quadrupole, the term $V_{22}V_{20}$ changes sign; its integral nearly vanishes. The term $(V_{20})^2$ is small compared to $(V_{22})^2$ and so can be ignored. This leaves only $(V_{22})^2$ as significant. The correcting fields that are then needed are V_{44}^{cor} and V_{40}^{cor}. The correcting fields reduce the emittance growth also by nearly a factor of two.

One more run was done where both corrections were done; the fourth order fields were cancelled and the energy effect was corrected. The plot of the emittance, figure 3, shows that there is no emittance growth and all of the essing in the phase space has disappeared.

The injector was also simulated with two other diode energies, 1 MV and 0.75 MV. As expected, the growth in both cases is much less than with the 0.5 MV diode. Application of the correcting fields reduced the emittance as in the 0.5 MV case.

SELF-CONSISTENT CORRECTIONS

The goal of these simulations is to reduce the emittance from the injector to an acceptable level. We have shown the cause of the growth in the above section and have demonstrated that it can fairly easily be removed. We now wish to remove the emittance by self-consistent means. This is done by changing the quadrupole's structure. Two changes that were made to the quadrupoles where to move all of the rods in or out to change the aperture, and to move two rods in and two out, maintaining the same average aperture. The voltages on the quadrupoles were also varied.

Moving all of the rods in or out and changing the voltages changed the focusing fields and therefore the envelope. Many runs were done, varying the apertures to reduce the maximum size of the envelope. The optimum apertures for the injector with the 1 MV diode were found. The emittance was reduced to nearly the source emittance, so no other adjustments are needed.

The injector with the 0.5 MV diode showed considerable emittance growth. By moving all of the rods in or out and changing the voltages, the emitance was reduced somewhat, though a growth of nearly a factor of two remained. To reduce the emittance further, the rods needed to be moved in and out, affecting only the fourth order fields. A sizeable number of runs have been made, doing both parameter scans and ad-hoc searches for minimal emittance. The emittance has to date been reduced down to one and a half times the inherent source emittance.

CONCLUSIONS

The methods described herein have allowed use of much larger grid cell sizes while still maintaining accurate fields from the conductors. This has made possible parameter scans and searches for minimal emittance.

The simulations of the ESQ injector have shown that much of the emittance growth is caused by fourth order fields. The speed of the code has allowed many runs to be done to study the system and to minimize the emittance. We find these three-dimensional simulations to be an effective design tool.

REFERENCES

[1] Alex Friedman, *et. al.*, these proceedings

[2] Debra Callahan, *et. al.*, these proceedings

[3] A. Friedman, D. P. Grote, and I. Haber, "Three-dimensional particle simulation of heavy-ion fusion beams," *Phys. Fluids B* **4**, 2203 (1992).

[4] T. Fessenden, R. Bangerter, D. Berners, J. Chew, S. Eylon, A. Faltens, W. Fawley, C. Fong, M. Fong, K. Hahn, E. Henestroza, D. Judd, E. Lee, C. Lionberger, S. Mukherjee, C. Peters, C. Pike, G. Raymond, L. Reginato, H. Rutkowski, P. Seidl, L. Smith, D. Vanecek, S. Yu, F. Deadrick, A. Friedman, L. Griffith, D. Hewett, M. Newton, and H. Shay, "ILSE, The Next Step toward a Heavy Ion Induction Accelerator for Inertial Fusion Energy," *Proc. 14th Int. Conf. on Plasma Physics and Controlled Nuclear Fusion Research*, IAEA, Wurzburg, Germany, Sept. 30-Oct. 7, 1992.

[5] Charles K. Birdsall and A. Bruce Langdon, *Plasma Physics via Computer Simulation*, McGraw-Hill Book Company, 1985

[6] D. P. Grote, A. Friedman, and I Haber, "3d Simulations of Axially Confined Heavy Ion Beams in Round and Square Pipes" *Particle Accelerators*, **37-38**, 141, (1992)

[7] William H. Press et. al., *Numerical Recipes: The Art of Scientific Computing* Cambridge University Press, 1986

LONGITUDINAL BEAM DYNAMICS FOR HEAVY ION FUSION USING WARPRZ*

Debra A. Callahan, A. Bruce Langdon, Alex Friedman
Lawrence Livermore National Laboratory, Livermore CA 94550

Irving Haber
Naval Research Laboratory, Washington DC 20375

ABSTRACT

WARPrz is a 2.5 dimensional, cylindrically symmetric, electrostatic, particle-in-cell code. It is part of the WARP family of codes which has been developed to study heavy ion fusion driver issues. WARPrz is being used to study the longitudinal dynamics of heavy ion beams including a longitudinal instability that is driven by the impedance of the LINAC accelerating modules. This instability is of concern because it can enhance longitudinal momentum spread; chromatic abhoration in the lens system restricts the amount of momentum spread allowed in the beam in the final focusing system. The impedance of the modules is modeled by a continuum of resistors and capacitors in parallel in WARPrz. We discuss simulations of this instability including the effect of finite temperature and reflection of perturbations off the beam ends. We also discuss intermittency of axial confining fields ("ears" fields) as a seed for this instability.

INTRODUCTION

Because of the large costs involved in building a full scale heavy ion fusion (HIF) accelerator, much effort has gone into simulating the physics of space charge dominated beams needed for HIF. These simulations have been coupled with experiments when possible. The WARP family of codes[1] has been developed to study driver issues. The code is made up of five major physics packages: a 3d particle-in-cell code in Cartesian geometry, a 3d electrostatic field solver, a cylindrically symmetric (r, z) particle-in-cell code, an r, z electrostatic field solver, and an envelope code. This family of codes is being used to study a variety of heavy ion fusion issues.[2,3]

The longitudinal dynamics of the beams in the induction linear accelerator for heavy ion fusion have been of concern for some time. Since we must focus the beams onto the target at the end of the driver, we cannot tolerate a large spread in longitudinal momentum. The longitudinal wall impedance instability can cause small errors launched at the beam head to grow to an unacceptable size. This instability has the same mechanism used in "resistive wall" amplifiers with the impedance coming from the accelerating modules.

In order to model the longitudinal dynamics of these beams, the r, z portion of the WARP code was developed. This code is a 2.5 dimensional, cylindrically symmetric particle-in-cell code. Field solution calculations are done in a window that moves with the beam. In this window, the fields are very close to purely electrostatic since the force due to magnetic fields is down by $(u/c)^2$ from the force due to the electric fields where u is the velocity in the beam frame. The beam frame velocity for a heavy ion fusion driver is much less than 1% of the speed of light.

---

*This work was preformed under the auspices of the U.S. D.O.E. by Lawrence Livermore National Laboratory under contract W-7405-ENG-48, and by the Naval Research Laboratory under contracts DE-AI05-92ER54177 and DE-AI05-83ER40112.

LONGITUDINAL INSTABILITY

The longitudinal wall impedance instability can be seen via a simple fluid model. If we consider an incompressible beam with radius a traveling down a pipe of radius r_{wall}, 1-d linear cold fluid theory shows that two waves will develop– a forward traveling wave and a backward traveling wave. These waves propagate with a phase velocity in the beam frame given by

$$v_{\text{phase}} = \sqrt{\frac{Ze\lambda g}{4\pi\epsilon_0 m}} \qquad (1)$$

where Z is the charge, λ is the line charge density (with units of charge/length), m is the mass, $g = \ln(r_{\text{wall}}^2/a^2)$. Adding a wall composed of a continuum of resistors and capacitors in parallel to this calculation results in the forward traveling wave decaying while the backward traveling wave grows. This growth is largest when the perturbation wavelength is large compared with the pipe radius. In a heavy ion fusion driver, the impedance that drives this instability comes from the induction acceleration modules.

In order to study the longitudinal instability, we added a model for a wall with a continuum of resistors and capacitors in parallel[4,5] to WARPrz. This approximation for the induction modules contains the relevant physics and also corresponds well with much of the analytic work being done. We calculate the resistive wall contribution to the electric field using the Poisson solve at the boundary. This is smoother and more physical than using the explicit beam current.

Our model assumes a continuity equation for the wall surface charge, σ which has units of charge/area.

$$\frac{\partial \sigma}{\partial t} + \frac{\partial K_z}{\partial z} = 0 \qquad (2)$$

where K_z is the surface current. We use Ohm's law for a resistor and capacitor in parallel

$$2\pi r_{\text{wall}} \eta K_z = E_z + \eta C \frac{dE_z}{dt} \qquad (3)$$

where η is the resistance per unit length and ηC is the "RC" time. Combining these equations and using the definition of the electrostatic potential gives

$$\frac{\partial \sigma}{\partial t} = \frac{1}{2\pi r_{\text{wall}}} \frac{\partial}{\partial z}\left[\frac{1}{\eta}\left(\frac{\partial \phi}{\partial z} + \eta C \frac{d}{dt}\frac{\partial \phi}{\partial z}\right)\right] \qquad (4)$$

For a constant η and ηC, we can Fourier transform with respect to z and use a finite difference approximation to the time derivatives. This then becomes the boundary condition for our Poisson solve routine. We also assume the radial electric field is zero outside the pipe wall. Since our computation mesh is a window moving with the beam, we need to advect the surface charge density and electrostatic potential backward when the mesh moves forward. This method has been tested in WARPrz by running in the linear, cold beam regime and the growth rates agree with the 1-d fluid theory to within 1.5%.

Our first set of realistic simulations included only a resistive component to the impedance. When resistance is turned on, we need to add an external electric field that keeps the beam from losing all of its energy to the wall. The field that we added was simply

$$E_{z,\text{external}}(z) = \eta I_0(z), \qquad (5)$$

where η is the resistance per unit length (ohms/meter) and $I_0(z)$ is the current profile at time zero. Since the axial electric field does not vary much across the beam radius, we applied this same external field at all radii.

Beam velocity	1/3 c
Beam current	3000 Amps
Pulse length	10 meters
Beam radius/Pipe radius	.4
Perpendicular temperature	10 keV
Parallel temperature	10 keV

Table 1: Beam parameters near the end of a HIF accelerator

We found that slight mismatches in the external field caused perturbations to be launched from the beam head and tail. Table 1 shows the parameters for the simulation; these are similar to those proposed for the end of the acceleration section of a HIF driver. These parameters were chosen because the growth rate for the longitudinal instability is the largest in this regime. The wall resistance was 100 Ω/m with no capacitance. Figure 1 shows the perturbation that has been launched from the beam head (right hand side of the plots) in the electrostatic potential on axis vs z and the z-v_z phase space at time 4.9 μs. At this time, the perturbation can be seen as a shallow dip the potential about 2.5 meters from the beam head. Figure 2 shows the same plots at time 17 μs. Growth is readily seen. The growth rate measured in this case is about 15% smaller than the cold beam theory predicts.

Electrostatic Potential on Axis vs z v_z vs z

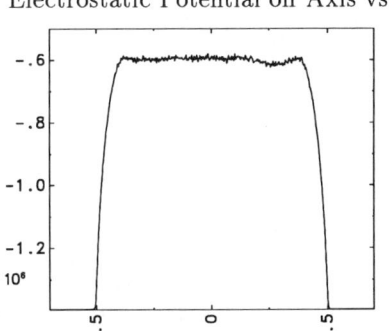 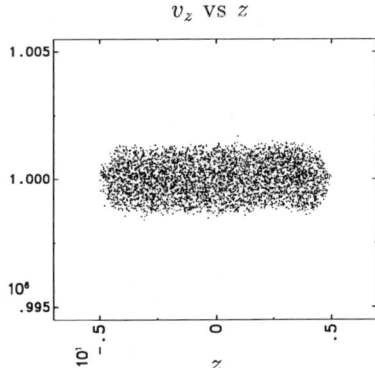

Figure 1: A perturbation is launched from the beam head

Continuing this simulation, we see that the perturbation reaches the beam tail and reflects, and begins moving towards the head. After reflection, the perturbation is seen to decay as is predicted by the cold beam theory. We do notice, however, that during reflection the perturbation width narrows. We believe this is a nonlinear effect since it has been observed to be more severe in larger perturbations.

The cold beam theory shows that capacitance has a partially stabilizing effect on this instability. When capacitance is added to the system, the external field

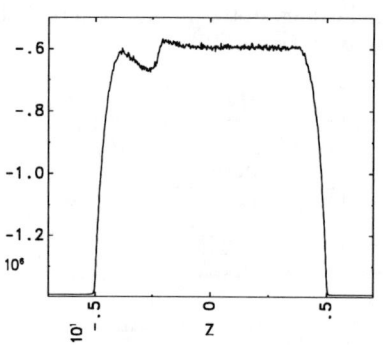 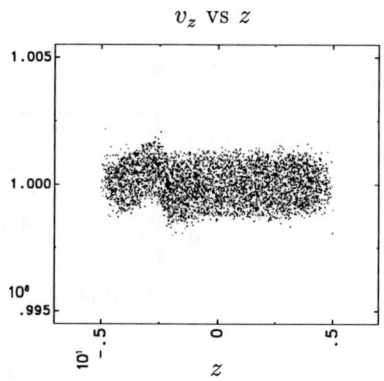

Figure 2: The perturbation grows as it travels from beam head to tail

added to keep the beam from losing its energy to the wall becomes:

$$\eta C \frac{dE_{z,\text{external}}}{dt} + E_{z,\text{external}} = \eta I_0(z) \qquad (6)$$

We found when we did simulations with $\eta = 100\ \Omega/m$ and $\eta C = 2.0 \times 10^{-8}$ seconds that very long wavelength perturbations were introduced because the equilibrium was not perfect. These perturbations did not undergo much growth, however. Video animation of the simulation showed that the perturbation sloshed back and forth from beam head to tail with little change in size.

INTERMITTENTLY APPLIED "EAR" FIELDS

To get a realistic look at the effects of the longitudinal instability, we need to look for sources of errors that will generate perturbations on the beam in an actual device. One source of such perturbations is the intermittently applied axial confining electric fields ("ear" fields). In the simulations in the previous section, we applied ear fields at each time step and these fields were designed to keep the beam from expanding or contracting. In an experiment, these fields will be applied at fixed locations along the accelerator and the beam will expand and contract between applications. The application of these fields can cause a train of perturbations to be launched from the beam head and these perturbations will be amplified by the longitudinal instability.

Each application of the ears has a period (τ_{ears}) made up of the following steps:

1. Let the beam expand from time 0 to $\tau_{\text{off}}/2$.

2. At time $\tau_{\text{off}}/2$, turn the ears on for a time τ_{on} and compress the beam.

3. At time $\tau_{\text{off}}/2 + \tau_{\text{on}}$, turn the ears off and let the beam expand until time $\tau_{\text{ears}} = \tau_{\text{on}} + \tau_{\text{off}}$. At this time, the beam should be back to its original length.

The waveform for the electric ears field was obtained by applying a force proportional to the integral over time of the force that was present during the free expansion.

$$E_z(z) = F(z)/q = -\hat{F} \int_0^{\tau_{\text{on}}} \frac{m}{q} \frac{du(z,t)}{dt} dt = -\hat{F} \frac{m}{q} u(z, \tau_{\text{on}}) \qquad (7)$$

where $u(z,t)$ is the average particle velocity as a function of z in the beam frame, m is the mass, and q is the charge. In evaluating the integral, we have assumed that the average velocity in the beam frame is zero at time zero. The proportionality constant $\hat{F}$ should be approximately equal to $2/\tau_{\text{on}}$ where the factor of two comes in because we need to not only reverse the expansion but also contract the beam so that it will expand back to its original size after the second $\tau_{\text{off}}/2$ of expansion time. In practice, we start with this value and then adjust $\hat{F}$ until we get good results from one ears application.

Since we are not accelerating the beam, we chose accelerator parameters like those near the end of the acceleration section of the HIF driver because the growth rate for the longitudinal instability is the largest there. In fact, the beam will not spend much time in this section of the machine, so we are looking at a worst-case scenario. The simulation parameters are summarized in table 1. This simulation had a wall resistance of 100 ohms/meter with no capacitance. The schedule for the ears was $\tau_{\text{off}} = 1$ μs (or 100 meters at $v_{\text{beam}} = c/3$) and $\tau_{\text{on}} = .1$ μs.

The first simulation in this series was an attempt to apply the intermittent ears with as little damage to the beam as possible ("perfect" ears). Figure 3 shows the electrostatic potential on axis vs axial position after 14 and 22 applications of the ears. By 14 applications, the perturbation launched at the beam head (the right hand side of the beam in the plots) has traveled about 7 meters and can be seen as a broad, shallow dip in the potential. After 22 applications, the beam looks similar. In this run, we do not see a train of perturbations being launched off the beam head and growing towards the tail. The beam remains very smooth.

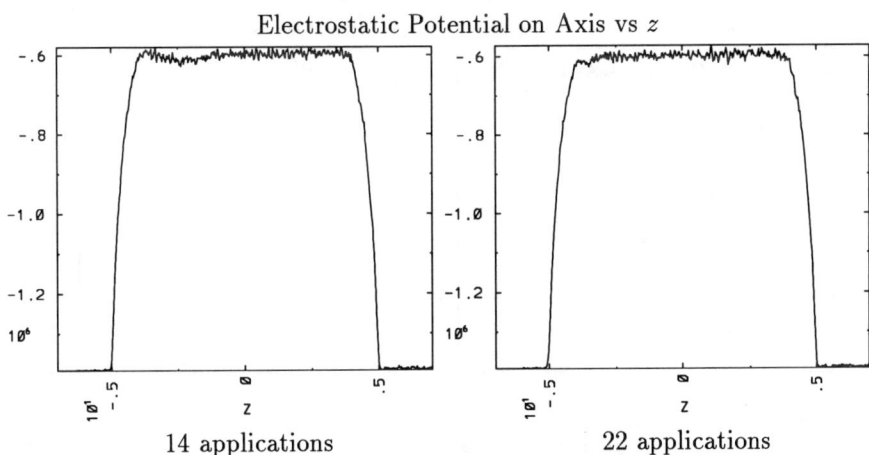

Figure 3: Applying "perfect" ears

The electric field that we are applying in this case is fairly large. Figure 4a shows the applied ear field as a function of z. The maximum electric field needed to contain the beam is on the order of 17.5 MV/m. This is about three times the size of the field that we would like to apply in the experiment and tells us that at the end of the accelerator, we will need to apply the ears more frequently than every 100 meters. It is comforting, however, to see that the beam remains in good shape even if we apply the ears as infrequently as 100 meters.

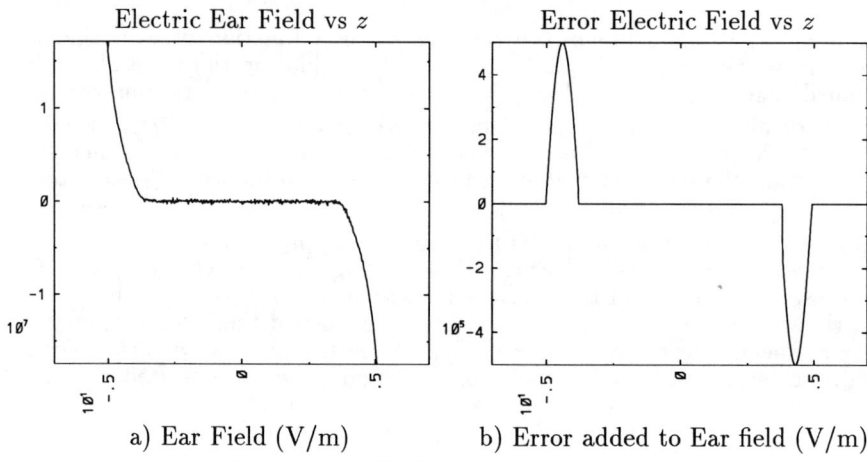

a) Ear Field (V/m)   b) Error added to Ear field (V/m)

Figure 4: Ear field and error introduced

Our first attempt at studying imperfect ears was to add a "bump" to the ear fields used in the case of perfect ears. Figure 4b shows the error added to the ears. The bump had the algebraic form of one half the period of a sine wave and magnitude of 5% of the local ear field. This bump has the effect of making the ears too large. We believed that by applying an error in the same direction every time, we would see a worst case because there was no way for the errors to cancel one another out. We found that this was not the case.

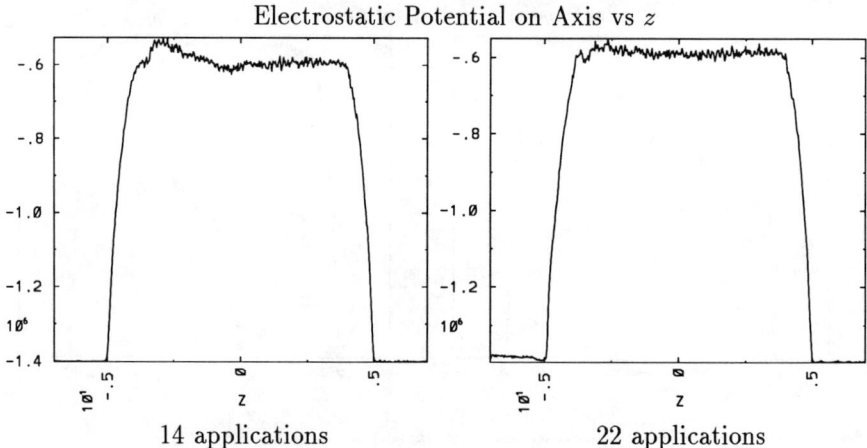

14 applications   22 applications

Figure 5: Applying ears that are always too large

When we applied these ears, we found that the first few applications caused a perturbation to be launched from the beam head. Figure 5 shows this perturbation after 14 applications as it nears the beam tail after undergoing amplification by the longitudinal instability. Interestingly, if we look after 22 applications, we do not see a trail of perturbations coming off the beam head as we expect. In fact, the beam has adjusted itself to the error in the ears.

This phenomenon has also been seen in experiments done by A. Faltens. These experiments were designed to test longitudinal bunch control in the beam tail on the SBTE at LBL. In this experiments, no attempt was made to match the waveform of the applied ear fields to the beam profile. Instead, fields of the form $[1 - \exp(-at)]$ were applied. In the experiment, mismatches in the ear fields caused waves to be launched from the beam tail in the early pulsers, but at later times the beam reached a new steady state configuration.

Once we found that the beam could adjust to a systematic error in the ears, we tried alternating the error. In this simulation, we applied the same size and shape error as in the previous run, but we alternated the sign of the error with each application. This amounted to applying ears that were too large on one application followed by ears that were too small on the next application.

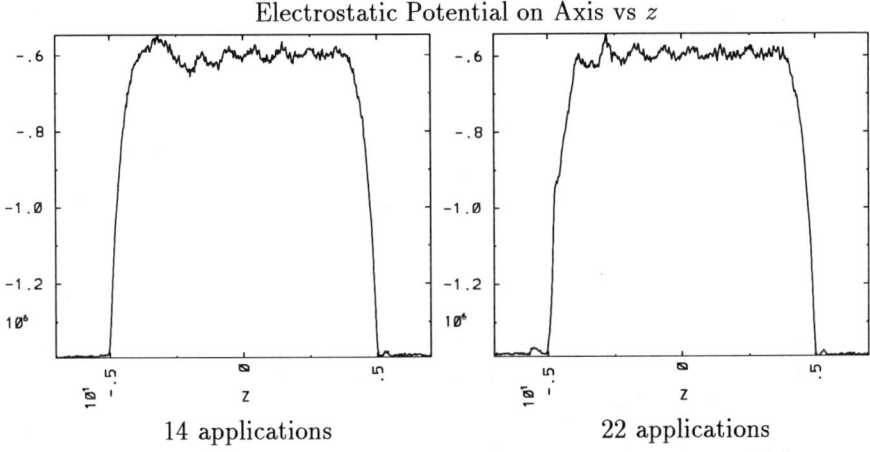

Figure 6: Applying ears that alternate too large and too small

As figure 6 shows, we generated the expected train of perturbations coming off the beam head and moving towards the tail. We expect a similar train of perturbations coming from the beam tail, but these perturbations decay as they travel towards the beam head and are not seen. The wavelength of the perturbations is measured to be about 1 meter. We calculate $2\tau_{ears}v_{phase} = 1.07$ meters. This would indicate the cycle is:

1. hit the beam too hard and cause a enhancement region

2. this region moves away from the beam head at the wave velocity (v_{phase}) in the time between ear applications

3. hit the beam too softly and cause a depletion region

4. this region moves away from the beam head at the wave velocity (v_{phase}) in the time between ear applications

5. repeat

Such a cycle implies a perturbation wavelength of $2\tau_{ears}v_{phase}$ which is the observed wavelength.

In the future, we will study further aspects of longitudinal beam dynamics for HIF with WARPrz. This will include extending the work presented here to cooler beams and studying other sources of beam perturbations such as errors in the accelerating fields. Since our beam travels at speeds much slower than the speed of light, it is possible to detect beam perturbations, send a signal down the accelerator, and correct the errors downstream. Such "feed-forward" schemes have been demonstrated in 1-d simulations[6] and we would like to further explore these ideas with WARPrz. We are also interested in how the beam emittance varies with longitudinal position in the beam ends in an equilibrium state.

CONCLUSIONS

We have modeled the longitudinal dynamics of space charge dominated heavy ion beams for heavy ion fusion. We have included a model of the impedance due to the induction acceleration modules as a continuum of resistors and capacitors in parallel in the WARPrz code. Using this code, we have modeled the longitudinal instability and seen waves launched from the beam head grow as they travel toward the tail, reflect off the beam tail, and decay as they travel forward. We have seen the partially stabilizing effect of the capacitive component of the impedance. We have looked at intermittently applying axial confining fields ("ear fields") as a seed for this instability. We have seen that the beam can adjust to systematic errors in the ear fields; however more random errors, such as errors which alternate sign, generate a train of perturbations on the beam.

REFERENCES

[1] A. Friedman, D. A. Callahan, D. P. Grote, A. B. Langdon, and I. Haber, Proc. of the Conference on Computer Codes and the Linear Accelerator Community, Los Alamos, NM, Jan 1990.

[2] A. Friedman, et. al. this meeting.

[3] D. P. Grote, et. al. this meeting.

[4] D. A. Callahan, A. B. Langdon, A. Friedman, D. P. Grote, and I. Haber, *Particle Accelerators*, **37-8**, 97, (1992).

[5] D. A. Callahan, A. B. Langdon, A. Friedman, D. P. Grote, and I. Haber, Proc. of the International Conference on the Numerical Simulations of Plasmas, Annapolis, MD, 1991.

[6] K. Hahn, Bul. Am. Phys. Soc., Seattle, WA, 1992.

HIGH POWER TRANSIT-TIME RF AMPLIFIER FOR SUPER ACCELERATORS[*]

Moe Joseph Arman [**]
PHILLIPS LABORATORY, Kirtland AFB, New Mexico, 87117
Donald J. Sullivan
MISSION RESEARCH CORPORATION Albuquerque, New Mexco.87106

ABSTRACT

When a charged particle beam passes axially through a cylindrical cavity, under proper conditions, it can excite and ultimately grow certain characteristic electromagnetic modes of the cavity. The effect, called transit-time effect, has been known since the 1930's. Transit-time effects can be used as an oscillator to generate microwave or as an amplifier for coherent amplification. Based on transit-time effects we have developed an rf amplifier with output power in the giga-watt range and an efficiency approaching 35%. With no foils to erode, the repetition rate can exceed 1000/sec. Furthermore, as a result of strong rf feedback on the structure, the emission into the diode gap is gated, leading to potentially very large efficiencies. In this device, called *acceletron*, a dc pulse is launched into a coaxial structure that is the diode, the buncher, and the oscillator all in one. Electrons are emitted from the surface of the inner conductor and as they accelerate axially towards the anode they interact with a selected characteristic mode of the structure, on the average losing kinetic energy to the mode. This process continues until the RF fields are strong enough to modify the transit time, leading to saturation. The source offers significant improvements in power, repetition rate, size, and efficiency. As is the case with all transit-time oscillators, the signal is stable and monochromatic. The device may be used as a buncher, as an oscillator, or as an amplifier. The RF may be extracted axially through a coaxial line, or radially through a radial waveguide perpendicular to the axis. The results of the numerical simulations will be presented.

1. INTRODUCTION

Large gradient rf fields have been used extensively in linear accelerators for accelerating charged particles. The rf source most frequently used for charge acceleration is the klystron. Though highly efficient and stable, the klystron is limited in power due to the high impedance intrinsic to this device. Furthermore, since the beam bunching in the klystron is the result of a spread in particle velocities, the drift region required for reasonable bunching becomes larger as the beam momentum approaches relativistic values. Also to maximize the beam cavity coupling in the klystron, a strong magnetic field is required to confine the beam very close to the cavity gaps involved.

As an alternate source of rf energy for particle acceleration, we have investigated the possibility of using a low-impedance (>100Ω) low energy (>500KV) electron beam in conjunction with transit-time instabilities to devise a narrow-band high efficiency rf amplifier with the output of approximately 1GW operating in the X-band of the rf spectrum.

Transit-time effect, like many other rf-plasma instabilities, is an interaction between a directed flow of charged particles and the characteristic modes of a conducting structure. As a flow of charged particles crosses a high-Q cavity supporting rf radiation, it is possible for the particles to be spending more time moving against the rf fields than the time they spend moving along the fields, and as such, on the average losing energy to the rf fields. The rf field could be seeded in the cavity via an input mechanism (amplifier) or grown out of noise by the beam (oscillator). In either case the rf grows as a result of this interaction while the particles lose some of the kinetic energy they gain in crossing the diode gap. This process, leading to a *coherent* radiation from a charged particle beam, is a powerful means of growing intense rf radiation.

* Work supported by the Department of Energy, SBIR Program.
** National Research Council Associate, on leave from Mission Research Corporation, Albuquerque, NM

Early efforts in using the transit-time effect for growing rf radiation failed because the growth rate was too slow and the saturation levels were too low. Efforts in increasing the beam power to boost the growth rate resulted in disruptive space charge effects and the idea of transit-time based rf radiation was soon abandoned. With the advent of intense *relativistic* electron beams, the problems associated with space charge effects could be partially circumvented but the pinching effects of an intense charged beam going through an anode foil, as is the case for non-accelerated (drifting) beam transit-time oscillator, would inhibit the maximum efficiency possible with these devices.

What is presented here is a new device called acceletron (*accele*rated elec*tron*) where we use the transit time effects of an accelerated flow of electrons to produce high power rf radiation. Figure 1 is a schematic drawing of the acceletron. It consists of a cylindrical solid cathode coaxial with a cylindrical anode. Electrons are emitted from the cathode face and as they move axially towards the anode, they interact with the TM_{020} mode of the cylindrical cavity which also defines the diode gap. If the transit time of the electrons is close to the period T of the mode selected, they will exchange energy with the rf fields of the that mode.

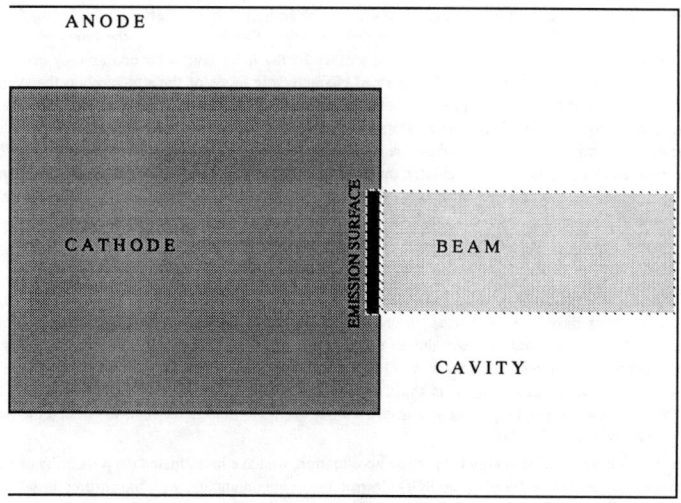

Figure 1. Axial section of the acceletron. The dc pulse is launched from the left. The beam is confined by a weak .5KG magnetic field. Drawing not to scale.

Since this device combines the diode, the drift region, and the buncher all into one cavity, there is no need for the anode foil thus eliminating the pinch. Furthermore with no foils to erode, the repetition rate can be very high (~1000Hz). The accelerated motion of the electrons reduces the space charge effects significantly thus allowing for a large growth rate and high saturation levels even for low energy(~300KV) electron beams. Furthermore, as a result of an interaction between the diode and the rf being grown in the diode, the emission becomes gradually modulated at the frequency of the radiated rf, further enhancing the radiation process. This feature is unique to the acceletron and may be used to further the on-going works in gated-emission research at several institutes.

In part 2. we present a brief description of the theory of transit-time oscillators for the TM_{0n0} modes. Part 3. is the results of sophisticated particle-in cell code simulations confirming the viability of this device as a source of high power rf radiation. Concluding remarks are presented in part 4.

2. ANALYTICAL CONSIDERATIONS

The non-relativistic theory of the transit time oscillator (TTO) for some simple cases such as the monotron and the Pierce diode was developed in the 1940's[1-3]. The theory is based on the calculation of the energy exchanged between an electron beam and the rf fields of the characteristic modes of the device. Depending on the physical dimensions of the cavity and the beam parameters involved, the energy exchanged can be either positive or negative leading to a growth or a damping of the rf fields. When conditions are right, the rf field growth continues until the non-linear effects become strong enough to modify the beam parameters, leading to saturation.

The non-accelerated(drifting) motion of *relativistic* beams in the rf fields has been investigated rather extensively for certain modes in the last few years[4-10]. The theories developed are all based on the assumption that the interaction is linear (small signal gain), and there is a well-defined direction associated with the flow of the beam electrons.

The energy exchange between the rf fields and the beam in a monotron, which is the non-accelerated version of the TM_{0n0} acceletron, is best described by

$$\frac{d\varepsilon}{dt} = \frac{1}{2} \text{Re} \int_{beam} (\underline{J}^* \cdot \underline{E}) \, dV \qquad (1)$$

where $\underline{J}^*$ is the complex conjugate of the current modulation produced by the gap electron field $\underline{E}$ and the integral is over the volume of the beam in the gap.

The reciprocal quality factor $1/Q_b$ is related to the energy exchange (eq. 1) according to

$$\frac{1}{Q_b} = \frac{-1}{\omega W} \frac{d\varepsilon}{dt} \qquad (2)$$

where ω is the angular frequency of the oscillation and W is the energy stored in the cavity. Using the Vlasov equation to find the perturbation on the electron momentum distribution, and the equivalent circuit model techniques to model the beam-gap interaction, it can be shown that Q_b is given by

$$\frac{1}{Q_b} = \frac{\nu}{2(\beta\gamma)^3} \frac{d^3|E|^2}{h_g\left(1 + \frac{\varepsilon_{ext}}{\varepsilon_g}\right)} f(\theta) \qquad (3)$$

where β and γ are the relativistic factors, ν is beam current, d is the gap length, E is the oscillating electric field, ε_{ext} is the energy stored in the resonator outside to the gap, ε_g is the energy stored in the gap, and $f(\theta)$ is given by

$$f(\theta) = \frac{2(1-\cos\theta) - \theta\sin\theta}{\theta^3} \qquad (4)$$

where $\theta = \omega\tau_0 = \omega d/v_0$ is the transit angle and τ_0 is the undisturbed transit time.

The equations 2-4 describe the dependence of the gain $1/Q_b$ on the transit angle θ. As the transit angle is increased, the function $f(\theta)$ displays oscillatory behavior. Where $f(\theta)$ is negative the rf can grow and where it is positive the rf is damped.

The accelerated version of the monotron (the acceletron) displays similar behavior with respect to the transit angle θ, although the analysis is considerably more complicated. Rather than drifting across the cavity, the particles are emitted from the cathode face on one side of the cavity by an applied dc electric field. The rf fields interact with the particles as they accelerate from nearly zero velocity to a final velocity roughly equal to the velocity corresponding to the applied potential. The particle trajectories and energy exchange with the rf fields depend not only on the applied voltage, but also on the details of the applied field profile. For instance, in a constant field, the particles spend less time at low energy than if the accelerating field is small near the cathode and ramps up across the cathode-anode gap. Furthermore, the rf field *modulates* the current emitted from the cathode which further complicates the analysis of the particle-field energy transfer.

The analytical study of the acceletron is not pursued any further here. Instead we have used an analytic-based particle trajectory code (ACCEL) capable of modeling the acceletron accurately. Independently, fully relativistic, fully electromagnetic Particle-In-Cell (PIC) codes ISIS and MAGIC have been used to model the device. The results of the analytic-based code and the PIC codes are in full agreement, although the analytic-based results are nor presented here.

3. SIMULATION RESULTS

Because of the axially symmetric structure of the acceletron a two-dimensional PIC code is all we need to model the device. Figure 2 shows a single-cavity acceletron with the particle trajectories as modeled by the code MAGIC. This figure shows the device at an early time t=5ns, long before the instability saturates. Because of the axial symmetry, only the upper half of the device is modeled. The input, a 750KV dc pulse is launched from the left. The beam is confined by a 0.5KG axial magnetic field. The beam is hollow near the axis to avoid

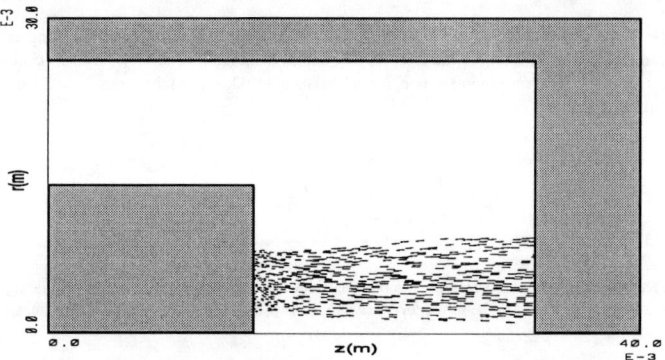

Figure 2. Single-cavity acceletron with particle trajectories at t=5ns.

enhanced numerical noise associated with the axis in cylindrical geometry. Figure 3. shows particle trajectories at t=20ns, after the instability has saturated. The particles are bunched near the cathode

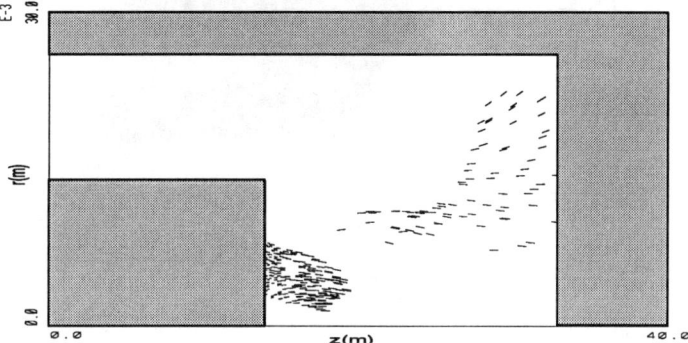

Figure 3. Particle trajectories after saturation showing strong bunching of the beam.

and the pattern typical of *gated emission* is evident in the distribution of the particles.

The cavity dimensions, the dc pulse voltage, and the diode impedance were chosen for the growth of the TM_{020} mode. This mode was chosen to maximize the overlap between a solid beam on the axis and the large values of the J_0 Bessel function for maximum coupling. This choice furthermore allows us to increase the gap on the input line without decreasing the cavity's quality factor Q below threshold. A large gap on the input line is desired to reduce chances of vacuum breakdown due to large dc electric fields.

Figure 4 is a perspective plot of the axial component of the electric field for the single-cavity acceletron. This plot shows the variations of E_z as a function of r, confirming the presence of the TM_{020} mode in the cavity. Figure 5a is a plot of the beam current density near the cathode as a function of time. It shows strong

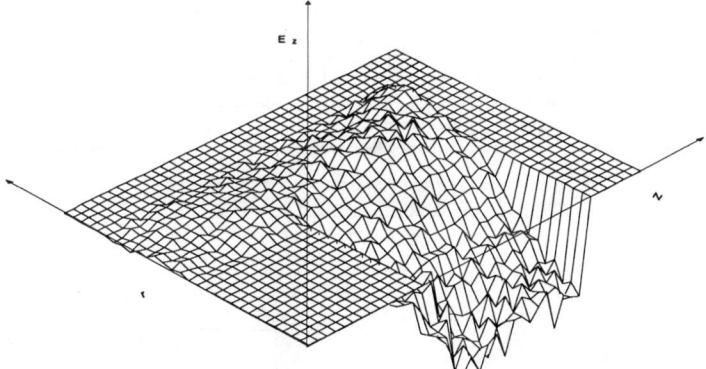

Figure 4. Perspective plot of the axial electric field indicates the presence of the TM_{020} mode.

bunching of the beam after saturation. Figure 5b is the fourier transform of 5a. It shows a monochromatic rf signal with no indication of mode competition in the cavity.

224 High Power Transit-Time RF Amplifier

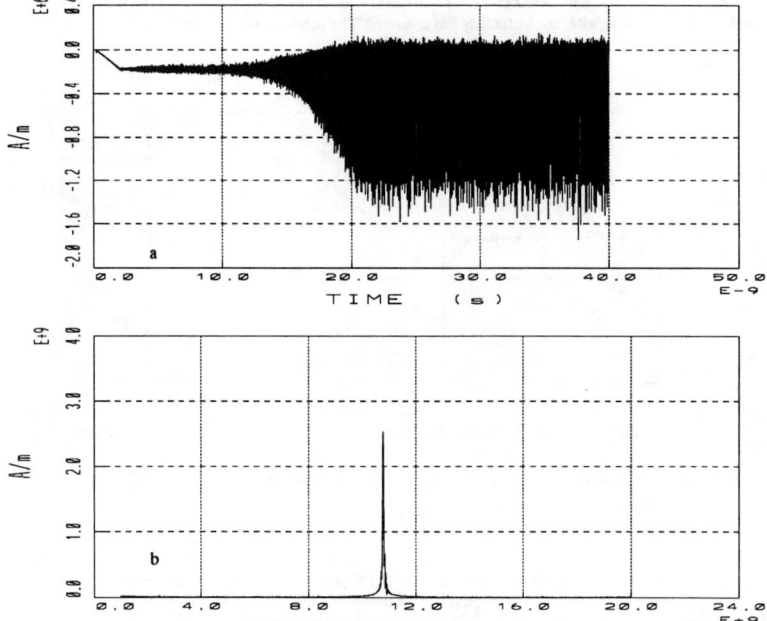

Figure 5 a. Time plot of the beam current density showing strong bunching in the beam.
b. Fourier transform of a. showing a monochromatic rf signal at ~ 11GHz.

In a single cavity acceletron, the beam reaches the end of the cavity with a portion of its kinetic energy intact. By letting the beam *drift* into a second cavity with properly chosen dimensions, a second transit-time instability can be grown. If the second instability is grown at a frequency matching that in the first cavity and the two cavities are electromagnetically connected, the effects will enhance each other leading to larger growth rates and saturation levels. Figure 6 is a plot of a two-cavity acceletron with the beam after saturation. The dimensions of the second cavity are chosen to grow a drifting-beam-initiated TM_{010} mode of matching frequency. The second cavity not only increases the growth rate, it also provides a convenient means of

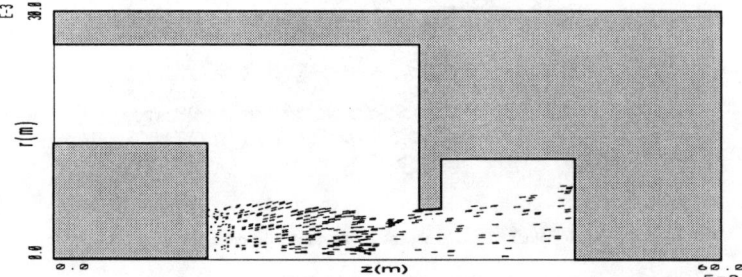

Figure 6. Axial section of the two-cavity acceletron with particle trajectories.

loading the system without reducing the Q of the first cavity. Figure 7 is a time plot of the beam current density for the two-cavity acceletron, showing a larger growth rate compared to that of the single cavity acceletron, figure 5.

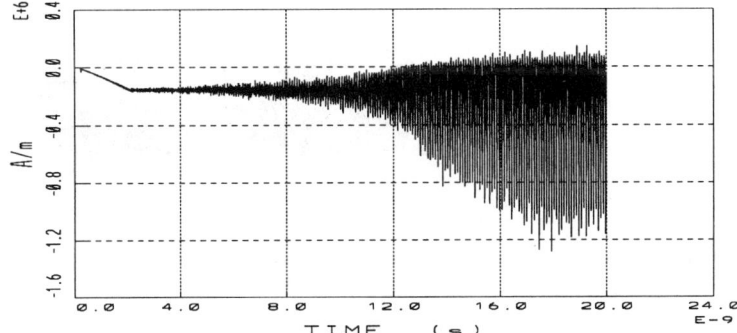

Figure 7. Time plot of the beam current density in the first cavity of a two-cavity acceletron. The instability saturates faster in a two-cavity system.

The loading of the acceletron can be achieved either radially through rectangular waveguides, or axially through a coaxial line along the axial direction. Figure 8 shows a plot of a preliminary design aimed at loading the device axially through the second cavity. This loading is unoptimized and is only used to study the loading characteristics of the system.

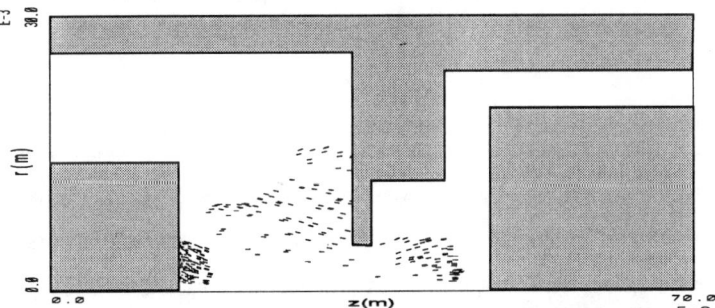

Figure 8. Axial section of the two-cavity acceletron with axial loading.

Figure 9 is a plot of the extracted power as a function of time. It shows a peak power of 3.0GW for an unoptimized load. Without Bragg reflectors on the input line, a comparable amount of power is lost through the input line. With optimized Bragg reflectors, the loss through the input line can be practically eliminated, thus doubling the output through the load. The rms efficiency with the reflectors is estimated at 32%. Based on related studies by the authors, switching to radial extraction will further increase the efficiency. The fourier transform of figure 9 (not shown) displays a strong monochromatic peak at twice the frequency of the rf, indicating very little or no dc power flow through the output line.

The acceletron *oscillator* can be turned into an acceletron *amplifier* by slightly detuning the device off the resonant conditions and launching a (low amplitude) rf signal of resonant frequency into the system. The most suitable parameters for detuning the device are the voltage of the dc pulse, the length, or the radius of the main cavity. Numerical simulations to verify the amplifier operation of the acceletron have been carried out and they confirm the operation as expected. The amplification gain has not been studied yet and the results of the amplification operation are not presented here.

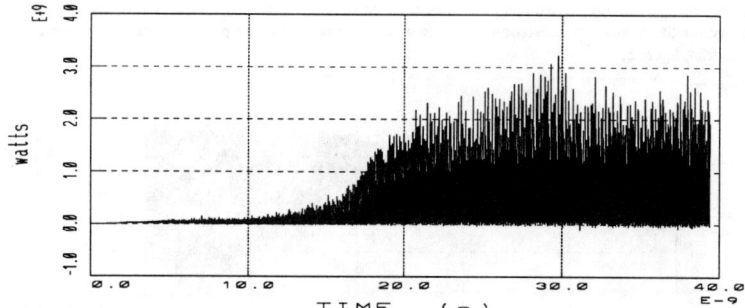

Figure 9. Time plot of the unoptimized extracted power through the coaxial line. No Bragg reflectors.

4. CONCLUDING REMARKS

The acceletron is an accelerated transit-time oscillator (or amplifier) capable of producing several gigawatts of rf power per single device. Because the electrons undergo acceleration while interacting with the rf, the space charge effects are suppressed, allowing the voltage to be dropped while the current is kept high. Furthermore, since there are no anode foils to erode, the device can be pulsed at a very high rate (~ 1000Hz). Also the confining magnetic field is low enough (0.5KG) that a permanent magnet can be used. The acceletron is compact, light, and easy to maintain. The efficiency, though not optimized yet, is potentially high because of very strong bunching in the beam. We plan to further study the loading, both radial and axial, and to optimize it for maximum efficiency.

REFERENCES

1. D. Marcuse, *"Principles of Quantum Electronics"*, pp. 125-144, Academic Press, New York (1980).
2. J. Marcum, *"Interchange of Energy between an Electron Beam and an Oscillating Electric Field"*, J. Appl. Phys. 17, 4 (1946).
3. J. R. Pierce, *"Limiting Stable Current in Electron Beams in the Presence of Ions"*, J. Appl. Phys. 15, 721 (1944).
4. B. B. Godfrey, *"Oscillatory Nonlinear Electron Flow in a Pierce Diode"*, Phys. Fluids 30, 1553 (1987)
5. B. B. Godfrey, D. J. Sullivan, M. J. Arman, T. C. Genoni, and J. E. Walsh, *"Linear Theory of Transvertron Microwave Sources"*, SPIE's OE/LAS 89 (1988).
6. M. J. Arman, D. J. Sullivan, B. B. Godfrey, R. E. Clark, and J. E. Walsh, *"Analytical Study and Numerical Simulation of High Power Vircator and Transvertron Microwave Sources"*, MRC/ABQ-R- 1029, Mission Research Corporation, Albuquerque (1988).
7. M. A. Mostrom, R. M. Clark, M. J. Arman, M. M. Campbell, B. B. Godfrey, D. J. Sullivan, and J. E. Walsh, *"Proton-Beam Transit-Time Oscillator for Producing High-Power Microwaves"*, MRC/ABQ- R-1030, Mission Research Corporation, Albuquerque (1988).
8. D. J. Sullivan, J. E. Walsh, M. J. Arman, and B. B. Godfrey, *"Simulation of Transvertron High Power Microwave Sources"*, SPIE's OE/LAS 89 (1988).
9. D. J. Sullivan, J. E. Walsh, M. J. Arman, and B. B. Godfrey, *"Small-Signal Gain and Numerical Simulation of Transvertron High Power Microwave Sources"*, Proc. IEEE 1989 Particle Accelerator Conf., (Chicago, March1989).
10. M. J. Arman, *"New Low Impedance HPM Source"*, Proc. Sixth National Conference on High Power Microwave Technology, (San Antonio, August 1992).

SELF-CONSISTENT SOLUTION OF MULTISPECIES CHARGED PARTICLE OPTICS PROBLEMS USING A HYBRID FLUID/RAYTRACE/POISSON-BOLTZMANN ALGORITHM*

J. L. Orthel
G. H. Gillespie Associates, Inc., P. O. Box 2961, Del Mar, CA 92014

ABSTRACT

The simulation of negative ion source physics should treat several charged species interacting with applied electric and magnetic fields, and also self fields resulting from ionic and electronic space charge, with complex boundary conditions imposed. Standard numerical codes exist to model these effects in positive ion source design problems. However in negative ion sources additional effects are important: ion, electron and neutral species can interact through collisions, charge exchange, stripping, and viscous effects, among others. A new multispecies hybrid fluid/ orbit integration/Poisson-Boltzmann algorithm is described, and applied to the LBL high brightness charge exchange (CXS) concept.

INTRODUCTION

The motivation of this work was to evaluate LBL High Brightness Charge Exchange Negative Ion Source concept. The original CXS concept[1] relies on the large cross section for electron transfer between a beam of fast (200-300 eV) high emittance negative ions produced conventionally, and a stream of cold neutrals. The dense stream of cold neutrals is produced in an RF dissociator, and the stream is directed through the perpendicular beam of fast negatives, where the neutrals pick up an electron and become cold negatives. If electric fields can be kept minimal in the production and extraction region, then no other processes are known that would increase emittance of the cold negative ions produced. The electric field E results from the charge density of the various interacting plasma species, electrons, and neutral along with the influence of conductor geometry. It should ideally be as uniform as possible and should be kept below E < 0.1 V/cm for the H⁻ temperature to be less than about 0.1 eV[2].

The rest of this paper discusses the relevant physics equations, numerical algorithms used to solve them, the computer implementation, and a preliminary application to a simplified CXS problem.

PHYSICAL MODEL AND NUMERICAL ALGORITHMS

The electric field is determined by the Poisson/Boltzmann equation and the boundary conditions imposed by conductor geometry[4],

---

* Supported in part by Nichols Research Corporation and the U.S. Army Strategic Defense Command under subcontract/contract numbers NRC-LE-91-0002 / DASG60-90-C-0132.

$$\nabla^2 \phi = (\sum_{ions}\rho_i + \rho_{eo}e^{-e\phi/T_e})/\varepsilon_o \qquad [1]$$

The source term is the sum of all the charged ions, plus the nonlinear Boltzmann term for the electron density. This equation is solved, given the source terms from the ions, by a standard succesive-over-relaxation (SOR) method, with a nonlinear Newton-Raphson[7] limit finder technique for the Boltzmann term on a finite difference grid. The source terms (excluding the term for the electrons) result from the particles whose motion is tracked and the charge density accumulated in the overall charge density grid matrix,

$$\rho_{ions} = \sum_{ions}\rho_i \; .$$

The force on the ions is from the electric field and viscosity due to collisions with other species (including neutrals)

$$\frac{d^2\mathbf{x}_i}{dt^2} = \frac{q_i}{m_i}(\nabla\phi + \mathbf{v}_i \times \mathbf{B}) + \sum_j v_{ij}n_j(\mathbf{v}_i - \mathbf{v}_j) \qquad [2]$$

The fluid description of the particles uses the coupled equations for number density n and the velocity $\mathbf{v}$,

$$\nabla \bullet (n_i\mathbf{v}_i) = \sum_{j,k}\chi_{ijk}n_j n_k \qquad [3]$$

(continuity) and the force equation,

$$\mathbf{v}_i \bullet \nabla\mathbf{v}_i = \frac{q_i}{m_i}\nabla\phi + \sum_j v_{ij}n_j(\mathbf{v}_i - \mathbf{v}_j) \qquad [4]$$

The coupled system of highly nonlinear equations [1]+[2], or alternatively equations [1]+[3]+[4] must be solved. The solution for ϕ can be accomplished if the positions of all the particles is known, (or alternatively n and $\mathbf{v}$). On the other hand, to solve [2] (or [3]+[4]) requires that the solution of [1] be accomplished.

For single and multiple species charged particle optics problems with no interactions other than with the smoothly varying externally applied and self fields, such as electron guns and positive ion sources, the raytrace/SOR method has been very successful[5-9]. This method iterates the solution of [1] with the solution of [2] using at each step values from previous iteration steps (see Fig. 1 below). Usually convergence is achieved after a modest number of steps. Orbits are integrated from starting surfaces or regions that emit particles, and a current is assigned to the ray based on the current density of the emitting surface and the number of rays starting from the surface. This current is used to deposit space charge on the numerical grid to be used as for the source term in the field calculation.

With multiple species, the numerical solution of the equations of motion for any one species requires knowledge of the solution for the other species, in addition to the electric field. Therefore the system of particle equations [2] (or [3]+[4]) must itself be iterated, in addition to the major iteration with the field solution (see Fig. 2).

The solution of [2] and [3] by a time-relaxation scheme using the $\partial n/\partial t$ and $\partial \mathbf{v}/\partial t$ terms added (and pseudo- diffusion/viscosity terms for stability), was tried and found to be inaccurate and computationally intensive (both time and memory). Furthermore, with boundary conditions from complex geometry and particle injection scenarios, it proved to be unstable.

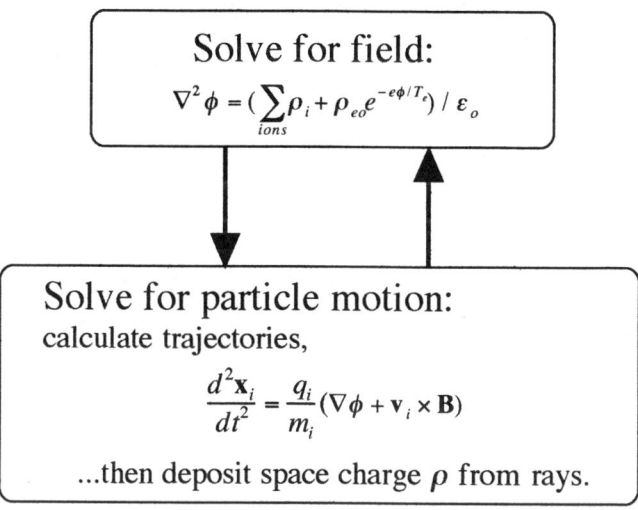

Fig. 1. Typical flow schematic of the standard iteration scheme[5-9] for the self-consistent solution of charged particle optics problems where the only inter-species interaction is through the electric field $\nabla\phi$ from the total space charge (and sometimes magnetic field **B** from total current). The fields ϕ (and **B** if used) are computed in the first step, and the rays (trajectories) computed and the resulting space charge field tabulated in the second step. The process iterated until convergence is reached.

A new algorithm, which may be called the "hybrid fluid/ray trace" method was formulated. At each iteration step, equation [1] is solved as usual, and the particle orbits are then integrated using the electric field based on the deposited charge from the previous iteration. The pressure and viscosity forces from the other particles are calculated from the right side of [4] using the n_j and $\mathbf{v}_j$ deposited from previous iteration steps, just as the total charge density was deposited in previous steps. As the ray traverses the numerical grid, it may experience current gain or loss due to the right side of [3], which is calculated from the fluid fields n_j and $\mathbf{v}_j$ deposited in previous iterations.

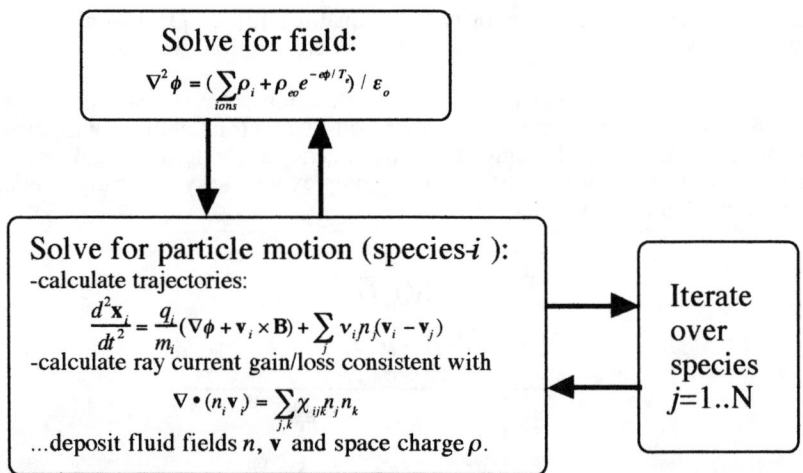

Fig. 2. Hybrid Fluid/Raytrace algorithm flow diagram. Inter-species interaction is not only through the electric field from the total space charge, but through fluid fields (n_j and $\mathbf{v}_j$) coupling via the fluid coupling coefficients v_{ij} and χ_{ij}.

SIMPLIFIED CXS PROBLEM

Fig. 3 shows a schematic of the initial model of the LBL charge exchange source concept. The four species used were positive ions, cold drifting negative ions, fast drifting negative ions, and cold neutrals. In a more refined scheme to improve symmetry and smoothness, the fast neutrals would be injected in counter streaming fashion from both top and bottom in Fig. 3.

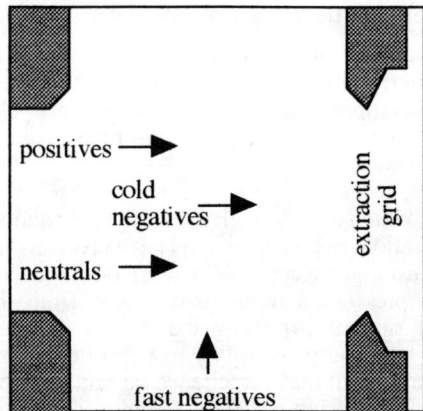

Fig. 3. Simplified LBL CXS problem. Slab geometry (x horizontal, y vertical) is used with dimensions 2 cm x 2 cm. A more optimized design would inject fast negatives from the top as well as bottom.

The neutrals can be integrated just as the ions using $q = 0$: this is shown in Fig. 4 and 5 below. However since they are expected to fan out from the surface that they are emitted from on the left, ray density becomes low on the right of the simulation box. The resulting low statistics in this area requires many additional rays. To get around this, a simple minded analytical approximation for the neutral flow was used instead. The model postulates that all neutrals are emitted from a vertical aperture surface on the left, and subsequently move in an expanding laminar flow.

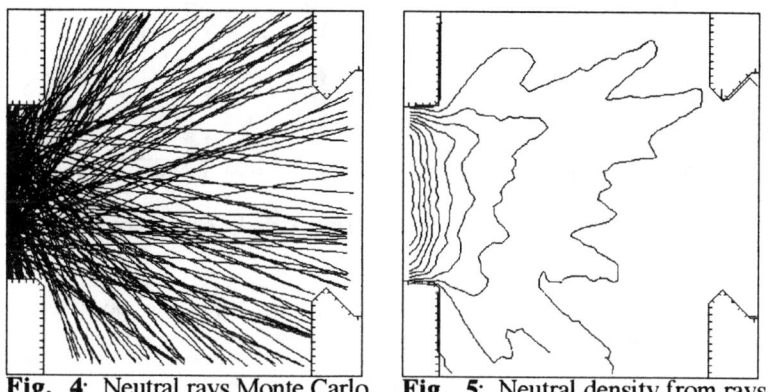

Fig. 4: Neutral rays Monte Carlo. **Fig. 5**: Neutral density from rays.

It can easily be shown that the velocity from a vertical aperture surface (along y) is given by

$$v_y = \frac{v_o}{L_y} \int_1^2 \frac{y' dy'}{\sqrt{x^2 + (y-y')^2}} = \frac{v_o}{L_y}(\sqrt{x^2 + (y-y_2)^2} - \sqrt{x^2 + (y-y_1)^2}) \quad [5]$$

where L_y is the length of the vertical surface emitting neutrals. The horizontal component is obtained from

$$v_x = \sqrt{v_o^2 - v_y^2}. \quad [6]$$

The density assumed in the reference[2,3] is of the form

$$n_o = n_o^o (1 - y^2/L_y^2)(1 - x^2/L_x^2) \quad [7]$$

Since equations [5], [6], and [7] are not necessarily consistent with the continuity equation [3], an alternative form consistent with the continuity equation and [6] was also tried,

$$n_o = \frac{n_o^o}{\pi}[\arctan(\frac{y-y_2}{x}) - \arctan(\frac{y-y_1}{x})]. \quad [7a]$$

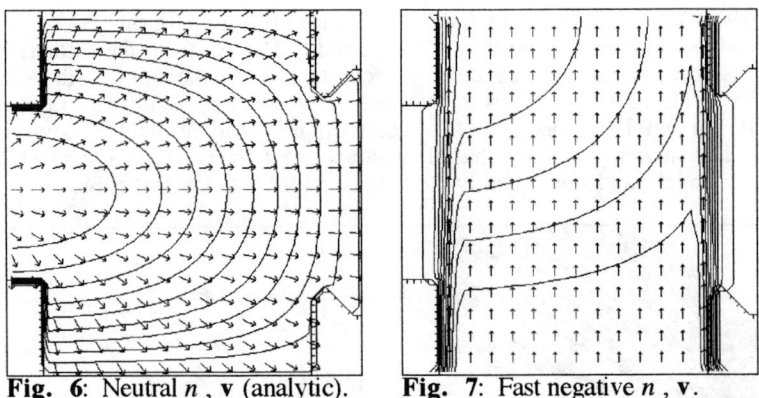

Fig. 6: Neutral n, $\mathbf{v}$ (analytic).   **Fig. 7**: Fast negative n, $\mathbf{v}$.

A combined contour plot of neutral density and vector plot of velocity from equations [5], [6] and [7] is shown in Fig. 6.

The continuity and viscosity coefficients used for the positives (p), cold negatives (c) and fast negatives (f) were[2,3],

$$\nabla \bullet (n_p \mathbf{v}_p) = -\sigma_{pc}\sqrt{\frac{T_p}{m_p}} n_c n_p \qquad [8]$$

$$\nabla \bullet (n_c \mathbf{v}_c) = \sigma_{fo} v_f n_o n_f - \sigma_{det}\sqrt{\frac{T_o}{m_o}} n_o n_c \qquad [9]$$

$$\nabla \bullet (n_f \mathbf{v}_f) = -\sigma_f v_f n_o \qquad [10]$$

$$\mathbf{v}_p \bullet \nabla \mathbf{v}_p = -\sigma_{cx} n_o \mathbf{v}_p \qquad [11]$$

$$\mathbf{v}_c \bullet \nabla \mathbf{v}_c = -(\sigma_{det} + \sigma_{cx}) n_o \mathbf{v}_c \qquad [12]$$

All other coupling terms were neglected.

The density and velocity plots for cold negatives and positives are shown below in Fig. 8 and 9. The initial (boundary) values assumed and the expressions and values for the coefficients are[2,3]:

$n_p = 10^{11}$ /cm^3	T_p (drift) = 0.1 eV	(on left)
$n_c = 0$	T_c (drift) = 0.01 eV	(on left)
$n_f = 10^9$ /cm^3	T_f (drift) = 200 eV	(on bottom)
$n_o = 10^{14}$ /cm^3	$T_o = 0.01$ eV	(on left).

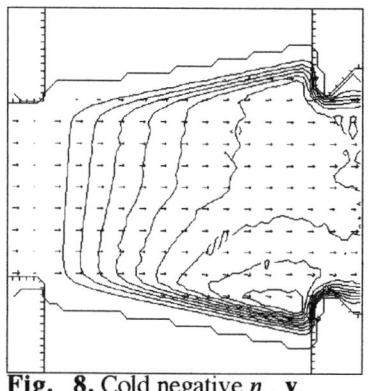
Fig. 8. Cold negative n, **v**.

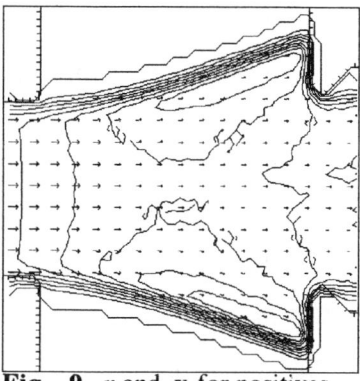
Fig. 9. n and **v** for positives.

The cold negatives start with zero density on the left and grow from charge exchange between fast negatives and neutrals to about 3×10^{10} /cm^3 on the right. Also note the peak at 5×10^{10} /cm^3 in the lower right corner. This production due to the product of fast negatives (Fig. 7) and neutrals (Fig. 6) densities. The cold negatives peak in the lower right because the fast negative density is highest there. Note also that the fast negatives are asymmetric in density because of losses from interactions with the fan of the neutrals (a good reason to have a counter-streaming fast negative beam from the top).

The fast negatives drop from 10^9 to about 6×10^8 /cm^3 by the time they cross the neutrals fan and reach the top. Note also that the positives and cold negatives fan out due to viscous force interaction with the neutrals. The positives also lose speed because of viscosity as they go from left to right (the arrows decrease in size from left to right in Fig. 9), and this causes their density to go up slightly from 10^{11} to 1.3×10^{11}, in spite of losses due to annihilation interactions with cold negatives.

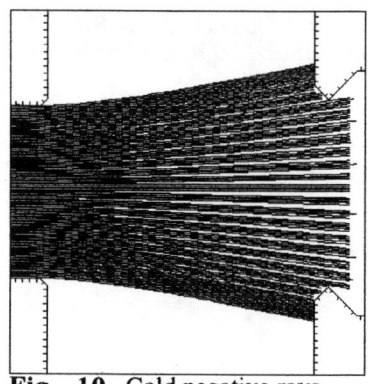
Fig. 10. Cold negative rays.

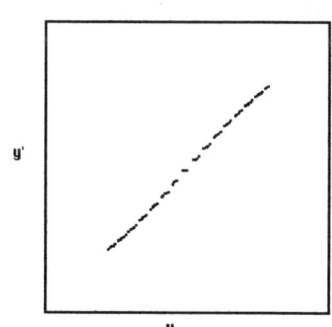
Fig. 11. Cold negative emittance.

Also shown is the ray plot for the cold negatives (Fig. 10), and their emittance diagram as they exit the simulation on the right (Fig. 11). A slight aberration in the diverging negative beam from the viscous interaction with the neutrals is evident.

What remains to be done is examine the full problem by turning on the Poisson/Boltzmann solver (and iterating with the fluid algorithm) to look for nonuniformity-induced voltage gradients in the production region as well as the extraction sheath, that might effect the emittance and brightness. Geometry, neutrals profile, counter-streaming fast negatives etc. can all be optimized to minimize any brightness degradation effects.

REFERENCES

1. W. Cooper and W. Kunkel, private communication (1991).

2. M.C. Vella, private communication.

3. M.C. Vella, Plasma Potential Scaling in the High Brightness Charge Exchange (CXS) H- source, LBID-1865.

4. S. A. Self; Exact Solution of the Collisionless Plasma-Sheath Equations; Phys. Fluids 6, 1762 (1963).

5. W. B. Hermannsfeldt; SLAC Electron Trajectory Program; SLAC 266, (1979), original SLAC 166, (1973).

6. W. S. Cooper, K. Halbach, S. B. Magyary; Computer Aided Extractor Design; Proceedings of the Second Symposium on Ion Sources and Formation of Ion Beams, Berkeley, LBL-3399, (1974).

7. See for example J. H. Whealton, and J. C. Whitson; Space Charge Ion Optics Including Extraction from a Plasma; Particle Accelerators 10, p. 235-251 (1979).

8. J. E. Boers; SNOW a Digital Computer Program for the Simulation of Ion Beam Devices; Sandia National Laboratory Report SAND 79-1127, (1979).

9. P. Spadtke, S. Wipf; KOBRA3, A Code for the Calculation of Space-Charge-Influenced Trajectories in 3-Dimensions, GSI-89-09 report, ISSN 0171-4546, June 1989.

DESIGN OF ION OPTICAL SYSTEMS BY COMPUTER SIMULATION

Bernard J. Hall
Charles Evans & Associates
301 Chesapeake Dr., Redwood City, CA 94063

ABSTRACT

The problem of mathematically describing the passage of ions through a series of field and field free regions has been addressed by many authors. TRIO, developed by Matsuo et al.[1], GIOSP, developed by Wollnik[2] and ION, developed at Charles Evans & Associates based upon the approach of Nakabushi et al.[3], are examples. These computer codes provide second and third order calculations of ion trajectories in a transfer matrix approach. If we expect spectrometer performance to match prediction it is necessary to develop simulations for the actual values of the individual ion optical parameters. The idealized parameters that are often used can lead to inaccurate total simulations. We expanded TRIO to incorporate minimization routines and monte carlo simulations and used existing code such as POISSON[4] in an integrated approach to the computer simulation of ion optical systems and specifically mass spectrometers. Comparison between simulation and experimental data using these principles shows very good agreement.

INTRODUCTION

Ion optical programs such as TRIO (developed by Matsuo et al[1]) GIOSP (developed by Wollnik[3]) and ION (developed at Charles Evans and Associates based on the approach of Nakabushi et al[2]) are examples of computer codes developed around mathematical models for the purpose of designing spectrometers. Our purpose here is to examine two groups of the many input parameters to these programs, the expansion coefficients of the electrostatic field used to describe electrostatic analyzers ('c-values') and the fringe field integrals for magnetic analyzers, and to demonstrate the advantage of computer simulation of these parameters. The simulation of relevant input parameters combined with the ability to perform multi-parameter minimizations and monte-carlo simulations of spectrometer performance provides a powerful tool for the development of ion optical systems and in mass spectrometers particular .

THE MATHEMATICAL APPROACH AND COMPUTER CODE

The passage of ions through a series of (sector) field and field free regions can be described using a transfer matrix approach[5]. In this approach each region has an associated transfer matrix with the combination of all regions being described by a single matrix, which is the product of all of the individual matrices. Each of the non-zero matrix elements is a function of parameters relevant to the particular region being described. Using this method, systems for focussing ion beams can be developed. In particular, double focussing mass spectrometers have been designed using this approach[6].

The calculation of these transfer matrices and the necessary matrix multiplication is quite tedious but easily performed by computer. TRIO, a set of computer sub-routines developed by Matsuo et al.[1] is one such example. Of course, the computation of the transfer matrices alone does not constitute a "computer simulation". Once the transfer matrices are computed for a desired geometry methods must be developed to simulate the performance of the system. Since the transfer matrices convert an input vector of position and angle (2-D), energy and mass (x,y,a,b,U,M) into an output vector, we can randomly generate many thousands of input vectors (monte-carlo simulation) and, using the total transfer matrix, produce output vectors for each of the input vectors. These output vectors can then be displayed as a scatter plot in X-Y space and a-b (angular) space or can be used to create simulated peak shapes. The incorporation of minimization routines to selectively zero or obtain target values of selected matrix elements by simultaneously varying one or more ion optical parameters is also desirable. This approach has been taken, with varying degrees of implementation, by Wollnik in GIOSP[3] and the author of this paper in TRIO-PC.

Problems do arise, however, when attempting to use computer simulations of this kind. The resulting transfer matrix and the simulated performance can be quite accurate, however the degree to which an actual device can be built according to the input parameters of the simulation turns out to be, in many cases, very poor. One option is to use the computer simulation as a rough estimate and design into the physical system enough variability (e.g. moveable entrance and exit slit positions) to compensate for errors in both the initial assumptions and errors in machining the ion optical components. The problem here is that although this method can ensure the desired first order performance, second and third order performance will almost certainly be compromised.

A second option is to use a more integrated approach to the computer simulation method. Rather than rely upon guess work to determine the individual ion optical parameters that go into the calculation of the transfer matrices one can simulate many of them before hand. Ion optical elements such as fringe field integrals and main electrostatic and magnetic fields are easily simulated using finite element programs such as POISSON. With the addition of some intermediary calculation the values obtained from POISSON can be translated into parameters needed in the transfer matrix calculations. Two advantages are immediately apparent in this approach. First, a better fit between theoretical and experimental design can be realized and second, the method forces the designer to consider the physical properties of the system early in the design stage, which can be of great benefit in the long run.

To illustrate this approach we will consider the simulation of two important ion optical parameters, the expansion coefficients of the electrostatic main field ("c-values") and the magnetic fringe field.

A BASIC SPECTROMETER DESIGN

To begin with, let us look at the design of a basic mass spectrometer composed of a drift space followed by a toroidal electrostatic analyzer then a drift space followed by a magnetic sector analyzer and a final drift space. The basic design is presented in Figure 1 which also includes the transfer matrices for each of the individual components and the full transfer matrix.

The necessary values for each of the parameters were determined by first providing approximate values and then allowing the program (TRIO-PC) to run a simplex minimization on appropriate matrix elements (e.g. x|a and x|d, which should be zero) by varying several parameters simultaneously (e.g. final drift length, "c-values" of the esa).

The performance of the spectrometer was simulated with the simulation of the ion beam cross section and peak shape at the end of the spectrometer presented in figure 2. Since the system given here is imaging, random input vectors (rays) from a 'grid' were used to highlight the performance.

Now, let us consider how to simulate the 'c-values' of the electrostatic analyzer portion of the spectrometer. The electrostatic potential of a toroidal electric field can be expanded as power series about the main path[7] where the potential U = 0 is given below in equation 1.

$$U(x,z) = -E_0(x - z^2/2r_0 + xz^2/2r^2_0 - x^2z^2/2r^2_0 + z^4/24r^3_0)$$

$$-E_0a_{20}((x^2-z^2)/2r_0 - xz^2/2r^2_0 + x^2z^2/2r^3_0 - z^4/24r^3_0)$$

$$-E_0a_{30}(x^3/6r^2_0 - xz^2/2r^2_0 - x^2z^2/4r^3_0 + z^4/12r^3_0)$$

$$-E_0a_{40}(x^4/24r^3_0 - x^2z^2/4r^3_0 + z^4/24r^3_0) \tag{1}$$

The 'c-values' used by TRIO, GIOSP etc. are given here in equations 2, 3 and 4.

$$c = -1 - a_{20} \tag{2}$$

$$c' = 1 + a^2_{20} - a_{30} \tag{3}$$

$$c'' = -2 - 2a^3_{20} + 3a_{20}a_{30} - a_{40} \tag{4}$$

After deciding upon a physical geometry (e.g. cylindrical electrodes terminated by matsuda plates[8]) we can input this geometry into POISSON and obtain a set of mesh points (an illustration of the process is seen in Figure 3).

Using the mesh points created by POISSON the values of the unknowns, E_0, a_{20}, a_{30} and a_{40} can be determined by using the method of least squares. As a test, experimental data obtained by Matsuda et al.[8] is compared with theoretical values obtained in this manner.

238 Design of Ion Optical Systems

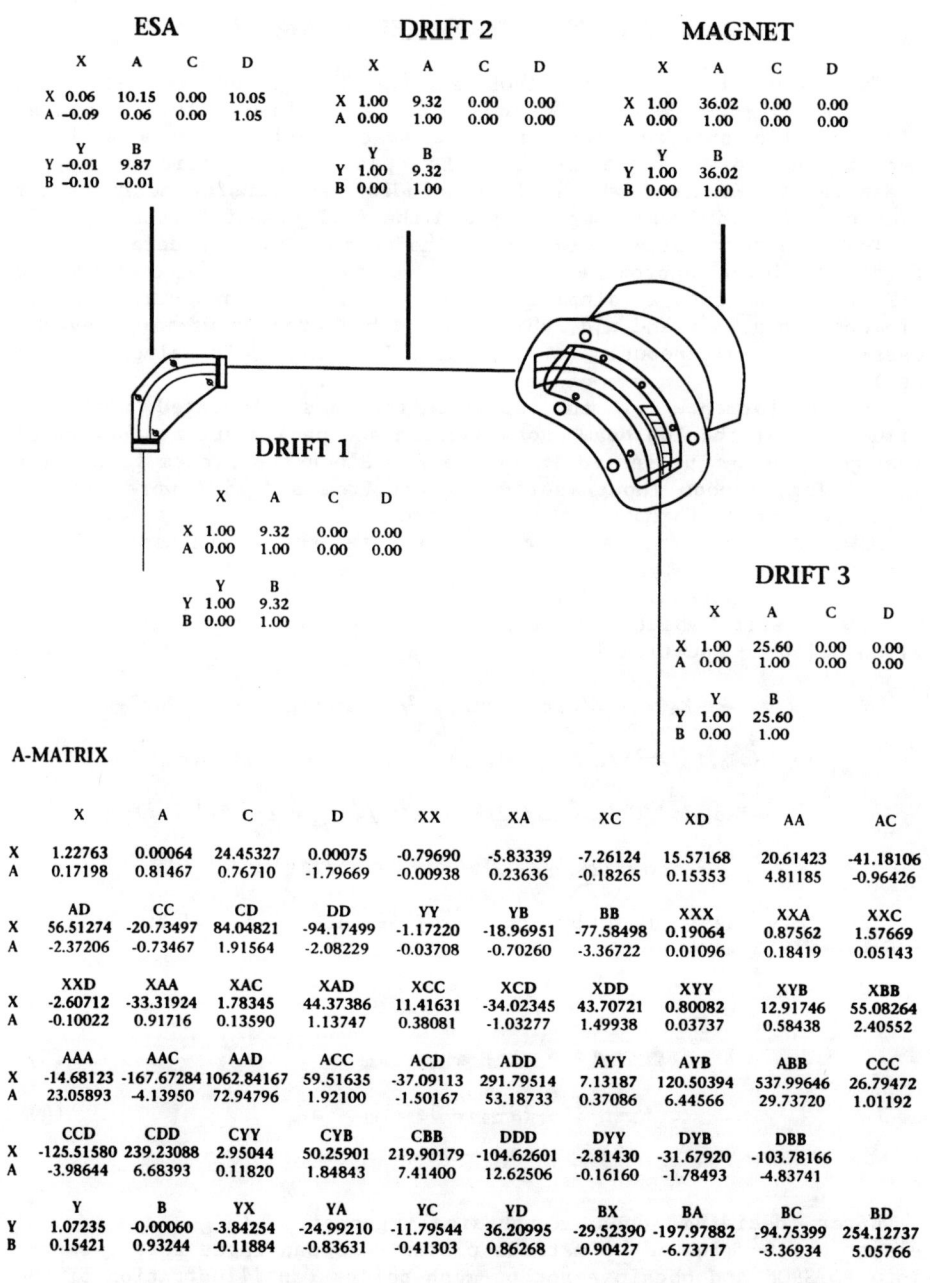

Figure 1
Basic Spectrometer Design and Full Transfer Matrix

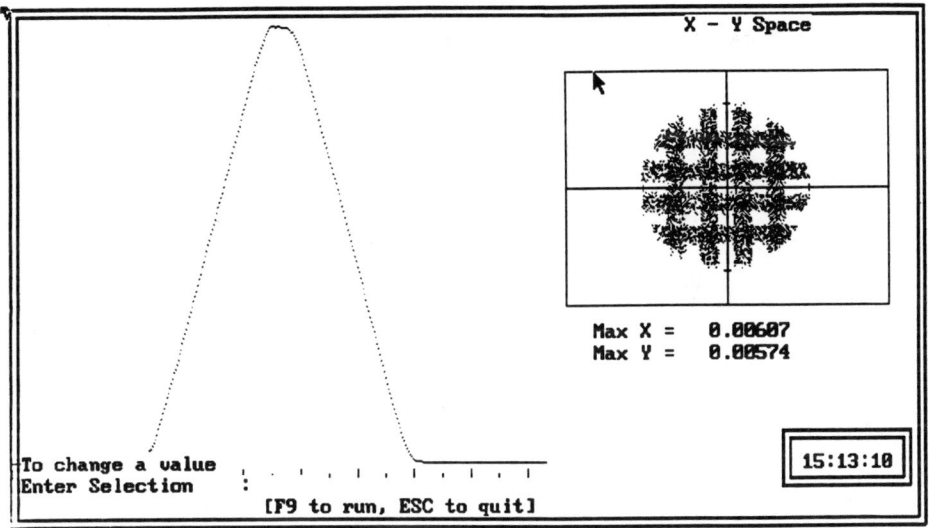

Figure 2
Simulated Spectrometer Performance

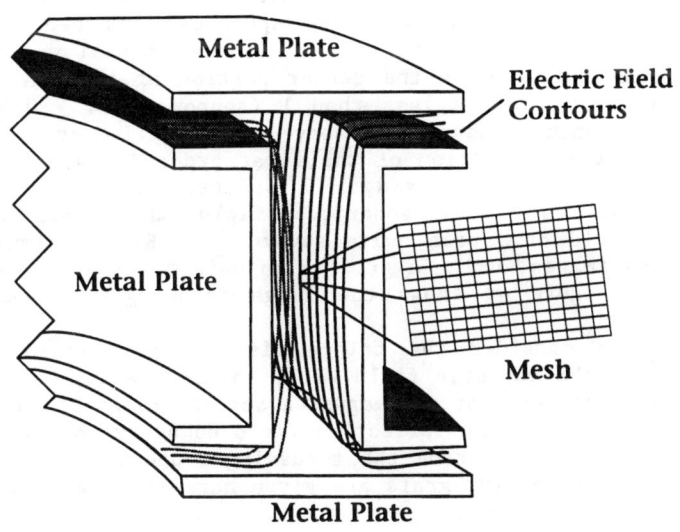

Figure 3
Cylindrical Electrodes Terminated by Matsuda Plates
POISSON produces Electric Field Values at each Mesh Point

The two compare extremely well as can be seen in Table I. (The "c's" are theoretical values and $B_1 = -1 + c$.)

Table I Theoretical vs Experimental 'c-values'

U_c	c	c'	c''	B_1(theory)	B_1(exp.)
0	-.06	-1.79	8	-1.06	-1
$-.25U_0$	0.32	-.86	-93	-0.68	-0.7
$+.25U_0$	-.44	-2.18	112	-1.44	-1.4
$+.50U_0$	-.82	-2.01	218	-1.82	-1.8
$-.50U_0$	0.69	0.59	-188	-0.31	-0.3

To illustrate the usefulness of the above simulation we can consider the example of a set of ESA plates recently constructed. The plates were machined to have spherical surfaces, a 2 cm total gap and were 7 cm high. It was assumed at the time of construction that since the electrode surfaces were machined to be spherical that the field close to the center would be spherical and hence have 'c-values' of $c = 1$, $c' = -1$ and $c'' = 2$. Experimental work on these plates showed that even in the center portion of the ESA the first order coefficient, c, was less than 1 (approx. 0.94 - 0.96). The simulation technique described above was applied to the given geometery and the prediction of the first order 'c-value' was 0.96. If these plates had been used in a spectrometer under the false initial assumption of a spherical field the resulting image abberations would have been quite severe. A TRIO-PC simulation of the resulting images, using a grid as an object, is shown in figures 4a and 4b. Figure 4a shows the image assuming c=1 and 4b with c=0.96.

Next let us consider the fringe field properties of a sector magnet. Since the magnetic field will, in most cases, extend beyond the physical boundaries of the magnet a set of fringe field integrals are used to determine the 'effective field boundary' of the magnetic field as well as to describe the focussing properties of the fringe field region[9]. These integrals are given here in equations 5, 6 and 7.

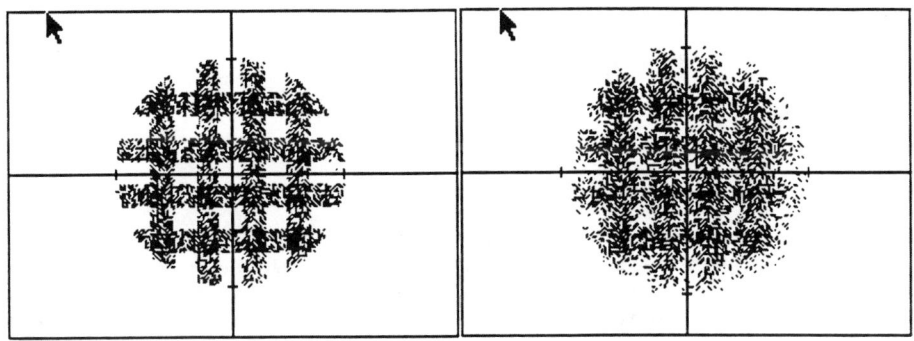

Figure 4a
Grid Image c = 1

Figure 4b
Grid Image c = 0.96

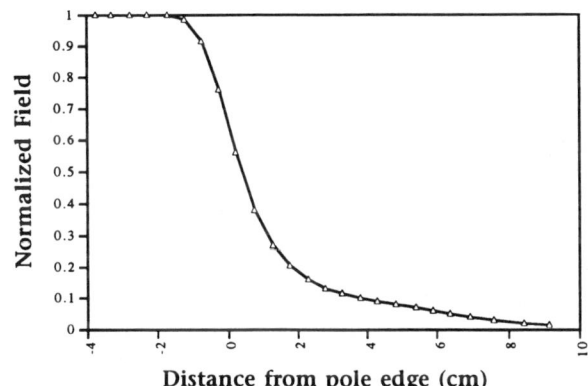

Figure 5

Poisson Simulation and Experimental Magnet Fringe Field.

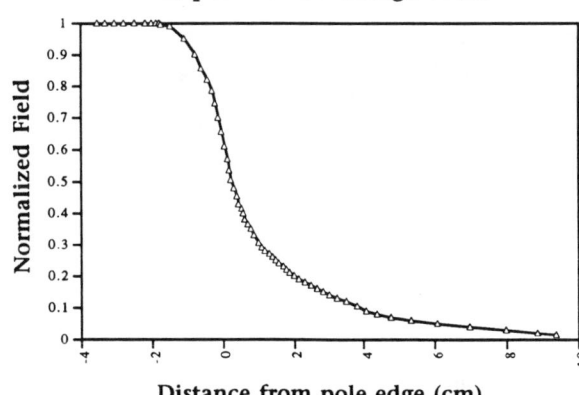

$$I_1 = B_0^{-1} \int_a^b \int B \, dg\,dg - 1/2 \, g_b^2 \qquad (5)$$

$$I_4 = B_0^{-2} \int_a^b B^2 \, dg - g_b \qquad (6)$$

$$I_{5m} = B_0^{-1} \int_a^0 gB \, dg \qquad (7)$$

Using a POISSON simulation to calculate the fringe field of a particular magnet design the values of these integrals are easily computed. These values are used directly in the transfer matrix calculations. A comparison between a POISSON prediction for a magnet fringe field and experimental data from the magnet are presented in Figure 5. The agreement is very good allowing us to conclude that fringe field integrals computed from this data will also be quite accurate. As an additional test of this simulation method a comparison against fringe field integral values predicted by Hu et al.[9] for both magnetic and electrostatic fields was made, again with very good agreement.

CONCLUSION

Using a combination of simulation techniques it is possible to design ion optical systems with a high degree of accuracy. The advantage of using such an approach allows the construction of systems with few or no moving parts which can be quite important if high performance and/or use of the system by many individuals is required.

REFERENCES

1. T.Matsuo,H.Matsuda,Y.Fujita,H.Wollnik, Mass Spec. V24 No.1, p.19,(1976).
2. H.Wollnik,J.Brezina,M.Berz,NIM,A258,p.408,(1987).
3. Nakabushi,Sakuri,H.Matsuda,IJMSIP,52,p.319,(1983).
4. K.Halbach,LLNL report,UCRL-17436,(1967).
5. H.Wollnik,NIM,52,p.250,(1967).
6. H.Matsuda,NIM,187,No.1,p127,(1981).
7. H.Matsuda,Y.Fujita,IJMSIP,16,p.395,(1975).
8. H.Matsuda,Rev.Sci.Inst.,32,No.7,p.850,(1961).
9. Z.Hu,T.Matsuo,H.Matsuda,IJMSIP,42,p.145,(1982).

NUMERICAL SIMULATIONS OF INPUT AND OUTPUT COUPLERS FOR LINEAR ACCELERATOR STRUCTURES*

C.-K. Ng and K. Ko
Stanford Linear Accelerator Center
Stanford University, Stanford, CA 94309

ABSTRACT

We present the numerical procedures involved in the design of coupler cavities for accelerator sections for linear colliders. The MAFIA code is used to simulate an X-band accelerator section with a symmetrical double-input coupler at each end. The transmission properties of the structure are calculated in the time domain and the dimensions of the coupler cavities are adjusted until the power coupling is optimized and frequency synchronism is obtained. We compare the performance of the symmetrical double-input design with that of the conventional single-input type by evaluating the field amplitude and phase asymmetries. We also evaluate the peak gradient in the coupler and discuss the implication of pulse rise time on dark current generation.

1. INTRODUCTION

At SLAC, we have an active program on the Next Linear Collider (NLC) R & D, and couplers are an important part of the accelerator structure work in this program. Indeed, the efficient delivery of power from RF sources such as klystrons to disk-loaded accelerator structures in linear colliders depends crucially on the coupler cavity. There are several requirements to be satisfied by such a cavity. First, it is to be well matched to the feeding waveguides (see Fig. 1) in order to couple the maximum amount of power into the structure to achieve the highest possible accelerating gradient. Second, it must be tuned to the synchronous frequency for the proper phase advance in the structure. Third, it must have minimal deleterious effect on the beam. Fourth, the couplers should ideally have surface fields no higher than the interior. These considerations, coupled with the fact that the geometry is intrinsically three-dimensional, make the design of the coupler cavity a nontrivial problem.

Previously, coupler cavities for disk-loaded accelerator structures have been designed following a set of procedures based on the Kyhl method[1]. It involves a sequence of experiments to determine the matching and tuning. Several iterations on actual prototypes may be needed before an optimal configuration can be obtained. The effort can be time-consuming and requires substantial empirical expertise. In this paper, we study an alternative approach by numerical simulation. We build a computer model that approximates closely the coupler cavity. Since changes in dimensions can be easily implemented on the computer, this approach offers a distinct advantage over cold tests in optimizing a design. Furthermore, valuable field information such as asymmetries and peak gradients, for example, is readily obtainable numerically, which otherwise would be difficult to measure experimentally. These advantages provide the motivation for our effort to develop an accurate and reliable computational procedure for matching and tuning this particular RF component.

In section 2, we describe the numerical model of the symmetrical double-input coupler using the MAFIA code[2]. Such a coupler has been incorporated into a 30-cavity section and operated up to 100 MV/m without evidence of breakdown. In section 3, the numerical procedure involved in matching and tuning the coupler is explained and the results are compared with experimental data in section 4. In section 5, we show the advantages of the symmetrical double-input design over the conventional single-input type in terms of field asymmetries. In section 6, we evaluate the peak gradients. The effect of pulse rise time is considered in section 7 where simulation results are presented to be corroborated with observations. We conclude with a summary in section 8.

---

* Work supported by Department of Energy, contract DE-AC03-76SF00515.

2. THE NUMERICAL MODEL

We model the symmetrical double-input coupler in the 30-cavity structure which was used in high power tests. Instead of all 30 cells, we simulate only a short section which is sufficient for matching the coupler and it is computationally more practical. Fig. 1 shows the mesh geometry we have constructed using MAFIA. It consists of two identical coupler cavities and two regular accelerator cavities. The coupler cavities are fed by WR90 rectangular waveguides through irises. Because the feeds are symmetrical, we only need to model one-quarter of the structure. The magnetic boundaries imposed at the two symmetry planes are consistent with the waveguide fields as well as the fields of the accelerating mode in the structure. The SLAC NLC operating frequency is chosen to be at X-band around 11.424 GHz. Accordingly, the dimensions of the regular cells in our model have been designed for that frequency at the $2\pi/3$ phase advance per cell. The dimensions of the couplers are different in order to fulfill the matching and tuning requirements described earlier.

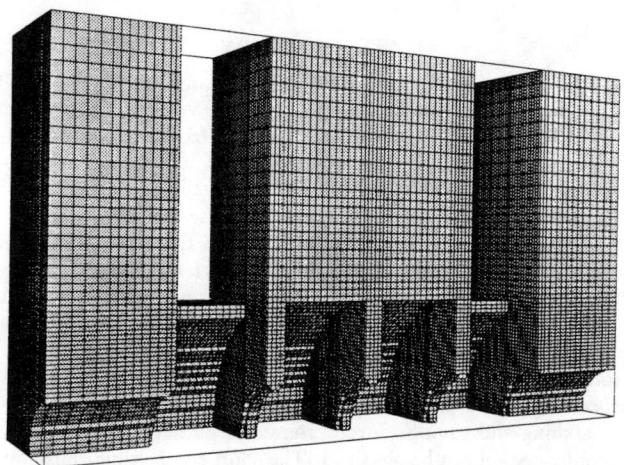

Fig. 1 MAFIA geometry for a 4-cell traveling wave section.

Given a coupler geometry, we perform a MAFIA simulation in the time domain. Power is fed continuously at the input waveguide port in the TE_{10} mode at a particular frequency, starting with a smooth initial rise and reaching 1 watt at flat-top. The input power couples to the accelerating mode via the irises, propagates through the section and exits by way of the output coupler. The simulation extends over many filling times of the section until a traveling wave at steady-state is reached. At the end of the run, the reflection coefficient S_{11} at the input waveguide port and the transmission coefficient S_{21} at the output waveguide port are evaluated. In addition, the electric field on axis and in designated regions of interest is recorded for subsequent post-processing.

3. MATCHING AND TUNING OF COUPLER CAVITY

The cross-section of the coupler cavity is shown in Fig. 2(a). There are three dimensions to be determined: the coupler diameter, the iris aperture and its thickness. Assuming that the iris thickness is held fixed, the design program is then to choose the two remaining dimensions in such a way that the matching and tuning are optimal. These conditions are assessed as follows. As far as matching is concerned, we look for the minimum VSWR for the section. In the simulation, this corresponds to the smallest reflection coefficient S_{11} at the input waveguide port. Fig. 3(a) shows the time history of S_{11} for a typical case when the coupler is matched. We see that the steady-state can be reached after several filling times and the amount of reflection is quite acceptable (VSWR = 1.023 in this case). This is in contrast to an unmatched case as

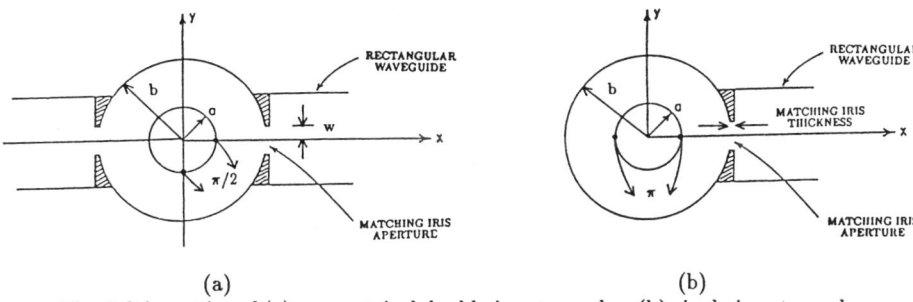

Fig. 2 Schematics of (a) symmetrical double-input coupler; (b) single-input coupler.

shown in Fig. 3(b) which corresponds to a different set of coupler diameter and iris aperture dimensions. The VSWR = 1.475 is large compared with the matched case.

To evaluate tuning, we examine the amplitude and phase variations of the electric field on axis. One can write

$$E_T(z) = |E_T(z)|e^{-i\theta_T(z)}, \qquad (1)$$

where the time variation has been left out. In the MAFIA run at steady-state, the electric field on axis along the structure is stored over several cycles which can be Fourier-analyzed to obtain E_T and θ_T. They are plotted in Figs. 4(a) and (b) for the same matched case mentioned above. The dashed lines mark the boundaries between cells. In both plots we see that the field is periodic in the regular cells. We also notice that it is symmetric about the center of the structure which should be the case when the couplers are nearly matched. In this case the fields look identical whether power is fed in at the input or output end. The phase advance in the two regular cells is 122°, close to the expected value of 120° at the driving frequency of 11.42 GHz. In the coupler cavities, the phase variation is zero across roughly half the cavity and totals to 62° for the whole cell. This suggests that the field in the half of the coupler cavity near the cut-off beam pipe is essentially a standing wave while the traveling wave in the other half advances by half the phase shift as compared to the regular cell. These results confirm earlier data from dielectric bead perturbation measurements[3]. As pointed out in that paper, the field amplitude and phase variations can provide a means by which the tuning of the coupler can be accurately determined. Numerically such a procedure is much easier to implement than in actual cold tests.

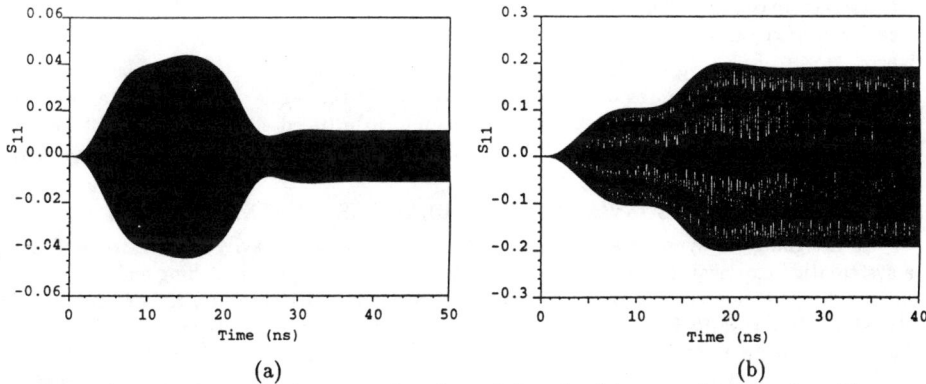

Fig. 3 The reflection coefficient as a function of time for (a) a matched coupler; (b) an unmatched coupler.

For comparison, we show an unmatched case in Figs. 5(a) and (b). Here the field ampli-

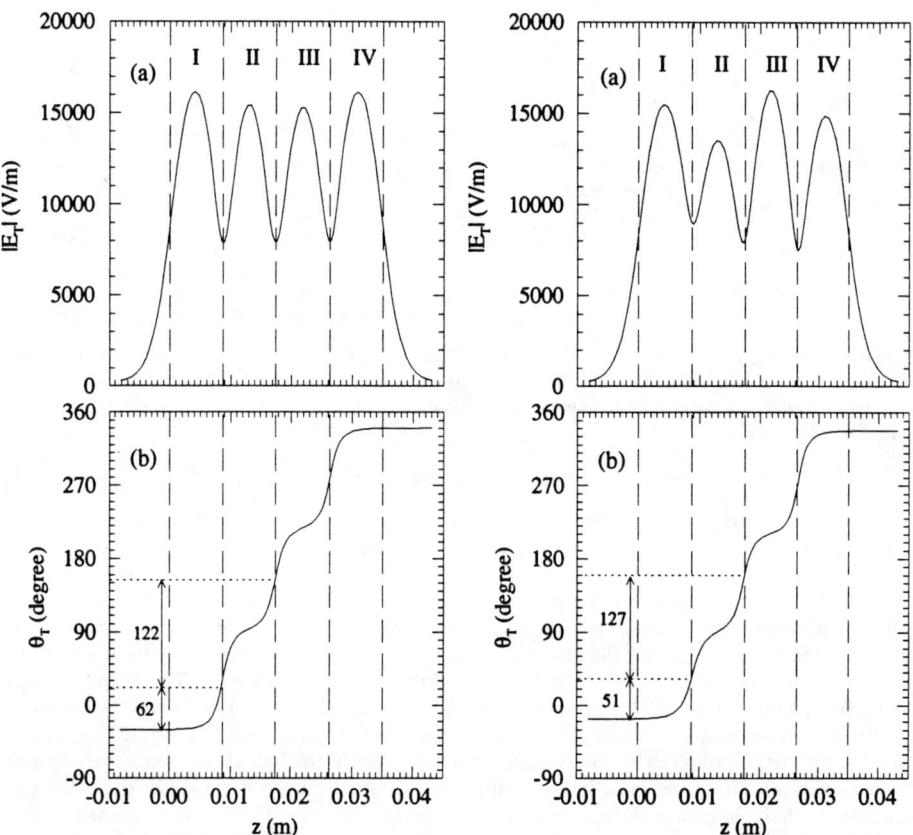

Fig. 4 Amplitude and phase variations for a matched coupler. Regions I, II & III, and IV are the input coupler cavity, the two structure cells and the output coupler cavity, respectively.

Fig. 5 Same as Fig. 4, but for an unmatched coupler with matching iris aperture increased by 0.006 inches.

tudes are irregular along the structure, quite different from the symmetric fields we saw earlier. Similarly, the phase advances in the coupler cavities and the regular cells deviate appreciably from the $\pi/3$ and the $2\pi/3$ values we expect to find in a tuned situation. The difference in dimensions in the coupler cavity from the matched case is an increase of 0.006 inches in the matching iris aperture.

4. COMPARISON BETWEEN SIMULATIONS AND EXPERIMENTS

In designing the symmetrical double-input coupler for the 30-cavity section, we performed a systematic numerical search for the optimal dimensions, using the matching and tuning conditions described above. We varied the cavity diameter and iris aperture, then calculated the VWSR and phase shifts in each iteration. Figs. 6(a) and (b) summarize the results. The driving frequency is again 11.42 GHz. The plots show, in effect, the sensitivity of matching and tuning to changes in coupler dimensions due to machining tolerances, for example. The perturbations (δb and δw) are measured with respect to the dimensions (b and w) for the matched case in which they are assumed to be zero. In Fig. 6(a) we find that the VSWR is much more sensitive to the coupler diameter than to the iris aperture. The same holds true for the phase shifts as

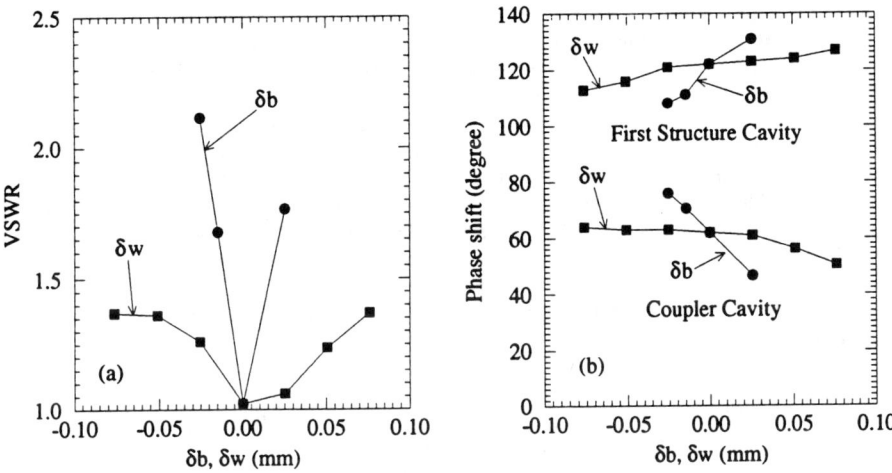

Fig. 6 The dependence of (a) VSWR; (b) phase shifts on the changes δb in coupler radius b, and δw in matching iris half aperture w.

shown in Fig. 6(b). We also note that about the optimum point, the dependence of the VSWR on slight changes in either dimensions is quadratic while that of the phase shifts is linear. Both are consistent with simple perturbation analysis.

With the coupler dimensions that we found for optimal matching and tuning, one can vary the frequency to explore the bandwidth. Fig. 7 shows a comparison of VSWR versus frequency between the MAFIA simulations and experiments[4]. Near the desired operating frequency of 11.424 GHz, the agreement is very good and the dimensions of the actual coupler are very close to those used in the MAFIA model. This is encouraging because it means that we can reasonably model the geometry for design purposes without expending an unrealistic amount of computational resources.

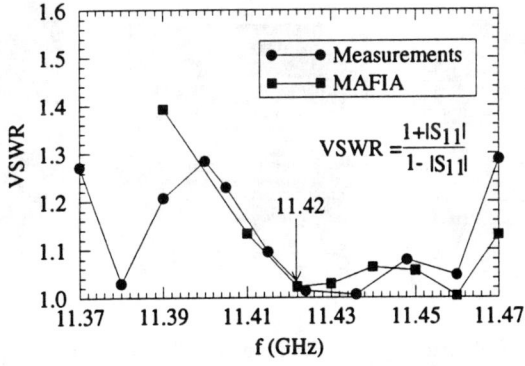

Fig. 7 VSWR versus frequency from MAFIA and measurements.

5. FIELD ASYMMETRIES IN COUPLERS

Conventional couplers are of the single-input type (see Fig. 2(b)) where power is fed in from a single waveguide. This configuration inherently introduces field asymmetries across the beam aperture in the form of a dipole component. The amplitude asymmetry leads to a shear force which spreads the bunch while the phase asymmetry results in a deflecting force on the bunch[5]. As discussed in Ref. 5, the amplitude asymmetry can be corrected by offsetting the cavity with respect to the beam axis. The effect of phase asymmetry on the beam can be reduced by tilting the coupler cavity or by feeding successive sections from opposite sides.

In the symmetrical double-input coupler, assuming that the fields in both feeds are equal in amplitude and have the same phase, the dipole component is eliminated by virtue of symmetry. The remaining asymmetries are then due to the quadrupole component which can be measured by comparing fields at points 90° around the beam aperture (see Figs. 2(a) and (b)). In Table 1, we list the asymmetries for the single-input coupler before and after offset, and for the double-input coupler. The data for the single-input coupler are taken from Ref. 5 while those for the double-input are obtained from the MAFIA simulation of the matched case. We conclude from the results that the field asymmetry should be negligible near the beam axis in the double-input coupler. This makes it a superior design over previous single-input types. Presently, the input couplers for the 75 cm and 1.8 m structures being planned at SLAC have incorporated this double-input feature.

Asymmetry	Single-input before offset	Single-input after offset	Double-input
$\frac{\Delta E}{E}\vert_\pi$	10%	0.1%	0
$\Delta\phi\vert_\pi$	1.5°	1.5°	0
$\frac{\Delta E}{E}\vert_{\frac{\pi}{2}}$			6%
$\Delta\phi\vert_{\frac{\pi}{2}}$			0.1°

Table 1. Amplitude and phase asymmetries for single-input and double-input couplers. The designations "$\frac{\pi}{2}$" and "π" correspond to asymmetries shown in Fig. 2(a) and (b) at $r = a$.

6. PEAK FIELD GRADIENTS

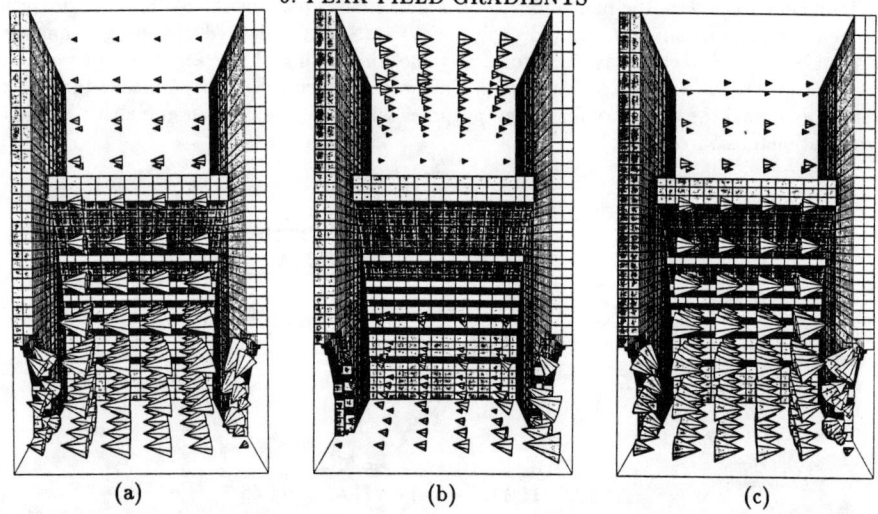

Fig. 8 Snapshots of electric fields in the input coupler at three different times. The left and right sides of the coupler are connected to the beam tube and the first structure cell, respectively. The peak gradients are (a) 235 MV/m; (b) 61 MV/m; (c) 239 MV/m.

One of the problems of common concern in accelerator structures is RF breakdown, which occurs when the peak electric field gradients reached in these structures exceed a certain critical value. At SLAC, an X-band 30-cavity accelerator section has been RF processed up to a stable accelerating field of about 100 MV/m, for a peak input power of 100 MW. While it is difficult to determine the locations of peak gradients experimentally, these can readily be obtained from our simulations.

Fig. 8 shows snapshots of the electric fields in the input coupler over approximately half a wave period at steady-state. We normalize the input power to 100 MW. The peak gradients are found to be 235, 61 and 239 MV/m for the three snapshots, respectively, and the maximum ($\sim$ 240 MV/m) is about a factor of 2.26 greater than the peak field along the axis. The maximum peak gradient occurs near the top part of the disk next to the first structure cell. Our result is in reasonable agreement with the measurement of the 30-cavity structure[6], where damage was seen near the top part of the coupler disk. Furthermore, the maximum peak gradient in the structure cell is found to be very close to that in the coupler from our suimulation.

7. PULSE RISE TIME EFFECT

A topic of great interest to accelerator designers is the generation of dark current by field emission at high gradient. Dark current observed in the high power tests of the 30-cavity traveling section was found to depend on the rise time of the RF pulse[7]. As the rise time got steeper, the resulting capture of electron increased. The effect was attributed to the upward frequency shift, caused by the rise time and resulting in a lower phase velocity. Here, we suggest a possible explanation by studying the space harmonic contents of the total field.

We have explored the effect of the pulse rise time with our MAFIA model. We scaled the rise times of the pulse to the filling time of our structure in the same ratios as those in the experiment, and hence chose 1.27 and 2.54 ns to correspond to the 10 and 20 ns cases that were measured. The fields in the half of the coupler which is traveling were Fourier-decomposed into their space harmonics. For a traveling wave, we can extend the analysis to a whole cell in the structure because of symmetry. Neglecting the time dependence, the peak electric field can be expressed as:

$$E_T(z) = \sum_{n=-\infty}^{\infty} C_n e^{-i\beta_n z}, \qquad (2)$$

where the propagation constant for the nth space harmonic is $\beta_n = \beta_0(1 + 3n)$. For a structure with spatial period D and operating at $2\pi/3$ mode, the synchronous component with phase velocity c has a propagation constant $\beta_0 = 2\pi/3D$. The complex field $E_T(z)$ can be expressed in terms of its amplitude $|E_T(z)|$ and phase θ_T as in Eq. (1). Hence, if $|E_T(z)|$ and $\theta_T(z)$ are known by measurement or by numerical means, then C_n can be determined as:

$$C_n = \frac{2}{D} \int_0^{\frac{D}{2}} |E_T(z)| \cos[\beta_n z - \theta_T(z)] \, dz. \qquad (3)$$

It is convenient to normalize $|E_T(z)|$ to the maximum peak field E_M. Comparing Eqs. (1) and (2), the contributions of each space harmonic to the total field are given by the individual terms in the sum of the following equation:

$$|E_T(z)| = E_M \sum_{n=-\infty}^{\infty} b_n e^{-i[\beta_n z - \theta_T(z)]}, \qquad (4)$$

where $b_n = C_n/E_M$. Furthermore, it can be shown easily that $\sum b_n = 1$. If the synchronous component has a phase velocity c, the phase velocities of the other space harmonics are given by $v_p = c/(1 + 3n)$, for $n = \pm 1, \pm 2, \ldots$. Therefore higher space harmonics have lower phase velocities.

In Fig. 9, we show the decomposition of the electric field along the axis at the end of the pulse into space harmonics across a cell for the two different rise times. The coefficients of $\{b_0, b_1, b_2, b_{-1}, b_{-2}\}$ for rise time 1.27 ns and 2.54 ns are $\{0.731, -0.076, 0.042, 0.369, -0.081\}$ and $\{0.741, -0.056, 0.026, 0.326, -0.051\}$, respectively. At the rise time of 1.27 ns, we see that the contributions of the higher order components ($n \neq 0$) are larger compared with those for the case with rise time of 2.54 ns. Therefore, electron capture by these lower phase velocity components is more important for short rise time. Of course, a realistic study of dark current

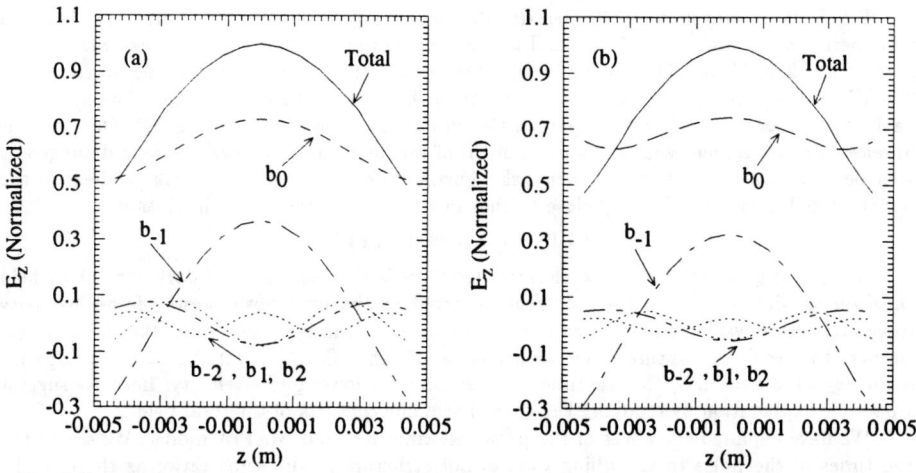

Fig. 9 Decomposition of the total field amplitude into space harmonics for rise time of (a) 1.27 ns; (b) 2.54 ns.

has to rely on particle simulation. Here, by simple Fourier analysis of the field amplitude, a qualitative understanding of dark current is obtained.

8. CONCLUSION

We have used MAFIA to simulate the symmetrical double-input coupler for an X-band accelerator structure. The numerical procedure for matching and tuning has been developed, and good agreement has been found between simulations and experiments. The advantage of the double-input over single-input geometry has been shown in terms of lower field amplitude and phase asymmetries, and the location of the peak gradient in the coupler cavity has been identified. Numerical results have also been presented to corroborate experimental observations on the dependence of captured dark current with RF pulse rise time. The present paper demonstrates that numerical simulations can provide very useful shortcut to cut-and-try prototyping in the design and analysis of linac coupler cavities.

ACKNOWLEDGEMENTS

We wish to thank H. Deruyter for the cold test data and many fruitful discussions. We are grateful to G. Loew for a critical reading of the manuscript and numerous valuable suggestions, and R. Miller for very useful comments. We also acknowledge H. Hoag, J. W. Wang, J. Haimson, N. Kroll, E. Nelson and W. Herrmannsfeldt for their interest in the problem.

REFERENCES

1. R. L. Kyhl, Impedance matching of disk loaded accelerator structures, unpublished (1976); E. Westbrook, Microwave impedance matching of feed waveguides to the disk-loaded accelerator structure operating in the $2\pi/3$ mode, SLAC-TN-63-103 (1963).
2. The MAFIA Collaboration, F. Ebeling et. al., MAFIA User Guide, 1992.
3. J. Haimson, Nucl. Instru. Meth., **39**, 13 (1966).
4. H. Deruyter et. al., Proceedings of Linac 92 Conference, p407, Ottawa, Canada, August, 1992.
5. G. A. Loew and R. B. Neal, Accelerator structures, in Linear Accelerators, p39-133, ed. P. M. Lapostolle and A. L. Septier, 1970.
6. J. W. Wang et. al., High-gradient studies on 11.4 GHz copper accelerator structures, SLAC-PUB-5900 (1992).
7. G. A. Loew, Review of studies on conventional linear colliders in the S- and X-band regime, SLAC-PUB-5844 (1992).

FAST SYSTEM DESIGN USING GUIs

Michael E. Kress[1]
Science Applications International Corporation Inc. (SAIC), Mc Lean, VA

Eunbong Sohn[1]
Science Applications International Corporation Inc. (SAIC), Mc Lean, VA

ABSTRACT

A compressible knowledge model is used to analyze contemporary design systems. Time compression techniques available with the onslaught of fast graphics processors and workstation are discussed and evaluated.

A brief history of interactive interface techniques is developed to set the stage for the dazzling Graphics User Interfaces, GUIs, now available on PCs and UNIX workstations. A hierarchy of GUI builders is presented. Examples are drawn from OSF/Motif widget sets and OPEN LOOK Intrinsics Toolkit widget sets. "Drag and drop" interactive interface builders are discussed and compared to a "cut and past" template technique using OSF/Motif's User Interface Language (UIL).

Examples are drawn from IDCFA, Interactive Design of Crossed Field Amplifiers.

COMPRESSIBLE KNOWLEDGE MODEL

A compressible knowledge model is used to analyze contemporary design systems. Time compression techniques available with the onslaught of fast graphics processors and workstation are used to minimize the time required to cycle through knowledge acquisition cycles.

We suggest the following set of governing equations for modeling the compression of knowledge flow into the design team:

$$\psi = F(X, t), \ u = \frac{d\psi}{dt}, \ \nabla \cdot u = \Gamma(\psi, U, V, S).$$

Here ψ is the knowledge base of the design team represented by the discrete variable X and t is time. u is the flow of knowledge into the system and throughout the team. U represents the user interface tools, V the visualization tools, and S the simulation codes.

This paper addresses the contribution of U to our design system. In

---

[1] Permanent address: College of Staten Island, CUNY, S. I., NY 10301

particular, we strive to maximize time compression effects of "user friendly" interaction systems by using the features of Graphics User Interfaces.

HISTORY OF INTERACTIVE INTERFACES

The first popular interactive user interfaces were text-based windows systems. In these interfaces, function keys, alphabetic and numeric choices, and mouse activation events are trapped and the program responds accordingly. One of the first such interfaces to gain notoriety was the IBM 3278 terminal. Here full screen display is used with the location of the cursor and function keys to determine the users request. The user can traverse layers of menus and execute commands and other programs from the command line. In spite of the claim of "user friendliness", the user can easily get lost in a quagmire of menus each of

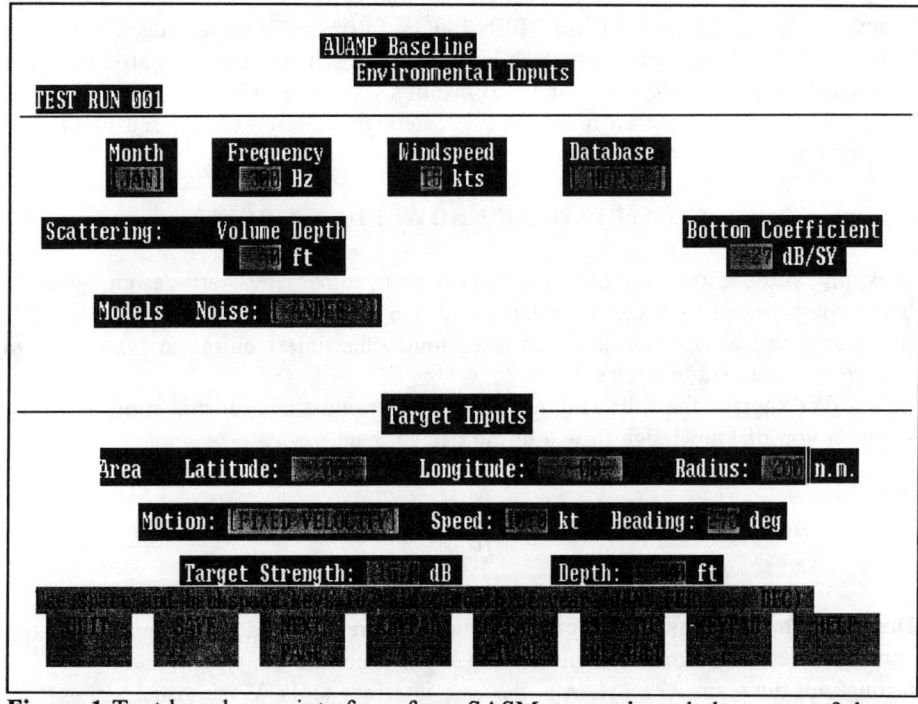

Figure 1 Text-based, user interface from SASM, to our knowledge, one of the earliest front-end interfaces to a numerical simulation.

which must be exited before certain tasks can be performed. Since multiple windows are not typically visible simultaneously, it is hard to remember: where in the program flow you are; how you got there; what is on the other forms; and how to get back. In spite of the versatility across systems and the speed of generating the text-based screens, the syntactic knowledge required for using the interface rarely fits our semantic knowledge concerning the application.[1] The resulting confusion is rarely user friendly.

The time compression of the knowledge is measurable but only after climbing a steep learning curve. Many users give up the climb and reverted back to the familiar, somewhat archaic, but reliable, line by line, command line oriented interface. In this latter case there is no time savings. The time spent attempting to learn the interface never pays off in terms of accelerating the interaction process.

Many text-based user interfaces have been built and used for countless applications including front-ends to physics codes. SAIC's Oceanographic Group

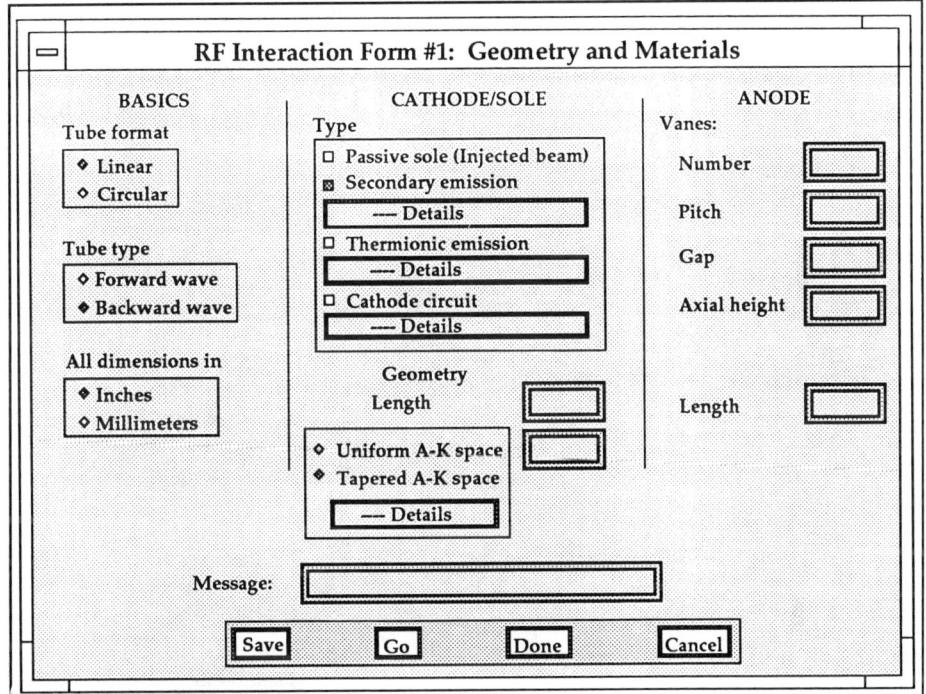

Figure 2 GUI from SAIC's Crossed-Field Amplifier Design System.

built SASM, a very nice text-based front-end to its Oceanographic Research Codes. Figure 1 shows a text-based SASM data entry form. The original version of the interface was built in 1986. Presently, a GUI version is under development.

With the availability of fast graphics processors the text-based constraint required for fast screen creation was removed. Bit map graphics can now be done quickly without loosing the user's attention. Fast graphics processing capabilities on today's PCs, Macs, and UNIX workstations have made possible user friendly GUIs. A GUI is bit map based. Typical GUIs permit multiple, visible windows that can be manipulated: re-sized, lowered, razed, and shrunk to an icon with the assistance of a window manager. Data can be entered on any window by simply moving or clicking the mouse in the window. Information can be cut from one window and pasted into another with the drag of the mouse and click of a button. Possible selections (the affordance) on a menu can be shown by crisply depicting the selection. Items not available can be hazed out while the selected item is highlighted. The graphical environment permits more information to be conveyed on the screen. Various fonts, colors, and icons can be used to convey information in a diverse spacial placement to lead the user through a sequence of possible choices. The buttons, boxes, menus, labels, and layout of the screen on a well organized display can couple the syntactic requirements of the application program with the users semantic knowledge of the computer concepts and the task concepts of the application resulting in a user friendly environment.[1]

To our knowledge, Apple Computer built the first popular GUI. Its fundamental operating system was graphics-based. The bit map requirement of the GUI was fundamental to the operating system. Pull-down menus, pop-up dialogue boxes, icon activated execution of programs are the essential building blocks of user interaction of the user friendly Apple Macintosh.

On the IBM PC compatible platform, Microsoft introduced Windows. This

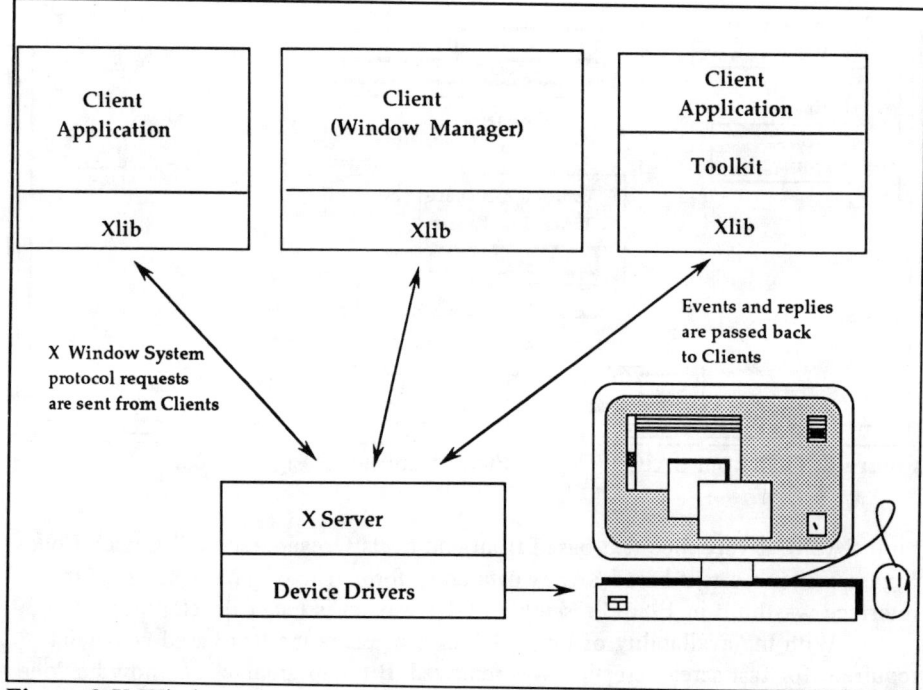

Figure 3 X Windows System's client server model handles inter-process communications.

GUI operating systems, followed Apple's look and feel and has become the operating system of choice for many PC users in spite of the heavy disk requirement of software applications.

On UNIX platforms, GUIs rule the user interaction domain with MIT's network based, client server model, user interaction system, The X Window System. The dazzling user friendly environment is currently used by SAIC to front-end to a number of their design codes. Figure 2 shows a typical data entry form used in SAIC's Crossed-Field Amplifier Design System.

ANATOMY OF A GUI

The fundamental building blocks of GUIs are a windowing system that permits creating graphical windows on the screen and trapping events including the position of the mouse pointer and type of mouse button click.

The "look and feel" of a GUI is the basic layout of the windows: the boarder type and titles, the fonts and styles, and the mouse interaction events. In

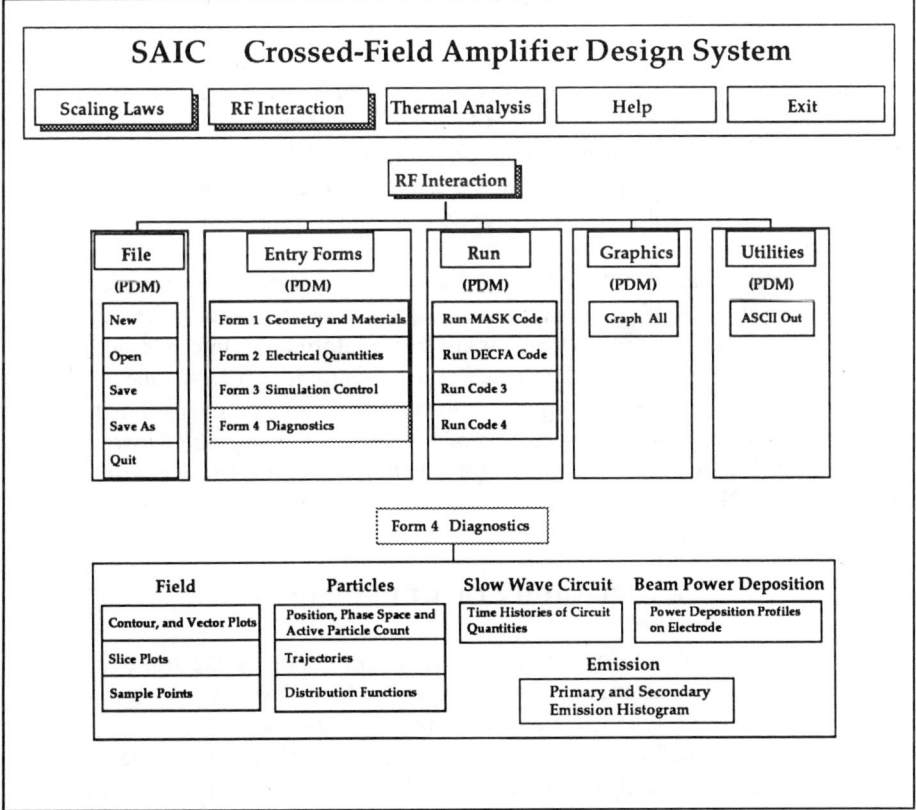

Figure 4 Menu map or menu tree from SAIC's GUI front-end shows the hierachy of the pull-down menus.

particular: How many mouse buttons are used? What event activates a window? How are windows lowered and raised? The "look and feel" of the GUI depends on the window system, the window manager, the toolkit, the widget set, and the design of the GUI. In addition, the X Windows's System supplies a Networking Library that handles the inter-process communications for the client server model of the window system. Figure 3 shows this client server model.

Various "looks and feels" are presented to the user on UNIX platforms with the following widget sets: Athena Xaw, Motif, DEC Windows, OPEN LOOK, and XView. Each widget set has an intrinsic library which serves as the framework for the windows and sits on the basic windowing system, Xlib (XLibrary). The widget sets add additional components like pop-up menus, scroll bars, dials, sliders, icons, file browsers, etc.

The front-end of the GUI is the section that meets the user's eye. The menus, forms, boxes, buttons, simple text widgets, and events trapped comprise the front-end of the GUI. The most appropriate language, we find, for building the front-end on a UNIX platform is the C-Language. (Visual Basic is quite good on a PC platform. On the Macintosh, we know of an impressive GUI built with Pascal.)

The details of the GUI's navigation and information prompted from the user are the essential components for determining what the application should do. The navigation through the menus and forms are specified in a menu tree or menu map. Figure 4 shows a menu map for SAIC's Crossed-Field Amplifier Design System GUI. The hierarchy of the menus as the programmer delves deeper into the application is depicted. Figure 3, from the same application, shows various types of widgets and a control panel on the bottom for navigating through the system that permits simultaneously visible forms.

Last, but certainly not least, is the back-end processor code of the GUI that does all the hard work of coordinating the program flow, processing the call back routines activated by the user's events, transferring information into and out of the widgets of the GUI, producing the name list decks and input files for the simulation codes, and spawning processes to run the simulation codes. For the most part, we prefer to write this section of the GUI in C-Language. However, when formatted input files are required for a FORTRAN program, we write the subroutines in FORTRAN. Logistically, we find writing the back-end code more time consuming than the menus, and forms.

A HIERARCHY OF GUI BUILDERS

The programming level a user chooses when developing the front-end of a GUI can vary, from an astonishing low level where he or she develops the windowing system in C-Language and a primitive graphics kernel, to a high level where a "drag and drop" GUI builder is used to generate C-Language code for the menus, menu hierarchy, call back structure, and forms. Figure 5 show the various levels a program can access Xlib routines when building a GUI based on the X Window System. Here the application code can access Xlib directly, use a widget set which in turn accesses Xlib and Xt Intrinsics, or use Xt Intrinsics directly.

Figure 6 show a hierarchy of what we perceive the levels of developing

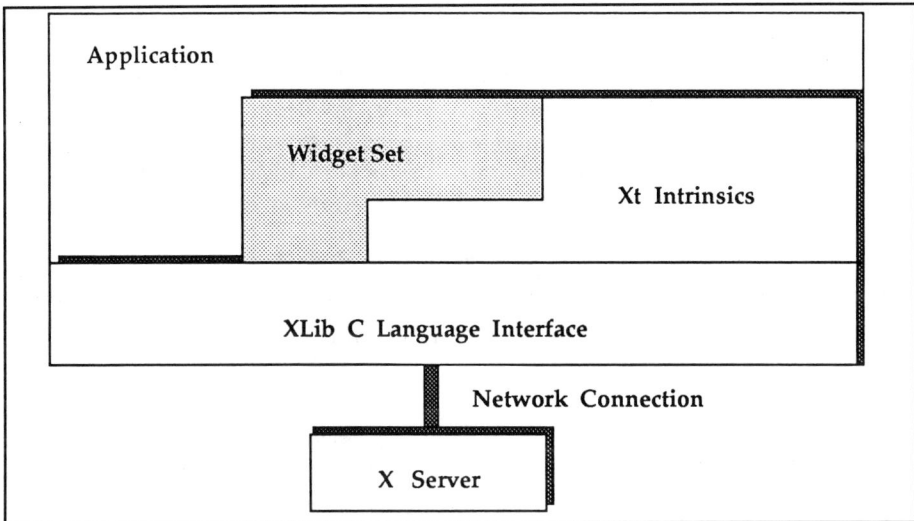

Figure 5 Various levels of interaction with Xlib gives the user maxixum flexibility in implementing GUI.

GUIs. Each level has benefits and draw backs. Levels are usually downwards

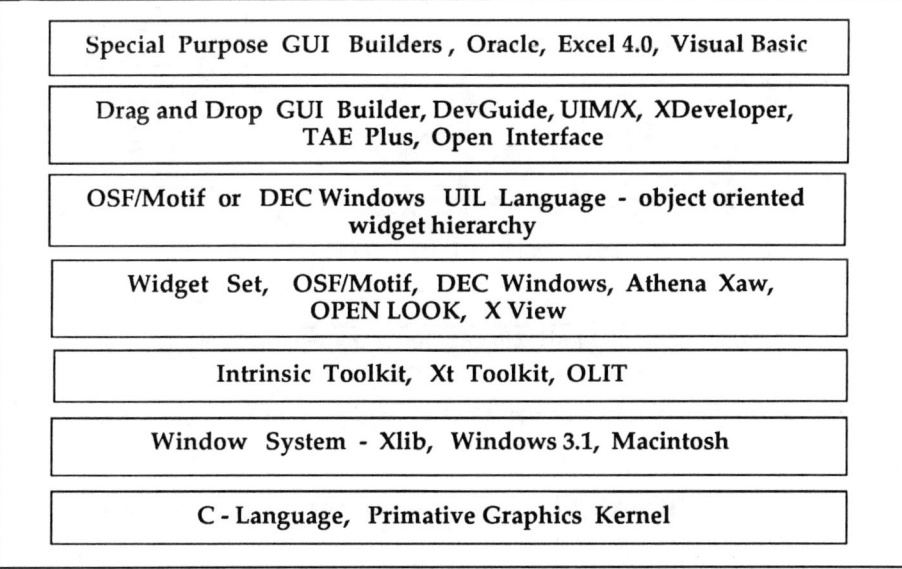

Figure 6 Hierachy of programming techniques for implementing GUIs

compatibly (lower level routines can be used by higher level routine).

OSF/MOTIF WIDGET SET

As an example of a widget set, we present Figure 7. It includes sliders, labels, file browsers, buttons, pull-down menues, and list boxes. These object oriented widgets save countless lines of code and "man years" of time.

Although the OPEN LOOK Intrinsics Toolkit widget set is similar to Motif the "look and feel" and style is not identical. Applications can be transformed from

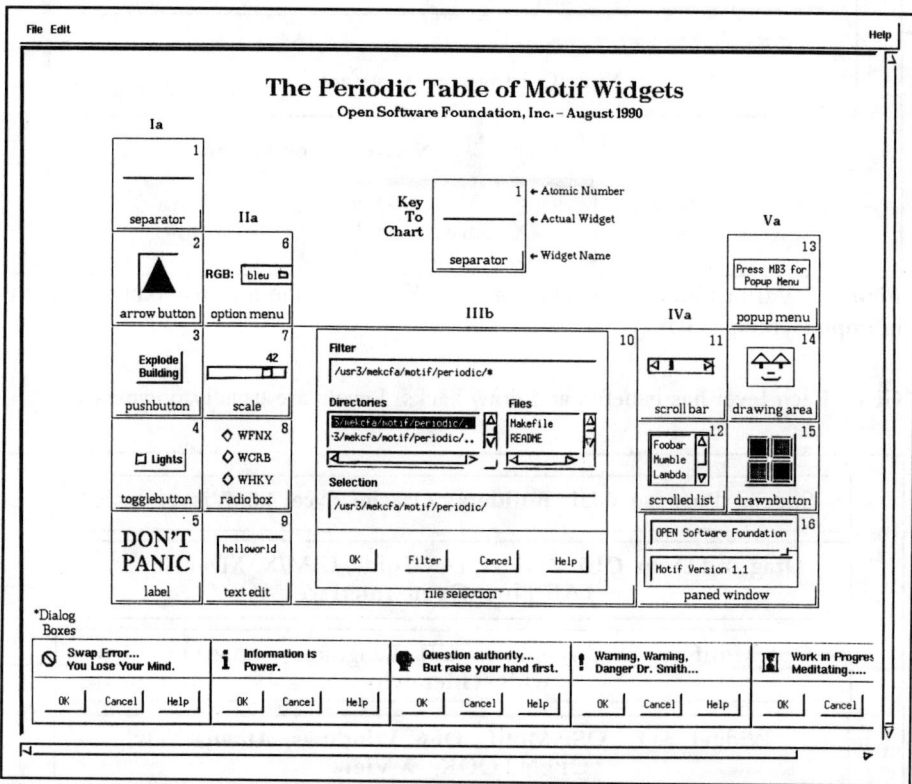

Figure 7 Periodic table of Motif widgets shows the various widgets in the set.

Motif to OPEN LOOK and vise verse but the widget transformations is not one to one.[2]

OPERATIONAL TECHNIQUES FOR BUILDING "GUI"S

Our first GUI project used a "cut and paste" editing technique from a UIL template file of widget objects, their attributes and associated hierarchy. We found this technique easy to learn and effective. The major drawback was the inability to see the positions and visual effects of the widgets while creating the forms. Each development iteration of menus and forms required recompilation of the UIL program and linking to the C Language back-end. This method was significantly faster than not using a UIL based GUI because the recompilation time of the UIL compiler is only a fraction of the that for the C Language compiler. Moreover, the conceptual separation of the front-end ,UIL program and the back-end C program has far reaching organizational benefits when developing a big GUI.

Lately, we have been experimenting with a number of "drag and drop" interactive GUI builders. Typical GUI builders permit the interactive creation of the front-end GUI using a pallet of widgets, a widget attribute specification form, a tree structure of menu hierarchy, a tool bar, and a view of the form. After the front-end is created, a code generator is used to produce C code (or BASIC, macro-language, etc. depending on the GUI builder), UIL code, or some combination of C and UIL. We worked with Devguide, XDesigner, and discussed TAE Plus version 5.1 with a satisfied user.[1] In general, our impression is that the learning curve for using these products was steeper than that for using UIL. However, there is a significant benefit of seeing the form interactively. On the PC platform we found Visual Basic exceptionally easy to use, fast to learn, and powerful. Figure 8 shows a representative widget pallet and display from a GUI builder.

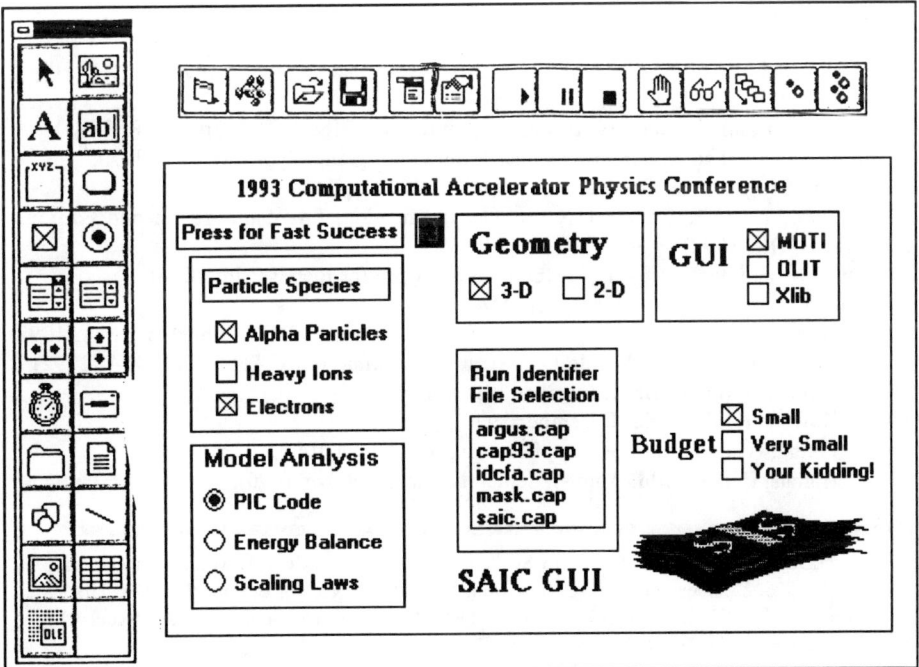

Figure 8 Visual Basic drag and drop GUI builder is representative of the GUI builders we have used.

REFERENCES

1. B. Shneiderman, Designing the User Interface, Strategies for Effective Human-Computer Interaction, Second Edition, Addison-Wesley Publishing Company, 1992.
2. Motif to OLIT Conversion Reference Paper, SunSoft 1990, Sun Microsystems, Inc.

BIBLIOGRAPHY

1. Open Software Foundation, OSF/Motif Programmer's Guide, (Prentice Hall, NJ, 1991).
2. Open Software Foundation, OSF/Motif Programmer's Reference, (Prentice Hall, NJ, 1991).
3. Open Software Foundation, OSF/Motif Style Guide Release 1.1, (Prentice Hall, NJ, 1991).
4. Open Software Foundation, OSF/Motif User's Guide, (Prentice Hall, NJ, 1991).
5. A. Nye and T. O'Reilly, X Protocol Reference Manual, Volume 0, (O'Reilly & Associates, Inc., MA, 1988).
6. A. Nye, Xlib Programming Manual, Vol. 1, (O'Reilly & Associates, Inc., MA, 1988).
7. A. Nye, editor, Xlib Reference Manual, Vol. 2, (O'Reilly & Associates, Inc., MA, 1988).
8. A. Nye and T. O'Reilly, X Window System User's Guide, OSF/Motif Edition, Volume 3, (O'Reilly & Associates, Inc., MA, 1991).
9. A. Nye and T. O'Reilly, X Toolkit Intrinsics Programming Manual, Volume 4, (O'Reilly & Associates, Inc., MA, 1990).
10. A. Nye and T. O'Reilly, X Toolkit Intrinsics Reference Manual, Volume 5, (O'Reilly & Associates, Inc., MA, 1990).
11. E. Johnson and K. Reichard, Power Programming... Motif, (MIS Press, OR, 1991).
12. D. A. Young, The X Window System Programming and Applications with Xt, OSF/Motif Edition, (Prentice Hall, NJ, 1990).
13. E. Johnson and K. Reichard, X Window Applications Programming, Second Edition, (MIS Press, OR, 1992).
15. Len Bass and Joelle Coutaz, Developing Software for the User Interface, (Addison-Wesley Publishing Company, MA, 1991.
16. M. Brian, Motif Programming: the essentials.. and more, (DEC Publishing, MA, 1992).
17. Scheifler and Gettys, X Windows System 3rd Edition, (DEC Publishing, MA, 1992).
18. K. Flowers, (UnixWorld, November 1990), p 135.
19. K. Flowers, (UnixWorld, December 1990), p 119.
20. D. Young, (UnixWorld, January, February, and March 1990).
21. A. Marcus, (UnixWorld, August, September, and October 1990).

---

[1]Devguide,Developers Guide, Sun Micro System. TAE Plus, COSMOS, University of Georgia, Athens, GA 30602. XDesigner, V.I. Corporation, MA. Object Vision, Borland Int., CA. UIM/X, Precision Visuals, NY. Open Interface, Neuron Data. ORACLE, Oracle Corp., CA. Excel 4.0, Microsoft Corp. Visual Basic, Micrsoft Corp.

GUIS FOR SCIENTIFIC CODE USAGE

N. Dionne
Raytheon Company, Equipment Division
190 Willow Street, Waltham, MA 02254

ABSTRACT

To achieve high-level functionality, an unadorned GUI based upon an enhanced version of the MIT X11 graphic routines has been integrated into SAIC's MASK code for keystroke-controlled, fully-interactive scientific application. Featured run-time capabilities include: a) buffered plot animation, b) mouse-driven data extraction, c) menu-driven parameter editing, d) postscript-based hard copy prints, e) run-state save, f) numerous plot display selections and g) optional GUI exit/return. A 400-line fortran-to-X-library interface (written in C) lies at the core of this utility, permitting either serial or concurrent interfacial keystroke control.

INTRODUCTION

Major PIC simulation codes have been traditionally employed in batch mode on large supercomputers. Plot information generally has been obtained using postprocessing graphics packages such as NCAR. With the availability of high performance UNIX-based work stations, graphical visualization systems such as AVS are often desirable for interactive control of PIC code applications, given the relatively long computation times incurred.

Since PIC codes often are applied to a wide variety of scientific problems, a customized setup of an interactive graphical user interface should be relatively quick and unencumbered with priority placed upon achieving a high degree of interactive functionality. A useful approach for accomplishing these objectives is summarized here for the specific case of SAIC's MASK code[1] adapted to a crossed-field amplifier simulation problem.

GUI BASED UPON MIT X11 ROUTINES

For simplicity and speed, a GUI approach based upon individual key press events has been adopted. Essentially, it is an extension of unsupported software based upon MIT's X11 routines which lies at the core of a 2-D graphics package called Graphic[2]. Using the X11 primitives, this brief code (written in C) performs a variety of operations such as opening and closing of the graphics windows, drawing lines and polygons, annotation, etc.. Among its most useful features are: 1) provision to wait for key press events and 2) function for returning both the cursor location and information needed in identifying the selected depressed key. Among several which have been added is a function for checking new key press events without waiting for them to occur. This collection of utilities enables a wide and powerful range of logical and graphical window operations under the control of the PIC code and user alike. A companion set of routines having similarly-named entry points generates postscript files corresponding to the graphics window displayed for run-time plotting requests.

GUI FUNCTIONALITY

From a scientific user's perspective, there are several desirable interactive capabilities which can significantly enhance the impact of a PIC code simulation. First, the ability to interrogate or edit various run control and key code variables for "mid-course" correction using a menu-driven editor is an obvious one. Second, mouse-driven extraction of 2-D function values via cursor location on any user-selected graphical display, circumvents the tedium of output data file perusal. Third, keystroke options to stop, continue or step the simulation time-iteration cycles promotes close-in scientific evaluation with options to save run state, print key plot displays, or effect early and orderly termination. Fourth, a particular useful feature for scientific visualization is a facility for making multiple plot displays in combination with synchronized animation having forward, reverse and freeze frame options for study and also extracting hard copy plots. Finally, keystroke option to terminate/ reactivate the GUI gives the user a selectable silent or interactive session. All of these feature have been incorporated into the interactive version of the MASK code along with a host of other convenient keystroke options.

INTERACTIVE PIC CODE MODALITIES

Depending upon the capabilities of the computer platform, there are at least two modalities in which keystroke control may be introduced. In the serial mode the keypress logic is placed directly into the PIC code, time iteration loop in the manner indicated on the left panel of Figure 1. In this case, keystroke control becomes active only when the keypress logic has been called at the completion of computational time step. For computationally-intensive simulations involving perhaps many charged particles, the delay in the action requested may be significant. For platforms supporting concurrent processing, this delay may be avoided by employing an alternate concurrent mode (see panel 2) in which the keypress monitoring is conducted in a separate parallel loop. Use of common blocks enables these forked processes to share pre-selected code variables for menu-driven editor access and for interprocess synchronization. In either mode of operation, the plotting control has been found to have negligible impact upon overall computation time.

DEMONSTRATION

Animation of power flow along a modeled crossed-field amplifier circuit followed by a dual display of both time-averaged rf power flow and corresponding particle plots have been recorded at actual run speed on a VCR tape, using a video converter. This recording served to demonstrate the "look and feel" of the basic GUI operating in a serial mode from a paused (inner loop) state.

REFERENCES

1. D. Chernin and N.Dionne,"Energy Exchange Instability In CFA Noise Generation",Technical Digest, IEDM92,p945.
2. J.Dannenhoffer,"Graphic Documentation, MIT X11 Software",1989.

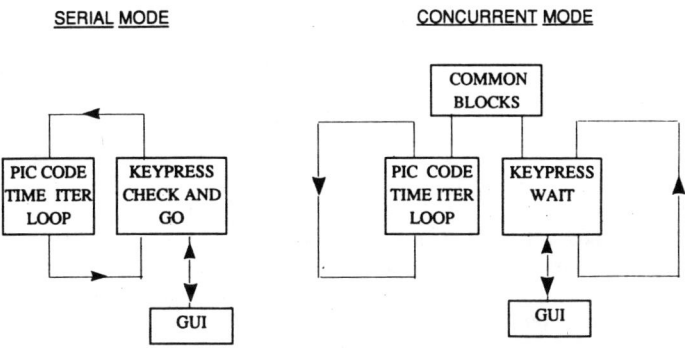

Fig. 1. Keypress Loop Modalities

Towards C++ Object Libraries for Accelerator Physics

Leo Michelotti
Fermi National Accelerator Laboratory

We must write robust, flexible, portable software that is easier to understand, maintain, modify, reuse, and extend. These attributes are more than mere buzzwords. They are important goals that computer scientists strive to achieve by refining their programming models and devising languages to support them. An important breakthrough was achieved by the introduction of object oriented programming (OOP) as the computing model behind such languages as Smalltalk, ADA, Eiffel, Objective-C, and C++ . OOP is *not* a "fad": it is arguably the most significant development in programming since the invention of FORTRAN, and it is the way that the best software will be written well into the next century. An "object" comprises structures of data, the functions that manipulate them, and rules for bringing them into and out of scope. OOP is a methodology for realizing and fully utilizing this abstract concept, an extension to programming of the basic technique that has advanced mathematics for centuries.

To avoid repeating the same material [1] too many times, I shall here merely state that two libraries of C++ classes (objects) have been under development for accelerator design and analysis: (a) MXYZPTLK, which implements the techniques of automatic differentiation and differential algebra, and (b) BEAMLINE, which provides objects for modelling beamline and accelerator components. The principal classes in the MXYZPTLK library are:

- `class DA : public dlist` ...is the fundamental object which carries data about functions and their derivatives and propagates them through arithmetic operations. The notation "`public dlist`" means that `DA` is derived from class `dlist`,[1] which is a doubly linked list. Each link of the list contains the index and value of a specific derivative.

- `class coord :  public DA` ...is the `DA` object which serves as the starting point for calculations; it is the `DA` implementation of a function which projects onto a coordinate. An important feature distinguishing it from a normal `DA` variable is that it cannot be changed by placing it to the left of an equal sign.

- `class LieOperator` ...implements vector field operators of the form $\underline{v}(\underline{z}) \cdot \partial/\partial \underline{z}$. Among other things, it is (to be) used in creating exponential maps.

while those in the BEAMLINE library include:

- `class bmlnElmnt` ...is the abstract base class containing data applicable to all possible beamline elements, including the geometric information determining its position and orientation in space. No `bmlnElmnt` declaration should ever appear in an application; rather, the beamline elements which do appear are all derived from this class.

- `class drift, quadrupole rbend, sector, ... : public bmlnElmnt` ...are the primitive beamline elements which can be instanced in application programs. Not all objects

---
[1] More accurately, it "inherits from" `dlist`.

appearing in MAD are presently available; for now, objects are created as needed by the author or upon request from others.

class beamline : public bmlnElmnt, public dlist ...is the composite object comprising a sequential, geometric arrangement of beamline elements. It inherits from **bmlnElmnt** as well as **dlist** to make it trivial to insert **beamlines** into **beamlines**, allowing for their hierarchical construction. This feature also facilitates certain recursive procedures in the class.

class proton, DAproton, bunchOfProtons, ...are the objects that can be propagated through beamline elements. Using a **proton** object performs tracking; using a DAproton object creates polynomial Poincaré maps, as its state is modelled with the DA variables from MXYZPTLK.

class beamlineImage, circuit, beamlineOverseer, ...are utility classes which, it is hoped, will make writing interactive programs easier. For example, **circuit** objects enable attributes of beamline elements to be changed in a correlated manner. The **beamlineOverseer** (not yet written) will perform communication between beamline elements, keep track of particles within the elements, and so forth. It will be useful for modelling such things as slow extraction of a bunch using septa or for connecting monitors to kickers in a stochastic cooling system.

Information describing a beamline element's location and orientation in space is contained in a "**geometry**" struct within the base class, **bmlnElmnt**. This struct comprises two frames, each consisting of a **threeVector** point of origin (yet another object) and three **threeVectors** which are unit vectors for a local right-handed coordinate system. The plane spanned by the first two unit vectors, and passing through the frame's point of origin, is a face of the beamline element; the two frames thus model the element's in-face and out-face. Every beamline element which inherits from **bmlnElmnt** possesses a member function called **.propagate** whose argument is a particle type (e.g, proton) and whose primary action is to change the state of the particle so as to reflect its passage from the in-face to the out-face. In this way, the beamline element classes implement the "lego" concept discussed by Forest. [2]

The methods employed by the **.propagate** function is contained in a collection of small files of C++ source code, the "physics files," one for each element type and particle class of interest. (The default provided by the base class, **bmlnElmnt**, is drift physics.) In fact, there will be a number of such collections reflecting varying levels of sophistication. Simple applications – such as calculating lattice functions for an uncoupled, design lattice – could use the most basic physics files, while more complicated applications – such as simulating stochastic cooling in a ring with alignment errors and using symplectic numerical integration through nonlinear focussing elements – would use the more elaborate physics files. (Of course, the latter could be used in the simpler calculations as well, but there is some value in doing simple calculations in a simple manner.) Because the physics will be isolated within collections of physics files, it will be easy to change from one to another, or even to modify them in order to try different ideas.

It was mentioned that the **.propagate** functions alter the state of the particle in passing from the in-face to the out-face of a beamline element. They can also change the element itself: for example, it is possible to write the functions so that a bunch of protons sets up a mode in an RF cavity which then modifies the cavity's action on subsequent protons. Using OOP, no further complexity need be added to the programming paradigm in order to achieve these things.

MXYZPTLK exists in a reasonably usable state; BEAMLINE is still in its infancy. The most ambitious goal is to put together a library of objects for use (and reuse) in programs at *all* levels of accelerator design and analysis, from the zeroth order task of placing dipoles so as to establish the correct design orbit to the problem of calculating normal forms of 137th degree polynomial maps. This goal is not something that can be accomplished by one person (at least, not *this* person) or in a short amount of time; although one can learn enough C++ to *use* objects fairly quickly, it takes years to become proficient at *creating* classes well. However, it is achievable provided that we start from the correct programming paradigm so that the library can evolve in a coherent, controllable manner ("organic programming") from its earliest stage all the way through, if desired, the development of complete, graphics-intensive accelerator CAD programs. I am proceeding under the beliefs that (a) object-oriented programming is that paradigm and (b) C++ is the correct language in which to implement it. Those who wish to play around with the MXYZPTLK and BEAMLINE libraries in whatever their current state of existence may obtain them via ftp from calvin.fnal.gov, in the subdirectories pub/outgoing/michelotti/beamline and pub/outgoing/michelotti/mxyzptlk. Regrettably, the only documentation included at this point is a three-year-old MXYZPTLK User's Guide, but a few simple demo programs are provided in the package.

References

[1] Leo Michelotti. Exploratory orbit analysis. In Floyd Bennett and Joyce Kopta, editors, *Proceedings of the 1989 IEEE Particle Accelerator Conference*. IEEE, March 20-23, 1989. IEEE Catalog Number 89CH2669-0.

— A C++ hacker's implementation of automatic differentiation. In *Automatic Differentiation of Algorithms: Theory, Implementation, and Application*. SIAM, Philadelphia, PA, 1991.

— MXYZPTLK: A practical, user-friendly C++ implementation of differential algebra: User's guide. Fermi Note FN-535, Fermilab, January 31, 1990.

— MXYZPTLK and BEAMLINE: C++ objects for beam physics. In *Advanced Beam Dynamics Workshop on Effects of Errors in Accelerators, their Diagnosis and Correction. (Corpus Christi, Texas. October 3-8, 1991)*. American Institute of Physics, 1992. Conference Proceedings No.255.

— Accelerator physics analysis with an integrated toolkit. Technical Report Fermilab-Conf-92/219, Fermilab, July 1992.

— A note on the automated differentiation of implicit functions. TM-1742, Fermilab, June, 1991.

— Automatic Differentiation of Limit Functions. Fermilab-Conf-93/135. To be published in the Proceedings of the 1993 U.S. Particle Accelerator Conference, Washington, D.C.

[2] E. Forest and K. Ohmi, KEK Report 92-14, September 1992.

NEW FEATURES IN COSY INFINITY

Martin Berz
Department of Physics and Astronomy, and
National Superconducting Cyclotron Laboratory
Michigan State University
East Lansing, MI 48824

ABSTRACT

COSY INFINITY is a DA-based code that allows the computation of high order maps including parameter dependencies for systems consisting of electric, magnetic, and circular elements, including their fringe fields. It is written in standard FORTRAN with dedicated versions for a variety of computers and supports several different graphics drivers. It has a very intuitive user interface based on a programming language with nonlinear optimization features built in and an interface to MAD input. One example for the power and generality of this interface is the analytical search for four cell third order achromats described in a companion paper. Currently the code is used by about 100 registered users in about 60 laboratories. Recently several new features have been added to the code, including the fast fringe field method and high precision fringe field descriptions described in companion papers. For the study of repetitive systems, the conventional DA - Lie algebraic normal form methods were replaced with a purely DA formulation, which is rather direct and also allows the analysis of non symplectic maps. Using the optimization features of COSY, these methods can readily be used for nonlinear optimization in general, and in particular the suppression of nonlinear resonances. Furthermore, the use of parameters allows a very direct and efficient computation of high-order chromaticities even without normal form theory. For the high-order correction of spectrographs, a new method of trajectory reconstruction allowing a computational correction of otherwise uncorrectable aberrations has been developed. This method will be used for the S800, a large acceptance high resolution magnetic spectrograph now under construction at NSCL.

INTRODUCTION

This paper gives the background and examples of some of the new features in COSY INFINITY[1], concentrating on tools relevant to the study of repetitive systems. Other features of the code are described in [2] and some less recent

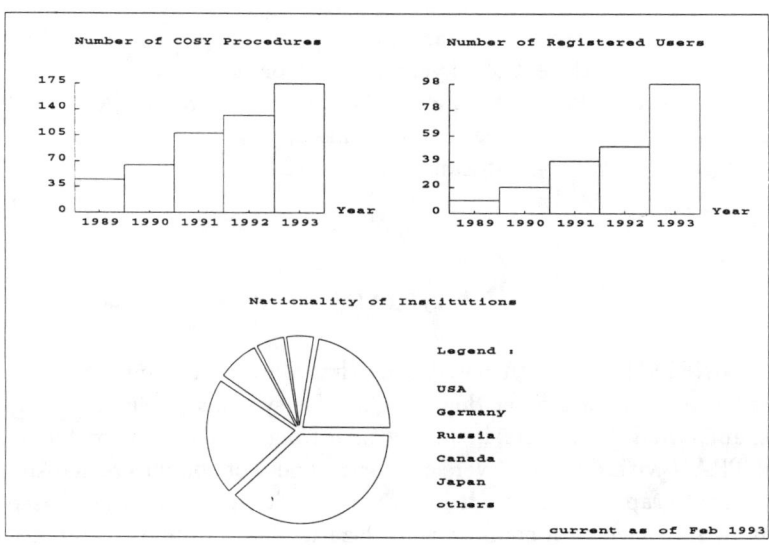

Figure 1: Some statistics about COSY INFINITY (picture generated in COSY's graphics environment)

papers [3, 4, 5]. We will discuss the use of various symplectification methods for the tracking in circular accelerators. Furthermore, a short overview over DA-based normal form techniques is presented and exemplified through one of its applications, the correction of nonlinear resonances.

Several other of the new features cannot be discussed here and are presented elsewhere; these include the new fast fringe field methods [6], the methods used for computational correction of aberrations in spectrographs by measurements in two planes [7, 8], the computation of maps from arbitrary measured fields, as well as the tools used for the design of third and higher order achromats [9, 10]. Besides these features based on new techniques, there are also a variety of other new tools of a more technical nature. These are connected to standard problems of accelerator design, to interactive graphics, as well as to several improvements of existing tools. A good overview over the key features of the code can be found in the extensive demo that is part of the COSY shipment.

THE CODE COSY

COSY INFINITY is a code for the simulation and design of particle optical systems. Since the first official version in 1989, a total of six versions with an increasing number of features have been released, and currently there are a total of about 100 registered users; more details are shown in figure 1. COSY is based on differential algebraic methods, which are described in detail elsewhere [11, 12, 13], and which lately have been widely used in most of the other newly emerging codes.

For the sake of portability, the code is based on standard FORTRAN 77. Despite the new powerful language environments which have become available recently, this workhorse of programming is still by far the most widespread language on the computers used for accelerator design and simulation, and it appears that this may only change slowly. In order to isolate COSY's programmers and users from the world of FORTRAN, the code employs its own programming language. The language is object oriented and has the flavor of PASCAL; its commands are as follows:

```
BEGIN ;        END ;                         { Begins and ends program              }
INCLUDE ;      SAVE ;                        { Includes and saves compiled code     }
VARIABLE ;                                   { Declares a local variable            }
PROCEDURE ;    ENDPROCEDURE ;                { Declares a local procedure           }
FUNCTION ;     ENDFUNCTION ;                 { Declares a local function            }
< assignments > ;                            { Sets value of variable               }
< procedure calls > ;                        { Calls previously defined  procedure  }
IF ;           ENDIF ;                       { Executed once if argument is true    }
WHILE ;        ENDWHILE ;                    { Executed while argument is true      }
LOOP ;         ENDLOOP ;                     { Stepping argument                    }
FIT ;          ENDFIT ;                      { varying arguments to fit conditions  }
```

Except for the last one, the flow control statements are rather standard. The FIT block is executed over and over again as long as the sum of squares of the objective variables listed in the ENDFIT statement can be reduced by modifying the values of the free variables listed in the FIT statement. The ENDFIT statement also contains the number of the optimizer to be used as well as the tolerance and the maximum number of iterations allowed.

The compiler for this language comprises about 5,000 of the approximately 30,000 lines of COSY's FORTRAN code. It has a completely rigorous syntax and error analysis and is comparable in speed to other compilers. The compiled code is stored in a machine independent integer-based meta format, which can either be saved for inclusion in later code or executed directly.

The object oriented features of the code allow a direct use of the differential algebraic operations contained in the 10,000 line DA package, the bulk of which was written in the middle of 1987, and whose interface was adjusted slightly to accommodate COSY's memory management.

SYMPLECTIC TRACKING

In this section we will briefly summarize the background behind the various symplectification methods in COSY as well as show some examples that show typical phenomena of symplectic tracking. These methods have also been described in greater detail elsewhere [14, 12, 11]. While symplectic tracking generally does not improve the mathematical accuracy of the tracking, it preserves the symplectic symmetry and thus in many cases hopefully allows a more detailed description of long term effects.

One important tool for symplectification as well as the normal form methods discussed below is the inversion of Taylor maps. Let $[A]_n$ be a map in the differential algebra[12, 11] $_nD_v^n$; we split the map into its linear and nonlinear parts $[A]_n = [A_1]_n + [A_2]_n$. Denoting the inverse to order n by $[M]_n$, we obtain

$$\begin{aligned}([A_1] + [A_2]_n) \circ [M]_n &= [E]_n \Rightarrow \\ [A_1] \circ [M]_n &= [E]_n - [A_2]_n \circ [M]_n \Rightarrow \\ [M]_n &= [A_1^{-1}] \circ ([E]_n - [A_2]_n \circ [M]_{n-1}).\end{aligned} \quad (1)$$

The composition of maps "∘", like many other operations for functions, transfers to our differential algebra because of the nilpotency of the map [12]. The necessary computation of A_1^{-1} is a linear matrix inversion. Equation (1) can now be used in a recursive manner to compute the $[M]_i$ order by order.

Symplectic tracking is usually based on representing maps by generating functions; these can be one turn maps, maps describing part of the ring, or maps from which the linear map (which is symplectic or can easily be symplectified) has been factored out. There are four possible types of generating functions that can be used:

$$\begin{aligned} F_1(\vec{q}_i, \vec{q}_f) \text{ satisfying } (\vec{p}_i, \vec{p}_f) &= (+\vec{\nabla}_{q_i} F_1, -\vec{\nabla}_{q_f} F_1) \\ F_2(\vec{q}_i, \vec{p}_f) \text{ satisfying } (\vec{p}_i, \vec{q}_f) &= (+\vec{\nabla}_{q_i} F_2, +\vec{\nabla}_{p_f} F_2) \\ F_3(\vec{p}_i, \vec{q}_f) \text{ satisfying } (\vec{q}_i, \vec{p}_f) &= (-\vec{\nabla}_{p_i} F_3, -\vec{\nabla}_{q_f} F_3) \\ F_4(\vec{p}_i, \vec{p}_f) \text{ satisfying } (\vec{q}_i, \vec{q}_f) &= (-\vec{\nabla}_{p_i} F_4, +\vec{\nabla}_{p_f} F_4), \end{aligned} \quad (2)$$

Tracking is then performed by first using the n-th order Taylor map to obtain a (usually very good) estimate of the final conditions of the particle, and then performing an iterative correction using the above conditions. In practice, in most cases one step of a Newton's method yields accuracy to machine precision.

Obtaining the "mixed" relations based on the gradient of the generating function simply requires inversion of appropriate maps. We denote with $\mathcal{M}_1$

the part of the transfer map describing the final positions, and with $\mathcal{M}_2$ the part describing the final momenta. Thus, we have $\mathcal{M} = (\mathcal{M}_1, \mathcal{M}_2)$. We do the same with the identity map: $\mathcal{E} = (\mathcal{E}_1, \mathcal{E}_2)$. In order to obtain the "mixed" relations $(\vec{q}, \vec{p}) = \mathcal{F}(\vec{q}, \vec{p})$ (the other cases are done analogously), we start by setting $\mathcal{N} = (\mathcal{E}_1, \mathcal{M}_2)$. Then,

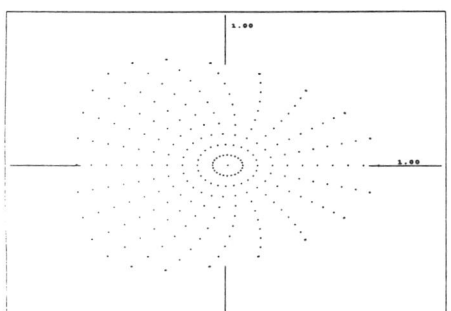 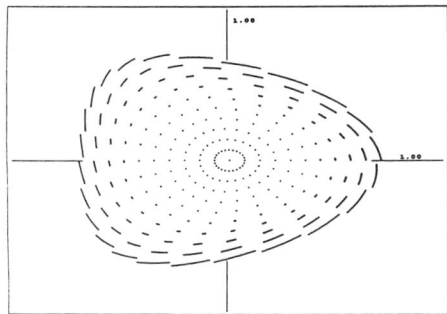

Figure 2: 1000 turn tracking through a 15 degree homogenous dipole, which is free of amplitude tune shifts and has exact tune of 1/24. The 19th order unsymplectified tracking (left) accurately describes the motion, while the fifth order unsymplectified tracking (right) introduces errors both in amplitude and in phase.

$$(\vec{q_i}, \vec{p_f}) = \mathcal{N}(\vec{q_i}, \vec{p_i}). \tag{3}$$

It turns out that the generating function exists if and only if $\mathcal{N}$ is invertible, and if it is we obtain

$$(\vec{q_i}, \vec{p_i}) = \mathcal{N}^{-1}(\vec{q_i}, \vec{p_f}). \tag{4}$$

Composing the map $(\mathcal{M}_1, \mathcal{E}_2)$ and the map $\mathcal{N}^{-1}$, we finally obtain the desired "mixed" relations:

$$(\vec{q_f}, \vec{p_i}) = \left((\mathcal{M}_1, \mathcal{E}_2) \circ \mathcal{N}^{-1}\right)(\vec{q_i}, \vec{p_f}) = \mathcal{F}(\vec{q_i}, \vec{p_f}). \tag{5}$$

Now going to the respective equivalence classes, it is again required that the transfer map $\mathcal{M}$ be origin preserving and hence the corresponding DA vectors nilpotent. Altogether, the whole process of obtaining the gradient of the generating function which allows symplectic tracking can be performed to arbitrary order using only composition and inversion of differential algebraic transfer maps.

In the following we want to analyze the behavior of various symplectic tracking methods for a case in which there are noticeable nonlinear effects, but in which there is a simple check for the accuracy of the predicted nonlinear behavior; a variety of other interesting studies related to symplectification methods can be found in [15]. In our example we choose the map of a fringe field less dipole magnet

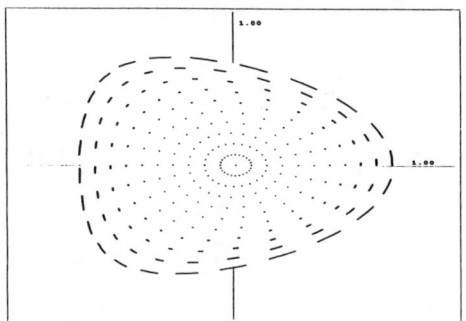 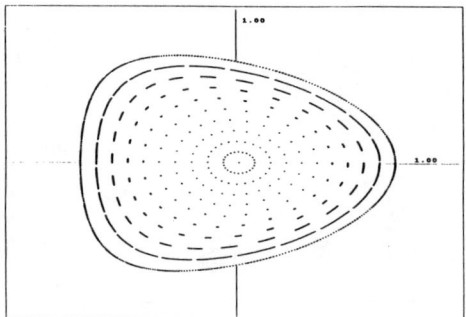

Figure 3: Tracking through dipole as in previous picture; shown are fifth order symplectic tracking using F_3 generating functions for the map (left) as well as for the nonlinear part only (right). While volume seems conserved, erroneous amplitude tune shifts prevail and even go in different directions in the two cases.

with a bending angle of 15 degrees. Over a large enough aperture, such a magnet exhibits sizable nonlinearities. Since the motion in uniform magnetic fields follows closed circles, we can conclude that the linear motion through the repeated dipole is stable with a tune is 1/24, and also that there are no amplitude dependent tune shifts. Because of this latter property, this element provides a good check for any tracking method as well as the ability of a code to describe the nonlinear effects of a dipole magnet properly.

The left phase plot in Figure 2 shows 1000 turn tracking through the dipole using a 19th order map generated with COSY INFINITY. Apparently over the whole shown aperture range of almost one meter, there is an exact repetition of the motion after 24 turns, and hence there are no noticeable tune shifts. The right phase plot shows the same tracking performed to fifth order, i.e. with less accuracy. Apparently, the repetitive behavior is predicted incorrectly in two respects; first of all, there are sizable artificial tune shifts, but also the total area of phase space seems to shrink.

Figure 3 shows the same fifth order tracking with two different symplectification methods, using an F_3 type generating function for the whole map (left), as well as for its nonlinear part only after the linear part has been factored out (right). In these cases, the errors in amplitude seem corrected, indicating that phase space volume is conserved due to symplecticity. On the other hand, the errors in the tune shifts prevail. While they have approximately the same magnitude as in the unsymplectified case, their signs are different in the two cases. While care is in order when trying to draw conclusions from examples only, these pictures seem to support the general notion that symplectification is not the cure of all evil, and that it is advantageous to start out with as accurate a map as possible.

DA NORMAL FORM THEORY

Normal form theory is a powerful technique to analyze the motion in nonlinear systems. It has been used in accelerator physics in various forms [16, 17, 18]; recently we showed that it is possible to phrase the method in a purely differential algebraic formalism which assumes a particularly elegant and simple form [19, 11]. We here sketch some of the features of this method. As a first step, we perform a transformation to the parameter-dependent fixed point of the map which satisfies

$$(\vec{z}_F, \vec{\delta}) = \mathcal{M}(\vec{z}_F, \vec{\delta})$$

which can also be written as $(\mathcal{M} - I_H)(\vec{z}_F, \vec{\delta}) = (\vec{0}, \vec{\delta})$, where the map I_H contains a unity map in the upper block describing the variables and zeros everywhere else. This form of the fixed point equation shows that the parameter dependent fixed point $\vec{z}_F$ can be obtained by inversion of the map $\mathcal{M} - I_H$, which can be readily performed in the DA picture as described above, and hence

$$(\vec{z}_F, \vec{\delta}) = (\mathcal{M} - I_H)^{-1}(\vec{0}, \vec{\delta}). \tag{6}$$

Interestingly, after the transformation to the fixed point has been performed, it is already possible to compute all parameter dependent tune shifts including chromaticities to arbitrary order directly without employing any normal form methods, resulting in a substantial gain in speed [20]. For our purposes, this also serves as an excellent cross check for the computation of tune shifts based on normal form theory. Table 1 shows the results of such a check, together with an independent test performed by Roger Servranckx using his code DIMAD. He compared numerically computed tune shifts with the DA results for the Saskatoon EROS ring. To obtain the complete agreement in the physics model used, a version of DIMAD that allows DA-based map extraction [21] based on the DA package and precompiler [22, 23] was used to compute the map, which was then read by COSY INFINITY and analyzed. As table 1 shows, agreement within the accuracy of the numerical methods used in DIMAD was obtained.

After the fixed point transformation has been performed, the first step in the process of the normal form transformation is the diagonalization of the linear map of the system, which is always possible if there are no resonances and linear stability, but also in various other cases[20]. After this step, the map has the form $\mathcal{M} = \mathcal{R} + \mathcal{S}_m$, where $\mathcal{R}$ is the linear diagonalized map. The main part of the actual normal form transformation consists of a sequence of transformations with maps of the form

Order	COSY Direct versus COSY Normal Form	COSY Direct versus DIMAD numerical
0	$1 \cdot 10^{-15}$	$1 \cdot 10^{-5}$
1	$1 \cdot 10^{-15}$	$1 \cdot 10^{-4}$
2	$5 \cdot 10^{-15}$	$5 \cdot 10^{-3}$
3	$1 \cdot 10^{-14}$	$1 \cdot 10^{-2}$
4	$1 \cdot 10^{-14}$	$4 \cdot 10^{-2}$

Table 1: Calculations of higher order chromaticities. Compared are the two analytical COSY modes based on a direct computation and normal form theory as well as the direct COSY mode with the numerical methods in DIMAD.

$$\mathcal{A}_m =_m \mathcal{E} + \mathcal{T}_m, \qquad (7)$$

where $\mathcal{T}_m$ vanishes to order $m - 1$. Because the linear part of $\mathcal{A}_m$ is the unity map, $\mathcal{A}_m$ is invertible. Moreover, inspection of the algorithm to invert transfer maps reveals that up to order m, we have $\mathcal{A}_m^{-1} =_m \mathcal{E} - \mathcal{T}_m$. To study the effect of the transformation, we now infer up to order m:

$$\begin{aligned}\mathcal{A} \circ \mathcal{M} \circ \mathcal{A}^{-1} &=_m (\mathcal{E} + \mathcal{T}_m) \circ (\mathcal{R} + \mathcal{S}_m) \circ (\mathcal{E} - \mathcal{T}_m) \\ &=_m (\mathcal{E} + \mathcal{T}_m) \circ (\mathcal{R} + \mathcal{S}_m - \mathcal{R} \circ \mathcal{T}_m) \\ &=_m \mathcal{R} + \mathcal{S}_m + (\mathcal{T}_m \circ \mathcal{R} - \mathcal{R} \circ \mathcal{T}_m) \end{aligned} \qquad (8)$$

For the first step, we have used $\mathcal{S}_m \circ (\mathcal{E} - \mathcal{T}_m) =_m \mathcal{S}_m$ which holds because $\mathcal{S}_m$ is nonlinear and $\mathcal{T}_m$ is of order m. In the second step we used $\mathcal{T}_m \circ (\mathcal{R} + \mathcal{S}_m - \mathcal{R} \circ \mathcal{T}_m) =_m \mathcal{T}_m \circ \mathcal{R}$ which holds because $\mathcal{T}_m$ is of exact order m and everything in the second term is nonlinear except $\mathcal{R}$. The last line now reveals that $\mathcal{S}_m$ can be simplified by choosing the commutator $\mathcal{C}_m = \{\mathcal{T}_m, \mathcal{R}\} = (\mathcal{T}_m \circ \mathcal{R} - \mathcal{R} \circ \mathcal{T}_m)$ suitably. Indeed, if the range of $\mathcal{C}_m$ is the full space, then $\mathcal{S}_m$ can be removed entirely. However, as we shall see, most of the time this is not the case. Let $(\mathcal{T}_{mj}^{\pm}|k_1^+, k_1^-, ..., k_n^+, k_n^-)$ be the Taylor expansion coefficient of $\mathcal{T}_{mj}$ with respect to

$(s_1^+)^{k_1^+}(s_1^-)^{k_1^-} \cdot \ldots \cdot (s_n^+)^{k_n^+}(s_n^-)^{k_n^-}$ in the j-th component pair of $\mathcal{T}_m$; so $\mathcal{T}_{mj}^\pm$ is written as

$$\mathcal{T}_{mj}^\pm = \sum (\mathcal{T}_{mj}^\pm | k_1^+, k_1^-, ..., k_n^+, k_n^-) \cdot (s_1^+)^{k_1^+}(s_1^-)^{k_1^-} \cdot \ldots \cdot (s_n^+)^{k_n^+}(s_n^-)^{k_n^-} \quad (9)$$

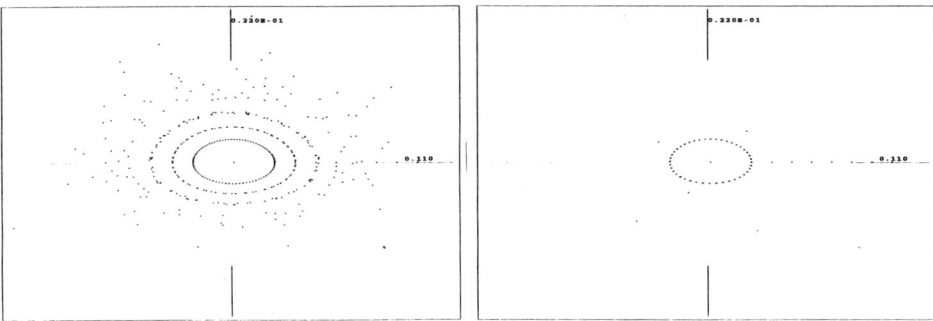

Figure 4: Tracking in ring without resonance correction; shown are horizontal phase diagrams without (left) and with (right) coupling of the vertical motion into horizontal motion.

Similarly we identify the coefficients of $\mathcal{C}$ by $(\mathcal{C}_j^\pm | k_1^+, k_1^-, ..., k_n^+, k_n^-)$. Because $\mathcal{R}$ is diagonal, it is easily possible to express the coefficients of $\mathcal{C}$ in terms of the ones of $\mathcal{T}$. One obtains

$$(\mathcal{C}_j^\pm | k_1, k_1^-, ..., k_n^+, k_n^-) = C_j^\pm(\vec{k}^+, \vec{k}^-) \cdot (\mathcal{T}_j^\pm | k_1^+, k_1^-, ..., k_n^+, k_n^-)$$
$$= \left((\prod_{l=1}^n r_l^{(k_l^+ + k_l^-)}) \cdot e^{i\vec{\mu} \cdot (\vec{k}^+ - \vec{k}^-)} - r_j \cdot e^{\pm i\mu_j} \right) \cdot (\mathcal{T}_j^\pm | k_1^+, k_1^-, ..., k_n^+, k_n^-) \quad (10)$$

Now it is apparent that a term in $\mathcal{S}_j^\pm$ can be removed if and only if the factor $C(\vec{k}^+, \vec{k}^-)$ is nonzero; if it is nonzero, then the required term in $\mathcal{T}_j^\pm$ is just the negative of the respective term in $\mathcal{S}_j^\pm$ divided by $C(\vec{k}^+, \vec{k}^-)$. It turns out [19] that if there are no resonances, the map resulting after the normal form transformation is just an amplitude dependent rotation. In case one is close to a resonance, some of the coefficients of the coordinate transformation map may become rather large and thus limit the quality of convergence of the normal form transformation. This fact shows that the coefficients of the transformation equation provide an excellent measure for the sensitivity of the system to resonances, and their reduction allows the improvement of the nonlinear motion.

The COSY environment provides a tool to compute these resonance strengths, and by using COSY's FIT and ENDFIT structure, it is easy to reduce these by adjusting certain system parameters accordingly. Figure 4 shows some tracking pictures for a small ring that is very sensitive to the $2\nu_x - \nu_y$ resonance. While the dynamical aperture without coupling is acceptable (left picture), as soon as particles also exhibit vertical motion, the dynamical aperture in the horizontal direction collapses (right picture). The ring was resonance corrected with two

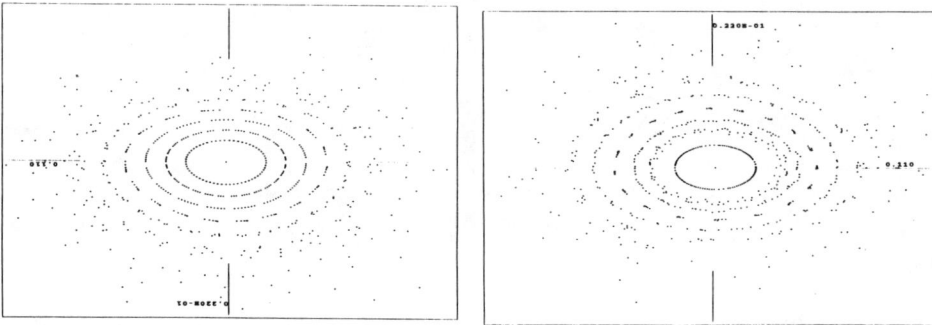

Figure 5: Tracking picture of horizontal motion after resonance correction. Again shown are phase diagrams without (left) and with (right) coupling of the vertical motion into the horizontal motion.

sextupoles using DA normal form methods. As figure 5 shows, after resonance correction, both the purely horizontal motion (left picture) as well as the horizontal motion under vertical coupling (right picture) show a significant improvement in stability.

In the meantime COSY's resonance suppression methods described here have been used for the SSC low energy booster [24] as well as a variety of other designs [21]; they are currently being applied to the PSR II ring at Los Alamos as well as the COSY synchrotron at KFA Jülich.

ACKNOWLEDGEMENTS

For financial support, we are grateful to the National Science Foundation as well as the Alfred P. Sloan foundation. For help with the pictures, I would like to thank Georg Hoffstätter. A lot of help with other parts of the code is also appreciated from Kyoko Fuchi, Ralf Degenhardt, Weishi Wan, and Meng Zhao.

References

[1] M. Berz. COSY INFINITY Version 6 reference manual. Technical Report MSUCL-869, National Superconducting Cyclotron Laboratory, Michigan State University, East Lansing, MI 48824, 1993.

[2] M. Berz. COSY INFINITY Version 6. *in: M. Berz, S. Martin and K. Ziegler (Eds.), Proc. Nonlinear Effects in Accelerators*, 1993.

[3] M. Berz. Computational aspects of design and simulation: COSY INFINITY. *Nuclear Instruments and Methods*, A298:473, 1990.

[4] M. Berz. COSY INFINITY. In *Proceedings 1991 Particle Accelerator Conference*, San Francisco, CA, 1991.

[5] M. Berz. COSY INFINITY, an arbitrary order general purpose optics code. *Computer Codes and the Linear Accelerator Community*, Los Alamos LA-11857-C:137, 1990.

[6] G. Hoffstätter and M. Berz. Efficient computation of fringe fields using symplectic scaling. In *Third Computational Accelerator Physics Conference*. AIP Conference Proceeings, 1993.

[7] M. Berz, K. Joh, J. A. Nolen, B. M. Sherrill, and A. F. Zeller. Reconstructive correction of aberrations in nuclear particle spectrographs. *Physical Review C*, 47,2:537, 1993.

[8] M. Berz, K. Joh, J. A. Nolen, B. M. Sherrill, and A. F. Zeller. On-line correction of residual aberrations in spectrographs. In *Proceedings 1991 Particle Accelerator Conference*, San Francisco, CA, 1991.

[9] E. Goldmann W. Wan and M. Berz. *in: M. Berz, S. Martin and K. Ziegler (Eds.), Proc. Nonlinear Effects in Accelerators*, 1993.

[10] W. Wan, E. Goldmann, and M. Berz. The design of a a four cell third order achromat. In *Third Computational Accelerator Physics Conference*. AIP Conference Proceeings, 1993.

[11] M. Berz. *High-Order Computation and Normal Form Analysis of Repetitive Systems, in: M. Month (Ed), Physics of Particle Accelerators*. American Institute of Physics, 1991.

[12] M. Berz. Arbitrary order description of arbitrary particle optical systems. *Nuclear Instruments and Methods*, A298:426, 1990.

[13] M. Berz. Differential algebraic description of beam dynamics to very high orders. *Particle Accelerators*, 24:109, 1989.

[14] M. Berz. *Symplectic Tracking in Circular Acceleratos with High Order Maps*, page 288. World Scientific, 1991.

[15] I. Gjaja and A. Dragt. In *M. Berz, S. Martin and K. Ziegler (Eds), Proceedings Workshop on Nonlinear Effects in Accelerators*. IOP Publishing, 1993.

[16] A. J. Dragt and J. M. Finn. Normal form for mirror machine Hamiltonians. *Journal of Mathematical Physics*, 20(12):2649, 1979.

[17] A. Bazzani, P. Mazzanti, G. Servizi, and G. Turchetti. Normal forms for Hamiltonian maps and nonlinear effects in a particle accelerator. *Il Nuovo Cimento*, 102 B, N.1:51, 1988.

[18] E. Forest, M. Berz, and J. Irwin. Normal form methods for complicated periodic systems: A complete solution using Differential algebra and Lie operators. *Particle Accelerators*, 24:91, 1989.

[19] M. Berz. Differential algebraic formulation of normal form theory. *in: M. Berz, S. Martin and K. Ziegler (Eds.), Proc. Nonlinear Effects in Accelerators*, 1993.

[20] M. Berz. Direct computation and correction of chromaticities and parameter tune shifts in circular accelerators. In *Proceedings XIII International Particle Accelerator Conference*, Dubna, 1992.

[21] Roger Servranckx. Private communication.

[22] M. Berz. The Differential algebra FORTRAN precompiler DAFOR. Technical Report AT-3:TN-87-32, Los Alamos National Laboratory, 1987.

[23] M. Berz. The DA precompiler DAFOR. Technical report, Lawrence Berkeley Laboratory, Berkeley, CA, 1990.

[24] R. Servranckx. Optics programs at TRIUMF. *in: M. Berz, S. Martin and K. Ziegler (Eds.), Proc. Nonlinear Effects in Accelerators*, 1993.

Zlib and Related Programs for Beam Dynamics Studies

Yiton T. Yan
Superconducting Super Collider Laboratory,* Dallas, TX 75237

Abstract

Zlib is a differential-algebraic and Lie-algebraic numerical library for subroutines that support beam dynamics studies. The source codes are written in Fortran. Hierarchical data structures are employed for speed optimization, particularly in vector computers (supercomputers). Dynamic memories are used for internal structural integer pointers and for required internal working memories. The use of Zlib is very much the same as the use of IMSL library except that a Zlib preparation subroutine should be called to set up the hierarchical structure before other Zlib subroutines can be called. There are currently about 200 subroutines in Zlib. Accompanied with Zlib are some specialized programs, such as one-turn-map extraction programs, nonlinear analysis programs, and symplectic one-turn-map tracking programs, for practical use in beam dynamics studies.

Natural properties of accelerators, of being perturbative for nonlinear effects and of being periodic for storage rings, have made it very interesting in applying differential algebras and Lie algebras for the study of single-particle beam dynamics.[1,2,3,4] However, such applications are very difficult or impossible without a systematic software that handles the algebras. Zlib is one among a few[5] that provide subroutines for such algebras. Currently, what is most needed for Zlib is an updated manual although an old manual,[6] which was written about three years ago, may still be somewhat useful.

Accompanied with Zlib are some programs for beam dynamics studies. Programs that use Zlib subroutines are Zmap: a map extraction program,[7] Zimaptrk: a symplectic implicit one-turn-map tracking program,[8] and some small specialized programs for nonlinear mapping analyses and post-tracking analyses. Implementation of Zlib subroutines in the program Teapot [9] is currently undergoing. Also, Zlib is used in the program SSCTRK for one-turn map extraction.[10]

Shown in the next page is a chart that shows how Zlib and its related programs are used. First a lattice design program, Synch[11] (not shown in the chart),

---

*Operated by the Universities Research Association, Inc., for the U.S. Department of Energy under Contract No. DE-AC35-89ER40486.

280 Zlib and Related Programs

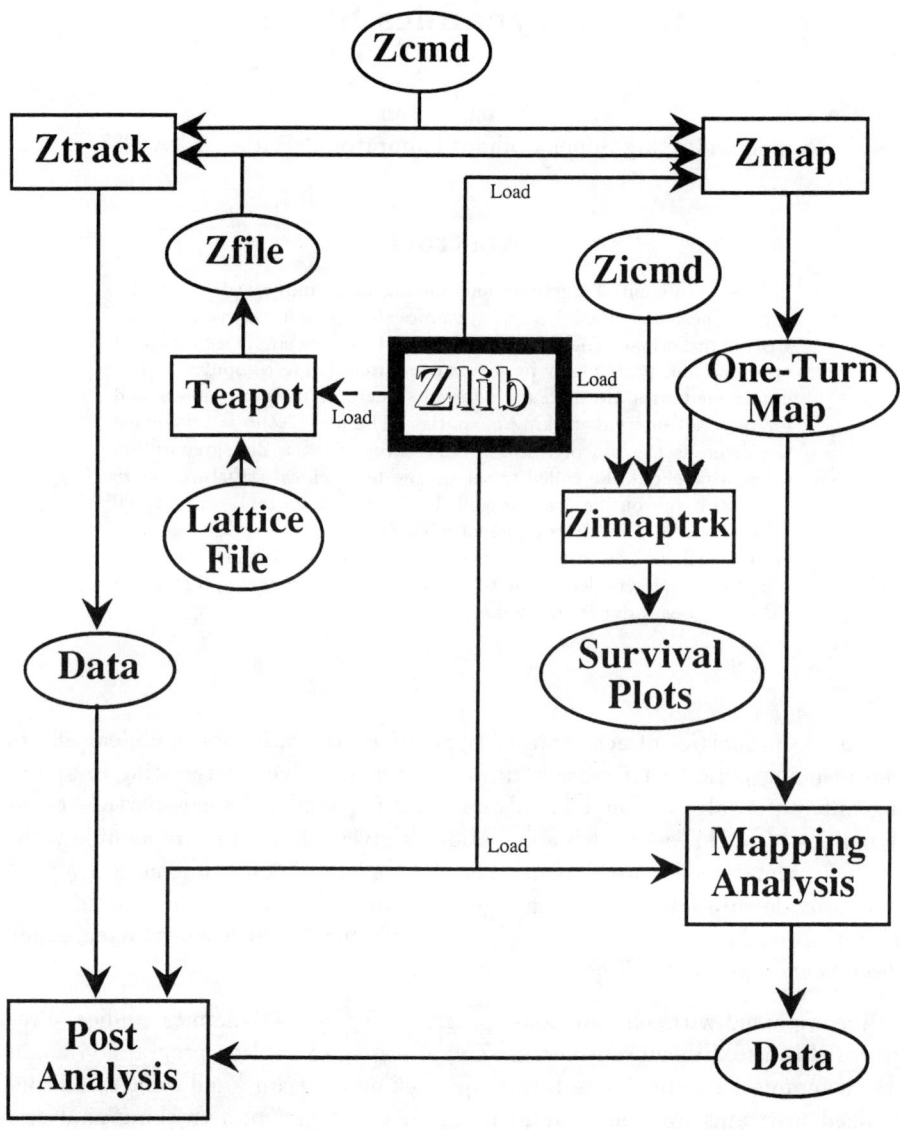

generates a linear lattice file which is then translated into a standard Mad[12] lattice file. After adding nonlinear error descriptions (for systematic and random multipole errors and random misalignment), the Mad lattice file serves as input of the program Teapot to generate Zfile, an element-by-element machine file. Accompanied with a command file, Zcmd, Zfile then serves as an input file of Ztrack for vectorized (and parallelized) element-by-element multi-particle tracking or as an input file for Zmap to extract one-turn maps. The one-turn maps are then analyzed by specialized mapping analysis programs or serve as an input of Zimaptrk for fast long-term tracking to obtain survival plots[13] for dynamic aperture studies. Occasionally post tracking data are analyzed with the aid of the information provided by the one-turn maps.

Since the start of its development in 1988, Zlib has made some contributions to practical studies, particularly on long-term stability studies. The development of Zmap made it possible for the extraction of one-turn high-order "order-by-order symplectic" Taylor maps for the SSC. Indeed, such maps were used to demonstrate, for the first time, that one-turn maps can be extracted for large static circular accelerators, that contains sufficient information for long-term stability studies.[14,15] Shown in Figure 1 is a survival plot for the SSC which shows that trackings up to a million turns with Ztrack (element-by-element tracking) and with the corresponding 11th-order Taylor-map tracking give the same dynamic aperture prediction. Straight-forward conversion (without explicitly using a generating function) of "order-by-order symplectic" Taylor maps into exactly symplectic implicit Taylor maps has concluded that one should be encouraged to use one-turn maps for long-term stability studies for large circular accelerators such as the SSC or the SSC High-Energy Booster (HEB).[8] For the same SSC nonlinear lattice as used for Figure 1, trackings with the corresponding 4th-order and 7th-order symplectic implicit Taylor maps show the same dynamic aperture as one can refer to the survival plots shown in Figure 1 of Ref. 8. Indeed, all trackings with the corresponding symplectic implicit Taylor maps with an order equal to or larger than 4 show very much the same dynamic aperture.

Shown in Appendix A is a small program that shows how the author used Zlib to test a Zlib subroutine for Lie transformation (Dragt-Finn factorization)[16] of "order-by-order symplectic" Taylor maps three years ago. For simplicity, only a fixed order "no" was used in this example program although different orders, equal to or lower than the maximum order "no" set up by the statement "call zpprep(nv,no,nm,0)", can be flexibly used in subroutine calls as provided by the subroutine parameters.

The author wishes to thank Robert Ryne and the program committee members for their invitation of his participation in this computational accelerator physics conference.

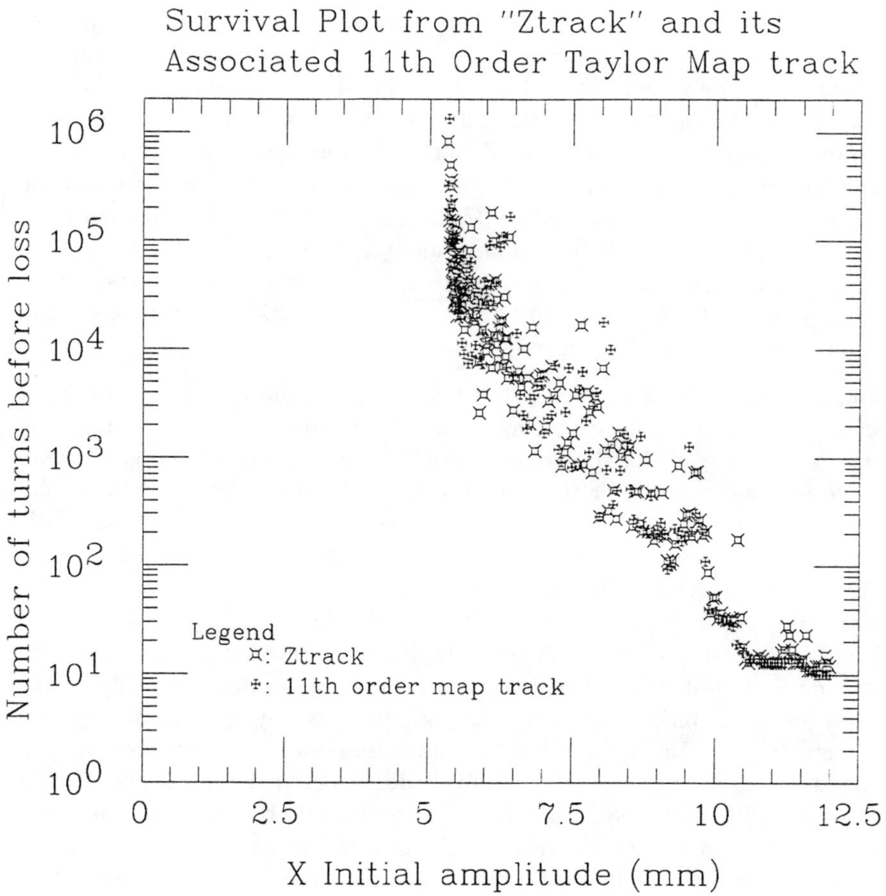

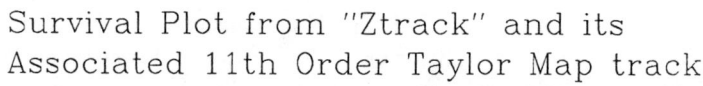

Figure 1. Million-turn survival plots for a 2-TeV injection lattice of the SSC, comparing the data from an 11th-order Taylor-map (extracted with "Zmap") tracking with the data from its associated "Ztrack" element-by-element tracking. The two sets of data match quite well, showing that: (a) round-off errors (64-bit precision) for long-term tracking in the SSC is of no concern since "Ztrack" element-by-element tracking is accomplished at an accuracy of 11 digits in one turn, while the 11th-order Taylor-map tracking is less accurate by 3 to 4 digits (7 to 8 digits of accuracy) in one turn for the selected region of interest, due to truncation of higher orders, and still generate the same result; and (b) Increasing the order of the Taylor map by one or two would allow reliable fast tracking up to 10 million turns for the SSC (the required proton-coasting time of the SSC injection lattice) since increasing one order higher in the Taylor map will enhance nearly 10 times as much accuracy (about one more digit accuracy) in one turn tracking. (Courtesy of Y. Yan et al., SSCL-301)

References

1. A. Dragt et al., "Lie Algebraic Treatment of Linear and Nonlinear Beam Dynamics," *Ann. Rev. Nucl. Part. Sci.*, **38**, 455 (1988).
2. M. Berz, "Differential Algebraic Description of Beam Dynamics to Very High Orders," *Particle Accel.* **24**, 109 (1989).
3. E. Forest, M. Berz, and J. Irwin, "Normal Form Methods for Complicated Periodic Systems," *Particle Accelerators*, **24**, 91 (1989).
4. Y. Yan, "Applications of Differential Algebra to Single-Particle Dynamics in Storage Rings," SSCL-500 (1991), in *Physics of Particle Accelerators*, M. Month and M. Dienes, eds., *AIP Conf. Proc.* No. 249, Vol. **1**, pp. 378–455 (1992).
5. See these proceedings.
6. Y. Yan and C. Yan, "Zlib—A Numerical Library for Differential Algebra," SSCL-300 (1990).
7. Y. Yan, "Zmap—A Differential Algebraic High-Order Map Extraction Program for Teapot Using Zlib," SSCL-299 (1990).
8. Y. Yan, P. Channell, and M. Syphers, "An Algorithm for Symplectic Implicit Taylor-Map Tracking," SSCL-Preprint-157 (1992); submitted to J. Comp. Phys.
9. L. Schachinger and R. Talman, "Teapot: A Thin-Element Accelerator Program for Optics and Tracking," *Particle Accel.* **22**, 35 (1987).
10. S.K. Kauffmann, D. Ritson, and Y. Yan, "Implementation of One-Turn Maps in SSCTRK using Zlib," SSC Laboratory Report SSCL-321 (1990).
11. A. Garren, A. Kenney, E. Courant, and M. Syphers, "A User's Guide to Synch," Fermi National Accelerator Laboratory Report FN-420 (1985).
12. D.C. Carey and F.C. Iselin, "A Standard Input Language for Particle Beam and Accelerator Computer Programs," *Proc. 1984 Summer Study on the Design and Utilization of the Supercollider*, Snowmass, CO (June 1984).
13. Y. Yan, "Supercomputing for the Superconducting Super Collider," *Energy Sciences Supercomputing 1990*, pp. 9–13 (1990), published by DOE National Energy Research Supercomputer Center, A. Mivin and G. Kaiper, eds.
14. Y. Yan et al., "Comment on Round-off Errors and on One-Turn Taylor Maps," SSCL-301 (1990); also appears in *Proc. Workshop on Nonlinear Problems in Future Particle Accelerators*, Capri, Italy (April 1990), p. 77, W. Scandale and G. Turehetti, eds., published by World Scientific (1990).
15. Y. Yan, "Brief Comment on One-turn Map for Long-term Tracking," *AIP Conf. Proc.*, No. 255, p. 305, A. Chao, ed. (1992).
16. A. Dragt and J. Finn, "Lie Series and Invariant Functions for Analytic Symplectic Maps," *J. Math Phys.* **17**, 2215 (1976).

Appendix A: a Testing Program for Lie Transformation

```
      program drgtst
      implicit double precision(a-h,o-z)
c---- This program has to be loaded with Zlib
      pointer (kuu,uu(1))
      pointer (kww,ww(1))
      pointer (kf,f(1))
      pointer (kar,ar(1))
      pointer (kam,am(1))
      pointer (kami,ami(1))
c---- input nv, no from the terminal
      write(6,*) 'input nv, no'
      read (5,*) nv, no
c---- set up zlib working environment
      call zpprep(nv,no,nm,0)
c---- nv = number of phase-space coordinates, input
c     no = the maximum order to be considered, input
c     nm = number of monomials, output
c---- nmnv coef's for nv-dimensional Taylor map
      nmnv=nm*nv
c---- nmover coefficients for f = f0 + f1 + f3 + ...
c     Should nmover be set as an output of zpprep?
      nmover=nm*(nv+no+1)/(no+1)
c---- generate users' memories
      call mcreator(kuu,nmnv)
      call mcreator(kww,nmnv)
      call mcreator(kf,nmover)
      na1=nv+1
      nss=na1*nv
      call mcreator(kar,nss)
      call mcreator(kam,nss)
      call mcreator(kami,nss)
c-1-- input an "order-by-order symplectic" Taylor map
      open (8,FILE='zpmap',STATUS='old')
      call rdmapzp(uu,nv,no,8)
c-2-- Lie transformation (Dragt-Finn factorization)
      call zpdragt(uu,nv,no,f,ar,am,ami,na1)
c-3-- Taylor expansion of the Lie transformation
      call zptylr(no,f,ar,am,ami,na1,nm,nv,ww,no)
c-4-- Inverse of the re-expanded Taylor map
      call zpmapinv(ww,no,ww,no)
c-5-- concatenation of the original Taylor map and the
c     inverse of the re-expanded Taylor map. the result
c     should be an identity Taylor map.
      call zpcnct(uu,no,ww,no,ww,no,nv)
c-6-- output the identity map for checking
      open (7,FILE='mapzp')
      call wrmapzp(ww,1,nv,no,7)
c==== Note that a lot of parameters in the calling
c     statements are to provide more flexibilities.
c     These flexibilities are not used in this program.
      stop
      end
```

NEW FEATURES IN THE DESIGN CODE TLIE *

Johannes van Zeijts
Continuous Electron Beam Accelerator Facility
12000 Jefferson Avenue, Newport News, VA 23606

ABSTRACT

We present features recently installed in the arbitrary-order accelerator design code TLIE. The code uses the MAD input language, and implements programmable extensions modelled after the C language that make it a powerful tool in a wide range of applications: from basic beamline design to high precision - high order design and even control room applications.

The basic quantities important in accelerator design are easily accessible from inside the control language. Entities like parameters in elements (strength, current), transfer maps (either in Taylor series or in Lie algebraic form), lines, and beams (either as sets of particles or as distributions) are among the type of variables available. These variables can be set, used as arguments in subroutines, or just typed out. The code is easily extensible with new datatypes.

INTRODUCTION

Here we give a short introduction to the physics and algorithms used in the program. We use the Hamiltonian formalism and canonical variables introduced in the map code field in the program MARYLIE 3.0[1]. The Hamiltonian describes in a compact way the dynamics of the particles in the full 6 dimensional phase space.

$$H = -\frac{P_\tau}{\beta} - \frac{A_z}{G} - \sqrt{1 - \frac{2P_\tau}{\beta} + P_\tau^2 - (P_x - \frac{A_x}{G})^2 - (P_y - \frac{A_y}{G})^2} , \quad (1)$$

G is the magnetic rigidity '$B\rho$' of particles on the design orbit. It is given by the relation

$$G = \frac{p_0}{q} , \quad (2)$$

and has units of tesla meters. Also, β and γ are the standard relativistic factors for the design orbit. They are related to p_0 (the design momentum) and $p_t^0 = -H|_{\text{design orbit}}$ by the equations

$$p_0 = \beta\gamma mc ,$$
$$p_t^0 = -\gamma mc^2 . \quad (3)$$

A_x, A_y, and A_z are the vector potentials; for example, the vector potential for a normal quadrupole with a gradient of b_2 tesla/meter is

$$A_z = -\frac{b_2}{2}(x^2 - y^2) \quad (4)$$

---

*Work supported by Department of Energy contract #DE-AC05-84ER40150

New physics is added by substituting more complex vector potentials. We provide vector potentials for all the simple elements and for realistic multipole magnets. In some of our multi-particle-effect applications the vector potential actually depends on the transported beam. However, the algorithms described below do not change no matter how complicated the vector potential becomes. The transfer maps for the 'time λ' flow are generated in a variety of ways from the Hamiltonian:

- Transfer map generation for the case where the Hamiltonian H does not depend on the independent variable is most readily done by direct exponentiation of the formal solution

$$T(z) = e^{-\lambda :H:}(z) = z - \lambda[H, z] + \frac{1}{2!}\lambda^2[H, [H, z]] + \cdots , \tag{5}$$

where $[,]$ is the Poisson bracket operator. This algorithm will eventually converge due to the $n!$ term in the denominator of the expansion of the exponential operator.

- Transfer map generation for the case where H does depend on the independent variable is implemented as:

$$T_{\lambda_1 \to \lambda_2}(z) = \int_{\lambda_2}^{\lambda_1} [H(z, \lambda), T(z)] \, d\lambda . \tag{6}$$

This algorithm is implemented in a particularly efficient way, which is the main reason we are able to generate high-order transfer maps for realistic systems in a time period practical for use in an optimization process (minutes for order 10). In practice we use a forward integration and calculate the inverse map $I = T^{-1}$

$$I_{\lambda_1 \to \lambda_2}(z) = \int_{\lambda_1}^{\lambda_2} [H(z, \lambda), I(z)] \, d\lambda . \tag{7}$$

The resulting Taylor series are converted into Lie algebraic form. This is necessary for the "concatenation" process and gives us a compact representation of the aberration coefficients. By default we split the Lie algebraic map up in homogeneous parts with the following standard 'Dragt-Finn' ordering

$$F(z) = e^{:f:}(z) = e^{:f_1:} M e^{:f_3:} \cdots e^{:f_n:}(z) , \tag{8}$$

where f_1 describes the misalignment part, f_3 the second order part, etc. The linear part M of the transfer map is carried as a symplectic matrix.

A number of other orderings of the Lie algebraic exponents are available and used in the program. The algorithms to translate between these different representations are most readily implemented to arbitrary order using Taylor series as intermediate steps. For instance to concatenate two maps we use the algorithm:

$$e^{:f:} e^{:g:}(z) = e^{:f:}(G(z)) = H(z) = e^{:h:}(z) , \tag{9}$$

where all steps in the process are uniquely determined as long as the matrix part of the maps are symplectic.

EXTENSIONS TO THE PHYSICS

We have added the capability to calculate transfer maps for cylindrical current sheet magnets, the current sheets may be stacked so as to produce overlapping multipole fields. Together with the fast integration algorithm this allows us to model realistic systems including fringe fields to high precision in a practical time period.

The vector potential off axis, for a given multipole symmetry, is determined from the appropriate magnetic field gradients and their longitudinal derivatives on axis. In the following we will write expressions only for the normal multipoles (for $m \neq 0$). Skew multipoles correspond to $cos(m\theta)$ terms in Eq.10. Given the Fourier expansion of the scalar potential

$$V(r, \theta, z) = \sum_{m=1}^{\infty} U_m(r, z) sin(m\theta), \quad (10)$$

a vector potential giving the same field is

$$\begin{aligned} A_z &= -\sum_{m=1}^{\infty} \frac{cos(m\theta)}{m} r \frac{\partial}{\partial r} U_m(r, z) \\ A_r &= \sum_{m=1}^{\infty} \frac{cos(m\theta)}{m} r \frac{\partial}{\partial z} U_m(r, z) \end{aligned} \quad (11)$$

Here we have chosen a gauge where $A_\theta = 0$. The scalar potential off-axis may be written

$$U_m(r, z) = r^m \sum_{l=1}^{\infty} \frac{(-1)^l (m-1)!}{l!(l+m)!} \left(\frac{r}{2}\right)^{2l} \left(\frac{\partial}{\partial z}\right)^{2l} g_m(z), \quad (12)$$

where

$$g_m(z) = \lim_{r \to 0} \frac{m U_m(r, z)}{r^m} \quad (13)$$

represents the profile of the mth multipole. This is a general solution to Maxwell equations order by order, for arbitrary $g_m(z)$. The problem is thus reduced to computing the generalized field gradients on axis for realistic magnet models. Subroutines to compute the required gradients are available for Halbach REC quadrupoles and for general multipoles, with the current distribution on a cylindrical surface specified by a shape function[3].

One particular powerful feature of the program is the ability to specify user profiles for the field gradient as functions of the independent variable s in the control language. These profiles can be arbitrarily constructed functions of s, and the expressions are automatically differentiated to the order needed in the integration process. We give several examples below and show how a user profile is given interactively to specify the octupole component of one multipole. Remember that any number of different multipoles can be layered on top of each other to represent arbitrarily complicated profiles.

```
f1(s) = 1/(1+E^(b1 + b2*(s/d)+ b3*(s/d)^2))
f2(s) = 1/(1+(s/d)^2)^(3/2)
f3(s) = B0*(tanh(s/d)-tanh((s-10)/d))

mult1: multipole, f1(), L = 1.0, m = 4, radius = 0.22
```

This feature for instance, allows users to install arbitrary fringe field profiles if he so chooses.

EXTENSIONS TO THE MAD LANGUAGE

We extend the MAD language in a variety of ways. First we introduce logic for loops, conditionals, subroutines, and functions returning real numbers. The control language is interpretive and fully programmable. It is based on a C like interpreter and written in the compiler generating language YACC[4].

Secondly we allow for element parameters to be easily accessible in the language; i.e., name[L] gives the length of a particular element, name[K] gives the quadrupole strength, etc. These parameters can be read and set interactively.

Next we introduce a transfer map datatype; i.e., map1 = namedline.F creates a new map variable from a given line element, in this case a Lie map variable. We also have map2 = namedline.T, which returns the Taylor map. Entries in map variables can be accessed as: map1[x,x] for entries in the matrix part, map1[x³], map1[x Pt²] for higher order entries in the Liemap variables, and map2[x,x²] etc. for Taylor map variables. Since maps are checked for dependencies on element parameters and brought up to date when they are used, it is easy to write a code segment that studies the effect of changing a parameter on transfer map entries,

```
for(quad1[K] = 0; quad1[K] <= 3.0; quad1[K] += 0.1) {
    type quad1[K], namedline.F[x,x]-map1[x,x],nameline.F[x Pt^2]
}
```

where namedline is a line which is dependent on the strength of quad1.

Transfer maps are constructed by patching maps for single element bodies together with coordinate rotations, fringe-fields, etc. We provide all the basic elements in the program, but it is also possible to specify new constructions interactively by using operations on maps. These Lie map expressions (Fexpr) can be:

- name.F

- Lie map variable

- Liemap(identity) , creates an identity liemap

- Fexpr + Fexpr

- Fexpr - Fexpr

- Fexpr / Fexpr, ignoring zero terms in denominator

- (Fexpr)
- Fexpr & Fexpr, result is the concatenated map.
- invert(Fexpr), returns the inverted Lie map.
- filter(Fexpr,expr), filters out entries less than expr
- prot(expr), the map for longitudinal reference plane rotation
- arot(expr), the map for transverse reference plane rotation
- monomial(Fexpr), the map in monomial factorized form
- reverse(Fexpr), returns a Lie map in reverse order
- flend(Fexpr), put the e^{f_1} term at the tail end of the map
- standard(Fexpr), to convert back to Dragt-Finn factorization;

i.e., to get the relative difference between two Lie maps we simple ask for

```
type (F1-F2)/F2
```

We allow for the evaluation of subroutines along any given line. These subroutines expect several parameters as their arguments: the longitudinal distance, coordinates and angles with respect to the floor, an element type, the transfer map up to the particular point in the line, and the name of the element as a string variable. An example of this can be as simple as typing the names and coordinates of a line

```
proc typeName(real s1, real x1, real y1, real z1, real angxz, real angyz, real
angxy, real eltype, liemap lie1, string name)
{
        type s1,name,z1,x1,y1
}
```

Since the transfer map is available we can evaluate any changing variable along the beam line. A useful procedure is to calculate lattice functions and, for instance, find the maximum value of a lattice function along a line:

```
proc findmaxBeta(liemap $9) {
  bX = betaxF($9,bx1,ax1)
  bY = betayF($9,by1,ay1)
  if (bX > maxX) { maxX = bX; maxsX = $1}
  if (bY > maxY) { maxY = bY; maxsY = $1}
  }
  layout(namedline) with findmaxBeta()
```

Here we see how Liemap variables can be used as arguments in functions. Arguments can be accessed by name or by the order they appear in. The layout with procedure command can be used inside any procedure. Hence we can use the optimizer to, for instance, minimize the maximum lattice functions

```
func maxBeta(line $1) {
    maxX = maxY = -1.0
    maxsX = maxsY = -1.0
    layout($1) with findmaxBeta()
    if(maxY > maxX) return(maxY) else return(maxX)
}
```

CONTROL ROOM APPLICATIONS

By the addition of a few new parameters and keywords we turn the code into a package useful for control room applications. We add the ability to set and read currents of magnets, and read the horizontal and vertical position from beam position monitors. A suitable control logic library,[5] and access to control room computers has to be available to be able to make use of these features.

The full power of the language is available to program experiments and data-analysis. For instance, a particularly simple application, to measure an entry in the transfer matrix between two points, is implemented as follows:

```
for(current1 = 0; current1 <= 3.0; current1 += 0.1) {
    kick1[I] = current1; sleep(1)
    type kick1[I], monitor1[x], monitor1[y]
}
```

where the `kick1` and `monitor1` keywords are declared as type kick resp. monitor. With suitable caution even the fitter and optimizer can be used for real-time control. This connection between a design code and a control code allows for measured data to be propagated using simulated transfer maps, which is useful in tuning algorithms. The interpretive nature of the code allows for many more applications to be implemented rapidly.

REFERENCES

1. A. J. Dragt, MaryLie 3.0, A Program for Charged Particle Beam Transport Based on Lie Algebraic Methods.

2. J. B. J. van Zeijts and F. Neri, The Arbitrary Order Design Code Tlie, presented at the Gosen High Order Optics Codes workshop, April 1992.

3. P. Walstrom, Filippo Neri and Tom Mottershead, High Order Optics of Multipole Magnets, LINAC Meeting Albuquerque 1990.

4. B.W. Kernighan and R. Pike, *The Unix Programming Environment*, (Prentice-Hall 1984).

5. M. Bickley, Star Documentation, CEBAF, May 1992.

RECENT ADVANCES AND APPLICATIONS OF THE MAFIA CODES

the MAFIA collaboration *
(presented by T. Weiland)
Technische Hochschule Darmstadt, Fachbereich 18, FG TEMF,
Schloßgartenstr. 8, 6100 Darmstadt, Germany

ABSTRACT

Over the last years MAFIA has grown to a more and more universal design tool for a vast range of applications not only in the field of accelerator physics. The currently distributed version 3.1 now includes a new solver module for time harmonic fields that enables the computation of eddy current distributions as well as the fields in driven rf systems. MAFIA 3.1 also includes static modules for electric and magnetic fields, 2D and 3D resonator solvers, 2D and 3D time domain solvers as well as 2.5D and 3D PIC modules. Thus MAFIA 3.1 now virtually covers the entire range of electromagnetic field problems. The fully menu driven user interface has been enhanced by implementation of macros, symbolic variables and language structures that makes MAFIA fully programmable. On the application side there are numerous highlights such as extremely fast and accurate computations of S-parameters, calculation of antennas including farfield patterns, non-destructive testing analysis for carbon fiber reinforced plastic as used as air plane material etc., to name only a few. In the accelerator physics area the new version added many enhancements on the calculation of impedances and wake fields with the possibility to simulate very short bunches without excessive need for memory. Version 3.2, scheduled for release in fall 1993, contains further new features such as fully lossy materials (complex fields), cylindrical coordinates for better cavity design, a possibility to add user-defined menus, various new 3D visualization tools, enhanced MAFIA language and an AUTOMESH option. The most important new module is an optimizer, called OO, which basically combines all MAFIA modules into one (big) program. OO allows fully automatic optimization of electromagnetic components such as waveguide transitions, cavities etc., according to user specified goal functions.

INTRODUCTION

The basic theory on which MAFIA is based is well documented in the literature [1] that it suffices to write down here Maxwell's Grid Equations without derivation.

---

*The MAFIA collaboration members are: S.G.Wipf, M.Marx, M.Dohlus from DESY, Hamburg, B.Steffen from KfA Jülich, U.Blell from GSI, Darmstadt, M.Bartsch, P.Hahne, A.Schulz, P.Schütt, .Wolter, T.Weiland from CST GmbH, Darmstadt, U.Becker, M.Dehler, X.Du, R.Klatt, A.Langstrof, Zhang Min T.Pröpper, U.van Rienen, D.Schmitt, P.Thoma, B.Wagner, from TH-Darmstadt

Maxwell's Grid-Equations

Integral Form	Matrix Representation
$\iint_A -\frac{\partial}{\partial t}\vec{B}\cdot d\vec{A} = \oint_{\partial A} \vec{E}\cdot d\vec{r}$	$-d/dt\ \mathbf{D_A b} = \mathbf{C\, D_s e}$
$\iint_{\partial V} \vec{B}\cdot d\vec{A} = 0$	$\mathbf{S\, D_A b} = 0$
$\iint_{\tilde{A}}\left(\vec{J}+\frac{\partial}{\partial t}\vec{D}\right)\cdot d\vec{\tilde{A}} = \oint_{\partial \tilde{A}} \vec{H}\cdot d\vec{r}$	$\mathbf{D_A}(\mathbf{i} + d/dt\ \mathbf{D_\varepsilon e}) = \tilde{\mathbf{C}}\, \mathbf{D_{\tilde{s}} D_\mu^{-1} b}$
$\int_{\partial \tilde{V}} \vec{D}\cdot d\vec{\tilde{A}} = \iiint_{\tilde{V}} \varrho\, d\tilde{V}$	$\tilde{\mathbf{S}}\, \mathbf{D_{\tilde{A}} D_\varepsilon e} = \mathbf{q}$
$\vec{D} = \varepsilon \vec{E}$	$\mathbf{d} = \mathbf{D_\varepsilon\, e}$
$\vec{B} = \mu \vec{H}$	$\mathbf{b} = \mathbf{D_\mu\, h}$
$\vec{J_L} = \kappa \vec{E}$	$\mathbf{i}_L = \mathbf{D_\kappa\, e}$

This set of matrix equations is a one-to-one translation of Maxwell's equations to a grid space doublet $G - \tilde{G}$ and represents the only known theory that not only allows practical solution on a computer but also maintains all analytical properties of electromagnetic fields when going to the grid space. Grid-fields represent not only a large bunch of numbers but also have a number of analytical, algebraic properties that ensure accurate numerical results and enable an algebraically exact self-testing of numerical results.

This solid theory is the basis for the MAFIA computer codes and resulted in an outstanding reliability of MAFIA results. MAFIA was released in three major versions. The current version 3.x, described in the next section, is the code with the broadest range of applicability including solvers for statics, rf fields, time domain fields and particle-in-cell problems.

With the enormous increase in computing power of nowadays workstations, MAFIA no longer needs super-computers but runs mainly on desktop workstations.

Due to the efficient memory management and last not least due to the exact linear rule of scaling for core size versus number of mesh cells, MAFIA can solve very big problems with millions of unknowns on workstations.

After one decade, in which the priority was focussed on getting an analysis of a given structure, we now enter the period where realistic big problems can be solved in a reasonable time, such as the analysis of a complete personal computer in terms of radiated electromagnetic fields.

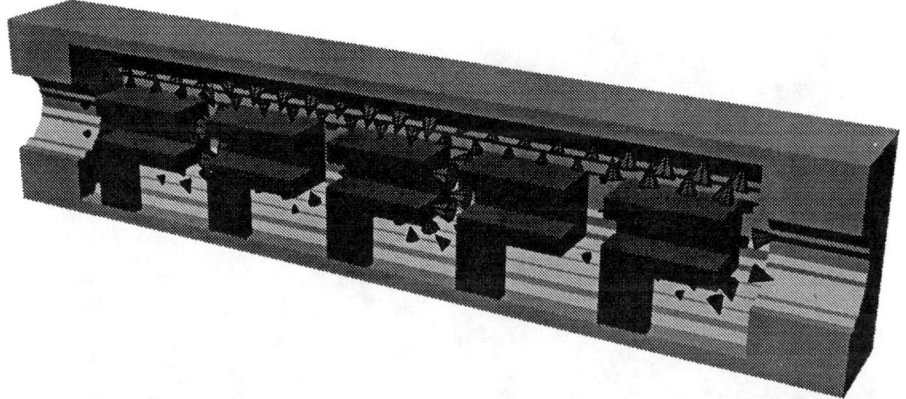

This example shows kicker tank built at GSI in Darmstadt. Besides the task of kicking the beam this kicker also represents a parasitic cavity with resonant fields having a sizeable shunt impedance. The plot shows the electric fields of one mode showing clearly that a strong coupling to the beam is present. The over all effect is reduced by the lossy ferrite material such that the effective impedance turned out to be tolerable. This example was solved by MAFIA-E with 150.000 real unknowns.

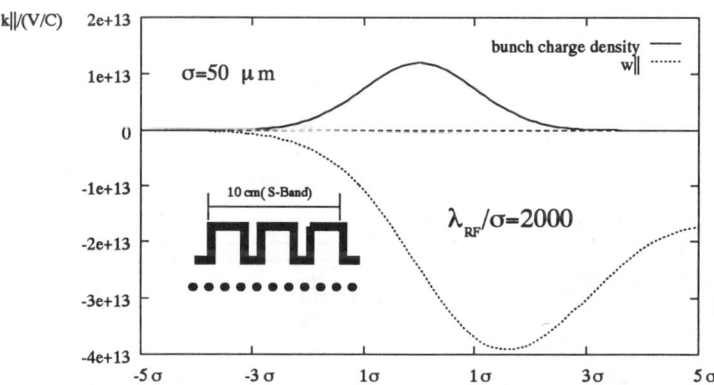

This example demonstrates the window option in MAFIA-T2 for calculating wake fields of very short bunches. Only the co-moving window is kept in core and thus very large meshes describing the geometry can be used without the needs for very large memory for the fields. On top of this option version 3.2 offers a unique ZOOM features that further increases the mesh density within the window by a given factor without blowing up all other arrays in core. The example here is a model of a SLAC-type of S-Band cavity used to study wakefields within the DESY/THD S-Band Linear Collider study[2]. The effective mesh size is as large as 200.000.000 cells cells with 5 micrometer step size using only 40 Mbytes of memory. The rms bunch length is as short as 50 micrometer and the structure length 11 cm. This example was solved by MAFIA-T2 within 50 hours cpu time on an HP apollo 730.

294   Applications of the MAFIA Codes

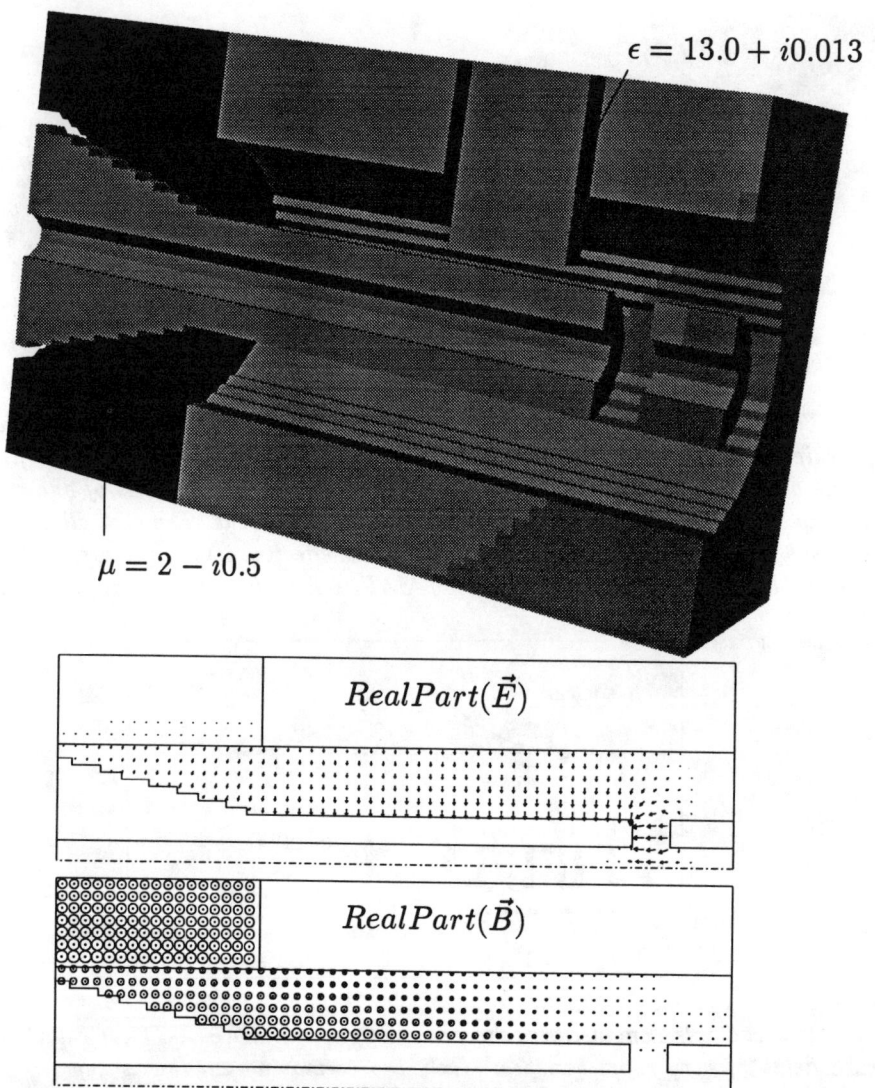

This example shows a ferrite loaded accelerating cavity with feed and ceramic window. The ferrite material is very lossy: $\mu = 2.0 - i0.5$. The top plot shows the geometry with the permeable lossy ring on the left and the lossy ceramic ring cealing the feed. The middle plots show the real part of the electric field and the magnetic flux in the middle plane. The lowest eigenfrequency was found as (56.5 + i 4.5) MHz. Note that MAFIA 3.2 cannot only solve for the (easy) case of small losses but also for the most complicated case of very heavy losses as shown here. This example was solved by the complex field option in MAFIA-E with 920.000 real unknowns (460.000 complex unknowns). The cpu time needed was about 7 hours on an IBM RISC/6000 model 550 using only 31 MBytes of core space.

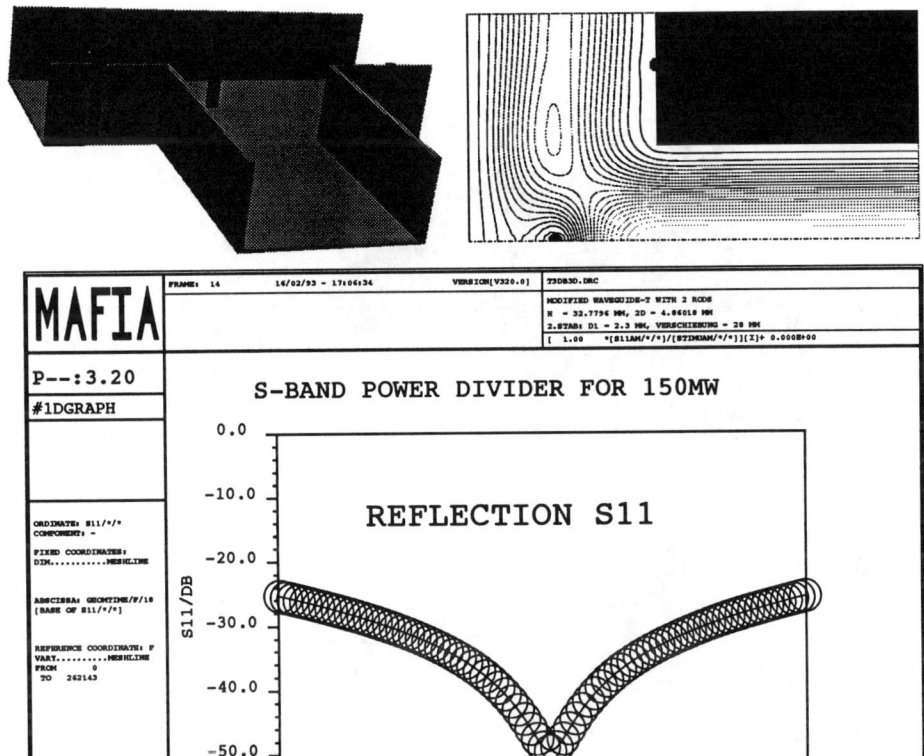

This example shows a 3dB waveguide-T as used for splitting rf power from one klystron into two waveguides. This component was developed for the S-Band Collider[2]. Due to the high input power (150MW) the critical point is the heating of the rods. The structure was optimized for a minimum reflection within a given bandwidth with the condition that the overall heating must stay within specified limits. The top-left picture shows the 3D geometry with the center rod and two half-rods on both exits. The top-right picture shows the electric field energy as contour plot. The center-plot finally shows the computed reflection coefficient versus frequency as calculated by MAFIA. Note that each circle represents one computed value and narrow peaks can easily be found by this technique. This example was solved by MAFIA-T3 with 19.000 cells using the new broad-band technique resulting in a frequency resolution of as high as 6MHz. The entire curve for S_{11} was obtained by one single run of MAFIA-T3 and took about 90 minutes cpu on HP730 (corresponding to roughly 1 minute cpu per frequency!!).

This example shows a 3D model of a device for non-destructive testing of carbon fiber reinforced plastic. This material is used in air plane technology and has rather unusual material properties such as anisotropic conductivity and permittivity. The task of the research done in this area was to find an optimimun frequency for field excitation. The eddy currents induced by the coil body are disturbed by material cracks and thus those material defects result in a change of the test coil impedance.

The upper plot shows the 3D geometry with the double coil body above the anisotropic multi layer material and the electric field vectors. The middle picture shows the 3D contour plot of one electric field component showing clearly the influence of the crack. The bottom plot shows an E-field component as 'hill'-plot.

This example was computed with MAFIA-W3 using 60.000 cells and 180.000 complex unknowns within less than 60 minutes on an IBM RISC/6000, model 550.

This example shows a 3D model of a photo gun built at CERN, similar to the BNL-type with a photo cathode and a 2-cell S-Band accelerating section. Rf is fed through the waveguide on top. Electron Bunches are ejected from the photo cathode (left hand side) and accelerated by the rf field along the cavity cells. The picture shows a significant emmittance growth due to space charge effects at the exit of the first cell. This example was solved by the MAFIA PIC module TS3 using 60000 cells (360.000 unknows) and pre-computed rf fields from E as initial fields in the time domain computation. The eigenmode and PIC simulation took less than 15 minutes cpu time on an IBM RISC//6000-550.

298   Applications of the MAFIA Codes

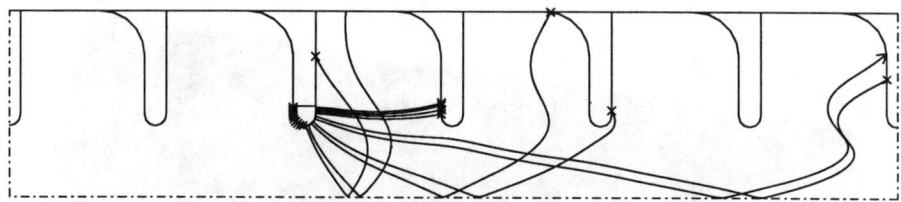

This plot shows an example of a dedicated trajectory version of MAFIA-T2/TS2. This module allows calculation of dark currents in high gradient accelerator and is currently being used to study the DESY/THD S-band collider[2].

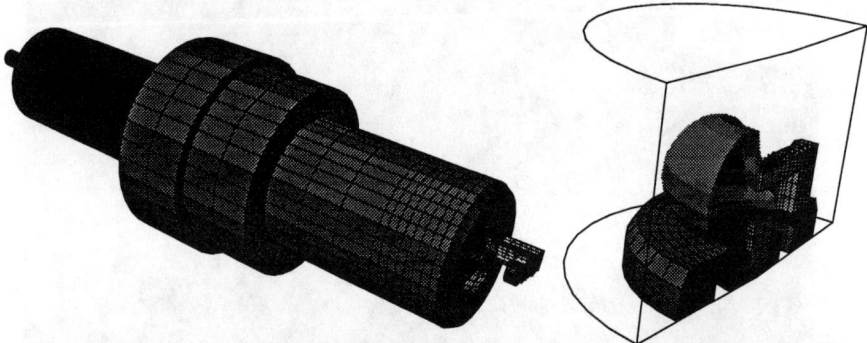

This plot shows an example of the new $r - \phi - z$ geometry option in MAFIA version 3.2. The left plot shows an ignition plug, the right hand side one a 3D iso-plot picture of an equipotential surface calculated with MAFIA-S.

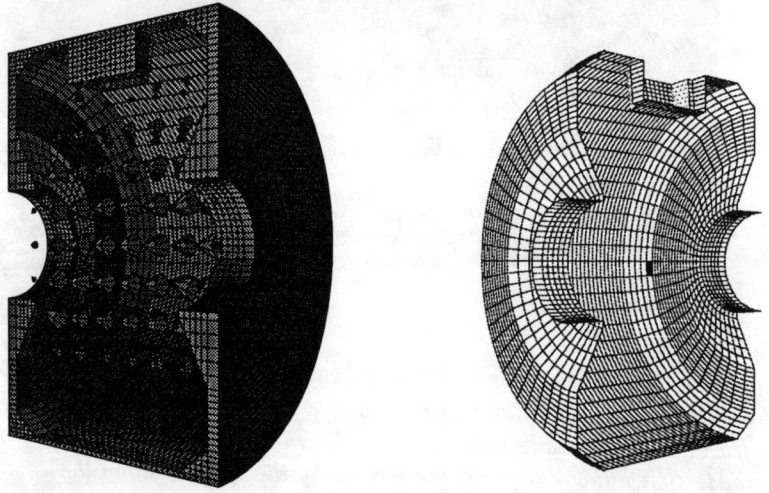

This plot shows an example of the new $r - \phi - z$ geometry option in MAFIA version 3.2. The plots show the mesh and the 3D field of the fundamental accelerating mode in a 500MHz accelerating cavity computed with MAFIA-E-3.2.

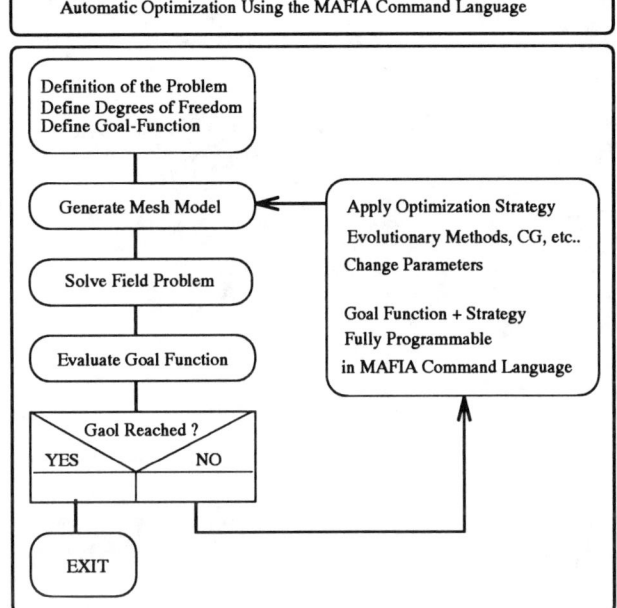

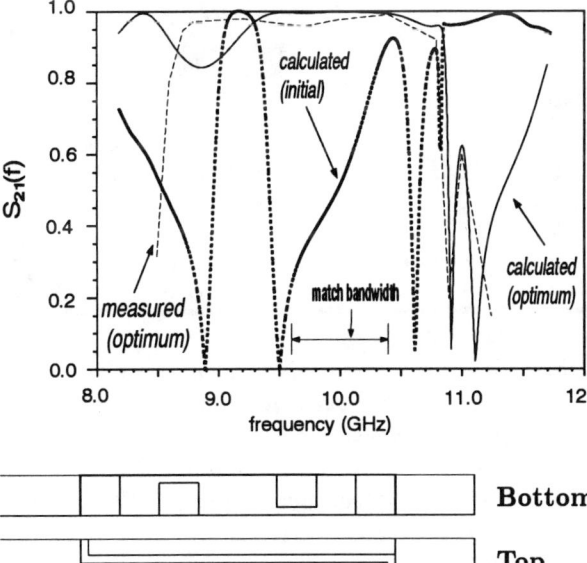

This page demonstrates the automatic optimizer module OO of MAFIA-3.2. This example is a transition from a rectangular waveguide to a micro strip line. To ease the measurement the transition was doubled and the micro strip is in fact in the center of the structure. The goal was to obtain a flat S_{21} for frequencies of 10GHz+/-0.4GHz. This example was solved by MAFIA-OO invoking the modules M, T3 and P and took about 4.3 days cpu on an IBM RISC/6000-550. During the optimization OO performed 350 runs optimizing with 30 degress of freedom for the shape of the bottom and top layer. The mesh model had 40.000 unknowns. Each of these 350 runs resulted in over 100 values for S_{21} in the desired frequency range of 9.6GHz to 10.4GHz. The middle plot shows S_{21} versus frequency for the initial shape, the optimized shape and the measured results.

The bottom plot shows the top and bottom layer shapes for the initial transition and the final optimized shape.

300 Applications of the MAFIA Codes

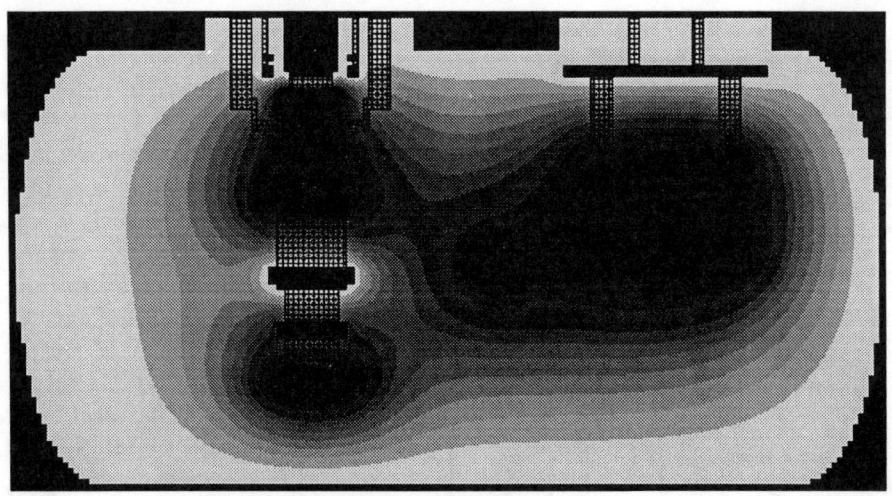

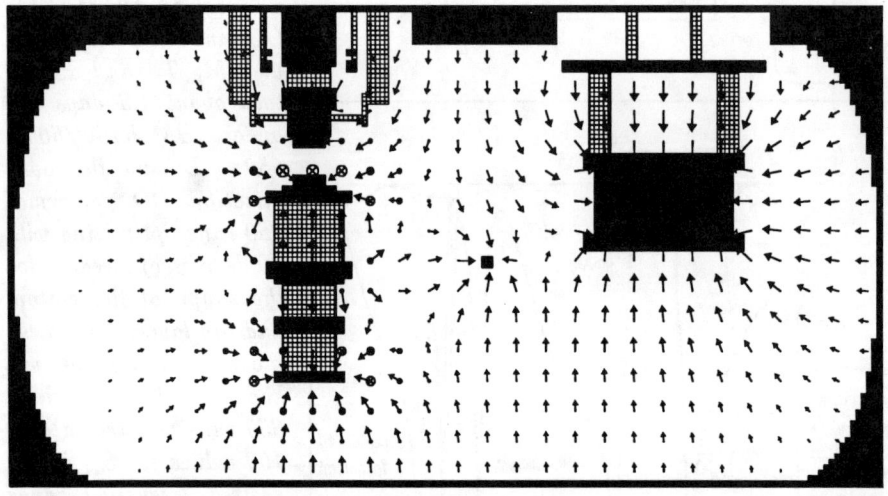

This example shows the three dimensional analysis of an oil-tank for a high voltage hard tube pulser. This device is being built at DESY as part of the DESY/THD S-Band Collider Study[2]. The upper picture shows a contour plot of the electric scalar potential. The lower plot shows the electric field distribution in logarithmic scaling. This example was calculated with MAFIA-S using 1.500.000 cells. The full 3D picture POSTSCRIPT file was too large to fit into my text system (13 MBytes!). The cpu time needed was less than 30 minutes on an IBM RISC/6000 model 550.

The next step after this, which is in reach already today, is to implement automatic optimization tools that run hundreds or thousands of field anaylsis cycles to reach without the user's interference an optimized component.

The best impression of what is possible today is obtained from pictures and examples and thus we concentrate here on demonstrating what MAFIA can do today and in the near future.

MAFIA - Version 3.x

MAFIA version 3.x is the successor of the well known version 2.x which is in use now since almost ten years. The step to version 3.x was more than an upgrade but was a complete redesign in view of future extensions and modernized user-interface. MAFIA 3.1 consists of the following modules

- **M - Mesh Generator** M is used to set up the mesh model. There are numerous predefined basic shapes such as arbitrary cylinders, spheres, cones etc. Version 3.2 also allows circular cylindrical coordinates ($r - \phi - z$) and has an AUTOMESH option. 2D and 3D visualization of the input geometry and the mesh model are supported as well as interactive re-meshing features via mouse clicks.

- **S - The Static Solver** S solves for electrostatic, magnetostatic and thermal fields. Nonlinear material properties are supported. MAFIA-S can handle millions of mesh cells by using a modern Multi-Grid solver.

- **W3 - Eddy Current Solver** W3 solves the inhomogeneous field equations including eddy currents and displacement currents and thus can be used in low and high frequency domain. Due to the incorporated lossy materials the solution vectors are complex. Numerical algorithms in use are COCG and similar derivatives.

- **E - Eigen Mode Solver** E solves for eigenmodes (2D and 3D) of cavities including lossy materials and complex solution vectors. Waveguide modes are also solved for with E.

- **T2 - 2D Time Domain Fields** T2 solves the transient (fast) fields in 2D r-z geometry for any azimuthal field order (monopole, dipole,..). The new ZOOM option is capable of computing wakes of extremely short bunches down to the 50 micrometer range for S-Band cavities.

- **T3 - 3D Time Domain Fields** T3 solves the 3D transient field problem including lossy materials and a very wide range of possible excitations. T3 covers antenna problems, has open boundaries and can use 2D waveguide fields as excitation. T3 is extremely fast in computing S-parameters of waveguide junctions using a new broad band technique. Not only that T3 is in such applications up to 200 times faster than usual frequency domain codes but it also is at the same time much more accurate.

- **TS2 - 2.5D Particle in Cell Simulator** TS2 simulates particle motion and field equations simultaneously for cylindrically symmetric structures such as klystrons, guns, etc. Any external field being a solution of MAFIA-S or MAFIA-E can be pre-loaded as starting field.

- **TS3 - 3D Particle in Cell Simulator** TS3 simulates particle motion and field equations simultaneously for three dimensional structures such as klystrons, guns, etc. Any external field being a solution of MAFIA-S or MAFIA-E can be pre-loaded as starting field.

- **P - Postprocessor** The post-processor P contains many visualization tools for vector fields and scalar fields in 2D and 3D. In MAFIA-P the user can evaluate many secondary results such as voltages, energies, wall losses, shunt-impedances etc.. MAFIA-P contains a special section that allows manifold manipulation of computed fields such as combination, mutliplication and operations such as **curl, grad** and **div**. The flexible MAFIA-Macro-Language enables the creation of user defined commands for complex post-processing such as evaluating shunt impedances, loss-factors, Q-values etc..

EXAMPLES

A sequence of examples is presented here to demonstrate the capabilities of MAFIA 3.1 and 3.2. All pictures shown here have been created by MAFIA-P version 3.2 which now supports many graphical output formats such as POSTSCRIPT, HPGL and others.

CONCLUSION

MAFIA 3.x is the only code available today that covers the full range of Maxwell's equations from static up to fully selfconsistent transient fields. Although MAFIA 3.x is a general purpose system it outperforms in many specialized areas dedicated codes such as S-parameter FE-programs by orders of magnitude. A big step forward is in reach now by the new optimization module that opens a new area in electromagnetic computation.

REFERENCES

1. T. Weiland, "Maxwell's Grid equations", URSI International Symposium on Electromagnetic Theory, Sydney, 17-20. August 1992, Proceedings and previous proceedings of this conference series.
2. T.Weiland (for the S-Band Collider Group), "Status of the DESY-THD 500 GeV S-Band Linear Collider Study" Talk at the LC92 in Garmisch Partenkirchen, to be published

APPLICATIONS OF THE ARGUS CODE IN ACCELERATOR PHYSICS*

J.J. Petillo, A. Mankofsky, W.A. Krueger, C. Kostas, A.A. Mondelli, and A.T. Drobot
Science Applications International Corporation, McLean, VA 22102

ABSTRACT

ARGUS is a three-dimensional, electromagnetic, particle-in-cell (PIC) simulation code that is being distributed to U.S. accelerator laboratories in collaboration between Science Applications International Corporation (SAIC™) and the Los Alamos Accelerator Code Group (LAACG). It uses a modular architecture that allows multiple physics modules to share common utilities for grid and structure input, memory management, disk I/O, and diagnostics. Physics modules are in place for electrostatic and electromagnetic field solutions, frequency-domain (eigenvalue) solutions, time-dependent PIC, and steady-state PIC simulations. All of the modules are implemented with a domain-decomposition architecture that allows large problems to be broken up into pieces that fit in core and that facilitates the adaptation of ARGUS for parallel processing. ARGUS operates on either Cray or workstation platforms, and a MOTIF-based user interface is available for X-windows terminals. Applications of ARGUS in accelerator physics and design are described in this paper.

DESCRIPTION OF ARGUS

The ARGUS code[1,2] has been under development at SAIC since 1983. It is a general-purpose three-dimensional simulation code. The code architecture is specifically designed to handle the problems associated with three-dimensional simulations. It uses sophisticated memory management and data handling techniques[3] to deal with the large volume of data that is generated in three-dimensional simulations. Recent work by Leabrook Computing, Ltd. shows that the domain decomposition technique used in ARGUS provides a suitable framework for coarse-grained parallelization of the code.

A modular architecture is employed so that ARGUS is in fact a system of three-dimensional codes (numerical modules) that utilize a common data structure and share utilities for structure input, grid generation, memory management, data handling, and diagnostics. The codes allow complicated geometrical structures to be represented on the computational grid. The grid can be nonuniform in all three dimensions. Cartesian and cylindrical coordinate systems have been implemented throughout ARGUS; some of the modules also support toroidal and mixed coordinates. Physics modules are in place to compute electrostatic and electromagnetic fields, the eigenmodes of rf structures, and PIC simulation in either a time-dependent mode or a steady-state mode. The PIC modules include multiple particle species, relativistic particle dynamics, and algorithms for the creation of particles by emission from material surfaces and by injection onto the grid. A plasma chemistry module allows species to be created or destroyed based on specified rate processes.

The structure input in ARGUS is carried out through combinatorial geometry. The code stores a library of basic three-dimensional objects (e.g., a rectangular solid, an elliptical cylinder, an ellipsoid, etc.). These objects are combined by the user with logical operations (to either *add* or *delete* the library object) to produce structures of arbitrary shape. The structures so specified are represented on the computational grid by a *structure mask* array, which stores the material and electrical properties of each cell on the grid.

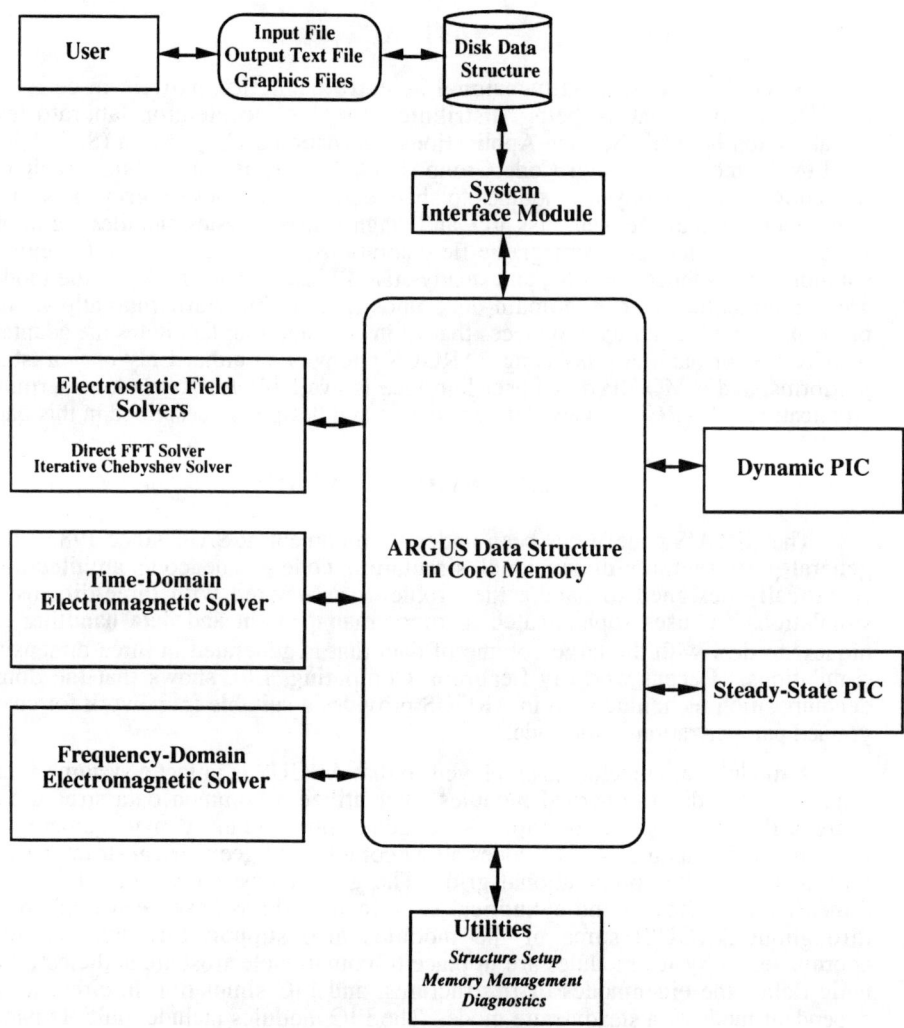

Figure 1. ARGUS modular code architecture.

Material properties can be associated with structures. The code allows perfectly-conducting materials, as well as materials with complex values of both permittivity and permeability; hence lossy materials are allowed. Furthermore, the permittivity and permeability may be specified as diagonal tensors to treat certain classes of non-isotropic materials.

ARGUS has a wide variety of diagnostic plots that are selected by user input and are available at run time. These include set-up graphics, contour and arrow plots of field quantities, and plots of particle trajectories and phase space. The code also will allow the user to create HDF files for exporting data to other visualization tools.

SOME EXAMPLES FROM THE ARGUS PRIMER

The current code release (ARGUS v.25) includes an ARGUS Primer, which provides a tutorial for new ARGUS users. It presents a set of simple example problems that illustrate how to use the features in ARGUS. The Primer will also be used in testing future ARGUS releases and ARGUS installations on new computing platforms. A subset of the examples featured in the Primer are presented here.

The "Tombstone" Cavity

An example in the MAFIA User's Guide consists of analyzing the normal modes of the structure shown in Figure 2. The structure is a cylindrical drift tube attached to a cavity made of a rectangular box topped with a half cylinder. MAFIA and ARGUS represent this structure in different ways. ARGUS uses a "stairstep" representation on the grid, while MAFIA allows triangular "half cells". (Half-cell masking will be included in a future ARGUS release.)

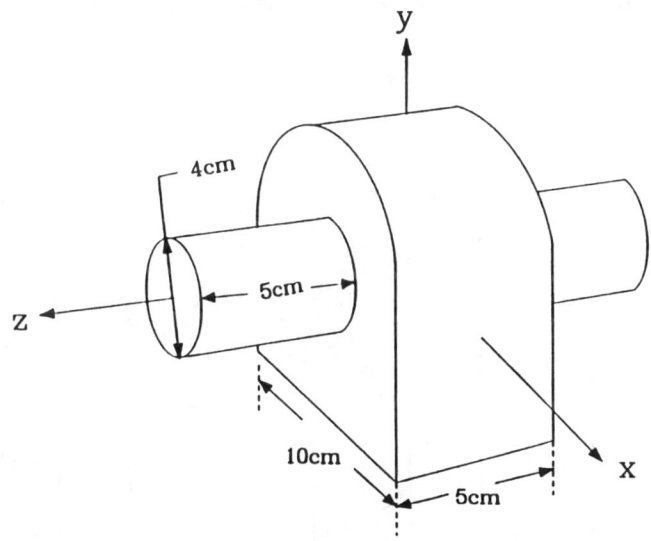

Figure 2. The "Tombstone" cavity.

Cold Test of the "Tombstone" Cavity

A comparison of ARGUS and MAFIA cold-test results for the "Tombstone" cavity are given in Figure 3. The ten lowest modes of the structure were computed. The discrepancy is generally in the 1-2% range, and is due to the differences in the structure representation by the two codes. This result is consistent with other comparisons of these codes on comparable meshes. With finer meshing, the codes agree more closely.

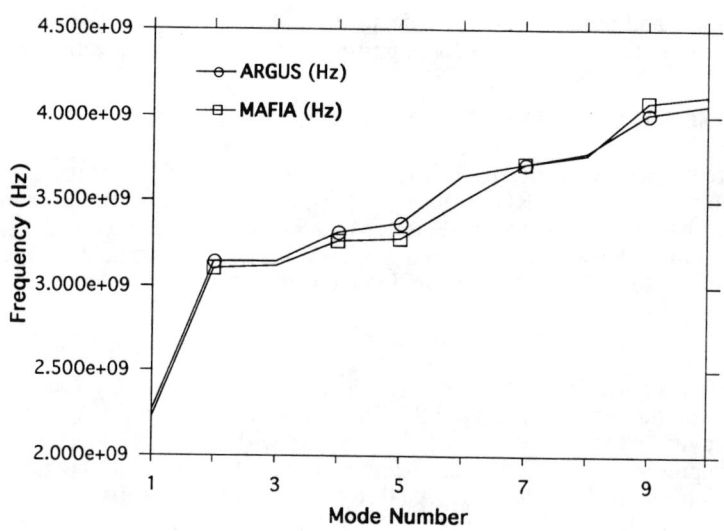

Figure 3. Comparison of ARGUS and MAFIA results for "Tombstone" cavity cold test.

Wake Fields in the "Tombstone" Cavity

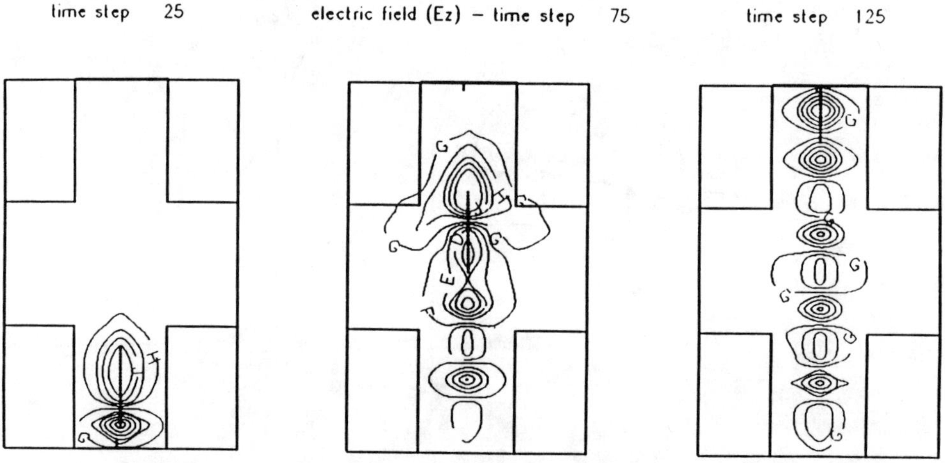

Figure 4. Wake fields due to a Gaussian bunch passing through the "Tombstone" cavity.

By initializing a Gaussian bunch of heavy particles in the time-dependent PIC module, and allowing them to drift through a structure, the user can cause ARGUS to create wake fields in the structure. This method was employed to display the wake fields in the "Tombstone" cavity. Figure 4 shows three snapshots of the particles and axial electric field as the bunch traverses the structure. The Gaussian bunch was placed on the axis of the cylindrical drift tube and launched into the tombstone cavity. The particles are visible in the figure as a line charge on axis. In the first frame, time step 25, the particles are just entering the tombstone cavity from the lower drift tube. In the second frame, time step 75, they are just leaving the tombstone cavity and entering the upper drift tube. In the third frame, time step 125, they are leaving the upper drift tube (by being absorbed on the upper boundary of the simulation).

Radiating Open Waveguide in the Time Domain

The ability of the ARGUS code to handle open, radiating boundary conditions in the time domain sets it apart from several other codes. Two separate utilities exist for carrying out this capability. The first is a "port" boundary condition, which treats a specified opening in the simulation boundary as though it were connected to a waveguide extending to infinity. The second is an implementation of the Lindman[4] algorithm for radiating boundaries. While the port condition only matches the boundary for radiation at a specified frequency, the Lindman condition is a general outgoing wave boundary condition.

This simulation provides a demonstration of the Lindman algorithm for the simple example of a rectangular waveguide that is driven at the lower end, and open at the upper end. Figure 5 shows the radiation fields at two time steps, selected to show the phase slippage at the top (open) boundary.

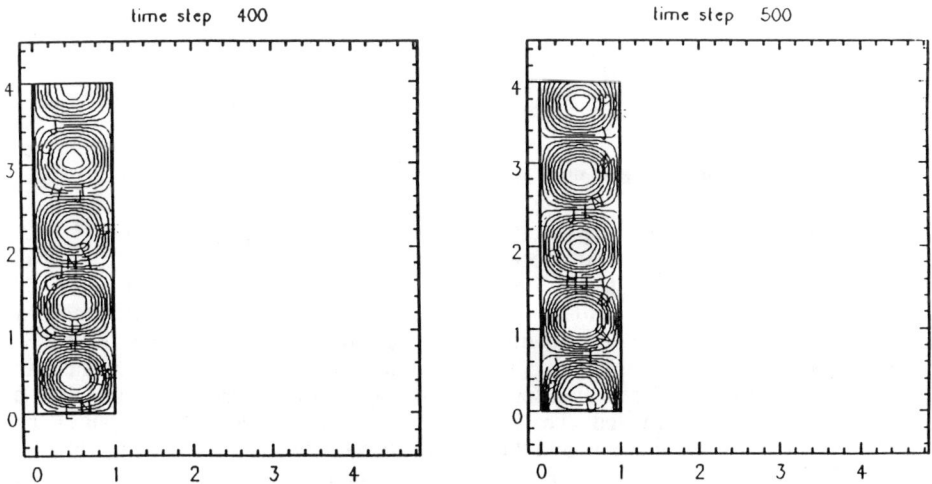

Figure 5. A radiating waveguide in the time domain.

WAVEGUIDE-CAVITY COUPLING

An important problem in the design of rf linacs is the coupling between the waveguide that feeds rf power to the accelerator and the cavity through which the beam is accelerated. The designer needs to know the coupling coefficient, the frequency shift, and the external Q due to the waveguide, as well as the fields in the aperture of the coupling iris. This problem is difficult for time-domain simulation codes because accelerator cavities often are very high-Q structures, and therefore require very many rf cycles to fill.

ARGUS has been employed in a collaboration between SAIC and AccSys Technology, Inc. to model the external Q of the drift-tube linac (DTL) cavities in the injector for the Superconducting Super-Collider (SSC). The intrinsic Q of the DTL cavities is approximately 40,000. Figure 6 shows the aperture in the waveguide cavity system, as represented in the ARGUS model. The drift-tube structures are represented by an "equivalent load," consisting of a dielectric rod.

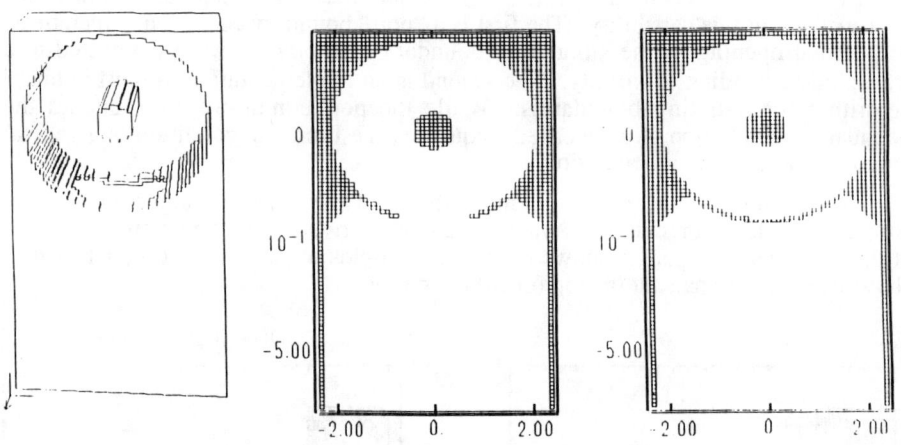

Figure 6. ARGUS representation of the DTL cavity/waveguide structure for the SSC injector.

The methods used for the solution are based on published techniques[5,6,7] for using electromagnetic frequency-domain algorithms to compute the external Q in high-Q cavity-waveguide systems. They require the eigenvalue solver to be exercised several times, with the waveguide shorted at different distances from the iris, thereby allowing a resonance curve of frequency vs. phase shift, like that shown in Figure 7, to be mapped out. The figure shows the resonance curve for a round iris, 12 cm in diameter. This analysis is repeated for different choices of the iris dimensions until a suitable design point is realized. The design is a trade-off between the external Q, the field strength in the iris aperture, and the cavity mode distortion due to the aperture.

AccSys has built an experimental test-stand which allowed the ARGUS results to be compared with measured data. The results of that test are shown in Figure 8, which presents the external Q and frequency shift as functions of the iris aperture length, for a fixed aperture width of 12 cm. The experimental data are plotted on the figure and agree with the simulations to within the experimental uncertainty.

AccSys Technology, Inc. is under contract to build the DTL cavities for the SSC Laboratory, and has used these simulations to verify the accuracy of their experimental test data prior to constructing full-scale cavities (which are approximately 10 m in length). Dr. Jim Potter of AccSys has collaborated with SAIC in this effort, and provided the experimental data which was compared with the ARGUS simulations in Figure 8.

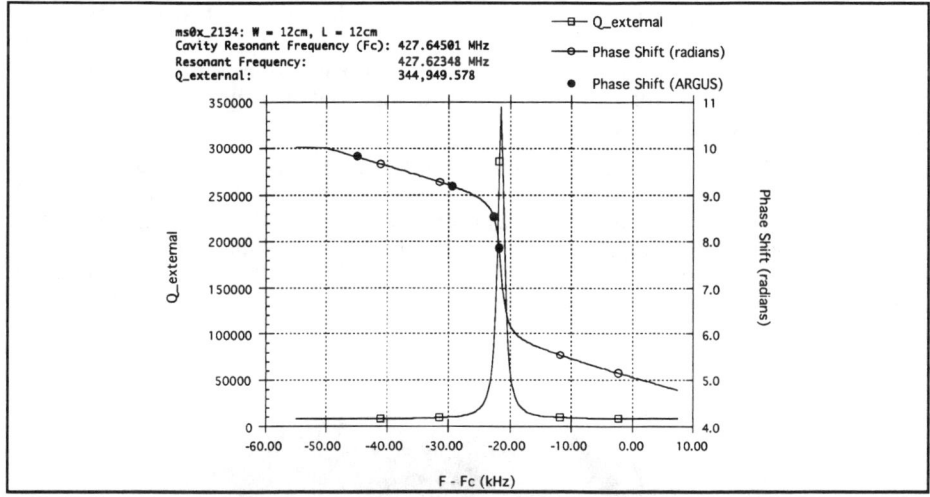

Figure 7. Phase shift vs. frequency for a 12 cm diameter iris aperture.

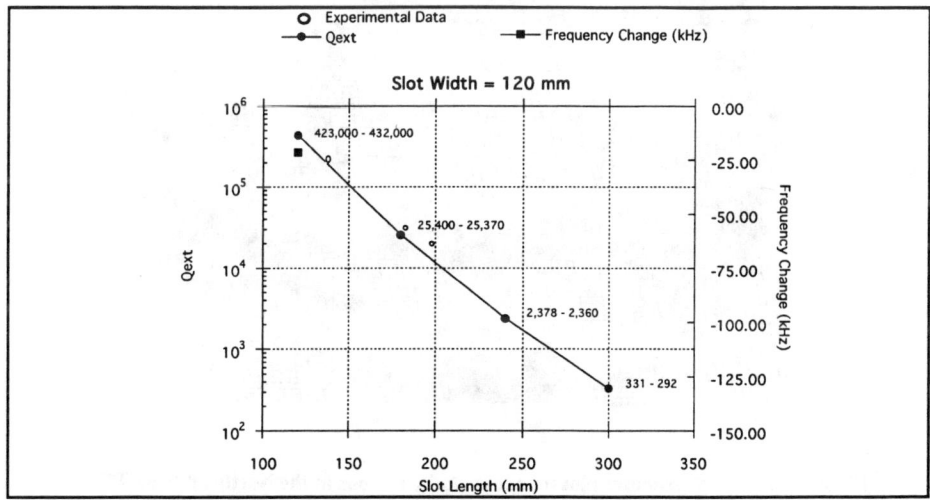

Figure 8. External Q and frequency shift vs. iris length for the DTL cavity/waveguide system. ARGUS simulation results are compared with experimental measurements (courtesy of J. Potter, AccSys Technology, Inc.)

HELIX TRAVELING-WAVE TUBE

The frequency-domain solver in ARGUS v.25 can treat complex fields, where the imaginary part allows the model to treat lossy materials and to handle periodic boundary conditions with sub-phase specification. This latter feature is particularly useful when attempting to compute the dispersion curves for long, periodic structures. The usual periodic boundary condition imposes a phase shift of 2π across the computational grid. If the user needs to impose a different phase shift across a symmetry period of the structure, he does so by modeling several structure periods to build up a phase shift of 2π (or π with an electric wall symmetry plane). The ARGUS v.25 algorithm allows the user to specify the phase shift to be imposed across the computational grid, thereby enabling him to map out a dispersion curve while modeling only a single structure period.

This module has been used to cold test a helix traveling-wave tube (TWT), developed by Northrop Corporation. Figure 9 shows the ARGUS representation of the helix TWT, including the metallic vanes that are used to short the fields between the helix supports. The supports (not shown in the figure) are dielectric rods that are attached to the helix and to the outer cylindrical wall. The ARGUS simulation model employed a $80 \times 80 \times 20$ grid, and requested six eigenvectors for each phase shift selected.

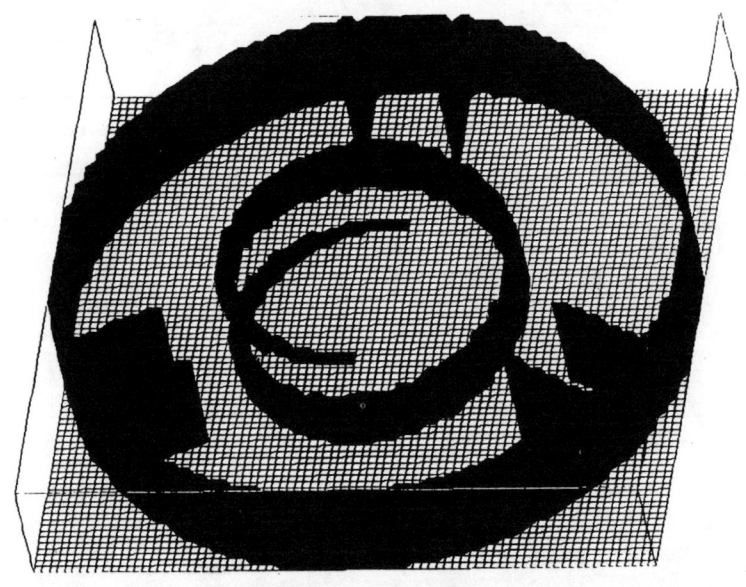

Figure 9. ARGUS structure plot for the metallic regions in the Northrop helix TWT.

The dispersion curve for the helix TWT is displayed in Figure 10. The figure shows the effect of the supports and the metallic vanes on the dispersion curve. The vanes are needed to flatten the dispersion curve (i.e., to reduce the dependence of the phase velocity on frequency). They are modeled in ARGUS as two-dimensional metallic sheets. Most three-dimensional electromagnetic codes are unable to include sub-cell features of this type, and therefore do not succeed in cold test analyses of devices like the helix TWT. Comparison with experimental data at Northrup indicate that the phase velocity corresponding to Figure 10 differs from the measured phase velocity by approximately 10%, but this difference can be largely accounted for by adjusting the contact area between the support rods and the helix. (The phase velocity of a helix TWT is known to be very sensitive to the area of contact.)

Dr. Gunther Doehler of Northrup Corporation has graciously provided a description of the helix TWT geometry, and has compared the ARGUS simulation results with his experimental data.

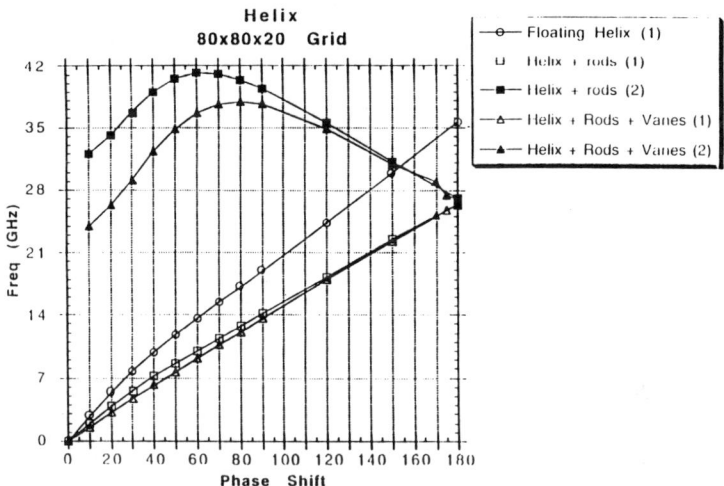

Figure 10. Dispersion curve for the Northrup helix TWT.

ACKNOWLEDGMENT

The authors gratefully acknowledge the collaboration of R. Ryne, R.K. Cooper, M.J. Browman, and G. Rodenz of the LAACG, J. Potter of AccSys Technology, Inc., and G. Doehler of Northrup Corporation in various portions of the material presented in this paper.

* Work supported by the US Department of Energy, Division of High Energy Physics under contract number DE-AC05-91ER40625.

REFERENCES

[1] A. Mankofsky, *Three-Dimensional Electromagnetic Particle Codes and Applications to Accelerators*, **Linear Accelerator and Beam Optics Codes**, C.R. Eminheizer, ed., A.I.P. Conf. Proc. No. 177 (American Institute of Physics, New York, 1988), p. 137ff.

[2] C.L. Chang, D. Chernin, A. Drobot, K. Ko, M. Kress, A. Mankofsky, A. Mondelli, and J. Petillo, *Three-Dimensional Modeling of Accelerators*, Proc. of the Conf. on Computer Codes and the Linear Accelerator Community, R.K. Cooper, ed. (Los Alamos National Laboratory, January 22-25, 1990, LA-11857-C), p. 27.

[3] A. Mankofsky, J.L. Seftor, C.L. Chang, K. Ko, A.A. Mondelli, A.T. Drobot, J. Moura, W. Aimonetti, S.T. Brandon, D.E. Nielsen, Jr., and K.M. Dyer, *Domain Decomposition and Particle Pushing for Multiprocessing Computers*, Comp. Phys. Commun. **48**, 155 (1988).

[4] E.L. Lindman, *"Free-Space" Boundary Conditions for the Time Dependent Wave Equation*, J. Comp. Phys. **18**, 66 (1975).

[5] N. Kroll and D. Yu, *Computer Determination of the External Q and Resonant Frequency of Waveguide-Loaded Cavities*, Part. Accel. **34**, 231 (1990).

[6] Y. Goren and D. Yu, *Computer-Aided Design of Three-Dimensional Waveguide-Loaded Cavities*, SLAC/AP-73 (1989).

[7] R.L. Gluckstern and R. Li, *Calculation of Cavity/Waveguide Coupling*, Proc. 1988 Linear Accel. Conf. (Newport News, VA, Oct. 3-7, 1988, CEBAF Report 89-001), p. 356.

ADVANCES/APPLICATIONS OF MAGIC AND SOS*

Gary Warren, Larry Ludeking, Khanh Nguyen, David Smithe, Bruce Goplen
Mission Research Corporation
8560 Cinderbed Rd, Suite 700
Newington, VA 22122

ABSTRACT

MAGIC and SOS have been applied to investigate a variety of accelerator-related devices. Examples include high brightness electron guns, beam-RF interactions in klystrons, cold-test modes in an RFQ and in RF sources, and a high-quality, flexible, electron gun with operating modes appropriate for gyrotrons, peniotrons, and other RF sources. Algorithmic improvements for PIC have been developed and added to MAGIC and SOS to facilitate these modeling efforts. Two new field algorithms allow improved control of computational numerical noise and selective control of harmonic modes in RF cavities. An axial filter in SOS accelerates simulations in cylindrical coordinates. The recent addition of an export/import feature now allows long devices to be modeled in sections. Interfaces have been added to receive electromagnetic field information from the Poisson group of codes and from EGUN and to send beam information to PARMELA for subsequent tracing of bunches through beam optics. Post-processors compute and display beam properties including geometric, normalized, and slice emittances, and phase-space parameters, and video. VMS, UNIX, and DOS versions are supported, with migration underway toward windows environments.

INTRODUCTION

MAGIC[1] and SOS[2] are self-consistent electromagnetic PIC codes that use finite-difference representations of the fields. MAGIC models two and one half-dimensional configurations; SOS models three-dimensional configurations. They each provide a variety of field and particle algorithms that can be selected to match the requirements of specific problems. Also, they provide a variety of boundary conditions. Both MAGIC and SOS solve problems in the time-domain; SOS also has frequency domain capabilities. POSTER[3] is a post analysis and graphics tool tuned to electromagnetics modeling needs. Examples of recent usages follow.

RF GUN MODELING WITH MAGIC[4,5,6]

MAGIC has been used by MRC and Grumman to examine the RF gun in use at Brookhaven National Laboratory, and a three and one half-cell high-brightness electron gun under development at Grumman. Nearly 200,000 grid cells are used to model the three and one half-cell gun using variable grid spacings. The MAGIC "DAMP" field algorithm is used to obtain single-mode fields in each cavity at the phase desired for that cavity (not all π apart). The resulting fields are saved to be used in many particle simulations. Particles are tracked self-consistently through the gun, using a short time step compared to the Courant condition to obtain correct phase velocities for the wakefields. Each particle simulation requires five hours on an IBM 6000/550 workstation.

Figure 1 shows the phase distribution of the particles at the gun exit (photo-emission-induced momentum fluctuations are not shown - they are added analytically in quadrature). The particle distribution depends on the launch phase, and the optimum launch phase depends on the subsequent optics. The distribution shown is phased for acceleration to 200 Mev. Particle data is output at multiple times to compute emittance and other beam properties at the output. A translator interfaces the MAGIC output to PARMELA for further tracking of the bunch through the beam optics.

KLYSTRON MODELING WITH MAGIC[7,8,9,10,11]

The design of efficient, high-power RF sources is a critical element of accelerator technology, and the MAGIC codes are presently being used extensively for this purpose. Because of the interest in klystrons, a template was developed for their simulation. (In essence, the template converts MAGIC from a general-purpose code to a device-specific (e.g., klystron) tool.) The klystron template provides useful, low-level design options such as the port approximation as well as algorithmic features essential to high-fidelity simulation. The high-Q filter field algorithm, for example, provides damping of undesirable particle noise associated with relativistic electron motion without an adverse effect upon the fundamental modes in high-Q cavities. (The field damping properties of the high-Q filter and the time-biased filter are compared in Figure 2.)

Figure 3 illustrates beam bunching and axial phase space for a klystron presently being investigated by SLAC. This 440 keV, 550 A, four-cavity device uses a traveling-wave structure for extraction. After 160 RF cycles in the "full-up" simulation, the device is approaching saturation, but the extraction efficiency is only about ten percent. Re-design of the traveling-wave structure can now be carried out with reasonable efficiency using the "export/import" capability in MAGIC.

DOUBLE CUSP GYRO-GUN[12]

MAGIC has been used in a design study for the Double Cusp Gyro-gun (DCG), for DARPA/NRL. This novel gun concept is a candidate beam-forming system for the second generation of mm-wave gyro-amplifiers. The DCG is comprised of three regions: an annular Pierce diode, a double-cusp focusing region, and an adiabatic compression region. By varying the magnetic field profile in these regions, it is possible to independently control the beam perpendicular to axial velocity ratio, guiding center radius, and velocity spread without changing the gun mechanical design. This makes the gun quite flexible, forgiving, and suitable to a very wide range of gyro-devices and linear beam devices. The design of the magnet system is accomplished via the use of the magnet design code, POISSON.[13] POISSON magnetic field results are transmitted to MAGIC for design of the Pierce diode and gun vacuum envelope. MAGIC results are further processed with POSTER to calculate beam parameters such as velocity spreads, guiding center spreads, and velocity ratios.

Figure 4 shows MAGIC results for the DCG in its large-orbit (axis-encircling) operation. The top and middle boxes show the beam perpendicular and axial beam momenta normalized by the speed of light as a function of axial distance. The bottom box shows the beam trajectory together with the gun vacuum envelope. As shown, the gun operates in the space-charge limited regime with a beam current of 4.4 amps. at a diode voltage of 60 kV. The beam final RMS axial velocity and guiding center

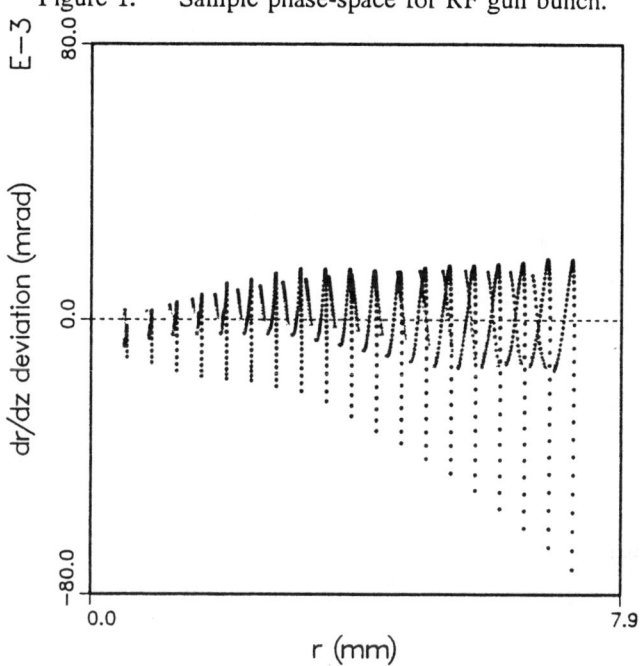

Figure 1. Sample phase-space for RF gun bunch.

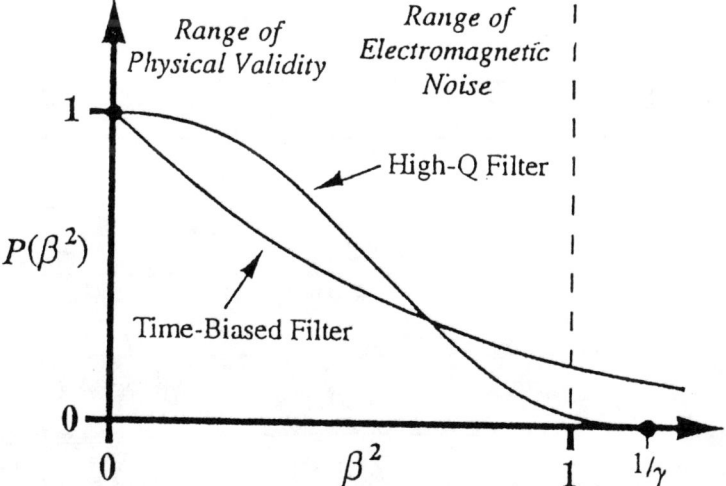

Figure 2. High-Q and time-biased filters.

Figure 3. Phase-space for SLAC klystron.

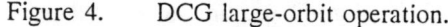

Figure 4. DCG large-orbit operation.

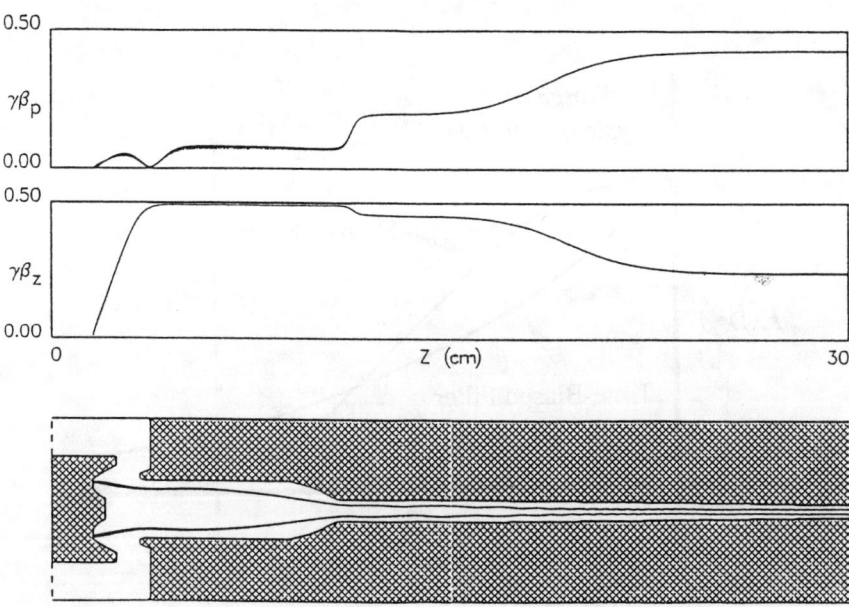

spreads are calculated by POSTER to be 0.7% and 6%, respectively, with a final velocity ratio of 1.73 in an 8.5-kG magnetic field.

PLASMA WAKEFIELD KLYSTRON[14]

MAGIC was recently applied to a novel microwave source concept called the Plasma Wakefield Klystron (PWK) to help evaluate its performance characteristics. The PWK operates in a manner similar to a klystron, except that beam modulation occurs in a plasma wakefield chamber instead of a buncher cavity. As the beam passes through the plasma chamber it excites plasma oscillations (wakefields) which maintain a constant phase with the beam and uniformly accelerate or decelerate beam electrons to create bunching. The PWK operates as an oscillator rather than an amplifier, since there is no input signal.

Both the startup and flat-top operating regimes of the PWK were simulated. Results provided estimates of flat-top RF power and efficiency estimates for the device. Figure 5 shows a schematic of the experiment for a two-cavity klystron design. The large cavity is the plasma-filled wakefield chamber. The first RF cavity serves the role of an intermediate cavity in a traditional klystron, the second cavity is the power extraction cavity. Figure 6 shows a snapshot of a simulation of a four-cavity PWK. This design used a single intermediate cavity, and three output cavities. The beam modulation is quite apparent in the picture. The ultimate power extracted from this design was estimated to be about 1 GWatt at near 1 GHz in frequency, corresponding to an RF efficiency of 35%.

MAGNETRONS

MIT has been using MAGIC to investigate a number of aspects for magnetrons. Figure 7 compares simulation and experimental results for the A6 magnetron for RF vs. magnetic field.[15,16,17]

SOS has begun to be used by MRC and Litton to examine a coaxial magnetron. Since cylindrical coordinates is the natural coordinate system for the magnetron, an axial field filter algorithm has been implemented in SOS to relax the on-axis Courant limitations thus accelerating frequency-domain simulations. Figure 8 shows the field vectors in the magnetron for a resonant mode, generated by SOS running in windows on a PC.

OTHER RECENT MODELING RESULTS

The 1992 Microwave Power Tube Conference included at least seven papers[10,11,12,18,19,20,21] presenting results from SOS and MAGIC. Subjects covered included multiple aspects of klystrons, TWT's, and the DCG.

RFQ results have been reported elsewhere by Grumman Corporation.[22]

DATA EXCHANGE

The MAGIC codes include a data storage and transfer facility. This software interfaces the MAGIC family of codes to each other, to the POISSON group, to PARMELA, and with some connections to HDE and spreadsheets. Some of the interfaces have been written by MRC, others by members of the MAGIC User's Group.

318   Advances/Applications of MAGIC and SOS

Figure 5.   Schematic of a single output cavity PWK.

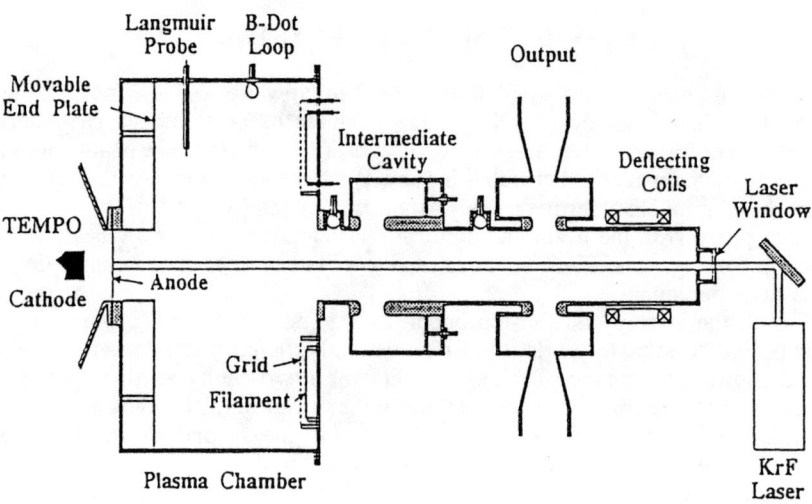

Figure 6.   MAGIC simulation of PWK.

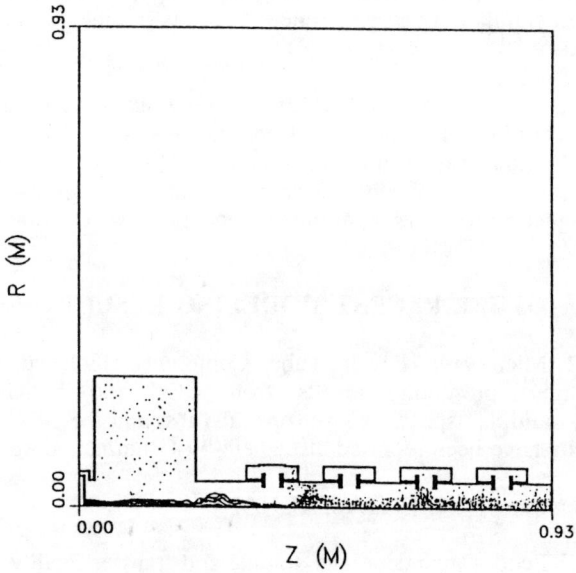

Figure 7. Magnetron RF power vs. magnetic field.

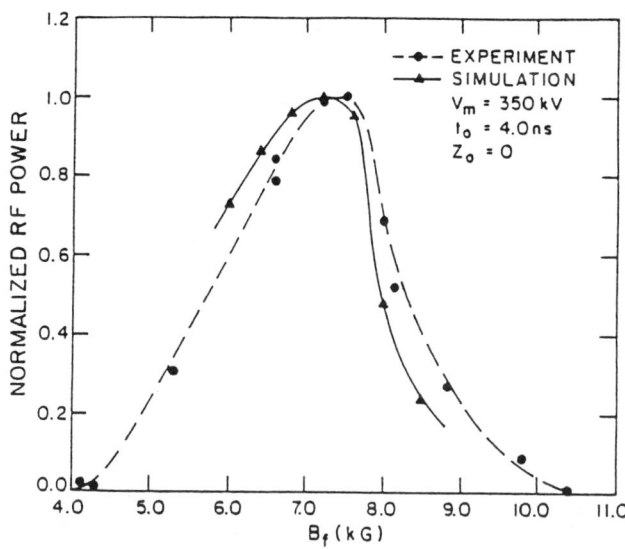

Figure 8. SOS running in windows.

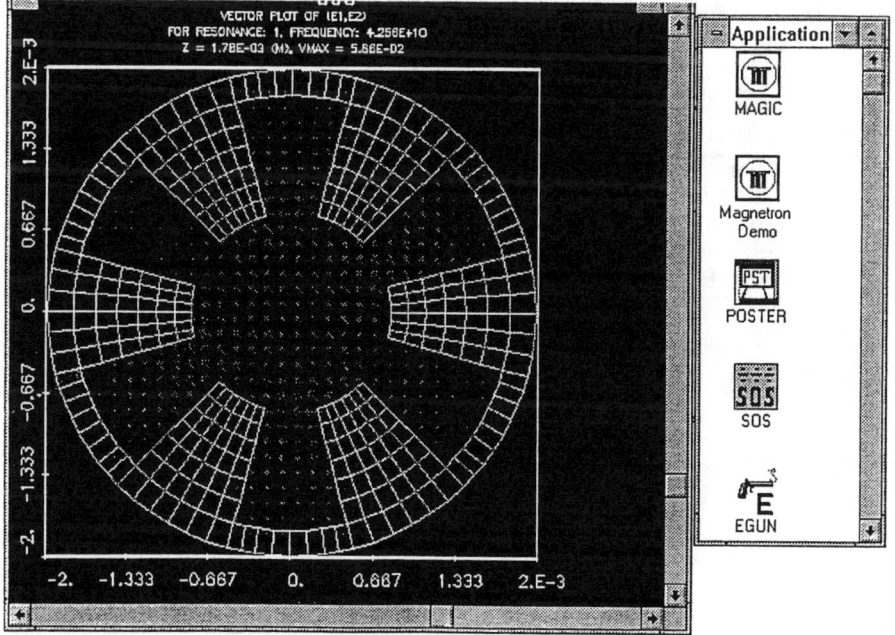

Under the MMACE program, Raytheon is currently connecting their UGUN code, and SAIC has indicated intent to connect the MASK code.

MAGIC USER'S GROUP

The MAGIC codes are made available through the MAGIC User's Group, which was formed by AFOSR to promote the use of electromagnetic particle-in-cell codes in plasma physics research. Initially consisting of five universities, the Group has expanded to more than thirty organizations, including universities, government agencies, and commercial firms.

The Group newsletter, now one year old, has become an effective means of communicating advances in simulation methods and research results. Group members now routinely contribute articles - the February 1993 issue will feature an article by Tim Davis at Cornell entitled "Traveling Wave Structure Interactions."

PLATFORMS

A PC version of MAGIC now runs under DOS, OS/2 and MS-Windows. Other new platforms include the HP750 workstation and the Silicon Graphics Iris workstation. The list of supported computers is shown in Table 1 along with timing results for the upcoming release of MAGIC. The benchmark is particle intensive and models magnetic insulation and field emission in pulsed power transport.

Table 1. Timing Benchmarks

COMPUTER	time hr:min:sec
CRAY-YMP	2:24
HP-550	9:18
IBM RS6000/550	10:02
SILICON GRAPHICS	10:20
IBM RS6000/320	18:12
VACCELERATOR	24:02
486-66MHz DOS (Gateway)	24:05
486-25MHz DOS (Gateway)	55:06
486-25MHz MS-WINDOWS	1:08:09
386-20MHz NOTEBOOK	5:03:16
Sun Sparc1,2,10	not available

---

*Work supported in part by Grumman Corp., AFOSR, NRL, DARPA/DSO, and members of the MAGIC User's Group.

REFERENCES

1. B. Goplen, L. Ludeking, D. Smithe, and G. Warren, "MAGIC User's Manual," (1991).

2. G. Warren, et al., Proceedings of the Conference on Computer Codes and the Linear Accelerator Community, January 22-25, 1990; Vol. 57.

3. L. Ludeking, G. Warren, and B. Goplen, Poster User's Manual, Mission Research Corporation Report, MRC/WDC-R-239, (1989).

4. I. Lehrman, I. Ben-Zvi, and G. Warren, "Design of a High Brightness High Duty Factor Photo-Cathode Electron Gun," presented at the 13th Intl. Conference, Santa Fe, NM, August 25-30, 1991.

5. G. Warren, et al., "Beam Dynamics in a High-Brightness High-Duty Factor Photocathode Gun," American Physical Society, Bulletin of the American Physical Society, Vol. 37, No. 2, (1992).

6. I. Lehrman, et al., "Design of a High-Brightness, High-Duty Factor Photocathode Electron Gun," Bulletin of the American Physical Society, Vol. 37, No. 2 (1992).

7. K. Nguyen, G. Warren, L. Ludeking, and B. Goplen, "Analysis of the 425-MHz Klystrode," IEEE Transactions on Electron Devices, Vol. 38, No. 10, October 1991.

8. K. Nguyen et al., "Numerical and Theoretical Analysis of the Klystrode Resonator and Collector Performance," IEDM Conference Proceedings (1989).

9. S. Begum et al., "Output Circuit for a 200-MW, 600-KV Relativistic Klystron," presented at 1992 Microwave Power Tube Conference, May 11-13, 1992.

10. F. Friedlander, "Analysis of a Waveguide-Loaded Klystron Cavity," presented at 1992 Microwave Power Tube Conference," May 11-13, 1992.

11. A. Young, W. Ansley, D. Reid, T. Larkin, "Beam Spread Characteristics in a Typical High-Power Klystron Collector," presented at 1992 Microwave Power Tube Conference, May 11-13, 1992.

12. K. Nguyen, D. Smithe, and L. Ludeking, "The Double Cusp Gyro-gun," Tech. Digest of Int. Elect. Dev. Meeting (1992).

13. Reference Manual for POISSON/Superfish Group of Codes, LANL Report LA-UR-87-126 (1987).

14. K. Nguyen, D. Smithe, and J. Pasour, "The Plasma Wakefield Klystron, A High-Power Microwave Generator," Mission Research Corporation Report, MRC/WDC-R-266, November 1991.

15. H. Chan, C. Chen, and R. Davidson, "Computer Simulation of Relativistic Multiresonator Cylindrical Magnetrons," Applied Physics Letters, Vol. 57, p. 12, September 1990.

16. C. Chen, H. Chan, and R. Davidson, "Parametric Simulation Studies and Injections Phase Locking of Relativistic Magnetrons," SPIE - The International Society for Optical Engineering, Vol. 1407, pp. 105-112, January 1991.

17. H. Chan, C. Chen, and R. Davidson, "Numerical Study of Relativistic Magnetrons," Journal of Applied Physics, December 1992.

18. K. Nguyen, D. Smithe, and L. Ludeking, "The Double-Cusp Gyro-Gun - A Flexible and High-Quality Beam-Forming System for Gyro-Amplifiers," presented at 1992 Microwave Power Tube Conference, May 11-13, 1992.

19. C. Kory and J. Wilson, "Simulated Cold-Test Characteristics of a TunneLadder Traveling-Wave Tube," presented at 1992 Microwave Power Tube Conference, May 11-13, 1992.

20. D. Schroeder and J. Wilson, "Micro-SOS Simulation of Coupled-Cavity Traveling-Wave Tube Cold-Test Characteristics," presented at 1992 Microwave Power Tube Conference, May 11-13, 1992.

21. S. Swanekamp, J. Holloway, T. Kammash, and R. Gilgenbach, "The Theory and Simulation of Relativistic Electron Beam Transport in the Ion-Focused Regime," Physics of Fluids, Vol. 4, No. 5, May 1992.

22. I. Lehrman and G. Warren, "Modeling of the End Regions of RFQs Using the 3-D SOS Code," Proceedings of the 1990 Accelerator Conference, Albuquerque, NM, September 10-14, 1990.

EXPLOITING PERIODICITY AND OTHER STRUCTURAL SYMMETRIES IN FIELD SOLVERS

E. M. Nelson*
Stanford Linear Accelerator Center
Stanford University, Stanford, CA 94309

ABSTRACT

Many RF structures have symmetries which can be exploited by field solvers with appropriate boundary conditions. These symmetries allow a reduced problem to be solved, which leads to faster and/or more accurate solutions. Of particular interest to the accelerator community are periodic structures. Quasi-periodic boundary conditions allow modes with any desired phase advance given a single cell of the periodic structure. For symmetric periodic structures there is a variation which requires only a half cell of the periodic structure. These boundary conditions can also be used for rotationally periodic structures, such as cross-field amplifiers and magnetrons. Boundary conditions for some other symmetries, such as reflection symmetry about a plane and about a point, will also be reviewed.

INTRODUCTION

Symmetries of the microwave structure can reduce the size of a problem. Reduced problems take less time and space to solve numerically. Alternatively, a finer mesh can be used with the reduced problem to obtain a more accurate solution. The symmetries described in the following sections can be exploited using appropriate boundary conditions while leaving the formulation for the interior of the structure unchanged. Other symmetries, such as axisymmetry, lead to different formulations for the interior as well as different boundary conditions.

The connection between Maxwell's equations and a symmetry is that Maxwell's equations, including the boundaries and any material properties, are invariant under the symmetry. Then fields can be found which are simultaneously eigenmodes of both Maxwell's equations and the symmetry operator, but with separate eigenvalues. In the language of quantum mechanics, Maxwell's equations (thought of as an operator) and the symmetry operator commute, so the operators are simultaneously diagonalizable. The size of the problem is reduced by constraining the solutions to be eigenmodes of the symmetry operator with a particular eigenvalue.

Examples of the exploitation of symmetries will be presented. These examples are based on the following formulation of the eigenmode problem: given the region Ω and its material properties ϵ and μ, find the eigenmode fields $\mathbf{E}$ and the corresponding eigenvalues ω^2/c^2 such that

$$\nabla \times (\mu^{-1} \nabla \times \mathbf{E}) = \frac{\omega^2}{c^2} \epsilon \mathbf{E} \quad \text{in } \Omega, \tag{1a}$$

$$\nabla \cdot (\epsilon \mathbf{E}) = 0 \quad \text{in } \Omega \tag{1b}$$

$$\text{and} \quad \hat{\mathbf{n}} \times \mathbf{E} = 0 \quad \text{on } \Gamma_{\text{metal}}. \tag{1c}$$

The region Ω represents the interior of the structure, and the surface Γ_{metal} represents the perfectly conducting walls which bound the structure. Specifying either the dirichlet condition $\hat{\mathbf{n}} \times \mathbf{E} = 0$ or the neumann condition $\hat{\mathbf{n}} \times (\mu^{-1}(\nabla \times \mathbf{E})) = 0$ on the boundary is sufficient for this type of problem. It should be made clear that despite the specific nature of these examples, the boundary conditions of the following sections are independent of the particular formulation for the interior of the structure.

The following section is a review of reflection symmetry about a plane. While this symmetry is well known and in common use, it will be helpful to introduce the operator notation and describe the steps leading to a reduced problem with this familiar case.

---

* Work supported by Department of Energy, contract DE-AC03-76SF00515

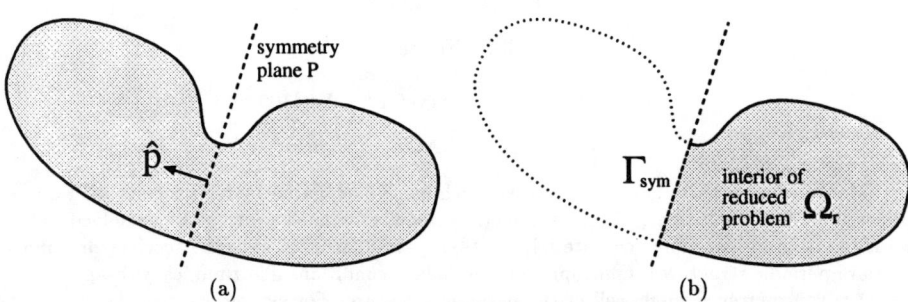

Fig. 1. (a) A structure with symmetry plane P. The shaded region is the interior Ω, and the normal to the symmetry plane is $\hat{p}$. (b) The region over which Maxwell's equations must be solved is reduced to Ω_r for a symmetric structure. The symmetry boundary is Γ_{sym}.

REFLECTION SYMMETRY ABOUT A PLANE

Consider a structure which is symmetric about a plane P, called the symmetry plane, defined by the equation $\mathbf{x}^T \hat{p} = p_o$, where $\hat{p}$ is a unit vector normal to the plane P and $|p_o|$ is its distance from the origin. An example is shown in figure 1a.

Let $\hat{P}$ be the reflection operator about the symmetry plane P. The reflection operator $\hat{P}$ can act on various types of objects. $\hat{P}$ acting on a point $\mathbf{x}$ yields the point

$$\hat{P}\mathbf{x} = \mathbf{P}(\mathbf{x} - \hat{p}p_o) + \hat{p}p_o = (\mathbf{I} - 2\hat{p}\hat{p}^T)(\mathbf{x} - \hat{p}p_o) + \hat{p}p_o, \tag{2}$$

where $\mathbf{P} = (\mathbf{I} - 2\hat{p}\hat{p}^T)$ is the reflection matrix and $\mathbf{I}$ is the identity matrix. The reflection operator acting on a vector field $\mathbf{E}$ is the vector field $\hat{P}\mathbf{E}$. Evaluating the reflected vector field $\hat{P}\mathbf{E}$ at a point $\mathbf{x}$ gives the vector

$$(\hat{P}\mathbf{E})(\mathbf{x}) = \mathbf{P}\,\mathbf{E}(\hat{P}^{-1}\mathbf{x}). \tag{3}$$

Finally, the reflection operator $\hat{P}$ acting on a pseudovector field $\mathbf{H}$ is $\hat{P}\mathbf{H}$. Evaluating the reflected pseudovector field at the point $\mathbf{x}$ gives the pseudovector

$$(\hat{P}\mathbf{H})(\mathbf{x}) = -\mathbf{P}\,\mathbf{H}(\hat{P}^{-1}\mathbf{x}). \tag{4}$$

The minus sign is present for pseudovectors because the determinant of the reflection matrix $\mathbf{P}$ is -1.

Consider an eigenmode of $\hat{P}$ with eigenvalue p. Denoting one of the fields ($\mathbf{E}$ or $\mathbf{H}$, for example) of the eigenmode by $\mathbf{A}$, then $\hat{P}\mathbf{A} = p\mathbf{A}$. Reflecting the mode twice gives the original mode, so

$$\hat{P}(\hat{P}\mathbf{A}) = \hat{P}(p\mathbf{A}) = p^2\mathbf{A} = \mathbf{A}. \tag{5}$$

Thus $p^2 = 1$, and the eigenvalues of $\hat{P}$ are $p = \pm 1$.

Let Γ_{sym} be the portion of the symmetry plane P in the region Ω. This will be a boundary of a reduced problem. Noting that

$$\hat{P}\mathbf{x} = \mathbf{x} \quad \text{for } \mathbf{x} \in \Gamma_{\text{sym}}, \tag{6}$$

then

$$\begin{aligned}(\hat{P}\mathbf{E})(\mathbf{x}) &= \mathbf{P}\,\mathbf{E}(\hat{P}^{-1}\mathbf{x}) = \mathbf{P}\,\mathbf{E}(\mathbf{x}) = p\mathbf{E}(\mathbf{x}) \\ (\hat{P}\mathbf{H})(\mathbf{x}) &= -\mathbf{P}\,\mathbf{H}(\hat{P}^{-1}\mathbf{x}) = -\mathbf{P}\,\mathbf{H}(\mathbf{x}) = p\mathbf{H}(\mathbf{x})\end{aligned} \quad \text{for } \mathbf{x} \in \Gamma_{\text{sym}} \tag{7}$$

and thus
$$\mathbf{PE} = p\mathbf{E} \quad \text{and} \quad -\mathbf{PH} = p\mathbf{H} \quad \text{on } \Gamma_{\text{sym}}. \tag{8}$$

Consider the action of the reflection matrix $\mathbf{P}$ on a vector $\mathbf{A}$. Let $A_n = \hat{\mathbf{p}} \cdot \mathbf{A}$ be the component of $\mathbf{A}$ normal to the symmetry plane P, and let $\mathbf{A}_t = (\mathbf{I} - \hat{\mathbf{p}}\hat{\mathbf{p}}^T)\mathbf{A}$ be the portion of $\mathbf{A}$ tangential to the symmetry plane P. Then

$$(\mathbf{PA})_t = \mathbf{A}_t \quad \text{and} \quad (\mathbf{PA})_n = -A_n. \tag{9}$$

That is, $\mathbf{P}$ reverses the normal component of the vector, but leaves the tangential component of a vector unchanged. Applying this to equation (8) yields

$$\mathbf{E}_t = p\mathbf{E}_t, \quad E_n = -pE_n, \quad \mathbf{H}_t = -p\mathbf{H}_t \quad \text{and} \quad H_n = pH_n \quad \text{on } \Gamma_{\text{sym}}. \tag{10}$$

The case $p = -1$ corresponds to a perfectly conducting boundary with boundary conditions

$$\text{perfectly conducting:} \quad \mathbf{E}_t = 0 \quad \text{and} \quad H_n = 0 \quad \text{on } \Gamma_{\text{sym}}, \tag{11}$$

and the case $p = 1$ corresponds to a perfectly insulating boundary with boundary conditions

$$\text{perfectly insulating:} \quad E_n = 0 \quad \text{and} \quad \mathbf{H}_t = 0 \quad \text{on } \Gamma_{\text{sym}}. \tag{12}$$

Similar conditions apply to $\mathbf{D}$ and $\mathbf{B}$ on Γ_{sym}. Like the boundary condition at perfectly conducting walls, the two boundary conditions at the symmetry plane are not independent. For example, given a solution to Maxwell's equations and the boundary condition $\mathbf{H}_t = 0$, then the other boundary condition $E_n = 0$ can be derived if time-varying fields are assumed.

Many field solvers call the perfectly conducting case a metal boundary condition and reserve the words symmetry boundary to mean only the perfectly insulating case. While this is reasonable for calculating the fields in an RF structure, some post-processing calculations, such as power loss due to the finite conductivity of the metal walls, need to distinguish between a real metal wall and the perfectly conducting case of a symmetry plane.

Here is a formulation for the eigenmode problem in a symmetric structure. The problem is reduced to a region Ω_r which is half of the original structure, as depicted in figure 1b. For the $p = -1$ (perfectly conducting) case: given the region Ω_r and its material properties ϵ and μ, find the eigenmode fields $\mathbf{E}$ and the corresponding eigenvalues ω^2/c^2 such that

$$\nabla \times (\mu^{-1} \nabla \times \mathbf{E}) = \frac{\omega^2}{c^2} \epsilon \mathbf{E} \quad \text{in } \Omega_r, \tag{13a}$$
$$\nabla \cdot (\epsilon \mathbf{E}) = 0 \quad \text{in } \Omega_r \tag{13b}$$
$$\text{and} \quad \hat{\mathbf{n}} \times \mathbf{E} = 0 \quad \text{on } \Gamma_{\text{metal}} \text{ and } \Gamma_{\text{sym}}. \tag{13c}$$

And for the $p = 1$ (perfectly insulating) case: given the region Ω_r and its material properties ϵ and μ, find the eigenmode fields $\mathbf{E}$ and the corresponding eigenvalues ω^2/c^2 such that

$$\nabla \times (\mu^{-1} \nabla \times \mathbf{E}) = \frac{\omega^2}{c^2} \epsilon \mathbf{E} \quad \text{in } \Omega_r, \tag{14a}$$
$$\nabla \cdot (\epsilon \mathbf{E}) = 0 \quad \text{in } \Omega_r, \tag{14b}$$
$$\hat{\mathbf{n}} \times \mathbf{E} = 0 \quad \text{on } \Gamma_{\text{metal}} \tag{14c}$$
$$\text{and} \quad \hat{\mathbf{n}} \times (\mu^{-1}(\nabla \times \mathbf{E})) = 0 \quad \text{on } \Gamma_{\text{sym}}. \tag{14d}$$

326 Exploiting Periodicity

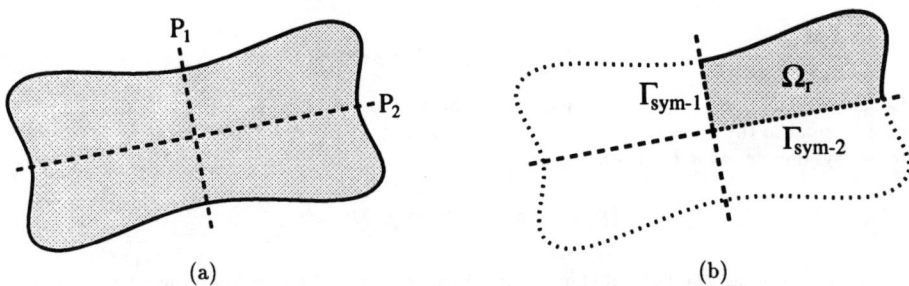

Fig. 2. (a) A structure with two symmetry planes, P_1 and P_2. (b) The region over which Maxwell's equations must be solved is reduced to Ω_r. The symmetry boundaries are $\Gamma_{\text{sym-1}}$ and $\Gamma_{\text{sym-2}}$.

It is possible for a structure to have more than one symmetry plane, in which case there is a symmetry operator for each symmetry plane. If the symmetry planes are perpendicular to each other then the symmetry operators commute and modes can be found which are simultaneously eigenmodes for all of the symmetry operators and of Maxwell's equations. An example with two symmetry planes is shown in figure 2. Let $\Gamma_{\text{sym(cond)}}$ be the symmetry planes for which the perfectly conducting ($p = 1$) case is chosen, and let $\Gamma_{\text{sym(ins)}}$ be the symmetry planes for which the perfectly insulating ($p = -1$) case is chosen. Then a formulation for the eigenmode problem is: given the region Ω_r and its material properties ϵ and μ, find the eigenmode fields $\mathbf{E}$ and the corresponding eigenvalues ω^2/c^2 such that

$$\nabla \times (\mu^{-1} \nabla \times \mathbf{E}) = \frac{\omega^2}{c^2} \epsilon \mathbf{E} \quad \text{in } \Omega_r, \tag{15a}$$

$$\nabla \cdot (\epsilon \mathbf{E}) = 0 \quad \text{in } \Omega_r, \tag{15b}$$

$$\hat{\mathbf{n}} \times \mathbf{E} = 0 \quad \text{on } \Gamma_{\text{metal}} \text{ and } \Gamma_{\text{sym(cond)}} \tag{15c}$$

$$\text{and} \quad \hat{\mathbf{n}} \times (\mu^{-1}(\nabla \times \mathbf{E})) = 0 \quad \text{on } \Gamma_{\text{sym(ins)}}. \tag{15d}$$

Plane symmetry may be combined with another type of symmetry, for example periodic symmetry. The guiding rule is that both symmetries can be used as long the corresponding symmetry operators commute.

REFLECTION SYMMETRY ABOUT A POINT

Consider a structure which is symmetric about a point $\mathbf{x}_o$, the center of the structure. An example of such a structure is shown in figure 3. Let $\hat{\mathbf{P}}$ be the reflection operator about the center. The reflection operator acting on a point $\mathbf{x}$ gives the point

$$\hat{\mathbf{P}}\mathbf{x} = -\mathbf{I}(\mathbf{x} - \mathbf{x}_o) + \mathbf{x}_o = -\mathbf{x} + 2\mathbf{x}_o. \tag{16}$$

The reflection operator acting on a vector field $\mathbf{E}$ and a pseudovector field $\mathbf{H}$ gives

$$(\hat{\mathbf{P}}\mathbf{E})(\mathbf{x}) = -\mathbf{E}(\hat{\mathbf{P}}^{-1}\mathbf{x}) \quad \text{and} \quad (\hat{\mathbf{P}}\mathbf{H})(\mathbf{x}) = \mathbf{H}(\hat{\mathbf{P}}^{-1}\mathbf{x}). \tag{17}$$

Let Ω_r be half of the interior Ω of the structure, and let the symmetry boundaries Γ_A, Γ_B and possibly the center $\mathbf{x}_o$ be the portion of the boundary of Ω_r which is in Ω. The boundaries Γ_A and Γ_B are chosen such that the symmetry operator $\hat{\mathbf{P}}$ maps Γ_A to Γ_B and vice versa. An example is shown in figure 3c.

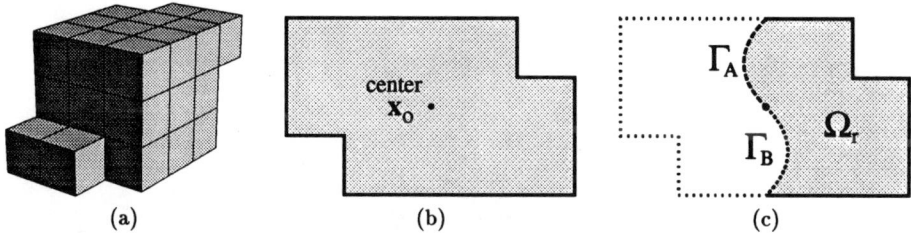

Fig. 3. (a) A structure with reflection symmetry about a point. (b) A 2d structure which is symmetric about its center x_o. The shaded region is the interior Ω. (c) The region over which Maxwell's equations must be solved is reduced to Ω_r for a symmetric structure. The symmetry boundaries are Γ_A and Γ_B.

As in the case of plane symmetry (equation (5)), reflecting an eigenmode of $\hat{P}$ twice gives the original mode, so the eigenvalues of $\hat{P}$ are $p = \pm 1$. The boundary conditions for the boundaries Γ_A and Γ_B are

$$(\hat{P}E)(x) = -E(\hat{P}^{-1}x) = pE(x) \quad \text{and} \quad (\hat{P}H)(x) = H(\hat{P}^{-1}x) = pH(x). \tag{18}$$

A special case occurs at the center x_o since $\hat{P}x_o = x_o$. At the center $-E(x_o) = pE(x_o)$ and $H(x_o) = pH(x_o)$, so

$$E(x_o) = 0 \quad \text{if } p = 1, \tag{19a}$$
$$H(x_o) = 0 \quad \text{if } p = -1. \tag{19b}$$

An example formulation for the $p = 1$ case is: given the region Ω_r and its material properties ϵ and μ, find the eigenmode fields E and the corresponding eigenvalues ω^2/c^2 such that

$$\nabla \times (\mu^{-1} \nabla \times E) = \frac{\omega^2}{c^2} \epsilon E \quad \text{in } \Omega, \tag{20a}$$
$$\nabla \cdot (\epsilon E) = 0 \quad \text{in } \Omega, \tag{20b}$$
$$\hat{n} \times E = 0 \quad \text{on } \Gamma_{\text{metal}}, \tag{20c}$$
$$E\big|_x = -E\big|_{\hat{P}x} \quad \text{for } x \in \Gamma_A \tag{20d}$$
$$\text{and} \quad E\big|_{x_o} = 0. \tag{20e}$$

Technically there is also a relation between the derivatives of the field at the boundaries Γ_A and Γ_B which is useful for proving various mathematical theorems, but in practice the relation isn't needed to exploit the symmetry in a field solver. Such relations will be neglected in this paper. The corresponding formulation for the $p = -1$ case is

$$\nabla \times (\mu^{-1} \nabla \times E) = \frac{\omega^2}{c^2} \epsilon E \quad \text{in } \Omega, \tag{21a}$$
$$\nabla \cdot (\epsilon E) = 0 \quad \text{in } \Omega, \tag{21b}$$
$$\hat{n} \times E = 0 \quad \text{on } \Gamma_{\text{metal}} \tag{21c}$$
$$\text{and} \quad E\big|_x = E\big|_{\hat{P}x} \quad \text{for } x \in \Gamma_A. \tag{21d}$$

To exploit this symmetry in a field solver, the field quantities at a point on one boundary, say $x \in \Gamma_A$, are constrained to be the same as the field quantities at the corresponding point on the other boundary, $\hat{P}x \in \Gamma_B$, times the eigenvalue factor p. This is easily accomplished if the mesh, or discretization, of the problem has the same symmetry. An example of a finite difference

328 Exploiting Periodicity

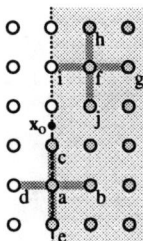

Fig. 4. An example of a finite difference mesh exploiting symmetry about a center $\mathbf{x}_o$. The shaded region is part of Ω_r and the dotted and dashed lines are the boundaries Γ_A and Γ_B, respectively. The filled circles are nodes where field values are computed. Dark gray lines indicate the field values necessary for a 5-point finite difference operator at nodes a and f. The field at node d is the field at node f times the eigenvalue p, while the field at point i is the field at node a times p. Note that the symmetry of the finite difference matrix $\mathbf{M}$ is preserved since $\mathbf{M}_{af} = \mathbf{M}_{fa}$.

mesh is shown in figure 4. The field quantities are computed for points in the region Ω_r and on the boundary Γ_B. Whenever a field quantity at a point $\mathbf{x}$ outside of the region Ω_r or the boundary Γ_B is required by the finite difference operator, the field quantity at the point $\hat{\mathbf{P}}\mathbf{x}$ (which is in the region Ω_r or the boundary Γ_B) multiplied by the eigenvalue p is used instead. A finite element formulation need only ensure that the global basis functions satisfy the boundary condition between Γ_A and Γ_B. The local basis functions of the finite elements remain the same.

PERIODIC STRUCTURES

A periodic structure has a symmetry operator $\hat{\mathbf{R}}$ which rigidly moves the structure by one period. The symmetry operator $\hat{\mathbf{R}}$ acting on a point $\mathbf{x}$ can be written generally as

$$\hat{\mathbf{R}}\mathbf{x} = \mathbf{R}\mathbf{x} + \mathbf{x}_o, \tag{22}$$

where the matrix $\mathbf{R}$ is an orthogonal matrix. That is, $\mathbf{R}^T\mathbf{R} = \mathbf{I}$. A common symmetry operation for periodic structures is translation by a cell length l along an axis, say $\hat{\mathbf{z}}$. In this case $\mathbf{x}_o = l\hat{\mathbf{z}}$ and $\mathbf{R} = \mathbf{I}$. Thus $\hat{\mathbf{R}}\mathbf{x} = \mathbf{x} + l\hat{\mathbf{z}}$. However, as indicated by the examples in figure 5, the general form of the operator $\hat{\mathbf{R}}$ allows more than translations. The symmetry operation can include a rotation as shown in figure 5b, and it can include a reflection as shown in figure 5c. A combination of translation and rotation describes helical structures, such as the example shown in figure 5d. These are all periodic structures and, as will be shown below, the modes of a periodic structure can be found by modelling a single period of the structure and using a boundary condition called the quasi-periodic boundary condition.

Periodicity differs from the previous symmetry operators in that $\hat{\mathbf{R}}^2$ is not the identity operator, so the eigenvalues of the symmetry operator are not simply ± 1. Floquet's theorem, described below, gives the eigenvalues allowed for the symmetry operator $\hat{\mathbf{R}}$ of a periodic structure.

Let $\mathbf{A}$ represent one of the fields, perhaps $\mathbf{E}$ or $\mathbf{H}$, of a mode which is a solution to the eigenmode problem in the periodic structure. Since Maxwell's equations are second order in space derivatives, there are two independent solutions to the eigenmode problem with frequency ω. For example, one solution could be $\mathbf{A}_1 \propto \cos(kz)$ while the other solution could be $\mathbf{A}_2 \propto \sin(kz)$. The fields $\hat{\mathbf{R}}\mathbf{A}_1$ and $\hat{\mathbf{R}}\mathbf{A}_2$ are also solutions to the eigenmode problem, and since any solution is a linear combination of $\mathbf{A}_1$ and $\mathbf{A}_2$,

$$\begin{pmatrix} \hat{\mathbf{R}}\mathbf{A}_1 \\ \hat{\mathbf{R}}\mathbf{A}_2 \end{pmatrix} = \begin{pmatrix} a_{11} & a_{12} \\ a_{21} & a_{22} \end{pmatrix} \begin{pmatrix} \mathbf{A}_1 \\ \mathbf{A}_2 \end{pmatrix}. \tag{23}$$

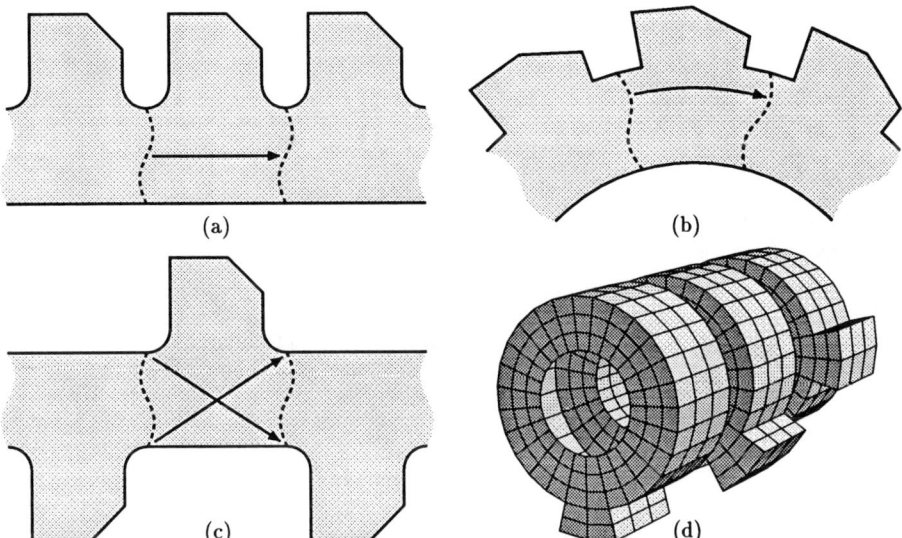

Fig. 5. Some examples of periodic structures. The dashed lines delimit one period, or cell, of the structure. The arrows indicate the action of the symmetry operations: (a) translation, (b) rotation and (c) glide reflection. The symmetry operation for a helical structure (d) is a combination of translation and rotation.

Let $\mathbf{A}_\alpha = A\mathbf{A}_1 + B\mathbf{A}_2$ be an eigenmode of $\hat{\mathbf{R}}$ with eigenvalue α. The eigenvalues α and the corresponding coefficients A and B are obtained from the eigenvalue problem

$$\begin{pmatrix} a_{11} - \alpha & a_{12} \\ a_{21} & a_{22} - \alpha \end{pmatrix} \begin{pmatrix} A \\ B \end{pmatrix} = 0. \tag{24}$$

It can be shown[1] that $a_{11}a_{22} - a_{12}a_{21} = 1$ for lossless structures. Then the characteristic equation is

$$(a_{11} - \alpha)(a_{22} - \alpha) - a_{12}a_{21} = \alpha^2 - (a_{11} + a_{22})\alpha + 1 = 0. \tag{25}$$

Letting $\beta = (a_{11} + a_{22})/2$, the eigenvalues are

$$\alpha = \beta \pm \sqrt{\beta^2 - 1}. \tag{26}$$

If $|\beta| > 1$ then the eigenvalues are real. One of the eigenvalues has $|\alpha| > 1$ and corresponds to a mode which grows geometrically along the structure. The other eigenvalue has $|\alpha| < 1$ and corresponds to a geometrically damped mode. These evanescent modes will not be considered further in this paper. If $|\beta| \leq 1$ then the eigenvalues are complex,

$$\alpha = \beta \pm i\sqrt{1 - \beta^2}. \tag{27}$$

Since $|\alpha| = 1$ then the two eigenvalues are $\alpha = e^{\pm i\psi}$ for some phase advance ψ. Furthermore, the field $\mathbf{A}$ is complex and represents a wave propagating along the structure. The real field $\mathbf{A}(\mathbf{x}, t)$ can be obtained from the complex field $\mathbf{A}(\mathbf{x})$ using

$$\mathbf{A}(\mathbf{x}, t) = Re\left\{\mathbf{A}(\mathbf{x})e^{-i\omega t}\right\}. \tag{28}$$

330 Exploiting Periodicity

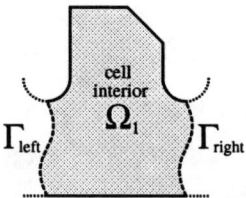

Fig. 6. One cell of the periodic structure of figure 5a. Γ_{left} and Γ_{right} are the quasi-periodic boundaries.

To exploit this symmetry the solutions are restricted to be eigenmodes of the symmetry operator $\hat{\mathbf{R}}$ with a particular eigenvalue $e^{i\psi}$. In other words, a phase advance ψ is selected and the vector field $\mathbf{E}$ must satisfy

$$\mathbf{E}(\hat{\mathbf{R}}\mathbf{x}) = \mathbf{R}\mathbf{E}(\mathbf{x})e^{i\psi} \tag{29}$$

and the pseudovector field $\mathbf{H}$ must satisfy

$$\mathbf{H}(\hat{\mathbf{R}}\mathbf{x}) = \pm\mathbf{R}\mathbf{H}(\mathbf{x})e^{i\psi}, \tag{30}$$

where the sign is the determinant of the matrix $\mathbf{R}$. The sign is positive except for the examples of figure 5 except for glide reflection, in which case the sign is negative.

Let the region Ω_1 be the interior of one period, or cell, of the periodic structure. The portion of the cell boundary in the interior Ω comprises the quasi-periodic boundaries Γ_{left} and Γ_{right}. The symmetry operator $\hat{\mathbf{R}}$ acting on the boundary Γ_{left} is the boundary Γ_{right}. An example is shown in figure 6. There is no unique choice for the cell and its boundaries. The boundaries Γ_{left} and Γ_{right} are usually planes, but in general they can be curved surfaces as shown in the example.

Here is a formulation for the eigenmode problem reduced to one cell Ω_1 of the periodic structure. Given the region Ω_1, its material properties ϵ and μ and a phase advance ψ, find the complex eigenmode fields $\mathbf{E}$ and the corresponding eigenvalues ω^2/c^2 such that

$$\nabla\times(\mu^{-1}\nabla\times\mathbf{E}) = \frac{\omega^2}{c^2}\epsilon\mathbf{E} \quad \text{in } \Omega_1, \tag{31a}$$

$$\nabla\cdot(\epsilon\mathbf{E}) = 0 \quad \text{in } \Omega_1, \tag{31b}$$

$$\hat{\mathbf{n}}\times\mathbf{E} = 0 \quad \text{on } \Gamma_{\text{metal}} \tag{31c}$$

$$\text{and} \quad \mathbf{E}\big|_{\hat{\mathbf{R}}\mathbf{x}} = \mathbf{R}\mathbf{E}\big|_{\mathbf{x}}e^{i\psi} \quad \text{for } \mathbf{x}\in\Gamma_{\text{left}}. \tag{31d}$$

SYMMETRIC PERIODIC STRUCTURES[2]

Now consider a structure which is both periodic and symmetric. In other words, the structure has two non-commuting symmetry planes. An example is shown in figure 7. Let $\hat{\mathbf{R}}$ be the rigid motion operator which moves the structure or field one period (equation (22)) and let $\hat{\mathbf{P}}$ be the reflection operator about a symmetry plane $\Gamma_{\text{sym,left}}$ (equation (2)). There is another symmetry plane $\Gamma_{\text{sym,right}} = \hat{\mathbf{R}}^{1/2}\Gamma_{\text{sym,left}}$, where $\hat{\mathbf{R}}^{1/2}$ is the rigid motion operator which moves the structure or field one half of a period. The reflection operator about the symmetry plane $\Gamma_{\text{sym,right}}$ is $\hat{\mathbf{R}}^{1/2}\hat{\mathbf{P}}\hat{\mathbf{R}}^{-1/2} = \hat{\mathbf{R}}\hat{\mathbf{P}}$.

According to Floquet's theorem, the fields can be decomposed into modes with phase advance ψ. Consider the electric field $\mathbf{E}$ of a mode with phase advance ψ satisfying $\mathbf{E}(\hat{\mathbf{R}}\mathbf{x}) = \mathbf{R}\mathbf{E}(\mathbf{x})e^{i\psi}$. The complex conjugate of the mode has the opposite phase advance,

$$\mathbf{E}^*(\hat{\mathbf{R}}\mathbf{x}) = \left(\mathbf{R}\mathbf{E}(\mathbf{x})e^{i\psi}\right)^* = \mathbf{R}\mathbf{E}^*(\mathbf{x})e^{-i\psi}. \tag{32}$$

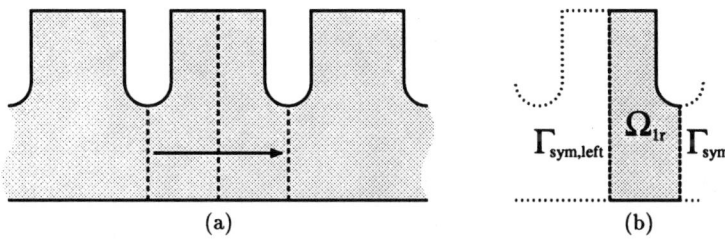

Fig. 7. (a) A symmetric periodic structure. The action of operator $\hat{\mathbf{R}}$ is indicated by the arrow. (b) The region over which Maxwell's equations must be solved is reduced to Ω_{1r}. The symmetry boundaries are $\Gamma_{\text{sym,left}}$ and $\Gamma_{\text{sym,right}}$.

$\hat{\mathbf{P}}\mathbf{E}$ is the mode reflected about the symmetry plane, and it also has the opposite phase advance,
$$(\hat{\mathbf{P}}\mathbf{E})(\hat{\mathbf{R}}\mathbf{x}) = \mathbf{R}(\hat{\mathbf{P}}\mathbf{E})(\mathbf{x})e^{-i\psi}. \tag{33}$$
Assuming $\mathbf{E}$ is a non-degenerate mode, equations (32) and (33) indicate that $\mathbf{E}^*$ and $\hat{\mathbf{P}}\mathbf{E}$ are the same mode. Let
$$\mathbf{E}^* = \alpha \hat{\mathbf{P}}\mathbf{E} \tag{34}$$
for some complex number α. Conjugating equation (34) and substituting for $\mathbf{E}^*$ with equation (34) gives
$$\mathbf{E} = \alpha^* \hat{\mathbf{P}}\mathbf{E}^* = \alpha^* \hat{\mathbf{P}}(\alpha \hat{\mathbf{P}}\mathbf{E}) = \alpha^* \alpha \hat{\mathbf{P}}^2 \mathbf{E} = |\alpha|^2 \mathbf{E} \tag{35}$$
which implies $|\alpha| = 1$. Without loss of generality choose $\alpha = 1$. A different α would just multiply the mode by an overall phase factor.

At the symmetry plane $\Gamma_{\text{sym,left}}$, $\hat{\mathbf{P}}\mathbf{x} = \mathbf{x}$ so
$$\mathbf{E}^*(\mathbf{x}) = (\hat{\mathbf{P}}\mathbf{E})(\mathbf{x}) = \mathbf{P}\mathbf{E}(\hat{\mathbf{P}}\mathbf{x}) = \mathbf{P}\mathbf{E}(\mathbf{x}) \quad \forall \mathbf{x} \in \Gamma_{\text{sym,left}}. \tag{36}$$
Let E_n be the component of $\mathbf{E}$ normal to $\Gamma_{\text{sym,left}}$ and let $\mathbf{E}_t$ be the vector tangential to $\Gamma_{\text{sym,left}}$. Then the above conditions are $E_n^* = -E_n$ and $\mathbf{E}_t^* = \mathbf{E}_t$, or
$$Re\, E_n = 0 \quad \text{and} \quad Im\, \mathbf{E}_t = 0 \quad \forall \mathbf{x} \in \Gamma_{\text{sym,left}}. \tag{37}$$
In other words, E_n is imaginary and $\mathbf{E}_t$ is real on $\Gamma_{\text{sym,left}}$.

At the other symmetry plane $\Gamma_{\text{sym,right}}$, $\mathbf{x} = \hat{\mathbf{R}}\hat{\mathbf{P}}\mathbf{x}$. Replacing $\mathbf{x}$ with $\hat{\mathbf{P}}\mathbf{x}$ in equation (32) and using equation (34) gives
$$\mathbf{E}^*(\mathbf{x}) = \mathbf{E}^*(\hat{\mathbf{R}}\hat{\mathbf{P}}\mathbf{x}) = \mathbf{R}\mathbf{E}^*(\hat{\mathbf{P}}\mathbf{x})e^{-i\psi}$$
$$= \mathbf{R}(\hat{\mathbf{P}}\mathbf{E})(\hat{\mathbf{P}}\mathbf{x})e^{-i\psi} = \mathbf{R}\mathbf{P}\mathbf{E}(\hat{\mathbf{P}}^2\mathbf{x})e^{-i\psi} = \mathbf{R}\mathbf{P}\mathbf{E}(\mathbf{x})e^{-i\psi} \quad \forall \mathbf{x} \in \Gamma_{\text{sym,right}}. \tag{38}$$
$\mathbf{R}\mathbf{P}$ is the reflection matrix about the symmetry plane $\Gamma_{\text{sym,right}}$, so let E_n be the component of $\mathbf{E}$ normal to $\Gamma_{\text{sym,right}}$ and $\mathbf{E}_t$ be the vector tangential to $\Gamma_{\text{sym,right}}$. Then the above conditions are $E_n^* = -E_n e^{-i\psi}$ and $\mathbf{E}_t^* = \mathbf{E}_t e^{-i\psi}$, or
$$Re\, E_n e^{-i\psi/2} = 0 \quad \text{and} \quad Im\, \mathbf{E}_t e^{-i\psi/2} = 0 \quad \forall \mathbf{x} \in \Gamma_{\text{sym,right}}. \tag{39}$$
In other words, $E_n \propto i e^{i\psi/2}$ and $\mathbf{E}_t \propto e^{i\psi/2}$ on $\Gamma_{\text{sym,right}}$.

The eigenmode problem reduced to one half cell Ω_{1r} is: given the region Ω_{1r}, material properties ϵ and μ and the phase advance ψ, find the eigenmode fields $\mathbf{E}$ and the corresponding eigenvalues ω^2/c^2 such that
$$\nabla \times (\mu^{-1} \nabla \times \mathbf{E}) = \frac{\omega^2}{c^2} \epsilon \mathbf{E} \quad \text{in } \Omega_{1r}, \tag{40a}$$
$$\nabla \cdot (\epsilon \mathbf{E}) = 0 \quad \text{in } \Omega_{1r}, \tag{40b}$$
$$\hat{\mathbf{n}} \times \mathbf{E} = 0 \quad \text{on } \Gamma_{\text{metal}}, \tag{40c}$$
$$Re\, E_n = 0 \text{ and } Im\, \mathbf{E}_t = 0 \quad \text{on } \Gamma_{\text{sym,left}}, \tag{40d}$$
$$\text{and} \quad Re\, E_n e^{-i\psi/2} = 0 \text{ and } Im\, \mathbf{E}_t e^{-i\psi/2} = 0 \quad \text{on } \Gamma_{\text{sym,right}}. \tag{40e}$$

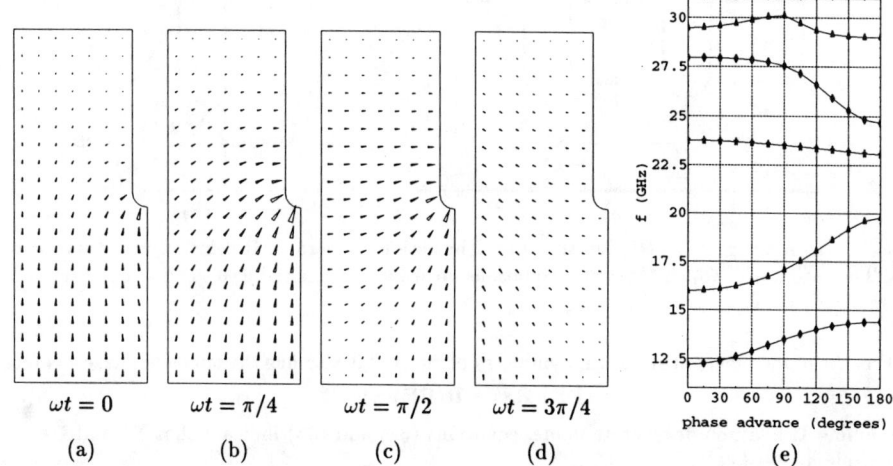

Fig. 8. (a-d) Snapshots of the electric field of the lowest dipole mode with phase advance $\psi = \pi/2$ for an X-band accelerator structure (disk loaded waveguide). The left and right sides are the symmetry planes. The size and direction of the arrows indicate the magnitude and direction of the fields. (e) Dispersion diagram for the five lowest dipole modes.

An example using symmetric quasi-periodic boundaries is shown in figure 8. Some dipole modes of an X-band accelerator structure were calculated using the finite element field solver YAP.[3] The first four figures illustrate the complex nature of a mode with phase advance $\psi = \pi/2$. The snapshot at $\omega t = 0$ is the real part of the calculated field and the snapshot at $\omega t = \pi/2$ is the imaginary part of the calculated field. Two intermediate snapshots are included to aid visualization of the wave travelling to the right. Since the modes can be calculated for an arbitrary phase advance using these boundary conditions, it is easy to calculate the dispersion diagram shown in figure 8e. If only metal and symmetry boundary conditions were available in the field solver then calculation of the dispersion diagram would be much more work for both the computer and the user.

CONCLUSION

Boundary conditions for exploiting reflection symmetry about a plane, reflection symmetry about a point, periodicity, and periodicity with a symmetry plane have been described. These conditions do not change the formulation for the interior of the structure and they are easy to implement in a field solver. Taking advantage of the symmetries of an RF structure reduces the size of the problem. This leads to less work for the computer and the user of the field solver.

REFERENCES

1. R. E. Collin, Field Theory of Guided Waves (IEEE Press, 1991), ch. 9.
2. A. G. Daikovsky, Yu. I. Partugalov and A. D. Ryabov, Particle Accelerators **20**, 23 (1986).
3. E. M. Nelson, SLAC-PUB-5881, 1992 Linear Accelerator Conf. Proc., p. 814.

CHARGED PARTICLE OPTICS WITHOUT DETAILED FIELD MAPS

David C. Carey
Fermi National Accelerator Laboratory*
P.O. Box 500, Batavia, Illinois 60510

ABSTRACT

For the initial design of a beam line or charged particle optical system, it is both useful and convenient to be able to describe the components in terms of a small number of parameters. These parameters are used in a calculation of a transfer map which represents the effect of the beam line on a particle trajectory. The transfer map is often expressed as some kind of series expansion. A calculation to first order requires the smallest number of descriptive parameters. Extension of the calculation to higher orders requires a greater number of parameters.

From our mathematical backgrounds we have come to have certain expectations as to the characteristics of a series expansion. These expectations may not always be commensurate with the physics of charged particle beam lines. The reconciliation of these expectations will be discussed.

The example used will be the program TRANSPORT and its extension to third order. The third-order expansion may represent the inherent limit of the series representation without numerical integration. We shall explain why we may have reached that limit.

INTRODUCTION

The effect of a beam line on the trajectory of a charged particle [1] is often described in terms of a transfer map. The transfer map can take the form of a multivariable Taylor series expansion [2] of TRANSPORT [3] or the symplectic maps of Alex Dragt and his collaborators [4]. In most of what follows, the exact nature of the map is not important. The details of the ensuing description will be described in terms of the multivariable Taylor series formalism, but most of the conclusions will be equally valid for other formalisms as well.

From our introductory courses in calculus that we took as undergraduates we came to have certain expectations about series expansions. The application of series expansions to charged particle optics gives us additional expectations. Finally, the desire that the mathematical formalism should produce useful results rounds out the set. A short list of the simplest of our expectations might be:

- the expansion converges rapidly

- The lower order coefficients should not depend on whether the higher orders are calculated

- The coefficents can be calculated analytically

The rapid convergence of the series expansion is a mathematically useful characteristic, since it means that the number of terms that must be calculated is relatively small. The use of the reference trajectory for the origin of the expansion facilitates rapid convergence. The helpful fact is that the origin of the series lies within the population of trajectories we wish to follow through the system. If, for example, the transformation of trajectories were to be expanded in absolute floor coordinates, the series would converge much less rapidly.

Rapid convergence of the series representation also describes a physical characteristic of the beam lines that physicists typically want to design. The elimination of nonlinearities is the most common use for correcting elements. A typical class of nonlinearities would be the chromatic aberrations. Sextupoles and octupoles can be used for correcting elements so that, as nearly as possible, all momenta can have the same focusing characteristics.

The invariance of the lower-order coefficients to whether the higher-order terms are calculated is a characteristic of expansions that we have come to take for granted. In elementary mathematics it is true of Taylor series and of expansions in orthogonal polynomials. It is not true of least squares fits. In the following discussions, we shall show that it is not necessarily an attainable ideal in charged particle optics.

That the coefficients can be calculated analytically is a fundamental prerequisite to the concept of transfer maps. If the coefficients have to be calculated numerically, then what we are doing is ray tracing and subsequently constructing transfer maps from the results of the ray tracing. Such a procedure may be useful in some instances. However, it is no longer the usual first step of describing a beam line in terms of a small number of parameters and then deriving a transfer map directly. It is more of a hybrid approach where the concept of the transfer map is used to characterize the effects of ray tracing.

For purposes of calculation, a charged particle optical system can be broken up into sections. These sections can be classified as being of two types:

1. translationally invariant fields

2. fringe fields

Below we shall show how other configurations are possible. We shall argue that inclusion of other types of field will violate our basic expectations for series expansions.

Our notation will be such that the two transverse directions are x and y, with y being vertical. The longitudinal direction is z if rectilinear and s if curvilinear. The complete six-component vector of trajectory coordinates is then:

$$(x, x', y, y', \ell, \delta)$$

The prime indicates differentiation with respect to s. The longitudinal coordinate is neither z nor s because here it represents the difference between the individual

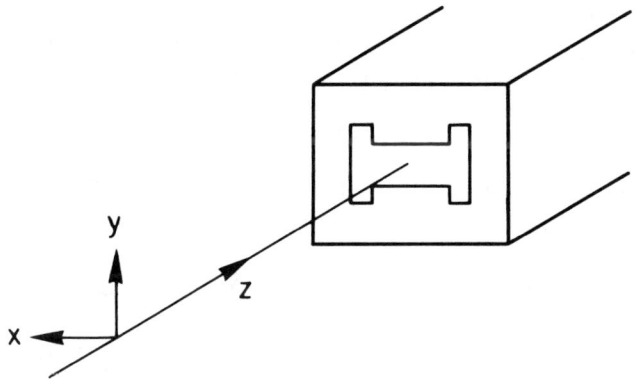

Figure 1: The trajectory coordinate system at the entrance to a magnet.

trajectory and the reference trajectory. The symbol ℓ represents the longitudinal separation, and δ the fractional momentum difference. The complete vector is sometimes denoted by the vector symbol **X**, and the individual components by x_i. The coordinate system at the entrance to a magnet is shown in figure 1.

With these essentials established, we begin our analysis of the different types of field region.

TRANSLATIONALLY INVARIANT FIELDS

Standard Beamline Components

The interior fields of the standard multipole components are translationally invariant along the reference trajectory. These components include the combined function bending magnet, the quadrupole, the sextupole, the octupole, and the solenoid. For the translationally invariant fields we can add another expectation to our list.

- The transfer matrix should segment longitudinally

This statement means that we can arbitrarily choose a point on the reference trajectory interior to the magnet. The portions of the magnet before and after this point constitute two shorter magnets. Transfer maps can be calculated separately for these shorter magnets. When combined the two transfer maps should then be identical to the transfer map for the entire magnet calculated as a whole. This statement will be true only to the order to which all the maps are calculated. If, for example, the transfer maps are all calculated to second order, the result of multiplying the two transfer maps will produce terms of third and fourth order.

These terms will not be found in the transfer map for the entire magnet considered as a whole, unless this transfer map is also calculated to third or fourth order.

The differential equations of motion of the particle trajectory coordinates take the form:

$$x_i'' + k^2 x_i = f_i \qquad (1)$$

where f is known as the driving term. The term f contains all the nonlinearities in the trajectory coordinates. It can be expanded to third order as:

$$f_i = \sum_j D_{ij} x_j + \sum_{jk} E_{ijk} x_j x_k + \sum_{jk\ell} F_{ijk\ell} x_j x_k x_\ell \qquad (2)$$

These equations are solved by iteration. A solution is first found to the homogeneous equation, where f_i is set to zero. Each iteration then raises by one the order of the solution. The solution found is substituted into the right side of equation (2) and solved again. At a particular stage, the solution takes the form:

$$x_i = \text{homogeneous solution} + \int G(t,\tau) f_i(\tau) d\tau \qquad (3)$$

The letter G denotes the Green's function, which is made up out of the solutions of the homogeneous equation.

The integrals can be evaluated analytically if k^2 and the coefficients D, E, F, and their higher-order equivalents are constants. Then, for any order, the terms inside the integral become sums and products of terms of lower order. These terms are, in turn, made up of solutions R_{ij} to the first-order equation and are nothing more than sums and products of trigonometric and hyperbolic functions. These last named functions can be evaluated analytically to give the higher-order transfer matrix elements T_{ijk} and $U_{ijk\ell}$.

The Acceleration Element

The accelerator element is not entirely translationally invariant. The field is translationally invariant, but since the element accelerates, the reference momentum is not constant. As a result, the wavelengths of the trigonometric and hyperbolic functions representing both the transverse and longitudinal motion change continuously as the element is traversed. For many years, the only representation of an acceleration element in TRANSPORT was for a massless particle. This made some kind of sense since TRANSPORT was originated at SLAC, which is an electron laboratory. The massless matrix element possessed the characteristic of longitudinal segmentation, in spite of its simple analytical form.

When this same formulation was applied to the massive particle, it failed miserably. The expressions did not agree at all with the results of numerical integration. The formula also did not come close to satisfying the segmentation test.

Fortunately, the mathematics of particle motion with continuously varying wavelength is already well developed because of quantum mechanics. The WKB method, applied to the acceleration cavity for massive particles, produces results of reasonably high accuracy [5]. Since the technique is perturbative, the expressions are not exact. However, they agree with the results of numerical integration to approximately the accuracy with which the transfer matrices are printed in TRANSPORT.

Magnet Mispowering and Violations of Midplane Symmetry

A bending magnet which is mispowered and/or has skew multipoles [6] in its field can still be translationally invariant. However, the path along which it is translationally invariant may no longer be the path that the reference particle would take in passing through the magnet. It is the path the reference particle would take if the magnet were correctly powered and the skew multipoles were absent.

It is now important to retain the field expansion about the path along which the translational invariance occurs. The integration will also be done along this path. Only then can the integrals described in the last section continue to be tractable analytically. Otherwise the field expansion will have to be paramaterized as a function of distance along the new trajectory. This parameterization will be an approximation, possibly to something which already has a known analytic form. The integrals will contain terms from this parameterization and possibly have to be evaluated numerically.

We will have lost one of the big advantages of the transfer map formalism. For this reason, even the skew dipole must be included in the skew field expansion. A combined function bending magnet with a skew dipole component cannot be rotated to eliminate that skew dipole component.

The coefficients D, E, and F, on the right side of equation (3) will have contributions from the skew components. The equations of motion will become considerably more complicated. Since sums and products of terms are used in making up higher-order terms, the number of terms in the expressions will considerably more than double. Since some of the terms are of first order (the skew dipole and quadrupole terms), iteration will continue to alter terms of all orders. It is clear that we need some criterion to terminate the iteration.

One criterion is that the matrix elements calculated be only linear in the skew fields. This is a reasonable approximation if the skew fields are small. By small we mean that the effect of quadratic or higher-order terms in the skew components is negligible for whatever purpose the beam line is being used for.

Smallness is indeed typically a characteristic of skew fields. Skew fields are present either through error of fabrication or for steering or correction purposes. These reasons all require small fields in well-designed beams. If the inherently skew fields need to be large, the beam designer should reconsider the beam configuration.

Finally, we should emphasize the difference between a magnet with skew multipoles and a rotated midplane-symmetric magnet. A magnet can often be rotated

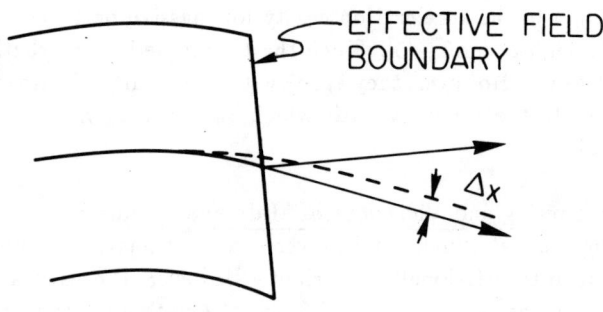

Figure 2: The parallel displacement of the reference trajectory in the fringe field of a dipole magnet

by a large angle. One possible purpose for rotating a set of quadrupoles might be to exchange horizontal and vertical phase space. A reason for rotating a dipole might be to provide vertical bending.

The field of such magnets, if analyzed in the external coordinate system, might be said to have large skew components. However, such magnets do have a magnet midplane about which the field is symmetric. About this midplane the field can then be expanded entirely in terms of symmetric multipoles.

A magnet where skew components are required for the field representation has no single plane of symmetry. In a multipole expansion about a reference curve, each multipole will have its own plane of symmetry. However, these planes for different multipoles will not necessarily coincide and there will be no necessary single magnetic midplane.

FRINGE FIELDS

The Reference Trajectory

In what is now a classic paper [7], Harald Enge derives the effect of an extended fringe field on the reference trajectory and on the first-order transfer matrices. His derivation is for a pure dipole field, where the interior of the magnet has no quadrupole or higher-order multipole terms. An equivalent sharply-cut-off field may be defined where the field integral is the same as for the extended fringe field case. The effect of the extended fringe field is to cause a parallel displacement of the reference trajectory compared to the case for a sharply cut off field. This parallel displacement is illustrated in figure 2.

The displacement is parallel because the angle by which the trajectory is deflected is proportional to the integral of the magnetic field traversed. Since the magnetic field is independent of the horizontal transverse coordinate the field integral does not depend on whether the field ends suddenly or is extended. The displacement of the reference trajectory is given by:

$$\Delta \xi = \sec^2 \beta \frac{g^2}{\rho} I_1 \qquad (4)$$

where β is the pole face rotation angle, g is the magnetic gap, and ρ is the radius of curvature of the reference trajectory in the interior field. The integral I_1 is sometimes called a "form factor". It is given by

$$I_1 = \int_{-z_1}^{\infty} \int_{-z_1}^{z} \frac{B_y^o - B_y}{g^2 B^o} dz' dz \qquad (5)$$

Matsuda et al. consider the case of a transformation through the fringe field of a combined-function bending magnet [8]. For a combined function bending magnet, the field strength is dependent on the horizontal transverse coordinate. The transverse displacement due to the extended fringe field causes the reference trajectory to get into a region where the field strength differs from the value on the reference trajectory for the sharply- cut-off field. The angle of deflection is then different, and the reference trajectory is not parallel to what it had been in the sharply-cut-off case.

In other words, there is no simple field integral which can be used to determine the deflection of the reference trajectory. Matsuda et al define what they call the "ideal field boundary for normal incidence." It is the boundary of a sharply-cut-off field that does give the same angular deflection as occurs in the reference trajectory with the extended fringe field. I believe that they have done as well as is possible in a difficult situation. However, the inevitable conclusion is that even the determination of the physical reference trajectory may require detailed ray tracing.

Form Factors

Getting back to Enge's work, he also derives the first-order transformation through an extended fringe field for a purely dipole interior field. In this case, one additional form factor is needed. It is given by:

$$I_2 = \int_{-z_1}^{\infty} \frac{B_y(B_o - B_y)}{gB_o^2} dz \qquad (6)$$

It is used in the modification of the vertical focusing strength for an extended fringe field. In the case of a sharply-cut-off field, the horizontal and vertical focusing strengths of the magnet entrance or exit are equal and opposite. In the case of an extended fringe field region they are not.

As the order of the calculation is increased, the number of form factors necessarily also increases. For a given order, the number of form factors also increases with the detail to which the calculation is done. The level of detail roughly corresponds to the number of times the equations of motion are iterated. We shall discuss this at greater length below.

The form factors, of course, depend on the detailed shape of the fringe field profile. They can all be derived from a simple field model to give order-of-magnitude estimates on the effect of the extension of the fringe field. However, as the order and accuracy of the calculation are increased the exact values of the form factors become more important. Eventually we reach the point where the paramaterization contains as much information as a detailed field map. The combined procedure of evaluating the form factors and transforming trajectories then becomes tantamount to numerical ray tracing.

Consequences of Maxwell's Equations

We have seen that a model of the extended fringe field is required in order to be physically realistic. We shall now investigate the requirements on the fringe field imposed by the mathematics of the situation. We shall, in this section, discuss only the case of the pure dipole field, where the interior field in the magnet is uniform.

From Maxwell's equations we can relate the derivatives of the magnetic field components:

$$\frac{\partial B_y}{\partial z} = \frac{\partial B_z}{\partial y} \tag{7}$$

The first term, $\frac{\partial B_y}{\partial z}$ is evaluated on the magnetic midplane. The field in the vertical direction B_y is the main bend field. It is the component which, when plotted against z, gives the drop off in the field as the magnet is exited.

The horizontal component of the magnetic field has the effect of vertical focusing. The field component B_z seems at first to be a longitudinal component. However, in this case, the z direction is taken perpendicular to the magnet pole face. If the magnet pole face is rotated, then B_z will have a component transverse to the beam. This component will be proportional to $\tan\beta$, where β is the angle by which the magnet pole face is rotated.

By symmetry the component B_z is zero on the magnetic midplane. The field value itself is given by

$$B_z = \Delta y \frac{\partial B_z}{\partial y} = \Delta y \frac{\partial B_y}{\partial z} \tag{8}$$

The deflection per unit length is proportional to the field strength. The total deflection is calculated by integrating longitudinally through the fringe field region. The derivative with respect to z integrates out and the deflection is simply proportional to the difference in field strength B_y in the interior and exterior. In the first iteration, at least, it does not matter if the field is sharply cut off.

To calculate the second-order fringe-field transmission characteristics, we need to determine the consequences of Maxwell's equation on the second derivatives of the field components:

$$\frac{\partial^2 B_y}{\partial z^2} = -\frac{\partial^2 B_y}{\partial y^2} \tag{9}$$

The longitudinal rate of fall off of the vertical dipole field is now related to the horizontal transverse rate of change of the same component. This component affects the horizontal focusing of the beam. The change in vertical strength due to the second derivative is:

$$\Delta B_y = \frac{1}{2}(\Delta y)^2 \frac{\partial^2 B_y}{\partial y^2} = \frac{1}{2}(\Delta y)^2 \frac{\partial^2 B_y}{\partial z^2} \qquad (10)$$

Integrating again longitudinally through the fringe field region, we get that the angular deflection due to this term is proportional to the difference in the first derivative $\frac{\partial B_y}{\partial z}$ inside and outside the magnet. Since the field is asymptotically constant both in the interior and exterior, the first derivative is zero in both regions. Hence, in second order a sharply-cut-off field can be used to calculate transfer matrix elements.

In third order, we must consider the field third derivatives. Maxwell's equations give us:

$$\frac{\partial^3 B_y}{\partial z^3} = -\frac{\partial^3 B_z}{\partial y^3} \qquad (11)$$

Here, as in first order, the influence is on the vertical focusing when the pole face is rotated. As with the two lower orders, we can integrate the field third derivative through the fringe field region to get the difference of two second derivatives. The two second derivatives are zero, and the effect on the focusing would seem to vanish.

However, in this case, the order of the partial derivative is so high that its effect does not entirely integrate out. In the course of traversing the fringe field region, the direction of the reference trajectory changes. This in turn is reflected by a change in angle between the field component B_z and the velocity vector. The combination of these terms gives a contribution which is proportional to the form factor

$$\frac{1}{B_o^2} \int_{-z_1}^{z_2} \left(\frac{dB_y}{dz}\right)^2 dz \qquad (12)$$

This form factor is not finite in the sharply-cut-off case. In order that the third-order matrix elements be finite, the function describing the vertical field component B_y must have a finite derivative almost everywhere. A linear fall off of the field strength in the fringe field region will produce finite third-order matrix elements.

As the order of the calculation is further increased, additional continuity conditions will be required for the higher-order transfer matrix elements to be finite. Ultimately the field will need to possess finite derivatives of all orders. The fact that the field model must become more refined as the order of the calculation is increased means that the lower orders will be affected also. Thus we are in seeming violation of one of our basic principles, namely that the lower order coefficients should not depend on whether the higher orders are calculated. The only solution

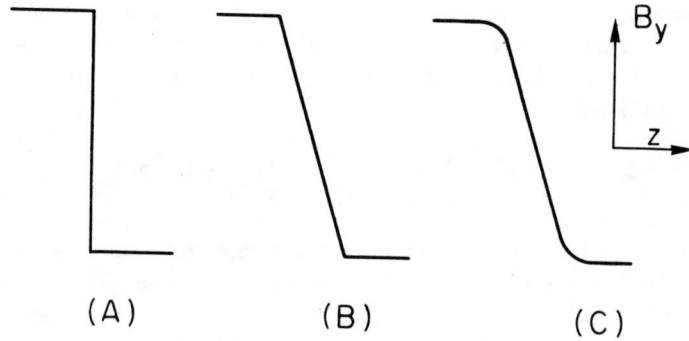

Figure 3: Three possible fringe field profiles: (A) A sharply-cut-off field where the field strength is discontinuous. (B) A linear fall off where the field strength is continuous. (C) A curved profile where the derivative of the field strength is continuous.

is to use at the outset a field representation which satisfies Maxwell's equations.

Relations Among Orders

There is still another reason why the calculation of a given order may affect lower orders. The equations of motion are solved by iteration. In the interior field of a magnet, at each iteration the lowest order affected is increased by one. Such a mathematical convenience does not hold for the fringe field. Here even the reference trajectory is determined by an iterative procedure.

In order to identify singularities and to better understand the iterative procedure, Sagalovsky [9] has developed a formalism based on an expansion in the quantity ϵ. The quantity ϵ is defined as the ratio of the separation g of magnetic poles to the interior radius of curvature ρ of the reference trajectory. The transverse coordinates are then all normalized to ρ while the longitudinal coordinate is normalized to g. The normalized coordinates are denoted by a bar over the letter, such as $\bar{z}$ for $\frac{z}{g}$. The effect of the iteration procedure is then such that the lowest order terms in ϵ that are changed increase by one with each iteration. The differential equations to be solved for the second-order transfer matrix elements are:

$$\frac{d}{d\bar{z}}T_{1ij} = \epsilon T_{2ij} \qquad (13)$$
$$\frac{d}{d\bar{z}}T_{2ij} = -3\epsilon\Gamma_1^2\Gamma_2 h T_{2ij} + f_{2ij}$$

$$\frac{d}{d\bar{z}}T_{3ij} = \epsilon T_{4ij}$$

$$\frac{d}{d\bar{z}}T_{4ij} = -\epsilon\Gamma_1^2\Gamma_2\frac{d}{d\bar{z}}(hT_{3ij}) + f_{4ij}$$

where

$$\Gamma_1 = (1+\Delta^2)^{\frac{1}{2}} \qquad (14)$$

$$\Gamma_2 = \frac{\Delta}{(1+\Delta^2)^{\frac{1}{2}}}$$

$$\Delta = -\frac{h}{\rho}(1+\Delta^2)^{\frac{3}{2}}$$

$$\Delta(z) = \epsilon z\tan\beta - \epsilon^2\sec^3\beta\int_{z_1}^{z}\int_{z_1}^{z'}h(z'')dz''dz'$$

Here h is the ratio of vertical magnetic field $B_y(z)$ at a given point in the fringe field to its value B_y^0 in the interior field. The terms denoted by the letter f are termed "driving terms" and are tabulated by Sagalovsky.

We want the solution to be at least of order zero in the parameter ϵ. The zero-order terms are those which do not depend on the spatial extent of the fringing field. The equations are solved by iteration. The driving terms f_{ijk} contain products of powers of the first-order transfer matrix elements R_{ij}. The second-order driving terms f_{ijk} contain an expression which is of order ϵ^{-1}. In order to solve the second-order equations to order ϵ^0, we need to include terms of order ϵ in the solution for the first order matrix elements R_{ij}.

The equations to be iterated to obtain the third-order matrix elements are similar to those used in second order. The driving terms are different and have also been tabulated by Sagalovsky.

$$\frac{d}{d\bar{z}}U_{1ij} = \epsilon U_{2ij} \qquad (15)$$

$$\frac{d}{d\bar{z}}U_{2ij} = -3\epsilon\Gamma_1^2\Gamma_2 hU_{2ij} + g_{2ij}$$

$$\frac{d}{d\bar{z}}U_{3ij} = \epsilon U_{4ij}$$

$$\frac{d}{d\bar{z}}U_{4ij} = -\epsilon\Gamma_1^2\Gamma_2\frac{d}{d\bar{z}}(hU_{3ij}) + g_{4ij}$$

These driving terms now contain a term of order ϵ^{-2}. The driving terms contain both the first- and second-order matrix elements. Again we wish to obtain a solution which is of at least zeroeth order in ϵ. Now we are required to include terms of order ϵ in the second-order matrix elements T_{ijk} and of order ϵ^2 in the first-order matrix elements R_{ij}.

We see that in general the extension of the calculation to each new order in the transverse trajectory coordinates requires an extension to a higher order in

ϵ for all the lower orders in the trajectory coordinates. We find again that the lower-order coefficients will depend on whether the higher orders are calculated. In this case, the preceding statement is true even if the fringe field representation is perfectly accurate from the outset of the calculation.

Multipole Equivalents

The final blow to our conventional notions about how mathematics can be used to solve physical problems comes from the consideration of multipole equivalents. A bending magnet fringe field can be configured so that optically it acts as a multipole element. If the bending magnet itself is combined function, so that it has multipole components in its interior field, then multipoles can also arise from the combined effect of the interior and fringing fields. The multipole components can be evaluated for a sharply-cut-off field, even when other parts of the transfer matrix would diverge.

An equivalent quadrupole component is produced by a flat but rotated pole face. The transformation does not depend on the characteristics of the interior field of the magnet. The calculation is performed in a coordinate system where two of the coordinates are in the face of the magnet and the third is perpendicular to it. There is no curvature to consider.

In second order, an equivalent sextupole component can be produced in either of two ways. The first and more straightforward is simply a curvature of the pole face. The effect of this curvature is independent of any nonuniformities of the interior field of the magnet. It also does not enter into the expressions for the first-order transformation. Its effect may be calculated by using a curvilinear coordinate system where one coordinate is along the curved pole face of the magnet.

The second sextupole equivalent arises from the combined effect of the transverse gradient of the interior field and the rotation angle of the pole face. The gradient of the interior field is measured in the curvilinear interior coordinate system, where one of the coordinates is along the reference trajectory. Thus we have to reconcile a curvilinear coordinate system with a pole face which is not perpendicular to the longitudinal coordinate at its intersection with the reference trajectory. Here we have the beginning of a bit of complication.

In third order, we have three ways of producing an octupole equivalent. The first, of course, is to give a cubic dependence to the pole face profile. Once again, the effect of this cubic dependence is independent of any inhomogeneities of the interior field of the bending magnet. It also does not affect any lower order. The system where one coordinate runs along the pole face now becomes a bit complicated, but is still analytically representable by a series expansion in rectilinear coordinates.

The second and third methods arise from combinations of pole face characteristics with those of the interior field. The second contribution of an octupole effect is due to the combined effect of the pole-face curvature and the linear dependence of the interior field. The third is due to the combined effect of the pole face rotation with the second transverse derivative of the interior field.

We shall first discuss the third method of producing an octupole equivalent since it is similar to one of the second order case. That second order case is the combined effect of the pole face rotation and the linear transverse variation of the interior field. The interior curvilinear system plus the pole face rotation again gives us a hint of impending complication.

The second contribution, mentioned above, to the effective octupole involves the reconciliation of two curvilinear coordinate systems. The effect of the pole face curvature falls off asymptotically in the interior or the magnet. However, there is no reason to think that the quadratic dependence of the interior field falls off asymptotically exterior to the magnet any more rapidly than does the dipole component. There is therefore no geometrical simplification for the field dependence. In third order the construction of a transfer map through an extended fringe field region inevitably points us in the direction of ray tracing through a detailed field map.

REFERENCES

[1] David C. Carey, *The Optics of Charged Particle Beams*, Harwood Academic Publishers, New York, 1987.

[2] Karl L. Brown, *A First- and Second-Order Matrix Theory for the Design of Beam Transport Systems and Charged Particle Spectrometers*, SLAC Report No. 75, Rev. 4, 1982.

[3] Karl L. Brown, Frank Rothacker, David C. Carey, and Christoph Iselin, *TRANSPORT, A Computer Program for Designing Charged Particle Beam Transport Systems*, SLAC Report No. 91, Rev. 2, 1977.

[4] Alex J. Dragt, "Lectures on Nonlinear Orbit Dynamics," Physics of High Energy Particle Accelerators, AIP Conference Proceedings No. 87, p. 147-313; Alex J. Dragt, Filippo Neri, Govindan Rangarajan, David R. Douglas, Liam M. Healy, and Robert D. Ryne, "Lie Algebraic Treatment of Linear and Nonlinear Beam Dynamics," Annual Review of Nuclear and Particle Science 38, 455-496 (1988).

[5] David C. Carey, "The WKB Approximation and the Travelling-Wave Acceleration Cavity," Record of the 1991 IEEE Particle Accelerator Conference, 1618.

[6] David C. Carey, "Two Dogmas of Charged Particle Optics," Proceedings of the Second International Conference on Charged Particle Optics, 417; David C. Carey, "Violation of Midplane Symmetry in Bending Magnets," IEEE Transactions on Nuclear Science, NS-32, 2255 (1985).

[7] Harald Enge, "Deflecting Systems," Focusing of Charged Particles, ed, Albert Septier, Academic Press, New York, 1967.

[8] H. Matsuda and H. Wollnik, "The Influence of an Inhomogeneous Magnetic Fringing Field on the Trajecories of Charged Particles in a Third Order Approximation," Nuclear Instruments and Methods 77, 40 (1970); H. Matsuda and H. Wollnik, "Third Order Transfer Matrices of the Fringing Field of an Inhomogeneous Magnet," Nuclear Instruments and Methods 77, 283 (1970); T. Sakurai, T. Matsuo and H. Matsuda, "The Vertical Component of an Ion Trajectory in a Homogeneous Magnet to a Third-Order Approximation," International Journal of Mass Spectroscopy and Ion Processes 91, 51 (1989).

[9] Leonid Sagalovsky, *Third-Order Charged Particle Beam Optics*, University of Illinois Doctoral Thesis, 1989.

*Operated by Universities Research Association, Inc., under contract with the United States Department of Energy.

OVERVIEW OF WARP, A PARTICLE CODE FOR HEAVY ION FUSION*

Alex Friedman, David P. Grote, Debra A. Callahan, and A. Bruce Langdon
Lawrence Livermore National Laboratory, Livermore CA 94550

Irving Haber
Naval Research Laboratory, Washington DC 20375

ABSTRACT

The beams in a Heavy Ion beam driven inertial Fusion (HIF) accelerator must be focused onto small spots at the fusion target, and so preservation of beam quality is crucial. The nonlinear self-fields of these space-charge-dominated beams can lead to emittance growth; thus a self-consistent field description is necessary. We have developed a multi-dimensional discrete-particle simulation code, WARP, and are using it to study the behavior of HIF beams. The code's 3d package combines features of an accelerator code and a particle-in-cell plasma simulation, and can efficiently track beams through many lattice elements and around bends. We have used the code to understand the physics of aggressive drift-compression in the MBE-4 experiment at Lawrence Berkeley Laboratory (LBL). We have applied it to LBL's planned ILSE experiments, to various "recirculator" configurations, and to the study of equilibria and equilibration processes. Applications of the 3d package to ESQ injectors,[1] and of the r, z package to longitudinal stability in driver beams,[2] are discussed in related papers.

INTRODUCTION

Heavy-ion particle accelerators are attractive candidates as drivers for inertial fusion energy applications.[3] However, in a fusion driver it is necessary to transport a much larger current than those which have been achieved in existing ion accelerators, and the physics of high-current beams is considerably more complicated than that of the beams in conventional ion accelerators. This is especially true for the recirculating induction accelerator being studied at Lawrence Livermore National Laboratory (LLNL) and at LBL as a lower-cost alternative to a linear driver for fusion energy,[4,5] because of the extra beam manipulations involved. A variety of numerical tools are employed in the study of HIF beams;[6] here we describe the WARP discrete-particle simulation effort.

WARP was developed specifically for the study of space-charge-dominated beams. In an HIF driver, such beams must be accelerated and transported over large distances, and undergo a number of manipulations, which may include: transport around bends (needed to enter the target chamber, or for recirculation); transport through imperfectly aligned focusing elements; non-steady acceleration; injection into rings; merging; and splitting.

---

*This work was performed under the auspices of the U.S. D.O.E. by Lawrence Livermore National Laboratory under contract W-7405-ENG-48, and by the Naval Research Laboratory under contracts DE-AI05-92ER54177 and DE-AI05-83ER40112.

In earlier stages of progress, the WARP code effort has been described in the *Proceedings of the International Symposium on Heavy Ion Inertial Fusion*, Dec. 3–6, 1990,[7–9] and elsewhere.[10–12] In this paper we briefly review the code concept, methods, and applications. These applications include studies of: beam drift-compression in a misaligned lattice of quadrupole focusing magnets; beam equilibria, and the approach to equilibrium; the MBE-4 experiment recently concluded at LBL;[13] and 3d simulations of bent-beam dynamics relevant to planned ILSE experiments.[14]

The code's newest capabilities include a model for an ESQ injector in 3d, using a beam formed "by injection," and an improved r,z package, incorporating a model of module impedances which can drive longitudinal instability. These are described in detail in other papers in these *Proceedings*.[1,2]

CODE OVERVIEW

The WARP code contains a number of distinct parts, including: a 3d PIC package, called WARP3d, which uses a "warped Cartesian" mesh in x,y,z to describe bends; an axisymmetric r,z PIC package, WARPrz, described elsewhere;[2,9] an envelope equation solver (used for loading a "matched" beam); and facilities for initialization, diagnostics, etc. The code uses Basis,[15] which provides a code development system that facilitates modular construction of programs, and a powerful interactive user interface.

WARP runs in single precision on computers with a 64-bit architecture (Cray C-90, Cray 2, etc.), and in double precision on 32-bit machines (currently Sun and IBM RS-6000; porting to others on which Basis runs will require very minimal effort). As the code runs, graphical output is generated using the NCAR graphics library; a high-level interface to this library provides for double-to-single precision conversion when necessary, as well as a convenient means of generating multiple plots per page, etc. The code can be run interactively, with the user calling (through the Basis interpreter) for advancement of the particles through one or more timesteps; generation of standard plots, or any other plots thought up on the spur of the moment by the user; production of data dumps (for future restarts or post-processing), and modification of physical and numerical parameters, as the run progresses—a full programming language is available to the user for these purposes. Alternatively, the code can be run to completion in a non-interactive "batch" mode for parameter studies; the same commands which a user might enter at the terminal can be included in one or more input files. Since run times (in 3-D) vary from as little as three minutes (some injector design runs) to tens of hours (runs which look at slow emittance growth phenomena, as for a recirculator), this flexibility is important.

The 3-D code's model accelerator "lattice" consists of a fully general set of finite-length (for the most part, sharp-edged) focusing and bending elements. The electric and magnetic fields of these elements (which have properties such as location, strength, etc. specified by the code's user) are computed algebraically at each particle location at each timestep. In combination with the self-fields, these applied fields are used in the Lorentz force law to advance the particle velocity timestep-by-timestep. Each multipole component (azimuthal harmonic) of the applied field is handled separately; for flexibility, different multipoles can overlap axially. The lattice can be made periodic, including periodic alignment errors, to simulate a recirculator or storage ring. Alternatively, the lattice can be periodic, but with aperiodic errors, to simulate a repetitive structure. For efficiency, we load a uniform 1d grid with lattice information (element starts, ends, strengths, etc.) at the beginning of each timestep. When advancing the particles we extract the necessary lattice data from this grid, rather than from the master lists of elements. A typical lattice, in this case one for a possible bend experiment on ILSE, is depicted in Fig. 1.

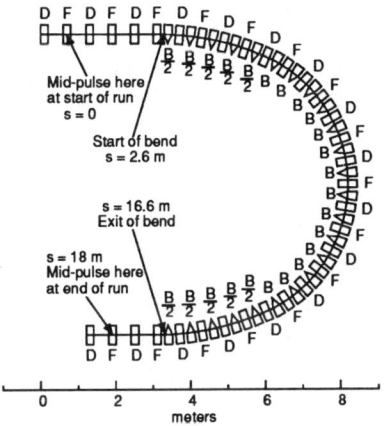

Figure 1: Lattice of focusing and bending elements in 180° ILSE bend

The simulation takes place in the laboratory frame. The computational mesh fills a moving window and is laid down anew at each timestep. The self-field is assumed electrostatic. The self-potential ϕ is obtained via vectorized fast Fourier transform (FFT) operations which assume periodicity in the axial coordinate z but use a Fourier-sine representation in each of the transverse coordinates x and y. These are implemented in such a way as to use negligible scratch space. Self-field boundary conditions are most naturally those of a square metal pipe at the transverse (x, y) edges of the mesh. A round pipe, or other shape independent of z, can be obtained by use of a 2d (transverse) capacity matrix applied independently to each axial Fourier mode. WARPrz's Poisson solver is also based on an FFT in z, with tridiagonal matrix inversion in r for each axial Fourier mode.

The fields from electrostatic quadrupoles were originally an idealization, with a perfect sharp-edged axial dependence and only quadrupole and dodecapole terms; this remains an option. Two other models for electrostatic quadrupoles are available. A set of rods, complete with self-field image effects, can be modeled; each group of four rods is handled by its own capacity matrix (the image coupling between rods at different axial locations is small). More recently, electrostatic quadrupoles and ESQ injector structures can be modeled in some generality, with plates attached to the quadrupole elements, holes in the plates through which the beam can pass, etc. This latter model uses an iterative (successive over-relaxation) solution. Most recently, the boundary conditions for this solver have been modified to allow subgrid-scale placement of the internal conductor boundaries. Specifically, Poisson's equation at all nodes overlapping a boundary is modified to enforce the condition that the interpolated potential at the intersections of the boundary with each of the three mesh axes has the desired value. This eliminates the "Lego-land" restriction common to particle codes, and allows much coarser meshes to be used.[1]

Usually the entire beam is loaded at the beginning of a run ($t = 0$), with guidance from the envelope solution for initial particle positions and velocities. For injector studies, the beam is formed by injection; particles are continually created along an equipotential surface as time advances.

To model driver-scale beams (which have speeds up to about $c/3$), we plan to use Lorentz transformations (at least in simple straight systems) to obtain the lab-frame self-**E** and **B** needed to advance the particles. We currently use $\mathbf{E}_{\text{self}}$ directly, a good approximation for the slower beams of near-term experiments.

The particle advance is based upon a leapfrog algorithm. However, when plots are to be made, particle moments collected, data dumped, etc., the velocities are advanced through a half-step so that the user sees only "synchronized" positions and velocities. In such a case the velocity advance on the next timestep is also through a half-step. Care is taken in computing particle moments for diagnostic purposes to assure that the transverse moments (*e.g.*, emittance) are not confounded by the phase advance which occurs over the finite-length "window" from within which particles contribute. Scratch copies of the particles' transverse positions and velocities are corrected to the values they would have at the midpoint of the window.

In a numerical calculation of particle trajectories, if a particle were to land within a sharp-edged focusing or bending element on four steps while its neighbor did so on only three, they would receive dramatically different impulses. Thus, the advance is modified to incorporate "residence corrections" for element forces; these corrections multiply the applied field by the fraction of the velocity advance step spent within the element. This technique allows much bigger computational steps than are otherwise possible.[10]

For efficiency on vector computers such as the Cray C-90, the particle advance is vectorized. Deposition of a particle's contribution to the charge density ρ on the computational mesh is also vectorized, but with a short vector of length eight, depositing "simultaneously" into the eight cells overlapped by each particle. No mesh arrays for the components of **E** are used; instead, values of ϕ are gathered from 32 cells in the neighborhood of each particle, and then differenced to obtain **E** at that particle's location. This saves the space of three 3d arrays.

Elongated zones (aspect ratios of order 10:1) have been found to work well provided the axial zone size is less than the beam radius. The timestep size Δt is chosen to resolve external field gradients, except at sharp transitions where residence corrections are employed. Relatively fine zoning in the transverse directions is necessary when the beam is strongly tune-depressed. To study slow emittance growth processes reliably, the transverse Debye length should be resolved by the mesh; more rapid, gross phenomena can often be studied using coarser meshes. The plasma period is long ($\mathcal{O}\ 100\Delta t$) and well-resolved.

We have developed a family of techniques for modeling accelerator bends. These are based upon following a particle's position and velocity in a sequence of rotated inertial Cartesian (laboratory) frames. This "warped" coordinate system is natural for the description of accelerators which include bends. The geometry is depicted in Fig. 2a.

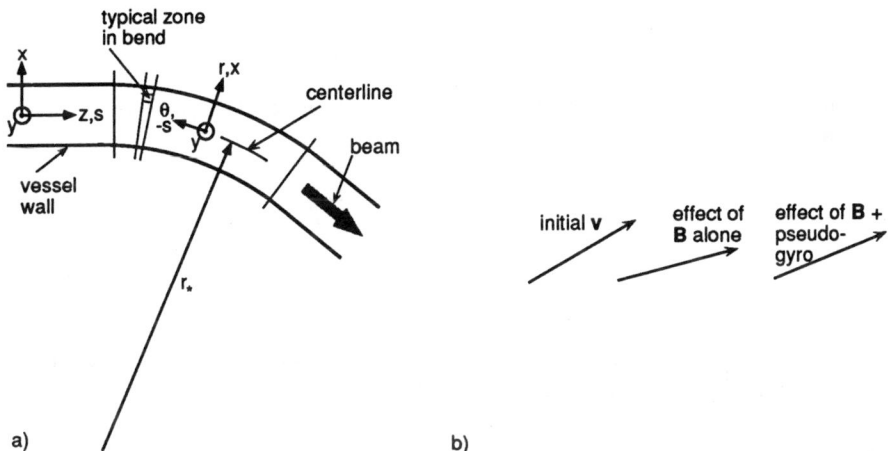

Figure 2: Geometry of the bent-beam algorithm. (a) Simulation domain and coordinate system; (b) effect of folding the pseudogyromotion associated with coordinate rotation into the bending field.

An "exact" method, which is symplectic and independent of aspect ratio, has been described previously, for both 3d and 2d (transverse) applications[12] We have tested that method in a single-particle code, but have not implemented it in WARP; however, a relativistic generalization of it has been successfully applied in the electron accelerator code ELBA.[16] In that method, a particle is advanced using (e.g.) a leapfrog advance in the coordinate system associated with the particle's location at the beginning of the step. Then, the particle's position and velocity are algebraically transformed into the rotated inertial coordinate system associated with its location at the end of the step. The scheme conserves phase-space area identically, and implies no large-aspect-ratio (gentle bend) expansion. Its numerical properties are those of the underlying difference scheme. The exact method has a non-negligible operation count, but is quite usable. Here, we describe a simpler approximate method now used in WARP3d which is both faster and sufficiently accurate for present purposes.[7,10]

The radius of curvature of the reference orbit (usually the vessel centerline) is $r_* \equiv h^{-1}$. Time is the independent variable for particle orbits. The conventional (for accelerator codes) independent variable s is in WARP a dependent variable for orbits, as are x, y. In straight sections, $s \equiv z$, while in bends, $s \equiv -r_*\theta$. The "radial" coordinate is $x \equiv r - r_*$; the unit vectors $\hat{x}$ and $\hat{s}$ evolve as a particle moves, and are different for each particle. The axial speed is $v_z = -r\dot{\theta}$ (we use subscripts z and s interchangeably). The axial position is advanced in time using:

$$ds/dt = -r_*\dot\theta = (r_*/r)v_z \ . \tag{1}$$

A particle's velocity vector rotates because of the rotation of the coordinate axes. Due to this alone, the rate of change of the velocity angle is:

$$\frac{d}{dt}\arctan\left(\frac{v_x}{v_z}\right) = -\dot\theta = \frac{v_z}{r_* + x} \ . \tag{4}$$

We thus need only augment the dipole (bending) field at each particle position with a "pseudo-gyrofrequency":

$$B_y \Leftarrow B_y - \frac{m}{q}\frac{v_z}{r_* + x} \tag{5}$$

where m is the particle's mass and q its charge. This folds the necessary back-rotation into existing coding. The net effect is shown in Fig. 2b. The algorithm is inexact because v_z and x change during the step, but is accurate enough for our needs; "residence corrections" on entry to and exit from bends are necessary.

Poisson's equation in "warped" coordinates is[17]:

$$\frac{1}{1+hx}\frac{\partial}{\partial x}\left((1+hx)\frac{\partial\phi}{\partial x}\right) + \frac{\partial^2\phi}{\partial y^2}$$
$$+ \frac{1}{1+hx}\frac{\partial}{\partial s}\left(\frac{1}{1+hx}\frac{\partial\phi}{\partial s}\right) = -4\pi\rho \ . \tag{6}$$

Expanding the derivatives, we solve this iteratively. At each iteration the 3d FFT Poisson solver inverts the dominant "Cartesian" second derivative terms. One term, proportional to $(\partial h/\partial s)(\partial\phi/\partial s)$, is included by a simple finite difference, assuming the change in h at bend entry/exit can be spread in s slightly. The iteration converges rapidly, in two or three passes. It is necessary to obtain the true charge density from the "conventional" ρ_c collected from the particles, using $\rho = \rho_c r_*/r$, since (in a bend) the "axial" separation of zones varies with x. Also, the axial field is $E_z = -(r_*/r)\partial\phi/\partial s$.

SUMMARY OF APPLICATIONS

Drift Compression: (current enhancement resulting from a head-to-tail velocity gradient or "tilt"): Relatively small misalignments of the focusing quadrupoles can lead to significant off-axis displacements. Image forces and fringing fields can then induce emittance growth. We seek to learn how fast and how much the beam may be compressed without unacceptable emittance degradation. The details of the errors, in a system of ILSE scale, are significant; different random-number seeds for the offsets lead to widely varying displacements.[11]

Equilibration: We are examining, in 3d and r,z, the transfer of thermal energy between transverse and longitudinal motions. For certain ranges of physical parameters, a beam initialized colder in z (axially) than in x,y (transversely) is observed to heat rapidly in z until T_z is a large fraction of $T_{x,y}$. This appears to be a collective process.[11]

Axial Confinement, Nature of Equilibria: To follow a finite-length beam for a long time, it is necessary to apply an axial confining force. This is done using shaped ends of the accelerating pulses, or "ears." We have modeled (in 3d) near-equilibrium beams that remain "quiescent" over runs as long as 175 lattice periods without significant mid-pulse emittance degradation in the simulation.[8] The emittance near the tips of the beam does grow. This suggests that the *ansatz* used in the initial loading, namely a parabolic falloff of line-charge density leading to a cigar-shaped beam with constant phase advance throughout the beam and its tips, is not a true equilibrium state. The beam appears to be evolving toward a constant-emittance or constant-$T_\perp$ configuration. We are currently exploring these issues further with WARPrz.

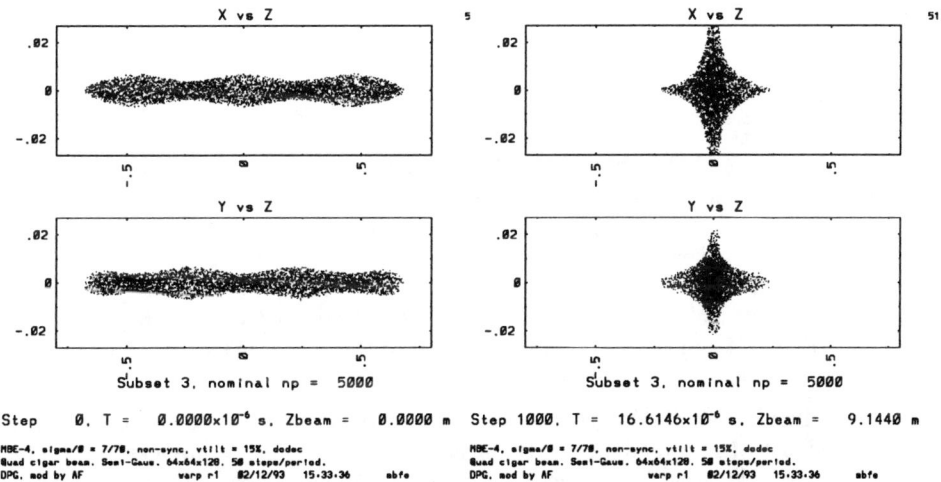

Figure 3: Views of MBE-4 beam: initial on left, at peak compression on right.

Simulations of the MBE-4 Experiment: In this LBL experiment, emittance growth has been observed to accompany aggressive drift compression. Using WARP, we have confirmed that this results from nonlinearities in the focusing fields which are sampled by the particles to a greater extent when the beam grows "fat" as a result of the compression. The dodecapole term is the chief culprit; however, a more realistic representation of the internal conductors as individual cylinders, using the capacity matrix method, showed less emittance growth in cases where the focusing potentials were at $\pm V_{\text{foc}}$ than in cases where they were at 0 and $2V_{\text{foc}}$. We have conjectured that this is a result of the beam particles "seeing" the pipe, which is at ground potential, while they are in the short drift regions between quadrupoles.

Figure 3 shows the initial beam in one such simulation, in top and side views. After aggressive drift compression has dramatically shortened the beam, it has the form shown in the right half of the figure. By this time some particles have

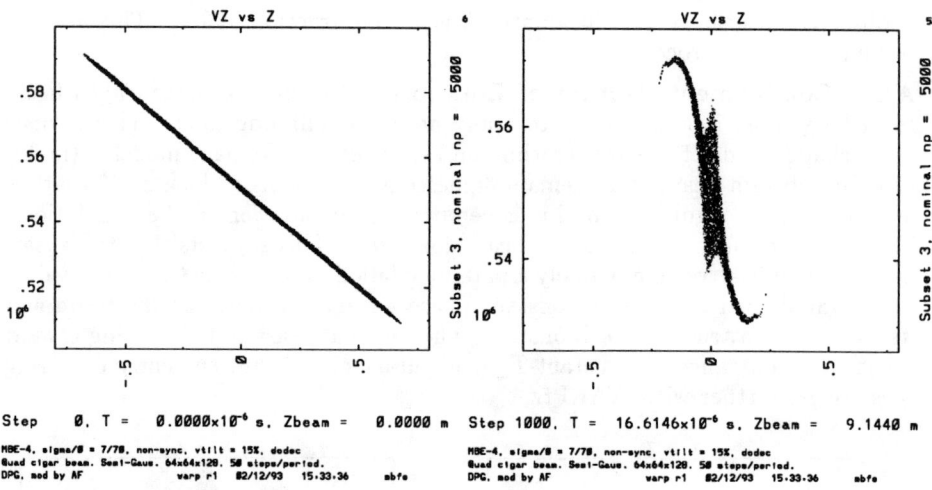

Figure 4: Axial velocity vs. position for MBE-4 beam.

scraped the walls of the chamber and thus have been lost. The actual computer output in diagnostics such as these uses color to indicate different initial axial segments of the beam. Plotted particles retain their color throughout, and so it is possible to discover where different parts of the initial phase space end up. In this run, the inward-directed (in the beam frame) velocity is reduced by the axial self-field, and particles from various axial locations are partly mixed by the end of the run. Figure 4 shows the (v_z, z) projection of the phase space at the beginning and end of the run.

Longitudinal Stability with Finite Gap Impedance: A known instability is associated with the impedance of the accelerating modules. While careful design and techniques such as feed-forward stabilization should afford suppression of the instability, it is important to be able to model (in a causal, self-consistent way) multidimensional effects such as wave reflection at the bunch ends and radial variations of the interaction between particles and modules. Such effects may be especially significant in the driver-relevant low growth rate regime. We are using WARPrz for these studies.[2,9]

Other applications of the (r, z) model include studies of equilibration processes, of the equilibrium axial dependence of the emittance at the beam ends, and of axial confinement using "ears."

Bent-beam dynamics: We have examined beam behavior in a variety of lattices which incorporate bends; these include models of the 180° bend planned as an ILSE experiment, and both round and racetrack-shaped recirculator configurations.[18]

A lattice we have considered, similar to one proposed for an ILSE experiment,[19] is shown in Fig. 1. (These runs simulated a somewhat different bend than is currently planned; modeling of the latest version is underway.) For this system, the phase advances per lattice period are $\sigma_0 = 72°$, $\sigma = 20°$, and dipoles

(20 cm) and quadrupoles (20 cm) alternate in a FOBODOBO lattice with full period 1.2 m. The first dipole begins at $z = 2.6$ m, the last ends at 16.6 m (after 180° of bending), and we ended the runs at 18 m (900 steps). We considered axially-cold and -hot ($T_z \sim T_\perp$) beams. The emittance of the axially-hot beam grows; that of the axially-cold beam does not. An axially-hot straight beam in a similar lattice without dipoles does not appear to suffer emittance growth, nor does an emittance-dominated beam in a bend. The beam centroid locations (away from mid-pulse) move radially during their transit of the bend because of the head-to-tail velocity "tilt;" nonetheless, they are re-injected nearly along the centerline of the straight section which follows, due to the "first-order achromat" design.

ESQ injector: We have recently begun modeling this class of systems in 3d.[1] One item of interest is an "energy effect" associated with the fact that the electrode potential differences are comparable to the beam energy, at least at the low-energy end. This leads to some emittance growth, which may be reduced through careful design.

REFERENCES

[1] D. P. Grote, A. Friedman, and S. S. Yu, these *Proceedings*.

[2] D. A. Callahan, A. B. Langdon, A. Friedman, and I. Haber, these *Proceedings*.

[3] T. J. Fessenden and A. Friedman, *Nucl. Fusion* **31**, 1567 (1991).

[4] J. J. Barnard, F. Deadrick, A. Friedman, D. P. Grote, L. V. Griffith, H. C. Kirbie, V. K. Neil, M. A. Newton, A. C. Paul, W. M. Sharp, H. D. Shay, R. O. Bangerter, A. Faltens, C. G. Fong, D. L. Judd, E. P. Lee, L. L. Reginato, S. S. Yu, and T. F. Godlove, "Recirculating Induction Accelerators as Drivers for Heavy Ion Fusion," submitted to *Physics of Fluids B*, December 1992.

[5] S. S. Yu, J. J. Barnard, G. J. Caporaso, A. Friedman, D. W. Hewett, H. Kirbie, M. A. Newton, V. K. Neil, A. C. Paul, L. L. Reginato, W. M. Sharp, T. F. Godlove, R. O. Bangerter, C. G. Fong, and D. L. Judd, *Particle Accelerators* **37-8**, 489 (1992).

[6] A. Friedman, J. J. Barnard, D. A. Callahan, G. J. Caporaso, Y.-J. Chen, J. F. DeFord, W. M. Fawley, D. P. Grote, I. Haber, K. D. Hahn, E. Henestroza, D. W. Hewett, D. D.-M. Ho, A. B. Langdon, E. P. Lee, A. N. Payne, C. C. Shang, W. M. Sharp, H. D. Shay, L. Smith, and S. S. Yu, "Simulation of Heavy Ion Fusion Beams," to appear in *Proc. IAEA Tech. Comm. Mtg. on Advances in Simulation and Modeling of Thermonuclear Plasmas*, Montreal, June 15–17, 1992; LLNL Report UCRL-JC-110193, 1992.

[7] A. Friedman, D. P. Grote, D. A. Callahan, A. Bruce Langdon, and I. Haber, *Particle Accelerators* **37-8**, 131 (1992).

[8] D. P. Grote, A. Friedman, and I Haber, *ibid.*, p. 141.

[9] D. A. Callahan, A. B. Langdon, A. Friedman, D. P. Grote, and I. Haber, *ibid.*, p. 97.

[10] A. Friedman, D. P. Grote, and I. Haber, *Phys. Fluids B* **4**, 2203 (1992).

[11] A. Friedman, R. O. Bangerter, D. A. Callahan, D. P. Grote, A. B. Langdon, and I. Haber, *Proc. 2nd European Particle Accelerator Conference*, Nice, June 12–16, 1990.

[12] A. Friedman, *Proc. 13th Conf. on Numerical Simulation of Plasmas*, Santa Fe NM, 1989; LANL, R. J. Mason, Ed.

[13] T. J. Fessenden et. al., *Proc. of the 1987 Particle Accelerator Conf.*, Pp. 898-902, IEEE Cat. No. 87CH2387-9 (1987).

[14] T. Fessenden, R. Bangerter, D. Berners, J. Chew, S. Eylon, A. Faltens, W. Fawley, C. Fong, M. Fong, K. Hahn, E. Henestroza, D. Judd, E. Lee, C. Lionberger, S. Mukherjee, C. Peters, C. Pike, G. Raymond, L. Reginato, H. Rutkowski, P. Siedl, L. Smith, D. Vanecek, S. Yu, F. Deadrick, A. Friedman, L. Griffith, D. Hewett, M. Newton, and H. Shay, "ILSE, The Next Step toward a Heavy Ion Induction Accelerator for Inertial Fusion Energy," *Proc. 14th Int. Conf. on Plasma Physics and Controlled Nuclear Fusion Research*, IAEA, Wurzburg, Germany, Sept. 30-Oct. 7, 1992.

[15] P. F. Dubois et. al., "The Basis System," LLNL Document M-225 (1988).

[16] S. Slinker, J. Krall, M. Lampe, and G. Joyce, in *Conference Record of the 1991 IEEE Particle Accelerator Conference*, San Francisco, edited by L. Lizama and J. Chew (IEEE, NJ, 1991), p. 242.

[17] K. L. Brown and R. V. Servranckx, in *Physics of High Energy Particle Accelerators*, AIP Conf. Proc. **127**, 75 (1983).

[18] J. J. Barnard, H. D. Shay, S. S. Yu, A. Friedman, and D. P. Grote, "Emittance Growth in Heavy Ion Recirculators," *Proc. 1992 Linear Accelerator Conference*, Ottawa, Ontario, August 24-8, 1992, 229 (AECL-10728, C. R. Hoffmann, Ed.).

[19] E. P. Lee, *Nucl. Inst. Meth. in Plasma Research* **A278**, 178 (1989).

MODELLING RF SOURCES USING 2-D PIC CODES[*]

Kenneth R. Eppley
Stanford Linear Accelerator Center, Stanford, CA 94309

ABSTRACT

In recent years, many types of RF sources have been successfully modelled using 2-D PIC codes. Both cross field devices (magnetrons, cross field amplifiers, etc.) and pencil beam devices (klystrons, gyrotrons, TWT's, lasertrons, etc.) have been simulated. All these devices involve the interaction of an electron beam with an RF circuit. For many applications, the RF structure may be approximated by an equivalent circuit, which appears in the simulation as a boundary condition on the electric field ("port approximation"). The drive term for the circuit is calculated from the energy transfer between beam and field in the drift space. For some applications it may be necessary to model the actual geometry of the structure, although this is more expensive. One problem not entirely solved is how to accurately model in 2-D the coupling to an external waveguide. Frequently this is approximated by a radial transmission line, but this sometimes yields incorrect results. We also discuss issues in modelling the cathode and injecting the beam into the PIC simulation.

INTRODUCTION

In recent years, 2-D PIC has been used to model a number of RF sources, e.g., klystrons, TWT's, lasertrons, magnetrons, CFA's, etc. We discuss a number of issues that arise in simulating RF devices, describe some successful solutions, and examine areas where difficulties still exist.

We separate the modelling process conceptually into several parts. The vacuum simulation involves the interior of the problem away from all boundaries. Beam injection requires modelling the cathode and transporting the beam into the interaction region. To model the RF structure one may either approximate the physical device with conducting boundaries, or else represent the circuit by imposing RF fields as a boundary condition to the vacuum region. Waveguides may also be modelled as RF boundary conditions, but they can present problems if the 3-D nature of the coupling invalidates the axisymmetric assumption. Surface physics must be understood to model secondary emission cathodes and multipactor.

VACUUM SIMULATION

Microwave tubes are well suited for PIC. The electron beams correspond to hot, low density plasmas, whose Debye length is of similar order to the scale of the RF structure. Although the internal temperature of the beams may be low,

---

[*] Work supported by the Department of Energy, contract DE–AC03–76SF00515.

the numerical heating rate is usually a function of the beam energy, which is high. The mesh size is generally determined by the RF structures rather than by the plasma properties. However, for relativistic beams, it may be necessary to reduce the mesh size to prevent unphysical emittance growth.

INJECTION

In pencil beam devices it is difficult to accurately simulate both the cathode region and the RF structure on a single mesh. Many Pierce guns have high convergence ratios. Thus the cathode radius is much larger than the drift tube. Also, the curved cathode surface cannot be modelled by stairsteps without grossly distorting the emission. For many guns, an electrostatic code such as EGUN gives an accurate calculation of the beam in the up to the RF region. Our klystron modelling uses EGUN, with a finer mesh and more radial trajectories than are used in the RF simulation, to generate a set of trajectories which are injected into the PIC simulation. Generally we reduce the number of radial trajectories by a factor of five to ten by averaging. It is important to properly model the space charge effects at the injection surface. Metal boundary conditions will cause a non-physical energy change. This is sometimes tolerable. However, for a long device it is often necessary to split the simulation into several pieces, injecting the output from one into the input of the next. In this situation a metal wall at the interface seriously distorts the bunching of the beam. We have found that using Neumann boundary conditions in the Poisson solver gives satisfactory results.

For an RF photocathode an electrostatic simulation would be inaccurate. When such devices have flat cathodes without radial compression, it is feasible to include the gun as part of the PIC calculation. One may model the actual electrode structure of the gun. Alternately, one may use EGUN to calculate the field distribution for the cathode region without space charge and impose this as a DC external field (either as a volume field or as a surface boundary condtion).

RF STRUCTURE

One way of simulating an RF structure is to approximate the physical boundaries with conducting surfaces. Alternately, one may replace the structure with an equivalent circuit and impose its fields as a boundary condition on the simulation region. For a structure with a short fill time and simple boundaries, such as disks or vanes, modelling of the physical geometry is practical. This approach has been used in magnetrons and disk-loaded structures. For high-Q cavities, especially with reentrant noses, the second method, replacing the cavity gap by a voltage boundary condtion ("port approximation") is highly useful. We describe this method in detail.

In the port approximation[1] to the modelling of RF cavities using an electromagnetic PIC code such as CONDOR[2], the cavities are simulated by imposing an RF voltage as a boundary condition across an opening or "port" in the outer wall of the drift tube (Figure 1). This method ignores the transient and looks only for the steady-state solution at a single operating frequency. (This is done for speed, but if desired, the port method could be used to calculate the transient as well.) By writing the equations for energy flow across the gap, one splits the problem into two simpler pieces:

$$V \cdot I_{ind} = \int E \cdot J dV \tag{1}$$

(Dot product involves the integration over RF cycle of the complex phase; for vector quantities it also subsumes the spatial dot product.)

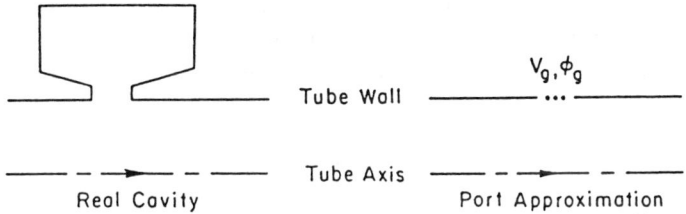

Figure 1. Simulation of a real cavity by a port boundary condition.

From the cavity side, in steady state the voltage across the gap and the current flowing in the walls uniquely determine the state of the cavity. From the drift tube side, the energy flow into or out of the beam is completely determined in steady state by the voltage and phase across the gap. The current flowing in the walls (the induced current) can be calculated simply in terms of the transform of the volume integral of $E \cdot J$. Of course, the current distribution is changed by the presence of the cavity voltage.

The voltage and phase must be chosen (by some means) to be consistent with the cavity impedance and with the RF current induced by the electron beam. Note that the induced current is not identical to the RF current flowing through the drift tube. The relation to cold cavity parameters comes through the relation

$$V = I_{ind} Z \tag{2}$$

It is straightforward to relate Z to cavity Q, ω, and R/Q (taking care to be consistent if voltages are measured on axis or across the gap), using the relationship:

$$Z = e^{j\psi}/\alpha \tag{3}$$

where

$$\alpha = I_{ind}/V = [Q_0^{-2} + 4(\Delta\omega/\omega)^2]^{1/2} \tag{4}$$

and

$$\psi = \phi_I - \phi_V = \tan^{-1}[-2Q_0\Delta\omega/\omega] \tag{5}$$

In steady state, the gap voltage should satisfy the condition

$$V_{ss} = I_{ss} \cdot Z \tag{6}$$

Here Z is the complex cavity impedance, and V_{ss} and I_{ss} are the Fourier components in steady state of the gap voltage and the induced current at the operating frequency. Now we assume a time dependence of the form:

$$V_t = V(t)e^{-j\omega t} \tag{7}$$

Here V_t is the instantaneous voltage across the port and $V(t)$ is an envelope which varies slowly in an RF cycle. Asymptotically, we want $V(t)$ to converge to V_{ss}. We can achieve this by making $V(t)$ satisfy a relaxation equation, i.e.,

$$dV(t)/dt = -k \cdot [V(t) - I(t) \cdot Z] \tag{8}$$

Thus $V(t)$ will adjust itself until the impedance relation is satisfied self-consistently. We compute the induced current at the operating frequency by keeping a running table of the volume integral of $E \cdot J$. The equation converges faster if one takes into account the beam loading, assuming that the change in induced current is a linear function of the change in voltage, i.e.

$$\Delta I = \alpha_l \cdot \Delta V \tag{9}$$

Then

$$\Delta V = -k\Delta t(V - IZ) \div (1 - \alpha_l Z) \tag{10}$$

The constant α_l depends on DC current, frequency, and drift-tube size but is insensitive to gap width and beam profile.

The port approximation has been used to calculate a number of klystrons at SLAC[3] and has agreed with experiments to within about five percent in peak efficiency and to about two dB in gain (Figure 2).

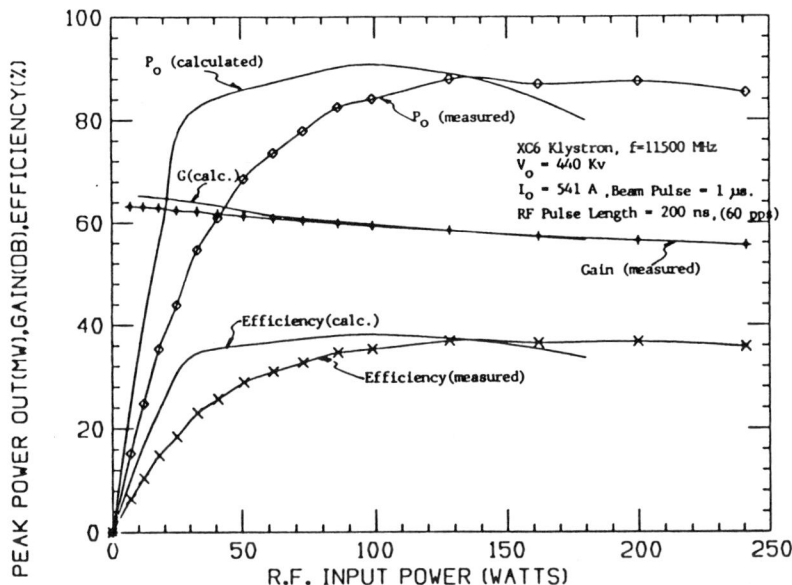

Figure 2. Comparison between CONDOR simulation and experimental results for the SLAC XC6 klystron.

An alternative to the port approximation is to model the cavity boundaries. This is feasible for low Q cavities or structures, but difficult for high Q gain cavities. An accurate gain calculation requires that the frequencies be correct to better than 0.1 percent. To avoid prohibitively fine zoning, it is necessary to "tweak" the cavity tuning by moving the boundaries a mesh point at a time. The MAGIC group has developed an algorithm so that their code can do this adjustment itself, rather than the user. This method works reasonably well, but it does require an approximate knowledge of the cavity dimensions beforehand. It is slower than the port approximation, because it must run until the transient decays, and also because the simulation region is larger. As mentioned, the port method could be modified to compute the trasient, if needed. The full cavity method is more accurate if the beam goes close to or intercepts the gap.

We have used full geometry modelling to simulate disk-loaded standing and travelling wave output structures (Figure 3). One can also use the port approximation to model some disk-loaded structures, using an impedance matrix to represent the coupling between the cells. This algorithm has been successful with slot-coupled standing wave double output cavities. If the inner radius of the disks is too large (e.g., if the drift tube is not cutoff to the second harmonic), the induced current calculation will be too inaccurate for the port approximation to be useable.

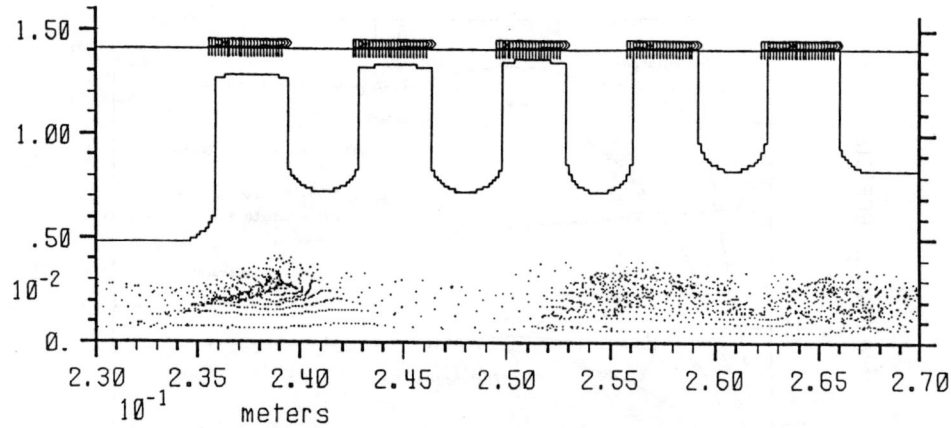

Figure 3. RF circuit and electron positions in a CONDOR simulation of an X-band disk-loaded output structure.

For narrow band structures, we have adopted the MAGIC idea of partially filling the outer mesh line for greater accuracy. We adjust the volume of metal to be the same as if the boundary were at its true phyical location. Without this method, the mesh required for accurate resolution would be prohibitively fine. For disks with rounded ends on their inner radii such an adjustment would be more complex, and we have not yet attempted it.

Helical TWT's have been modelled using MAGIC[4], representing the 3-D helix with an array of conducting wires and dielectric support rods (Figure 4). Although the simulation model was quite different geometrically from the physical helix, its RF properties were similar. The output was represented by a completely radiating freespace boundary. This model agreed well with experimental coldtest results and hottest efficiency, except at low current.

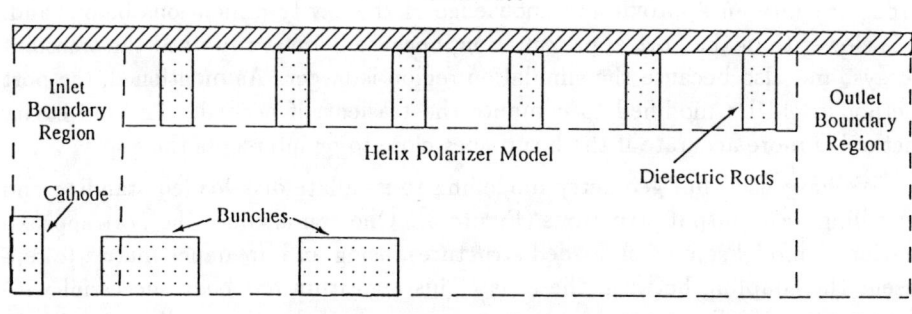

Figure 4. Simulation geometry used to model a helical travelling wave tube using MAGIC.

Modelling the geometry in magnetrons is relatively simple, since the devices have low Q and the structures are usually vanes or slots. Periodicity allows one to model a single section with periodic boundary conditions. A full 360 degree simulation is sometimes necessary, if the output coupling is not periodic (Figure 5). There are also modes which only show up when the full circumference is included. CONDOR calculations of Varian phase-locked magnetrons[5] agreed with experiment to about five percent in efficiency and to a few percent for the V-I characteristic (Figure 6).

Originally we believed that it was essential to use a capacitance matrix to properly impose the DC voltage on the anode. We have now found a simpler method to be satisfactory. We model the anode with conducting blocks and impose a voltage on the upper surface. Image charges on the conductors will force the field inside the metal to be zero.

Cross field amplifiers may be too long to model in their entirety. The RF sever makes this device non-periodic. We have used a variation on the port approximation to model such tubes. Rather than model each cell as a port, we

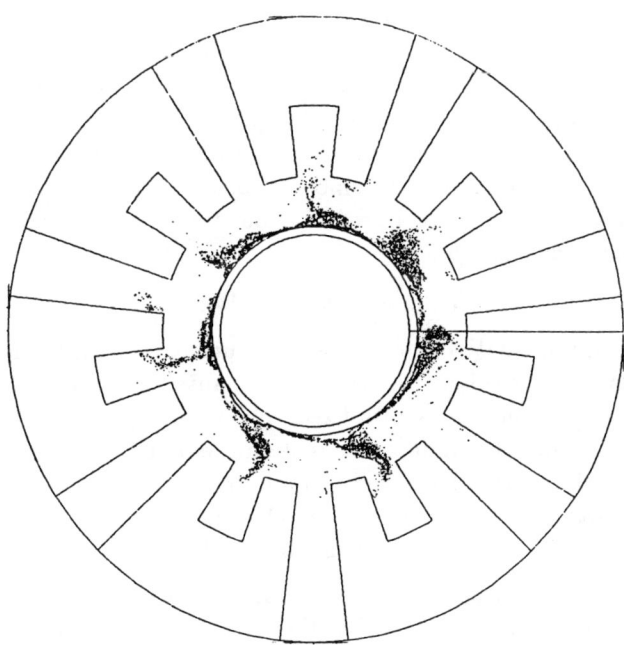

Figure 5. RF structure and electron distribution for a CONDOR simulation of a rising sun magnetron.

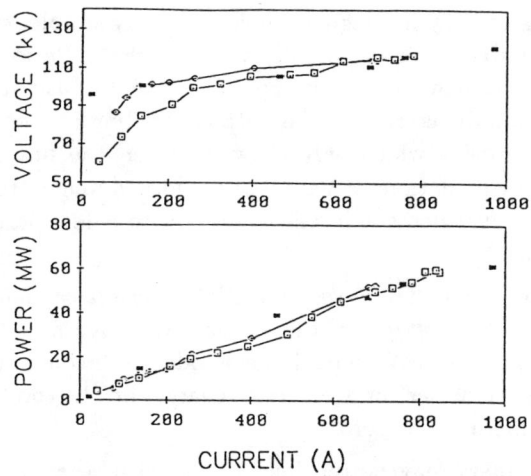

Figure 6. Comparison between simulation and experimental results for rising sun magnetrons. The open squares are experimental data. The closed squares are simulation results.

replaced the entire RF circuit with a traveling wave voltage imposed across the top boundary[6]. We again calculated the beam-circuit interaction from the energy balance equations. The simulation region corresponded to one travelling wave wavelength which moved with the electron spoke. In forward wave CFA's the power was quite variable, depending on how much bunching was maintained across the sever. Averaging over a number of passes gave reasonable agreement for a Varian CFA. The V-I curve agreed to about three percent and the efficiency to about fourteen percent.

COUPLING TO EXTERNAL WAVEGUIDES

In the port approximation, coupling to an output waveguide merely changes the Q of the cavity, with otherwise identical formalism. For the input waveguide, we make an initial calculation with a fixed voltage, independent of the specified drive level. We calculate the induced current on the input cavity and the voltages on the small signal cavities. Then we calculate the relation between input voltage and available power, and make a second calculation with the desired drive, scaling the small signal voltages linearly with the drive voltage.

For standing wave disk-loaded structures, coupling to a waveguide may be modelled by a port on the outer wall of the cell. The port behaves like a radial transmission line. We adjust the transmission coefficient of the port to produce the desired Q. Equivalently, we calculate the surface current from the magnetic field, then set the port voltage to equal $I \cdot Z$. We can vary Z to adjust the Q. If we need a complex impedance, we can Fourier transform the current to find the phase (for a single frequency). Another method, used in MAGIC, is to insert a resistive element behind the port and adjust the resistivity for a desired Q.

For a travelling wave structure the output coupling may be more complicated. Generally, these structures are adjusted experimentally to produce a good cold match. This adjustment may involve not only the waveguide geometry, but also the geometry of the cell previous to the waveguide. If we model the cells of a disk-loaded structure which was cold-matched in the lab, and adjust the impedance of the port (in general complex) for maximum transmission, then we get rough agreement between the 2-D model and the experimental device. Two such tubes built at SLAC agreed to ten percent in peak efficiency in such post-hoc simulations. We cannot always go the other way. A 2-D design that was fairly well matched in simulation was poorly matched in the lab. When the lab structure was adjusted to give a good match, its RF properties were very different from the originally simulated structure.

We are still studying this problem. We expect that if we could match the coldtest voltages and phases in the simulation with coldtest voltages and phases in the lab, then the hottest behavior would be similar. 3-D codes, which give good agreement with the lab, are still too slow to for extensive parameter studies. We hope that by iterating between 2-D and 3-D calculations we will be able to obtain accurate answers in a reasonable timeframe.

For magnetrons we can also adjust the port impedance or transmission factor to get a desired Q. For a phase-locked magnetron, modelling the input cavity is somewhat more complex. This is because the input voltage does not remain constant in time, but will drop as a result of beam loading. Many simulations of magnetrons have ignored this effect. Including the beam loading effect, the voltage across the port is given by:

$$V_{port} = (1 + Z_-/Z_+) \cdot V_{input} - I_{wall} \cdot Z_- \qquad (11)$$

where V_{input} is the incoming voltage on the waveguide, I_{wall} is the current flowing in the wall (calculated from the magnetic field), and Z_+ and Z_- are the forward and backward impedances of the waveguide. The relative sign of V and I is defined so that positive power corresponds to energy going into the tube. For a matched load, $Z_{load} = Z_+ = Z_-$, so $I = V_{port}/Z_{load}$, $V_{port} = V_{input}$, and there is no reflected wave.

SURFACE PHYSICS

Magnetrons and CFA's may use secondary emission to produce the electron beam. In CONDOR we used a universal curve with two parameters, peak yield and energy at the peak. We have used this model to simulate CFA's and magnetrons with platinum and oxide coatings. The oxide results generally agreed with a space charge limited current model, and to experimental data. At high current densities, the secondary emission model did not agree as well with the data as the space charge limited model. The secondary model predicted a peak current above which the current sheath was too depleted to replenish itself. Experimentally, higher currents were obtainable. The discrepancy might be due to the experimental coating having higher yields than assumed in the model, to 3-D variations in the spoke allowing the sheath to replenish itself, to a higher number of very low energy electrons being produced than were assumed in the model, or to some combination of these.

REFERENCES

1. K. Eppley, "Algorithms for the Self-Consistent Simulation of High Power Klystrons," *SLAC PUB 4622* (Stanford Linear Accelerator Center, May 1988).

2. B. Aiminetti, S. Brandon, K. Dyer, J. Moura, and D. Nielsen, Jr., *CONDOR User's Manual,* Livermore Computing Systems Document (Lawrence Livermore National Laboratory, Livermore, CA., April, 1988).

3. K. Eppley, A. Drobot, W. Herrmannsfeldt, H. Hanerfeld, D. Nielsen, S. Brandon, R. Malendez, "Results of Simulations of High Power Klystrons," *Proceedings of the Particle Accelerator Conference* (Vancouver, British Columbia, May, 1985).

4. B. Goplen, D, Smithe, K. Nguyen, M. Kodis, and N. Vanderplaats, "MAGIC Simulations and Experimental Measurements from the Emission Gated Amplifier Experiments," *Proceedings of the IEDM Meeting* (San Francisco, CA., December, 1992).

5. T. Treado, P. Brown, R. Bolton, T. Hansen, and K. Eppley, "High Power, High Energy, and High Efficiency Phase-Locked Magnetron Studies," *Proceedings of the IEDM Meeting* (Washington, D.C., December, 1991).

6. K. Eppley, "Numerical Simulation of Cross Field Amplifiers," *Proceedings of the Conference on Computer Codes and the Linear Accelerator Community* (Los Alamos, N.M., January, 1990).

Numerical Simulation of Relativistic Klystrons[*]

Giulia M. Fiorentini[†]
Lawrence Berkeley Laboratory,
University of California, Berkeley,CA 94720

ABSTRACT

The Microwave Source Facility at the Lawrence Livermore National Laboratory is a facility for testing high power microwave devices for possible linear collider applications. In order to study the feasibility of a Relativistic Klystron/Two Beam Accelerator concepts, we are planning a series of experiments that involve reacceleration of a modulated beam alternating with power extraction. In support of these planned experiment, we have performed numerical simulations using the two-dimensional, time-dependent code RKS2. We describe the main features of RKS2 and present simulation results including analysis of beam transport, modulation, phase space and power extraction.

I. Introduction

In a relativistic klystron we extract energy from a relativistic beam of electrons and feed the resonant mode in the RF band for which the klystron is built. The RKS2 code studies the interaction of a charged particle beam with an electromagnetic wave in a relativistic klystron.

This code models only cylindrically symmetric problems. Only two coordinates (r,z) are necessary to describe forces since they are assumed to be cylindrical symmetric. Each particle is a charged point with longitudinal position, radius and an angle. The angular distribution is uniform. The source code is designed to be able to model several different combinations of standing wave (SW) cavities, traveling wave (TW) tubes, drift tubes, potential gaps, transverse kick cavities and others components.

In section II we describe the physics of the code, while in section III a particular application is presented.

II. The Physics of the Code

The code solves a system of 2nd order differential equations to compute the current density from the fields and the fields from the current density.

---

[*] This work was supported by the Director, Office of Energy Research, Office of Basic Energy Sciences Division, of the U.S. Department of Energy under Contract No. DE-AC03-76SF00098.

[†] Permanent address: Universita' Statale Milano, Via Celoria 18 , Milano, Italy.

The system of equations comes from the waves equation

$$\nabla^2 \vec{E}_n - \frac{1}{c^2}\frac{\partial^2 \vec{E}_n}{\partial t^2} = \frac{4\pi}{c^2}\vec{J} \qquad (II.1)$$

where n is an index for which cavity is considered, and from the motion equations as function of z:

$$\frac{dx}{dz} = v_x/v_z, \qquad \frac{dy}{dz} = v_y/v_z, \qquad \frac{dy}{dz} = v_y/v_z$$

$$\frac{d\psi}{dz} = \omega/v_z, \qquad \frac{dv_x}{dz} = \frac{qm}{\gamma}\cdot[\vec{E}+\vec{v}\times\vec{B}]_x, \qquad \frac{dv_y}{dz} = \frac{qm}{\gamma}\cdot[\vec{E}+\vec{v}\times\vec{B}]_y. \qquad (II.2)$$

The code computes the Lorentz force from all the electromagnetic fields acting on the particles: space charge, microwave and focusing fields, and also longitudinal accelerating fields, transverse kicks and so on.

1. THE CIRCUIT EQUATION

Assuming that a single cavity mode is dominant, the fields ringing in a cavity can be expressed as the product of a slowly varying time dependent term, a position **r** dependent term and a phase term:

$$\vec{E}_n(r,t) = f_n(t)\cdot\vec{E}_n(r)\cdot\exp(-j\omega t) \qquad (II.3)$$

n is an index for the cavity, ω is the frequency of the driving field or current depending on whether the cavities are fed by an electromagnetic wave or by an RF current, and usually is equal to the resonant frequency (but it only needs to be close to the resonance frequency) and $\vec{E}_n(r)$ is the normalized eigenmode that has been excited. The eigenvalue and the eigenmode depend on the geometry of the cavity and are part of the input data for the RK codes. $f_n(t)$ is the dynamic amplitude of the field that is solved through the circuit equation, obtained by substituting eq. (II.3) into (II.1) and integrating over the volume of the nth cavity. Here below the result:

$$\ddot{f}_n + \left(\frac{\omega_n}{Q_n} - 2j\omega_n\right)\dot{f}_n + \left(\omega_n^2 - \omega^2 - \frac{j\omega\omega_n}{Q_n}\right)f_n -$$
$$(K_n^{n-1} f_{n-1} + K_n^{n+1} f_{n+1}) = \frac{j\omega}{\varepsilon_0}\int dV_n\cdot \vec{E}_n^*\cdot\vec{J} \qquad (II.4)$$

with

$$\frac{\omega_n^2}{c^2} = \int dV_n |\vec{\nabla}\times E_n(r,z)|^2, \qquad (II.5)$$

and

$$K_n^{n-1} = c^2 \int_{S_{n-1/n}} d\vec{S} \cdot \left[\vec{E}_n^* \times \left[\vec{\nabla} \times \vec{E}_{n-1}\right]\right] . \qquad (II.6)$$

The code solves for the value of the term $f_n(t)$ (n=1,...N), as a function of the current density J, the amplitude $\vec{E}_n(\mathbf{r})$; the code need to be supplied with the value of Q_n (the quality factor of the cavity n), $K_{n,n-1}$ and $K_{n,n+1}$ (the coupling terms between the cavity n and the cavity n-1 and n+1: they are zero in case n represents a SW cavity), and ω_n that comes out to be the resonant frequency for SW cavity, while coincides with the frequency of the $\pi/2$ mode for a traveling wave cavity.

2. THE PREPROCESSOR TO CONVERT SW-MODES INTO TW-MODES

Before running RKS2 we use Superfish to determine the resonant frequency and the vector potential of the mode with which we are concerned. By knowing the vector potential the code solves for the electric field and the magnetic field. For TW structures another step is required before to running RKS2. A preprocessor turns SW modes into TW modes[1]. Furthermore it furnishes the group velocity from computing the rate between the power flux through one surface and the energy stored per unit length:

$$v_g = \frac{P}{W_{TW}} = \frac{\frac{1}{2}\int E_r \cdot H_\phi \, ds}{\int_{\text{unit length}} \varepsilon \frac{E^2}{2} dV + \int_{\text{unit length}} \mu \frac{H^2}{2}} \qquad (II.7)$$

3. HOW TO COMPUTE ω_n AND THE COUPLING FACTORS.

For a SW cavity at steady state, with no beam, the circuit eq. becomes

$$\left(\omega_n^2 - \omega^2 - \frac{j\omega\omega_n}{Q_n}\right) f_n = 0 \qquad (II.8)$$

If Q_n is very large we can neglect the term $\frac{j\omega\omega_n}{Q_n}$ in eq. (II.8) and ω_n coincides with the resonant frequency. For a TW cavity to determine the value of ω_n let us consider the circuit equation at steady state and with driving current J=0. The equation becomes

$$\left(\omega_n^2 - \omega^2 - \frac{i\omega\omega_n}{Q_n}\right) f_n - \left(K_n^{n-1} f_{n-1} + K_n^{n+1} f_{n+1} \right) = 0. \qquad (II.9)$$

As above we will omit $\frac{j\omega\omega_n}{Q_n}$ with respect to $\omega_n^2-\omega^2$.
We obtain

$$(\omega_n^2-\omega^2)f_n - (K_n^{n-1} f_{n-1} + K_n^{n+1} f_{n+1})=0 \qquad (II.10)$$

If we consider an infinite structure the coupling terms are all the same, and for Flouquet theorem the fields only differ by a factor of phase. So we can write:

$$(\omega_n^2 - \omega^2)f_n - (K\, f_n \cdot e^{i\delta \cdot \ell} + K\, f_n \cdot e^{-i\delta \cdot \ell})=0, \qquad (II.11)$$

and

$$\omega_n^2 - \omega^2 = 2 \cdot K\cos(\delta\, \ell), \qquad (II.12)$$

where the formula (II.12) is a simple version of the dispersion relation. Moreover it says that for $\omega=\omega_n$ the phase advance is $\pi/2$.
After a differentiation, we obtain

$$\frac{d\omega}{d\delta} = \frac{K}{\omega} \cdot \ell \sin(\delta\, \ell), \qquad (II.13)$$

where the left hand size is the group velocity.

Since from Superfish we know the frequency ω, phase advance δ for a certain mode, the length ℓ of the cavity and from the preprocessor we know the group velocity, equation (II.13), can be used to calculate the coupling factor K. Substituting ω, δ, K in (II.12) we find out ω_n to input to the code.

4. THE FOCUSING MAGNETIC FIELD AND THE CONSERVATION OF CANONICAL MOMENTUM.

There are two different ways to specify the magnetic field along the klystron: reading an input data file or giving the geometry of the solenoids. The code also takes into account what magnetic path the beam takes as it goes from the cathode to the entrance of the structure. It is well know that if a beam goes through a non uniform magnetic field by the time it reaches the entrance it has gotten a certain angular momentum (from the conservation of the canonical angular momentum). This helps the beam to be constrained.
This law is also expressed in the envelope equation:

$$\ddot{R} + K_\beta^2 R - \frac{K}{2R} - \frac{\varepsilon^2 + (q\psi_0/p)^2}{R^3} = 0. \qquad (II.14)$$

where R=rms radius=$\sqrt{<r^2>}$, $K_\beta = \dfrac{eB_z}{2\gamma\beta mc}$, K=perveance=$\dfrac{2I/I_0}{(\beta\gamma)^3}$,

I_0=17000amp, ψ_0=magnetic flux enclosed within the beam at the source.

If the beam at the source is subject to no magnetic field, ψ_0 is zero in the eq.(II.13). If the beam at the source is subject to a non zero magnetic field, ψ_0 is non zero and in the envelope equation we need to add to the real emittance term the additional $\dfrac{(q\psi_0/p)^2}{R^3}$ to get the effective emittance of the beam. This is one of the features the code supplies if we input the value of the magnetic field at the cathode.

III. An Application: The Reacceleration Experiment

A collaboration between the Lawrence Livermore National Laboratory and the Lawrence Berkeley Laboratory has been studying microwave sources which could be suitable drivers for a future TeV linear e$^+$e$^-$ collider. The Choppertron[2], a high-power microwave generator which uses transverse modulation of the drive beam, has been successfully tested at the Microwave Source Facility[3]. Although the Choppertron has demonstrated high-power pulses, >150 MW per output at 11.424 GHz with stable phase and amplitude and >400 MW total peak power, the conversion efficiency of beam energy to microwaves is only about 30%. The efficiency could be significantly improved by adding up more and more TW structures for the power extraction and reaccelerating the beam after each extraction unit. Besides the energy of the beam has been increased (5Mev) to reduce the space charge effect. The application of this concept to a linear collider is referred to as the Relativistic Klystron Two-Beam Accelerator (RK-TBA)[4].

Fig. 1 shows a layout of the proposed reacceleration experiment. The major experimental components, except for the induction accelerator that generates the drive beam, are shown in the layout. These components include the Choppertron beam modulator, traveling-wave microwave extraction structures, and induction cells for reacceleration.

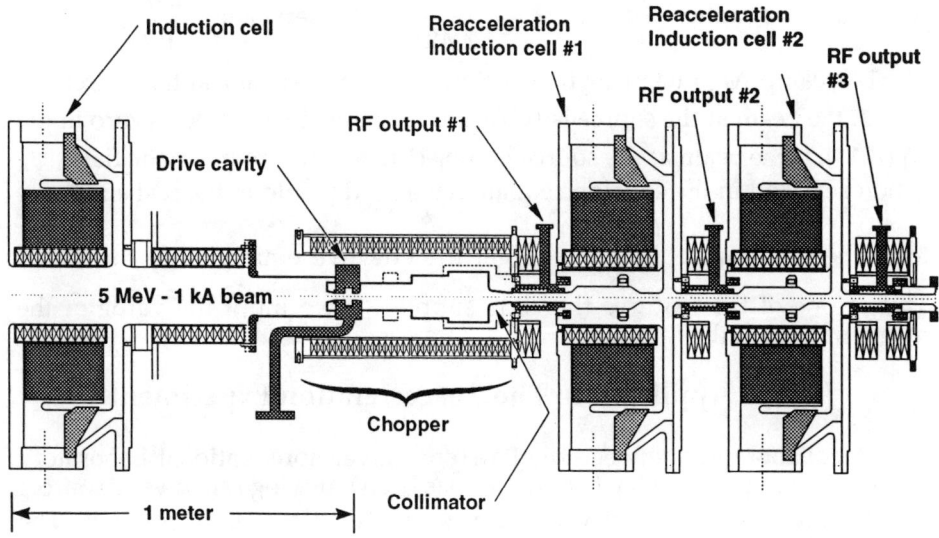

Figure 1. Schematic of proposed reacceleration experiment.

The RKS2 code has been used to simulate the beam transport and the field dynamics. As mentioned before it only deals with cylindrically symmetric fields, and transverse instabilities has been studied separately.
We shall not describe the experiment here , as that has been described elsewhere[5,6] but simply show how RKS2 has been helpful. we shall separately talk about the modulation section and the extraction section.

1. THE MODULATOR SECTION

The choppertron modulator has been analyzed elsewhere[5,6]. We will just introduce the geometry and the way it works: In a dipole cavity an alternated kick (f=5.712Ghz) is given to the beam. The beam is allowed to oscillate in a drift region of uniform magnetic field. After a length=$\lambda_\beta/4$, where

$$\lambda_\beta = \frac{4\pi\gamma\beta mc}{eB_z} \qquad (III.1)$$

is the betatron wavelength, the beam has reached a maximum deflection from the axis. At this point a collimator chops the beam resulting in a current modulation at twice the kick frequency. The condition

drift length=$\lambda_\beta/4$,

furnishes the best rf/dc ratio. Since the drift length is fixed eq.(III.1) determines the magnetic field that optimizes the deflection. For a 5Mev beam, the value of Bz that satisfies eq.(III.1) for our geometry is $B_z=1kG$. The choppertron was designed for 1kA of dc current and a normalized emittance $\varepsilon_N=30\pi$ cm mrad.

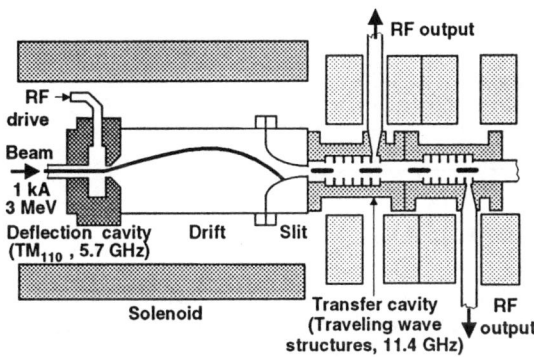

Figure 2. Schematic of the original Choppertron.

From the envelope equation (II.14) we set $\ddot{R}=0$ to determine the radius of the matched beam. The design envelope radio was R=4mm. In the real experiment the beam actual emittance was different,
$\varepsilon_N\approx100\pi$ cm mmrad. With this value of emittance it was not possible to match the beam (for an acceptable R value) keeping $B_Z=1KG$.

2. OPTIMIZATION OF PARAMETERS IN THE MODULATOR.

The use of RKS2 allowed us to determine the optimum choice of the following parameters , I, B_z, ε_n, R, to get the maximum rf current coming out of the modulator. The value of emittance depends upon the value of dc current. The higher is the dc current, the larger is the emittance. As a consequence beyond a certain value of dc current (1≈kA) the benefit of having an intense beam is canceled by the disadvantage of having a larger emittance. From the value of magnetic field two parameters depend through inverse proportionality: the deflection and the radius of the envelope. From the deflection the rf current depends according to direct proportionality. From the envelope radius R the rf current depends according to some inverse proportionality law. There must be a middle value that optimizes the rf current.

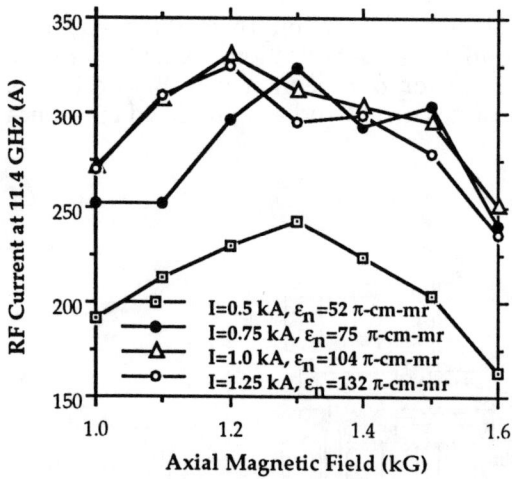

Figure 3. Predicted rf current for different beam emittances and dc currents, and fixed drive level = 1.2 MW.

In Fig.3 we show the result of series of simulations. Each curve has been obtained running RKS2 up to the end of the collimator and for different values of initial current. Each point of these curves has been obtained by running the code with different magnetic fields. The code produces in an output file the value of the dc current and the first two harmonics as a function of the longitudinal position. Besides for the analysis of the transient time it is possible to plot these variables at a fixed longitudinal position as a function of time. We chose to work with I=1.0kA, consequent ε_N=100π cm mmrad, B_z=1.2kG, R=0.568cm.

3. BEAM TRANSPORT IN THE GAIN SECTION.

The RKS2 analyzes the beam transport. Plots of the current harmonics versus z, plots of the profile of the beam, plots of the radius of the envelope, plots of the longitudinal phase space were available. In Fig.6 we show the power extracted by TW#1, TW#2, TW#3. The big particle loss that is fingered by the most left arrow in Fig.4 has been attributed to a defocusing effect of the TW transverse fields. The beam profile plotted in Fig.5&7 show at different z how the later particles of each bucket receive a defocusing kick and how the earlier particles receive a focousing kick inside the first TW structure.

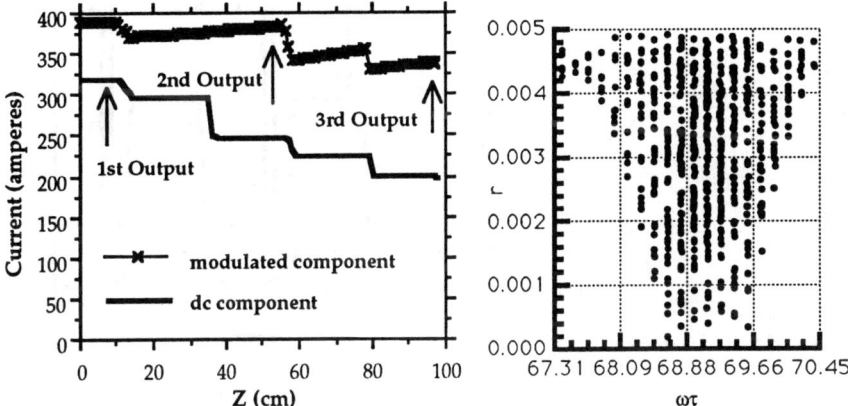

Figure 4. RF and DC current versus z.

Figure 5. Profile r(m) of 1 bucket at the entrance of the first TW tube.

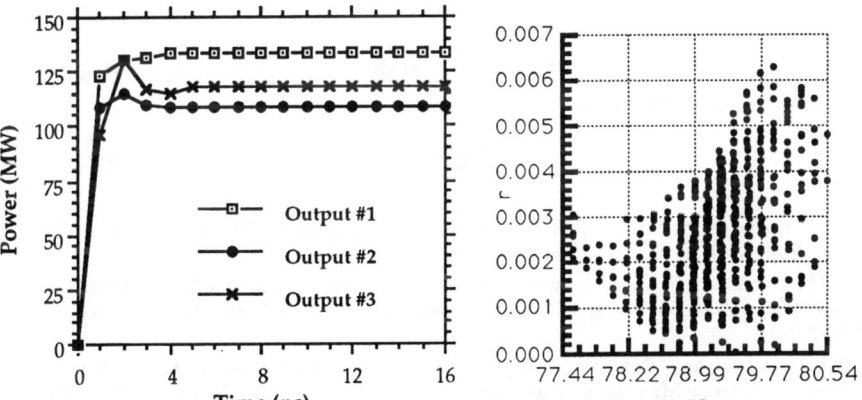

Figure 6. Power out versus time in the extraction cavities.

Figure 7. Profile r(m) of 1 bucket at the exit of the first TW tube.

In fig.8 as a verification of this effect we show the beam radius behavior versus z in the first TW structure and what it would be if we replaced the TW tube with a simple drift tube. In the second case the radius would reduce. In Fig.9 finally we show the rms radius expansion inside the pipe.

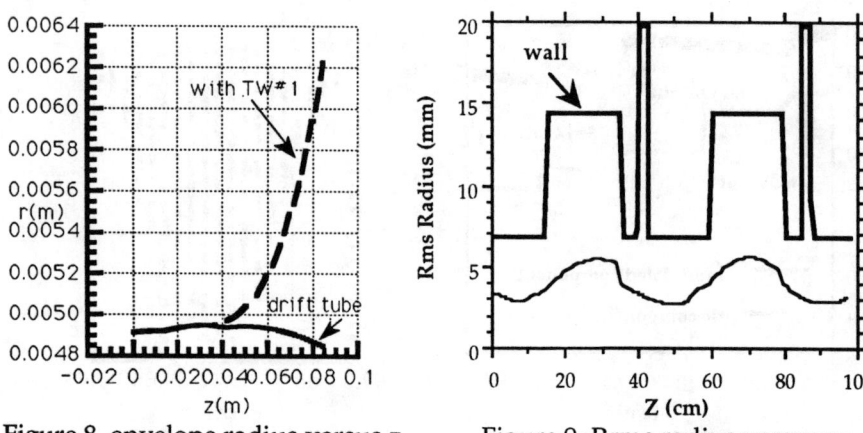

Figure 8. envelope radius versus z.   Figure 9. Rrms radius versus z.

IV. Acknowledgments

I would like to thank Tim Houck, Rob Ryne, and Andrew Sessler for their helpful suggestions and comments.

VI. References

[1] G.A. Loew, R.H. Miller, R.A., "Computer Calculatios Of Travelling-Wave Periodic Structure Properties"

[2] J. Haimson and B. Mecklenburg, "Design and Construction of a Chopper Driven 11.4 Ghz Traveling Wave RF Generator", Proceedings of the 1989 IEEE Particle Accel. Conf.,pp243-245.

[3] T.L.Houck et al., "Relativistic Klystron Research for Two-Beam Accelerators," SPIE Symposium onIntense Microwave and Particle Beams II Proceedings Vol 1629-47 (1992).

[4] A.M.Sessler and S.S. Yu, "Relativistic Klystron Version of the Two-Beam Accelerator,"Phys.Rev.Lett.,58,2439,1987.

[5] T.L.Houck and G.A. Westnenskow, " Status of an Induction Accelerator Driven, High Power Microwave Generator at Livermore," Spie Proceedings Vol 1872-16,1993

[6] G. M. Fiorentini, T. L. Houck, and C. Wang,"Design of a Reacceleration Experiment " SPIE proceedings 1993, Vol 1872-17,1993

LIDOS - UNCONVENTIONAL HELPER FOR LINAC BEAM DESIGNING

B.I.Bondarev, A.P.Durkin, B.P.Murin,
G.T.Nikolaishvili, O.Yu.Shlygin
Moscow Radiotechnical Institute
Warshavskoe shosse, 132, 113519, Moscow, Russia

Abstract

The new standpoint on linac beam dynamics computer studies is proposed. Codes LIDOS (Linac Ion Dynamics Optimization and Simulation) is a realization of this method. At every investigation stage long before the beam dynamics simulation will be running LIDOS information are well suited to find linac accelerating/focusing channel structure satisfactory for further studies. The LIDOS includes RFQ, DTL, High-Beta Linac, multi ion transport systems etc. LIDOS permits users to make choice of channel optimal parameters and tolerances, to carry out matching of different channel parts and to simulate beam dynamics with space charge as well. Unlike the other existing codes LIDOS is an expert system, so it can not only to solve equations of ion motion, but give intellectual advice and help users to come to the optimal decision.

The advancement of computer engineering is a contributory factor for creating the more sophisticated mathematical models of physical processes. In computational accelerator physics such problem's typical example are concurrent calculation of intense beam motion through complex electrodynamic structure. The computer-based tools for solving the problem of such class involves as rule the calculation of multi dimension meshfunction and the numerical integration of many macroparticle motion equations. The solving process is memory and time consumed and is oriented to supercomputers.

Unfortunately, a knowledge of *"What is happened"* and not a knowledge of *"Why it is happened"* or *"What is the best way"* is a result of complex beam dynamics calculation. Such information can be received if a knowledge-based part (at a later time it is called as *"Adviser"*) arises in computer tool. The knowledge-based part helps the designer to put out the information (in the most convenient form) about received results quality and about the best way for this quality growth. The quality properties may be both parameters combination and images of visualized mathematical equations solution. (A beam phase portrait serves as example of Image). The image-based codes user must be a specialist for which used images have the physical meanings.

Adviser operates on-line and that is why it must base on the simplest models and covers only the main features of physical process.

The LIDOS Adviser (LIDOS1) is the first work of the authors in the image-based computer codes direction. LIDOS tools package includes **LIDOS1** and is designed for ion linac accelerating/focusing channel calculation and optimization.

The visual, graphical and numerial information having been obtained by user helps him to find the best way toward linac

configuration and region of linac parameter space the most likely satisfy his design requirements. In addition LIDOS1 must:
1) permit of on-line operation,
2) give all possible information about final results long before these results have been obtained.
3) represent the input data and output results in convenient form easy to understanding,
4) use the mathematical methods of optimization.

LIDOS1 consists of several unconnected parts corresponding different linac types. Each part includes set of tools connected by special shell codes (see Part 2 and Part 3 of *LIDOS User Manual* [1]) creating the calculation scenario.

The channel optimization process lies in investigating the array of accelerating/focusing field variation curves and fast search of the best ones.

LIDOS1 is adapted for IBM PC-486. The designer during one run may choose desirable linac version or must assure himself that the input data are invalid.

Let us consider several Images examples.

If $E(n)$ is RF-field amplitude and $\varphi_s(n)$ is synchronous phase dependence on n - accelerating period number, then the accelerating channel parameter optimization consists in selection of $E(n)$ and $\varphi_s(n)$ curves form in order to maximize accelerated ion current. The main parameter governing a beam quality is input phase interval in which the stability accelerating particle was located. However this result presentation as the input beam phase width is a "blind" one because it does not pointed the way to the more desirable version. The current phase portrait as well as beam losses distribution can be viewing but it is lack of understanding in this case too because the designer can not locate the effects which is a cause of particle loss.

The Image which can present the real cause of beam losses is a phase portrait on the background of separatrix of periodic channel starting from the current period. The particle exits from stability domain means that they have been lost. Viewing the curve which has described particle proportion within the separatrix on the background of $E(n)$ and $\varphi_s(n)$ variation curves the designer understands how these curves must be corrected in order the beam phase portrait to be will within the separatrix. This correction would give the designer the chance to increase the output current or accelerating rate.

The visual information example is presented at Fig.1. In the upper right side of screen the designer sees the beam longitudinal phase portrait on the background of periodic channel separatrix. In down right side the proportion of particles inside separatrix is indicated on the background of chosen Curves $E(n)$ and $\varphi_s(n)$.

The Image which give a similar presentation about focusing channel is a presenting point trajectory on stability diagram. At Fig.2 the Mathie diagram for RFQ linac is presented. The generalized coordinates are used. The isolines of transverse oscillation frequencies, matching beam transverse sizes and beam envelope modulations are indicated as well as lines of simple and parametric resonances. The equilibrium crossection transverse size along the channel is uniquely determined by presenting point trajectory (Fig.3) with resonances no taking into account. Together, information given by these two picture is large enough for focusing channel better parameter determination. The further information gives beam current stability diagram (Fig.4) specified on generalized coordinates one of them (the horizontal) is proportional to the current phase density and other (the vertical) - to focusing force gradient. The current diagram allows to estimate the accelerating bunch limit intensity.

All Image information much alleviates the focusing channel optimization problem because the beam simulation is not needed for trajectory plotting on the stability diagram.

At high energy linac the focusing channel quality largely depends on sensitivity with respect to random errors taking place during manufacture, installation and alignment. The channel sensitivity is a combination of its geometric and magnetic parameters [2-4]. A knowledge of different channels sensitivity and corresponding tolerance values helps the designer to choose the focusing channel optimal structure (Fig.5,6).

The main information about LIDOS and LIDOS1.Adviser is considered below.

The computer codes LIDOS has been developed for ion linac design and analysis. The codes can be used for accelerating/focusing channel calculation and helps user in choosing the best channel version.

The seven accelerator components have been successfully designed using the LIDOS:
 - RFQ linac,
 - drift tube linac (DTL) with magnet focusing,
 - DTL with RF focusing,
 - high-beta linac (HBL),
 - beam transport,
 - beam matching and separating,
 - beam transverse position correction.

The LIDOS consists of the three levels codes as well as linac designing process has three main stages.

First level codes LIDOS1 are the new generation of linac codes. LIDOS1 helps the designer find the region of linac parameter space most likely to satisfy his design requirements. The process is fast because the beam simulation has not used for large arrays of possible configuration. LIDOS1 is an expert system, so it can not only solve ion equations of motion, but gives also the user an intellectual advise and helps him to come to the best channel version by the shortest way. LIDOS1 has major advantages over the more conventional codes because it involves not only computing but also optimizing codes. The new language GIA for shell codes are used. The meta-language GIA gives a simple and convenient way of adapting the codes for user's requirements.

The first level codes are based on simple models in which the electric field is constant in accelerating gaps and equal zero outside of them (so called "rectangular" approximation). The same distribution has magnet field in focusing elements.

For computation of the beam space charge field the LIDOS1

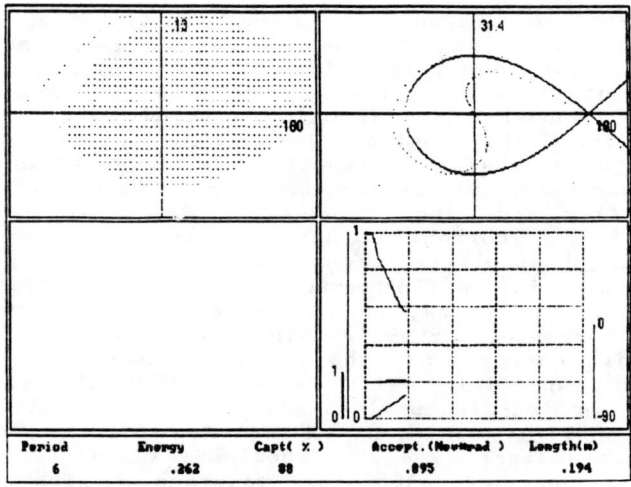

Fig.1. Current Separatrix and Phase Portrait

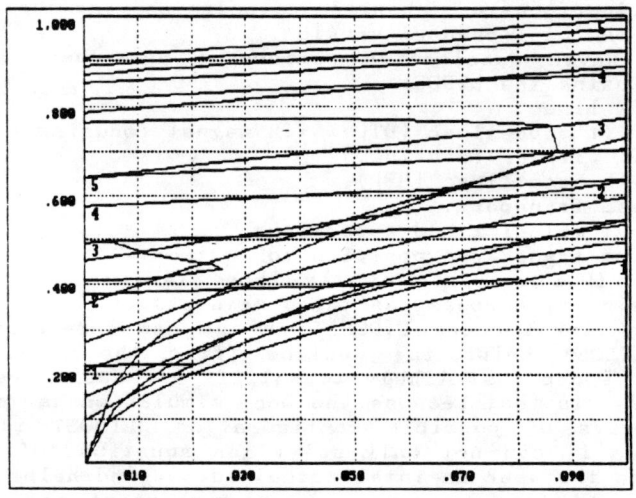

Fig.2. Representing point trajectory on stability diagram

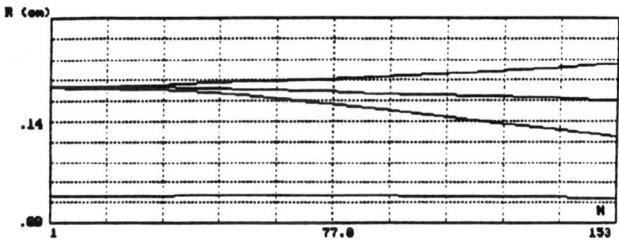

Fig.3. Transverse size of equilibrium crossection

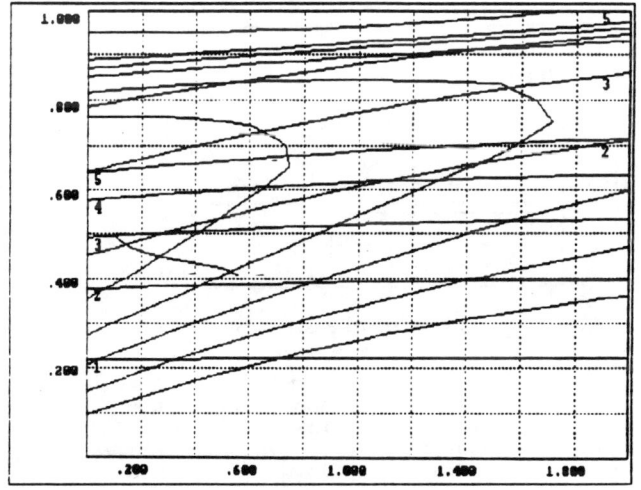

Fig.4.
Representing point trajectory on stability diagram with current

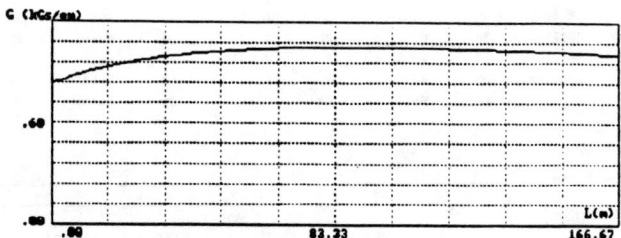

For dH/H= .13% the probablity of Rout/Rint being less than 1.10 is .90
For dφ = .13° the probablity of Rout/Rint being less than 1.15 is .92
With the same probability Rout/Rint is less than 1.13
For dX = 33μm the probablity of Xout being less than 4.0 mm is .80

Fig.5. Sensitivity of HBL focusing channel

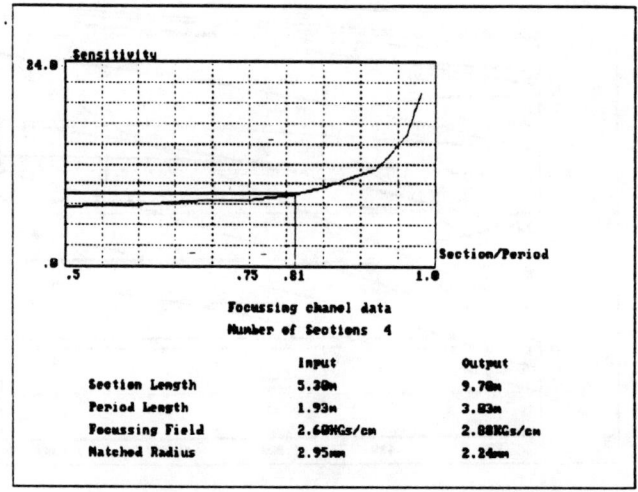

Fig.6. Tolerances for the long focusing channel

codes use the models which do not require the Poisson's equation solution in the three-dimensional space (x,y,z). Within the limits of these models the beam is represented by a uniformly charged cylinder and the beam phase volume is represented by an ellipsoid in the four-dimensional phase space. The LIDOS1 mathematical models do not require a big number of particles for beam simulation but they describe well enough the major factors influencing on the dynamics of accelerated or transported beam.

The input data are introduced into LIDOS1 shell by the keyboard in the form of decimal numbers with floating point or as integers. With the aid of setting the parameter distribution along linac length the user may enter the tables containing the arguments (period number or energy) and parameter value corresponding to those arguments. All the introduced points except the first two ones are connected together either with a straight line or with the parabola.

The LIDOS1 computation carries out in an on-line regime according to pre-arranged schedule. The visualization of the results allows the user to change the channel parameters most likely satisfy the output beam characteristics without the beam simulation.

It is expedient to use the LIDOS1 at the initial stage of linac designing, for example, for choosing a type of focusing and focusing period structure, for choosing a synchronous phase distribution and so on. Experience accumulated with the LIDOS1 shows that although the models it is based on do not make allowance for all nonlinearities, the latter in most cases do not require to change the structure of the choosing channels. In particular, focusing channels optimized by LIDOS1 give the least beam effective emittance growth.

LIDOS1 is the intellectual part of the LIDOS package.

LIDOS2 is the central tool for the linac design. Starting from LIDOS1 output files it generate detailed descriptions of the linac geometry taking into account the real distribution of RF and magnetic fields. The stochastic simulation codes using for estimation of random perturbation on the beam motion also belong to the LIDOS2.

LIDOS3 generates detailed descriptions of beam dynamic performance. It provides the designer with information about the beam motion in the already designed channel using the more sophisticated beam dynamics programs. They are based on various modifications of the large particles method and determines the influence of various nonlinearities upon the output beam including the beam emittance growth.

The second and third level codes LIDOS2 and LIDOS3 are based on the traditional tasks and methods of their solution.

All **LIDOS** parts are adapted to IBM PC (desirable, for its powerful modifications). The LIDOS1 system associates with the user on a real time scale, while LIDOS2 and LIDOS3 are time consuming. They may be adopted to supercomputers.

The **LIDOS** structure is easily adaptable to the user's needs.

The **LIDOS1** is the system open for adding by tools oriented to linac other structure.

The **LIDOS1** using are given in reports [6,7].

REFERENCES

1. LIDOS User Manual (*Version 3.0*), S&T Center ELEON, Moscow, 1993.
2. B.P.Murin et al, Particle Accelerators v.6, 27(1974)
3. B.I.Bondarev et al, Journal of Technical Physics v.49, 1662(1979) (in Russian)
4. Ion Linear Accelerators (Atomizdat, Moscow, 1978), p.264 (in Russian)
5. B.P.Murin et al, 1992 Linear Accelerators Conference Proceedings v.2, 731(1992)
6. B.P.Murin et al, 1992 Linear Accelerators Conference Proceedings v.2, 734(1992)
7. B.I.Bondarev, A.P.Durkin, G.T.Nicolaishvili, O.Yu.Shlygin "Multi-ion Transport System", this Conference

NUMERICAL SIMULATIONS AT CEBAF USING PARMELA*

H. Liu

Continuous Electron Beam Accelerator Facility
12000 Jefferson Avenue, Newport News, VA 23606

ABSTRACT

PARMELA has been used at CEBAF for numerical modeling of the nuclear physics injector chopping system, a possible FEL laser gun injector, and the rf steering and focusing effects of the standard CEBAF SRF cavities. These applications call for the code to input field data consistently from SUPERFISH, POISSON, and MAFIA, to properly treat a focusing solenoidal lens having an actual field profile either individually or together with its adjacent rf cavity, to deal with the space charge forces, to model the longitudinal phase space matching required for bunching electrons using a phase-compressor chicane, etc. In this paper, we describe in detail these issues of general interest.

INTRODUCTION

PARMELA[1] is a versatile multi-particle computer code that can be used for studying the effects of electromagnetic fields on the dynamics of electrons. It has achieved great popularity because of its power in designing electron injectors and accelerators.

In this paper, we present our use of the code at CEBAF for numerical modeling of the nuclear physics injector chopping system, a possible FEL laser gun injector, and the rf steering and focusing effects of the standard CEBAF SRF cavities on electron beams. The emphasis will be put on how to input field data consistently from SUPERFISH, POISSON, and MAFIA, to properly treat a focusing solenoidal lens having an actual field profile either individually or together with its adjacent rf cavity, to deal with the space charge forces, to use the code effectively for longitudinal phase space matching required for bunching electrons using a phase-compressor chicane, etc.

THE CHOPPER SYSTEM

As is shown in Fig. 1, the initial part of the CEBAF nuclear physics injector consists of a 100-kV thermionic gun, a pair of apertures (A_1 and A_2) for limiting

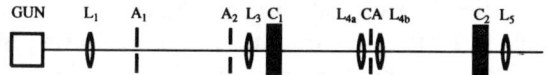

Fig. 1 Chopper system of the CEBAF nuclear physics injector

the initial emittance of the beam, and a pair of chopper cavities (C_1 and C_2) to chop a cw beam through an aperture (CA). The first lens (L_1) focuses the beam

---

* Supported by D.O.E. contract #DE-AC05-84ER40150

to a waist at the first aperture; the third lens (L_3) makes an image-to-image transform from A_1 to CA; and the lens pair L_{4a}-L_{4b} makes an image-to-image transform between the centers of C_1 and C_2. The chopper system is symmetric with respect to the chopper aperture.

Each chopper cavity is a square box operating presently at a fundamental frequency of 1500 MHz on the two modes: TM_{210} and TM_{120}. When an electron beam moves through the first cavity, it is deflected radially outward and gradually turned into a hollow beam by the two orthogonal modes. Then it is chopped at the chopper aperture. The second identical cavity is used to compensate for the radial momentum introduced by the first cavity, and completes the chopping process. See Fig. 2.

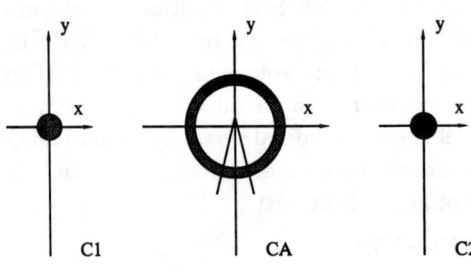

Fig. 2 Chopping process from C1 to C2

The conventional version of PARMELA assumes a hard-edge field profile for a solenoidal lens. It additionally requires that the length of the hard-edge field profile be the same as that of the rf element when the two elements overlap each other. The chopping process is treated using a zero-length transform. All these assumptions fail to apply to most of the actual cases. For example, in our case, the field profiles of all the solenoidal lenses can not be well approximated by the square ones. In addition, there exists a mutual penetration between the solenoidal field of $L_{3(5)}$ and rf field of $C_{1(2)}$. More importantly, we needed to model the energy spread performance of the chopper system, and a zero-length transform can not provide the needed information.

We modified the code to meet the requirements for our modeling. First we fitted the actual on-axis axial solenoidal field profiles using Glaser and Gaussian distributions. A paraxial approximation is applied to the transverse magnetic field components. The lens L_1 is treated as an individual one. We modified the card $CELL$ and the subroutine $CELLIMP$ to accommodate both the rf fields from $C_{1(2)}$ and the solenoidal fields from $L_{3(5)}$ in one numerical element. The rf fields in the chopper cavities are calculated using MAFIA. The approach adopted to calculate the rf fields in PARMELA from those obtained using MAFIA is the one developed by Z. Li[2] and will be introduced in a later section.

The purpose for modeling the chopper system is to clarify its energy spread performance. For each slice of the beam, an energy spread is induced when it passes through the first cavity. As the beam passes through the second cavity properly in phase, its emittance introduced by the first cavity is undone. We needed to clarify whether the energy spread is undone simultaneously with the emittance.

Numerically we found that the phase difference between the two orthogonal modes must be 90° in both cavities, instead of +90° in one cavity whereas −90° in the other. This finding is consistent with the experiment. Then we found that the relative phase difference between the two identical modes in the two cavities controls the cancellation of emittance, and the energy spread follows exactly the same process as for the emittance. See Fig. 3. The underlying mechanism is that an electron is flipped 180° in the transverse plane by the lens pair L_{4a}-L_{4b}, therefore it experiences an acceleration or deceleration process which is opposite to that occured in the first cavity. The details will be presented elsewhere[3]. This numerical calculation has been confirmed with the experiments carefully designed and conducted by M. Tiefenback et al.[4]

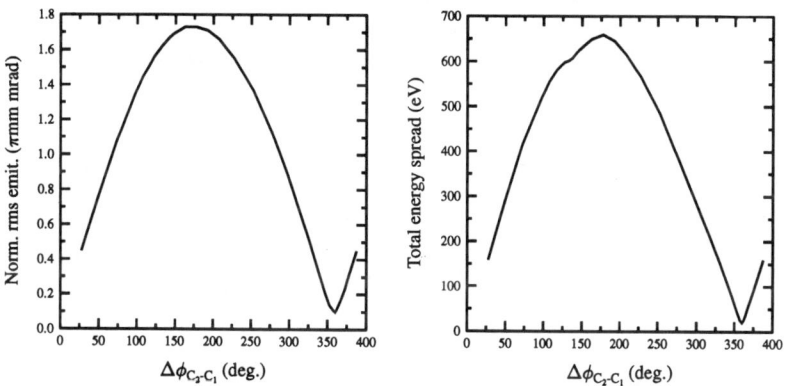

Fig. 3 Emittance and energy spread of the chopper system versus the rf phase difference between the two identical modes in the two cavities

THE FEL INJECTOR

CEBAF has been studying an IR FEL and a UV FEL utilizing the superconducting accelerator technology that has been developed at CEBAF, aimed at industrial applications and fundamental scientific research[5-8]. An FEL injector, consisting of a photocathode DC gun, a prebuncher, a cryounit containing two standard CEBAF SRF cavities, and a phase-compressor chicane, would be used as a high-brightness cw source. The specifications for the FEL injector are summarized in Table 1. $4\sigma_t$ and $4\sigma_E$ are used to represent the bunch length and

Table 1 FEL Injector Specifications

Energy	10 MeV
Charge per bunch	120 pC
Bunch length ($4\sigma_t$)	2 ps
Energy spread ($4\sigma_E$)	400 keV
Normalized rms emittance (ϵ_n)	15 mm mrad
Average beam current	900 µA
Repetition frequency	7.677 MHz

bunch energy spread. For ideal Gaussian distributions, they correspond to 95% particles.

Based on the previous calculation[8], the performance of the FEL injector has been optimized and thoroughly investigated using time-consuming but accurate integrated numerical modeling[9]. The beam dynamics is calculated using a version of PARMELA with the point-by-point method for space charge treatment[10,11]. The code POISSON was used to generate the DC electric field in the photocathode gun, and the code SUPERFISH was used to generate the 2-D RF field distributions in the prebuncher and two standard CEBAF SRF cavities in the cryounit. In each integrated simulation ($\sim$ 10 cpu-hours on an HP 9000/730 UNIX workstation), the same electrons are followed from emission at the photocathode through the gun, the prebuncher, the cryounit and the chicane. The results for the baseline design are listed in Table 2. The various distributions are shown in Fig. 4.

Table 2 Baseline Design and its Performance

Gun parameters:	
Voltage (V_0)	500 kV
Laser pulse length ($4\sigma_l$)	100 ps
Cathode diameter (d_0)	3 mm
Field gradient (E_0)	10 MV/m
System characteristics:	
Prebuncher:	two-cell scheme
Solenoidal lenses:	two rotating lenses
Quadrupoles:	one triplet
Chicane:	$R_{56}=-0.085$ cm/%
System performance:	
Bunch charge	120 pC
Bunch length ($4\sigma_t$)	0.96 ps
Energy spread ($4\sigma_E$)	290 keV
Mean energy (E_m)	9.492 MeV
Norm. rms emittance ($\epsilon_{nx}/\epsilon_{ny}$)	4.44/4.73 (mm mrad)

The conventional version of PARMELA has no input for an electrostatic element. We deal with the DC fields in the gun calculated using the code POISSON as if they were rf fields but with $\sin(\omega t)$ and $\cos(\omega t)$ excluded. We used 14 Fourier coefficients calculated using the code SUPERFISH for the rf fields in the prebuncher and the SRF cavities, which is a standard use of the code. We found that the key point for our injector simulation is to accurately match the electron bunches in the longitudinal phase space from the second SRF cavity to the chicane to achieve the shortest possible bunch length and hence the highest possible FEL optical gain.

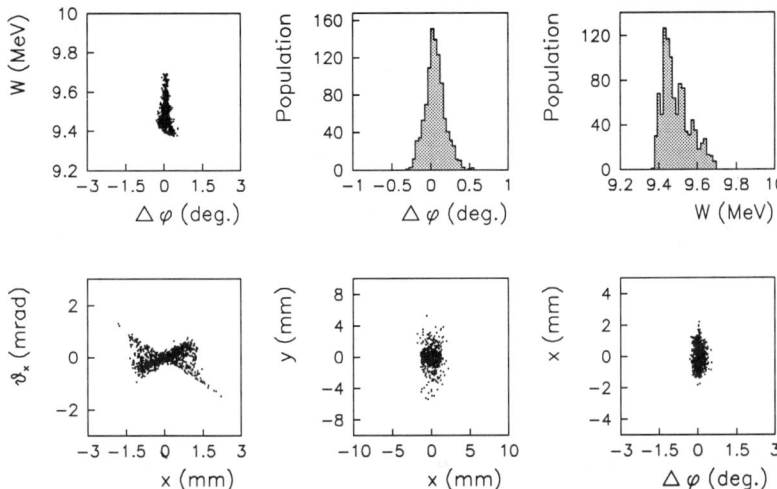

Fig. 4 Various distributions of 1000 superparticles at the exit of the chicane, showing the optimized baseline design performance listed in Table 2. *Upper left*: longitudinal phase space distribution (W - energy; $\Delta\phi$ - relative phase); *Upper middle*: phase profile; *Upper right*: energy profile; *Lower left*: horizontal trace space distribution (x - horizontal position; θ_x - horizontal divergence angle); *Lower middle*: cross-sectional distribution; *Lower right*: horizontal snap-shot.

The electrons are bunched at first for suitable injection into the cryounit using a prebuncher. Then two SRF cavities and a chicane are used for further bunching, as shown in Fig. 5. Using the σ matrix representation[12], we have

$$\sigma_{55}(1) = \sigma_{55}(0)(1 - R_{56}/f_{56})^2 + R_{56}^2 \sigma_{66}(-1), \tag{1}$$

where $\sqrt{\sigma_{55}(0)}$ is the bunch length at the entrance of the chicane, $\sqrt{\sigma_{55}(1)}$ the bunch length at the exit of the chicane, $\sqrt{\sigma_{66}(-1)}$ the momentum spread at the entrance of the second SRF cavity, $R_{56} = \delta l/(\delta p/p)$ the parameter of the bunching property of the chicane, δl the path difference between electrons having an energy spread of $\delta p/p$, $f_{56} = -\sigma_{55}(0)/\sigma_{56}(0)$ the tilt of the longitudinal phase space distribution of the bunch at the exit of the second SRF cavity. It is seen that when $f_{56} \simeq R_{56}$, the final bunch length depends only on the product of the momentum spread and R_{56} of the chicane. We call the above condition the *conditioning for final bunching*, which is a term borrowed from Ref. 13. Eq. (1) has been incorporated into PARMELA so that the matching can be predicted accurately after the second SRF cavity. This has turned out to be an indispensable means for optimizing the design.

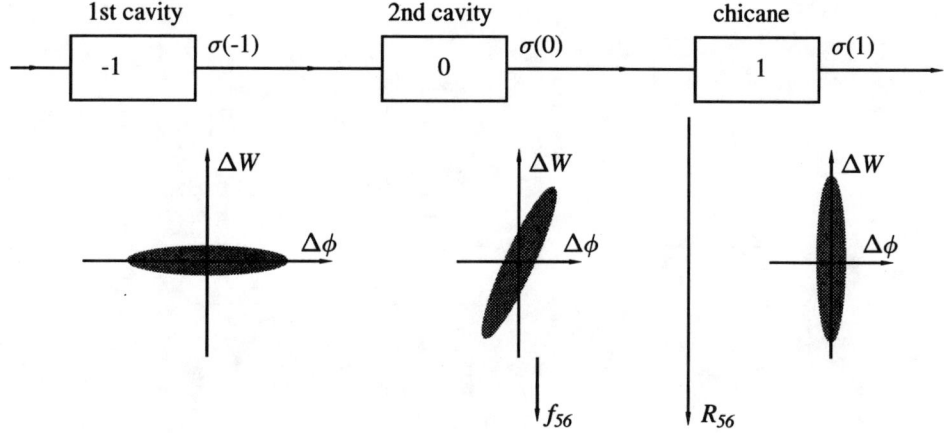

Fig. 5 Bunching process using the two SRF cavities and the chicane.

Using such a version of PARMELA equipped with Eq. (1) and the point-by-point method for space charge calculations, the following investigations have been made:

(1) the robustness of the baseline design against the laser intensity fluctuation and so the bunch charge fluctuation;

(2) the sensitivity of the baseline design on the basis of $\delta\phi = \pm 2°$ for rf phase fluctuations and $\delta E/E = \pm 2\%$ for rf amplitude fluctuations in the prebuncher and the two rf cavities in the cryounit;

(3) the maximum operational flexibilities under various different gun operating conditions;

(4) the fluctuation of emission phases of electron bunches.

With these integrated simulations, the bunch length, energy spread, emittance, bunch-to-bunch centroid energy shift and bunch-to-bunch centroid phase shift have been fully examined. It has been demonstrated that the design will perform beyond the specifications over a quite wide range of operation conditions. In addition, some potentialities have been found for making a better compromise between the optimum performance and minimum cost. These possibilities include: (1) to employ the one-cell prebuncher scheme instead of the two-cell scheme; (2) to increase the value of the matrix element R_{56} of the chicane to allow an easier longitudinal phase space matching and thus to reduce the sensitivity in electrons bunch length and centroid phase shift. These features are under consideration for further elaborate designs.

CAVITY STEERING AND FOCUSING

The transport properties of the standard CEBAF SRF cavities have been carefully studied by Z. Li using PARMELA and MAFIA[2]. Fig. 6 shows the 5-cell cavity configuration with two couplers for 3-D rf field calculations using MAFIA.

MAFIA modeling gives the six field components for each mesh point. The six components are not defined right on that mesh point, as shown in Fig. 7. The

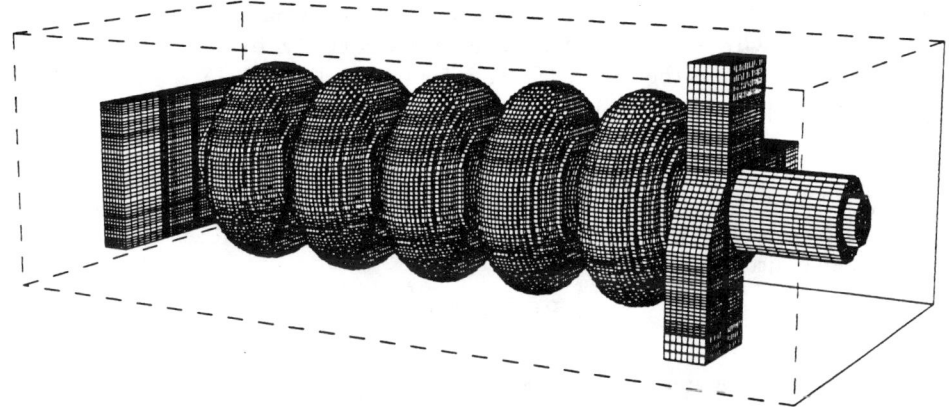

Fig. 6 The standard CEBAF SRF cavity configuration

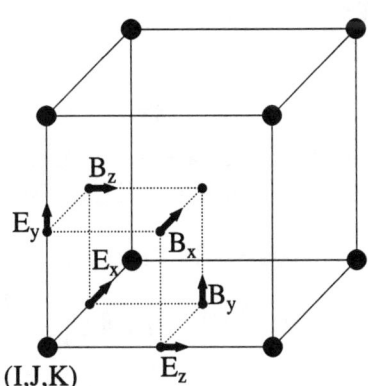

Fig. 7 Field components defined in MAFIA calculation

E_x, E_y and E_z components are defined on the grid axis with half mesh size off the grid in the x, y and z directions. The B_x, B_y and B_z components are defined at the central points of the mesh areas in the (y, z), (x, z) and (x, y) planes. Besides, the two couplers are arranged alternatively for different cavities, so the parity of a cavity must be taken into account. Z. Li has considered all these details carefully and used a 3-D interpolation subrountine $Q3DVL$ from $IMSL$ for calculating the 3-D rf fields in PARMELA based on the field data from MAFIA. The Fourier transform has been used to analyze the steering and focusing effects of the cavities.

SUMMARY

We have introduced our use of PARMELA at CEBAF for numerical modeling of the nuclear physics injector chopping system, a possible FEL laser gun injector, and the rf focusing and steering effects of the standard CEBAF SRF cavities on the beam. The emphasis has been put on how to modify and use the code consistently; for example, to input field data from SUPERFISH, POISSON, and MAFIA, to properly treat a focusing solenoidal lens having an actual field profile either individually or together with its adjacent rf cavity, to deal with the space charge forces, to use the code effectively for longitudinal phase space matching required for bunching electrons using a phase-compressor chicane, etc.

ACKNOWLEDGEMENTS

I wish to thank J. Bisognano and C. Sinclair for support and enlightening discussions. I also thank S. Benson, Z. Li, P. Liger, M. Tiefenback and B. Yunn for valuable discussions.

REFERENCES

1. The code was originally developed by K. R. Crandall.
2. Z. Li, private communication.
3. H. Liu et al., to be presented at the 1993 PAC, May, Washington, DC.
4. M. Tiefenback et al., ibid.
5. J. Bisognano et al., NIM A318, 216(1992).
6. G. R. Neil et al., ibid., p. 212.
7. C. K. Sinclair, ibid., p. 410.
8. P. Liger et al., ibid., p. 290.
9. H. Liu et al., to be presented at the 1993 PAC, May, Washington, DC.
10. K. T. McDonald, IEEE Trans. ED, 35, 2052 (1988).
11. H. Liu, this conference.
12. T. Raubenheimer, Workshop on Fourth Generation Light Sources, Feb., SSRL, 263(1992).
13. A. M. Sessler et al., Phys. Rev. Lett. 68, 309 (1992).

NEW FEATURES AND APPLICATIONS OF ABCI

Y. H. Chin
Lawrence Berkeley Laboratory, Berkeley, CA 94720

ABSTRACT

ABCI is a computer program which solves the Maxwell equations directly in the time domain when a Gaussian beam goes through an axi-symmetrical structure on or off axis. Many new features have been implemented in the new version of ABCI (presently version 6.2.4), including the "moving mesh" and Napoly's method of calculation of wake potentials. The mesh is now generated only for the part of the structure inside a window, and moves together with the window frame. This moving mesh option reduces the number of mesh points considerably, and very fine meshes can be used. Napoly's integration method makes it possible to compute wake potentials in a structure such as a collimator, where parts of the cavity material are at smaller radii than that of the beam pipes, in such a way that the contribution from the beam pipes vanishes. For the monopole wake potential, ABCI can be applied even to structures with unequal beam pipe radii. Furthermore, the radial mesh size can be varied over the structure, permitting to use a fine mesh only where actually needed. With these improvements, the program allows computation of wake fields for structures far too complicated for older codes. Its usefulness is illustrated by showing some numerical examples. A newly installed mesh generator performs automatic circular and elliptical connections of input points. In addition to the conventional method, ABCI permits the input of the structure geometry by giving the increments of coordinates from the previous positions. In this method, one can use the repetition commands to repeat input blocks when the same structure repeats many times. Plots of a cavity shape and wake potentials can be obtained in the form of a Top Drawer file.

INTRODUCTION

The first version (version 2.0) of ABCI (Azimuthal Beam Cavity Interaction)[1] was written in 1984, however, its manual was published only in 1988. It was a computer program which solved the Maxwell equations directly in the time domain when a Gaussian beam passed through an axi-symmetrical structure on or off axis. It used the FIT method[2] to discretize the Maxwell equations, similar to TBCI.[3] However, in addition to some internal differences, it was preferable to TBCI mainly due to capability to change dimensions of arrays to make a larger mesh if necessary (which could not be done in TBCI which was only distributed in the compiled form), and the possibility of different mesh sizes in r- and z-directions. Furthermore, one could input the mesh sizes rather than the number of mesh lines, and could use `CONTINUE` cards to calculate with different bunch lengths and/or mode numbers (m=0 or 1) in a single job. In this program, the beam was assumed to be hollow, with surface charges azimuthally distributed either in an uniform or sinusoidal way. In the first version of ABCI, the radius of the hollow beam was always chosen to be equal to that of a beam pipe so that no fields were brought with it into the structure of concern. The wake fields were integrated at the radius of the beam pipe, which left the integration across the cavity gap as the only contribution to the wake potentials and thus made long beam pipes unnecessary. The program was compact, and simply structured so that users could easily change important parameters such as an array size for the number of mesh points, and modify the program for their special needs. Since the main body of the program was small, relatively large arrays could be allocated to mesh points in a limited memory space. Furthermore, permitting unequal mesh sizes in the axial and radial directions helped to reduce the number of mesh points.

However, if one tried to apply the program to long structures and/or very short bunches, the total number of mesh points easily becomes of the order of many hundred thousands or more. For example, the recently proposed "stagger-tuned" structure for the NLC of SLAC[4] consists of a disc-loaded waveguide with a large number cells with slightly different dimensions of the order of μm or less. In order to correctly represent such tiny differences, many million mesh points would be needed.

A. Moving Mesh

Not all of these mesh points are simultaneously necessary at each time step for the calculation of fields. If we are only interested in the wake potentials not too far behind the beam, the fields need to be calculated only in the area called, "window". The window is defined by the area of the structure which starts at the head of the bunch and ends at the last longitudinal coordinate in the bunch frame (which is often the tail of the bunch) up to which we want to know the wake potentials. The fields in front of the bunch are always zero. The fields behind the window can never catch up with the window, which is moving forward with the speed of light, and thus do not affect the fields inside the window. Since the calculation is confined to the area inside the window, the "mesh" is needed only for this frame and moves together with it. One of main new features of ABCI is the implementation of this "moving mesh" in lieu of the conventional static mesh. Since the window is usually much smaller than the total structure, the number of mesh points can be drastically reduced. In addition, since the window length is determined only by the last longitudinal coordinate of the wake potentials, the number of mesh points does not change as the structure length increases.

B. Napoly's Integration Method

Another main new feature of ABCI is the implementation of "Napoly's integration method" of fields to calculate wake potentials.[5,6] The conventional integration method at the radius of the beam pipe breaks down when a part of the structure comes down below it, or when the radii of the two beam pipes at both ends are unequal. The only alternative was to integrate over a straight line at an allowable radius and with beam pipes long enough to allow the fields to catch up with the beam far behind the structure. Napoly's integration method is a solution to this classical problem (the integration along the structure surface was already described by Gluckstern and Neri[7] in 1985). It relies on the expression of the wake potentials, at any multiple order, as an integral of electromagnetic fields along any one dimensional conture spanning the structure longitudinally. For the particular case of the contures parallel to the r- and z-axes, the integration is considerably simplified.[6] For this reason, ABCI has an option which uses a path of integration ("Napoly-Zotter conture") that starts as usual along the beam pipe, then descends radially to pass underneath the smallest material structure radius. It then rises again to the radius of the outgoing beam pipe and moves along it to the end of the structure (Napoly's method and a proper integration conture are actually automatically chosen as soon as a material point has a radius smaller than the beam pipe). This path is shown in Fig. 1 by the broken curve.

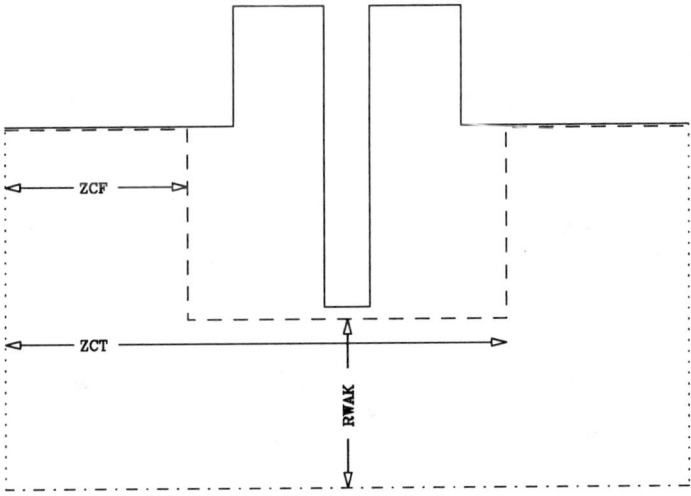

Fig. 1. Napoly-Zotter integration conture for computation of wake potentials.

The first axial coordinate where the path descends, the radius to which it goes, and the second axial coordinate where it rises again can also be chosen as input commands. In particular for structures with a complicated boundary extending to the inside of the beam pipes, this technique leads to a considerable saving in computing time. For the monopole (longitudinal) wake potential case, this method permits a structure with unequal beam radii at both ends. For the dipole (transverse and longitudinal) wake potential case, the beam pipe radii must be equal.

C. Other New Features

In addition to these two new main features mentioned above, ABCI has a completely new mesh generator, which permits circular and elliptical inputs just as TBCI. The program allows variable radial mesh sizes for different radial intervals for the better fitting of mesh and reducing the total number of mesh points by permitting to use a fine mesh only where actually needed.. In addition to the conventional method of inputting the shape of the structure by giving the absolute coordinates of points, users can now input the structure by giving the increments of coordinates from the previous positions (incremental input). In this method, one can use repetition commands to repeat input blocks which saves time and labor when the same structure repeats many times. The new ABCI also has better plotting facilities. It can show on a separate page each the input and actual shape of a cavity used for calculation, the wake potentials, and finally the Fourier transforms of the wake potentials. ABCI creates a "Top Drawer" file[8] for the corresponding figures. By this method, ABCI's graphical output becomes independent of computers and graphic devices. One can easily import/export the graphical output to other computers, and/or edit it if desired.

APPLICATIONS

I this section, we show a number of examples of structures which have so far been evaluated using the new version of ABCI.

A. Collimator

A Saclay collimator shown in Fig. 2 is a simple constriction of a beam pipe, which can be computed easily with the new version of ABCI using Napoly's method. The beam pipes at both sides have 5 cm length, long enough to allow wake fields to propagate in both z-directions like plane waves (this is the essential assumption for the engaged "open" boundary condition). The integration conture used is shown by the broken curve. The rms bunch length is chosen to be 0.5mm. The longitudinal loss factor was then found to be -1.755×10^{13} V/C. For comparison, longitudinal loss factors were also computed by the integration along a straight line at the inner radius of the collimator and subtracting the contribution of the beam pipe from it (similar to the "WAKCOR" option in TBCI). The results are shown by the solid curve in Fig. 3 as a function of the beam pipe length L at both sides. The dotted line denotes the loss factor obtained by Napoly's method. One can see that a quite long beam pipe ($\gtrsim$30cm) compared to the beam pipe radius of 1 cm is needed for the WAKCOR method to saturate to the correct result.

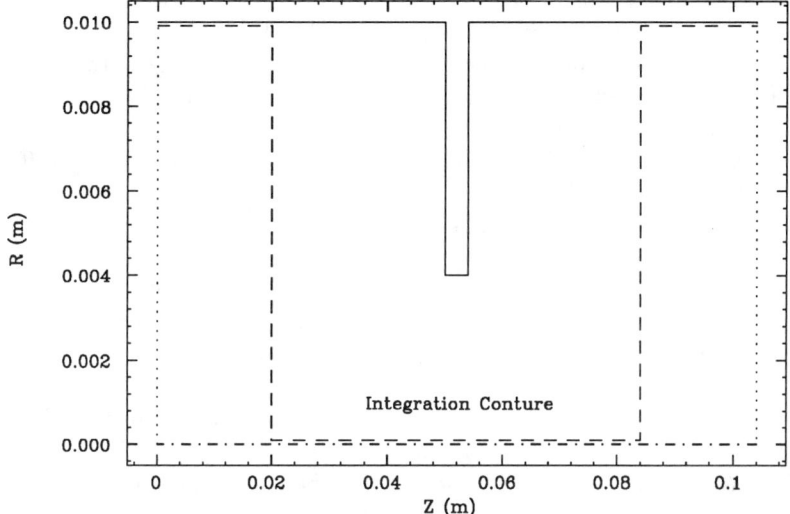

Fig. 2. Saclay collimator.

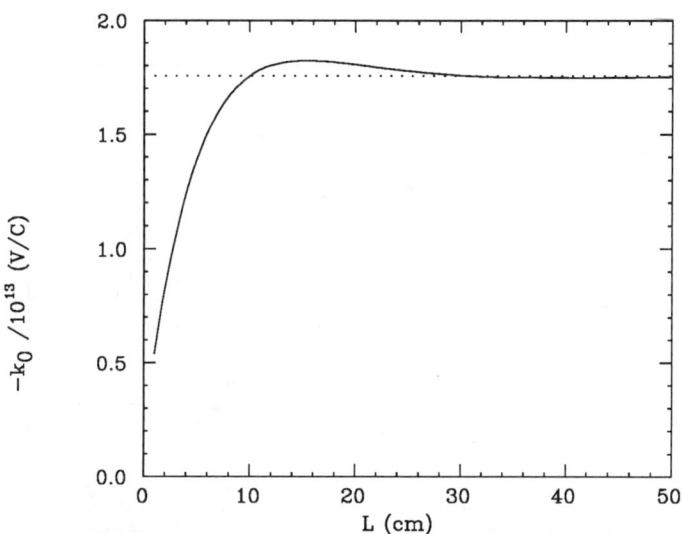

Fig. 3. Comparison of the longitudinal loss factor obtained by Napoly's method (dotted line) with that obtained by the "WAKCOR" method (solid line) as a function of beam pipe length L at both sides.

B. Step-in and Step-out Structures

Next, we show computations of loss factors of a tube with a step to a smaller radius (step-in) and to a larger one (step-out). Figure 4 shows the step-in structure used in this example. The integration path is denoted by the broken curve. The step-out structure for comparison is a mirror symmetrical image of the step-in structure. The bunch length is chosen to be 1cm. Figure 5 shows the (normalized) longitudinal wake potentials for the step-in structure. The broken and solid curves express the wake potentials with and without the potential energy difference term [5,6], respectively. The positive value of the broken curve predicts that the bunch will be accelerated after passing though the step-in structure. The resulting longitudinal loss factors for the step-in and step-out structures without the potential energy difference terms are -3.804 $\times 10^{11}$ V/C and -3.806 $\times 10^{11}$ V/C, respectively. They are equivalent within the limit of the computer accuracy. This result supports the analytical prediction.[9] The sum of those loss factors are compared with the one for a double-step geometry (pill box) shown in Fig. 6 obtained by integrating along a straight line at constant radius of the beam pipes. The result is summarized in Fig. 7. The horizontal axis L is the distance between the two steps in the pill box. The dotted line denotes the sum of loss factors for the two steps previously obtained, while the solid curve shows the loss factor for the pill box. One can see that the loss factor of the pill box converges to the sum for the two steps if they are sufficiently separated, and thus their interference becomes negligible. It is thus possible to split computation of long geometries in smaller parts and thus save memory and cpu time.

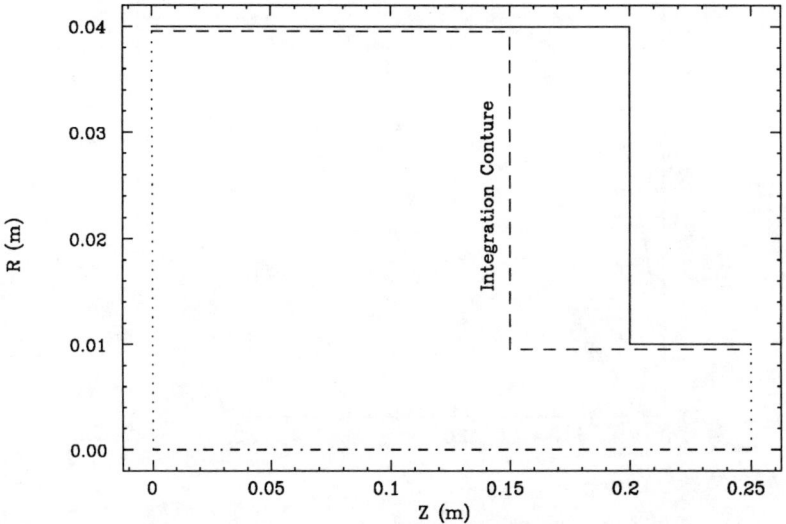

Fig. 4. Step-in structure.

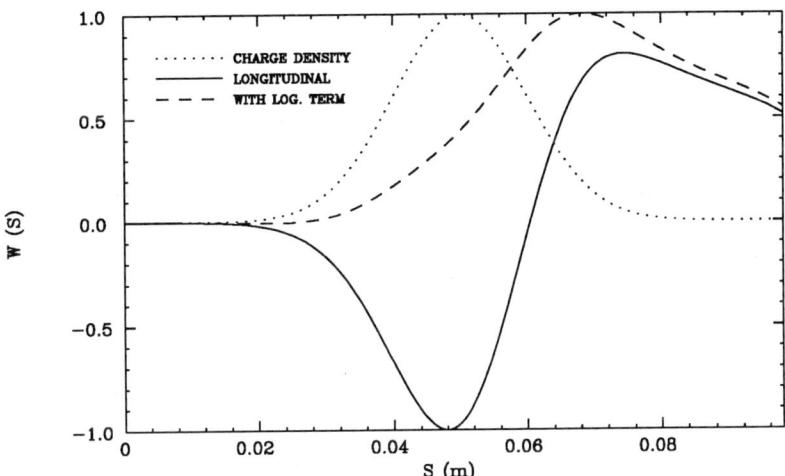

Fig. 5. Normalized longitudinal wake potentials for the step-in structure.

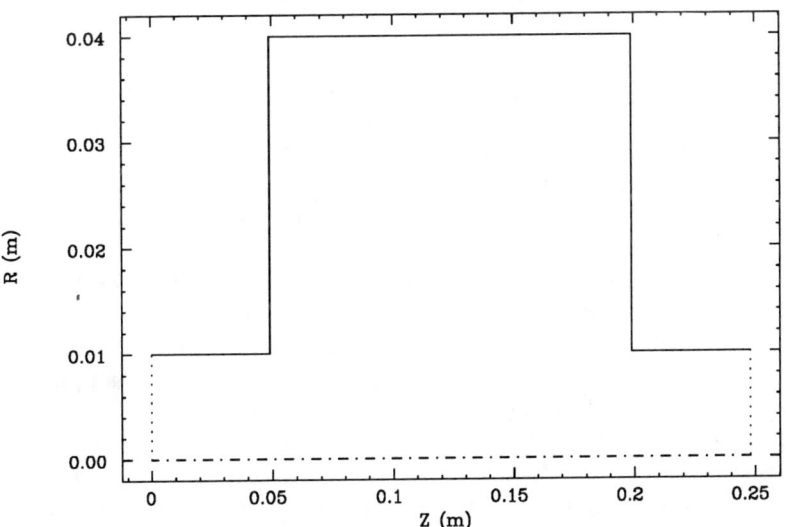

Fig. 6. Pill box structure.

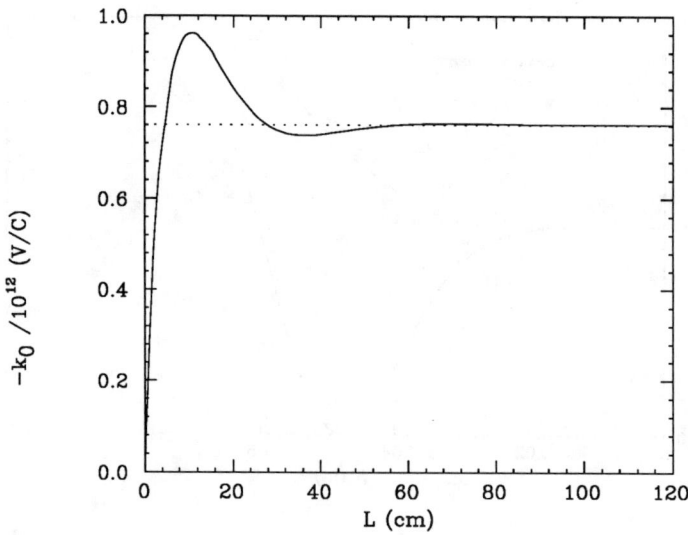

Fig. 7. Comparison of the sum of the longitudinal loss factors for the step-in and step-out structures (dotted line) with that for the pill box (solid line) as a function of the distance L between the two steps in the pill box.

C. CLIC stagger-tuned disk-loaded waveguide

A "stagger-tuned" structure of the CLIC (CERN Linear Collider)[10] is a disc-loaded waveguide composed of many cells with slightly different dimensions in such a way that the mode frequencies of each cell are distributed around the average values. Then, wake fields from each cell are expected to cancel each other so that the total wake fields will damp away rather quickly. Figure 8 shows an example of the CLIC stagger-tuned structure with 20 cells. The computed (normalized) transverse wake potential is plotted in Fig. 9 up to 1cm behind the head of the bunch (the bunch length in this case is only 0.17 mm). A clear damping of the transverse wake potential can be seen.

If TBCI is used instead in this example, it would have required about 4.6 million mesh points of uniform mesh size for required mesh sizes of 10 μm and for almost 7.2cm long structure of over 6.4mm radius. That would probably not fit any computer. With the moving, variable and unequal meshes, ABCI requires only 84 thousands mesh points, by factor $\sim$ 60 less than TBCI does. This example vividly illustrates that the new ABCI can open up a new possibility for computation of wake fields for structures which have been thought be too complicated to be dealt with.

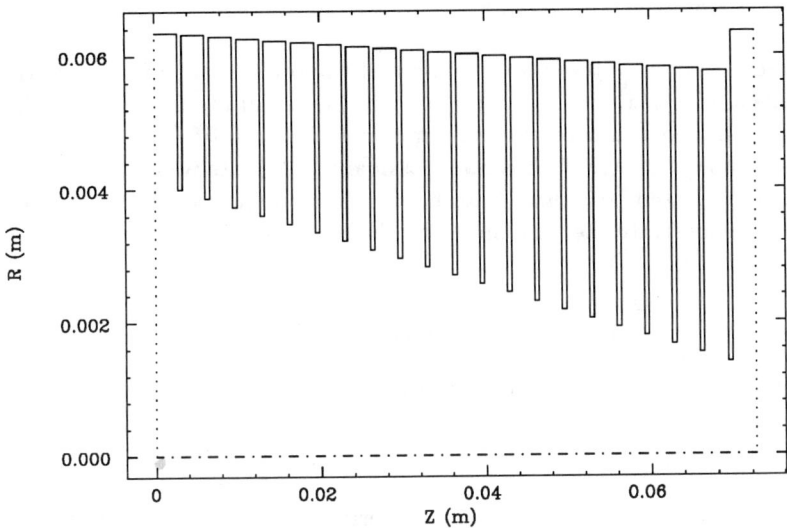

Fig. 8. Stagger-tuned disk-loaded waveguide of CLIC.

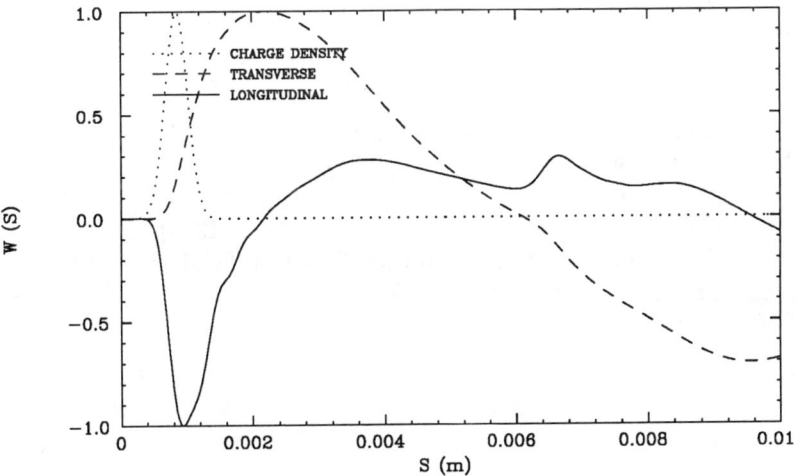

Fig. 9. Normalized transverse wake potential for CLIC stagger-tuned structure up to 1cm behind the head of bunch. The rms bunch length is 0.17mm.

CONCLUSIONS

The implementation of the moving mesh and Napoly's method for computing wake potentials, together with the option of variable radial mesh sizes, permits a large saving in memory and computing time, and thus drastically enhances the computational power of ABCI. It is now possible to compute wake potentials in much more complicated structures than before. The numerical examples shown in this paper demonstrate the usefulness and the remarkable advances in the new version of ABCI.

ACKNOWLEDGEMENTS

The author would like to thank B. Zotter and O. Napoly for helpful discussions and providing important calculation results.

This work was supported by the Director, Office of Energy Research, Office of High Energy and Nuclear Physics, High Energy Physics Division, of the U.S. Department of Energy under Contract No. DE-AC03-76SF00098.

REFERENCES

1. Y. H. Chin, CERN LEP-TH/88-3, 1988.
2. K. S. Yee, IEEE Trans. Antennas Propagat., Vol. AP-14, 302 (1966).
3. T. Weiland, DESY 82-015, 1982.
4. K. L. F. Bane, and R. L. Gluckstern, SLAC-PUB-5783, 1992.
5. O. Napoly, Part. Accelerators, 36, 15 (1991).
6. O. Napoly, Y. H. Chin and B. Zotter, DAPNIA/SEA/93-01, 1993.
7. R. L. Gluckstern, and F. Neri, IEEE NS-32, 5, 2403 (1985).
8. Top Drawer Manual, SLAC Computation Group, CTGM-189, 1980.
9. S. Heifets, SLAC/AP-81, ABC-14, 1990.
10. B. Zotter, private communications.

SYNCH–STATUS AND RECENT USE AT SSCL[*]

A. A. Garren and A. S. Kenney
Lawrence Berkeley Laboratory, Berkeley, CA 94720

E. D. Courant
Brookhaven National Laboratory, Upton, NY 11973

A. D. Russell
Fermi National Accelerator Laboratory, Batavia, IL 60510

M. J. Syphers, T. Sen, and T. Barts
Superconducting Super Collider Laboratory, Dallas, TX 75237

ABSTRACT

SYNCH is a computer program for use in the design and analysis of synchrotrons, storage rings, and beamlines. The first accelerator program to organize its input in the form of a language, **SYNCH** provides a natural way to describe and design lattices and to analyze them through calculations of betatron functions, dispersion, beam envelopes, emittances, and closed orbits. In addition, particle tracking, non-linear transformations, misalignments, electron integrals, and orbit corrections may be treated.

A brief summary of some features of the program and recent applications for designing the Interaction Regions (IR) of the SSC will given.

INTRODUCTION

The **SYNCH** program was first used in 1964-5 for the 200 BeV Design Study, and at present is programmed in FORTRAN 77.[1] The CMS code management system[2] is used to generate equivalent versions for different platforms. The latest version incorporates the ANSI/ISO GKS standard[3] for plotting and currently executes on VAX/VMS systems and Sun/UNIX Workstations with Sun GKS. A new Users Guide[4] will be available this year.

A table from the Users Guide summarizing commands available in **SYNCH** is included, as well as a discussion of how **SYNCH** has been used in recent design work for the Interaction Region (IR) of the Superconducting Super Collider (SSC).

CALCULATIONS PERFORMED BY **SYNCH**

The **SYNCH** program is used for the design and analysis of circular accelerators and beamlines. The program uses input commands to describe a lattice for the structure being studied. This lattice consists of basic elements such as drift spaces, dipole, quadrupole and sextupole magnets, and other non-linear elements. These elements are combined in beamline descriptions which can be manipulated to build up realistic accelerator lattices composed of thousands of elements.

---

[*]Work supported by the Universities Research Association, Inc., for the U.S. Department of Energy under Contract No. DE-AC35-89ER40486.

Table 1. **SYNCH** Program Commands, by Topic

Topic					
Program Control	ACT	DEACT	RUN	SIZE	STOP
SYNCH Subroutines	CALL SUB	END VPAR	INCR	MESH	REPL
Mathematical Operations	CALC PARA	(SIN RAND	SQRT SUM	1/X VAR	etc.) =
Beamlines	BML	LIST			
Element Definitions	DEQ MAP	DRF NPOL	KICK ROTZ	MAG SOL	MAGV SXTP
Operations on Transfer Matrices	EQU REF	INV ROT	INV2 ROTZ	MMM **	MXV
General Matrix/Vector Definitions	MAT	MAT3	VEC		
Betatron Function Calculations	BETA TRKE	CYA TRKM	CYC	IBET	TRKB
Particle Beam Calculations	BVAL	CYAE	CYEM		
Fitting Routines	FITB SOLV	FITQ	FITR	FITV	SMIN
Closed Orbit Determination and Particle Tracking	FXPT	PVEC	TRK		
Element Misalignments	BMIS SHF7	EMIS	MAGS	MOVE	SHF
Orbit Correction	ORBC				
Plotting	BEP	BEST	TRKB		
Files	IOUT	KEEP	OPEN	UPDAT	SELCT
Output Statements	C PBML PTAB	ECHO PCYC REM	NECHO PRNT WBE	P PRTV WMA	PAGE PRTV7

NEW FEATURES IN SYNCH

New features incorporated in this 1993 version of **SYNCH** include calculations of transverse coupling parameters using the formalism of Edwards and Teng;[5] generalizing the calculation of the electron integrals, rf output, and electron emittance factors to include the effects of vertical bending; and optionally creating a new input file with updated parameters calculated by the fitting routines.

The new version of **SYNCH** is substantially coded in FORTRAN 77 so that the program is easily transportable to many operating systems and computer architectures. Also, portions of of the program have been re-coded for enhanced performance. Plots of the β (amplitude) and η (dispersion) functions through a beamline (with a schematic diagram of the magnet layout) are programmed with a GKS interface for increased portability (see Figures 1 and 2.) It is very probable that the UNIX and possibly the VAX/VMS versions of the code will be available as a complete package with the GKS platform included; testing is currently being done on a public domain version of GKS.

Final testing of the new draft versions of **SYNCH** is being done on a VAX/VMS and on Sun SPARCstations at the Superconducting Super Collider Laboratory.

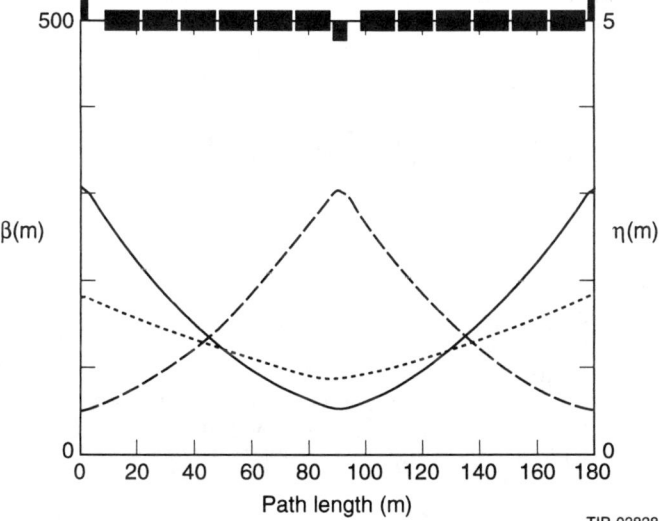

Figure 1. In this sample betatron-function plot, β_x is a solid line, β_y is a chain dash line, η_x is a dash line, and η_y is a chain dot line. Dipole magnets straddle the top horizontal line, focussing quadrupoles are above the line and the de-focussing quadrupole is below the line.

INTERACTION REGION DESIGN AT THE SSC

Most of the design effort on the SSC Interaction Regions has been done using **SYNCH**. The present status of the design will be reported at the 1993 IEEE Particle Accelerator Conference in May.[6] The most significant change since the 1990 Conceptual Design[7] is the procedure for the beta squeeze. Six families of independently powered quadrupole magnets change the optics from injection to collision while the strengths of the final-focus triplet quadrupoles are held constant. The phase advance from the Interaction Point (IP) to the arc quadrupoles was optimized to provide more effective positions of the local quadrupole and sextupole correctors for the IR. Two secondary focii where the IP is imaged have been incorporated symmetrically on both sides of the IR for beam diagnostics.

Other modifications include a significant reduction in the peak beta-function values at injection, optimized configuration of the M=-1 section, minimized number of different quadrupole lengths, standard field for the vertical bends and reduced total length of the magnets. The present design allows for a range of values between 40 to 180 meters for the space allocated to a detector at each IP, and for a range of β^* values from 7 to 0.25 meters for 40 meters detector space.

For each value of β^*, the gradients of the six variable strength quadrupoles were calculated to match the lattice functions into the arcs. The phase advance in each plane from the IP to the arc was also held constant. There were six conditions on the variables $\beta_x, \alpha_x, \eta_x, \beta_y, \alpha_y$, and η_y which were satisfied by using the fitting routine, **SOLV** on the six quadrupole gradients as parameters. Figure 2 is a plot of half an IR showing the amplitude and dispersion functions.

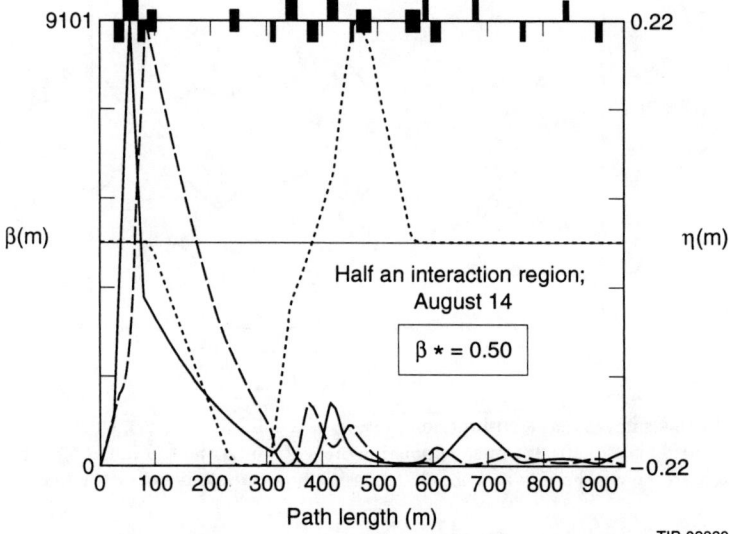

Figure 2. Betatron plot of one half of the IR with the schematic diagram of magnets at the top.

REFERENCES

1. FORTRAN (ANSI, N.Y., 1978).
2. VAX DEC/Code Management System Version 3.4, Digital Equipment Corporation, 1990.
3. Graphic Kernel System (GKS), (ANSI, N.Y., 1985).
4. A. A. Garren, A. S. Kenney, E. D. Courant, A. D. Russell, M. J. Syphers, **SYNCH** Users Guide, (SSCL) 1993 [in publication].
5. D. A. Edwards, L. C. Teng, Parameterization of Linear Coupled Motion in Periodic Structures, (IEEE Trans. Nucl. Sci, 20, No.3, 85, 1973) IEEE.
6. Y. Nosochkov, E. Courant, A. Garren, D. Ritson, T. Sen, R. Stiening, Current Design of the SSC Interaction Regions, [to be presented at the IEEE Particle Accelerator Conference, Washington, 1993].
7. J. R. Sanford, D. M. Mathews, Eds., SITE-Specific Conceptual Design, SSCL Document SSCL-SR-1056, 1990.

CODE TRANSPORTABILITY AND SIMPLE MAINTENANCE TOOLS

H. Grote, CERN, SL Div.
CH-1211 Geneva 23, Switzerland

ABSTRACT

General accelerator programs such as MAD have to run on many different configurations and require therefore certain coding standards, and tools to maintain and ship versions for different computers in the simplest possible way.

The transportability of Fortran-77 code will be discussed in the light of experiences with several systems. In particular, certain non-ANSI extensions are now widely available.

Two programs have been written by the author to maintain and extract versions for different computers. The first, called BASTA allows to keep successive versions in one single library, similar to the CDC program UPDATE.

The second, called ASTUCE is shipped together with MAD and allows to extract versions for different computers from one single "source" file containing the MAD code. This avoids sending one or several specific Fortran files to a given receiver, giving the latter the choice of all possible versions. Finally, a derivative of ASTUCE called ASTAP allows Unix-specific file extraction for the convenient creation of binary libraries.

INTRODUCTION

The accelerator simulation program MAD [1] is a fairly good example of a general purpose program in this field, in that it fulfills the needs of a great variety of users, with a wide spectrum of problems to be solved. Consequently, it has encountered considerable success world-wide and now runs on an (estimated) 80 computers with an unknown number of users. At CERN, the program is installed on eight different types of computers, running under two brands of Apollo/Aegis, Cray-Unicos, DEC/VMS, IBM/VM, IBM RISC/6000, HP-UX, and SUN Unix.

It is crucial that such a program be transportable, and easy to develop and to maintain if the developer does not want to spend most of his time upgrading and shipping a great number of versions for the different systems: although ANSI Fortran 77 has made this task easier, it is far from allowing to have only one single version of the code for all possible computer systems.

The transportability is achieved (as far as it goes) by adhering strictly to the ANSI Fortran 77 standard, with minor exceptions in FORMAT statements. Based on experience with the CDC program UPDATE (which works only on CDC mainframes) and with the CERN product Patchy [2] specific maintenance tools for MAD were developed with the following aims:

1. Keep only one "master file" with all the code for all systems, and use it for development and maintenance alike.

2. Allow temporary modifications of the master file for the test of new or modified code, without disturbing the consolidated "production" version. Once the modifications are tested they can be implemented, but remain "traceable" and may again be removed later should the necessity arise.

3. Ship a special form of the master file together with an extraction tool which allows the implementer of the program to create the version or versions he needs, since at many laboratories more than one type of computers is available.

4. Use the same special form of the master file to create single file Fortran decks for the creation of Unix libraries at CERN.

TRANSPORTABILITY OF ANSI FORTRAN 77

One may say that after some initial problems with the interpretation of the standard (certain software suppliers interpret more than others), ANSI Fortran 77 has been well transportable over different systems for many years now. This includes the somewhat more thorny I/O of sequential and random access files, apart from minor differences in file name conventions, and in unit numbers available, and some divergence of views on how to interpret an empty carriage return. What is missing from the standard (and therefore widely different on the different systems) is any kind of error condition handling, except for trivial I/O occurrences.

Nowadays there are even many extensions available that work on most or all systems we have encountered so far: FORMAT statements in the style of the previous standard (all), code in lower case (most), double precision complex COMPLEX*16 (most), and quadruple precision real REAL*16 (most). Of these, MAD employs only the somewhat more convenient FORMAT style.

A word of caution: many compilers these days make use of the possibility to share memory for temporary variables and common blocks. This is very often a cause of trouble when a program is transported to a different computer system, and normally extremely hard to locate. Fortunately, all compilers seem to offer an automatic SAVE of all variables and commons, albeit at the expense of a slightly downgraded performance. MAD uses SAVE statements for all common blocks and for critical variables, and will normally work when compiled with the highest optimization level of a given compiler.

THE SOURCE FILE

Initially, it is convenient to keep a program in the form of a plain coded file, and one may even do this for the whole development phase, depending on its size. The form of this file is the SOURCE form, inspired to equal parts by Patchy and

UPDATE. The source file contains the Fortran code in its usual form (except for common blocks and certain other declarations), and in addition some annotation, special "control" statements with a "+" in column one. These are of two types:

- +DECK and +COMDECK statements to organize the code in reasonable pieces. Normally, a deck coincides with a subroutine, and a comdeck with a common block declaration, a group of parameter statements, or the like. Comdecks are inserted in the places where a +CALL statement with their name is found. This works therefore like the Fortran INCLUDE (which is not ANSI) and has the advantage of keeping all comdecks in one file.

- Statements of the form +IF .. +ELSE .. +ENDIF which allow conditional inclusion or exclusion of code; they can be nested. It is with the help of these constructs that all computer-specific features are implemented.

The source file is edited during code development. Existing Fortran code can be added after having passed through a tool that inserts +DECK statements, extracts common blocks and replaces them by +CALL statements, and prints warnings if common blocks with the same name are not identical (this practice of the early sixties can still be found). Before compilation, the program ASTUCE creates the Fortran from the source file. This is very fast and does not represent an overhead, and is particularly convenient when Unix "make" is used.

ASTUCE

ASTUCE is written in Fortran 77 and is shipped together with the MAD source file. The MAD implementer has then first to install ASTUCE, i.e. compile and load it, and can then extract the MAD version he wants. The advantages are:

- The source file is shorter than the Fortran file. The MAD source file, for example consists of 60000 lines, a MAD Fortran file of 85000 lines.

- The user may want to extract versions for different computers, but as well different versions for the same computer, e.g. with or without GKS plotting interface.

- If the user ever has to modify the MAD memory pool size, there is just one line to change.

ASTUCE receives instructions specifying the selection of flags and decks. It then reads the source file and writes the Fortran file. A special version of ASTUCE exists that creates one file per deck, plus the three Unix shell scripts to compile all these files (about one thousand for MAD), to enter them in a relocatable library, and finally to delete all intermediate files.

BASTA

Eventually the program reaches its "production phase", i.e. it is installed on one or several computers, users are admitted (!), and these users want to live with a relatively stable version of the program. On the other hand, development goes on, of course, and one single source file is no longer sufficient.

Rather than using now two or three or more source files (for the "old", the "production", the "new", and the development version), the source file is converted into a random access binary file under the control of the program BASTA. This program is currently used at CERN for the maintenance of MAD. It is similar to the CDC program UPDATE in that it gives each line a unique identifier which remains constant in time, i.e. which is given when the line is read in, and never changed. Lines can then be accessed via their identifier through so-called "correction sets" forming the input of BASTA. However, the user does not have to create these correction sets by hand, they are automatically created by BASTA itself. BASTA allows to extract the following files from the (binary and modified) "source" file:

1. The current (including all changes) source file, either in full, or only selected comdecks and decks, with or without line identifers.

2. The Fortran file corresponding to the selection flags set, in full or part, with or without line identifiers, and with or without extra Fortran comment lines for automatic correction set creation.

3. The original source file, with all correction sets appended. This is the only way of removing correction sets entirely. Normally it is sufficient to correct an existing correction set with another one, and so forth.

A programmer wanting to modify a program kept under BASTA would then normally proceed as follows:

1. Extract the decks to modify. This would be in most cases in the Fortran form with extra annotations, except when common decks have to be modified, or conditional code has to be added which is done using an extracted source file.

2. Modify the code; if it is in Fortran form, compile and run it, and possibly iterate until the code is correct.

3. Run BASTA with the modified code as input to create a correction set containing only the changes.

4. Run BASTA with the correction set to extract the modified code again, this time always in Fortran form. If the code has already been tested, the correction set may be implemented at the same time.

5. Compile and run the modified code.

6. If not already done, run BASTA to implement the correction set.

BASTA is written in Fortran. In order to keep the overhead small, it keeps a complete cross reference list between all decks, comdecks, and correction sets. In this way, if only one deck is extracted from a large program, only the comdecks and correction sets needed for this deck have to be read from the binary file. The execution time is thus more or less proportional to the amount of code extracted, which is not always the case with programs of this type.

FINAL REMARK

The programs ASTUCE and BASTA may of course be used to maintain any type of file, and not just Fortran, since they only distinguish between control statements, and other input. For example, one could imagine to keep large data files in source form, the way this is done with Patchy at CERN for certain applications. However, the way in which C or C++ are implemented under Unix calls for a different type of tool to maintain such programs.

References

[1] H. Grote and F.C. Iselin, CERN-SL/90-13 (1990), Rev. 3, Jan. 1993.

[2] J. Zoll, CERN Program Library Long Write-up L400, CERN 1972, rev. 1977

FULL-TURN SYMPLECTIC MAP
FROM A GENERATOR IN A FOURIER-SPLINE BASIS *

J. S. Berg, R. L. Warnock, and R. D. Ruth
Stanford Linear Accelerator Center, Stanford University, Stanford, CA 94309

É. Forest
Lawrence Berkeley Laboratory, University of California

ABSTRACT

Given an arbitrary symplectic tracking code, one can construct a full-turn symplectic map that approximates the result of the code to high accuracy. The map is defined implicitly by a mixed-variable generating function. The implicit definition is no great drawback in practice, thanks to an efficient use of Newton's method to solve for the explicit map at each iteration. The generator is represented by a Fourier series in angle variables, with coefficients given as B-spline functions of action variables. It is constructed by using results of single-turn tracking from many initial conditions. The method has been applied to a realistic model of the SSC in three degrees of freedom. Orbits can be mapped symplectically for 10^7 turns on an IBM RS6000 model 320 workstation, in a run of about one day.

INTRODUCTION

Long term stability of orbits in circular accelerators is usually studied by tracking codes, which integrate the equations of motion through the lattice by some symplectic integration algorithm, proceeding element-by-element. There have been various attempts to summarize the full-turn evolution defined by a tracking code in an analytic formula, a *full-turn map*. If the map represented the code to sufficient accuracy, and could be evaluated in substantially less time than the time for tracking one turn, it could be used for economical studies of long-term evolution.

The method of automatic differentiation [1] allows one to differentiate the tracking algorithm, so as to generate a large number of Taylor coefficients of the corresponding map. The resulting map, given as a truncated Taylor series, cannot be exactly symplectic. In a region of phase space close to the dynamic aperture, the failure of symplecticity may be so large as to raise doubt about the usefulness of the map. This is the case for the highest order Taylor maps generated for the SSC (Superconducting Super Collider).

One possibility is to symplectify the map by producing a mixed- variable generating function that induces an exactly symplectic map that closely approximates the underlying map. This can be done by using formal power developments in Cartesian coordinates to solve the nonlinear equations that define the generator in terms of the map. This method was proposed and carried out long ago [2]. Because of convergence difficulties it proved not to be very useful for some accelerators (for instance the Berkeley Advanced Light Source and the Tevatron), but recently Yan, Channell, and Syphers have reported some success with an application to the SSC [3].

We describe a different way to construct a symplectic full-turn map from a tracking code or other "source map". We again define the map through a mixed-variable generating function, but given as a function of action-angle coordinates rather than Cartesian coordinates. We avoid the use of Taylor series in favor of methods based on Fourier developments and spline interpolation. We believe that these methods are more appropriate at large amplitudes, since they use information on the function to be represented at many points in the region of interest. By contrast, the Taylor method uses information on the function and its derivatives at one

---
* Work supported by Department of Energy contracts DE-AC03-76-SF-00515 and DE-AC03-76-SF-00098.

point far from the region of interest, and tries to extrapolate from that point. A singularity at a complex point can cause divergence of the Taylor expansion, but does not necessarily spoil our Fourier-spline representation. An added bonus of the Fourier method is that a large fraction of the low Fourier modes (and all sufficiently high ones) prove to be negligible, and can be discarded without affecting symplecticity. This aids in fast evaluation of the map.

This paper is a brief summary of our mapping method. Details and associated references can be found in [4].

CONSTRUCTING THE MAP

The map is defined to be a transformation from the "old" variables $(\mathbf{I}, \mathbf{\Phi})$ to the "new" variables $(\mathbf{I'}, \mathbf{\Phi'})$. The generating function in this case will be in terms of old action and new angle variables:

$$G(\mathbf{I}, \mathbf{\Phi'}) = \sum_{\mathbf{m}} g_{\mathbf{m}}(\mathbf{I}) e^{i\mathbf{m}\cdot\mathbf{\Phi'}}. \tag{1}$$

The transformation equations are then

$$\mathbf{I'} = \mathbf{I} + G_{\mathbf{\Phi'}}(\mathbf{I}, \mathbf{\Phi'}), \qquad \mathbf{\Phi} = \mathbf{\Phi'} + G_{\mathbf{I}}(\mathbf{I}, \mathbf{\Phi'}). \tag{2a, b}$$

We start with a "source map," which gives the final variables as an explicit function of the initial variables:

$$\mathbf{I'} = \mathbf{I} + \mathbf{R}(\mathbf{I}, \mathbf{\Phi}), \qquad \mathbf{\Phi'} = \mathbf{\Phi} + \mathbf{\Theta}(\mathbf{I}, \mathbf{\Phi}). \tag{3a, b}$$

This map will usually be defined as the result of tracking over one turn, but in the numerical work reported here it was a 12th order Taylor series map.

The Fourier coefficients are obtained from (2a) and (3a) as

$$\begin{aligned} g_{\mathbf{m}}(\mathbf{I}) &= \frac{1}{(2\pi)^d i m_\alpha} \int_0^{2\pi} d\mathbf{\Phi'} G_{\Phi'_\alpha}(\mathbf{I}, \mathbf{\Phi'}) e^{-i\mathbf{m}\cdot\mathbf{\Phi'}} \\ &= \frac{1}{(2\pi)^d i m_\alpha} \int_0^{2\pi} d\mathbf{\Phi'} R_\alpha(\mathbf{I}, \mathbf{\Phi}(\mathbf{I}, \mathbf{\Phi'})) e^{-i\mathbf{m}\cdot\mathbf{\Phi'}}. \end{aligned} \tag{4}$$

Since we do not know $\mathbf{R}$ as a function of $\mathbf{\Phi'}$, we perform a change of variables in the integral to get an integral over $\mathbf{\Phi}$:

$$g_{\mathbf{m}}(\mathbf{I}) = \frac{1}{(2\pi)^d i m_\alpha} \int_0^{2\pi} d\mathbf{\Phi} R_\alpha(\mathbf{I}, \mathbf{\Phi}) e^{-i\mathbf{m}\cdot\mathbf{\Phi}} e^{-i\mathbf{m}\cdot\mathbf{\Theta}(\mathbf{I}, \mathbf{\Phi})} \det(1 + \Theta_{\mathbf{\Phi}}(\mathbf{I}, \mathbf{\Phi})). \tag{5}$$

The integral is then discretized to obtain

$$g_{\mathbf{m}}(\mathbf{I}) = \frac{1}{i m_\alpha \prod_\beta J_\beta} \sum_{\mathbf{j}} R_\alpha(\mathbf{I}, \mathbf{\Phi_j}) e^{-i\mathbf{m}\cdot\mathbf{\Phi_j}} e^{-i\mathbf{m}\cdot\mathbf{\Theta}(\mathbf{I}, \mathbf{\Phi_j})} \det(1 + \Theta_{\mathbf{\Phi}}(\mathbf{I}, \mathbf{\Phi_j})), \tag{6}$$

where J_β is the number of Φ_β mesh points in the β dimension, and the summation is over integer vectors $\mathbf{j}$ such that $j_\beta \in \{0, \ldots, J_\beta - 1\}$.

The $\mathbf{m} = 0$ mode must be handled differently. We instead must use Θ values. The resulting summation is

$$g_0(\mathbf{I}) = -\frac{1}{\prod_\beta J_\beta} \sum_{\mathbf{j}} \Theta(\mathbf{I}, \mathbf{\Phi_j}) \det(1 + \Theta_{\mathbf{\Phi}}(\mathbf{I}, \mathbf{\Phi_j})). \tag{7}$$

To increase the speed of evaluation of the map, Fourier modes that are smaller than the expected or desired accuracy of the map can be removed from the generating function.

We obtain values of $g_\mathbf{m}(\mathbf{I})$ for values on a mesh in $\mathbf{I}$. We then choose a set of basis functions $B_j^{(\alpha)}(I)$ to use in interpolating the coefficients such that

$$g_\mathbf{m}(\mathbf{I}) = \sum_j g_{\mathbf{m},\mathbf{j}} \prod_\alpha B_{j_\alpha}^{(\alpha)}\left(I^{(\alpha)}\right). \tag{8}$$

The index α labels the different degrees of freedom. For the $\mathbf{m} \neq 0$ modes, the interpolation is straightforward. For the $\mathbf{m} = 0$ mode, one must be careful to consider the fact that the derivatives of the basis functions are linearly dependent. Details of this can be found in [4]. It is advantageous to choose B-splines for the basis functions. Because they have a small region where they are nonzero, their use greatly increases the speed of evaluation of the map.

EVALUATING THE MAP

The map is evaluated by performing a Newton iteration to obtain $\boldsymbol{\Phi}'$ and then substituting into (2a) to get $\mathbf{I}'$. An initial guess for the Newton iteration is provided by an explicit map with a small number of modes retained.

THREE DIMENSIONS

The method can be used in any number of dimensions. In a three dimensional accelerator problem, however, it is not advantageous to do the third dimension in action-angle variables. Instead, note that most of an accelerator ring is time independent. One can construct a map for the time independent part that has the energy deviation as an additional parameter, which is treated on equal footing with the actions. The time-dependent parts (usually r.f. cavities) can then be treated separately as the user chooses. Time-of-flight information is obtained by taking a derivative of the generating function with respect to energy deviation.

PRECONDITIONING THE SOURCE MAP

Finally, note that since one wants to perform the action interpolation over a finite domain that does not include the origin in each phase space plane, the plain source map is sometimes not well-suited for direct application of this method. This can be overcome by performing a preliminary canonical transformation on the source map so as to have the new source map take an annulus of initial conditions into a similar (larger) annulus. This can be done easily by a linear transformation or a low-order Taylor series mixed-variable generating function.

RESULTS

As an example, we take the source map to be a 12th order Taylor series map for a realistic model of the SSC. Results for accuracy (agreement with the source map) and iteration time are shown in figures 1 through 4. The "mode cutoff" is a measure of the maximum size of the Fourier modes that are being removed from the generating function. The number of actions indicates the number of mesh points in each dimension of action interpolation. The order refers to the order of B-splines used in action interpolation. The curves have approximately slope 1 when the error is dominated by the number of Fourier modes being thrown away. They begin to level off when the error is dominated by the action interpolation (low actions) or failure of symplecticity of the source map (high actions).

We have constructed maps at amplitudes near the dynamic aperature, and have found that we can track stable trajectories for 10^7 terms in about a half a day in 2 dimensions and about a day in three dimensions. Times are on an IBM RS6000 320H workstation.

CONCLUSIONS

A method has been devised that will allow the construction of exactly symplectic maps. These maps are highly accurate and can be evaluated in a very short time, sufficient to perform long-term tracking in a reasonable time.

REFERENCES

[1] M. Berz, Particle Accelerators **24**, 109 (1989).
[2] D. R. Douglas and A. J. Dragt, *Proc. 12th International Accelerator Conf.*, edited by F. T. Cole and R. Donaldson, p. 139 (FNAL, Batavia, 1983).
[3] Y. Yan, P. Channell, M. Syphers, SSCL-Preprint-157 (1992).
[4] J. S. Berg, R. L. Warnock, R. D. Ruth, É. Forest, Construction of Symplectic Maps for Nonlinear Motion of Particles in Accelerators, SLAC-PUB-6037 (1993), submitted for publication.

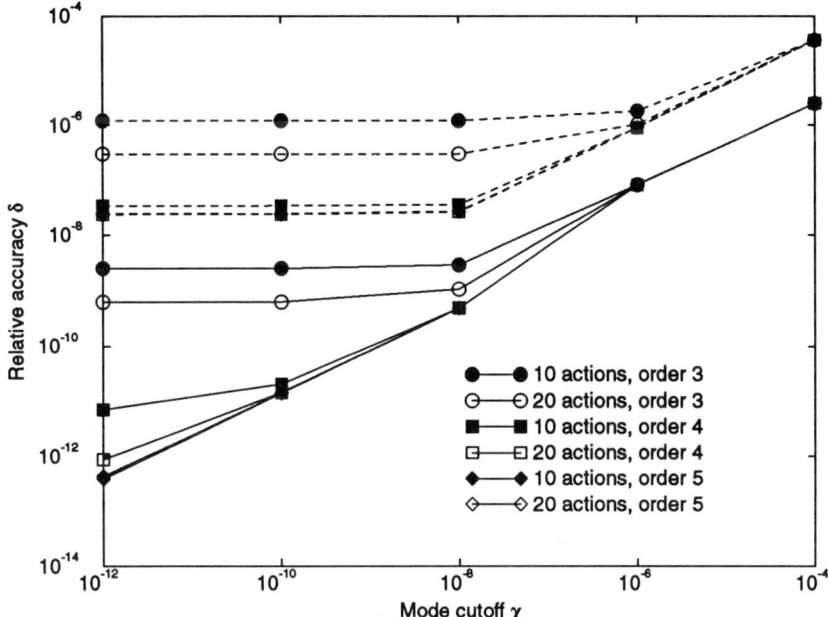

Figure 1: Relative accuracy of 2-D map. Solid lines are low action $\mathbf{I} = (0.1, 0.1)$, dashed lines are high action $\mathbf{I} = (3.0, 3.0)$.

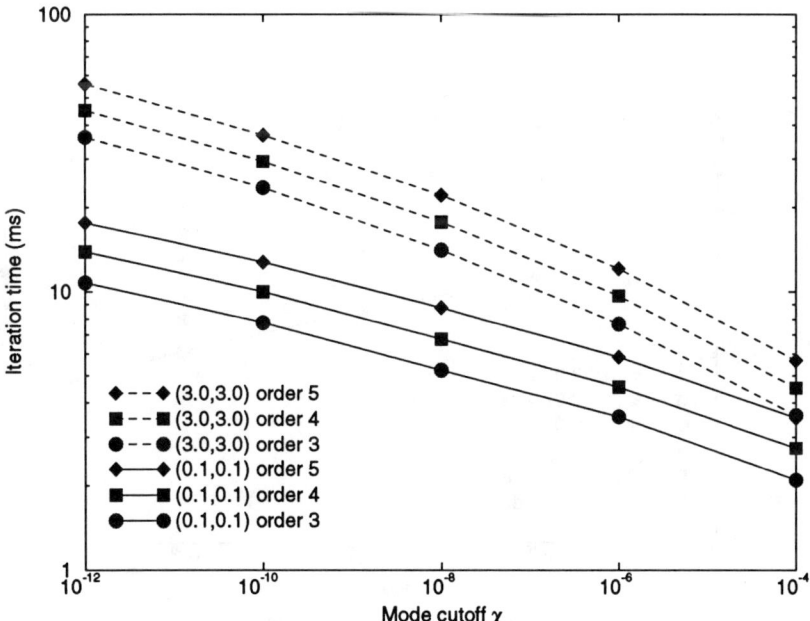

Figure 2: Iteration time of 2-D map. Iteration time is independent of number of action mesh points.

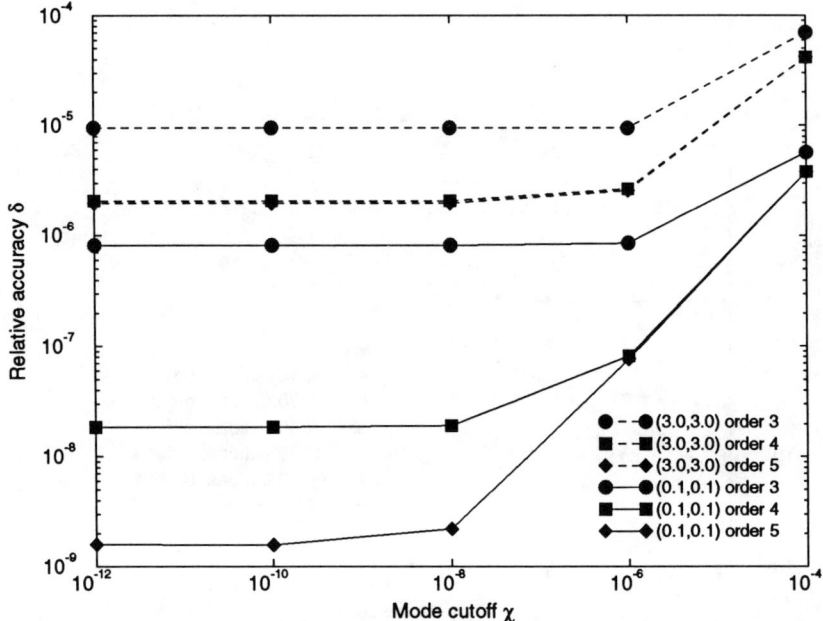

Figure 3: Relative accuracy of 3-D map

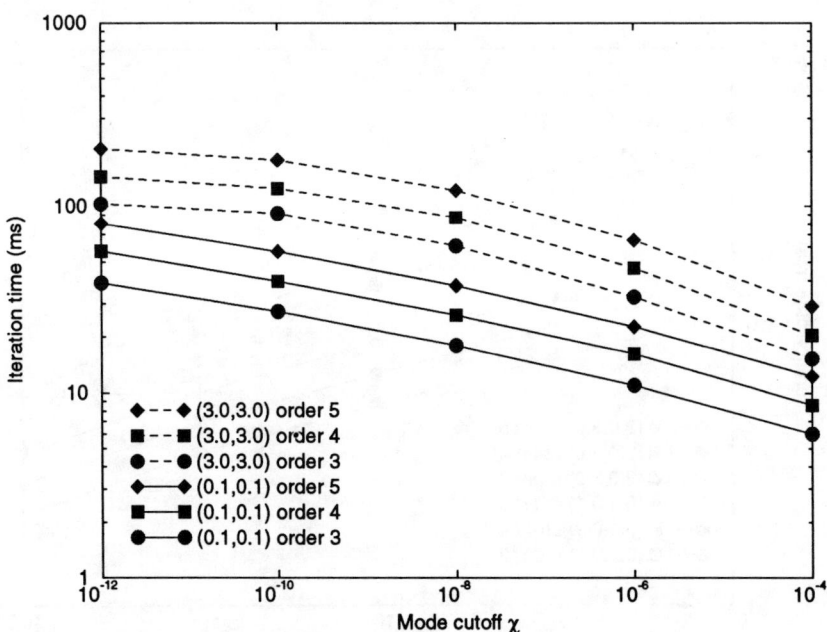

Figure 4: Iteration time of 3-D map.

PARTICLE DISTRIBUTION GENERATOR IN 4D PHASE SPACE

Y.K.Batygin
Electrophysics Department
Moscow Engineering Physics Institute
115409, Moscow, Russia

ABSTRACT

A random number generator for particle distributions with elliptical symmetry in four-dimensional phase space is presented. Distribution function $f(x,\dot{x},y,\dot{y})$ depends on a parameter which describes the hyperellipsoid surface in 4D phase space. The values of root-mean-square beam emittances and orientation of elliptical beam projections on phase planes are arbitrary. The accuracy of representation of the second moments of the distribution function for 1000 points is better than 1%.

INTRODUCTION

Many particle-in-cell beam simulations in linear accelerators require multidimensional phase space distribution generators to provide for elliptical projections of the beam on phase planes $(x,\dot{x})$, $(y,\dot{y})$ with arbitrary tilts of ellipses, semiaxes ratio and areas. A study of RF ribbon beam accelerator[1] to increase the limited linac beam current under fixed space charge density of the beam is an example to that. The traditional measurements of beam emittance include slit-slit method which gives projection of the distribution on phase plane. Generation of particle distribution for PIC simulation is an inverse problem: we have to reestablish the position of the particles in multidimensional phase space using information about 2D projections of this distribution.

PHASE SPACE DISTRIBUTIONS

Consider the distributions in 4D phase space[2] $x,\dot{x},y,\dot{y}$ in which the distribution function $f(x,\dot{x},y,\dot{y}) = dN/(dx\,d\dot{x}\,dy\,d\dot{y})$ depends on parameter F:

$$f(x,\dot{x},y,\dot{y}) = f(F)$$

$$F = A_x^2 + \varepsilon A_y^2 \qquad (1)$$

where[3]

$$A_x^2 = (a_x\dot{x} - \mathring{a}_x x)^2 + \left(\frac{x}{a_x}\right)^2$$

$$A_y^2 = (a_y\dot{y} - \mathring{a}_y y)^2 + \left(\frac{y}{a_y}\right)^2 \qquad (2)$$

The equation F = const describes a hyperellipsoid surface in phase space $x,\dot{x},y,\dot{y}$. It follows from eq.(1) that phase space density is constant on a hyperellipsoid surface and varies from one surface to another. The distribution function is normalized under the following condition:

$$\int\int\int\int_{-\infty}^{+\infty} f(x,\dot{x},y,\dot{y})\, dx\, d\dot{x}\, dy\, d\dot{y} = 1 \tag{3}$$

The projections of distributions defined by eq.(1) on phase planes $(x,\dot{x})$, $(y,\dot{y})$ are ellipses described by eq.(2) with different areas:

$$A_x^2 = F$$
$$A_y^2 = \frac{F}{\varepsilon} \tag{4}$$

Beams with the same root-mean-square (RMS) values [4,5] are considered for different phase space distributions

$$\langle x^2 \rangle = \int\int\int\int_{-\infty}^{+\infty} x^2\, f(x,\dot{x},y,\dot{y})\, dx\, d\dot{x}\, dy\, d\dot{y} \tag{5}$$

with analogous relations for $\langle \dot{x}^2 \rangle$, $\langle x\dot{x} \rangle$, $\langle y^2 \rangle$, etc. The projections of such beams on phase plane $(x,\dot{x})$ can be described by RMS ellipse (see fig. 1):

$$\left(\frac{4\langle \dot{x}^2 \rangle}{E_{RMS}^{(x)}}\right)x^2 - 2\left(\frac{4\langle x\dot{x} \rangle}{E_{RMS}^{(x)}}\right)x\dot{x} + \left(\frac{4\langle x^2 \rangle}{E_{RMS}^{(x)}}\right)\dot{x}^2 = E_{RMS}^{(x)} \tag{6}$$

with analogous expression for the plane $(y,\dot{y})$. The area of the RMS ellipse is proportional to RMS beam emittance

$$E_{RMS}^{(x)} = 4\left(\langle x^2 \rangle \langle \dot{x}^2 \rangle - \langle x\dot{x} \rangle^2\right)^{\frac{1}{2}} \tag{7}$$

Coefficients a_x, $\dot{a}_x$, a_y, $\dot{a}_y$ are defined by the values of semiaxes X1, X2, Y1, Y2 and tilts of the beam ellipses on phase planes [3]:

$$a_x = \left(\frac{X1}{X2}\cos^2\alpha_x + \frac{X2}{X1}\sin^2\alpha_x\right)^{\frac{1}{2}}$$

$$\dot{a}_x = \frac{1}{2a_x}\left(\frac{X1}{X2} - \frac{X2}{X1}\right)\sin 2\alpha_x \tag{8}$$

with analogous expressions for a_y, $\dot{a}_y$.

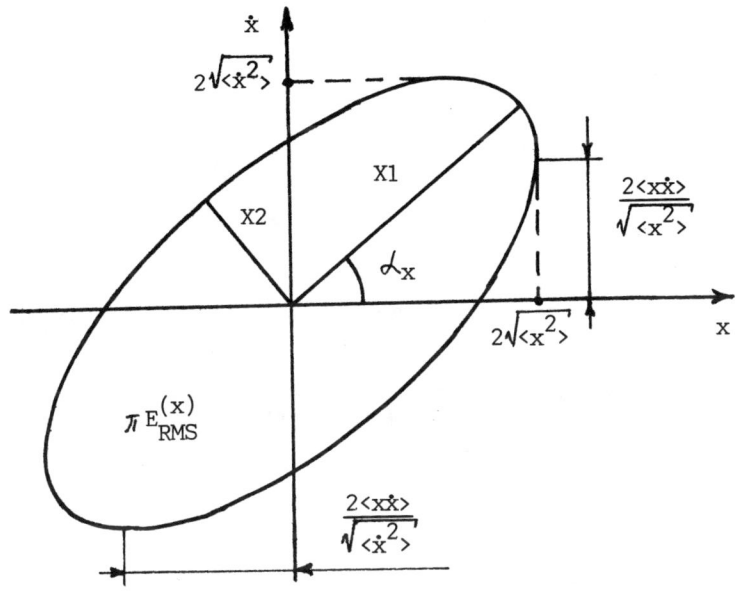

Fig. 1. The RMS ellipse for the $(x,\dot{x})$ plane.

NUMERICAL ALGORITHM

The problem of particle phase space distribution generation is to find 4N values of x_i, $\dot{x}_i$, y_i, $\dot{y}_i$, $i = 1,2,\ldots N$ which correspond to given function $f(\bar{F})$ and given RMS ellipses on the phase planes $(x,\dot{x})$, $(y,\dot{y})$. The distribution of values F

$$g(F) = \frac{dN(F)}{dF} \tag{9}$$

is necessary to implement the algorithm. Let us substitute new variables F, θ, φ, ψ ($0 \leq \theta \leq 2\pi$, $0 \leq \varphi \leq 2\pi$, $0 \leq \psi \leq \pi/2$) for $x, \dot{x}, y, \dot{y}$:

$$a_x \dot{x} - \dot{a}_x x = F^{\frac{1}{2}} \sin\psi \sin\varphi$$

$$\frac{x}{a_x} = F^{\frac{1}{2}} \sin\psi \cos\varphi \tag{10}$$

$$\varepsilon^{\frac{1}{2}}(a_y \dot{y} - \dot{a}_y y) = F^{\frac{1}{2}} \cos\psi \cos\theta$$

$$\varepsilon^{\frac{1}{2}} \frac{y}{a_y} = F^{\frac{1}{2}} \cos\psi \sin\theta$$

The phase space element is transformed as

$$dx\, d\dot{x}\, dy\, d\dot{y} = \frac{1}{4\varepsilon} F\, dF\, \sin 2\psi\, d\psi\, d\varphi\, d\theta \tag{11}$$

It follows from eq.(11)

$$g(F) = \frac{\pi^2}{\varepsilon} f(F)\, F \tag{12}$$

The distribution generation algorithm is based on the above equations and is divided into three steps.

1. To calculate coefficients a_x, $\dot{a}_x$, a_y, $\dot{a}_y$ according to eq.(8) using given parameters of RMS ellipses on the phase planes $(x,\dot{x})$, $(y,\dot{y})$.

2. To simulate distribution $g(F)$ using inverse function method i.e. to take integral distribution

$$G(F) = \int_0^F g(F')\, dF' \tag{13}$$

to find inverse function $F = F(G)$ under assumption that the values of G are uniformly distributed within the interval $[0,1)$.

3. To take for each value of F two random numbers A_x, A_y which satisfy equation (1) to describe ellipses from eq.(2). On each ellipse an arbitrary point is taken with the following coordinates:

$$\begin{aligned}
x &= A_x\, a_x\, \cos\beta_x \\
\dot{x} &= A_x\left(\dot{a}_x \cos\beta_x - \frac{\sin\beta_x}{a_x}\right) \\
y &= A_y\, a_y\, \cos\beta_y \\
\dot{y} &= A_y\left(\dot{a}_y \cos\beta_y - \frac{\sin\beta_y}{a_y}\right)
\end{aligned} \tag{14}$$

where the values of β_x, β_y are randomly distributed within the intervals

$$\begin{aligned}
0 &\leq \beta_x \leq 2\pi \\
0 &\leq \beta_y \leq 2\pi
\end{aligned} \tag{15}$$

The values obtained from eq.(14) give the point in 4D phase space which belongs to distribution f(F). The steps 2, 3 are repeated N times intil the N points cover the phase space volume.

NUMERICAL RESULTS

The above method was used in SUBROUTINE PDS to simulate distributions which are commonly used in linear accelerator PIC codes[6,7]: microcanonical (KV), "water bag", parabolic, Gaussian. Definitions and some characteristics of such distributions are given in Table 1. The input parameters for the subroutine are the following: type of distribution, number of points, values of semi-axes and tilts of RMS beam ellipses at phase planes $(x,\dot{x})$, $(y,\dot{y})$. As the inverse function method defined by the eq.(13) is too slow for Gaussian distribution simulation the appropriate random number generator described by Forsythe et. al.[8] was used.

The results of simulation of different distributions are presented in fig. 2,3. The accuracy of RMS emittances representation resulted from SUBROUTINE PDS better than 1% is achieved when N = 1000 and more.

Table 1. Characteristics of Different Phase Space Distributions

	Distributions			
	KV	Water Bag	Parabolic	Gaussian
Definition	$\dfrac{\varepsilon}{\pi^2 F_0} \delta(F-F_0)$	$\dfrac{2\varepsilon}{\pi^2 F_0^2}$	$\dfrac{6\varepsilon}{\pi^2 F_0^2}(1-\dfrac{F}{F_0})$	$\dfrac{\varepsilon}{\pi^2 F_0^2}\exp(-\dfrac{F}{F_0})$
Projection $(x,\dot{x})$	$\dfrac{1}{\pi F_0}$	$\dfrac{2}{\pi F_0}(1-\dfrac{A_x^2}{F_0})$	$\dfrac{3}{\pi F_0}(1-\dfrac{A_x^2}{F_0})^2$	$\dfrac{1}{\pi F_0}\exp(-\dfrac{A_x^2}{F_0})$
Projection $(y,\dot{y})$	$\dfrac{\varepsilon}{\pi F_0}$	$\dfrac{2\varepsilon}{\pi F_0}(1-\dfrac{\varepsilon A_y^2}{F_0})$	$\dfrac{3\varepsilon}{\pi F_0}(1-\dfrac{\varepsilon A_y^2}{F_0})^2$	$\dfrac{\varepsilon}{\pi F_0}\exp(-\dfrac{\varepsilon A_y^2}{F_0})$
Eq. (13)	$F=F_0$	$F=F_0 G^{\frac{1}{2}}$	$2F^3-3F_0 F^2 +GF_0^3 = 0$	$(1-G)-(1+\dfrac{F}{F_0})\cdot \exp(-\dfrac{F}{F_0}) = 0$

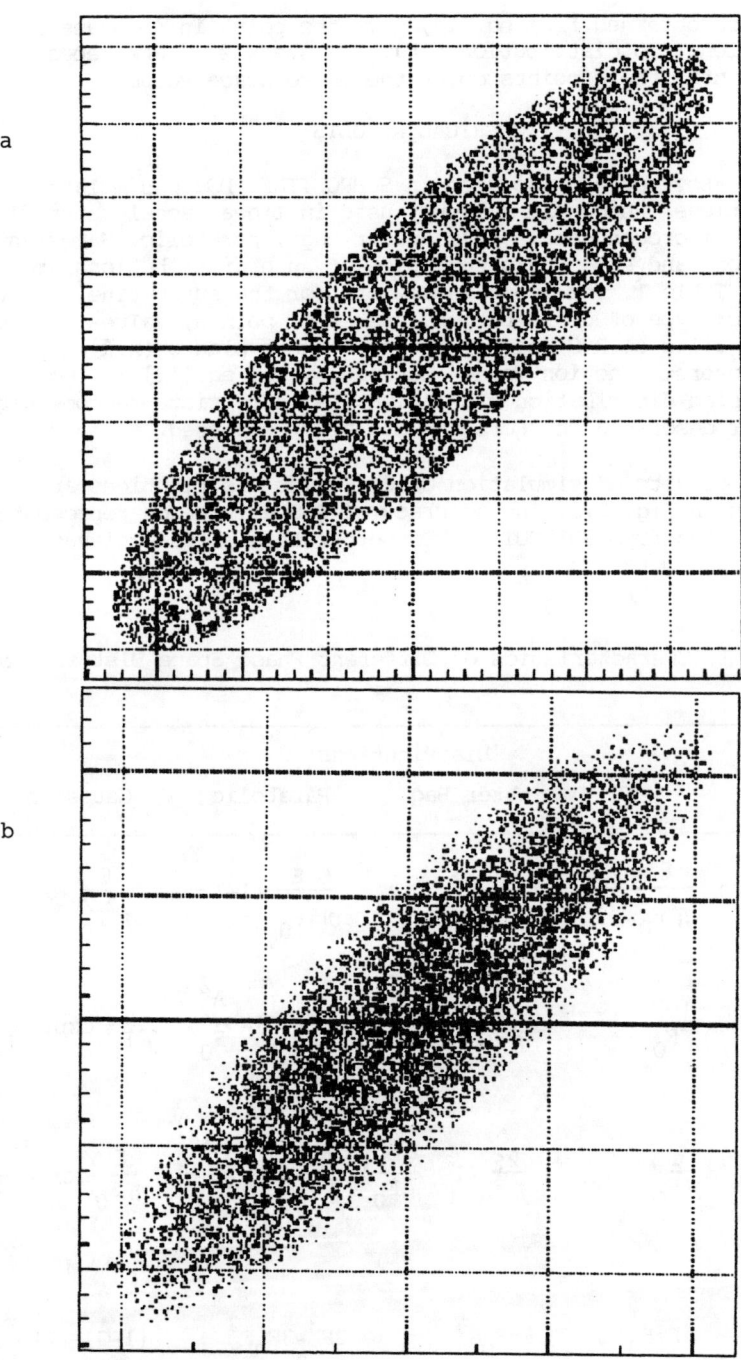

Fig. 2. Projections of 4D phase space distributions: a) KV, b) water bag.

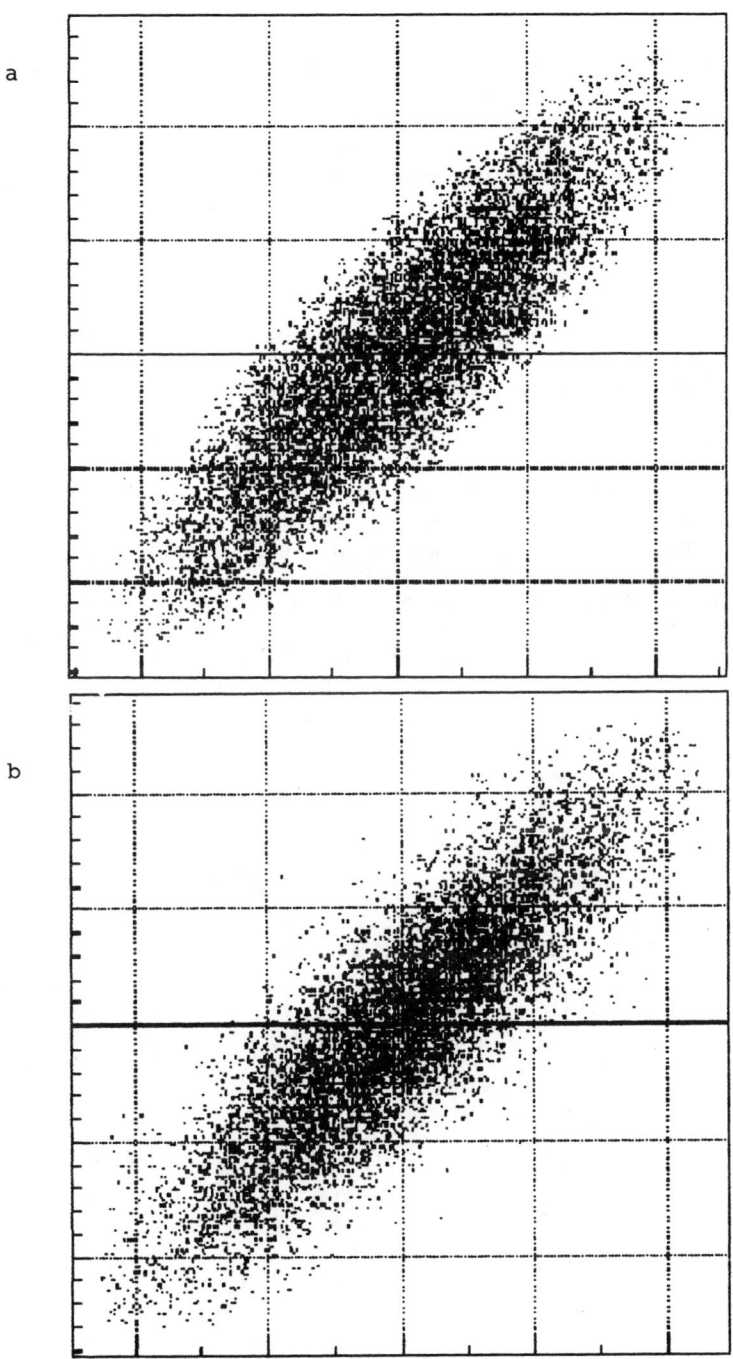

Fig. 3. Projections of 4D phase space distributions: a) parabolic, b) Gaussian.

CONCLUSIONS

The method for particle distributions generation in 4D phase space is described. The numerical results obtained from the method are given. The subprogram PDS developed according to the above method is a part of the program library BEAMPATH[9] for simulation of a wide range of problems connected with linear accelerators design.

REFERENCES

1. V.D.Danilov, A.A.Iliin, Y.K.Batygin, Proc. 3rd Europ. Part. Accel. Conference, Berlin, 1992, p.569.
2. J.Struckmeier, J.Klabunde and M.Reiser, Particle Accelerators, Vol. 15, 1984, p.47.
3. I.M.Kapchinsky, Theory of Linear Resonance Accelerators, Moscow, Energoizdat, 1982.
4. P.M.Lapostolle, IEEE Trans. NS, 18, 1971, p.1101.
5. F.J.Sacherer, IEEE Trans. NS, 18, 1971, p.1105.
6. G.P.Boicourt, Proc. Workshop on linear accelerator and beam optics codes, San Diego, 1988, p.1.
7. K.R.Crandall and T.P.Wangler, ibid., p.22.
8. G.E.Forsythe, M.A.Malcolm and C.B.Moler, Computer Methods for Mathematical Computations (Prentice-Hall, N.J., 1977).
9. Y.K.Batygin, Proc. 3rd Europ. Part. Accel. Conference, Berlin, 1992, p.822.

WIDER AVAILABILITY OF PARMILA AND RECENT IMPROVEMENTS TO PARMILA [a]

Jean L. Merson, AT-7 and Lawrence J. Rybarcyk, MP-6
Los Alamos National Laboratory
P.O. Box 1663
Los Alamos, NM 87545

ABSTRACT

PARMILA (Phase And Radial Motion in Ion Linear Accelerators) is a drift-tube linac (DTL) ion-beam dynamics code. Over its long life, many versions have developed. The Los Alamos Accelerator Code Group distributes a version, for which a manual[1] is available. Unless otherwise specified, all mentions of PARMILA in this document refer to that LAACG-distributed version. Until recently, this documented and distributed version functioned only under CTSS. Users who wished to run on a different operating system needed to convert the code themselves. PARMILA now operates under UNICOS, a much more widely available CRAY operating system, and under VAX/VMS. This paper describes some new features of the code, and gives directions for obtaining the manual and the UNICOS and VMS versions of the code.

INTRODUCTION

Reference 1, the PARMILA manual, describes the major features of PARMILA as it is distributed by the LAACG. Unless otherwise specified, all mentions of PARMILA in this document refer to the LAACG-distributed version. This paper emphasizes features that have been added since the first version of that report was published in 1990, and those that are not well documented. The 1992 revision to the manual was minor. The new features include the ability to scale data from SUPERFISH, availability of a second linac generation subroutine, two new output files giving emittances and power requirements, and the ability to handle comments in the input file.

We also describe a post-processor for PARMILA, named NBEAM6.

---

[a] Work supported by U S Department of Energy, Office of Energy Research: Office of High Energy and Nuclear Physics, Office of Basic Energy Sciences, Office of Fusion Energy, Office of Superconducting Super Collider, and Scientific Computing Staff.

SCALED SFDATA INFORMATION

A table of information generated by running the cavity code SUPERFISH[2,3] provides the basis for generating a linac. PARMILA now has the ability to scale information in an SFDATA table based on SUPERFISH analysis of cells having one resonate frequency to a different design frequency[4]. In order to utilize this feature, the user must specify the reference frequency for which the SUPERFISH runs were done, FREQREF. The fifth data element after the LINAC label is set equal to the value of FREQREF, in megahertz. Then PARMILA multiplies each shunt impedance, Z, in the SFDATA table by the square root of the ratio of the design frequency to the reference frequency. (The design frequency is the third data element following the LINAC label, as described in the manual.) If no value for FREQREF is given, PARMILA assumes it to be equal to the design frequency. This scaling is done in the main program, and is independent of the linac generation routine selected for use (see next section).

CHOICE OF ROUTINE TO GENERATE LINAC

PARMILA now contains two routines that the user can select to generate the linac. The two routines have somewhat different capabilities, and even for the same input data, they generate slightly different linacs. The new routine, which is named GENLAT1 in PARMILA, is similar to the linac-generating routine used in the AT-1 PC version of PARMILA. It is the routine that is called by default when all the cells for which SFDATA information is provided are symmetric.

The old routine, GENLIN2, can deal with asymmetric cells, and is called automatically if any asymmetric cells are represented in the SFDATA input. If the SFDATA is for a symmetric cell, entry number 8 equals zero, and entries 9 and 10 are ignored but must be present for spacing. For an asymmetric cell, entry number 8 is nonzero, and it, along with entries 9 and 10, gives the characteristics of the second half of the cell. See reference 1 for details. GENLIN2 also generates a file of power requirements, described below.

In order to use GENLIN2 when all cells are actually symmetric, an artificial value for SFDATA entry 8 is used. Set entry number 8 in the SFDATA information for at least one cell equal to -9000.0. PARMILA will reset it to zero but will set a flag to cause GENLIN2 to be called.

The new routine, GENLAT1, allows an automatic ramp of phase, either with or without ramping E0[5] . To ramp phase only, set vv(30) [b] following a TANK label to 1, and vv(5) following the TANK label to less than 0.1e10. To ramp both phase and E0, set vv(5) equal to 1.0e10 and VV(30) to 1. For ramp of E0 only, see the manual. (The E0 only ramp capabilities are similar in GENLAT1 and GENLIN2.) A 10-cm delay in the phase ramp is programmed into GENLAT1. It can be changed by changing the numeric value in one "if" statement that checks

---

[b]vv(n) is the nth data element following its associated label.

"(gnlen .le. 10.00)" and recompiling. Similarly, the phase limit can be changed by changing the value assigned to SPMINV in the routine and recompiling. This assignment occurs immediately after the comment "generate linac".

NEW OUTPUT FILES

Two additional output files not documented in the manual are now available. The file EMITT contains 100%, 90%, and rms emittances, as well as α and β in the x-xprime, y-yprime and phase-energy phase space planes at each cell, and the number of particles that have not been lost to that point. The file is written by subroutine EMIT.

POWFILE contains the power requirement in megawatts for each cell and the cumulative power requirement to that cell in the tank. It is written by subroutine GENLIN2.

COMMENTS

Data on a line that starts with a COMMENT label will now be ignored. This permits comments to be included in input files.

POST-PROCESSOR

A post-processor named NBEAM6 has been locally available for use with the version of PARMILA but has not been documented. It is now also available for use with UNICOS. It provides graphical and tabular descriptions of a beam from PARMILA, taking the beam particle coordinates from a file named by the user. Histogram plots of the beam particle distributions in x, xprime, y, yprime, phase, and energy are available, either with or without Gaussian curves fitted to the distributions. Figure 1 shows an example histogram plot with fitted Gaussian.

Scatter plots in each of the phase-space planes are available. An example is given in figure 2. Tabular descriptions of either a set of nine percentages of the beam (a "full range analysis") or of a single percentage of the beam specified by the user are presented in the output file named FORT.4. The values of the step, mean, standard deviation, and third moment associated with the fitted Gaussian curves are given in a file named OUTPUT. The program makes use of the proprietary graphics software, DISSPLA.

Use of the program is interactive, and the user is prompted for input. An example interactive session follows.

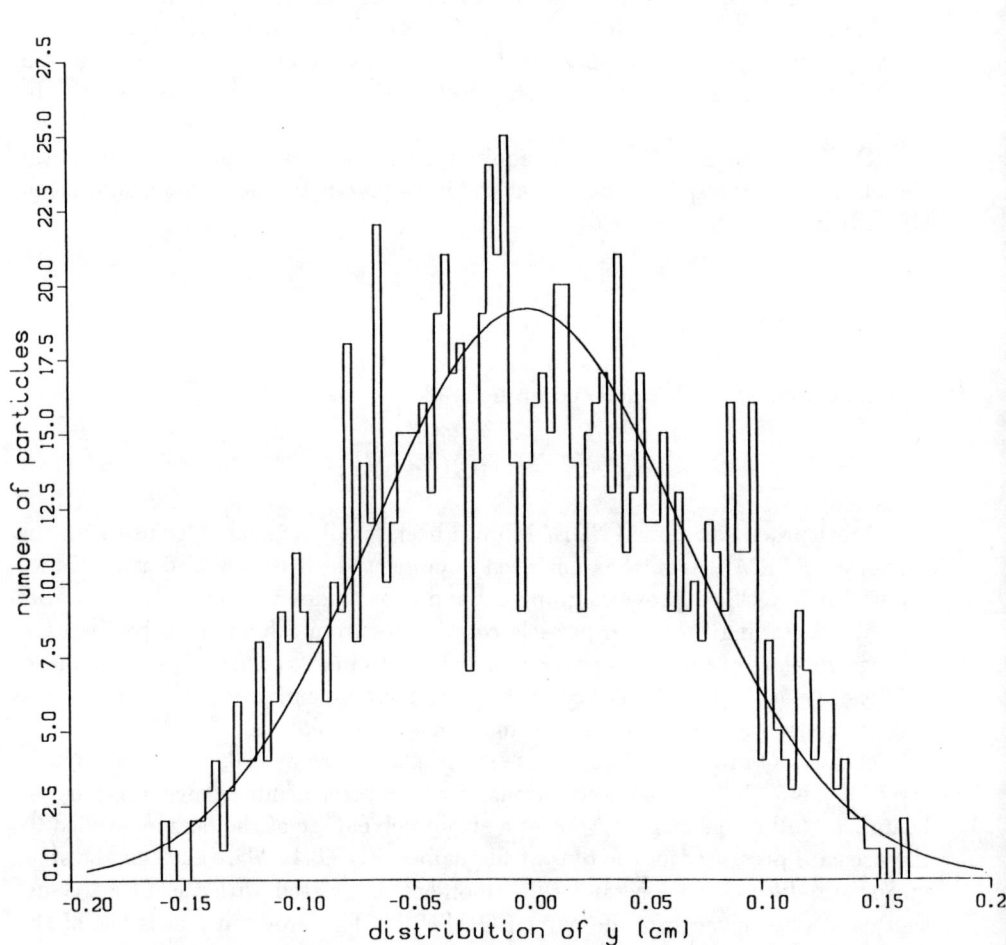

Fig. 1. Histogram plot of particle distribution in y, with fitted Gaussian.

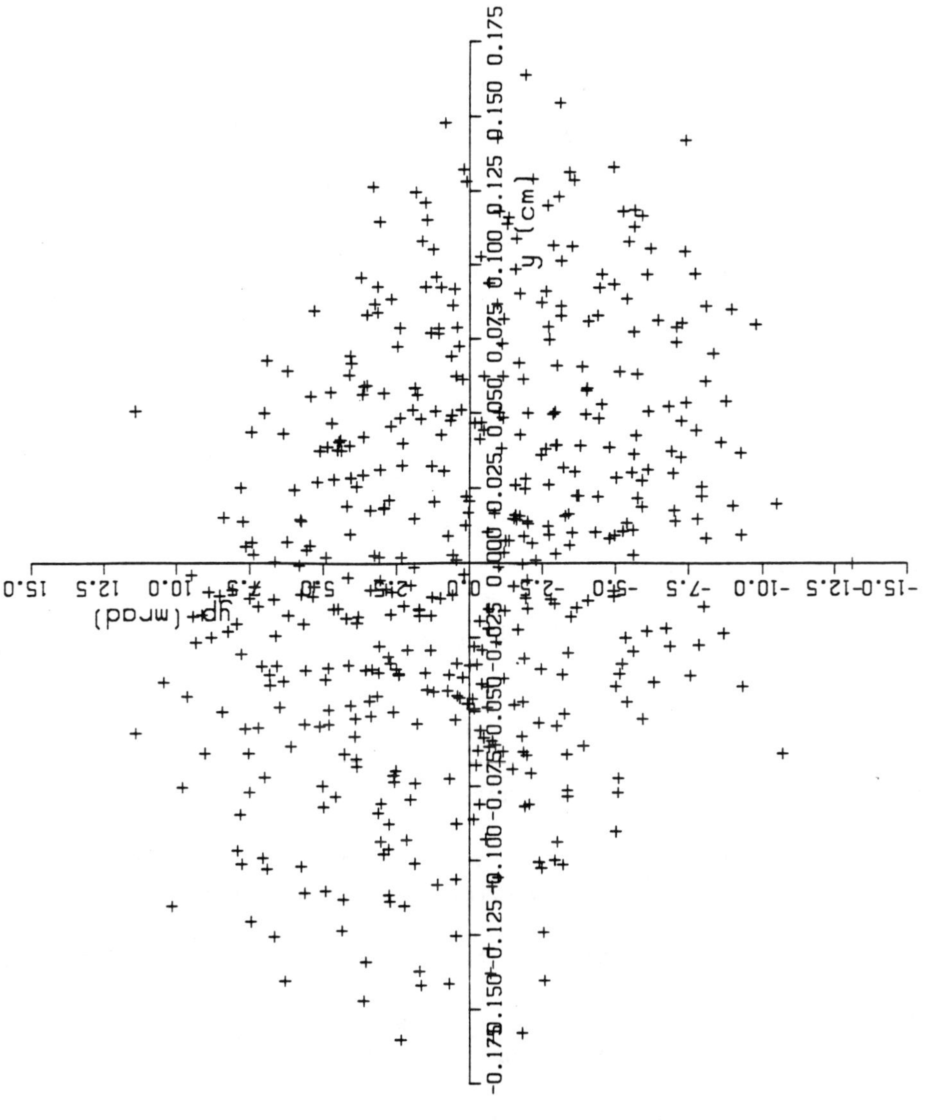

Fig. 2. Phase-space distribution in y vs. y-prime.

```
  give coordinate input file name.
tape25
  what is dimension of cord on input file?
6
  give first four letters of type of particle.
try <prot>,<deut>,or <hmin> or <stop> to exit code.
hmin
  want to use synch energy-synch phase <y> <cr>?
  if <cr>, will use wbar-pbar calculated in this code.
y

    read from tape5      synch energy       5.0504 mev
                         synch phase       -0.6981 radians

do you want a full range analysis? <y> or <cr>
y

parmila short input file

  do you want plot output ? <y> or <cr>
y

do you want histograms plotted? <y> or <cr>
y
  do you want plots of fitted gaussian distribution? <y> or <cr>
y
to change distribution limits enter any or all xl,xpl,yl,ypl,phimin,phimax
wlow,whigh,npar. namelist format. namelist name=limits
  $limits$
  do you want sample emittance plots? <y> or <cr>)
y
to change plot limits enter any or all xmx,xpmx,ymx,ypmx,wmx,pmx
use namelist format. namelist name is maxs
  $maxs$
  plot    done.  pages =   11.  words =     10517
  graphics cl = u
END OF DISSPLA 11.0 9003, DRIVERS 9003 -- 20519 VECTORS IN 10
PLOTS.
RUN ON 1/12/93 USING SERIAL NUMBER 2545 AT LOS ALAMOS NATIONAL
LABORATORY
PROPRIETARY SOFTWARE PRODUCT OF COMPUTER ASSOCIATES, INC.
9763 VIRTUAL STORAGE REFERENCES; 7 READS; 0 WRITES.
```

CONCLUSIONS

The manual, the UNICOS version of PARMILA, or the VAX/VMS version can be obtained by sending a request to The Los Alamos Accelerator Code Group by one of the means given below.

Mail request to:
>Los Alamos Accelerator Code Group (LAACG)
>Mail Stop H825
>Los Alamos National Laboratory
>Los Alamos, New Mexico 87545
>USA

Phone:
>(505) 667-9131

E-Mail:
>laacg@lanl.gov

Please include the following information with your request: code and/or documentation requested, your name, organization, address, phone number, fax number, and E-mail address, and the computer and operating system on which you intend to use the program.

The LAACG is restricted in its dealings with persons from countries on the U.S. Department of Energy's sensitive countries list. The code group can send such persons documentation, but they must request software from the Energy Science and Technology Software Center (ESTSC). For those who must request software from the ESTSC, the center can be reached as follows:

>Phone: (615) 576-2606
>Fax: (615) 576-2865
>Mail: Energy Science and Technology Software Center
>P.O. Box 1020
>Oak Ridge, TN 37831
>USA

The Los Alamos Accelerator Code Group is able to send the manual to requestors in sensitive countries.

References

[1] G. Boicourt and J. Merson, "PARMILA Users and Reference Manual," Los Alamos National Laboratory report LA-UR-90-127 (January 10, 1990, revised September 25, 1992).

[2] M.T. Menzel and H.K. Stokes, "User's Guide for the POISSON/SUPERFISH Group of Codes," Los Alamos National Laboratory report LA-UR-87-115 (January 1987).

[3] Los Alamos Accelerator Code Group,"POISSON/SUPERFISH Reference Manual," Los Alamos National Laboratory report LA-UR-87-126 (January 1987)

[4] Thomas P. Wangler, private communication, October 1991

[5] George H. Neuschaefer, private communication, December 21, 1992

ADLIB - A Simple Database Framework for Beamline Codes[*]

C. Thomas Mottershead
AT-3, MS H808, Los Alamos National Laboratory,
Los Alamos, NM 87545
(505) 667-9730
email: motters@atdiv.lanl.gov

Summary

There are many well developed codes available for beamline design and analysis. A significant fraction of each of these codes is devoted to processing its own unique input language for describing the problem. None of these large, complex, and powerful codes does everything. Adding a new bit of specialized physics can be a difficult task whose successful completion makes the code even larger and more complex. This paper describes an attempt to move in the opposite direction, toward a family of small, simple, single purpose physics and utility modules, linked by an open, portable, public domain database framework. These small specialized physics codes begin with the beamline parameters already loaded in the database, and accessible via the handful of subroutines that constitute ADLIB. Such codes are easier to write, and inherently organized in a manner suitable for incorporation in model based control system algorithms. Examples include programs for analyzing beamline misalignment sensitivities, for simulating and fitting beam steering data, and for translating among MARYLIE, TRANSPORT, and TRACE3D formats.

The Problem With Big Codes

There are many independent and well developed beam optics codes, which typically are large, self-contained, batch-oriented programs with uniquely formatted input and output files. The input languages vary from simple and rather limited to very complicated and powerful. The input parser is the code that reads this input file, and stores the commands and machine parameters in some kind of data structure, usually FORTRAN common blocks, where they can be accessed by the computational routines. Depending on its power and complexity, the input parser itself can be a substantial piece of code, sometimes a significant fraction of the whole code.

A substantial investment of time is required to develop the expertise necessary to use any of these codes effectively. But none of them have all the desired capabilities: some do space charge to first order, some do high order without space charge, some track particles, while others deal with misalignments, etc. The complete design process often requires all of these capabilities and more. We therefore often need to either add a missing capability, or move the problem to another code that already has it. Error studies can be particularly tedious, requiring many runs of the code, interleaved with minor editing of the input file and tabulation of results from the output file. The process cries out for automation, but modification of these big codes is difficult, often even for the author, with the risk of damaging existing features in the effort to introduce new ones. The desired new functionality may even be impossible, if it clashes too much with the internal structure of the code. If the major code does not allow private USER routines, successful modification results in a private version with features that are wiped out by the next official distribution.

---

[*] Work supported and funded by the US Department of Defense, Army Strategic Defense Command, under the auspices of the US Department of Energy

The Virtues of Small Codes

The alternative is to write small, simple, single purpose beamline physics and utility programs to fill in the gaps between the major optics codes. The big virtue of such small codes is that they may be understood whole, without the burden of complex, but extraneous, functionality. They are easier to write, and may be readily cloned and modified to create a new small code that serves another special simple purpose, but shares the same data structures, and perhaps many subroutines.

This paper describes an attempt to move in this direction, by providing a prototype open, portable, public domain database framework that facilitates the writing of such small codes. The database mechanism allows coding to begin with the machine parameters already available in memory, and provides the common link that allows the power of the whole family of these small utility codes to grow with each new module in the inventory. Examples from the current inventory are discussed below.

Internal Data Structure

The database itself is the backbone of this entire development. Its implementation must be simple, practical, portable, flexible, and efficient, with well-defined units and variables, capable of describing a general beamline and a general beam. Generality means to be inclusive enough to support all the commonly used formulations of accelerator physics. That is, while the physics modules themselves may use very different techniques, the machine and beam parameters they need should all be present, at least implicitly, in the common data structure. Of course it is difficult to anticipate every future need. So the formulation reported here is to be regarded as experimental, and is expected to evolve in the course of working on actual problems, hopefully toward the most convenient and generally useful definitions. A set of database service routines is provided to make data access easy, and to isolate the implementation details of the database from the application codes that use it. This facilitates upgrades, and enhances portability, because only the service routines would need to be rewritten to implement this database in other languages and operating systems.

Modularization

This modularization of the physics codes, while valuable in its own right, also produces code that is inherently organized in a manner suitable for incorporation in a scheme of the sort illustrated in Figure 1., which shows how optics codes could contribute to, and benefit from, a larger design, simulation, and operational control system for a real machine. The database mechanism allows separation of the physics part of the optics codes from both their inputs and their outputs. The optics modules, which may in general be thought of as transforming the initial beam model Bi into the final beam model Bf, draw their parameters from, and store their results in, the database.

Modern User Interfaces

Other modules put parameters into, and display results out of, the database. A modern graphic user interface, of the kind being written these days to allow the use of workstations as control system consoles, is one way to put parameters into the database. These use multiple windows, icons, pointing devices, colors, and interactive graphics to display and reset machine parameters via a set of database access routines. The control system in turn, runs the actual hardware from the contents of the data base, shown in the diagram as the link to the "real machine". Some measures of the resultant changes in the "real beam" are then reported by the diagnostics, fit to some beam model and shown to the operator by the beam display module. The interactive design interface shown in the figure is similar to the operator console, but programmed more flexibly to allow the machine configuration to change. Some previous experiments in linking accelerator codes into fancy user interfaces (or even AI systems)[1] have been done by generating the input file, running a large code as a black box, and parsing the output file to extract parameters for the next step. Small beamline physics codes written in the database style to begin with are much more suited to this purpose.

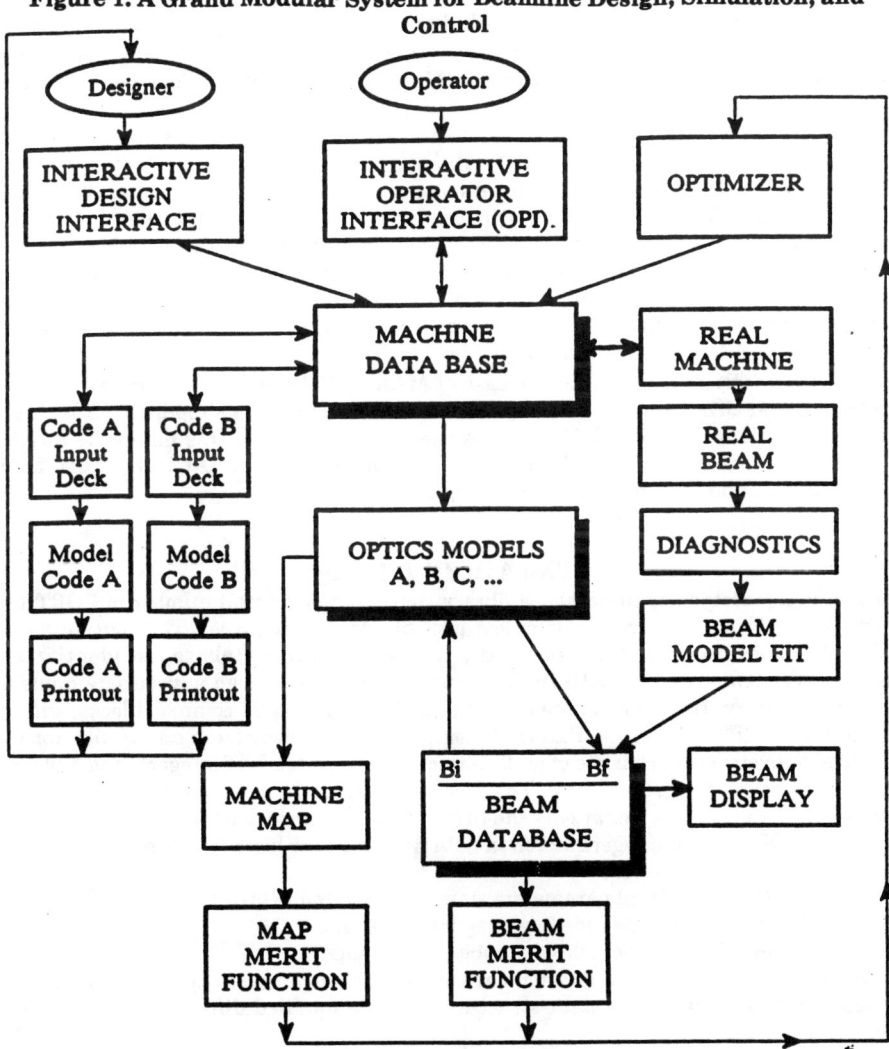

Figure 1: A Grand Modular System for Beamline Design, Simulation, and Control

External Optimizers

The second way to change the machine parameters is via the optimizer module, which contains subroutines that recommend new values for selected machine parameters, deduced from the empirical behaviour of selected merit functions, whose values are received as input. The new machine parameters are then loaded into the database for use in computing the next iteration of the merit functions. Most available optimizers and equation solvers are written to be in control, calling the function(s) to be solved or minimized as often as they please. The emphasis is on maximizing their internal efficiency. For our purposes, this is backwards: a good optimizer or fitter should be callable, leaving the external program in control, and extremely frugal with function calls, because these can each be a very expensive model computation, or even an experiment. We have written three such routines: AMDII [2] is an efficient non-linear system solver using a new adaptive multidimensional inverse interpolation algorithm; NLS is a similar non-linear least squares routine; and QSO [3] is a new quadratic step optimizer that uses quadratic fitting of the Hessian to the best previous points to speed convergence.

Translators

The third way to load machine parameters into the database is from an external file such as the input deck for your favorite major code, shown in the figure as "code A". The input parser from code A could be removed surgically, and used to initialize the database. Another program must then be written to write the current contents of the database back into the format of code A's input deck. As other operational codes receive this treatment, the set of programs to map their various input file formats to and from the database will, in itself, constitute a valuable translation facility. The separation of the input parser may well be a non-trivial exercise. In the case of MARYLIE,[4] it has so far proven easier to write a special USER routine to dump the current beamline to a simplified ASCII file. A short stand alone program (DMPIN) was then written to read this dump file into the database. At this point, code for translation between TRACE3D, MARYLIE, TRANSPORT, and TeX tables is partially operational.

The ADLIB Library

ADLIB is a prototype open database library consisting of a small number of FORTRAN routines to store and fetch the simplest possible data structures. The routine names (almost) all begin with the prefix AD, for Accelerator Database, to identify and distinguish them. The term ADLIB, meaning "to improvise", is an appropriate name for the whole library. Internal storage is in standard FORTRAN 77 common blocks, without computer specific extensions. External storage is in a direct access binary file, for fast I/O. The data structure consists of an 8 character name, a block of integers, and a block of double precision real parameters. The block of integers is meant to hold various type codes, flags, and indices. So far only the first four integers are defined in general:
  1. NI = Number of integers stored in data item. NI is at least 4, but may be as large as necessary.
  2. NR = Number of real parameters stored in data item. Unrestricted.
  3. MAJOR = Major type code (e.g. magnetic element).
  4. MINOR = Minor type code (e.g. subtype in MAJOR).

Beyond this, ADLIB itself doesn't care what the parameters mean, and in other languages and systems, the data structure could be handled differently. The routines defined so far are:

ADINIT: This subroutine initializes (clears) the internal common blocks to zero or blank, for fresh starts.

ADREAD(lun,fname,info): This subroutine uses logical unit **lun** to load the internal commons from the direct access binary disk file named **fname**. The integer **info** will be an error message. Any code that begins with a call to **adread** will have access to the parameters stored in the specified copy of the database.

ADSAVE(lun,fname,info): This subroutine reverses the process, using **lun** to write the contents of the internal commons back to binary file **fname.** Any code that ends with a call to **adsave** will store its results in the specified copy of the database.

ADNEW(ni,nr,iib,kxb,nuslot): This subroutine creates a new storage record in the database containing room for **ni** integers and **nr** real parameters. The slot number assigned to the new item is **nuslot.** This is normally the next available location in the database, but if the database is full, it is returned negative as an error message. The integer block offset **iib,** and real block offset **kxb** are returned only for the convenience of users who want to manipulate the commons directly, without the overhead of a subroutine call. This is an example of the open, permissive, philosophy that aims to put the user in complete control of his own code.

ADPUT(nslot,cname,iblok,rblok,info): This subroutine stores the complete record consisting of the character*8 name **cname,** the block of integers **(iblok(j),j=1,ni),** and the block of double precision reals **(rblok(j),j=1,nr)** into database slot **nslot.** The error flag **info** is returned to indicate the success of the operation.

ADGET(nslot,cname,iblok,rblok,info): This subroutine is the inverse of **ADPUT.** It retrieves the complete record consisting of **cname,** the **ni** integers in **iblok,** and the **nr** double precision reals in **rblok** from database slot **nslot.**

ADMAX(maxi,maxr) This subroutine allows control of dimension overruns by setting the maximum length of the **iblok** and **rblok** arrays that will be returned on subsequent calls to **ADGET.**

ADPOKE(nslot,nvar,rval,info): This subroutine stores a single real value **rval** into location **nvar** of database slot **nslot.**

ADVAL(nslot,nvar): This double precision function returns the **nvar**th real parameter stored in database slot **nslot.** It is the inverse of **ADPOKE.**

NUMAD(nslot,nvar): This integer function returns the **nvar**-th integer stored in database slot **nslot.** It is useful for rapid scans of the database for a particular integer flag, but has no inverse because the stored integers are generally type codes and dimensions that should not be easy to change.

ADNAME(nslot,cname,info) retrieves only the name '**cname**' of the specified database slot.

ADLIB Utility Programs:
The following programs are available for manipulating the database itself, which is normally stored as a binary file for efficient I/O.

ADOUT writes the database to an ASCII file suitable for human reading and modification with a text editor.

ADIN loads the above human readable ASCII file into a copy of the database.

ADUMP writes the database to a concise and complete ASCII file which, if not particularly human readable, is at least suitable for long term storage and email.

ADEDIT is still under development, but will allow direct editing of the binary database, including adding and deleting beamline elements, and defining indirect parameter linkages representing constraints.

Type Codes

As in all optics codes, the meaning of the parameters is specified by type code flags, and is of course crucial to their use in the physics programs. Here we have allowed both a **major** type code in **iblok(3)**, and a **minor** type code in **iblok(4)**. But since the application programs are small and specialized, they do not necessarily need to cope with the general case. Ease of use is paramount. A program that does not use a particular data type can be written to simply ignore it. Control of this rests with the particular program, not the database library. A need may be forseen for the the following major type codes, for which tentative definitions are under development. New type codes may therefore be defined as needed, hopefully in an orderly and backward compatible manner.

Simple Magnetic Elements: MAJOR=1, MINOR specifies element type:
 drift, marker, bend, quad, sextupole, etc. For compactness and
 simplicity, the real block contains only the aperture, length, and
 strength in MKS units.
Complex Magnetic Elements: MAJOR=2, MINOR specifies the
 maximum order of the normal and skew multipole components
 stored in the real block, which also contains aperture, length, and
 misalignment parameters. Magnet type and fringe field
 specifications are also needed.
RF and Electric Elements: MAJOR=3 Details to be determined
Lines: MAJOR = 4, The integer array is just a list of the slot numbers
 containing elements of the line (Easy to use, but effort to define.)
Beam Moments: MAJOR=5, MINOR specifies the storage mode and
 format, either in the database, or an external file (e.g.
 cname.mom) The usual beam sigma matrices are the second
 moments of the beam distribution. For higher orders, the
 Giorgelli [5] monomial indexing scheme used in both MARYLIE[4] and
 BEDLAM[6] is natural for the database as well.
Beam Particle Sets, MAJOR = 6, represented by arrays of canonical
 MKS particle coordinates, stored internally, or in external ASCII
 or binary files.

ADLIB Physics Applications:

About 20 ADLIB application programs have been written to date, mostly by modifying a predecessor program, sometimes extensively. The point is not so much what particular little job these small codes do. It is that they are small, were written relatively quickly, and could be readily cloned and modified to do something else. Thus the capability of the entire family is expected to grow, filling in the gaps between the major design codes, and perhaps eventually becoming a serious contender themselves. The following are some representative examples:

DMPIN loads a MARYLIE beamline into the database using a simple ASCII dump file written by a MARYLIE user routine.

MATRIX interactively displays the R-matrix between any two selected points of the beamline in the database.

ADVIBS does a vibration sensitivity analysis of the entire beamline by computing the final beam steering angle induced by misalignment of each quad in the line. The sensitivity coefficients are written in a TeX table that may be formatted and printed using TeX, without further editing.

SIMS simulates beam steering due to quadupole rolls and displacements, and reports expected beam position monitor readouts at all indicated locations in the beamline. All quadrupole gradients and misalignments, along with input beam centroids, can be interactively changed.

STEX writes a ready to print TeX table containing steering sensitivity coefficients, at all the diagnostics, to movements of two steering quadrupoles in a particular beamline.

GENFIT uses the non-linear system solver AMDII to adjust selected database parameters to fit measure steering data for the above system.

Conclusion

None of the major optics codes provides all of the capabilities needed for complete design and simulation. Consequently, there is often the need to translate beamline descriptions between codes, as well as to write small specialized adlib codes to fill in a missing functionality. In addition, there is both an opportunity and a need to work toward a broader synthesis of accelerator design, simulation, and control systems, with prospective benefits on all sides. Full implementation of this grand design is a large undertaking, but plans can be laid so that small practical steps accumulate in that direction. This paper reports one of the most basic of these steps, namely an experimental prototype of the machine data base.

References

1. R. R. Silbar, ``*An Interactive Interface to the Beam Optics Code TRANSPORT,*'' in Linear Accelerator and Neam Optics Codes, AIP Conference Proceedings No. 177, C.R. Eminhizer, ed, 1988

2. L. Schweitzer and T. Mottershead, *A Multidimensional Inverse Interpolation Algorithm for Accelerator Design,*, Los Alamos National Laboratory technical note AT-6:ATN-86-28, August 1986.

3. H. K. Overley and C. T. Mottershead, *A Quadratic Form Fitting Algorithm for Unconstrained Minimization,* Los Alamos National Laboratory technical note, AT-3:TN-87-26, August 1987.

4. A. J. Dragt, et.al, *MARYLIE, A Program for Charged-Particle Beam Transport Based on Lie Algebraic Methods.* Univ of Maryland Technical Report, Mar 1991.

5. Georgelli, Comp. Phys. Comm. 16, 331, 1979

6..W. P Lysenko, "*Status of the BEDLAM Optics Code*", this conference

DILUTE: A CODE FOR STUDYING BEAM EVOLUTION UNDER RF NOISE*

H.-J. Shih
Superconducting Super Collider Laboratory, Dallas, TX 75237

J. A. Ellison
University of New Mexico, Albuquerque, NM 87131

W. E. Schiesser
Lehigh University, Bethlehem, PA 18015

ABSTRACT

Longitudinal beam dynamics under rf noise has been modeled by Dôme, Krinsky and Wang using a diffusion-in-action PDE. If the primary interest is the evolution of the beam in action, it is much simpler to integrate the model PDE than to undertake tracking simulations. Here we describe the code that we developed to solve the model PDE using the numerical Method of Lines. Features of the code include (1) computation of the distribution in action for the initial beam from a Gaussian or user-supplied distribution in longitudinal phase space, (2) computation of the diffusion coefficient for white noise or from a user-supplied spectral density for non-white noise, (3) discretization of the model PDE using finite-difference or Galerkin finite-element approximations with a uniform or non-uniform grid, and (4) integration of the system of ODEs in time by the solver RKF45 or a user-supplied ODE solver.

THE DIFFUSION EQUATION

When there is noise in the rf system, the voltage across the gap of the rf cavity at time T is

$$V(T) = V_0(1 + a(T))\sin(2\pi f_{rf}T + \psi(T)) \quad (1)$$

where $a(T)$ is the amplitude noise, $\psi(T)$ the phase noise, V_0 the rf peak voltage, and f_{rf} the rf frequency. The longitudinal dynamics of a single particle in the presence of rf noise is described by the Hamiltonian[1]

$$H(P, \phi, T) = H_0(P, \phi) + H_1(P, \phi, T) \quad (2a)$$

$$H_0(P, \phi) = \frac{P^2}{2} + U(\phi) \quad (2b)$$

---

* Work supported by the Superconducting Super Collider Laboratory which is operated by the Universities Research Association, Inc., for the U.S. Department of Energy under Contract No. DE-AC35-89ER40486.

$$H_1(P, \phi, T) = a(T)U(\phi) + P\dot{\psi}(T) \qquad (2c)$$

where $U(\phi) = \Omega^2(1 - \cos\phi)$, $P = 2\pi h\eta\delta/T_0$, ϕ is the phase deviation from the synchronous phase, Ω the angular frequency of the small amplitude synchrotron oscillation, h the harmonic number, η the slip factor (related to the momentum compaction factor α_c by $\eta = \alpha_c - \frac{1}{\gamma}$ where γ is the relativistic factor), δ the relative momentum deviation from the synchronous momentum p_s, and T_0 the revolution period. Without rf noise, particles move on the closed orbits in the P-ϕ phase space as shown in Fig. 1. Under the perturbation of rf noise, particles gradually move onto bigger closed orbits, eventually cross over the separatrix, and get lost.

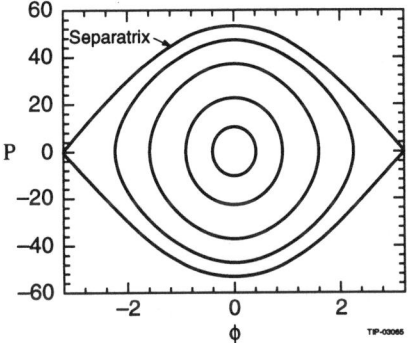

Fig. 1. Unperturbed particle orbits in longitudinal phase space.

If we define 2π times the action J to be the area in the P-ϕ phase space enclosed by the unperturbed closed orbit, i.e.,

$$J = \frac{1}{2\pi} \oint P d\phi \qquad (3)$$

then the effect of rf noise is to induce a diffusion of particles in action. In the Dôme-Krinsky-Wang (DKW) theory,[1,2] this diffusion process is approximated by a Markov process, and the evolution of the probability density in action, $p(J,T)$, is governed by the following diffusion equation:

$$\frac{\partial p}{\partial T} = \frac{\partial}{\partial J}\left(\mathcal{D}(J)\frac{\partial p}{\partial J}\right). \qquad (4)$$

Here $\mathcal{D}(J)$ is the diffusion coefficient determined from the noise spectral density. The solution of the diffusion equation (4) is uniquely defined by specifying

$$p(J_b, T) = 0, \qquad (5)$$
$$p(J, 0) = p_0(J), \qquad (6)$$

where $p_0(J)$ is the initial action density. Equation (5) is an absorbing boundary condition at $J = J_b \lesssim J_s$ where $J_s = 8\Omega/\pi$ is the action at the separatrix. Thus particles are lost once they are diffused to the orbit at J_b near the separatrix. Since $\mathcal{D}(0) = 0$ according to the theory, no boundary condition at $J = 0$ is needed. We have made extensive comparisons between the DKW diffusion theory and tracking simulations, and found good agreement. See Ref. 3 for details.

It is convenient to introduce the non-dimensional variables,

$$x = \frac{J}{J_b}, \qquad t = \frac{T}{T_1}$$

where T_1 is some convenient time scale. In terms of these variables, the probability density in x, $\rho(x,t) = J_b p(J,t)$, satisfies

$$\frac{\partial}{\partial x}\rho(x,t) = \frac{\partial}{\partial x}\left(D(x)\frac{\partial}{\partial x}\rho(x,t)\right), \qquad (7)$$

$$\rho(1,t) = 0, \qquad (8)$$

$$\rho(x,0) = J_b\, p_0(J_b x). \qquad (9)$$

Here $D(x)$ is the scaled diffusion coefficient related to $\mathcal{D}(J)$ by

$$D(x) = \frac{T_1}{J_b^2}\mathcal{D}(J_b x). \qquad (10)$$

The code DILUTE was developed to solve the well posed PDE problem formed by Eqs. (7), (8) and (9). Before we solve it using the Method of Lines, we must compute the diffusion coefficients and the initial action density.

THE DIFFUSION COEFFICIENTS

In general, the diffusion coefficient is given by $\mathcal{D}(J) = \mathcal{D}_a(J) + \mathcal{D}_\psi(J)$ where[1]

$$\mathcal{D}_a(J) = 4\sum_{m=2,4,\ldots}^{\infty}\frac{(m\omega_s)^4}{\sinh^2(mv)}S_a(m\omega_s), \qquad (11)$$

$$\mathcal{D}_\psi(J) = 4\sum_{m=1,3,\ldots}^{\infty}\frac{(m\omega_s)^4}{\cosh^2(mv)}S_\psi(m\omega_s). \qquad (12)$$

Here S_a and S_ψ are the spectral densities for amplitude and phase noise, respectively. The quantities J, $\omega_s(J)$, and $v(J)$ are easily defined through the intermediate variable k, $0 \leq k < 1$, by

$$J = \left(\frac{8\Omega}{\pi}\right)k^2 B(k), \qquad (13)$$

$\omega_s = \Omega(\pi/2K(k))$, and $v = (\pi/2)K(\sqrt{1-k^2})/K(k)$, where K is the complete elliptic integral of the first kind and $B(k) = \int_0^{\pi/2}\cos^2 x\,dx/\sqrt{1-k^2\sin^2 x}$. The variable k is related to the unperturbed energy $h = H_0(p,\phi)$ by $h = 2k^2\Omega^2$. It is also the relative momentum amplitude of the unperturbed orbit with respect to the separatrix, i.e., $k = \delta_{\max}/\delta_h$ where δ_h is the bucket half-height.

For white phase and amplitude noise, the following expressions for $\mathcal{D}$ can be derived:[1]

$$\mathcal{D}_a(J) = \frac{32}{\pi^2}\Omega^4 S_a k^4 K(k) B_1(k), \qquad (14)$$

$$\mathcal{D}_\psi(J) = \frac{8}{\pi^2}\Omega^4 S_\psi k^2 K(k)(B(k) - 4k^2 B_1(k)), \tag{15}$$

where $B_1(k) = \int_0^{\pi/2} \sin^2 x \cos^2 x \sqrt{1 - k^2 \sin^2 x}\, dx$. The functions $B(k)$ and $B_1(k)$ are evaluated by quadrature. As we compute $\mathcal{D}$ and J as functions of k, we spline fit $\mathcal{D}(k)$ vs. $J(k)$ to obtain $\mathcal{D}(J)$.

THE INITIAL ACTION DENSITY

If the initial distribution of the beam in the P-ϕ phase space is bi-Gaussian, i.e.,

$$p(P, \phi) = \frac{1}{\sqrt{2\pi}\sigma_p} e^{-P^2/2\sigma_p^2} \frac{1}{\sqrt{2\pi}\sigma_\phi} e^{-\phi^2/2\sigma_\phi^2}, \tag{16}$$

the initial action density $p_0(J)$ is calculated as follows. Let $p_H(h)$ be the probability density in energy $h = 2k^2\Omega^2$, then

$$\int_0^h p_H(x) dx = 4 \int_0^{\phi_h} \left(\int_0^{\sqrt{2(h-U(\phi))}} p(P, \phi) dP \right) d\phi,$$

where ϕ_h is the maximum phase of the unperturbed orbit, i.e., $U(\phi_h) = h$. Differentiating the equation above with respect to h gives

$$p_H(h) = 4 \int_0^{\phi_h} [p(P_m(\phi), \phi)/P_m(\phi)]\, d\phi, \tag{17}$$

where $P_m(\phi) = \sqrt{2(h - U(\phi))}$. The action density $p_0(J)$ is then related to $p_H(h)$ by

$$p_0(J) = p_H(h)\frac{dh}{dJ} = p_H(h)\omega_s. \tag{18}$$

The standard deviations σ_p and σ_ϕ in Eq. (16) are determined by specifying the longitudinal emittance $\epsilon_L = \sigma_{\Delta E}\sigma_{\Delta t}$ and using the relationship $\sigma_{\Delta t} = (\eta/\Omega\beta^2 E_s)\sigma_{\Delta E}$, where ΔE is the energy deviation from the synchronous energy E_s, $\Delta t = T_0(\phi/2\pi h)$, and $\beta = p_s c/E_s$. The integral in Eq. (17) is evaluated by quadrature. Again, p_H, ω_s and J are computed as functions of k, and the spline fit between $\rho_H \cdot \omega_s$ and J is made to obtain $p_0(J)$.

THE METHOD OF LINES SOLUTIONS

Our PDE problem is solved numerically using the Method of Lines. That is, we 1) set up a uniform or variable grid: $x_1 = 0, x_1, x_2, \ldots x_{N-1}, x_N = 1$, 2) obtain from the PDE an ODE in time at each grid point by some discretization method, and 3) integrate the system of ODEs by an ODE solver. The finite-difference and Galerkin finite-element methods have been used for discretization. Reliability of our numerical solutions has been checked. See Ref. 4 for details.

THE FINITE-DIFFERENCE METHOD

In this method, some finite-difference approximation is employed to evaluate the right-hand side of Eq. (7) (and thus to obtain an ODE in time) at each grid point.

In the case of a uniform grid, the following three-point, second-order approximation is used

$$f_{x,i \neq 1,N} = \frac{1}{2\Delta x}(f_{i+1} - f_{i-1}) \tag{19a}$$

$$f_{x,i=1} = \frac{1}{2\Delta x}(-3f_1 + 4f_2 - f_3) \tag{19b}$$

$$f_{x,i=N} = \frac{1}{2\Delta x}(f_{N-2} - 4f_{N-1} + 3f_N) \tag{19c}$$

in evaluating $\frac{\partial \rho}{\partial x}$ and $\frac{\partial}{\partial x}\left(D\frac{\partial \rho}{\partial x}\right)$. Here $\Delta x = 1/(N-1)$ is the grid spacing.

In the case of a variable grid, the following approximation is used

$$\rho_{t,i \neq 1,N} \cong \frac{2}{\Delta x_i + \Delta x_{i-1}}\left(D_{i+\frac{1}{2}}\frac{\rho_{i+1} - \rho_i}{\Delta x_i} - D_{i-\frac{1}{2}}\frac{\rho_i - \rho_{i-1}}{\Delta x_{i-1}}\right) \tag{20a}$$

$$\rho_{t,i=1} \cong \frac{2}{\Delta x_1}\left(D_{\frac{3}{2}}\frac{\rho_2 - \rho_1}{\Delta x_1} - 0\right) \tag{20b}$$

$$\rho_{t,i=N} = 0 \tag{20c}$$

where $\Delta x_i = x_{i+1} - x_i$ and $D_{i+\frac{1}{2}} = (D_i + D_{i+1})/2$.

THE GALERKIN FINITE-ELEMENT METHOD

We approximate $\rho(x,t)$ by a sum of basis functions $\Phi_i(x)$:

$$\rho_G(x,t) = \sum_{i=1}^{N} C_i(t)\Phi_i(x). \tag{21}$$

The basis functions $\Phi_i(x)$ are piecewise linear with $\Phi_i(x_i) = 1$ and $\Phi_i(x_{i-1}) = \Phi_i(x_{i+1}) = 0$, as shown in Fig. 2. Thus $\rho_G(x_j,t) = C_j(t)$. Since $\rho(1,t) = 0$, $C_N(t) = 0$ and we only need to determine $C_j(t)$, $j = 1, 2, \ldots, N-1$.

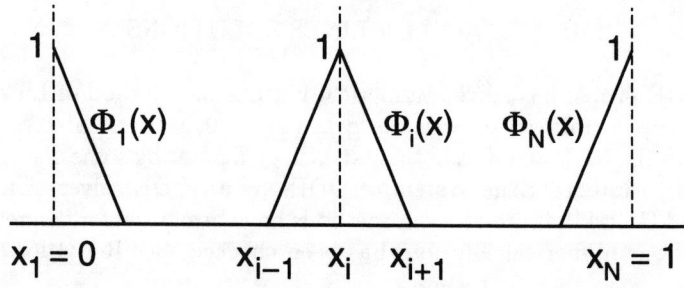

Fig. 2. Basis functions in the Galerkin approximation.

To obtain the conditions on $C_j(t)$, we define the residual functions $R(x,t)$,

$$R(x,t) = \frac{\partial \rho_G}{\partial t} - \frac{\partial}{\partial x}\left(D(x)\frac{\partial \rho_G}{\partial x}\right) \tag{22}$$

and apply the Galerkin condition,

$$\int_0^1 \Phi_j(x)R(x,t)dx = 0, \quad j = 1, 2, \ldots, N-1. \tag{23}$$

The conditions on $C_j(t)$ are given by

$$\mathbf{C}'(t) = (\mathbf{A}^{-1}\mathbf{B})\mathbf{C}(t), \tag{24}$$

where $\mathbf{C}(t)$ is the column vector $(C_1(t), C_2(t), \ldots, C_{N-1}(t))$ and the $N-1$ by $N-1$ matrices $\mathbf{A}$ and $\mathbf{B}$ are

$$\mathbf{A} = \begin{pmatrix} \frac{\Delta x_1}{3} & \frac{\Delta x_1}{6} & 0 & 0 & \cdots & 0 \\ \frac{\Delta x_1}{6} & \frac{\Delta x_1+\Delta x_2}{3} & \frac{\Delta x_2}{6} & 0 & \cdots & 0 \\ 0 & \frac{\Delta x_2}{6} & \frac{\Delta x_2+\Delta x_3}{3} & \frac{\Delta x_3}{6} & \cdots & 0 \\ \vdots & & & & & \vdots \\ 0 & \cdots & 0 & \frac{\Delta x_{N-3}}{6} & \frac{\Delta x_{N-3}+\Delta x_{N-2}}{3} & \frac{\Delta x_{N-2}}{6} \\ 0 & \cdots & 0 & 0 & \frac{\Delta x_{N-2}}{6} & \frac{\Delta x_{N-2}+\Delta x_{N-1}}{3} \end{pmatrix}$$

$$\tag{25}$$

$$\mathbf{B} = \begin{pmatrix} -\frac{d_1}{\Delta x_1^2} & \frac{d_1}{\Delta x_1^2} & 0 & 0 & \cdots & 0 \\ \frac{d_1}{\Delta x_1^2} & -\frac{d_1}{\Delta x_1^2} - \frac{d_2}{\Delta x_2^2} & \frac{d_2}{\Delta x_2^2} & 0 & \cdots & 0 \\ 0 & \frac{d_2}{\Delta x_2^2} & -\frac{d_2}{\Delta x_2^2} - \frac{d_3}{\Delta x_3^2} & \frac{d_3}{\Delta x_3^2} & \cdots & 0 \\ \vdots & & & & & \vdots \\ 0 & \cdots & 0 & \frac{d_{N-3}}{\Delta x_{N-3}^2} & -\frac{d_{N-3}}{\Delta x_{N-3}^2} - \frac{d_{N-2}}{\Delta x_{N-2}^2} & \frac{d_{N-2}}{\Delta x_{N-2}^2} \\ 0 & \cdots & 0 & 0 & \frac{d_{N-2}}{\Delta x_{N-2}^2} & -\frac{d_{N-2}}{\Delta x_{N-2}^2} - \frac{d_{N-1}}{\Delta x_{N-1}^2} \end{pmatrix}$$

$$\tag{26}$$

Here $d_i = \int_{x_i}^{x_{i+1}} D(x)dx$. See Ref. 4 for the derivation of Eq. (24) for the case of a uniform grid. The extension to the case of a variable grid is straightforward.

INTEGRATION

The resulting system of ODEs in time is integrated by RKF45, a non-stiff ODE solver written by H.A. Watts and L.F. Shampine.[5] However, when the diffusion coefficient or the action density varies rapidly, the integration with RKF45 can be very slow. In this case, a stiff ODE solver, for example, LSODES written by A.C. Hindmarsh,[6] is recommended to speed up the integration.

THE CODE USAGE

The code is written in Fortran 77. It makes use of the IMSL library for evaluating some special functions and integration by quadrature.

INPUT PARAMETERS

The input parameters are defined in the subroutine INIT. They are listed below in four categories:

1) Beam/machine parameters:
   - CS — C_s, the machine circumference (m).
   - ALC — α_c, the momentum compaction factor.
   - EV0 — V_0, the rf peak voltage (GeV).
   - P0 — E_s, the total energy of synchronous particle (GeV).
   - HN0 — h, the harmonic number.
   - AM — m_s, the rest energy of synchronous particle (GeV).
   - EPSL — ϵ_L, the longitudinal rms emittance (GeV-sec).
   - AKB — k_b, the critical boundary in k. It is defined by $J_b = J_s k_b^2 B(k_b)$.
   - IB — flag for beam type: 0 for a Gaussian beam and 1 for a user-supplied beam.

2) RF noise parameters:
   - IW — flag for noise type: 0 for non-white noise and 1 for white noise.
   - IF — flag for including phase noise: 0 for no and 1 for yes.
   - IA — flag for including amplitude noise: 0 for no and 1 for yes.
   - T1 — T_1, the convenient time scale (sec).
   - SPHI — spectral density for white phase noise to be specified when IW = 1 and IF = 1.
   - SA — spectral density for white amplitude noise to be specified when IW = 1 and IA = 1.

3) Discretization parameters:
   - N — total number of grid points.
   - IV — flag for grid type: 0 for uniform grid and 1 for variable grid.
   - ID — flag for discretization method: 0 for finite-difference method and 1 for Galerkin finite-element method.

4) Integration parameters:
   - T0 — initial time of integration.
   - TF — final time of integration.
   - TP — integration time step.
   - IS — flag for integrator type: 0 for the integrator RKF45 and 1 for user-supplied integrator.
   - ABSERR — absolute error tolerance in integration.
   - RELERR — relative error tolerance in integration.

USER-SUPPLIED FUNCTIONS OR SUBROUTINES

In addition to the subroutine INIT, the following functions or subroutines must be supplied when the indicated flag is selected:

1) Subroutine BEAM:
   supplied when IB = 1. Here the user defines the longitudinal variables δ and l for up to 5000 particles, where l is the longitudinal displacement from the center of the bunch (related to ϕ by $\phi = 2\pi h l/C_s$).
2) Function SPECA:
   supplied when IW = 0 and IA = 1. Here the user defines the spectral density for amplitude noise as a function of angular frequency.
3) Function SPECF:
   supplied when IW = 0 and IF = 1. Here the user defines the spectral density for phase noise as a function of angular frequency.
4) Subroutine VGRID:
   supplied when IV = 1. Here the user defines the number and coordinates of the grid points for a variable grid.
5) Subroutine SOLVE:
   supplied when IS = 1. Here the user calls his or her own ODE integrator.

OUTPUT

The integration results are written out to the file DILUTE.OUT at the end of each integration step in the following format:

```
      WRITE(1,9000)T
      WRITE(1,9000)(U(I),I=1,N)
9000  FORMAT(1X,11E12.5)
```

where T is the final time of the integration step, N the number of grid points and U(I) the density at grid point i, $\rho(x_i, t)$.

REFERENCES

1. G. Dôme, "Diffusion Due to RF Noise," in *CERN Advanced Accelerator School on Advanced Accelerator Physics*, CERN Report No. 87-03, 1987 pp. 370–401.
2. S. Krinsky and J.M. Wang, "Bunch Diffusion Due to RF Noise," Part. Accel. 12, 107 (1982).
3. H.-J. Shih, J. Ellison, B. Newberger, and R. Cogburn, "Longitudinal Beam Dynamics with RF Noise," SSC Laboratory Report No. 578; submitted for publication.
4. H.-J. Shih, J. Ellison and W. Schiesser, "Reliability of Numerical Solutions of a Diffusion Equation Modelling RF Noise-Induced Dilution in Particle Beams" in *Advances in Computer Methods for Partial Differential Equations VII*, R. Vichnevetsky, D. Knight and G. Richter, eds. (IMACS, New Brunswick, NJ, 1992) p. 663.
5. G.E. Forsythe, M.A. Malcolm, and C.B. Moler, *Computer Methods for Mathematical Computations* (Prentice-Hall, Englewood Cliffs, NJ, 1977).
6. G.D. Byrne and A.C. Hindmarsh, "Stiff ODE Solvers: A Review of Current and Coming Attractions," J. Comput. Phys., 70, 1 (1987).

XORBIT — AN X-WINDOWS ACCELERATOR SIMULATION*

Kenneth Evans, Jr.
Advanced Photon Source
Argonne National Laboratory, Argonne, Illinois 60439

ABSTRACT

Xorbit is an accelerator physics code that tracks particle orbits. Its two distinguishing features are a rich graphical interface and the ability to connect to and be controlled by external programs such as Mathematica or control-system software. The design goal is to have a code that can be controlled in much the same way as a real machine is controlled. This allows the testing of control algorithms before the real machine is commissioned or without disturbing the real machine at any time. The graphical interface provides a means of changing magnet parameters easily with immediate visual feedback on the resulting orbit changes. There are a number of features including interactive plotting of orbits and Twiss parameters; the ability to display error positions, monitor readings, or the full orbit; the ability to display true or difference orbits; as well as the ability to find closed orbits, track from given initial conditions, or apply a variety of correction methods. There is a Design Mode in which element strengths and positions can be changed with the mouse with continuous display of the results. All of these operations are fast and intuitive.

INTRODUCTION

Xorbit is a Unix computer code designed to help in the design and operation of circular accelerators, storage rings, and beam lines. It is a lattice design code that calculates the orbit primarily by using 2 x 2 matrices. (Nonlinear elements are simulated with kicks, and random element displacements are supported. Dispersion is obtained from the 2 x 2 matrices.) It has been designed to be fast and easy to use and to provide immediate feedback when changes are made in the lattice.

This ease-of-use and immediate-feedback are mostly provided through a graphical user interface (GUI) which is based on Motif. It is especially easy to make changes to individual elements in the lattice and then see what happens with a minimum of typing input files and reading output files. In fact, after the initial input file is made, most of the input consists of clicking with a mouse on options, and most of the output is either automatically displayed or available by clicking with the mouse.

The second important feature of Xorbit is its Channel Access capability. This feature was originally implemented to simulate the way the Advanced Photon Source (APS) control system accesses (or will access) the devices in the APS rings and beam lines over the Ethernet, so that Xorbit could be used as a simulation for those devices to test control algorithms and procedures. The mechanism, using Unix Interprocess Communication (IPC), by which Xorbit mimics Channel Access turns out to also provide a powerful means of connecting programs, such as Mathematica,[1] or spreadsheets like Excel,[2] to the orbit calculations in Xorbit. It is felt that the ability to easily use such powerful programs, which contain a multitude of flexible tools, is perhaps a better approach than providing specialized (and to some extent inflexible) tools in the program itself. It is possible, in this case, to keep the orbit code small and clean, and to leave the development of mathematical tools to companies who can afford to spend more time developing and refining them. These tools also have the advantage that many people are already familiar with them and use them.

Xorbit is not necessarily a replacement for large, well-established, accelerator codes such as Mad.[3] It is, however, an accelerator code that provides good visual feedback and which can tap the capabilities of existing mathematical analysis programs to provide even more power than conventional codes in a flexible way. It can be effectively used alongside codes like Mad. Xorbit is an accelerator code for the '90's.

Xorbit is written with a combination of Fortran and C. The physics is mostly Fortran, and the interface is mostly C. The code has been developed on Sun workstations and should work well in that environment. It has been used with the Motif window manager, Mwm, and with the X Windows window manager, Twm, as well as with the Open Windows window manager, Olwm. Because it is an X-Windows code, it can be run on one platform, and the graphical results can be displayed on any terminal or computer that has an X server. The mixed-language programming and the use of IPC, which is not implemented in a standard way, may make porting the code to other platforms difficult, however.

The lattice is specified in an input file. There are means provided to convert Mad input to that used by Xorbit and back. In the interactive mode the remaining input is through the interface. Xorbit can also be operated as a program that reads the input file and produces several output files without the interface and without further user intervention.

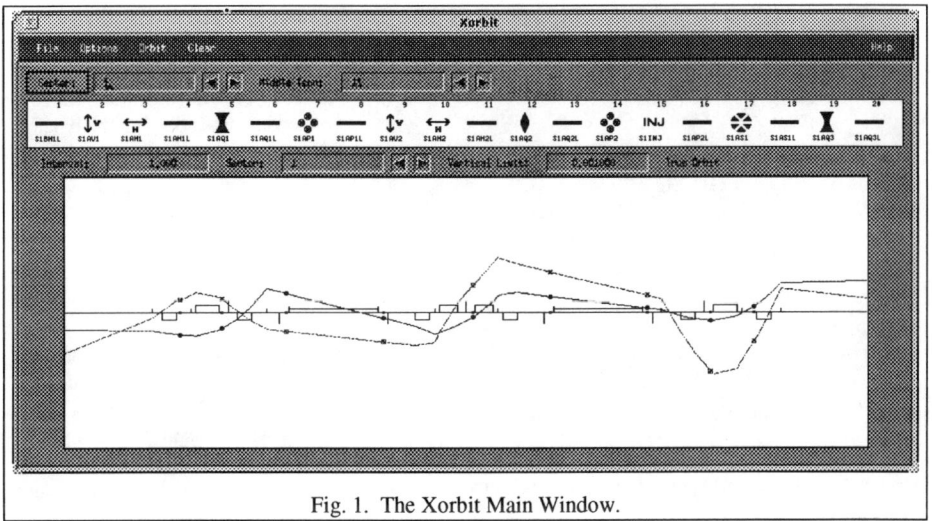

Fig. 1. The Xorbit Main Window.

MAIN WINDOW

Most of the work in Xorbit is done in its Main Window shown in Fig. 1. The Menu Bar is at the top of the main window, with buttons labeled File, Options, Orbit, Clear, and Help. Below the Menu Bar is the Icon Control Area. This area contains the controls for manipulating the Icon Area, which is just below it. The Icon Area displays icons representing the magnet elements. Below the Icon Area is the Graph Control Area. This area contains controls for adjusting and scrolling the Graph Area, the bottom area, in which the orbits are displayed

The Icon Area contains icons for each of the elements in the lattice. Both the index of the element in the lattice and its name are shown. The display can be scrolled in several ways. If an element is clicked with Button 1 in the Graph Area, the icon for that element will appear at the middle of the Icon Area display. This is the easiest way to pick icons.

The main reason for the Icon Area is that if you click on an icon in the Icon Area, a dialog box like that shown in Fig. 2 will pop up giving information about that element. This information is

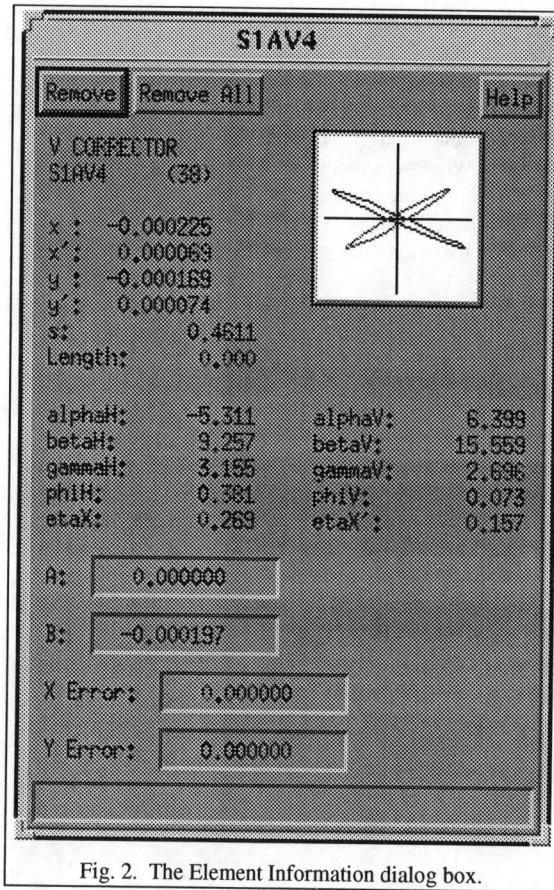

Fig. 2. The Element Information dialog box.

current at the time the icon is clicked and will not be updated. Clicking at a later time on the icon for the same element will display another box with the then-current information. A fairly large number of boxes can be displayed at once. They can be dragged around the screen and placed as desired. They can be used as a means of keeping track of parameters as changes are made in the orbit. In addition, the items that appear in text boxes can be changed, and this is the primary means of changing lattice parameters via the interface. In the Element Information dialog box the quantities A and B typically represent the element's normal and skew strengths, respectively. The meaning of these quantities varies somewhat depending on the type of element, however.

The Graph Area contains the orbit display. The horizontal axis is s, the longitudinal distance along the lattice, and the vertical axis is x or y, the transverse coordinates. Exactly what is displayed can be controlled via the Orbit Options item in the Options pull-down menu, and the area can be cleared via items in the Clear pull-down menu. In addition to the menus, there are controls in the graph control area to size and scroll the display, and there are mouse operations.

There are three operations that can be performed with the mouse in the Graph Area.

1. Clicking Button 1 moves the icon for the element at the mouse pointer to the middle of the Icon Area, where it can be clicked to display information about the element.

2. Clicking Button 2 moves the position which is at the mouse pointer to the center of the Graph Area. The Graph Area display can be scrolled discontinuously by repeatedly clicking Button 2.

3. Pressing and holding Button 3 scrolls the lattice continuously. The speed depends on how far you are from the center, and the direction depends on which side of center.

There are a number of options to control just what is displayed in the Graph Area. These are available through the Orbit Options dialog box actuated from the Options Menu. They include which coordinates to display, whether to display symbols for beam position monitors (BPM's) and element displacements, whether to show the orbit at all elements or only at BPM's, the initial values (x_0, x'_0,

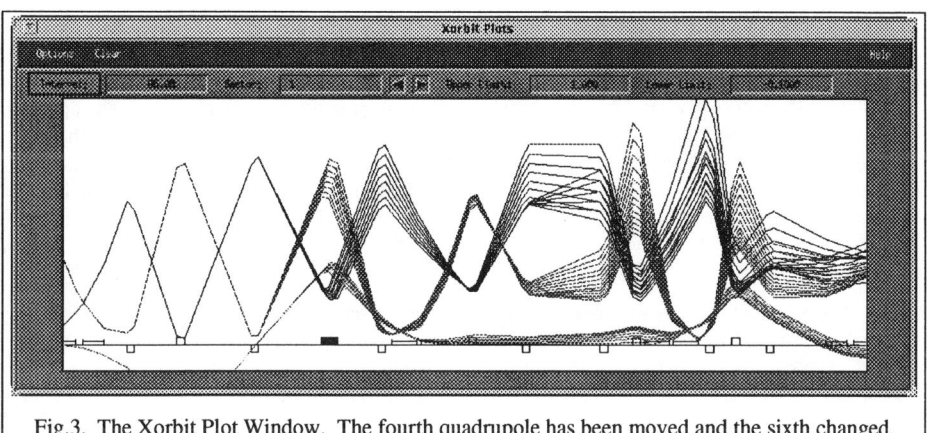

Fig.3. The Xorbit Plot Window. The fourth quadrupole has been moved and the sixth changed.

y_0, y'_0) for the orbit, and the number of orbit turns. In addition, there are many menu options, which are summarized below.

PLOT WINDOW

Plots of α_h, α_v, β_h, β_v, γ_h, γ_v, η_x, η'_x, x, y, x', and y' *vs.* s can be displayed in the Plot Graph Area in the Plot Window, shown in Fig. 3. This window is selected from the Main Options Menu. What is displayed and how it is displayed are controlled by switches in the input file and by the Plot Options dialog box, available from the Options Menu in the Plot Window Menu Bar. The same mouse operations as for the Main Graph Area are available, as are similar controls for sizing and scrolling the display area.

The real advantage of the Plot Window is that there is a Design Mode that allows the element strengths and positions to be changed by clicking with the mouse or by entering values in a

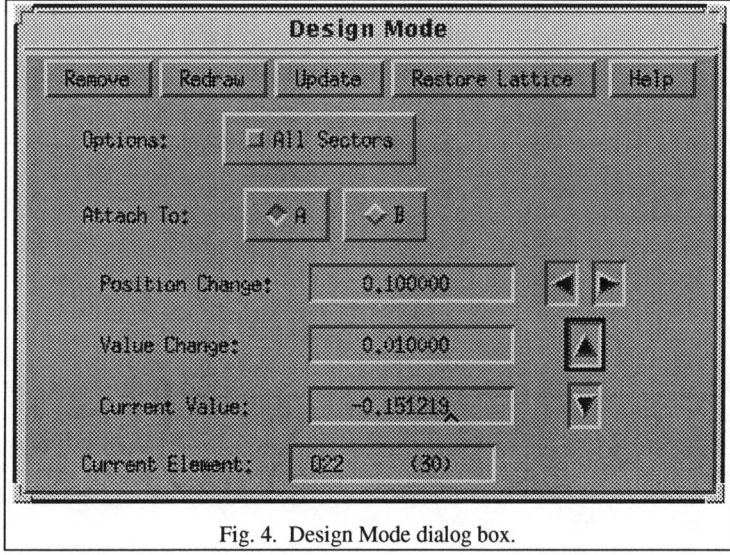

Fig. 4. Design Mode dialog box.

text box. The plots in the Plot Window change as the elements are changed, allowing immediate visual feedback of the results of the changes on any of the above parameters. The Design Mode is controlled by the Design Mode dialog box, shown in Fig. 4. An element is selected the same as it is selected for the Icon Area, most conveniently by clicking on it in the Plot Graph Area.. The up and down arrows increment the chosen element's normal (A) strength or its skew (B) strength. The left and right arrows move the element sideways. Normally, each new set of curves is displayed along with the previous ones as the chosen element is changed. The graph can be redrawn with the latest curves only, and the lattice can be restored to its original values. It is easier to do than to explain.

MENU OPTIONS

This section gives a summary of the operations in Xorbit that can be accessed via the menus. Since these are most of the operations, it thus serves as a summary of the features available with Xorbit. There are two menu bars: one in the Main Window and one in the Plot Window. The Clear Menu and the Help Menu appear in both menu bars and have similar options.

File Menu

The File Menu contains operations for saving and restoring lattice files and files with other orbit information, as well as the Quit button.

Options Menu

The plot window is initiated from this menu. A dialog box with the tunes can be displayed, and the lattice can be restored to its original values. Displacements and correctors can be saved, restored, and zeroed, and new random numbers for displacements can be calculated. A variety of plots, including the Twiss parameters, x, x', y, y', and the horizontal and vertical corrector settings, can be displayed and manipulated using Xvgr,[4] a public-domain, X-Windows plotting program. An XTerm window can be started, and programs such as Mathematica can be started to access Xorbit.

Orbit Menu

A dialog box with the orbit options can be popped up from the Orbit Menu. The orbit can be started and continued for a number of turns (chosen in the Orbit Options dialog box). A new closed orbit can be calculated. A dialog box with orbit statistics (average, rms, and maximum values for the orbit, the correctors, and the displacements) can be displayed. Several (currently five) orbits can be saved and later displayed. It is also possible to display difference orbits with respect to any one of the saved orbits. Several correction schemes can be applied.

Clear Menu

The graph area can be cleared, updated with the current orbit, or redrawn with the current orbit only. On startup all new orbits are drawn without clearing the old ones. This can be changed to clear the graph area before the orbit is plotted.

Help Menu

Xorbit has a hypertext Help package, HyperHelp,[5] which is a stand-alone Motif application developed by Bristol Technologies. It is very similar to the help provided in Microsoft Windows. The HyperHelp viewer is necessary to utilize this Xorbit feature, but it may be distributed with Xorbit, along with the Xorbit help file. The help compiler, which is necessary to make the Xorbit help file in the first place may not. Figure 5 shows the viewer positioned at the contents screen, one of over 30 screens of Xorbit help. The menu items allow bringing up HyperHelp with the contents screen, the overview screen, or screens with help on HyperHelp itself. The user can subsequently jump to other places by clicking with the mouse or by using the search or browse modes. In addition, dialog boxes which do not rely on HyperHelp can be displayed with a help summary and version information.

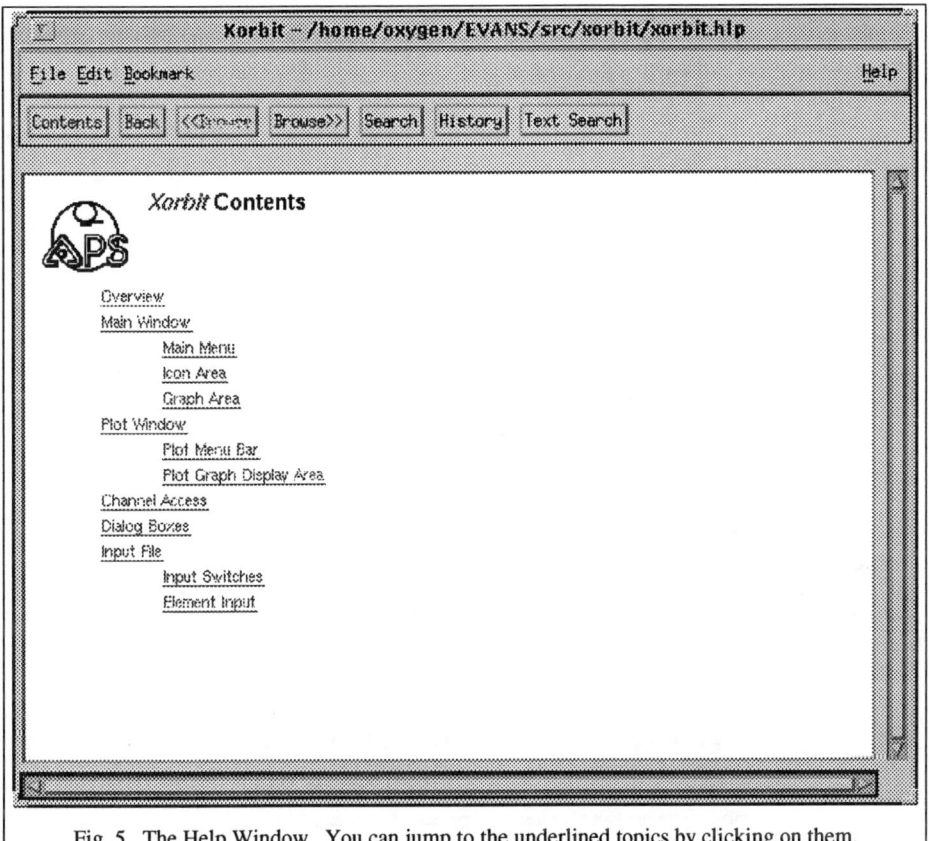

Fig. 5. The Help Window. You can jump to the underlined topics by clicking on them.

Plot Window Option Menu

The Option Menu in the Plot Window can be used to bring up a dialog box which can be used to change what is plotted in the plot window, how each plot is scaled, and what colors are used. The Design Mode dialog box, described above, is initiated from this menu. There are also Remove and Quit buttons to remove the Plot Window and quit Xorbit, respectively.

CHANNEL ACCESS

Xorbit provides a Channel Access capability which emulates that used in the APS control system. The real (Epics[6]) Channel Access communicates with real devices over the Ethernet. Xorbit Channel Access communicates with Xorbit, which simulates many aspects of the real devices, via Unix IPC, primarily via a message queue.

Channel-access interfaces developed at APS to such programs as Mathematica, Devtest,[7] WingZ,[8] and Excel work in just the same way with Xorbit as they do with the real machine. The only necessary change is that they need to be linked with the Xorbit Channel Access routines rather than the Epics Channel Access routines. The Xorbit routines have the same names and arguments and, in fact, use the same header file, cadef.h.

Apart from emulating the real Channel Access, these routines provide a powerful and easily modified way to perform complex procedures, such as orbit correction, on the lattice in Xorbit. They also make the extensive features, such as plotting or matrix manipulation, in programs like Mathematica available for analyzing orbit dynamics.

Devices are read or set via their process variable names. The process variable names used in Xorbit are of the form:

elementname.variable

elementname is the full name of the element in Xorbit, e.g., S34AQ1. The possible values for *variable* include most of the element and orbit parameters in Xorbit, including variables such as the tunes and Twiss parameters.

To read the x displacement of S34AQ1 one would use the process variable name S34AQ1.xerr, and to set the strength of horizontal corrector S1AH2 to, say, 0.1 mrad, one would use the process variable name S1AH2.a. In Mathematica, for example, this might be done by

xdisp=CaGet["S34AQ1.xerr"]

and

newstrength=CaSet["S1AH2.a",.0001],

respectively. A Mathematica procedure to measure a response matrix is shown in Fig. 6. This procedure simulates a real-life accelerator operation. One could, in addition, use the sophisticated

```
getcvres:=(
        vnum=Length[vnames];
        pnum=Length[pnames];
        cvres=Table[0,{j,1,vnum},{i,1,pnum}];
        vbnames=makenames[vnames,"b"];
        vb0=CaGet[vbnames];
        pbpmynames=makenames[pnames,"bpmy"];
        pbpmy0=CaGet[pbpmynames];
        For[j=1,j <= vnum,j++,
                {vb}=CaSet[vbnames[[j]],vb0[[j]]+delc];
                pause[pausetime];
                pbpmy=CaGet[pbpmynames];
                delc1=vb-vb0[[j]];
                cvres[[j]]=(vx-vx0)/delc1,
                CaSet[vbnames[[j]],vb0[[j]]];
                pause[pausetime];
        ];
        cvres=Transpose[cvres];
        Print["cvres[",pnum,",",vnum,"] has been measured"];
)
```

Fig. 6. A Mathematica procedure for measuring a response matrix. The variables, **vnames** and **pnames**, are previously-defined lists of vertical corrector and position monitor names. The matrix is **cvres**. The short procedure, **makenames** (defined elsewhere), makes lists of the required process-variable names for the vertical correctors and the position monitors from **vnames** and **pnames**, respectively. The old corrector values are obtained from Channel Access, and the correctors are then changed through Channel Access in turn by **delc**, getting the list of the resulting y monitor readings from Channel Access. Then the corrector settings are restored through Channel Access. The pause of 1 - 2 seconds is to wait for Xorbit to calculate the closed orbit.

graphics and matrix routines in Mathematica to, say, invert the response matrix, plot corrector strengths or monitor readings, *etc*. In general, one has access to all the orbit information in Xorbit and can read it, change it, or use it.

In a spreadsheet Channel Access might be done by selecting the cell or cells with the desired name or names and indicating the cells for the input and output values before running an appropriate macro. An example of using Excel is shown in Fig. 7. This Excel is running on a PC and is communicating over the Ethernet to the workstation on which Xorbit is running.

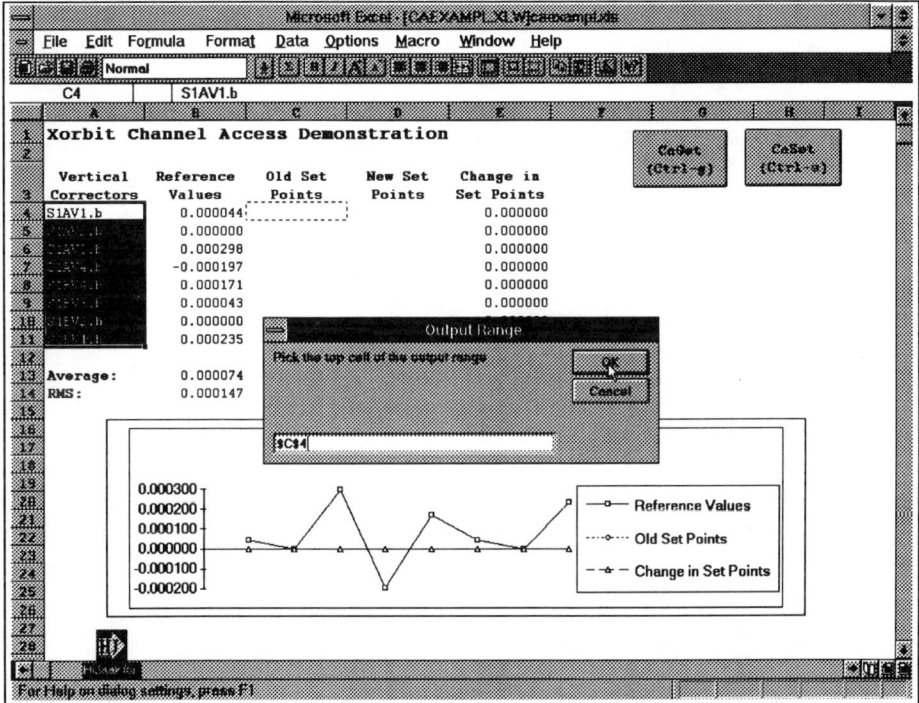

Fig. 7. Xorbit Channel Access from Excel. The device names have been selected, and the CaGet macro has been started by clicking on the button or by typing Ctrl-g. There is a dialog box prompting for the start of the output range, which has been chosen to be where the dotted box is. After the OK button is clicked, the current settings will be placed in the output range, and the spreadsheet, including the plot, will be updated.

SUMMARY

This paper has given an overview of the capabilities and features of Xorbit. More detailed information is available in a reference manual and through the on-line help. The major features of Xorbit are an easy-to-use, intuitive, interactive interface and the ability to communicate with existing, well-known, mathematical analysis programs as well as the ability to act as a simulation for a control system.

REFERENCES

* Work supported by U.S. Department of Energy, Office of Basic Energy Sciences under Contract No. W-31-109-ENG-38.

[1] S. Wolfram et. al., *Mathematica* (Addison-Wesley, Redwood City, CA,1991).

[2] Computer code *Excel* (Microsoft Corp., Redmon, WA, 1993).

[3] H. Grote and F.C. Iselin, "The MAD Program," CERN Report No. CERN/SL/90-13(AP), 1991.

[4] P. J. Turner, computer code *Xvgr* (Oregon Graduate Institute of Science and Technology, Beaverton, OR, 1992).

[5] Computer code *HyperHelp* (Bristol Technology, Inc., Ridgefield, CT, 1992).

[6] J. O. Hill, "Channel Access: A Software Bus for the LAACS," in *Accelerator and Large Experimental Physics Control Systems*, edited by D. P. Gurd and M. Crowley-Milling (ICALEPCS, Vancouver, British Columbia,1989), pp. 288-291.

[7] Y. N. Yang and J. D. Smith, "Devtest — An Interpreter," Brookhaven NSLS Technical Note No. 356A, 1990.

[8] Computer program *Wingz* (Informix Software, Inc., Lenexa, KS, 1992).

THE SIMPSONS PROGRAM
6-D PHASE SPACE TRACKING WITH ACCELERATION*

S. Machida
Superconducting Super Collider Laboratory,[†] Dallas, TX 75237

ABSTRACT

A particle tracking code, Simpsons, in 6-D phase space including energy ramping has been developed to model proton synchrotrons and storage rings. We take time as the independent variable to change machine parameters and diagnose beam quality in a quite similar way as real machines, unlike existing tracking codes for synchrotrons which advance a particle element by element. Arbitrary energy ramping and rf voltage curves as a function of time are read as an input file for defining a machine cycle. The code is used to study beam dynamics with time dependent parameters. Some of the examples from simulations of the Superconducting Super Collider (SSC) boosters are shown.

INTRODUCTION

Particle tracking codes developed so far mostly deal with either transverse or longitudinal phase space coordinates separately, or at most, track both of them but without acceleration. In a rapid cycling proton synchrotron, such as the Low Energy Booster (LEB) of the Superconducting Super Collider (SSC) which accelerates protons from 1.22 GeV/c to 12 GeV/c within 50 msec (about 26,000 turns, total), the synchrotron tune changes quite rapidly and sometimes becomes large (~ 0.1). Furthermore, multiturn injection, adiabatic capture and bunching all occur in a short time period right after the injection, usually with strong space-charge force in all directions. The Medium Energy Booster (MEB) of the SSC has the transition crossing where it is essential to observe the particle behavior in 6-D with acceleration. Coupling effects between transverse and longitudinal planes are expected to play a significant role in these machines thus, a more sophisticated tracking model with acceleration is necessary.

From a practical point of view, for example to simulate a commissioning procedure, it is desirable to have a machine simulation model which can handle time-dependent parameters such as rf frequency errors and magnet strength (including multipole errors), as arbitrary functions. We developed the code Simpsons to track multi-particles in full 6-D phase space with acceleration. The code reads external tables which specify time-dependent parameters when they are necessary. First, we will explain the detail of the simulation method. Secondly, we will show some results of the LEB and MEB of the SSC.

THE CODE SIMPSONS[1]

The code Simpsons is a multi-particle tracking program in 6-D phase space with acceleration. It takes time as the independent variable. Phase space coordinates of macro-particles are updated by a fixed time interval. Choice of the time interval depends on the physics one is studying. When the internal force among particles, such as space-charge effects, is not negligible, one needs to choose a small enough time interval because the force is a function of the coordinates themselves and self-consistency of the particle distribution and the force should be kept. When the external force, namely the force from magnetic lattice elements and rf cavities, is dominant, each macro-particle behaves independently so that a large time interval can be taken.

---

* To be presented at the Computational Accelerator Physics Conference February, 1993.

† Operated by the Universities Research Association, Inc., for the U.S. Department of Energy under Contract No. DE-AC35-89ER40486.

© 1994 American Institute of Physics

The tracking is done for one cycle of a machine (it does not necessarily mean from injection to extraction), and some machine parameters have to be specified throughout a cycle in addition to the geometric lattice configuration. When the synchronous momentum and the rf voltage are constant throughout a cycle, for example the simulation of an injection porch in the MEB and the HEB, values of these constants and total real time of a cycle (or total turn numbers) are sufficient to specify a cycle. For a simulation of a machine with acceleration, one has to prepare tables which specify the strength of a bending field $B\varrho(t)$, and the rf voltage $V(t)$ as a function of time. In the code, these tables are interpolated to get values in a time interval. Figures 1 and 2 show these functions for the LEB.

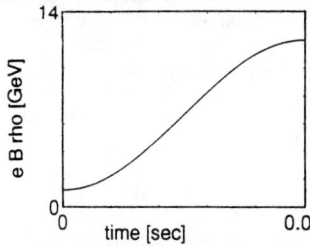

Fig. 1. $eB\varrho$ of a LEB Cycle.

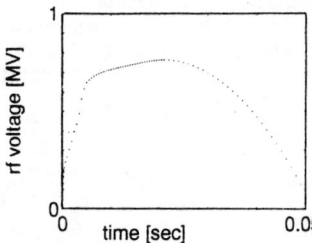

Fig. 2.   Rf Voltage of a LEB Cycle.

The Simpsons itself does not have the interpretation part of the Standard Input and Command Language[2] for describing and tuning an accelerator lattice. Instead, we adopted the TEAPOT[3] as a preprogram. A lattice description file in the Standard Language is processed by the TEAPOT beforehand and replaced by a thin lens machine file. Furthermore, transverse tune and chromaticity are adjusted, multipole and misalignment errors in lattice elements are introduced, and lattice functions; β, α, η and closed orbit distortions are calculated by the TEAPOT. In this way, a thin lens machine file and a lattice function file are ready for the Simpsons as input files.

Normally, a simulation starts from creating initial 6-D phase space coordinates; s, p_s, x, p_x, y, p_y for all the macro-particles. The Gaussian distribution with a certain cut of a tail for both longitudinal and transverse directions is commonly assumed, but a parabolic and a uniform distribution for longitudinal, and a waterbag and a K-V distribution for transverse can be also selected.

Within one time interval, a particle either remains in drift space or receives one or more momentum kicks due to magnets and/or rf cavities. If there are no lattice elements in the time interval, as is the case only with drift space, the positions are updated,

$$s_{new} = s_{old} + \frac{p_s c}{\sqrt{(m_0 c^2)^2 + (pc)^2}} \Delta t , \qquad (1)$$

$$x_{new} = x_{old} + \frac{p_x c}{\sqrt{(m_0 c^2)^2 + (pc)^2}} \Delta t , \qquad (2)$$

$$y_{new} = y_{old} + \frac{p_y c}{\sqrt{(m_0 c^2)^2 + (pc)^2}} \Delta t , \qquad (3)$$

where p is the total momentum of the particle,

$$p = \sqrt{p_s^2 + p_x^2 + p_y^2} , \qquad (4)$$

m_0 is the mass of proton, c is the speed of light, and Δt is the time interval. The momenta remain constant.

At the location of rf cavities, a particle gains or loses its energy as

$$\Delta E = eV(t) \sin \phi_{rf}(t) , \qquad (5)$$

where $V(t)$ is the rf voltage, e is the unit charge, and $\phi_{rf}(t)$ is the rf phase at the time a particle is passing through the rf cavity. Without acceleration, the rf phase can be simply replaced by

$$\phi_{rf}(t) = \omega_{rf} \cdot t + \phi_{rf}(0) , \qquad (6)$$

where ω_{rf} is the constant rf frequency, the product of the revolution frequency of the synchronous particle and the harmonic number h,

$$\omega_{rf} = h \cdot \omega_{rev} . \qquad (7)$$

That is, the rf phase is a linear function of time.

The rf phase is no longer a simple linear function of time once a particle is being accelerated. With acceleration, the rf phase at time t is

$$\phi_{rf}(t) = \int_0^t \omega_{rf}(t) dt + \phi_{rf}(0) , \qquad (8)$$

where $\omega_{rf}(t)$ is the "instantaneous" rf frequency and as a function of the bending field,

$$\omega_{rf}(t) = h \cdot \omega_{rev}(t) = h \cdot \frac{2\pi c \beta(t)}{C}$$

$$= h \cdot \frac{2\pi c \left(m_0 c^2 / e B \varrho(t)\right)^2}{C \sqrt{1 + \left(m_0 c^2 / e B \varrho(t)\right)^2}} , \qquad (9)$$

where $\beta(t)$ is the Lorentz factor and C is the machine circumference. When a beam goes through the transition energy, the rf phase must be shifted by a proper value. In the code, we add the phase shift $\Delta \phi_{rf}$ to the rf phase at specified time instantaneously.

In either case, with and without acceleration, the total energy of a particle increases, or decreases at an rf cavity, by the amount of the Eq. (5)

$$E_{new} = E_{old} + \Delta E , \qquad (10)$$

and the new total momentum is

$$p_{new} = \sqrt{E_{new}^2 - (m_0 c^2)^2} . \qquad (11)$$

Since only the longitudinal momentum should change at the rf cavity,

$$p_{s,new} = \sqrt{p_{new}^2 - p_{x,old}^2 - p_{y,old}^2} . \qquad (12)$$

In summary, the most different feature of the longitudinal dynamics of the Simpsons is that it does not solve differential (or difference) equations with respect to the synchronous phase. Instead, the rf frequency is integrated to keep track of the rf phase all the time so that one can apply proper voltage when a particle is passing through an rf cavity.

If there are magnets in a time interval, first the positions are updated to the magnet location with the Eqs. (1) to (3) (Δt should be substituted by its appropriate fraction), then the transverse momenta are changed in the same way as the TEAPOT,

$$p_{x,new} = p_{x,old} - \frac{1}{1+\delta} \left(\frac{B_y L}{p_0/e} \right) p , \qquad (13)$$

$$p_{y,new} = p_{y,old} + \frac{1}{1+\delta}\left(\frac{B_y L}{p_0/e}\right)p , \qquad (14)$$

where p_0 is the synchronous momentum at that particular time, δ is the momentum deviation from the synchronous one

$$\delta = \frac{p - p_0}{p_0} , \qquad (15)$$

L is the magnet length, and B_x and B_y are the magnetic fields at the particle position as the summation over multipoles.

$$(B_y + iB_x)L = (p_0/e)\sum_{n=0}^{M}(\tilde{b}_n + i\tilde{a}_n)(x+iy)^n , \qquad (16)$$

where $\tilde{b}_n$ and $\tilde{a}_n$ are the normal and skew multipole strengths, respectively, in the TEAPOT notation, which are related to the MAD[4] k_n by

$$\tilde{b}_n = \frac{k_n L}{n!} . \qquad (17)$$

Since the magnitude of the total momentum should not change by magnets,

$$p_{new} = p_{old} , \qquad (18)$$

and the longitudinal momentum is adjusted as

$$p_{s,new} = \sqrt{p_{new}^2 - p_{x,new}^2 - p_{y,new}^2} . \qquad (19)$$

In the real machines, designed values never stay perfect and noise of all kinds of frequencies affects machine operations. Besides, one may intentionally vary some machine parameters as a function of time. For example, in low energy proton synchrotrons the operation point is set with a large fractional tune at the beginning of a cycle to accommodate the large space-charge tune shift and it is gradually pushed down for the suppression of other instabilities such as resistive wall instability. In the Simpsons, the deviation from the perfect value of the magnetic strength, the rf frequency and so on, are read as an external file tabulating values as a function of time.

EXAMPLES

We have been using the Simpsons to study several beam dynamics issues in the LEB, MEB, and HEB of the SSC. In the following section, some tracking results of the LEB and MEB will be shown to demonstrate potential capability of the code. The application to the HEB can be found elsewhere.[5]

Synchro–beta Coupling

The synchrotron tune of the LEB becomes as high as 0.05 a few milliseconds after injection and decreases slowly toward the end of a cycle. The betatron tune is around 11.85 right after the injection to accommodate the large space-charge tune shift. We have looked at the emittance growth due to synchro-beta coupling effects which are excited by rf cavities at a finite dispersion position. As a matter of fact, all the rf cavities in the LEB are planned to be located in one straight section where dispersion functions of both planes are ideally zero. We examined, however, some lattices where some of rf cavities are in an arc section with finite dispersion. In all the examined lattices, the total rf voltage is kept same.

The 100 macro-particles are tracked in a whole cycle of the LEB (50 msec). The rms $\Delta p/p$ of 100 particles is shown in Figure 3 and the estimated synchrotron tune in Figure 4. The horizontal emittance stays almost constant throughout a cycle when all four rf cavities are put in a dispersionless (except residual ones due to errors) straight section as designed, Figure 5(a). When we put one rf cavity at the maximum horizontal dispersion position of about 3.5 m, the horizontal emittance becomes as twice as much, which mostly occurs when the synchrotron tune has the maximum value of 0.05, Figure 5(b). In both cases, the vertical emittance stays constant.

An arc section consists of four identical modules. Horizontal phase advance between each module is fixed to $0.75 \times 2\pi$. The lattice functions including dispersion repeat in each module. Instead of putting one rf cavity in one module, we tested two rf cavities in two modules. First, the two rf cavities were separated by phase advance $0.75 \times 2\pi$ (one module apart). The horizontal emittance growth becomes smaller but still exists, Figure 5(c). Second, the two rf cavities were separated by phase advance $1.50 \times 2\pi$ (two modules apart). The result shows the cancellation of synchro-beta coupling effects by means of proper phase advance, Figure 5(d).

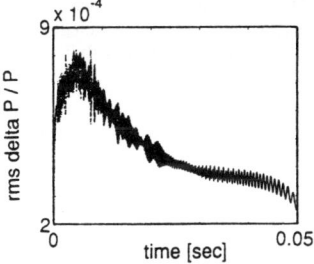

Fig. 3. Rms $\Delta p/p$ of 100 Macro-particles.

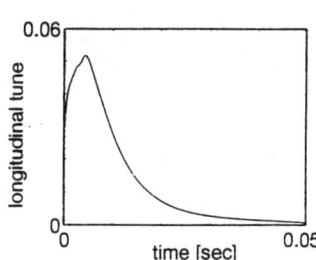

Fig. 4. Synchrotron Tune.

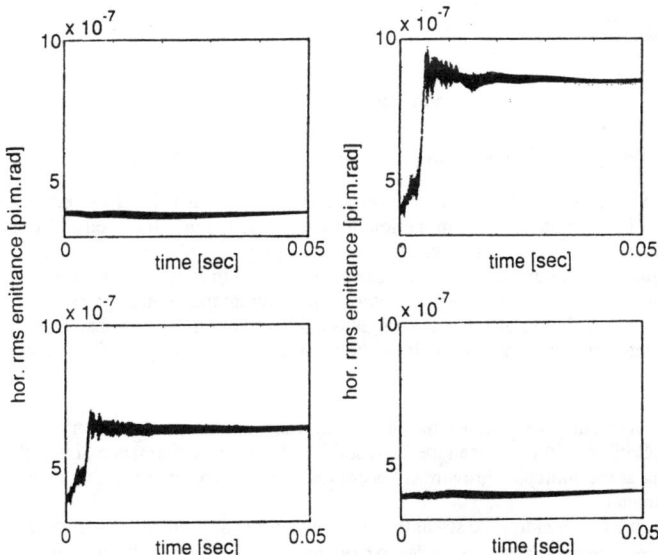

Fig. 5. Horizontal Rms Emittance; (a) 4 Rf Cavities in Dispersionless Section, (b) 1 Rf Cavity at Large Dispersion, (c) 2 Rf Cavities at Large Dispersion with Phase Difference of $0.75 \times 2\pi$, and (d) 2 Rf Cavities at Large Dispersion with Phase Difference of $1.50 \times 2\pi$.

Transition Crossing

The transition energy is defined by a lattice itself unlike a tracking program of longitudinal phase space only, in which it should be given as an input parameter. That is also true for higher order coefficients of $\Delta p/p$. Therefore, a simulation of transition crossing starts from finding transition energy itself. The revolution frequency shift is expanded as

$$\frac{\Delta f}{f} = c_0 \frac{\Delta p}{p} + c_1 \left(\frac{\Delta p}{p}\right)^2 + \dots \quad . \tag{20}$$

The first coefficient c_0 is

$$c_0 = \frac{1}{\gamma^2} - \alpha_0 , \tag{21}$$

where γ is the Lorentz factor and α_0 is the momentum-compaction. At the transition, c_0 becomes zero and second order of $\Delta p/p$ is dominated.

To find out the transition energy of the MEB, off-momentum particles were tracked for 100 turns by scanning the synchronous momentum. When the synchronous momentum is not equal to the transition momentum, the trace of off-momentum particles draws an asymmetric curve, Figure 6(a) because the first term of Eq. (20) is not zero. At the transition energy, off-momentum particles behave symmetrically, Figure 6(b). In that MEB lattice, it turns out that a particle reaches the transition energy 0.39237 sec after acceleration starts.

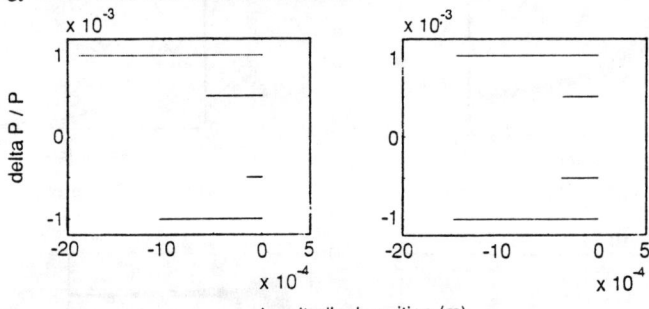

Fig. 6. Phase Space Plot of Off-momentum Particles for 100 Turns. (a) On-momentum Particle is not at the Transition Energy. (b) On-momentum Particle is at the Transition Energy.

We started a simulation 0.380 sec after acceleration starts (about 1000 turns before the transition) with 1000 macro-particles. The initial longitudinal distribution is matched to the rf bucket at that time. We set the bending field increases quadratically and the rf voltage does linearly. The bunch length and $\Delta p/p$ throughout the transition crossing are shown in Figures 7 and 8. Because of unavoidable mismatch at the transition crossing, namely the different momentum particle has the different transition energy, Johnsen effects[6], the longitudinal emittance growth is observed in Figure 9 while the transverse emittance stays almost constant as shown in Figure 10 as expected.

Resonance Correction

The way of correcting betatron resonances is well-known and several harmonic correctors are designed for the SSC boosters. We used the Simpsons to check the performance of these correctors by looking at the emittance growth and beam loss due to resonance crossing taking bandwidth as a parameter.

In the LEB, the trim quadrupole strength is linearly shifted as a function of time to simulate resonance crossing. We used 100 macro-particles to calculate the evolution of the emittance. In the Simpsons, a particle loss happens if the radial amplitude of the particle is more than the beam pipe radius, which is 35 mm in the LEB. The resonance crossing simulation of the sum resonance; $v_x + v_y = 23$ is shown in Figure 11. Keeping the vertical tune constant at 11.60, the horizontal tune is scanned from 11.41 to 11.39 in 2.5 msec so that a beam crosses the sum resonance about halfway. The 50% correction of the bandwidth is not enough and a large emittance growth occurs. Figure 11(b), however, indicates the decrease of the beam loss even with the 50% correction. For the second order normal resonance; $2v_x = 23$, the horizontal tune is scanned from 11.51 to 11.49 in 2.5 msec and it shows that the 50% correction works well in Figure 12. There is no beam loss.

S. Machida 465

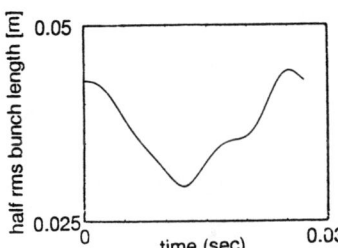

Fig. 7. Half Rms Bunch Length Around the Transition Energy Which Is At $t = 0.01237$ s.

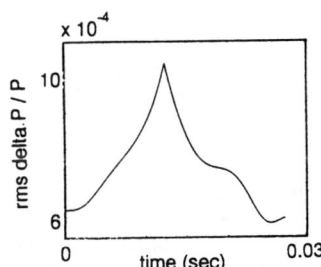

Fig. 8. Rms $\Delta p/p$ Around the Transition Energy.

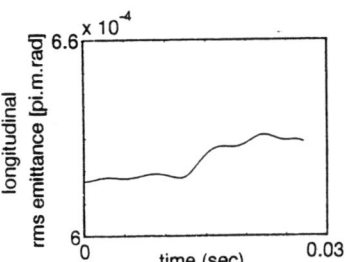

Fig. 9. Longitudinal Rms Emittance Around the Transition Energy.

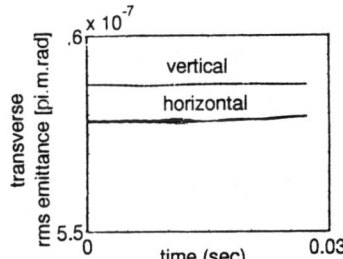

Fig. 10. Transverse Rms Emittance Around the Transition Energy.

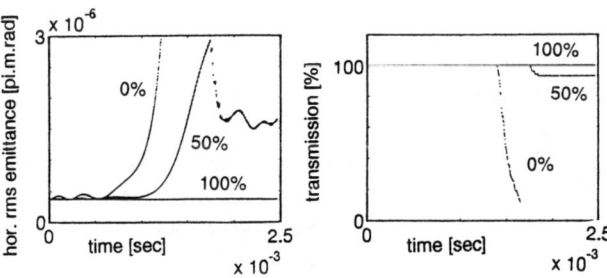

Fig. 11. (a) Horizontal Rms Emittance Growth Due To Resonance Crossing of $\nu_x + \nu_y = 23$ Which Occurs at About $t = 0.1$ milliseconds. (b) Beam Loss Due To Resonance Crossing.

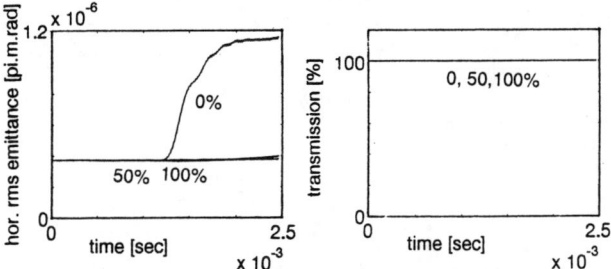

Fig. 12. (a) Horizontal Rms Emittance Growth Due To Resonance Crossing of $2\nu_x = 23$ Which Occurs at About $t = 0.1$ milliseconds. (b) Beam Loss Due To Resonance Crossing.

SPACE-CHARGE EFFECTS

There are two methods of calculating space-charge effects in the Simpsons. One employs macro-particles for calculation of charge distribution and emittance. The electromagnetic space-charge field is solved self-consistently in 3-D configuration space according to macro-particle distribution which is updated every time interval. A more detailed calculation method and some results can be found elsewhere.[7] The other, which is a simplified version, uses macro-particles only for emittance calculation and the space-charge field is fixed throughout a simulation. An initial charge distribution is assumed to be Gaussian in 3-D configuration space and we use the analytical formula of the transverse space-charge field for that distribution.[8] Nevertheless, an evolution of the bunching factor due to the rf voltage envelope, adiabatic damping of transverse beam size, and cancellation of an electric field by a magnetic field depending on the Lorentz factor γ are included. Once the distribution evolves into non-Gaussian or the emittance grows significantly, the approximation which assumes constant distribution function and emittance is not appropriate. In that sense, the latter for space-charge calculation is only appropriate to see the growth rate in a quick way, not to predict asymptotic emittance.

Space-charge effects are one of the crucial beam dynamics issues for the emittance preservation especially in the LEB and MEB. In each stage of the tracking studies we have mentioned above, we should include the effects in a proper way and that is certainly one of the goals of the code Simpsons. Some other methods to incorporate space-charge effects are also under development.[9] Although the simulation of the LEB with space-charge effects becomes almost possible in an established way, the transition crossing simulation with space-charge effects, especially of the transverse direction, seems to need further development.

REFERENCES

1. More detailed description and test results of the code can be found in S. Machida, SSCL-570, SSC Laboratory, 1992.
2. E. Keil, *Physics of Particle Accelerators*, AIP Conference Proceedings 153, p. 83, 1987.
3. L. Schachinger and R. Talman, *Particle Accelerators*, Vol. 22, pp. 35–56, 1987.
4. H. Grote and F. C. Iselin, CERN/SL/90-13(AP), Revision 2, CERN, 1991.
5. M. Li, P. Zhang, and S. Machida, *this conference*.
6. K. Johnsen, Proc. of CERN Symposium on High Energy Accelerators and Pion Physics, Vol. 1, p. 106, 1956.
7. S. Machida, *et al.*, to be published as Proc. of XVth International Conference of High Energy Accelerators, 1992.
8. M. Furman, SSC-N-312, SSC Laboratory, 1987.
9. J. Koga, T. Tajima and S. Machida, *this conference*.

AN EFFICIENT SYMPLECTIC APPROXIMATION FOR FRINGE–FIELD MAPS

G. H. Hoffstätter and M. Berz
Department of Physics and Astronomy, and
National Superconducting Cyclotron Laboratory
Michigan State University, East Lansing, MI 48824

ABSTRACT

The fringe fields of particle optical elements have a strong effect on optical properties. In particular higher order aberrations are often dominated by fringe-field effects. So far their transfer maps can only be calculated accurately using numerical integrators, which is rather time consuming. Any alternative or approximate calculation scheme should be symplectic because of the importance of the symplectic symmetry for long term behaviour. We introduce a method to approximate fringe-field maps of magnetic elements in a symplectic fashion which works extremely quickly and accurately. It is based on differential algebra (DA) techniques and was implemented in COSY INFINITY. The approximation exploits the advantages of Lie transformations, generating functions, scaling of the map with field strength and aperture, and the dependence of transfer maps on the ratio of magnetic rigidity to magnetic field strength. The results are compared to numerical integration and to the approximation via fringe-field integrals. The quality of the approximation will be illustrated on some examples including linear design, high order effects, and long term tracking.

INTRODUCTION

The fringe–field map of a static particle optical element is defined as the concatenation of an inverse drift from the effective field boundary to a point outside the field, the map through the fringe field, and the inverse map of the main field back to the effective field boundary[1,2,3]. This is illustrated in figure 1.

So far high order transfer maps of fringe fields can only be calculated accurately using numerical integrators[4], which is very time consuming. Figure 2 shows the ratio of the time used for the computation of the main–field map to the time used for the fringe–field maps of a typical quadrupole and dipole for different expansion orders. (The quadrupole used in the examples of this paper has: length 41.9cm, a pole–tip field of 2T, and an aperture of 2.54cm. The wedge dipole has radius 2m, an angle of 30°, and an aperture of 2.54cm. The ion chosen is $^{16}O^{3+}$ with an energy of 25MeV per nucleon. The fringe fields used are those of the Enge model[5,6,7].) The accurate consideration of fringe fields slows the computation down by orders of magnitude.

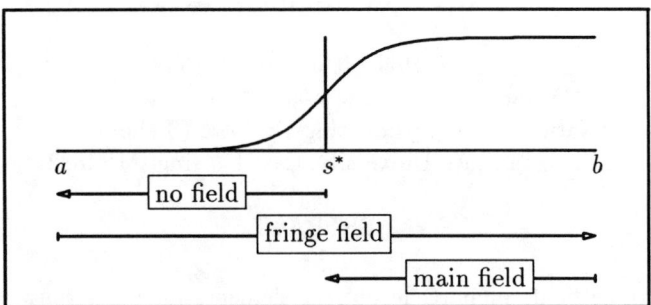

Figure 1: Definition of fringe–field maps.

The computation of main–field maps for multipoles is so speedy because no numerical integration is needed[8]. The map $\mathcal{M}(\vec{z})$ that transforms the canonical variables of motion $\vec{z}$ of a particle before the main field into those variables behind the main–field region can be calculated directly using the Lie derivative $L_f = \vec{f} \cdot \nabla + \partial_s$ that governs the motion $d\vec{z}/ds = \vec{f}(\vec{z})$ in the main field:

$$\frac{dg(\vec{z})}{ds} = L_f g(\vec{z}) \quad , \quad \mathcal{M}(\vec{z}) = e^{l_0 L_f} \vec{z} \tag{1}$$

where g is an differentiable function, s is the path length of the reference trajectory, and l_0 is the length of this trajectory in the main–field region. The first and second order is particularly fast since the equation of motion is evaluated analytically up to second order for multipoles. For dipoles, the equation of motion is solved analytically to all orders by geometrical reasoning, including edge angles and edge curvatures. Furthermore, the main–field map is accurate to machine precision (10^{-16}), whereas COSY uses a Runge–Kutta of eighth order which for the sake of speed is usually set to an accuracy of 10^{n-9} for order n. This integrator is thought to be most efficient for the differential equations of particle dynamics[9]. The righthand picture in figure 2 shows the normalized average relative difference for map coefficients computed with and without fringe fields for different orders of the map. Note especially that this average $< \frac{|a-b|}{|a|+|b|} > / \ln(2)$ would approach 1 for totally randomly distributed numbers a and b.

The simplest approximation is to ignore fringe fields, which is often referred to as the sharp cut off fringe field (SCOFF) approximation. This approximation is very inaccurate. Firstly, a field raising rapidly from zero to the main–field value does not satisfy Laplace's equation. In second order this problem was solved with the impulse approximation[10] which is used in the second–order particle optics code TRANSPORT and related codes[11]. Secondly, the detailed fringe-field shape has a strong effect on optical properties[1,2]. The results of computations with fringe

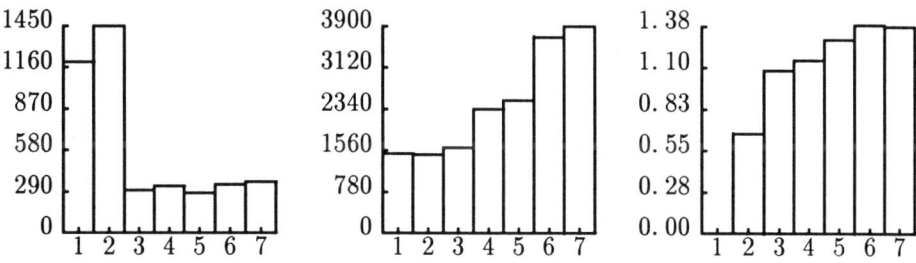

Figure 2: The ratio of computation speed for the two fringe-field maps and the main-filed map using COSY INFINITY. Left: quadrupole, middle: dipole.
Right: Average relative difference of map coefficients for a dipole with and without fringe field.

fields differ significantly from the results of the SCOFF approximation.

Often the effect of fringe fields is approximated by fringe-field integrals. This was done in third order for GIOS[12,13] and attempts are being made to extend this to fifth order for some particle optical elements[1,2]. The approximation by fringe-field integrals, however, has some serious disadvantages:

- It is nonsymplectic and therefore not especially suited for cases where symplectic tracking can be advantageous, for example circular machines.

- It represents the fringe effect well only if the region of the fringe field is not much bigger than the dimension of the beam diameter.

- It is limited to low orders and usually not very accurate.

In order to speed up the fringe-field calculation in the arbitrary order code COSY INFINITY, we searched for a method that does not have those drawbacks, works fast, and to all orders.

SYMPLECTIC SCALING

In the following we will use TRANSPORT notation for the map[14] which means the variables of motion are the cartesian coordinates and slopes x, x', y, y', the path length difference l, and the relative momentum deviation from a reference momentum δ_p. Those six quantities form the vector $\vec{z}$. The transfer map describes the transformation of an initial $\vec{z}_i$ into a final $\vec{z}_f$ by means of an optical element:

$$\vec{z}_f = \mathcal{M}^P(\vec{z}_i) . \qquad (2)$$

The index P indicates that the map $\mathcal{M}$ depends on certain parameters like the momentum p, mass m, and charge z of the reference particle and the aperture A of the optical element.

From now on we will restrict ourselves to magnetostatic elements, although parts of the procedure are applicable to electrostatic elements, too. The bending radius of the path of a particle with momentum p at field B is

$$R = \frac{p}{zB_\perp} \tag{3}$$

where $B_\perp$ denotes the field component perpendicular to the momentum of the particle. All maps that describe particles with equivalent bending radii along their path are identical. If the map for a specific beam is known as a function of the field B at the pole tip, the transfer map for all other beams can be computed:

$$\mathcal{M}^{p^*,m^*,z^*,B^*} = \mathcal{M}^{p,m,z}(B)\big|_{B=B^*\frac{z^*p}{p^*z}} \, . \tag{4}$$

The map on the right hand side is known as a function of the field B, whereas the left hand side describes a map that is calculated at a certain field B^*. This is just another way of saying that the map depends only on the ratio of field to magnetic rigidity.

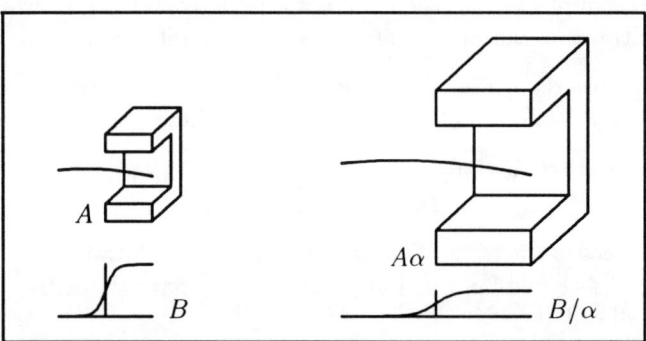

Figure 3: The coordinates of the particle trajectories in two elements scale with the factor α if the elements scale with the factor α and the fields scale with the factor $1/\alpha$.

Let us now consider two similar magnetostatic elements that differ only by a scaling factor α. If the bending radii also differ by a factor of α, the maps are similar. Equation (3) shows that this is the case whenever the increase in size by a factor α is accompanied by a decrease in the field strength by the same factor. After scaling the coordinates x, y, l, we obtain

$$\begin{pmatrix} \mathcal{M}_x^{B/\alpha,A\alpha} \\ \mathcal{M}_{x'}^{B/\alpha,A\alpha} \\ \mathcal{M}_l^{B/\alpha,A\alpha} \\ \mathcal{M}_{\delta_p}^{B/\alpha,A\alpha} \end{pmatrix}_{(x_i,x_i',l,\delta_p)} = \begin{pmatrix} \alpha \mathcal{M}_x^{B,A} \\ \mathcal{M}_{x'}^{B,A} \\ \alpha \mathcal{M}_l^{B,A} \\ \mathcal{M}_{\delta_p}^{B,A} \end{pmatrix}_{(x_i/\alpha,x_i',l/\alpha,\delta_p)}. \tag{5}$$

The second dimension (y, y') is not mentioned because it has the same properties as (x, x').

To conclude, we state that the knowledge of the transfer map as a function of the field strength at the pole tip for particles with a specific magnetic rigidity and for an element with a specific aperture is sufficient to know all transfer maps of similar elements for all energies, masses, and charges. In fact the map does not even have to be known as a function of the field because the dependence on the field can be obtained by equation (4) from the dependence of the map on the momentum. In general it is useful and customary to work with the canonical coordinates[4] $(x, p_x/p_0, y, p_y/p_0, (t_0 - t)E_0/p_0, \delta_E)$ where the subscript 0 indicates quantities of the reference particle that defines the reference trajectory. For this purpose one has to transform between TRANSPORT and COSY notation.

COSY INFINITY can readily compute the map of a fringe field of certain aperture $\mathcal{M}^{E,m,z,A}$, with δ_E being the sixth variable. The map $\mathcal{M}^{E,m,z,A}$ of a certain fringe field for a certain beam has to be stored once as a function of B in order to compute maps of similar fields and all kind of beams. The functional dependence on B can be approximated by a Taylor expansion, the coefficients of which are computed by COSY INFINITY automatically. The accuracy of this expansion depends on the chosen order and on the relative difference of the ratio of bending radius to aperture in the scaled and the saved element. Unfortunately the symplectic structure of the map would not be conserved in this process. Symplecticity, however, is an intrinsic symmetry of canonical motion that arises from the special structure of Hamilton's equations. It should not be violated, especially when long term behaviour is of interest[15,16,17].

This drawback can be eliminated by storing the reference map $\mathcal{M}^{E,m,z,A}$ in a symplectic representation, either in the form of a generating function[18], or in the form of a Lie exponent

$$\vec{z}_f = L(B)e^{:P(B):}\vec{z}_i. \tag{6}$$

In higher orders the representation via generating functions is slow because a map inversion is required[8]. The Lie representation has the disadvantage that the matrix $L(B)$ can only be approximated and is not exactly symplectic. A combination of both methods is most efficient: We represent the nonlinear part by $P(B)$ and the linear part by the generating function $F(B)$ that is most accurate for the given matrix $L(B)$.

EXAMPLES

In this section, we will illustrate the profitable use of the method with several examples. In order to evaluate speed and accuracy of the proposed approximation, we study a certain aberration coefficient of a quadrupole. Figure 4 shows the dependence of the expansion coefficient $(x|xxa)$ as a function of the field B at the pole tip. Because functions like this can be closely approximated by polynomials, symplectic scaling (SYSCA) is quite accurate. Even at the border of the

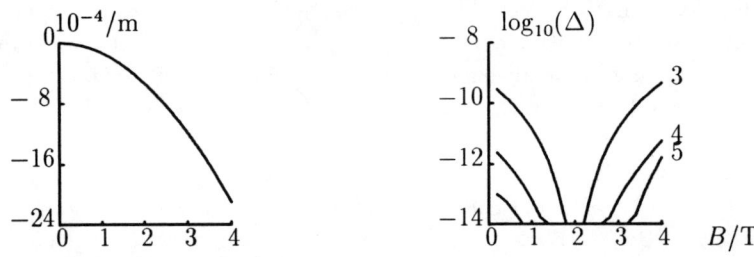

Figure 4: left: $(x|xxa)$ for a quadrupole as a function of the field at the pole tip. right: Error Δ of the approximation of $(x|xxa)$ with different expansion orders for the reference representation at $B=2T$.

range in figure 4 the presented method is more accurate than the COSY standard integrator. Close to the value with which the reference file was produced, the accuracy increases drastically. The results in figure 4 were obtained by evaluating the symplectic reference representation to third, fourth, and fifth order. The accuracy can be further improved by increasing this order which of course increases the computation time that has to be invested for creating the reference map in advance. This investment can be very much rewarding, especially when beamlines or spectrometers are being fitted or when system errors are analyzed so that maps of similar fringe fields are needed over and over again with only slightly different parameters. The SYSCA approximation is especially helpful in the design of a realistic system after approximate parameters of the elements have been obtained by neglecting fringe fields. These values can be used to create a reference file for symplectic scaling. In this way, a very high accuracy almost equivalent to accurate but time intensive numerical integration can be obtained. The time advantage of this method is illustrated in figure 5.

Fringe fields do have noticeable effects already in first order. In the example of the A1200[19] isotope separator at the NSCL, the effect of the fringe fields on

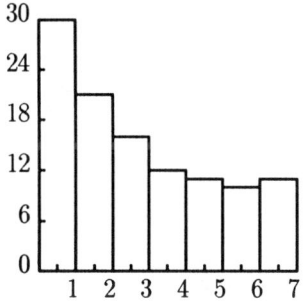
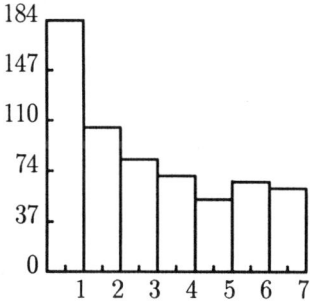

Figure 5: Factor of time advantage of SYSCA to numerical integration with accuracy of 10^{n-9} as a function of the expansion order. Left: Quadrupole, Right: Dipole.

Θ and C_0 with SCOFF approximation	80.8840°	−65.96m
Θ and C_0 with dipole fringe fields only	81.1696°	−65.96m
Θ and C_0 with quad fringe fields only	81.2694°	−682.68m
Θ and C_0 with SYSCA approximation	81.2701°	−687.10m
Θ and C_0 with actual fringe fields	81.2702°	−687.10m

Table 1: Tilt angle and opening aberration for various fringe–filed models.

the calculated setting of the field strength is shown in figure 6. The fringe fields were described by Enge functions, and the Enge coefficients had been fitted to measured field data. Here the time advantage of the proposed approximation in the fit is three minutes versus two hours. As a measure of accuracy, we study the tilt angle Θ of the dispersive image plane and the opening aberration C_0 for various approximation methods. In the discussed device the coefficient $(x|aa)$ vanishes because of symmetry of the axial ray and anti symmetry of the dipole fields; therefore $(x|aaa)$ is the relevant opening aberration,

$$\Theta = -\frac{(x|a\delta)}{(a|a)(x|\delta)} \quad , \quad C_0 = (x|aaa) \ . \tag{7}$$

Table 1 shows Θ and C_o for various finge–field models. The values of Θ with and without fringe fields differ by 0.5% for the first dispersive image plane in the A1200; the third order aberration, however, is completely wrong if fringe fields are disregarded. This comparison also shows that quadrupole fringe fields, although often disregarded, can have effects which dominate over dipole fringe fields. Nonlinear effects can be seen by sending a cone of particles through the 7th order A1200 map. The images with SCOFF and SYSCA approximation are shown in figure 7. The maximum angle used is 15mrad.

An Efficient Symplectic Approximation

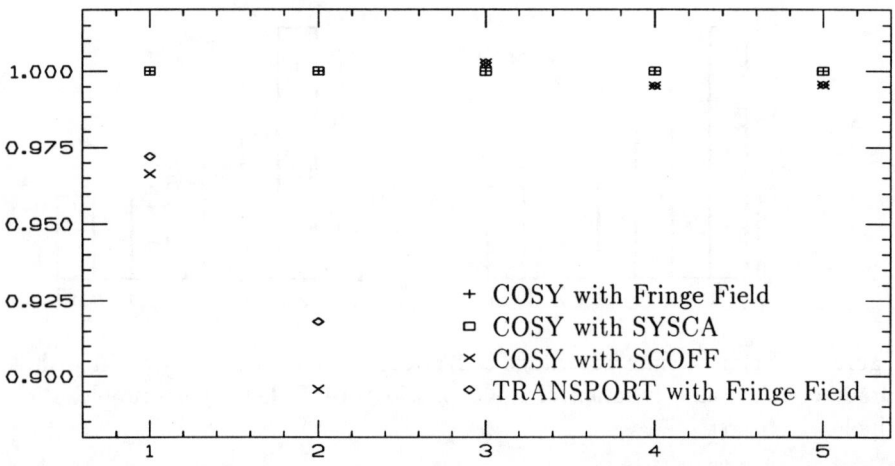

Figure 6: Relative deviation of predicted field settings with SCOFF and SYSCA from the correct settings for five quadrupoles. The standard fringe field approximation of TRANSPORT is given as a reference; the deviation is mainly due to the neglect of quadrupole fringe fields.

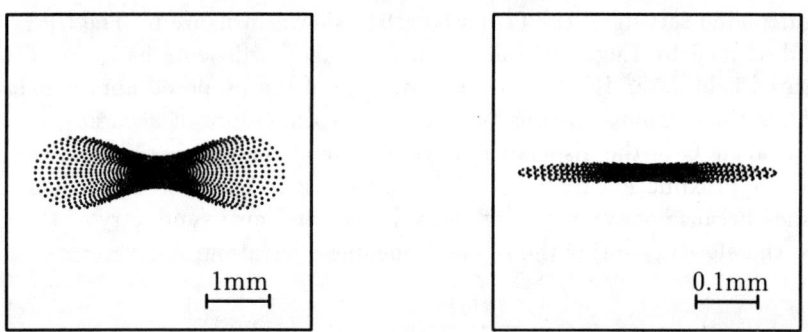

Figure 7: Beam spots with SYSCA (left) and SCOFF (right) approximation. The plot produced with the exact fringe fields can not be distinguished from the plot produced with SYSCA.

The effort involved in generating a symplectic approximation is rewarded when repetitive tracking is being performed. The example lattice of choice is the proposed PSR II Ring. The 9th order 5000 turn tracking pictures are displayed in figure 8. The tracking was performed with the described standard numerical

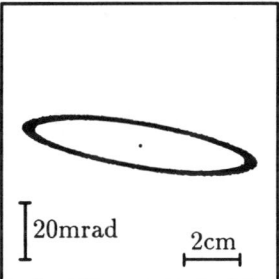 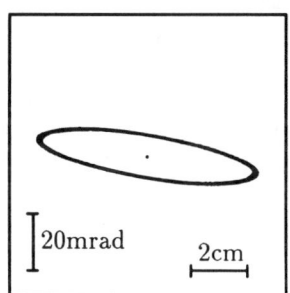 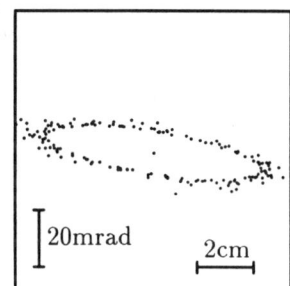

Figure 8: 5000 turn tracking with fringe fields obtained by numerical integration (left), SYSCA (middle), and a nonsymplectic fringe field approximation (right). The initial position of the particle is $(x, y) = (3cm, 3cm)$ with no initial inclination.

integration, SYSCA, and a nonsymplectic fringe–field approximation obtained by low accuracy numerical integration. Nonsymplectic tracking rapidly destroys the phase space. SYSCA yields more stable results than the numerical integration since the limited accuracy of the numerical integrator slightly violates symplecticity. The corresponding 9th order maps were produced with the SYSCA mode in COSY INFINITY in 30 minutes, whereas the standard numerical integration took 15 hours, and the nonsymplectic approximation took 44 minutes on a VAX 4000–90 computer.

REFERENCE

1. B. Hartmann, M. Berz and H. Wollnik, Nucl. Instr. and Meth. A208, 343 (1990)
2. B. Hartmann, H. Irnich and H. Wollnik, Workshop on Nonlin. Effects in Accel. Phys., Berlin, IOP Publishing (1992)
3. H. Wollnik, Nucl. Instr. and Meth. 38, 56 (1965)
4. M. Berz, Nucl. Instr. and Meth. A298, 473 (1990)
5. M. Berz, COSY INFINITY User's Manual, MSUCL-869 (1993)
6. S. Kowalski and H. A. Enge, RAYTRACE User's Manual, Dept. of Physics MIT (1987)
7. K. L. Brown IEEE Transact. on Nucl. Science, NS-28, 3, 2568 (1981)
8. M. Berz, Nucl. Instr. and Meth. A298, 426 (1990)
9. Ingolf Kübler, Master's thesis, JLU Gießen, Germany (1987)
10. R. H. Helm, SLAC Technical Report 24 (1963)
11. R. Servranckx, private communication

12. H. Wollnik, B. Hartmann, and M. Berz, AIP Conference Proceedings, 177, 74 (1988)
13. H. Wollnik, GIOS User's Manual, JLU Gießen, Germany (1992)
14. K. L. Brown, F. Rothacker, D. C. Carey, and Ch. Iselin, TRANSPORT User's Manual, SLAC-91 (1977)
15. I. Gjaja, A. J. Dragt, D. T. Abell, Workshop on Nonlin. Effects in Accel. Phys., Berlin, IOP Publishing (1992)
16. R. Kleiss, F. Schmidt, Y. Yan, F. Zimmermann CERN SL/92-02
17. R. Kleiss, F Schmidt, F. Zimmermann CERN SL/92-31(AP)
18. H. Goldstein, Classical Mechanics (1980)
19. B. M. Sherrill et al. The First Intern. Conf. on Radioactive Nucl. Beams, Berkeley, World Scientific Publishing (1990)

SIMULATING DARK CURRENT EFFECTS IN LINEAR COLLIDER STRUCTURES

U. Becker, M. Dehler, T. Weiland
Technische Hochschule Darmstadt, Fachbereich 18, FG TEMF,
Schloßgartenstr. 8, 6100 Darmstadt, Germany

ABSTRACT

With the requirement for particle accelerators getting higher and higher in energy and accelerating gradients many phenomena become more and more dominant. One of them is the dark current, resulted from collective movement of electrons which by quantum effects leave cavity walls and cause additional thermal heatup as well as many undesired effects in the linac.

In this paper we present the simulation of two multicell cavities for the DESY/THD S-band collider and the TESLA collider project. The electrons are emitted at different phases of the accelerating fields. The trajectories are calculated for several accelerating gradients with no space charge effects taken into account. From this the deposited thermal energy inside the cavity walls as well as the dark current rate are determined.

INTRODUCTION

The development of linear colliders with kinetic particle energies up to 1 TeV requires high field gradients of the accelerating mode. Thus the probability of electron emissions from the metallic boundings of cavities increases. Therefore it is necessary to simulate the motion of emitted charges to get information about their energies and trajectories.

Pertaining to this subject the following points should be remarked. The ratio of captured electrons to emitted electrons has to be known. The captured electrons are accelerated with the beam up to relativistic velocities. The accelerating electromagnetic field is weakened by the energy extracted by those particles.

There is an other reason why they should be prevented from existing in an accelearting structure, namely the distortion of beam monitors.

Another effect is the thermal load of cavity walls caused by hitting electrons. It must be examined if an enlargement of the cooling capacity of cavities is necessary.

All these reasons make it necessary to examine electron currents, which are emitted by metal surface of an accelerating tube.

PROCEDURE

For the examination of rotationally symmetric accelerating tubes the cylindrical coordinate system (r, φ, z) without azimuthal dependance is suitable. In that coordinate system the Maxwell's equations are approximated with the FIT-method (Finite Integral Technique). The rz-plane is divided in a non-equidistant grid. The allocation of field components at the grid satisfies implicitly the continuity conditions of electric and magnetic fields. The Maxwell's equations are transformed to matrix equations with corresponding properties[1]. The fields are computed in the time domain step by step with a leap-frog-scheme[2]:

$$b^n = b^{n-1} - \Delta t \, D_A^{-1} C D_s e^{n-1/2} \tag{1}$$

$$e^{n+1/2} = e^{n-1/2} - \Delta t \, \tilde{D}_\epsilon^{-1}(\tilde{D}_A^{-1} \tilde{C} \tilde{D}_s D_\mu^{-1} b^n) \tag{2}$$

where $e^{n+1/2}$ and b^n represent the electric and magnetic fields in the grid, the other symbols are special differential operators, derived by FIT-method. The equations of motion

$$\frac{d\vec{r}}{dt} = \frac{c}{\gamma}\vec{u} \tag{3}$$

$$\frac{d\vec{u}}{dt} = \frac{q}{m_0 c}(\vec{E} + \vec{v} \times \vec{B}) \tag{4}$$

with $\vec{u} = \frac{\vec{p}}{m_0 c} = \gamma \vec{\beta}$ are discretized with the same timestep Δt as the discrete Maxwell's equations. They are approximated by a similar leap-frog-scheme:

$$\vec{r}^{n+1/2} = \vec{r}^{n-1/2} + \Delta t \frac{c}{\gamma} \vec{u}^n \tag{5}$$

$$\vec{u}^{n+1} = \vec{u}^n + \Delta t \frac{q}{m_0 c} \vec{E}^{n+1/2} + \Delta t \frac{q}{m_0 \gamma^{n+1/2}} \vec{u}^{n+1/2} \times \vec{B}^{n+1/2} \tag{6}$$

The advantage of combining these two algorithms is that the calculated fields, positions and moments have to be stored only at current timesteps.

The forces on charged particles are calculated using the field components, which are situated next to the particles' location (nearest gridpoint weighting). The number of considered electrons is limited. Therefore in the simulation the space charge effects can be neglected, namely, the charged particles have no retroaction on the electromagnetic fields. Thus the needed cpu-time decreases considerably with comparison to a particle-in-cell code.

The initialization of particles is independant of field, that means the emission is not really simulated. At a given initial phase of the oscillating electromagnetic wave a fixed number of particles is simulated. Relative depositions can be made, particularly there can be decided if any dark current is possible for special field phases and accelerating gradients.

RESULTS FOR THE S-BAND LINEAR COLLIDER

The DESY/THD-collaboration is engaged with a feasibility study of a 500 GeV center of mass linear collider. In S-band Linear Collider cavities a travelling $2\pi/3$-mode of the frequency 3 GHz is used as accelerating mode. This mode is shown in different phases in Fig. 4 and 6. To simulate this wave in the time domain module a periodic boundary condition is needed. Therefore first the eigenmodes in the structure are calculated separately with magnetic and electric boundary condition in z-direction with MAFIA. Combining these solutions as initial fields in the time domain with a phase shift of 90 degrees yields the desired travelling $2\pi/3$-mode.

The amplitude of the fields is given by the accelerating voltage V for the electrons flying on the z-axis with velocity of light. That voltage can be calculated by

$$V = \int_0^{z_{max}} E_z(z) e^{j\omega z/c}\, dz \tag{7}$$

The average accelerating gradient is given by V/z_{max}. The aim is to use a gradient of 17 MV/m.

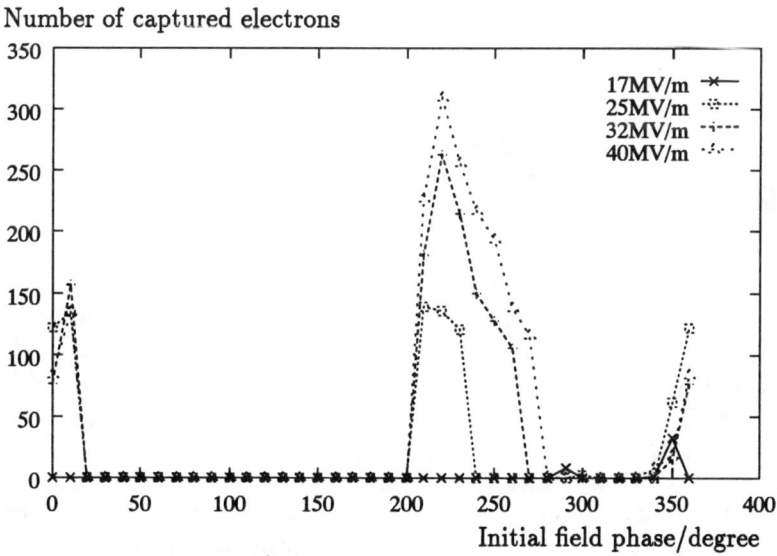

Figure 1: *The number of captured electrons (Emission of 1000 electrons) for the accelerating gradients 17MV/m, 25MV/m, 32MV/m and 40MV/m.*

480 Simulating Dark Current Effects

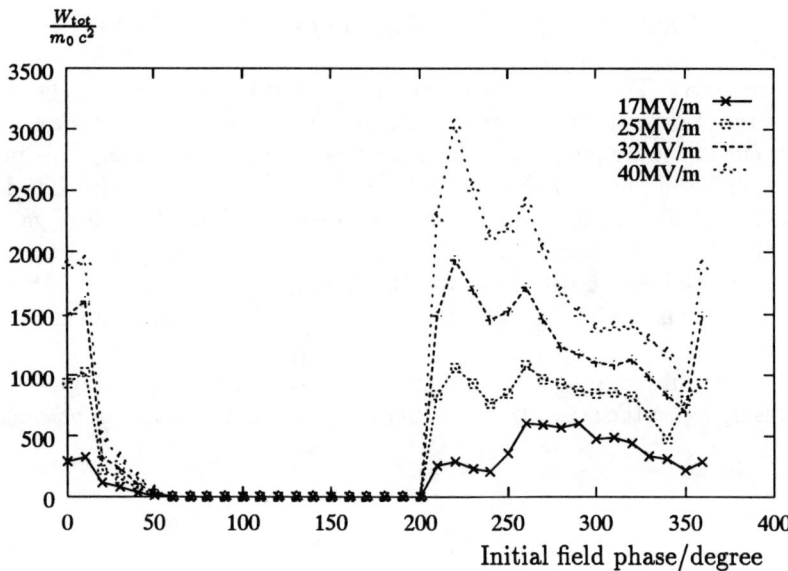

Figure 2: *The total energy extracted by 1000 electrons from electromagnetic fields for the accelerating gradients 17MV/m, 25MV/m, 32MV/m and 40MV/m.*

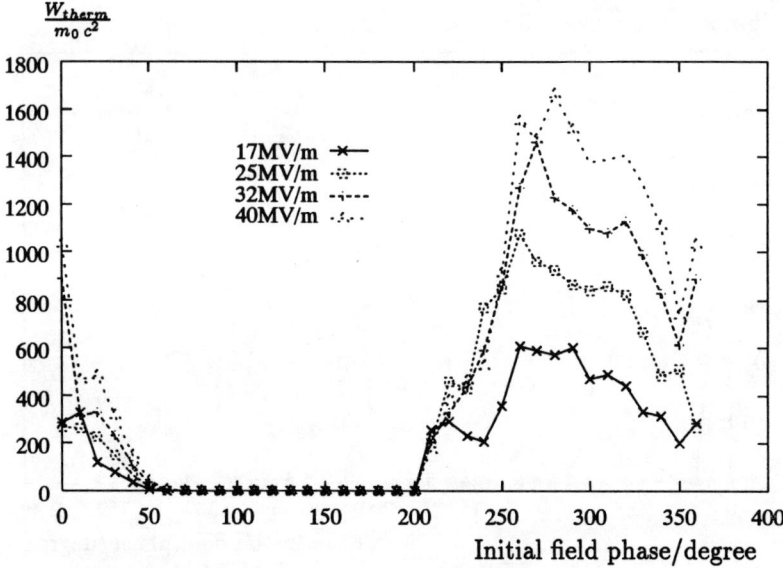

Figure 3: *The thermal load of cavities, caused by 1000 emitted electrons, for the accelerating gradients 17MV/m, 25MV/m, 32MV/m and 40MV/m.*

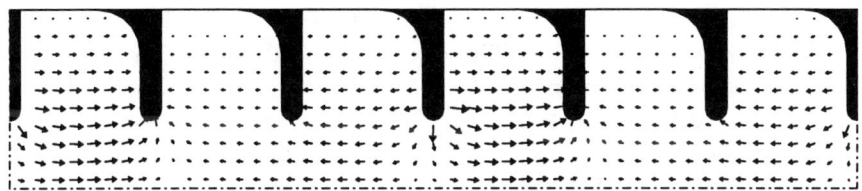

Figure 4: *The electric field of $2\pi/3$-mode in 6 cells of S-band collider (phase 240°).*

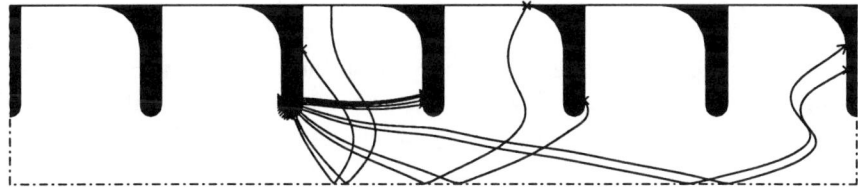

Figure 5: *The trajectories of 20 electrons, calculated for an accelerating gradient 17MV/m and initial field phase 240°.*

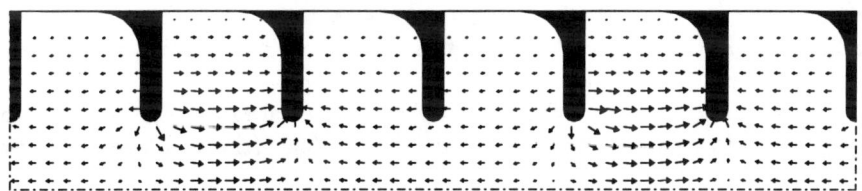

Figure 6: *The electric field of $2\pi/3$-mode in 6 cells of S-band collider (phase 350°).*

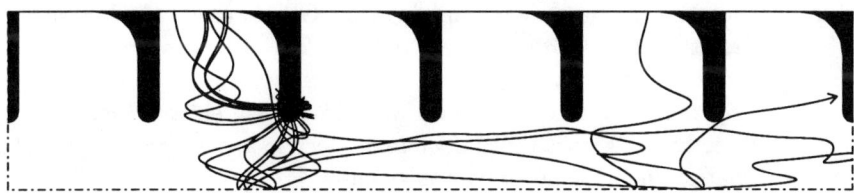

Figure 7: *The trajectories of 20 electrons, calculated for an accelerating gradient 17MV/m and initial field phase 350°.*

The trajectories of electrons which are emitted at the third iris are calculated in a part of six cells of the travelling wave tube. The simulation was done for different initial field phases and gradients. The number of simulated electrons is constant set to 1000 (in Fig. 5 and Fig. 7 only 20 trajectories are shown).

The decision whether an electron is captured or not is taken at the right side of 6 cell structure where the electron leave the calculaton area. The asymptotic behaviour is described by the following equation:

$$\sin \Theta_\infty = \sin \Theta_0 + \frac{2\pi m_0 c^2}{e E_0 \lambda} \left(\gamma - \sqrt{\gamma^2 - 1} \right) \qquad (8)$$

Θ is the relative phase between electron flying on the axis and the minimum value of z-component electric travelling field (maximum acceleration of electrons). The electron is captured if Θ_∞ exists, namely $\sin \Theta_\infty \leq 1$ (initial conditions Θ_0, γ at right bounding of simulated region). Otherwise the electron will only distribute to thermal heat but not to the dark current.

The diagram 1 shows that for a field gradient of 17MV/m dark current is only possible around the phases 290° and 350°. By increasing the voltage the number of captured electrons rises especially at the inital phase ranges 200° to 280° and 340° to 100°. Obviously there exists an threshold value of accelerating gradient, where the relative rate of captured electrons rises. This value is greater than 17MV/m.

Concerning the energy the critical phase ranges are from 220° to 260° (total energy, caused by captured electrons) and 260° to 320° (thermal load of cavities).

RESULTS FOR THE TESLA LINEAR COLLIDER

The TESLA acceleration tube consists of superconducting cavities. As acceleration mode the standing wave π-mode of 1.3 GHz is used, which is shown in Fig. 8. In this structure at two successive irises 200 electrons are emitted and their trajectories are calculated.

The π-eigenmode is calculated with MAFIA-E with electric boundary condition in z-direction. The used initial value of electric and magnetic field at the start of calculation in the time domain is given by:

$$\vec{E}(t=0) = \vec{E}_r \cos \phi \qquad (9)$$

$$\vec{B}(t=0) = \vec{B}_r \sin \phi \qquad (10)$$

The initial electric field and thus the probability of emission reach the maximum at the phases $\phi = 0°$ and $\phi = 180°$. So the results obtained in phase ranges near $\phi = 90°$ or $\phi = 270°$ can be ignored because only few electrons will be extracted out of cavity walls if the electric field is small.

Because of that the phases $\phi = 30°$ and $\phi = 200°$ are most critical concerning dark current and thermal load of cavities.

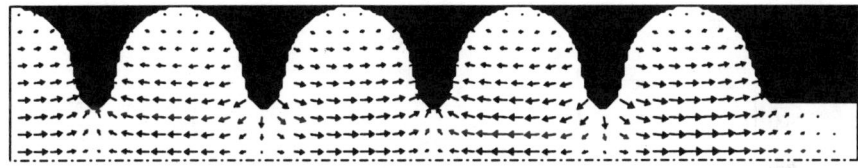

Figure 8: *The electric field of π-mode in TESLA acceleration tube (phase $0°$).*

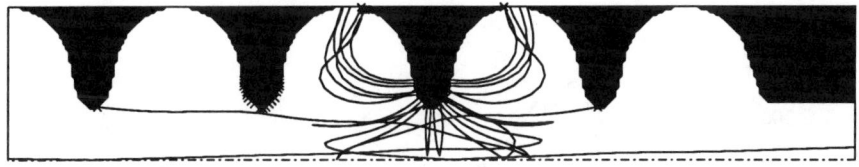

Figure 9: *The trajectories of 20 electrons per iris, calculated with accelerating gradient 15MV/m and initial field phase $0°$.*

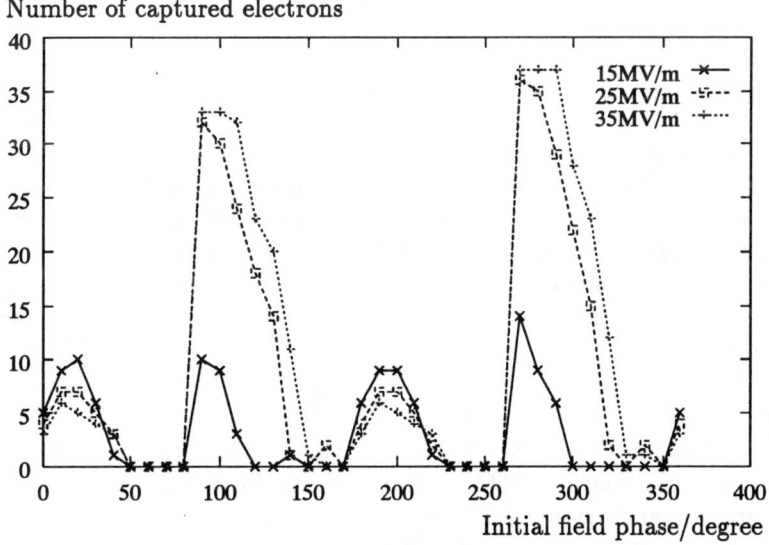

Figure 10: *The number of captured electrons (Emission of 100 electrons per iris) for the accelerating gradients 15MV/m, 25MV/m and 35MV/m in TESLA acceleration tube.*

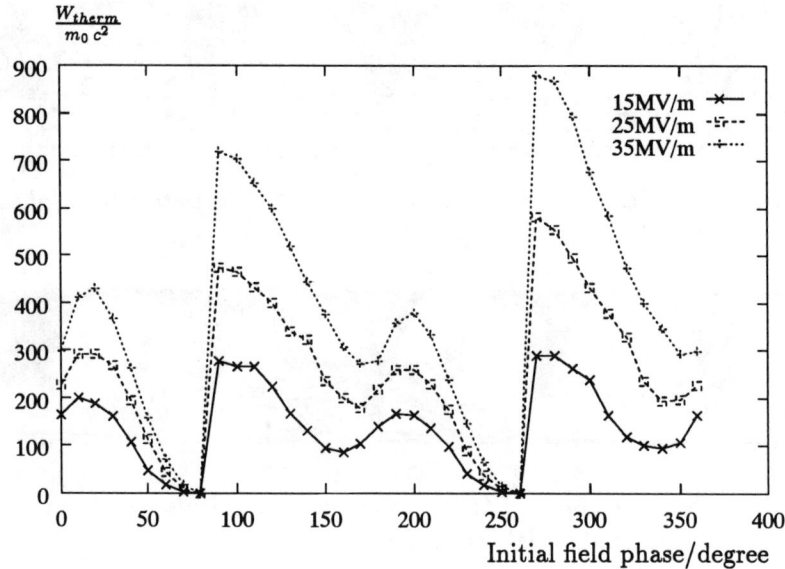

Figure 11: *The thermal load of cavities in TESLA acceleration tube, caused by 100 emitted electrons per iris, for the accelerating gradients 15MV/m, 25MV/m and 35MV/m.*

CONCLUSION

In this paper we presented the simulation of effects caused by emitted electrons in Linear Collider structures. The calculation has been done with a travelling wave tube of the DESY/THD S-band Linear Collider Study. The dark current is uncritical by operating with an accelerating gradient of 17 MV/m.

An equivalent simulation of the TESLA acceleration tube yields the result that even an accelerating gradient of 15 MV/m may produce dark current at different phases.

REFERENCES

1. Th. Weiland, *On the Numerical Solution of Maxwell's equations and Applications in the Field of Accelerator Physics*, Particle Accelerators, Vol.15, pp. 245-292, 1984
2. K. S. Yee, *Numerical Solution of Initial Boundary Value Problems Involving Maxwell's equations in Isotropic Media*, IEEE AP-14 (1966)
3. C. K. Birdsall and A. B. Langdon, *Plasma Physics via Computer Simulation*, McGraw-Hill, 1985
4. Th. Weiland, *Status Report of a 500 Gev S-Band Linear Collider Study* DESY 91-153, December 1991

A HIGH-ORDER MOMENT SIMULATION MODEL*

K. T. Tsang, C. Kostas, and A. Mondelli

Science Applications International Corporation
1710 Goodridge Dr., McLean, VA 22102

Abstract

A new transport code has been developed in which characteristics of the beam are described by transverse averages of moments of the particle distribution in four-momentum space. The set of moments satisfies a system of fluid-like equations derived from the relativistic Vlasov equation. The model is applicable to relativistic electron and ion beams, and is suitable for modeling nonlinear effects in the beam transport for heavy-ion fusion. Preliminary numerical results are presented here.

Introduction

The physical length of accelerator systems presents computational difficulties for three-dimensional discrete-particle simulations of the dynamics and emittance growth in beam transport. An alternative to the discrete-particle approach is a high-order moment equation model, which is an extension of envelope equations formulated first by Kapchinskij and Vladimirskij[1] and later adapted by Lee, Close and Smith[2].

Chernin[3] has derived a beam envelope equation from the equations of motion for a mono-energetic particle beam in a magnetic field that is linear in the transverse coordinates. In this work we are interested in beams with a spread in energy, and transport systems with a nonlinear transverse dependence of the magnetic field. For this type of system the spatial coordinate along the beam motion cannot be used as the time variable; faster particles will overtake the slower particles as time evolves. We present relativistically covariant moment equations for modeling beam transport, based on the work of Newcomb[4] and Amendt and Weitzner[5]. The beam is described by a set of partial differential moment equations, instead of a set of ordinary differential envelope equations. Our formalism is based on transverse averages of the moment equations obtained from the relativistic Vlasov equation. The spatial coordinate along the beam motion and time are the only independent variables. A similar formulation of moment equations by Channel and co-workers[6] used spatial averaging in all three coordinates to model a bunched beam with time as the only independent variable. For bunch lengths that are large compared to the betatron wavelength, it is impractical to carry sufficiently high longitudinal moments to model the oscillations within the bunch.

In the following we detail the formulation of our approach. A nonlinear space-charge model, similar to the one used in the BEDLAM code[6], has been developed and implemented. The moment equations are closed by setting higher order correlation functions to zero. Preliminary numerical results of the High-Order Moment (HOM) code are presented in the last Section.

In the current design of the heavy ion driver for inertial confinement fusion, space-charge-dominated beams are to be accelerated and transported over a large distance through recirculating bends. To increase the transportable current, magnets with a large bore-to-length ratio are used[7]. Consequently end magnetic fields become important compared with the main field, and effects of high-order nonlinear field components and moments are significant. The HOM code developed here will be an ideal tool to study beam behavior under such condition.

Relativistic Formuation

We denote the time t and local Cartesian space coordinates (x^1, x^2, x^3), where x^3 is measured along the beam motion direction and x^1 and x^2 are the transverse directions. We define $x^4 = ct$, where c is the speed of light, so that space-time is parametrized by $x^\mu, \mu = 1, 2, 3, 4$. We use a summation convention, and we assume that Latin subscripts and superscripts, i, j, k, l, are summed from one to three, while Greek subscripts and superscripts are summed from one to four. The space-time metric $(ds)^2 = dx^i dx^i - c^2(dt)^2$ becomes $(ds)^2 = dx^\mu dx^\nu g_{\mu\nu}$, where the non-zero elements of the metric tensor $g_{\mu\nu}$ are $g_{ij} = \delta_{ij}$ and $g_{44} = -1$. The metric tensors $g_{\mu\nu}$ and $g^{\mu\nu}$, which are defined so that $g_{\mu\nu} g^{\nu\lambda} = \delta_\mu^\lambda$, may be used to raise and lower indices covariantly. The usual three velocity v^i may be extended to a relativistic covariant four-velocity u^μ by the definitions $\gamma^{-2} = 1 - v^i v^i / c^2$ and $u^i = \gamma v^i, u^4 = \gamma c$ so that $u^\mu u_\mu = -c^2$.

The electromagnetic field tensor $F_{\mu\nu}$ is antisymmetric and is given by

$$E_i = cF_{i4} = -cF_{4i}, \quad B_1 = F_{23} = -F_{32}, \quad B_2 = F_{31} = -F_{13}, \quad B_3 = F_{12} = -F_{21};$$

while the Lorentz force on a particle of charge q is $q(\vec{E} + \vec{v} \times \vec{B})_i = qF^{i\mu} u_\mu / \gamma$. The general form of the external magnetic field of interest can be expressed as

$$B_i = B_{i0} + B_{i1} x^1 + B_{i2} x^2 + B_{i11} x^1 x^1 + B_{i12} x^1 x^2 + B_{i22} x^2 x^2,$$

where all the coefficients $B_{i0}, B_{i1}, B_{i2}, B_{i11}, ..., B_{i22}$ are functions of x^3, with B_{i0} the dipole, B_{ij} the quadrupole, and B_{ijk} the sextupole components. Higher order nonlinear field components can be included if desired. The beam distribution function, $f(x^\mu, u^i)$, satisfies the relativistic Vlasov equation;

$$\left[\frac{\partial}{\partial t} + \vec{v}\cdot\vec{\nabla} + \frac{q}{m}\left(\vec{E} + \vec{v}\times\vec{B}\right)\cdot\frac{\partial}{\partial \vec{u}}\right]f = 0 , \quad (1)$$

which can be rewritten in a covariant form as:

$$\left(u^\mu \frac{\partial}{\partial x^\mu} + \frac{q}{m} F^{i\mu} u_\mu \frac{\partial}{\partial u^i}\right) f = 0 \quad (2)$$

where m is the particle mass. The volume element $d\omega = du^1 du^2 du^3/\gamma$ in the four-momentum space is invariant under a Lorentz transformation. Since the transverse coordinates, x^1 and x^2, are invariant under a Lorentz transformation, we define an invariant phase space volume element under a Lorentz transformation to be $d\Omega = dx^1 dx^2 du^1 du^2 du^3/\gamma$, and a phase space average

$$\langle X\rangle = h^{-1} \int X f d\Omega, \quad (3)$$

with $h = \int f d\Omega$. We also define the following second order correlation functions,

$$[u^\mu u^\nu] = h^{-1} \int f(u^\mu - \langle u^\mu\rangle)(u^\nu - \langle u^\nu\rangle)d\Omega, \quad (4)$$

and similar definitions for the third order correlation functions. The lowest moment of the Vlasov equation (Eq. 2) gives

$$\frac{\partial}{\partial x^3} h\langle u^3\rangle + \frac{\partial}{\partial x^4} h\langle u^4\rangle = 0. \quad (5)$$

With Eq. 2 multiplied by u^ν and $u^\nu u^\lambda$ then integrated over $d\Omega$, we have

$$\frac{\partial}{\partial x^3} h\langle u^3 u^\nu\rangle + \frac{\partial}{\partial x^4} h\langle u^4 u^\nu\rangle = \frac{q}{m} h\langle F^{\nu\mu} u_\mu\rangle \quad \text{and} \quad (6)$$

$$\frac{\partial}{\partial x^3} h\langle u^3 u^\nu u^\lambda\rangle + \frac{\partial}{\partial x^4} h\langle u^4 u^\nu u^\lambda\rangle = \frac{q}{m} h(\langle F^{\nu\mu} u_\mu u^\lambda\rangle + \langle F^{\lambda\mu} u_\mu u^\nu\rangle), \quad (7)$$

respectively. There are four independent equations represented in Eqs. 6 and ten in Eqs. 7. Equations 5 to 7 are basically the same as the fluid equations of Newcomb[4] and Amendt and Weitzner[5], with the additional averaging over the transverse coordinates. If $F^{\mu\nu}$ is independent of the transverse coordinates, then Eqs. 5 to 7 can be reduced to a closed system by assuming the third order correlations are negligible, which is the standard approximation used in truncating most fluid equations. Since $F^{\mu\nu}$ depends on the transverse coordinates, Eqs. 6 and 7 cannot be closed without introducing the spatial moment equations:

$$\frac{\partial}{\partial x^3} h\langle u^3 x^i \rangle + \frac{\partial}{\partial x^4} h\langle u^4 x^i \rangle = h\langle u^i \rangle; \tag{8}$$

$$\frac{\partial}{\partial x^3} h\langle u^3 u^\nu x^i \rangle + \frac{\partial}{\partial x^4} h\langle u^4 u^\nu x^i \rangle = h\langle u^\nu u^i \rangle + \frac{q}{m} h\langle F^{\nu\mu} u_\mu x^i \rangle; \text{ and} \tag{9}$$

$$\frac{\partial}{\partial x^3} h\langle u^3 x^i x^j \rangle + \frac{\partial}{\partial x^4} h\langle u^4 x^i x^j \rangle = h\langle x^j u^i \rangle + h\langle x^i u^j \rangle \quad \text{for } i, j = 1, 2 \text{ only.} \tag{10}$$

To more concisely represent and to allow easier numerical solution, the system of partial differential equations for the second order model (described above) and higher order models should be reorganize to combined both fluid and spatial moment equations. Consider a new general variable y^λ such that $y^1 = 1$, $y^2 = x^1$, $y^3 = x^2$, $y^4 = u^1$, $y^5 = u^2$, $y^6 = u^3$, and $y^7 = u^4$. The twenty-eight equations represented in Eq. 5 thru 10 can be obtained by multiplying Eq. 2 by $y^\lambda y^\nu$ and integrating over $d\Omega$. ($\lambda = 1$ to $7; \nu = \lambda$ to 7.)

$$\frac{\partial}{\partial x^3} h\langle u^3 y^\lambda y^\nu \rangle + \frac{\partial}{\partial x^4} h\langle u^4 y^\lambda y^\nu \rangle = \Gamma(\lambda, \nu) + \Gamma(\nu, \lambda) \tag{11}$$

where

$$\Gamma(\lambda, \nu) = \begin{cases} 0, & \lambda = 1; \\ \int y^\nu u^{\lambda-1} f d\Omega, & \lambda = 2 \text{ or } 3; \\ \frac{q}{m} \int y^\nu F^{\lambda-3,\mu} u_\mu f d\Omega, & \lambda = 4 \text{ thru } 7. \end{cases} \tag{12}$$

To close the second order system of equations we assume that correlations above second order are zero. This still allows third order moments to be nonzero. The closing condition creates several equivalent families of independent variables: $\{h\langle y^\lambda y^\nu u^3\rangle\}$ and $\{h\langle y^\lambda y^\nu u^4\rangle\}$ where $\lambda = 1$ to 7 and $\nu = \lambda$ to 7, $\{h, \langle y^\lambda y^\nu \rangle\}$ where $\lambda = 2$ to 7 and $\nu = \lambda$ to 7, or $\{h, \langle y^\lambda \rangle, [y^\lambda y^\nu]\}$ where $\lambda = 2$ to 7 and $\nu = \lambda + 1$ to 7. Note that each set has twenty-eight elements. The natural set of independent variables to advance in time is $\{h\langle y^\lambda y^\nu u^4\rangle\}$. At each cell in x^3 these 28 variables describe the moments of f when correlations above second order are zero. To solve this system of equations we need to calculate $h\langle u^3 y^\lambda y^\nu \rangle$ and $\Gamma(\lambda, \nu)$ after each time advance. We need to relate $\{h\langle y^\lambda y^\nu u^4\rangle\}$ to the 84 non-zero third order moments, $\{h, \langle y^\lambda y^\nu y^\alpha\rangle\}$. Define $[\![\lambda, \nu]\!] \equiv h\langle y^\lambda y^\nu u^4\rangle$. From the expansion of the third order correlation functions,

$$[\![\lambda, \nu]\!] = h[y^\lambda y^\nu u^4] + [\![1,1]\!]\langle y^\lambda y^\nu\rangle + [\![1, \nu]\!]\langle y^\lambda\rangle + [\![1, \lambda]\!]\langle y^\nu\rangle - 2[\![1,1]\!]\langle y^\lambda\rangle\langle y^\nu\rangle. \quad (13)$$

With this we can derive a mapping from $[\![\lambda, \nu]\!]$ to $\{h, \langle y^\lambda y^\nu y^\alpha\rangle\}$.

$$\begin{aligned}
\langle u^4\rangle &= (3[\![1,7]\!] + \sqrt{9[\![1,7]\!]^2 - 8[\![1,1]\!][\![7,7]\!]})/(4[\![1,1]\!]); \\
h &= [\![1,1]\!]/\langle u^4\rangle; \\
\langle y^\lambda\rangle &= ([\![\lambda,7]\!] - 2[\![1,\lambda]\!]\langle u^4\rangle)/([\![1,7]\!] - 2[\![1,1]\!]\langle u^4\rangle); \quad (14)\\
\langle y^\lambda y^\nu\rangle &= ([\![\lambda, \nu]\!] + 2[\![1,1]\!]\langle y^\lambda\rangle\langle y^\nu\rangle - [\![1,\nu]\!]\langle y^\lambda\rangle - [\![1,\lambda]\!]\langle y^\nu\rangle)/[\![1,1]\!]; \\
\langle y^\lambda y^\nu y^\alpha\rangle &= \langle y^\lambda\rangle\langle y^\nu y^\alpha\rangle + \langle y^\nu\rangle\langle y^\lambda y^\alpha\rangle + \langle y^\alpha\rangle\langle y^\lambda y^\nu\rangle - 2\langle y^\lambda\rangle\langle y^\nu\rangle\langle y^\alpha\rangle.
\end{aligned}$$

For order n, Eq. 2 may be multiplied by $y^{\lambda_1} y^{\lambda_2} \cdots y^{\lambda_n}$ and integrated over $d\Omega$. The system of partial differential equations is ($\lambda_1 = 1$ to $7; \lambda_2 = \lambda_1$ to $7;\ldots;\lambda_n = \lambda_{n-1}$ to 7)

$$\frac{\partial}{\partial x^3} h\langle u^3 y^{\lambda_1} y^{\lambda_2} \cdots y^{\lambda_n}\rangle + \frac{\partial}{\partial x^4} h\langle u^4 y^{\lambda_1} y^{\lambda_2} \cdots y^{\lambda_n}\rangle = \Gamma(\lambda_1, \lambda_2, \ldots, \lambda_n)$$
$$+ \Gamma(\lambda_2, \lambda_1, \lambda_3, \ldots, \lambda_n) + \cdots + \Gamma(\lambda_n, \lambda_1, \ldots, \lambda_{n-1}) \quad (15)$$

where

$$\Gamma(\lambda_1, \lambda_2, \ldots, \lambda_n) = \begin{cases} 0, & \lambda_1 = 1; \\ \int y^{\lambda_2} \cdots y^{\lambda_n} u^{\lambda_1 - 1} f d\Omega, & \lambda_1 = 2 \text{ or } 3; \\ \frac{q}{m}\int y^{\lambda_2} \cdots y^{\lambda_n} F^{\lambda_1 - 3, \mu} u_\mu f d\Omega, & \lambda_1 = 4 \text{ thru } 7. \end{cases} \quad (16)$$

To close the n'th order system of equations we assume that correlations above order n are zero. Each order has a different closing condition, thus each order has a different mapping from $\{h\langle y^{\lambda_1} y^{\lambda_2} \cdots y^{\lambda_n}\rangle\}$ to $\{h, \langle y^{\lambda_1} y^{\lambda_2} \cdots y^{\lambda_{n+1}}\rangle\}$.

For second and third order systems, the systems have 28 and 84 equations respectively. Currently our computer model allows a fourth order system, which has 210 equations. The general expression for the total number of equations in an n-th order system can be expressed as:

$$1 + \sum_{m=1}^{n} \sum_{i=1}^{min(6,m)} C_i^6 C_{i-1}^{m-1}.$$

This gives 462 for the fifth order, and 924 for the sixth order systems. The advantage of Eqs.(11)-(14) is that they can be easily manipulated symbolically.

Space Charge Models

A space charge model has been implemented to include the image charges of the metallic boundary and the longitudinal component of the space charge fields. The model assumes that the charge density, ρ, can be approximated by a two-dimensional distribution of charged rods inside a cylindrical metallic pipe;

$$\rho = \sum_{i=1}^{N} q_i(x_3, t) g(\vec{x} - \vec{x}_i), \tag{17}$$

where g is the spatial distribution function for the finite size charged rods whose location are independent of x_3. Note that g depends on x_1, and x_2, while q_i depends only on x_3. The charges on the rods, q_i, are chosen to be consistent with the spatial moments. In the second order moment system there are six spatial moments, and therefore we have $N = 6$ and a matrix equation to relate the coefficients q_i with the spatial moments.

$$hX = MQ, \tag{18}$$

where M is a 6×6 matrix whose elements are of the form $\int x_k^m x_l^n g(\vec{x} - \vec{x}_i) dx_1 dx_2$, with $k, l = 1, 2$ and $m + n \leq 2$, X and Q are column matrices such that $X^T = (1, \langle x_1 \rangle, \langle x_2 \rangle, \langle x_1 x_1 \rangle, \langle x_2 x_2 \rangle, \langle x_1 x_2 \rangle)$ and $Q^T = (q_1, ..., q_6)$. Equation (18) can be easily inverted to express q_i in terms of the spatial moments.

The image charges of these rods are easy to determine. For a charge rod located at $\vec{x}_i$ with a charge q_i the image is located at $\vec{x}'_i$ with charge $-q_i$, where $\vec{x}'_i$ is along the same direction outside the metallic cylinder with a magnitude $a^2/|\vec{x}_i|$, and a is the radius of the cylinder. The electric field due to the charged rods inside the cylinder in the beam frame can be written as

$$\vec{E}(\vec{x}) = \sum_{i=1}^{N} q_i \frac{\vec{x} - \vec{x}_i}{|\vec{x} - \vec{x}_i|} - \sum_{i=1}^{N} q_i \frac{\vec{x} - \vec{x}'_i}{|\vec{x} - \vec{x}'_i|}. \tag{19}$$

This result can be transformed back to the laboratory frame. With this simplified model, the self-field contribution to moments involving $F^{\nu\mu}$ can be readily expressed in terms of other moments retained in the system and a close set of equations can be achieved.

Special Cases

The number of moment equations increases rapidly with the order. Eventhough the model resembles a one-dimensional fluid system, at high order there will be a large system of equations for numerical solution. In some applications, an approximate model can be used to obtain much faster simulations. Two simple cases deserve special attention in this regard: (1) the steady state model and (2) the single slice model.

In the steady state model, all moments and parameters are independent of time. The system is then reduced to a set of ordinary differential equations with the beam profile independent of time but the longitudinal variation to be determined. In the single slice model, the beam profile is uniform in the longitudinal direction and its time dependence has to be determined. Both of these models will yield substantial savings in computer time and memory requirements.

Results

The High-Order Moment (HOM) model agrees identically with SAIC's "ABBY"[3] code, a linear envelope model for the steady state evolution of a monoenergetic beam. To test the effect of energy spread in the HOM model, we injected two monoenergetic circular beams at $z = 0$ with γ's of 3 and 3.1, respectively, into a mismatched constant guide field. Each beam will independently exhibit betatron oscillations in the beams radius as a function of the distance of propagation, z. The space charge model has been turned off to eliminate any interaction between the two beams; the addition of the results from an envelope equation will be the exact solution. Figure 1a shows the expected beat pattern in the $\langle x^1 x^1 \rangle$ moment caused by the slight difference in frequencies of the betatron oscillations. The HOM model, when retaining up to third order correlations, does not capture this energy mixing. The average of the two betatron frequencies developes in time as shown in figures 1b and 1c. To capture the effects of this energy spread fourth order correlations must be retained (see fig. 1d).

* Work supported partially by DARPA/DSO

References
1 I. M. Kapchinskij and V. V. Vladirmirskij, in *Proceedings of the International Conference High Energy Accelerators* (CERN, Geneva, 1959), P. 274.

2. E. P. Lee, E. Close, and L. Smith, in *Proceedings of the 1987 IEEE Particle Accelerator Conference*, edited by E. R. Lindstrom and L. S. Taylor (IEEE, New Jersey, 1987), p. 1126.
3. D. Chernin, *Particle Accelerators*, *24*, 29 (1988).
4. W. A. Newcomb, *Phys. Fluids*, *25*, 846 (1982).
5. P. Amendt and H. Weitzner, *Phys. Fluids*, *28*, 949 (1985).
6. P. J. Channell, *IEEE Trans. Nucl. Sci.*, *30*, 2607 (1983); P. J. Channell, L. M. Healy, and W. P. Lysenko, *IEEE Trans. Nucl. Sci.*, *32*, 2565 (1985).
7. A. Faltens, S. Mukherjee, and V. Brady, *Particle Accelerators*, *37-38*, 123 (1992).

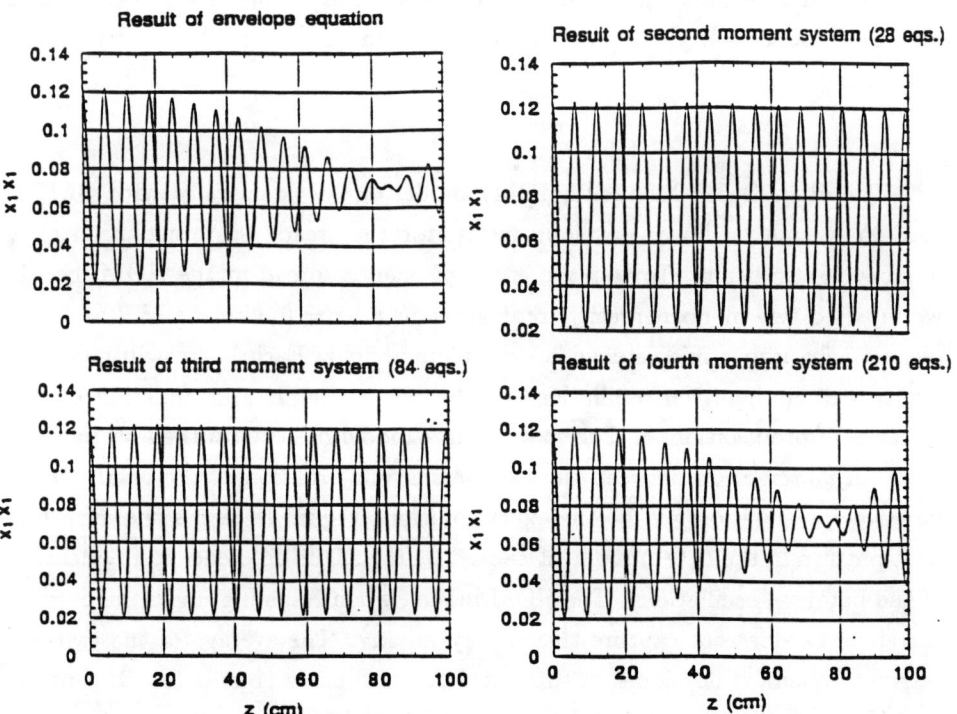

Fig. 1: a) top left; b) top right; c) bottom left; d) bottom right

A GENERAL PURPOSE RELATIVISTIC BEAM DYNAMICS CODE

Richard True

Litton Systems, Electron Devices Division
San Carlos, California 94070

ABSTRACT

DEMEOS is a finite-element based deformable triangular mesh gun code which has been used for almost two decades in the microwave tube field in the design of gridded and diode electron guns, uniform and periodic focussing systems, single and multi-stage depressed collectors. This paper describes a fully-relativistic version of the code, RDEMEOS, which has been used in the design of superpower and high impulse klystrons, gyrobeam devices, linacs, and other high power microwave devices. The variable mesh allows beam optical simulations all the way from gun to collector in practical devices. Recently, another version of the basic (nonrelativistic) code has be used to model Spindt cathodes.

CODE DESCRIPTION

The fully-relativistic version of DEMEOS can be illustrated by means of an example. Figure 1 is a simulation of a one megavolt Pierce gun and magnetically focussed beam.[1] In this case, flux threads the cathode resulting in space-charge balanced flow. Computed microperveance for the gun is 1.825 and average current density in the focussed beam is approximately 710 a/cm^2 (a very dense beam).

The gun of Fig. 1 has three grading electrodes which provide a means of attaining higher voltage operation, or longer pulse length operation.[1,2] The mesh on which the problem was solved is shown in Fig. 2. Mesh points lie along boundaries in the deformable triangular mesh which is an obvious advantage in solving guns with grading electrodes.

Basic principles behind DEMEOS are summarized in Ref. 3, and modifications to make the code fully-relativistic are summarized in Ref. 1. Referring to Fig's 3 and 4, basically, the code uses the method of self-consistent fields (Poisson's equation solved in alternation with particle trajectory tracing and deposition of space-charge). RDEMEOS integrates the equations of motion expressed in terms of pseudovelocities which are proportional to momentum.[4,5] Canonical angular momentum is assumed to be conserved. The relativistic mass factor, γ, is obtained from the potential within the region for solution stability at very high energy levels.

Ampere's circuital law is used to compute the self-magnetic field.[6] Current carried by rays enclosed by the traced ray is added to 1/2 that carried by the traced ray

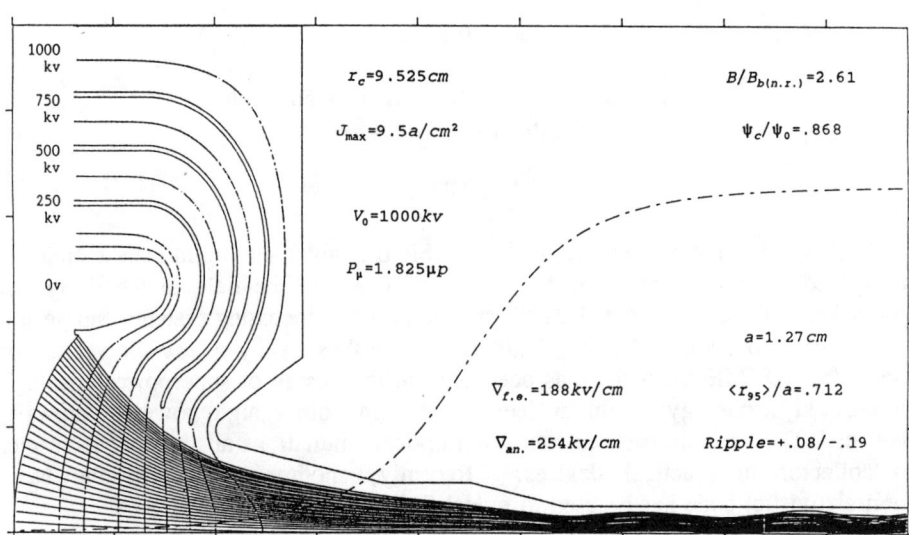

Fig. 1. Computer simulation of 1 megavolt 1825 ampere electron gun and magnetically focussed beam over a portion of the drift tube (axial scale compressed). This gun includes 3 grading electrodes and has been designed to provide a 9.13 kilojoule per pulse beam for use in high power microwave (HPM) klystrons. (From Ref. 1 ©1991 IEEE)

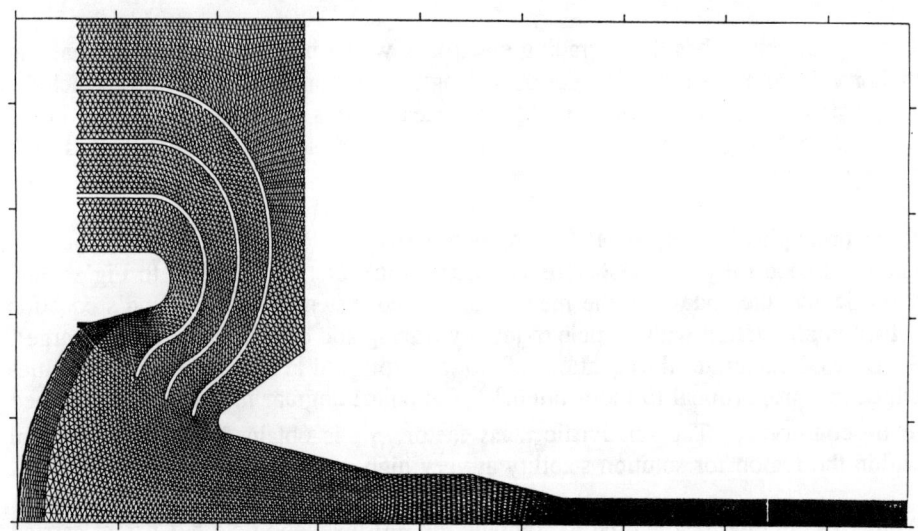

Fig. 2. Deformed triangular mesh generated by RDEMEOS which is a fully-relativistic finite element electron beam dynamics code (axial scale compressed). This code was used to produce the simulation of Fig. 1. (From Ref. 1 ©1991 IEEE)

$$\ddot{\theta} = \frac{1}{\gamma r^2} (\eta\, r\, A_\theta - \eta\, r_c\, A_{\theta c})$$

$$\frac{d}{dt}(\gamma \dot{r}) = \eta V_r - \eta (r\dot{\theta}) B_z + \eta \dot{z} B_\theta + \gamma r \dot{\theta}^2$$

$$\frac{d}{dt}(\gamma \dot{z}) = \eta V_z - \eta \dot{r} B_\theta + \eta (r\dot{\theta}) B_r$$

$$\gamma = 1 + V_p / E_0$$

WHERE V_p IS VOLTAGE AT ANY POINT ALONG TRAJECTORY FROM POTENTIAL MATRIX (2-D TAYLOR SERIES EXPANSION)

CENTER OF GRAVITY APPROACH

$$B_\theta = -\frac{\mu_0}{2\pi} \frac{|I_{encl}|}{r}$$

$$B_\theta = -\frac{\mu_0}{2\pi} \left(\sum_{i=1}^{l-1} I_i + \frac{1}{2} I_j \right) / r_p$$

DIVISION BY r_p IN PREDICTOR-CORRECTOR LOOP IN RAY TRACING INTEGRATION ROUTINE.

Fig. 4. Basic relativistic trajectory equations used in code RDEMEOS. (From Ref. 1 ©1992 IEEE)

```
┌─────────────────────┐
│ INPUT PROBLEM DATA  │
└──────────┬──────────┘
           │
┌──────────▼──────────┐
│    GENERATE MESH    │
└──────────┬──────────┘
           │
┌──────────▼──────────────────┐
│  SOLVE POISSON'S EQUATION   │◄─────┐
│  FOR POTENTIALS WITHIN REGION│     │
└──────────┬──────────────────┘     │
           │                         │
      ╱ CONVERGENCE? ╲ ──YES──┐      │
      ╲              ╱        │      │
           │ NO                │      │
┌──────────▼──────────────────┐│     │
│ COMPUTE SOURCE TERMS FOR    ││     │
│ POISSON'S EQUATION          ││     │
│ (INTEGRATE EQUATIONS OF     ││     │
│ MOTION AND DISTRIBUTE       ││     │
│ SPACE-CHARGE)               ││     │
└──────────┬──────────────────┘│     │
           │                    │     │
┌──────────▼──────────────────┐│     │
│ COMPUTE WEIGHTED ENCLOSED   │└─────┘
│ CURRENT MATRIX USED IN SELF-│
│ MAGNETIC FIELD CALCULATION  │
└──────────┬──────────────────┘
           │
           ▼
┌─────────────────────┐
│   OUTPUT RESULTS    │
└─────────────────────┘
```

Fig. 3. RDEMEOS Flow chart. (From Ref. 1 ©1991 IEEE)

itself. Weighted enclosed current values are computed once each iteration of the self-consistent fields loop and are stored in a matrix corresponding to every trajectory step of each ray. A predictor/corrector method of integration is used in ray tracing, whence the radial divisor used in computation of the self-magnetic field is taken to be the radius at each time step. A solution is reached when all variables and matrix elements are substantially identical for two consecutive iterations of the self-consistent fields loop.

Magnetic field components and the vector potential throughout the region are typically obtained from second-order (first in the case of B_r) off-axis expansions to decrease numerical sensitivity. Alternatively, these components can be obtained to a high degree of accuracy from POISSON (or PANDIRA)[7].

TEST CASES AND APPLICATIONS

In order to test the code, numerous theoretical and practical problems have been solved. One shown in Ref. 1 is a one megavolt, 1829 amp electron beam injected into a drift tube for three different magnetic field conditions, and in another test case, a 10000 amp 2.5 megavolt beam was launched into a drift tube. In all cases, results are in excellent agreement with theory.

RDEMEOS can be used to analyze and design very high convergence guns. An example of an extreme case is a gun designed in Russia for a magnicon experiment.[8] Another high convergence gun which has been run is one of the gun designs used in the SLAC 100 megawatt X-band klystron[9].

Design of collectors which can handle relativistic beams represents a challenge insofar as ion effects can cause self-magnetic field induced beam pinch which can melt a hole through the back of the collector.[1] The relativistic version of DEMEOS can handle partial or full ion neutralization of space-charge to simulate this effect. The code is also capable of simulating other specialized problems associated with relativistic beams.

At Litton, we have applied various versions of DEMEOS to design or analyze the beam optical systems in most of our travelling wave tube and klystron products which are used in high-performance microwave radars, electronic countermeasure systems, communication, and missile systems. The relativistic version of the code is particularly useful in the design of beam optical systems for our high power klystron line which includes superpower klystrons, and broadband klystrons such as the Clustered Cavity Klystron™ [10]. Many of our klystrons are used to power linear accelerators such as the L-5859 used in the Fermilab linac upgrade, and the L-5868 used in commercial medical accelerator linacs. Since DEMEOS is a generalized code, it has been used to design a host of other devices including injectors for linacs and free-electron lasers, cathode ray tubes, microfocus x-ray tubes, magnetron injection gun switch tubes (Litton Injectron™), and numerous other devices.

Further, in the research area at Litton, we have used versions of DEMEOS to study Pierce guns with grading electrodes, a microperveance 2 shadow gridded gun for a 5 megawatt klystron, beam dynamics in various specialized magnetic focussing systems, high power collectors, a novel ion expeller to prevent relativistic beam collapse within them, and the Litton Etatron™ which has promise of being a highly efficient UHF accelerator driver tube or high perveance switch tube. Finally, the code has been used to design gyrotron guns and other magnetron injection guns, and the new advanced centerpost gun for the Litton harmonic gyroklystron[11], and the NRL gyropeniotron experiment[12].

SPINDT CATHODES

Recently, a version of the basic nonrelativistic code has been applied to the analysis and design of field emitter array cathodes.[13] A Fowler-Nordheim emission model[14] was installed to enable modelling of Spindt cathodes[15] which are used in vacuum microelectronic devices. Emitters having tip radii as small as 50 angstroms have been successfully modelled in a single computer run. Results of a sample case are shown in Fig's 5-7.

CONCLUSIONS

This paper has described a fully-relativistic deformable triangular mesh gun code which has been applied to the design of very high power guns, magnetic focussing systems, and collectors. A high perveance one megavolt Pierce gun having three grading electrodes was shown which should be capable of providing a 1.83 gigawatt 9.13 kJ per pulse beam. Recent work has extended the basic nonrelativistic code to handle Spindt and silicon gated field emitter array cathodes.

ACKNOWLEDGEMENTS

The author wishes to thank his co-workers at Litton for their help and support especially G.R. Good, G.P. Scheitrum, R.S. Symons and J.F. Middaugh. The author also would like to acknowledge the following individuals in the high power microwave area: Z.G. Guiragossian, TRW; A.E. Vlieks, SLAC; S.H. Gold, NRL; R.J. Temkin, MIT; and in the field emitter array area: I. Brodie, A. Rosengreen, and C.A. Spindt of SRI; G. Jones of FED; and C.T. Sune of MCNC.

REFERENCES

1. R. True, "Beam Optics Calculations in Very High Power Microwave Tubes," IEDM Tech. Dig., pp. 403-406, Dec. 1991.
2. R. True, "Formation and Focussing of Very High Power Beams," IEEE Microwave Power Tube Conf., Monterey, CA, May 1990.
3. R. True, "The Deformable Relaxation Mesh Technique for Solution of Electron Optics Problems," IEDM Tech. Dig., pp. 257-260, Dec. 1975.

4. M. Caplan and C. Thorington, "Improved Computer Modelling of Magnetron Injection Guns for Gyrotrons," Int. J. Electronics, Vol. 51, No. 4, pp. 415-426, 1981.
5. C. Thorington, "Particle Simulation of Electron Beams with Self-Consistent Magnetic Fields," IEEE Trans. Electron Devices, Vol. ED-33, No. 11, pp. 1883-1889, Nov. 1986.
6. W.B. Herrmannsfeldt, "Electron Ray Tracing Programs for Gun Design and Beam Transport," Workshop on Linear Accelerator and Beam Optics Codes, AIP Conf. Proc. 177, pp. 45-58, 1988.
7. R.K. Cooper, et al., "Poisson/Superfish Reference Manual," Los Alamos Accelerator Code Group (MS H829) LA-UR-87-126, Los Alamos National Lab., Jan. 1987.
8. Y.B. Baryshev, et al., "Electron Optic System for Forming 100 MW Beam with High Current Density and Microsecond Pulse Duration for X-Band Magnicon," Inst. Nucl. Phys., Novosibirsk, 1990.
9. A.E. Vlieks, et al., "100 MW Klystron Development at SLAC," 1991 IEEE Particle Accelerator Conf., SLAC-PUB-5480, May 1991.
10. R.S. Symons, et al., "An Experimental Clustered-Cavity Klystron," IEDM Tech. Dig., pp. 153-156, Dec. 1987.
11. G.P. Scheitrum, R.S. Symons and R.B. True, "Low Velocity Spread Axis Encircling Electron Beam System," IEDM Tech. Dig., pp. 743-746, Dec. 1989.
12. A.K. Ganguly, S. Ahn and S.Y. Park, "Three Dimensional Nonlinear Theory of the Gyropeniotron Amplifier," Int. J. Electronics, Vol. 65, No. 3, pp. 597-618, 1988.
13. R. True, "Simulation of Thin Film Field Emitter Array Cathodes," IEDM Tech. Dig., pp. 379-382, Dec. 1992.
14. R.H. Fowler and L. Nordheim, "Electron Emission in Intense Electric Fields," Proc. Royal Soc. London, vol. 119a, pp. 173-181, 1928.
15. C.A. Spindt, et al., "Physical Properties of Thin-Film Field Emission Cathodes with Molybdenum Cones," J. Appl. Phys., vol. 47, no. 12, Dec. 1976.

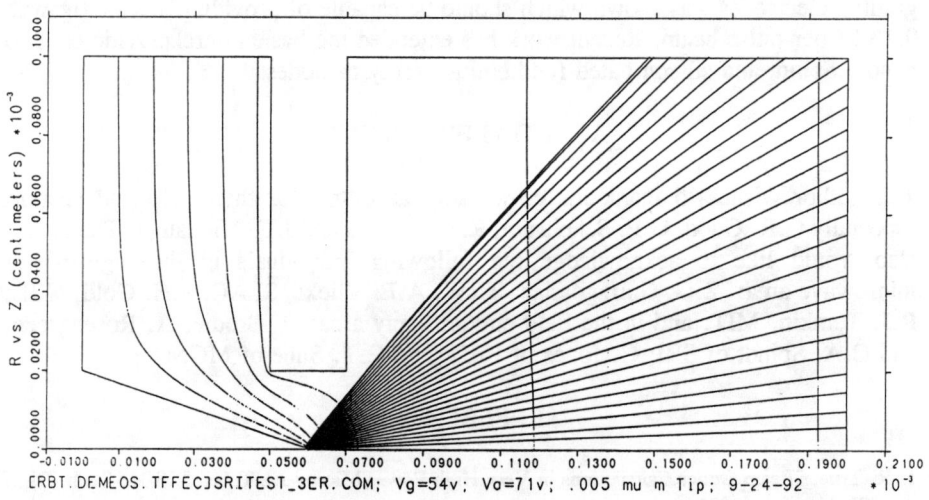

Fig. 5. Rays and equipotentials in gate of field emitter array having 50 angstrom radius moly tip. For the voltages shown, computed current equals 6.92 microamps. (From Ref. 13 ©1992 IEEE)

R. True 499

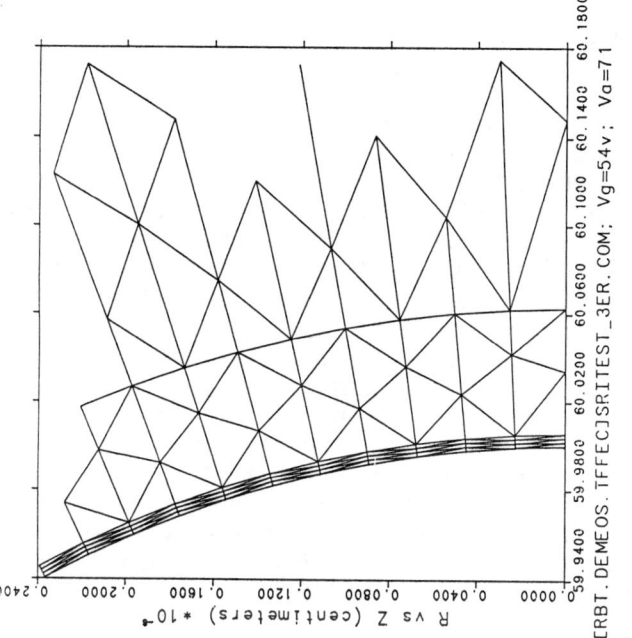

Fig. 7. Higher magnification zoom showing the mesh near the tip in the simulation of Fig. 5.

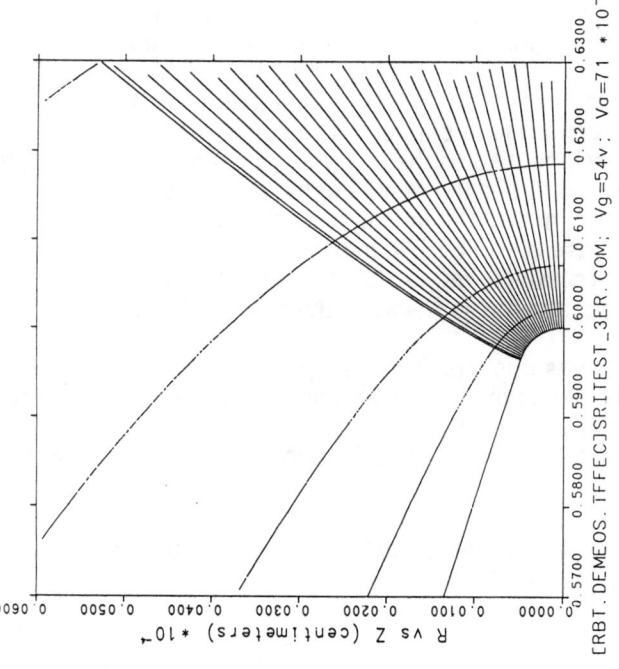

Fig. 6. Zoom view near the tip in the simulation of Fig. 5. (From Ref. 13 ©1992 IEEE)

GRAPHICAL USER INTERFACE FOR AMOS AND POISSON [*]

T.L. Swatloski
Lawrence Livermore National Laboratory
Livermore, CA 94550

ABSTRACT

A graphical user interface (GUI) exists for building model geometry for the time-domain field code, AMOS. This GUI has recently been modified to build models and display the results of the Poisson electrostatic solver maintained by the Los Alamos Accelerator Code Group called POISSON. Included in the GUI is a 2-D graphic editor allowing interactive construction of the model geometry. Polygons may be created by entering points with the mouse, with text input, or by reading coordinates from a file. Circular arcs have recently been added. Once polygons are entered, points may be inserted, moved, or deleted. Materials can be assigned to polygons, and are represented by different colors. The unit scale may be adjusted as well as the viewport. A rectangular mesh may be generated for AMOS or a triangular mesh for POISSON. Potentials from POISSON are represented with a contour plot and the designer is able to mouse click anywhere on the model to display the potential value at that location. This was developed under the X windowing system using the Motif look and feel.

INTRODUCTION

Part of the value of an experimental physics computer code is related to the number of runs the designer can perform, or the length of time it takes for a run to complete. Traditionally this has been addressed by using more and more powerful hardware, and better software algorithms, Graphical user interfaces can also increase the number of runs a designer can perform by minimizing the turn-around time of a run. The concept of turn-around time as used here is the total time it takes the designer to define the problem, to run the modeling code, and to see the results of the run. The purpose of this GUI, named Dragon, is to minimize the time it takes the designer to define a problem and in the case of POISSON, minimize the time to display the results to the designer, both of which happen to be where the designer's time is spent; generally a more valuable resource than computer time in today's world.

The interface was designed to be simple to use, promoting experimentation in building the model, and giving as much feedback as possible in the process. The GUI uses intuitive techniques such as pull-down menus, multiple choice wherever possible, and use of the mouse to select or

---

[*] Work performed under the auspices of the U.S. Department of Energy by LLNL under contract W-7405-ENG-48.

inquire about an object or location. There is a help menu that gives short descriptions of all fields in the GUI including parameters for the AMOS input file, and parameters to POISSON so the designer need not go to the user's manual for simple questions.

GRAPHIC EDITOR

The File menu (Fig. 1) allows the designer to open of a set of previously saved geometry information, open a file containing simple x-y coordinates of points to be added, save a set of geometry information, save an input file for AMOS, print the geometry model, or quit the Dragon. Filters and display of directories and existing files make retrieving and saving of files more convenient.

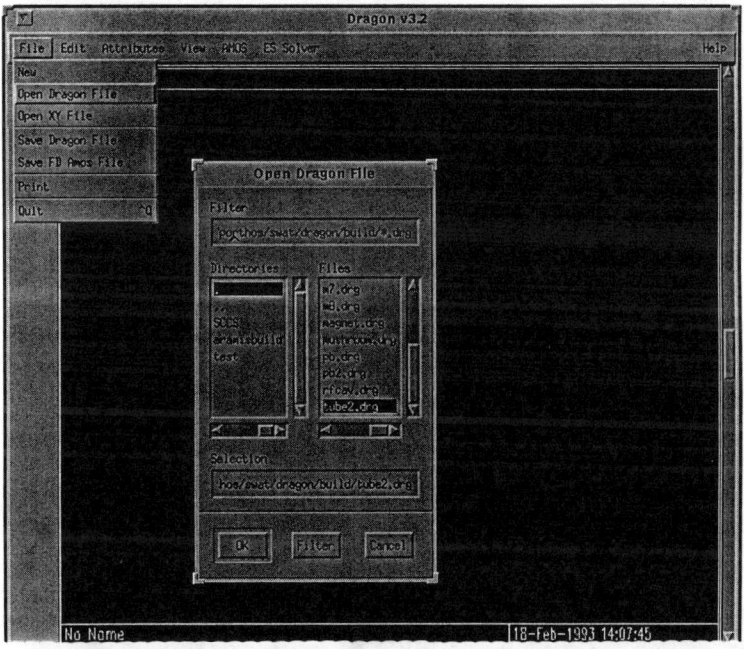

Fig. 1. Opening a Dragon file from the File Menu.

The Edit menu allows the designer to manipulate points and lines and a newly added feature is the ability to add an arc. Most edit functions are performed with the mouse.

The designer can change attributes of the model such as scale factor, materials, and grid size through the Attributes menu as shown in Figs. 2 and 3.

The View menu allows the designer to manipulate the display of the model, such as changing what is visible in the viewport, hiding or showing the material fill, and hiding or showing the grid.

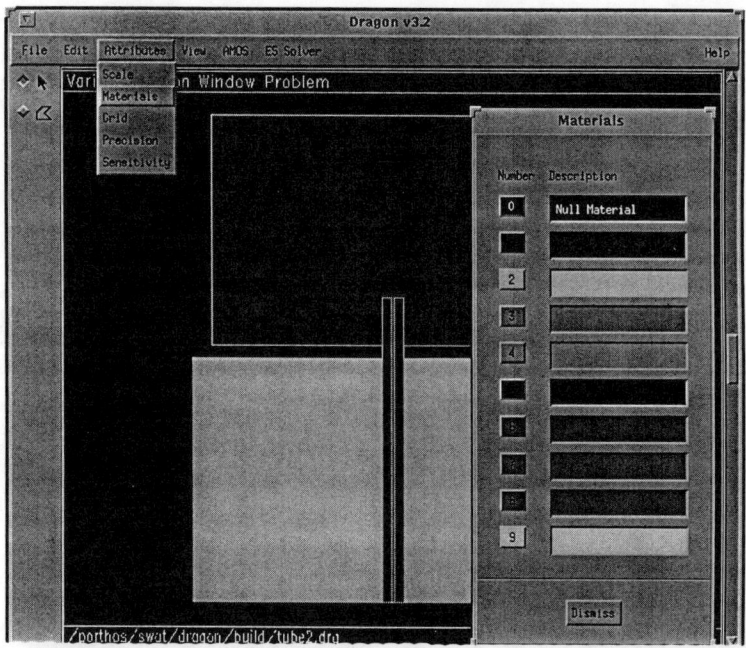

Fig. 2. Assigning materials to regions from the Attributes menu.

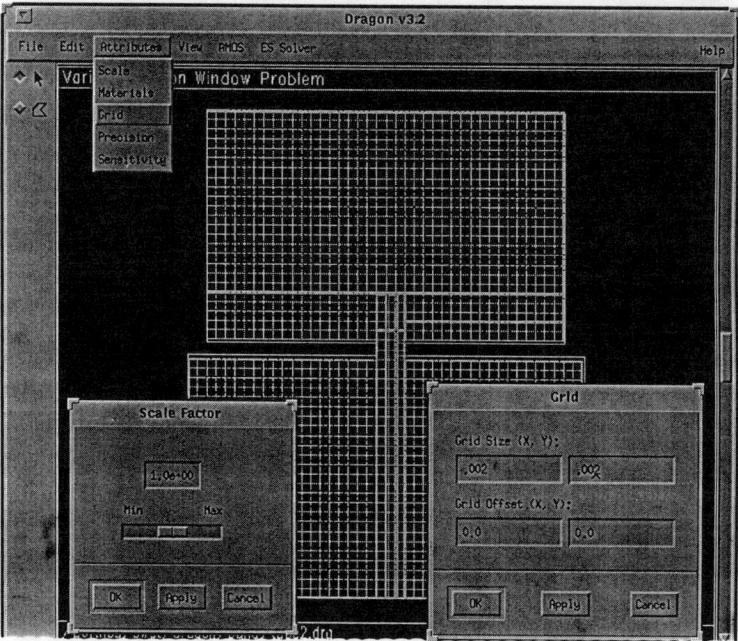

Fig. 3. Specifying the mesh and scale factor from the Attributes menu.

AMOS

AMOS (Azimuthal Mode Simulator) is a physics simulation code solving Maxwell's time dependent curl equations for axisymmetric bodies (2-1/2 D). The azimuthal mode number, m, entered by the designer allows solution of the fields of the form $e^{jm\phi}$. The code models lossy dielectrics, ferrites, and has up to five pole fit for dispersive media modeling. The code is used to identify and suppress modes leading to beam instabilities in accelerator cells, to study the rf in microwave structures and to model scattering problems.[1]

The AMOS menu allows the designer to easily build additional information for the AMOS input file by using mouse clicks on the model to fill in fields that require model coordinates, and giving lists of values from which to choose. An example of this is shown in Fig 4.

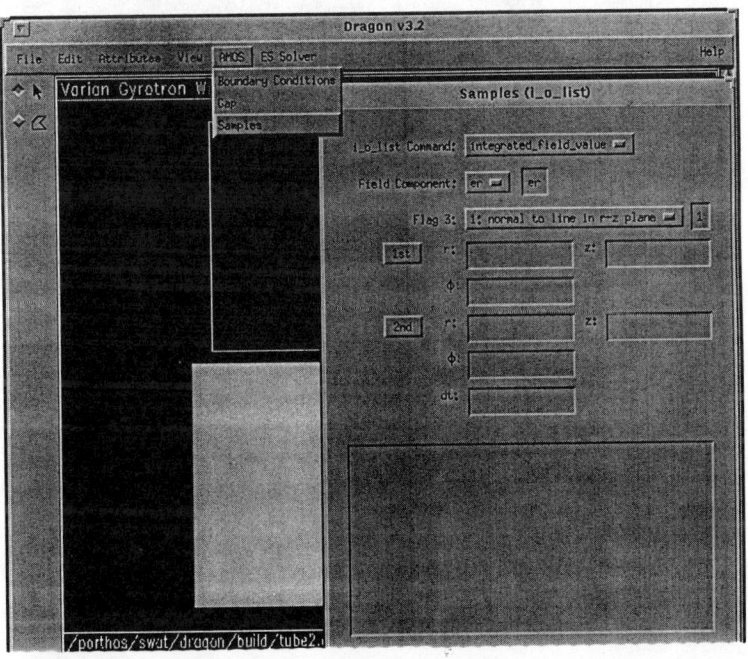

Fig. 4. Entering sample information in building the input file for AMOS.

POISSON

POISSON solves Maxwell's magnetostatic (electrostatic) equations for the vector (scalar) potential with nonlinear, isotropic iron (dielectric) and electric current (charge) distribution for two-dimensional Cartesian or three-dimensional cylindrical symmetry. It calculates the derivatives of the potential, namely, the fields and their gradients, calculates the stored energy, and performs harmonic (multipole) analysis of the potential. The code uses successive over-relaxation algorithm and an iterative scheme that steps successively throughout the mesh points.[2]

Fig. 5 shows the drawn model of an example in the POISSON reference manual for an H-shaped dipole magnet.[2] Fig. 6 shows the window for entering parameters to the mesh generation routines, AUTOMESH and LATTICE. Specific region information is in a separate window which can be scrolled to view each region's parameters. Fig. 7 shows the visual output from AUTOMESH and LATTICE, and the input window for POISSON. Fig. 8 shows the results of applying the contour plot and a window that allows the designer to change the number of intervals for the plot or the interval width.

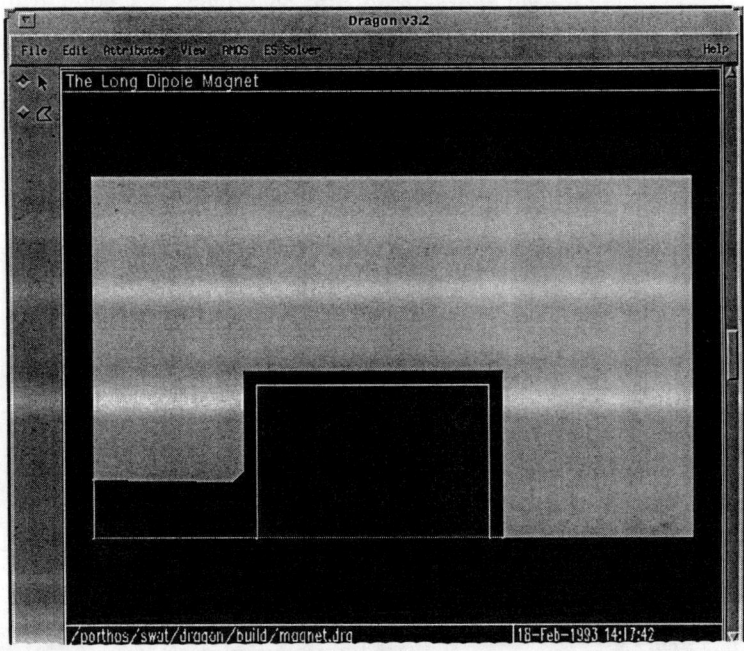

Fig. 5. H-shaped dipole magnet example.

FUTURE WORK

Dragon's graphic editor is somewhat limited and does not behave in accordance with currently accepted standards for graphic editors. A more modern graphic editor has been recently written and implemented in another application. (See Fig. 9.) It was written with the intention of being a generic 2D graphic editor, so that it may be used with other applications as well. This graphic editor will be used in any future 2D modeling development efforts in our area. Plans are to integrate this new graphic editor with Dragon in the near future, possibly by the end of this year (1993).

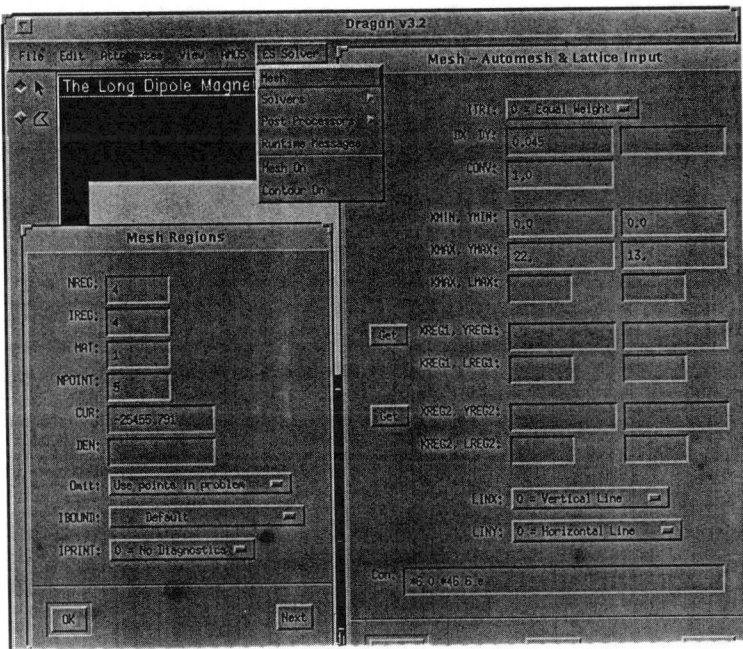

Fig. 6. Entering mesh information for AUTOMESH and LATTICE.

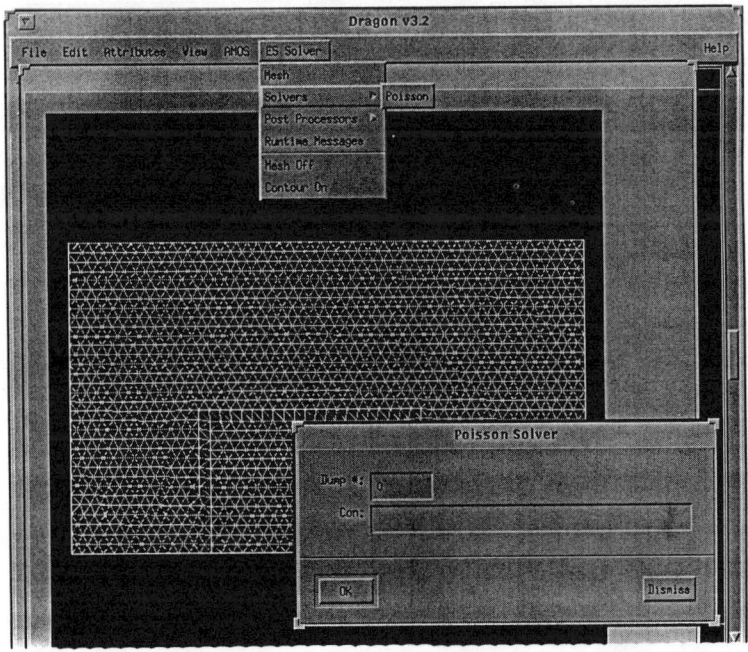

Fig. 7. Display of mesh and entering parameters for POISSON.

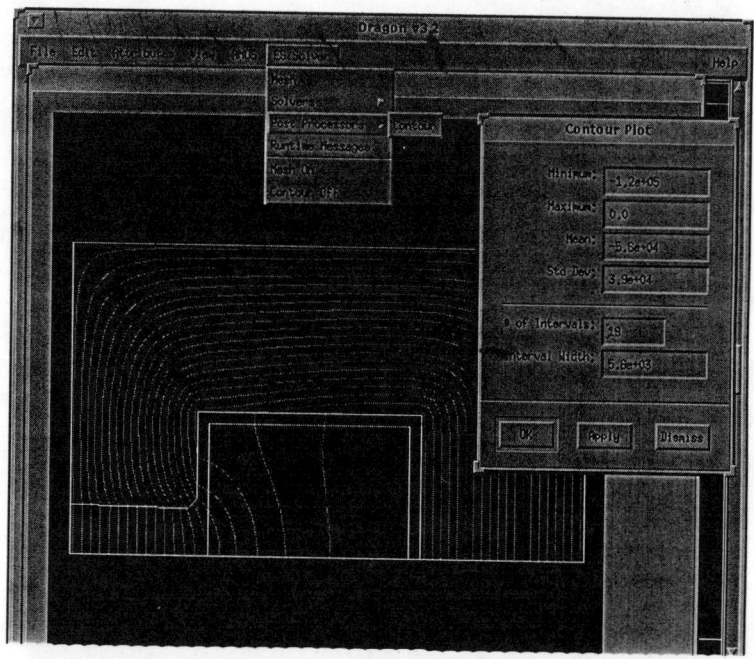

Fig. 8. Viewing default parameters for the contour plot.

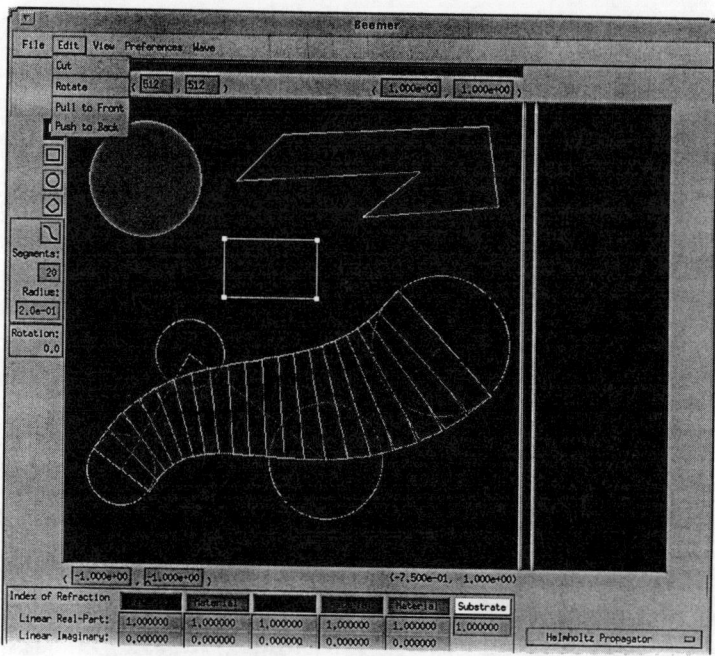

Fig. 9. Future graphic editor for Dragon.

This new graphic editor allows more types of objects to be drawn such as rectangles, circles, free-form polygons, and Bezier splines. Objects are drawn only with the mouse, no text input. It also allows better and easier manipulation of the objects with such functions as cut, copy and paste, grouping, rotation, and dynamic update display of dimensions and angles as the object is being drawn or manipulated.

The new editor uses ANSI C, Motif, and X-window graphic primitives, making it fully portable to most UNIX-based systems, with the executable easily distributed because there are no licensing issues.

ACKNOWLEDGMENTS

We wish to thank C.C. Shang and J.F. DeFord (LLNL) for developing the AMOS modeling code, giving reason for the existence of the GUI. Also, thanks to R.K. Cooper (LANL) for his help in implementing the GUI for the POISSON group of codes. And, thanks to M.E. King and R.R. McLeod for developing the original interface for AMOS.

REFERENCES

1. C.C. Shang (Livermore, CA, 1992)
2. POISSON/SUPERFISH Reference Manual, LA-UR-87-126 (Los Alamos National Laboratory, NM, Jan. 1987) part A.1.I & part B.2

ANALYSIS OF SPACE CHARGE CALCULATION IN PARMELA AND ITS APPLICATION TO THE CEBAF FEL INJECTOR DESIGN*

H. Liu

Continuous Electron Beam Accelerator Facility
12000 Jefferson Avenue, Newport News, VA 23606

ABSTRACT

The space charge calculation in PARMELA is analyzed in detail. Two different methods, the 2-D mesh method and the 3-D point-by-point method, are compared based on a cylinder model. Mesh dividing and choice of screening factor for alleviating the numerical noise are discussed and clarified. The analysis is applied to the CEBAF FEL injector design.

INTRODUCTION

Space charge is one of the most intractable issues in designing high intensity charged particle injectors and accelerators. In this respect, it has turned out that PARMELA is suitable for handling an electron bunch of several nC in several ps[1], given that space charge has been calculated properly and all the missing physics have been included.

It seems that the details of the methods for calculating space charge forces in the code are interesting to many users. In this paper, two different methods, the 2-D mesh method and the 3-D point-by-point method[2] are analyzed in detail, and compared based on a cylinder model. Several issues like mesh dividing, choice of screening factor for alleviating the numerical noise, etc. are discussed and clarified. The analysis is applied to the CEBAF FEL injector design.

CYLINDER MODEL

Fig. 1 shows a uniformly charged cylinder in a long metal tube. The cylinder has a radius r_0, a length L, and a charge Q; and the tube has a radius a. The normalized field components e_r and e_z at a point (r, z) inside the cylinder are[3]

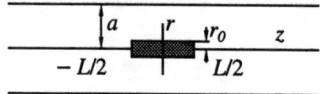

Fig. 1 Cylinder bunch model

$$e_r(u,v) = \frac{2}{\pi} \int_0^\infty S(\alpha\xi) \cos(2\alpha\xi v) I_1(u\xi)[K_1(\xi) + I_1(\xi)K_a] d\xi, \qquad (1)$$

$$e_z(u,v) = \frac{2}{\pi} \int_0^\infty S(\alpha\xi) \sin(2\alpha\xi v)\{1 - \xi I_0(u\xi)[K_1(\xi) + I_1(\xi)K_a]\}\xi^{-1} d\xi, \qquad (2)$$

where $u = r/r_0$ and $v = z/L$, $\alpha = L/2r_0$ the aspect ratio, $K_a = K_0(ka)/I_0(ka)$ the image charge contribution with $k = \xi/r_0$, $S(x) = \sin(x)/x$ the sampling function, and I_n and K_n the nth-order modified Bessel functions. The electric field strength

---

*Supported by D.O.E. contract #DE-AC05-84ER40150

$E_0 = \sigma_s/2\epsilon_0$ is used for normalization, where ϵ_0 is the vacuum dielectric constant, and $\sigma_s = Q/\pi r_0^2$ the surface charge density. The integral form instead of the series expansion is chosen for its fast convergence.

2-D MESH METHOD

The 2-D mesh method is virtually of a PIC (Particle-In-Cell) scheme. It assumes cylindrical symmetry and the bunch is embedded in an r-z mesh in the rest frame. The mesh is specified by Rmesh Zmesh Nr Nz and Frm, where Rmesh and Zmesh represent the maximum radial and longitudinal dimensions of the mesh, Nr and Nz the numbers of radial and longitudinal intervals and Frm the factor for enlarging the longitudinal dimension as the bunch is accelerated.

The mesh is designated by two 1-D arrays: Rm and Zm. Then, as shown in Fig. 2, through integrating over a *finite-size* charged ring corresponding to a bin of the mesh, a background table is established, containing

$$E_n^{(1)}(r_k, z_l) = \sum_{i=1}^{n_2} \sum_{j=1}^{n_1} \frac{R_{mi}}{\bar{R}_m} \lambda_{mi}^{(n2)} \lambda_{mj}^{(n1)} E_{kl}^{(0)}(r_k, z_l; r_{mi}, z_{mj}), \quad \begin{array}{l}(k=1,...,Nr+1)\\(l=1,...,Nz)\end{array} \quad (3)$$

where $E_n^{(1)}$ is E_r or E_z at the node (r_k, z_l), $n=(m-1)(Nr+1)Nz+(l-1)(Nr+1)+k$ the sequence number of the field data, n_1 and n_2 the radial and longitudinal numbers of zero-size rings in the bin used for integration over $\Delta S = \Delta R \Delta Z$, $\lambda_{mi}^{(n2)}$ and $\lambda_{mj}^{(n1)}$ the 2-D Gaussian integral coefficients, r_{mi} and z_{mj} the location of the zero-size ring within the mth bin, $E_{kl}^{(0)}$ the space charge fields at the node (r_k, z_l) produced by a ring located at (r_{mi}, z_{mj}) in the bin, R_{mi} and $\bar{R}_m$ the radius of the ring (i, j) and the average radius of the corresponding finite-size ring. For a given bin, this integration is done $(Nr+1)Nz$ times for all the nodes in the mesh, and then another bin is picked up and the integration is repeated till the last bin at the first column of the mesh (no more than this is necessary because of the symmetry). Therefore Nr layers of meshes are established with each layer representing $(Nr+1)Nz$ field data at all nodes. Note that n_1 and n_2 are related to the parameter Opt on the space charge card *Scheff*.

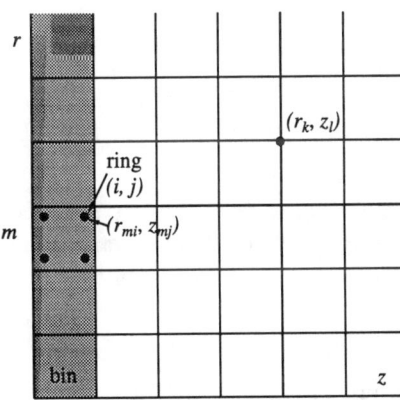

Fig. 2   2-D r-z mesh

The second issue is the charge assignment. By shifting a half interval relative to the Rm-Zm mesh, another mesh designated by Rs and Zs has been formed in advance for the use of charge assignment. See Fig. 3, where solid lines correspond to Rm-Zm arrays, and dotted lines Rs-Zs arrays. The macroparticles are located in the bins in the Rs-Zs mesh. Recall that the space charge fields at all nodes have been calculated previously by assuming a unit and uniform charge distribution in a bin. Now if the charge density in each bin can be found, the space charge fields at all nodes can be calculated by summing up the contributions from all the bins rated by the charge densities assigned to each of them.

Suppose a macroparticle (MP) is located in a bin, as shown in Fig. 3. It occupies an area of $\Delta S = \Delta R \Delta Z$ in the Rs-Zs mesh. It can be divided into four parts, A, B, C, and D. Then part A is assigned to the bin (i, j), B to $(i\text{-}1, j\text{-}1)$, C to $(i, j\text{-}1)$ and D to $(i\text{-}1, j)$, which constitute the area weighting coefficients for these four bins in the Rm-Zm mesh. The assignment is done for all particles. Now the space charge fields at all nodes can be calculated according to

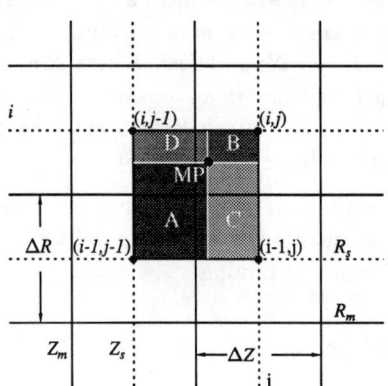

Fig. 3 Charge assignment (MP: MacroParticle)

$$E_n^{(2)}(r_k, z_l) = \sum_{j=1}^{M} A_j E_{nj}^{(1)}(r_k, z_l), \quad \begin{matrix}(k=1,...,Nr+1)\\(l=1,...,Nz)\end{matrix} \tag{4}$$

where $M = Nr \times Nz$ is the total number of bins in the mesh, A_j the charge density weighting coefficient of the jth bin, $E_{nj}^{(1)}$ the contribution to the field at that node from the jth bin, which has been calculated previously according to Eq. (3). The summation goes over all the bins in the Rm-Zm mesh.

A space charge field table based on the actual charge distribution has thus been established. Using linear interpolation, the space charge fields at any point possibly occupied by a macroparticle in the mesh can be obtained. Then the fields are transformed to the laboratory frame.

3-D POINT-BY-POINT METHOD

The 3-D point-by-point method is simple. The space charge fields produced by a moving charge Q are calculated in the laboratory frame as follows

$$\mathbf{E} = Q\mathbf{r}/\gamma^2 s^3, \qquad \mathbf{B} = \boldsymbol{\beta} \times \mathbf{E}, \tag{5}$$

where $s = r(1 - \beta^2 \sin^2 \theta)^{1/2}$, $\mathbf{r}$ the vector from the source to the observer, $\beta = \mathbf{v}/c$ the normalized velocity, θ the angle between $\mathbf{r}$ and $\boldsymbol{\beta}$, and $\gamma = 1/(1-\beta^2)^{1/2}$.

This method does not require any symmetry, which is a great release with respect to the mesh method. However, when a macroparticle is found inside another one, the one applying forces must be screened properly to avoid artificial close-encounter or numerical noise. The screening strongly depends on how the *size* of a macroparticle is defined. All the macroparticles have the same rms size which is defined by

$$\sigma_{x,y,zrms}^{(MP)} = \sigma_{x,y,zrms}^{(bunch)}/N_p^{1/3}, \qquad (6)$$

where $\sigma_{x,y,zrms}^{(bunch)}$ is the rms size of the total bunch in a specific dimension, and N_P the number of simulated macroparticles. The half width of a macroparticle, e.g., in the x dimension, is correlated with its rms size in the form

$$L_x^{(MP)} = f \sigma_{xrms}^{(MP)}, \qquad (7)$$

where f is the screening factor. A close-encounter is defined by $\Delta x < L_x^{(MP)}$, $\Delta y < L_y^{(MP)}$ and $\Delta z < L_z^{(MP)}$, where Δx, Δy and Δz represent the relative distances between two particles in the x, y and z dimensions. When a close-encounter happens, the macroparticle charge is reduced according to

$$Q^* = Q \frac{\Delta x \Delta y \Delta z}{f^3 \sigma_{xrms} \sigma_{yrms} \sigma_{zrms}} < Q. \qquad (8)$$

Table 1 Screening Factor

Profile	Screening factor
point	0
ring	1
disk	$\sqrt{2}$
square	$\sqrt{3}$
Gaussian	2
parabolic	$\sqrt{5}$

We found that the screening factor may vary from 0 to $\sqrt{5}$, depending on the density distribution assumed for a macroparticle. See Table 1. In McDonald's version of the code[2], a uniform density profile is assumed. Therefore a screening factor of $\sqrt{3}$ we found here explains the empirical screening factor of 1.75 in Ref. 2.

COMPARISON

The space charge field profiles calculated using PARMELA have been compared with the exact ones from Eqs. (1) and (2) for three different aspect ratios of $\alpha = 10$, 1 and 0.1. In all cases, the bunch has a charge of 6.681 nC, and remains 1 cm in diameter. The bunch length is changed from 10 cm to 1 cm, and in the last case, to 1 mm, to be representative and complete. 10^4 macroparticles are generated uniformly and randomly in the x-y plane but deterministically in the z-dimension. See Fig. 4.

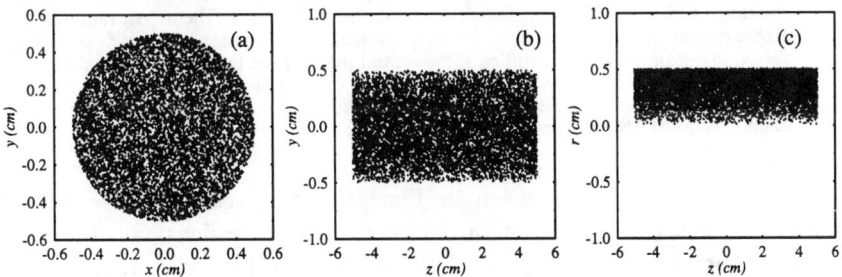

Fig. 4 Bunch model (α=10, Q=6.681 nC, N_p=10^4, (b) & (c) not to scale)

The radial and axial field profiles for $\alpha = 10$ are shown in Fig. 5. They are obtained by sending a test particle through a specified trace and recording its field response. The abscissas are normalized so that $u = 1$ represents the radial boundary, and $v=\pm 0.5$ the two end planes of the cylinder. The solid smooth curves represent the exact field profiles. The radial field profiles both at the central plane ($v = 0$, solid triangles) and at the end planes ($v = \pm 0.5$, solid and open squares) are demonstrated from (a) to (c). The on-axis axial field profiles ($u=0$, dots) are indicated from (e) - (f).

From Fig. 5, several interesting points are revealed: (1) the fields are noisier at the central part than at the edges of the bunch; (2) with the screening factor increased from $\sqrt{3}$ to 10 (which is a huge step), the noise can be substantially reduced, but in the meanwhile the particles are over-screened at the edges; (3) the mesh method gives a perfect agreement except at the central part of the bunch with mesh dividing of 40×50.

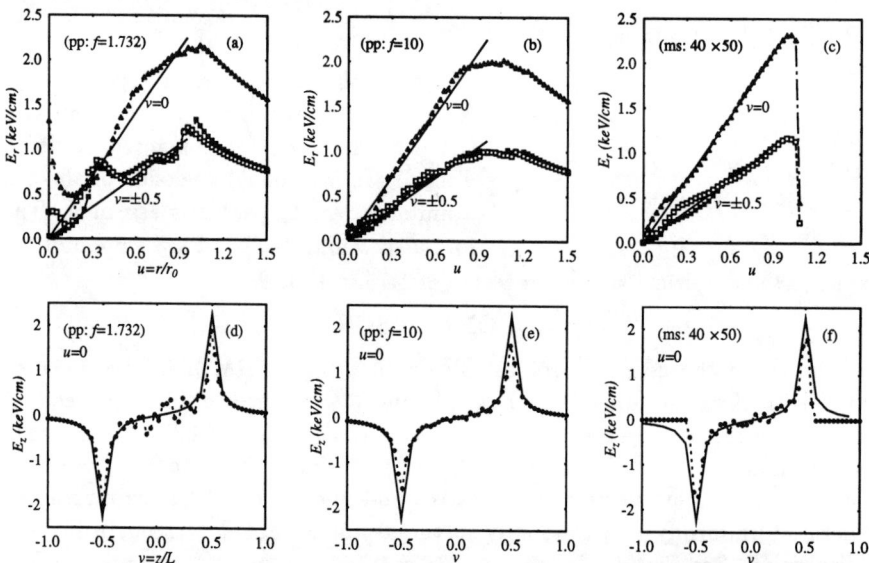

Fig. 5 Comparison between two different methods of space charge calculations in PARMELA. The space charge bunch model is shown in Fig. 4. The aspect ratio $\alpha = 10$. Abbreviations in the figure: pp - point-by-point method; ms - mesh method; f - screening factor. (a) - (c): radial field profiles at the central plane of the bunch ($v=0$) and at the end planes of the bunch ($v=\pm 0.5$). (e) - (f): on axis ($u=0$) axial space charge field profiles. Solid smooth curves represent the exact field profiles from Eqs. (1) and (2) with image charge omitted. Solid triangles: $v=0$; solid squares: $v=-0.5$; open squares: $v=0.5$.

The results for $\alpha=1$ are shown in Fig. 6. It is seen that the point-by-point method agrees completely with the exact field expressions out of the bunch ($v=\pm 1$). The agreement for the axial field profile is excellent with a screening factor of $\sqrt{3}$, as shown in (d). It is clear that a smaller screening factor under-screens at the central part but seems precise for the edges, whereas a larger screening factor over-screens at the edges but seems precise for the central part. We emphasize that it is the edge part of a bunch that needs to be treated more accurately, for that is the region where various nonlinear effects are acting.

The last case is for a very short bunch of 1 mm corresponding to $\alpha = 0.1$. It was claimed that the mesh method seems less suitable for very short bunches[2]

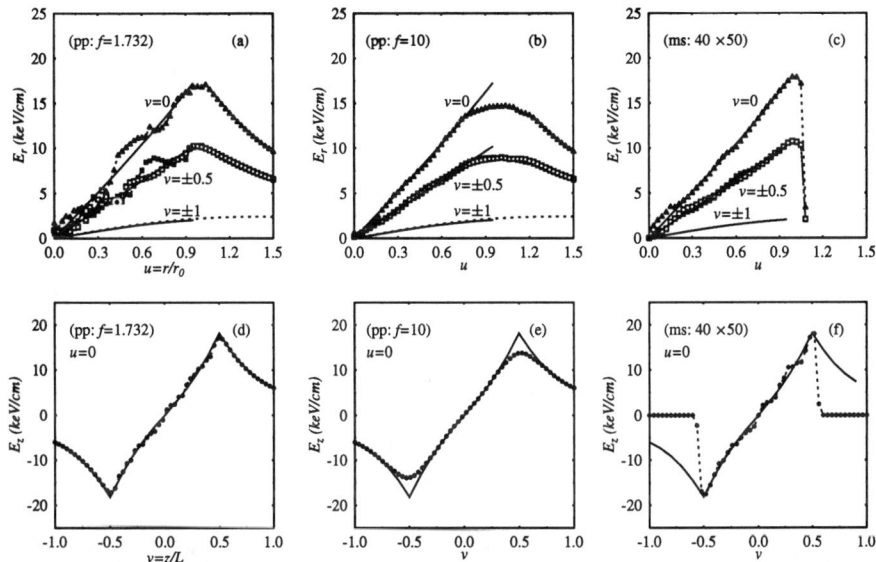

Fig. 6 Same as Fig. 5 except that: (1) the aspect ratio $\alpha = 1.0$; (2) the radial field profiles out of the bunch ($v=\pm 1$) are added. Solid smooth curves represent the exact field profiles from Eqs. (1) and (2) with image charge omitted.

and therefore the point-by-point method was developed. Usually, one believes that the aspect ratio is the watershed between the two methods. It was also claimed that the point-by-point method fails because of the artificially large collisions that occur[4]. However, we found that both methods remain accurate for a highly charged short bunch. See Fig. 7.

IMAGE CHARGE

In the mesh method, the image charge is treated based on the ring-model as well. It could be easily proved that[3] a charged ring inside a pipe induces a continuous charge density distribution along the pipe wall as

$$\sigma_z(Z) = \frac{1}{\pi} \int_0^\infty \frac{I_0(k\rho)}{I_0(ka)} \cos(k\Delta Z) dk, \quad (9)$$

where $\Delta Z = Z - Z'$, Z' is the location of the ring, I_0 the zeroth-order modified Bessel function, ρ the radius of the ring, and a the radius of the pipe. This density distribution is divided into a series of rings along the inside wall of the pipe,

514 Analysis of Space Charge Calculation

applying forces on other charges. This treatment holds as long as a cylindrical symmetry exists for the bunch.

The point-by-point method deals with the image charge simply as it is. When a particle is emitted from a cathode, it has an image on the other side of the cathode plane; when a particle is inside a metal pipe, it has an image outside the pipe. The image of a macroparticle applies forces on others as if it was a member at a distance from the ensemble.

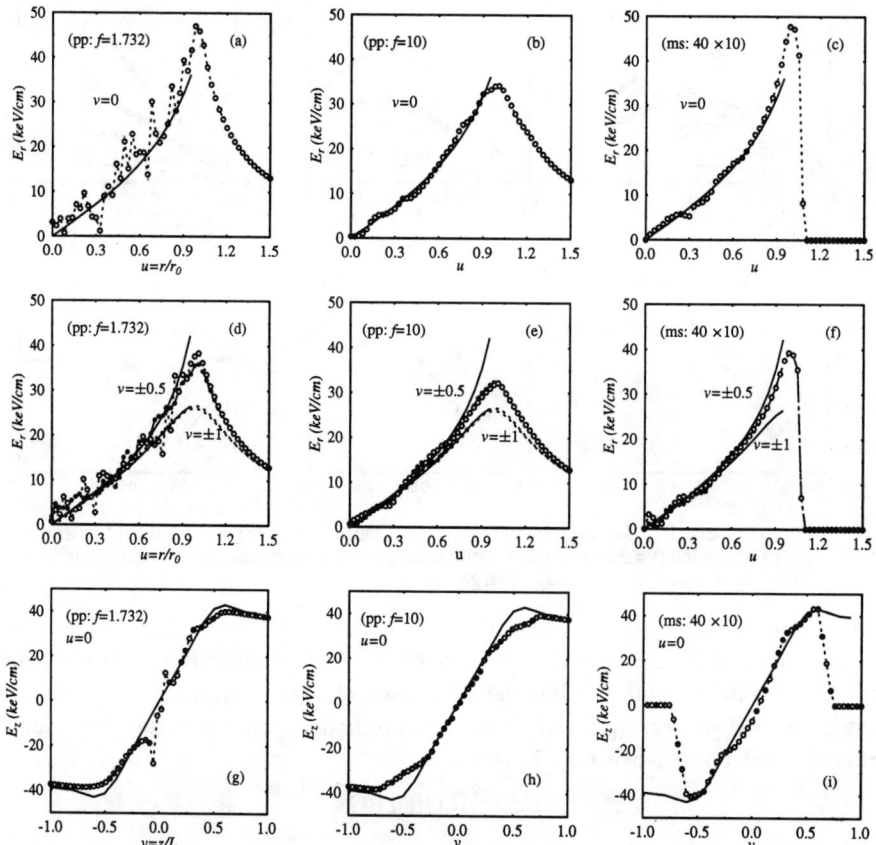

Fig. 7 Same as Fig. 5 except that: (1) the aspect ratio $\alpha = 0.1$; (2) the radial field profiles out of the bunch ($\nu = \pm 1$) are added; (3) mesh dividing changed from 40×50 to 40×10. Solid smooth curves represent the exact field profiles from Eqs. (1) and (2) with image charge omitted.

APPLICATION

PARMELA has been used for a free-electron laser (FEL) injector design at CEBAF for which the space charge effect is important. CEBAF proposes to build an IR FEL and a UV FEL utilizing the superconducting accelerator technology that has been developed at CEBAF[5,6]. The FEL injector consists of a photocathode DC gun, a prebuncher, a cryounit containing two standard CEBAF SRF cavities, and a phase-compressor chicane. The DC laser gun will be operated at ~ 500 kV and generate a cw train of bunches having a charge of 120 pC and a length of ~ 100 ps from a $\sim$3-mm-diameter photocathode[7].

Based on the previous calculation[8], extensive integrated numerical simulations have been conducted with the point-by-point method for space charge treatment from the cathode all the way down to the exit of the chicane. The injector performance has been fully investigated based on an optimized baseline design for its robustness, sensitivity, and operational flexibilities. The effects of space charge on the phase spread, energy spread and emittance of the electron bunches have been closely examined. It is shown that the design will perform beyond the specifications. The results will be presented elsewhere[9].

SUMMARY

It has been revealed that both methods remain suitable for very short bunches as long as a proper screening factor is introduced for the point-by-point method and a proper mesh is divided for the mesh method. We suggest to set 40×50 as the limit for mesh dividing. The reason is that the space charge is always overestimated at the central part, which could be alleviated if a larger radial nodal number is allowed.

It is emphasized that the number of macroparticles per cell must be significantly larger than unity for both methods. For the mesh method, this means $N_p/N_r N_z \gg 1$. For the point-by-point method, it means that the condition $N_p \gg f^3$ must be satisfied. We suggest to choose $f \in \{\sqrt{3}, 0.5 N_p^{1/3}\}$.

Finally, it is worthwhile to mention that the method introduced in Ref. 4 might be most promising for accurate but less time-consuming 3-D space charge calculations. It would be helpful if a comparison could be made between that method and the point-by-point method on the basis of an appropriate screening factor using a 3-D bunch model.

ACKNOWLEDGEMENT

I wish to thank R. Li for helpful discussions.

REFERENCES

1. B. E. Carlsten *et al.*, IEEE J. of QE, <u>27</u>, 2580 (1991).
2. K. T. McDonald, IEEE Trans. ED, <u>35</u>, 2052 (1988).
3. H. Liu, 1989, unpublished.
4. R. W. Garnet and T. P. Wangler, IEEE PAC, <u>1</u>, 330(1991).
5. J. Bisognano *et al.*, NIM <u>A318</u>, 216(1992).
6. G. R. Neil *et al.*, *ibid.*, p. 212.
7. C. K. Sinclair, *ibid.*, p. 410.
8. P. Liger *et al.*, *ibid.*, p. 290.
9. H. Liu, this conference.

STATUS OF THE BEDLAM OPTICS CODE*

W. P. Lysenko

AT-3 MS H808, Los Alamos National Laboratory, Los Alamos, NM 87545

ABSTRACT.

The BEDLAM simulation code (BEam Dynamics in Linear Accelerators by Moments) represents the beam by its phase-space moments. Here, we present a status report on our work in developing this approach, which is an efficient way to compute 3-D beam motion to high order. The space-charge algorithm, which extends Sacherer's results to higher order, now works well for some test cases but is inaccurate for others. We found that for high-brightness beams, higher-order motion is qualitatively different from that predicted by the linear model. While this means BEDLAM is modeling real-beam behavior not seen in the linear model, it makes it difficult to compare BEDLAM to linear codes. We have also verified the feasibility of using BEDLAM to do nonlinear matching.

INTRODUCTION

In the BEDLAM code,[1-3] we represent the beam by moments of its phase-space distribution. Moments are averages of monomials in the phase-space variables. For example, the moment $<x^2>$ is the average value of x^2. At present, the BEDLAM code considers moments up to fourth order (force to third order). Given a set of initial moments, BEDLAM numerically integrates the moment-evolution equations using a Lie-Poisson integrator, which preserves the Hamiltonian structure. The integrator has proved to be stable as expected. The space-charge effects are computed using an extension of Sacherer's ideas to higher order. In this paper, we review these principles and present the results of recent tests of the space-charge model. We also compare BEDLAM to second-moment (linear force) codes like TRACE3D. There are fundamental differences in the dynamics because BEDLAM includes nonlinear forces. The final topic concerns nonlinear matching, a new capability made possible by the moment approach.

SIMULATION BY MOMENTS

There are at least three advantages to describing a beam by its moments. First, this description deals with distributions rather than with single-particle motion. This is important in studying matching. For example, if a beam is matched to a periodic focusing system, then all the moments are periodic. This is a much cleaner situation than in a particle description in which the individual particle motions contain betatron-frequency components that average to zero only when the whole distribution is considered. Second, the moments are closely related to laboratory quantities like beam sizes and divergences. Finally, using moments, the computational cost for 3-D simulations is little more than for 2-D simulations of the same accuracy. For example, for fourth-order simulations, the number of moments in 3-D is only about 4 times that in 2-D. In a mesh method (particle-in-cell code), the amount of work increases by the number of of mesh points in the third direction, which could be considerable. For this reason, we believe the BEDLAM code will be useful in studying high-order, 3-D space-charge effects in high-brightness beams.

The equation for the evolution of the $<xp_x>$ moment is derived as follows.

$$\frac{d}{dt}<xp_x> \;=\; <(\frac{d}{dt}x)p_x> + <x(\frac{d}{dt}p_x)>$$
$$=\; \frac{1}{m}<p_x^2> + <xF_x>$$

---

*This work supported and funded by U.S. Department of Defense, Army Strategic Defense Command, under the auspices of the US Department of Energy.

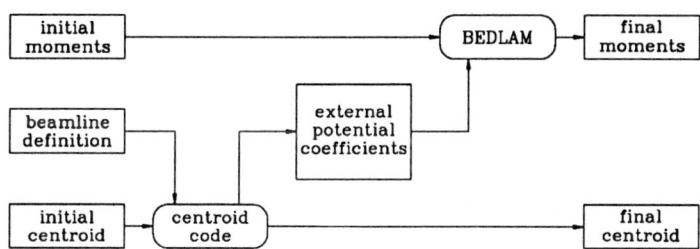

Fig. 1. Data flow for a BEDLAM simulation. The rounded boxes are processes that transform data.

$$= \frac{1}{m}<p_x^2> + <x(a_0 + a_1 x + a_2 y + a_3 z + a_4 x^2 + a_5 xy + \cdots)> \quad (1)$$
$$= \frac{1}{m}<p_x^2> + a_0<x> + a_1<x^2> + a_2<xy> + a_3<xz> + a_4<x^3> + a_5<x^2 y> + \cdots$$

We have replaced the force F by a polynomial approximation. This turns the integral on the right side into a sum of moments. The other moment equations are similar. If we truncate the moments at some order, we get a system of ordinary differential equations for the moments. The present version of BEDLAM truncates after fourth order so there are 203 moments and a corresponding number of evolution equations for these moments. (BEDLAM treats the six first moments separately because of limitations imposed by our integrator.) BEDLAM numerically integrates the truncated system starting with a set of initial moments.

The moment-evolution equations are not in the form of Hamilton's equations (the moments cannot be paired into coordinates and conjugate momenta). This means we cannot use symplectic integrators to solve the differential equations. However, the moment equations do have a Lie-Poisson structure, which is related to the Poisson bracket. We have developed Lie-Poisson integrators[4] that preserve this bracket structure exactly. This provides the numerical stability analogous to that provided by symplectic integrators for single-particle motion.

Figure 1 shows how a BEDLAM simulation works. The first moments (beam centroids) are treated separately. The inputs to the BEDLAM code itself are the 203 second through fourth initial moments and the external-potential-expansion coefficients, evaluated at the centroid positions. The BEDLAM code internally computes the space-charge effects from the beam moments. The output of BEDLAM is the vector of moments at the end of the time step.

SPACE-CHARGE ALGORITHM

At each time step in the simulation, we need to compute the 31 coefficients to a fourth-degree polynomial that approximates the space-charge potential.

$$U(\vec{x}) = u_1 x^2 + u_2 xy + u_3 xz + u_4 y^2 + \cdots + u_{31} z^4 \quad (2)$$

These coefficients are computed from the 31 purely-spatial moments that make up the 203 second through fourth moments. Our algorithm uses a generalization of Sacherer's results of 1971.[5]

Sacherer's Results

Sacherer's results, which relate to the evolution of the second moments, are

1. Moment evolution depends on the nine integrals $<x_i F_j>$.

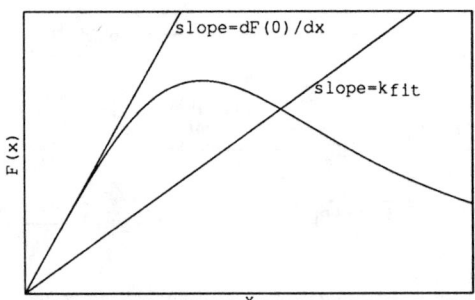

Fig. 2. The curve shows a typical space-charge force as a function of distance x. The fitted linear force has a slope less than that of the actual force's slope at the origin.

2. These integrals are almost independent of distribution details (higher moments).

We have seen in Eq. 1 how the first result arises. The second result was obtained by computing the $<x_i F_j>$ integrals for a variety of distributions having the same values of the second spatial moments ($<x_i x_j>$). For a second-order truncation, the space-charge force must be replaced by its linear approximation.

$$F_x = k_{\text{fit}} x, \qquad (3)$$

where the fitted force constant k_{fit} is chosen by

$$k_{\text{fit}} <x^2> \, = \, <x F_x> \qquad (4)$$

Sacherer's second result says that we can use any convenient model to compute the integral $<x F_x>$. Figure 2 shows a typical space-charge force field, which is linear near the origin and drops off as r^{-2} at infinity. Since the fitted force constant takes into account the whole beam, its value is less than the slope of the force field at the origin. Equation 4 is actually the condition that minimizes the least-squares error in $k_{\text{fit}} x - F_x$.

Extension of Sacherer's Results to Higher Order

In the case of fourth-order moments, the evolution is determined by 57 generalized Sacherer integrals, which are $<m_i F_j>$, where the m_i are the 19 monomials of degrees one through three. Sacherer's second results holds when extended to higher order. Table I shows some of the values of the integrals $<x E_x>$ and $<x^3 E_x>$ for a spherically-symmetric beam with various radial charge distributions. All distributions have the same value of the second moment $<x^2>$. (The integrals are normalized to the uniform-beam case.) The values of $<x E_x>$ for the first four distributions are the same as in Sacherer's paper. We see that this integral is nearly constant, regardless of the distribution. The surprising fact is that $<x^3 E_x>$ is also nearly constant, even though we have not fixed the fourth moments. So we have even more than is required (that $<x E_x>$ and $<x^3 E_x>$ depend only on moments of order fourth and lower). We have developed an algorithm that determines the 31 space-charge potential coefficients of Eq. 2 from the 31 spatial moments. We can simplify the computation if we augment the vector of the 31 spatial moments with the fixed zeroth and first moments to form a 35-component vector of moments π_i. At each time step, the computation of the 31 space-charge potential coefficients U_i is given by a matrix multiply

$$U_i = T_{ij} \pi_j, \qquad (5)$$

where T_{ij} is a 31×35 matrix. This matrix does not have to recomputed at each time step but only after the beam has undergone a substantial change in size in any direction.

Table I. Sacherer integrals for various spherically-symmetric charge distributions. For the Gaussian and hollow distributions, there is a cutoff at 5σ. All distributions have the same second moments.

distribution	$<xE_x>$	$<x^3E_x>$
uniform	1.00	1.00
parabolic $1-r^2$	1.01	1.02
Gaussian $e^{-r^2/2\sigma^2}$	1.05	1.03
hollow $r^2 \times e^{-r^2/2\sigma^2}$	1.02	1.02
r^4	1.02	0.96
r^8	1.04	0.94

The matrix T_{ij} is derived in two steps. First, we compute the 35 charges in a charge model consisting of 35 Gaussian charges. These charges are chosen by making the spatial moments of the 35-Gaussian model match those of the beam being simulated. Now assuming this charge model, we obtain the 31 potential coefficients by requiring the field of the model to match the field of the polynomial approximation at N "evaluation points." If the evaluation points are distributed like the actual beam, then this procedure minimizes the error in the 57 generalized Sacherer integrals $<x_i F_j>$, which determine the evolution of the moments. The 31 coefficients are passed to the integrator (after adding the external-forces part) and the integrator builds the Sacherer integrals.

TEST OF SPACE-CHARGE ALGORITHM

To test the space-charge algorithm, we considered a 3-D example we could compute analytically: a charge distribution consisting of two spherical Gaussians. To judge the accuracy of the algorithm, we compared the exact value of the following Sacherer integrals (which are the largest of the 57) to the BEDLAM result.

$$<xF_x> \quad <yF_y> \quad <zF_z>$$
$$<x^3F_x> \quad <y^3F_y> \quad <z^3F_z>$$

We used the BEDLAM values of u_i in

$$F_x = -\frac{\partial}{\partial x}[u_1 x^2 + \cdots + u_{31} z^4] \qquad (6)$$

to compute numerically the Sacherer integrals corresponding to the BEDLAM simulation. For our examples, we have 75% of the charge in one Gaussian and 25% in the other. Table II shows the results for various cases in which we varied the separation of the Gaussians and their widths (σ values).

Table II. Maximum of errors in the six Sacherer integrals for five cases of double-Gaussian distributions.

case	separation of centers	σ_1	σ_2	max. error in Sacherer int.
1	1	0.5	0.5	2 %
2	1	1	1	1
3	1	2	2	1
4	2.15	0.5	0.5	20
5	1	0.9	0.5	15

The BEDLAM calculations for Cases 1—3, which had equal widths for the two Gaussians, were very accurate. The Sacherer integrals, between cases, varied by more than a factor of three for the various widths but the algorithm computed them to within one or two percent. We also found that the results were not very sensitive to the overall size of the beam, which is evidence that the recalculation of the T_{ij} matrix of Eq. 5 would not have to be done often.

Case 4 also had equal widths but the separation of the Gaussians was large compared to the widths. This case was not accurately computed by BEDLAM, which is not surprising. This level of approximation (fourth order moments) is not good enough to model a charge distribution consisting of two separated blobs. Case 5 shows that we also get poor performance if the widths of the two Gaussians are different. This is an unexpected result. We plan further tests to understand the problem.

NONLINEAR SPACE-CHARGE DYNAMICS

TRACE3D is a second-order moment code (linear forces) that implements Sacherer's results by using a uniformly-charged ellipsoid space-charge model. To compare TRACE3D to BEDLAM, we simulated one period of a FODO channel of Halbach[6] PMQs transporting a high-brightness beam. In our example, the tune depression was $\sigma/\sigma_o = 0.53$, which corresponds to a space-charge parameter value of $\mu = 1 - (\sigma/\sigma_o)^2 = 0.72$. We found poor agreement between the codes for this high-brightness beam, although the agreement was excellent with space charge turned off. We found poor agreement even when we truncated the BEDLAM force expansion to turn off nonlinearities.

Does this indicate a problem with BEDLAM? To see that it does not, consider the following. (For simplicity, we assume 1-D here, although both TRACE3D and BEDLAM are actually 3-D codes.) In TRACE3D, the space-charge force is linear

$$eE_x = k_{11}x, \qquad (7)$$

where the space-charge force constant k_{11} is determined from the uniform-ellipsoid charge model using Eq. 4. In BEDLAM, we have a cubic-force model

$$eE_x = k_{31}x + k_{32}x^2 + k_{33}x^3, \qquad (8)$$

where the constants k_{31}, k_{32}, and k_{33} are chosen by

$$\begin{aligned}
<x^2>k_{31} + <x^3>k_{32} + <x^4>k_{33} &= e<xE_x> \\
<x^3>k_{31} + <x^4>k_{32} + <x^5>k_{33} &= e<x^2E_x> \\
<x^4>k_{31} + <x^5>k_{32} + <x^6>k_{33} &= e<x^3E_x>,
\end{aligned} \qquad (9)$$

which makes Eq. 8 a least-squares fit to the actual force eE_x. The cubic model is, of course, a better approximation to the actual force than the linear model. This is illustrated schematically in Fig. 3. The linear part of the cubic fit is larger than that for the linear fit, $k_{31} > k_{11}$. This difference can be a large effect.

When we want to characterize the strength of the space-charge effect by a single number, we use the tune depression factor σ/σ_o or the space-charge parameter μ. Let k_{ext} be the effective focusing-force constant. For an alternating-gradient focusing system, this means k_{ext} is the value of the time-independent force constant that produces the same phase advance as the actual alternating-gradient force. (This is often called the smooth approximation.) Then μ is given by

$$\mu = -\frac{k_{sc}}{k_{ext}}, \qquad (10)$$

where k_{sc} is the effective (smooth) space-charge force constant, and can be the smooth equivalent of k_{11}, k_{31}, or the value of slope at the origin of the actual force. In 3-D, we can still describe the space-charge strength by a single number, even though there are three tune depressions, by using (smooth) potentials

$$\mu = -\left.\frac{\nabla^2 \phi_{sc}}{\nabla^2 \phi_{ext}}\right|_o. \qquad (11)$$

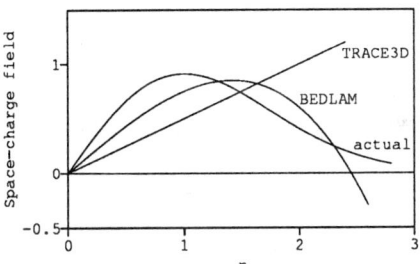 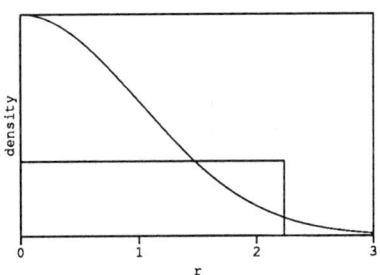

Fig. 3. For a typical charge distribution, BEDLAM's cubic-force model is more like the actual space-charge force than the linear model in TRACE3D.

Fig. 4. Gaussian and uniform charge distribution in 3-D. Both charge distributions have the same total charge and second moments ($\sigma = 1$) but very different central densities.

This means μ is proportional to the central charge density. Nonuniform beams can have much greater central charge densities than uniform beams having the same total charge and second moments. Figure 4 shows charge density ρ as a function of r for a uniform beam and a Gaussian beam with the same total charge and width ($x_{rms} = y_{rms} = z_{rms} = 1$). The graph is to scale. The central density of the Gaussian is about three times that of the uniform beam.

Thus we see that for a situation with a moderately high μ value in a linear code like TRACE3D, the μ value using the cubic model (or the actual space-charge force) can be greater than unity. This does make sense. Near the center of the beam the space charge force can be larger than the external focusing force. Even though μ is larger than unity, the motion can be stable, consistent with the smaller-than-unity value of μ given by the linear model, in which the whole beam is modeled by a different, smaller effective space-charge force constant. Figure 5 shows the phase-space trajectories for this situation. For small $|x|$, the motion looks unstable; the orbits are locally like hyperbolas. However, the overall motion is stable; the phase-space orbits circle the origin. We see that for bright beams, higher-order space-charge effects introduce qualitatively different behavior. We need to further study how well BEDLAM is modeling this new behavior.

MATCHING WITH BEDLAM

The moment approach makes it easy to do nonlinear matching. For a periodic channel, a matched beam has all moments periodic. We have tested the idea of computing matched beams by writing a matching code. This code is separate from the BEDLAM code. It reads the final moments computed by BEDLAM and generates a new file of input moments, which can be used by BEDLAM. (At the present time, the matcher code simply averages the input and output moments of the BEDLAM run to get the new input moments. Since this simple algorithm does converge, we did not attempt to develop a more sophisticated method.) Thus to compute a matched beam, we just execute BEDLAM and the matching code alternately until the moments stop changing. Figure 6 shows this process. This approach of using a separate code to do the matching is practical because because the amount of data is small (203 numbers). For our example, we took a PMQ channel without space charge. To enhance nonlinear effects, we used short magnets (length=bore) with small bores (bore=$2 \times x_{rms}$). With the nonlinearity of the magnet fringe fields temporarily turned off, we computed the matched beam. For this matched beam, all 203 moments final moments were the same as the initial moments. The transverse Courant-Snyder parameters for this linearly-matched beam are shown in Table IV, where it is called Beam A.

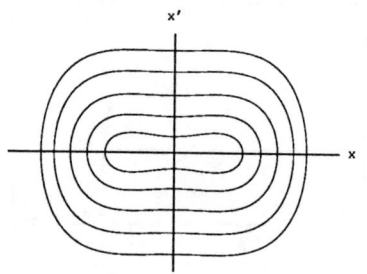

Fig. 5. Motion in phase space for nonuniform beams like Gaussians when $\mu > 1$. Motion for small $|x|$ appears unstable (locally hyperbolic) even though overall motion is stable (circles origin of phase space).

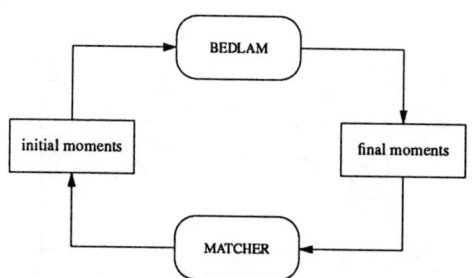

Fig. 6. Data flow for matching with BEDLAM. The BEDLAM code transports initial moments to their final values. The MATCHER code computes a suggested initial set of moments for BEDLAM to try. This process is repeated until the moments stop changing.

Table IV. Courant-Snyder Parameters for Matching Study.

	A. Linear match.		B. Final beam using beam A as input.		C. Matched to nonlinearities.	
	α	β	α	β	α	β
x	-1.4461	1.3475	-1.1985	1.0949	-1.3818	1.2261
y	1.4464	1.3478	1.5204	1.2447	1.3822	1.2270

When we turned the fringe-field nonlinearities on and ran Beam A through the channel, the final beam was not the same as the initial beam. This final beam is shown in Table IV as Beam B. Besides not having a match anymore, we found that the rms emittances decreased by 2%. This is not surprising. A mismatched beam in a nonlinear system is not expected to preserve emittance. Next we rematched by runing the matcher-BEDLAM iteration cycle several times. We found a new matched beam. (The input and output were again identical.) This beam is shown in Table IV as Beam C. The emittance was preserved for this matched-to-the-nonlinearity case. Thus nonlinear matching works as expected. It computes a beam matched to a nonlinear system by adjusting all moments up to fourth order. The matched beam, which differs from the one matched to the linear system, preserves emittance.

Beams matched to periodic systems preserve all moments. The familiar Courant-Snyder parameters are a complete description only for uncoupled motion for second-order moments. These parameters are obtained from the second moments as follows:

$$\begin{aligned} \alpha_x &= -<xp_x>/\epsilon_x \\ \beta_x &= <x^2>/\epsilon_x \\ \epsilon_x &= (<x^2><p_x>^2 - <xp_x>^2)^{1/2} \end{aligned} \quad (12)$$

DISCUSSION AND CONCLUSIONS

Our 3-D space-charge tests with the two-Gaussian charge distribution obtained excellent results for the cases with both Gaussian widths equal, over a wide range of widths. The accurate determination of all Sacherer integrals means that we have access to high-order effects of space charge. However, the cases with different widths produced poor results that we cannot yet explain. We need to understand this behavior before we can predict new phenomena with BEDLAM.

With nonlinear codes (moment order > 2), space-charge models must be nonuniform, just as real beams are. As we have seen, this introduces the possibility of having space-charge μ values larger than unity for beams of even moderate brightness. The reason this happens is the nonorthogonal nature of the potential expansion. The linear part of a linear fit is different from the linear part of a cubic fit, for example. This behavior makes it hard to compare a high-order code like BEDLAM to linear codes like TRACE3D. The effect is real, however. A bright beam does have motion near the spatial origin that looks unstable, so a μ value that is greater than unity is reasonable. The higher-order motion is qualitatively different from the linear motion.

Although Sacherer's results appear to still apply at higher orders, we need to further develop the theory. The convergence of the truncated moment-evolution equations can be tested independently, using other codes. We can do this by using moment invariants,[7] which are certain functions of moments that are conserved for linear motion. For truncated moment systems (as in BEDLAM), some of these linear invariants are conserved for the general (nonlinear) case. (I call these invariants BEDLAM invariants.) We can test the validity of the convergence of the truncated moment equations, using even nonmoment codes, by seeing how well the BEDLAM invariants are conserved.

As expected, nonlinear matching has been verified as a new capability. Since the matcher code is separate from BEDLAM itself, we have a flexible system in which various tools can be used together without linking them all beforehand into one huge code. This capability is possible because of the small amount of data that has to be handled in the moment approach. It now appears that an optimizer that runs as a process separate from BEDLAM will also be feasible.

ACKNOWLEDGEMENTS

Paul Channel, who originated the BEDLAM approach, has also contributed to some of the newer developments described here.

REFERENCES

1. P. J. Channell, "The Moment Approach to Charged Particle Beam Dynamics," IEEE Trans. Nucl. Sci., **30** (4), August 1983, 2607.

2. P. J. Channell, L. M. Healy, and W. P. Lysenko, "The Moment Code BEDLAM," IEEE Trans. Nucl. Sci., **32** (5), October 1985, 2565.

3. W. P. Lysenko and P. J. Channell, "New BEDLAM," Proceedings of the Conference on Compute Codes and the Linear Accelerator Community, Los Alamos, January 22–25, 1990, Los Alamos National Laboratory Report LA-11857-C (July 1990).

4. P. J. Channell and J. C. Scovel, "Integrators for Lie-Poisson Dynamical Systems," Physica D **50** 80 (1991).

5. F. J. Sacherer, "RMS Envelope Equations with Space Charge," IEEE Trans. Nucl. Sci., **18** 1105 (1971).

6. K. Halbach, "Physical and Optical Properties of REC Magnets," Nucl. Instrum. Meth. **187**, 109–117 (1981).

7. D. D. Holm, W. P. Lysenko, and J. C. Scovel, "Moment invariants for the Vlasov equation," J. Math. Phys. **31** 1610 (1990).

HIGH-ORDER OPTICS WITH SPACE CHARGE: THE TOPKARK CODE*

David L. Bruhwiler and Michael F. Reusch
Grumman Corporate Research Center, Princeton NJ 08540-6620

ABSTRACT

TOPKARK is a three-dimensional high-order optics code that incorporates a simple space charge model. This code uses the differential algebra library DA to generate an arbitrary-order Taylor map describing a given lattice, then the Lie algebra library LIELIB is used to obtain the Dragt-Finn factorization of the corresponding Lie polynomial. The Lie polynomial generated by TOPKARK without space charge has been successfully benchmarked through third order against MARYLIE 3.0 and through fifth-order (for a single example) against TLIE. With space charge turned on, TOPKARK generates a linear map that agrees well with TRACE 3-D. We describe here the various algorithms employed by TOPKARK and address the issue of verifying the high-order map when space charge is turned on.

INTRODUCTION

TOPKARK is a six-dimensional high-order relativistic beam optics code which exists in two separate but parallel versions: a particle tracking code and a mapping code. Here we discuss only the mapping version of TOPKARK, in which a fully three-dimensional linear space charge model has been implemented.

TOPKARK employs a fourth-order, adaptive-step-size, Runge-Kutta integration scheme[1] which provides good accuracy and reasonable computational speed. The differential algebra library DA[2] is used to generate a high-order Taylor map expansion of the dynamical variables about the design trajectory (in practice up to fifth order has been used) step by numerical step along the length of the lattice. This Taylor map is used to propagate the spatial moments of the (assumed) initial particle distribution from one integration step to the next.

At each integration step, a 3-D uniformly-filled ellipsoid (possibly tilted) is constructed according to the calculated spatial moments. The exact linear electric fields associated with this ellipsoid are calculated and, in combination with any magnetic fields, are used to advance to the next step. At the end of the lattice, the final Taylor map is used to calculate the emittance and Twiss parameters of the final distribution. The Lie algebra library LIELIB[3] is used to obtain the Dragt-Finn[4] factorization of the Lie polynomial corresponding to this final Taylor map.

GENERAL FEATURES

TOPKARK currently implements a number of "hard edge" or uniform-field magnet elements, including a dipole (i.e. a normal entry and exit sector bend) and quadrupole through duodecapole. Also available are "thin fringe" elements for dipole and quadrupole magnets. All of these elements, including the fringe fields, have been successfully benchmarked against MARYLIE 3.0[5] through third order. The fringe field models, although calculated independently, were based on ideas developed previously by Forest.[6]

---

* This work was supported by the Independent Research and Development funds of Grumman Aerospace Corporation

TOPKARK also employs one extended-fringe magnet model. This is a line-dipole model for large-bore magnets constructed from a cylindrical array of magnetized rods, including quadrupole, octupole and duodecapole configurations. TOPKARK was successfully benchmarked against TLIE[7] through fifth-order in a single test-case where these extended-fringe quadrupole and octupole models were used. New element types are easily added to the list above.

The code includes two optimizing routines, one based on the downhill simplex method[1] and another based on Powell's method.[1] Either of these algorithms can be used by a matching routine that sets the final transverse Twiss parameters to specified values by modifying any four of the lattice parameters. This matching routine has been successfully used both with and without space charge.

Another type of matching routine, which can also use either of the optimizing algorithms (when necessary), is used to zero specified terms in the Lie polynomial, sometimes while simultaneously satisfying other imposed constraints. For example, TOPKARK can determine the required strengths of three (or more) octupoles in order to eliminate third-order geometric aberrations.

UNITS, NOTATION AND EQUATIONS OF MOTION

The code uses MKS units throughout, although all momenta are normalized to the design longitudinal momentum: $p_0 \equiv \gamma_0 \beta_0 mc$ (and thus are dimensionless). The lab-frame electric field is $\mathbf{E}_L(x,y,z)$ in volts per meter; the magnetic field is $\mathbf{B}(x,y,z)$ in Tesla. The normalized momentum is $\mathbf{p} = \gamma m v/p_0$; the relativistic beta is $\beta = v/c$. The normalized time coordinate is $\tau = ct$ in meters; the time coordinate with respect to the design trajectory's time is $\delta\tau = c(t - t_0)$. We work with the normalized *negative* energy $p_\tau = -E/p_0 c$; the energy difference with respect to the design trajectory is $\delta p_\tau = (E_0 - E)/p_0 c$. This convention for the particle energy is an artifact from the test-particle tracking version of TOPKARK, which uses Hamiltonian formalism and symplectic integration techniques. The magnetic rigidity is $B\rho = q/p_0$ in tesla per meter, and the independent variable is z, the distance along the axis of each lattice element.

In a straight lattice, the equations of motion are:

$$\frac{dx}{dz} = \frac{\beta_x}{\beta_z} \; ; \qquad \frac{dp_x}{dz} = \frac{1}{B\rho}\left(\frac{\beta_y}{\beta_z} B_z - B_y + \frac{1}{\beta_z c} E_{xL} \right) ; \qquad (1a,b)$$

$$\frac{dy}{dz} = \frac{\beta_y}{\beta_z} \; ; \qquad \frac{dp_y}{dz} = \frac{1}{B\rho}\left(-\frac{\beta_x}{\beta_z} B_z + B_x + \frac{1}{\beta_z c} E_{yL} \right) ; \qquad (1c,d)$$

$$\frac{d\delta\tau}{dz} = \frac{1}{\beta_z} - \frac{1}{\beta_0} ; \qquad \frac{d\delta p_\tau}{dz} = -\frac{1}{B\rho c}\left(\frac{\beta_x}{\beta_z} E_{xL} + \frac{\beta_y}{\beta_z} E_{yL} + E_{zL} \right). \qquad (1e,f)$$

In a (constant field) sector bend, the equations of motion are:

$$\frac{dx}{ds} = \left(1 + \frac{x}{\rho_0}\right)\frac{\beta_x}{\beta_s}; \quad \frac{dy}{ds} = \left(1 + \frac{x}{\rho_0}\right)\frac{\beta_y}{\beta_s}; \quad \frac{d\delta\tau}{ds} = \left(1 + \frac{x}{\rho_0}\right)\frac{1}{\beta_z} - \frac{1}{\beta_0}; \qquad (2a\text{-}c)$$

$$\frac{dp_x}{ds} = \frac{1}{\rho_0}(p_s/p_0 + A_s/B_\rho) + \left(1 + \frac{x}{\rho_0}\right)\frac{1}{B_\rho}\left\{\frac{\beta_y}{\beta_s}B_s - B_y + \right.$$
$$\left. + \frac{1}{\beta_s c}\left[E_{xL}\cos((s-s_0)/\rho_0) + E_{sL}\sin((s-s_0)/\rho_0)\right]\right\}; \quad (2d)$$

$$\frac{dp_y}{ds} = \left(1 + \frac{x}{\rho_0}\right)\frac{1}{B_\rho}\left(-\frac{\beta_x}{\beta_s}B_s + B_x + \frac{1}{\beta_s c}E_{yL}\right); \quad (2e)$$

$$\frac{d\delta p_\tau}{ds} = -\left(1 + \frac{x}{\rho_0}\right)\frac{1}{B_\rho c}\left\{\frac{\beta_x}{\beta_s}\left[E_{xL}\cos((s-s_0)/\rho_0) + E_{sL}\sin((s-s_0)/\rho_0)\right]\right.$$
$$\left. + \frac{\beta_y}{\beta_s}E_{yL} + \left[E_{sL}\cos((s-s_0)/\rho_0) - E_{xL}\sin((s-s_0)/\rho_0)\right]\right\}; \quad (2f)$$

where the s-direction is locally tangent to the (curving) design trajectory, and ρ_0 is the radius of curvature of the design trajectory.

CALCULATION OF THE ELECTRIC FIELD

Electric fields are calculated in the bunch frame, then relativistically transformed to the laboratory frame. The length of the bunch as observed in the lab frame is shortened due to relativistic length contraction, so the distribution is first stretched out in the z-direction before calculating the fields: $\delta z_B = \gamma_0 \delta z_L$. This reduces the particle density: $n_B = \gamma_0^{-1} n_L$, so E_{xB}, E_{yB}, E_{zB} are all reduced by a factor γ_0 from what one would naively calculate in the lab.

However, the fact that the distribution is stretched out in z effectively increases the value of E_{zB} by γ_0 at the position of each "particle", thus negating the decrease in E_{zB} noted above. This stretching of the bunch also alters the geometry of the distribution, which affects the values of all three components of $\mathbf{E}_B$ accordingly.

The current and self-magnetic field are neglected in the bunch frame. Thus, the Lorentz transformations yield:

$$\mathbf{E}_L^{\parallel} = \mathbf{E}_B^{\parallel} \; ; \qquad\qquad \mathbf{E}_L^{\perp} = \mathbf{E}_B^{\perp} \; ; \qquad (3a,b)$$

$$\mathbf{B}_L^{\parallel} = \mathbf{B}_B^{\parallel} = 0 \; ; \qquad\qquad \mathbf{B}_L^{\perp} = \gamma_0\, \mathbf{B}_B \times \mathbf{E}_B/c \; . \qquad (3c,d)$$

The Lorentz force equation is:

$$\frac{1}{q}\mathbf{F}_L = \mathbf{E}_L + \mathbf{v} \times \mathbf{B}_L \; . \qquad (3e)$$

Combining these results yields an *effective* electric field in the lab frame:

$$\mathbf{E}_{L\,\text{eff}} = \mathbf{E}_B^{\parallel} + \mathbf{E}_B^{\perp}/\gamma_0 \; . \qquad (3f)$$

This is the quantity used to advance the particles: the longitudinal field is altered by geometric effects only, while the transverse fields are *also* reduced by a factor of γ_0^2.

The bunch being propagated along the accelerator lattice by TOPKARK is an ensemble with every "particle" having the z of the design trajectory. Therefore, we must first expand the bunch in z so it has a finite volume. Then we find the frame (if necessary) that eliminates any coupling between x, y and z (the electric field is subsequently rotated back to the original frame at the end of the calculation). We then obtain the spatial second moments from the Taylor map (and the assumed initial distribution).

We use the above information to construct a uniformly-filled 3-D ellipsoid, and calculate the electric fields of this object as follows:[8]

$$E_{xB} = x \left(\frac{3}{8\pi 5^{3/2}} \frac{Q_{tot}}{\varepsilon_0} \right) \int_0^\infty \frac{du}{(<x^2>+u)^{3/2} (<y^2>+u)^{1/2} (<z^2>+u)^{1/2}}, \quad (4)$$

with analogous equations for E_{yB} and E_{zB}. These integrals have been evaluated in terms of complete and incomplete elliptic integrals, with the resulting formulas depending upon the relative sizes of the spatial moments $<x^2>$, $<y^2>$ and $<z^2>$.

PROPAGATION OF THE SPATIAL MOMENTS

TOPKARK works with a Taylor-series expansion about the design trajectory, so the obvious question arises: How then does one propagate a particle distribution down the beamline? The answer is explained in this section.

The current dynamical variables at any value of z -- x', p'$_x$, y', p'$_y$, $\delta\tau$', δp_τ' (denoted below as w_i' with i ranging from 1 to 6) -- are known in terms of a Taylor-series expansion in the initial variables, w_i:

$$w_i' = \sum_{n=1}^{6} R_{in} w_n + \sum_{n\,m=1}^{6} T_{inm} w_n w_m + \cdots . \quad (5a)$$

Therefore, we may calculate the moments of our current distribution as a function of the *initial* moments as follows:

$$<w_i'w_j'> = \sum_{n\,m=1}^{6} R_{in} R_{im} <w_n w_m> + \sum_{k\,l\,m\,n=1}^{6} T_{inm} T_{ikl} <w_k w_l w_m w_n> + \cdots . \quad (5b)$$

In addition to letting us propagate the spatial moments down the beam line, this algorithm allows for detailed examination of any second moment of the distribution at the end of the lattice. We call this technique the Moment Evolution (ME) algorithm.

Of course, we must assume a *convenient* initial distribution function, and one that is consistent with our assumption of a uniformly-filled 3-D ellipsoid in space:

$$F(x, p_x, y, p_y, \delta\tau, \delta p_\tau) = \frac{3}{4\pi 5^{3/2} x_{rms} y_{rms} \delta\tau_{rms}} g_x(x, p_x) g_y(y, p_y) g_\tau(\delta\tau, \delta p_\tau) . \quad (6a)$$

Equation (6a) holds only when the condition below is satisfied:

$$\frac{x^2}{5\langle x^2\rangle} + \frac{y^2}{5\langle y^2\rangle} + \frac{(\delta\tau)^2}{5\langle\delta\tau^2\rangle} \le 1 \; ; \tag{6b}$$

otherwise, $F(x,p_x,y,p_y,\delta\tau,\delta p_\tau) = 0$. We also impose the condition that

$$\int_{-\infty}^{\infty} dp_x \; g_x(x,p_x) = 1 \,, \tag{6c}$$

with analogous conditions imposed on g_y and g_τ. In other words, the x in $g_x(x,p_x)$ appears only to account for any x-p_x coupling.

The resulting spatial distribution function is:

$$f(x,y,\delta\tau) \equiv \int_{-\infty}^{\infty} dp_x \int_{-\infty}^{\infty} dp_y \int_{-\infty}^{\infty} d\delta p_\tau \; F(x,p_x,y,p_y,\delta\tau,\delta p_\tau) = \frac{3}{4\pi 5^{3/2} x_{rms} y_{rms} \delta\tau_{rms}}, \tag{7a}$$

where again the criterion given by Eq. (6b) must be satisfied. If we integrate further, we obtain the distribution along the x-axis:

$$f_x(x) = \int_{-r_x y_{rms}}^{r_x y_{rms}} dy \int_{-r_{xy}\delta\tau_{rms}}^{r_{xy}\delta\tau_{rms}} d(\delta\tau) \; f(x,y,\delta\tau) = \frac{3}{4\sqrt{5} x_{rms}} \left(1 - \frac{x^2}{5\langle x^2\rangle}\right), \tag{7b}$$

where we must require that $|x| \le \sqrt{5}\, x_{rms}$, or else $f_x = 0$. Also, we have defined the quantities $r_x \equiv \sqrt{5 - x^2/\langle x^2\rangle}$ and $r_{xy} \equiv \sqrt{5 - x^2/\langle x^2\rangle - y^2/\langle y^2\rangle}$.

We choose Gaussian distributions in momentum that yield the appropriate emittance and Twiss parameters when the entire distribution is projected onto any of the three phase planes:

$$g_x(p_x) = (2\pi\varepsilon_x/\beta_x)^{-1/2} \exp\left[-\frac{\beta_x}{2\varepsilon_x}\left(p_x + \frac{\alpha_x}{\beta_x} x\right)^2\right], \tag{8a}$$

with analogous equations for g_y and g_τ. Thus the initial distribution projected onto the x-p_x phase plane is:

$$f_{xp}(x,p_x) = \frac{3}{4\sqrt{10\pi}\,\varepsilon_x}\left(1 - \frac{x^2}{5\varepsilon_x\beta_x}\right) \exp\left[-\frac{\beta_x}{2\varepsilon_x}\left(p_x + \frac{\alpha_x}{\beta_x} x\right)^2\right], \tag{8b}$$

for $x^2 \le 5\varepsilon_x\beta_x$ (otherwise, $f_{xp} = 0$). We emphasize that we are using RMS Twiss parameters, which means that $\langle x^2\rangle \equiv \varepsilon_x\beta_x$, $\langle xp_x\rangle \equiv -\varepsilon_x\alpha_x$, and $\langle p_x^2\rangle \equiv \varepsilon_x\gamma_x$, with $\gamma_x = (1+\alpha_x^2)/\beta_x$ (and similarly for y and $\delta\tau$).

In order to apply the ME algorithm, we need to be able to calculate arbitrary moments of this projected distribution:

$$<x^n p_x^m> = \frac{3}{4(10\pi)^{1/2}\varepsilon_x} \int_{-x_{max}}^{x_{max}} dx \, x^n \left(1 - \frac{x^2}{5\varepsilon_x \beta_x}\right) \int_{-\infty}^{+\infty} dp_x \, p_x^m \exp\left[-\frac{\beta_x}{2\varepsilon_x}\left(p_x + \frac{\alpha_x}{\beta_x}x\right)^2\right] \quad (9a)$$

$$= \frac{3}{4}\sqrt{\frac{\beta_x}{2\pi\varepsilon_x}} (5\beta_x\varepsilon_x)^{m/2} \int_{-\infty}^{+\infty} du \, e^{-bu^2} \int_{-1}^{+1} dv \, (1-v^2) \, v^m \, (u-av)^n \quad , \quad (9b)$$

where $a \equiv \alpha_x\sqrt{5\varepsilon_x/\beta_x}$, $b \equiv \beta_x/2\varepsilon_x$, and $x_{max} = \sqrt{5\varepsilon_x\beta_x}$. Macsyma[9] has been used to evaluate the double integral above for all values of m and n such that $m+n \leq 10$.

SIMPLE EXAMPLE: A POINT-TO-POINT TRANSFER LINE

In order to demonstrate TOPKARK's matching capability and its close agreement (in the linear limit) with TRACE 3-D[10], we consider the simple example of a point-to-point transfer line. This example also demonstrates how second- and third-order aberrations cause emittance growth and degrade the final focus. TOPKARK (to third order) and PARMILA[11] both show these effects with reasonable agreement.

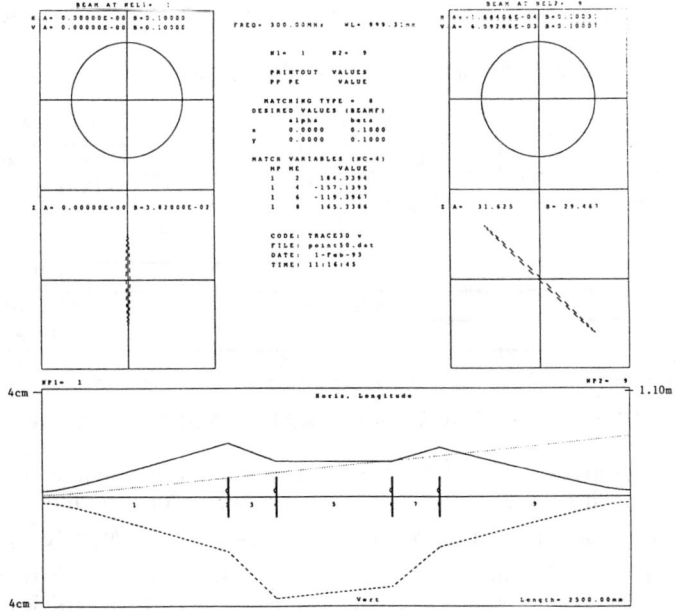

Fig. 1. TRACE 3-D plot of the Point-to-Point Transfer Line, using 50 mA of 4.73 MeV protons and quad strengths found by TOPKARK: $\beta_0=0.1$; $\gamma_0=1.005$; $B_\rho=0.315$

Table I. Code Comparison using 50.0 mA

Code	$\varepsilon_{xf}/\varepsilon_{xi}$	α_{xf}	β_{xf}	γ_{xf}	x_{rms}	$P_{x\ rms}$
TRACE3D	1.000	4.03e-3	0.100	10.0	1.00e-3	1.00e-2
TOPKARK (linear)	1.000	7.48e-4	0.100	10.0	1.00e-3	1.00e-2
PARMILA	1.026	-3.05e-2	0.101	9.92	1.02e-3	1.01e-2
TOPKARK (3rd order)	1.034	-8.35e-3	0.104	9.65	1.04e-3	0.99e-2

Code	$\varepsilon_{yf}/\varepsilon_{yi}$	α_{yf}	β_{yf}	γ_{yf}	y_{rms}	$P_{y\ rms}$
TRACE3D	1.000	-4.10e-3	0.100	10.0	1.00e-3	1.00e-2
TOPKARK (linear)	1.000	5.92e-3	0.100	10.0	1.00e-3	1.00e-2
PARMILA	1.209	-8.00e-2	0.118	8.51	1.20e-3	1.01e-2
TOPKARK (3rd Order)	1.248	-1.43e-1	0.124	8.24	1.24e-3	1.01e-2

Table I presents the final distribution obtained from TRACE 3-D, PARMILA, and from TOPKARK (both in the linear limit and to third order). Comparison shows that TOKARK agrees very well with TRACE 3-D in the linear limit, and that reasonable agreement is obtained with PARMILA when higher-order effects are included.

Table II demonstrates how TOPKARK can separate high-order aberrations into chromatic and geometric contributions, as well as distinguishing them order by order. This capability reveals that second-order geometric aberrations (which are prohibited by symmetry in the zero-current limit) do in fact occur with 3-D space charge forces.

Table II. Breakdown of Contributions to $\langle y^2 \rangle_f$:

$\langle y^2 \rangle_f$ Contributions	0 mA	50 mA
Total	1.336 e-06	1.546 e-06
Linear	1.000 e-06	1.000 e-06
Chromatic	0.341 e-06	0.544 e-06
2nd Order	0.342 e-06	0.542 e-06
3rd Order	-0.001 e-06	0.002 e-06
Geometric	-0.005 e-06	0.002 e-06
2nd Order	0	0.009 e-06
3rd Order	-0.005 e-06	-0.007 e-06

SPACE CHARGE -- SOME GENERAL CONSIDERATIONS

The algorithms of TOPKARK *cannot* follow the development of any structure *within* a bunch, such as might arise due to plasma waves or instabilities. For this reason,[12] we must impose the criterion: $\omega_p^{-1} \gg T_{transit}$, where $\omega_p = (q^2 n/\varepsilon_0 m \gamma_0)^{1/2}$ and $T_{transit} = L/\beta_0 c$. Furthermore, because we impose a *smooth* form on the distribution, we must be in a regime where the combined field of the particles is smooth and individual collisions are a *secondary* effect. For this reason,[12] we must impose the criterion: $n\lambda_D^3 \gg 1$, where $\lambda_D = v_{thermal}/\omega_p$ and $v_{thermal} \sim v_{rms}$.

Simple calculations show that $n\lambda_D^3 \sim 1\times 10^7$ throughout, so our second criterion is satisfied. Because ω_p varies by an order of magnitude, we assumed a simple model for its z-dependence, then integrated over the first half of the transport line to find:

$$T_{transit} \int_{z=0m}^{z=1m} dz\, \omega_p(z) \sim 2\; . \qquad (10)$$

To satisfy our first validity criterion, the above integral should be much less than 1; however, the fact that it is of order unity indicates that plasma effects will not dominate.

CONCLUSIONS

TOPKARK is a tested and reliable high-order beam optics code in which a TRACE3D-like space charge model has been successfully implemented. In a simple example, TOPKARK demonstrates close agreement (in the linear limit) with TRACE 3-D and reasonable agreement (to third order) with PARMILA, showing comparable transverse emittance growth.

Three-dimensional space charge effects introduce an asymmetry of the variable $\delta\tau$ into the equations of motion, thus allowing for the existence of *new* second-order geometric aberrations. This result is contrary to the physical intuition that has been developed through the use of test-particle codes, 2-D space charge codes, and 3-D *linear* space charge codes.

ACKNOWLEDGEMENTS

The authors wish to acknowledge many fruitful discussions with Alan Todd of the Grumman Corporate Research Center and with Tom Mottershead, Filippo Neri, Walter Lysenko and Paul Channell of Los Alamos National Laboratory.

REFERENCES

1. W. H. Press, B. P. Flannery, S. A. Teukolsky and W. T. Vetterling, Numerical Recipes (Cambridge University Press, 1990).
2. M. Berz, Part. Accel. **24**, 109 (1989).
3. E. Forest, M. Berz and J. Irwin, Part. Accel. **24**, 91 (1989).
4. A. J. Dragt and J. M. Finn, J. Math. Phys. **17**, 2215 (1976).
5. A. J. Dragt, L. M. Healy, F. Neri and R. Ryne, IEEE Trans. Nucl. Sci. **5**, 2311 (1985).
6. E. Forest, "The Absolute Bare Minimum for Tracking in Small Ring (Minus Radiation), unpublished report (1989).
7. J. van Zeijts and F. Neri, in these proceedings.
8. O. D. Kellogg, Foundations of Potential Theory, (Dover Publications, 1954).
9. MACSYMA™ User's Guide, (Symbolics Inc., 1987).
10. K. R. Crandall and D. P. Rusthoi, TRACE 3-D Documentation, (LAACG, 1990).
11. Much-used but little-documented particle code -- for an overview: G. P. Boicourt, AIP Conference Proceedings **177**, ed. C. R. Eminhizer (AIP, NY, 1988), p. 1.
12. J. D. Lawson, Applied Charged Particle Optics, ed. A. Septier, (Academic Press, 1983).

PIC Space-Charge Emission with Finite Δt and Δz *

Dennis W. Hewett and Yu-Jiuan Chen
Lawrence Livermore National Laboratory
University of California
Livermore, California 94550

ABSTRACT

A new algorithm for space charge emission has been developed to provide the correct (to a few percent) Child-Langmuir steady-state current limits as the number of mesh points in the voltage gap drops to O(10). Further, the transient behavior of such flows compares well with idealized, analytic cases, lending confidence as we extend these algorithms into full RZ geometry with curved emitting surfaces to investigate transient characteristics of realistic injector designs.

INTRODUCTION

Charged particle source physics has a wide range of applications. Of particular interest are new applications in plasma-aided manufacturing where ions essential to the process are extracted from a plasma. This same physics is applicable to the heavy ion source, a crucial part of Heavy Ion Fusion HIF. The common theme is the extraction of the desired ion species from a region of neutral or quasi-neutral material by an external voltage. The dividing line between the beam and quasineutral regions can be viewed as a **Space-Charge-Emitting** Surface (SCES) that can be modeled in a simulation as a boundary condition. We hold this surface at a fixed potential and assume it to be able to deliver all the particles or charge necessary to reduce the surface normal electric field to zero. The numerical boundary must emit these particles such that they may be extracted with little transverse temperature and thus quickly achieve the anisotropic velocity profiles that are an essential feature of many applications. In practice, building such a numerical boundary condition is not difficult if we use of order 100-200 grid points in the gap between the anode and cathode (called

---

* Work performed under the auspices of the U.S. Department of Energy by Lawrence Livermore National Laboratory under contract W-7405-ENG-48.

the A-K gap) and a sufficiently small time step so that the fastest particles do not travel more than a cell in Δt. As may be surmised, we do not have the luxury of carefully resolving the SCES in codes that are to be used frequently as design tools. We are required by computational demands to use very large space steps Δz and large time steps Δt.

Electrode design in such sources is also critical to the overall performance. Steady state codes, such as EGUN[1], can provide excellent predictive capabilities for the steady-state flow characteristics of a given configuration. However, in short pulse operation in which the control of beam transient behavior is essential, time-dependent acceleration and focusing voltages add yet another dimension to the problem. Only time-dependent simulation can provide the detailed information needed to assess the effect of voltage waveforms on the beam bunch shape. This information is essential to the design of such low-emittance, short-pulse source electrode configurations and to the design of the associated pulse power supplies.

In this paper we describe such a time-dependent code and concentrate on the numerical treatment of the SCES. We utilize the external and internal electrode structure specification capabilities of the time-dependent, axisymmetric RZ PIC code, GYMNOS[2]. Since we will not generally have the luxury of carefully resolving the SCES region, We have developed a space-charge-emission algorithm that allows us to use large space and time steps and satisfies three criteria. First, it must provide steady-state currents to within a few percent of *both* the well-proven EGUN result *and* the experimental measurement even as the resolution becomes marginal. Second, agreement is also required between GYMNOS, EGUN, and the experiment in steady-state normalized emittance values. Finally, we expect GYMNOS to provide detailed agreement with the few beam transient test cases that can be found analytically.

GEOMETRIC PROPERTIES OF GYMNOS

GYMNOS stores all quantities on the corners of a regular uniform RZ mesh. The boundaries of all structures, internal or external, also are assumed to lie on these corners. In a simple electrostatic case, a charge density ρ is accumulated from the PIC ion representation on all mesh points — taking care to use the correct reduced volume when finding ρ at a mesh point that is on a structure boundary. Given the instantaneous ρ, we then solve for the consistent electrostatic potential ϕ on *all* the mesh points, including those that represent the space-charge-emitting SCES. On the SCES we specify ϕ (typically $\phi = 0$). At most points **E** is obtained by central differencing. On the SCES boundary, since we know ϕ and ρ at all points, we can reconstruct the potential value ϕ_{in} just inside the SCES structure from the ϕ value *on* the SECS, the one just outside the SCES, and the ρ at the point in question. Finite differencing for **E** at this point using ϕ_{in} gives a second-order approximation for the electric field on the SCES.

IMPLEMENTATION OF THE SCES BOUNDARY

We now add the additional physics that characterizes a SCES; charge is emitted until the normal **E** is also zero. We use this condition to determine how much charge would have to be emitted each time step to bring the surface field described above back to the physics-required zero. The condition is simple: the surface normal **E** is equated to an induced surface charge σ that, multiplied by the surface area represented by that node, just gives the space charge that should be emitted this time step.

We now describe our implementation of this straightforward algorithm together with numerical tests that have allowed us to tune the algorithm so that it provides remarkable agreement with very coarse finite difference representations. Given the above prescription for the surface normal **E** and thus the charge induced from the surface for emission, we fill a reservoir at each mesh point and emit particles from the reservoir until it no longer exceeds the charge carried by each simulation particle. (Other approaches use variable particle weights so as to emit constant *numbers* of particles/cell/Δt; so far our results seem more than adequate using uniform weighting.)

In the limit of a small time step and many spacial grid points between anode and cathode, many straightforward algorithms can provide adequate results. We now present the test results that show that one algorithm continues to work well as resoluation degrades and seems to be robust enough to also work as the SCES is generalized to a curved surface. The method is to place randomly as many particles as can be extracted from the reservoir randomly in the first half cell outside the SCES, and to give these particles a normal velocity $\mathbf{u}_{norm} = 2\Delta t\, q\, \mathbf{E}_{norm}/m$. This expression for u_{norm} is derived from the condition that the normal force times rate at which the particle gains kinetic energy, i.e.,

$$qE_{\text{norm}}u_{\text{norm}} = mu_{\text{norm}}^2/2\Delta t \quad , \qquad (1)$$

where the electric field is evaluated one-half cell outside of the SCES. Empirically, we have found that a very good choice is to evaluate the electric field one-half cell outside the SCES.

NUMERICAL TESTS OF THE SCES BOUNDARY

The simulation tests that lead us to these choices were an 1-D potassium (A=39) diode studies with a gap distance of 1.6 cm and a voltage of -6.56 kV in which we varied the number of mesh points in the gap. The Child-Langmuir current for a such diode is 0.057 mA. Shown in Fig. 1 are 1D cases with 240 mesh points in the A-K gap in Fig 1a and the corresponding run with only 8 mesh points in Fig. 1b. We have shown the steady-state v_z vs. z phase space in which the random loading in the first half cell outside of z_{min} is very apparent in the coarse mesh case. Nonetheless the total current for both the 240 mesh point case and the 8 mesh point case is 0.057 mA, within PIC noise, as predicted by the Child-Langmuir law. The time step is a relatively small $\Delta t = 0.5$ ns in both cases.

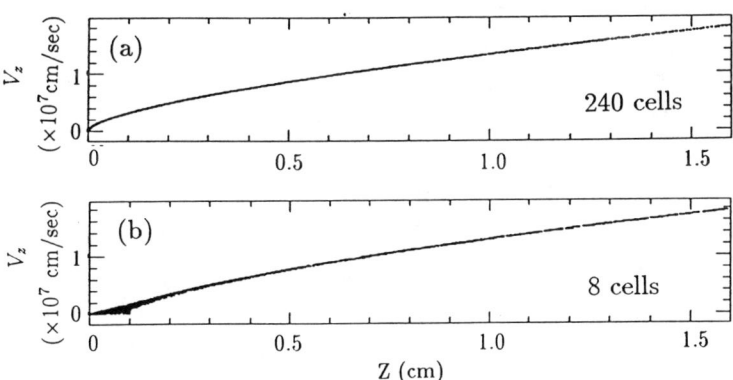

Fig. 1 The v_z vs. z phase spaces for the 1-D diode with (a) 240 and (b) 8 mesh points in the A-K gap.

Since one of the purposes of the time-dependent approach is to study the effect of transients, we show in Fig. 2 the simulation results of the same 1-D potassium diode using the A-K voltage waveform

$$\phi(t) = \left[\frac{4}{3}\frac{t}{t_{\text{rise}}} - \frac{1}{3}\left(\frac{t}{t_{\text{rise}}}\right)^4\right]\phi_0, \qquad t \leq t_{\text{rise}} ,$$
$$\phi(t) = \phi_0, \qquad\qquad\qquad\qquad\qquad t > t_{\text{rise}} . \qquad (2)$$

Using the same time step as in Fig. 1 and 8 mesh points in the A-K gap, the steady current values in both Figs. 2a and 2b are the same 0.057 mA since the voltages and gap distances are the same as that of the Fig. 1 tests. For the case in Fig. 2a, the waveform used in the simulation has the rise time t_{rise} equal to the transit time, t_{trans}, for an ion to cross the A-K gap, i.e., the so-called Lampel and Tiefenbach voltage waveform[3]. By using this voltage waveform, we obtained the predicted constant current profile for the front end and the flat-top of the beam pulse. When $t_{\text{rise}} < t_{\text{trans}}$, we expect the same asymptotic Child-Langmuir current at the flat-top portion of the beam pulse led by a higher current during the rise time. In the case $t_{\text{rise}} = 150$ ns, the current during the rise time is estimated to be roughly 0.08 mA. Our result for this situation is shown in Fig. 2b. The degree to which our simulation results agree with analytic predictions in both cases gives us confidence that we can achieve useful results in more complicated geometries where we are forced to work with limited spatial resolution.

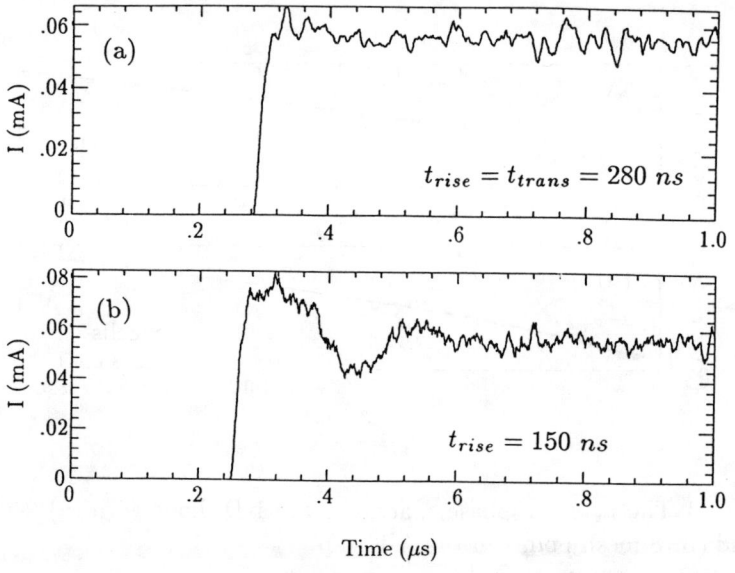

Fig. 2 The current profile calculated by GYMNOS when the A-K gap voltage waveform's rise time is (a) equal to and (b) less than the ion transit time, respectively.

We have used GYMNOS to study the axisymmetric LBL HIF electrostatic column injector. In Fig. 3 we show a case with a current valve mesh[4,5], used to control the beam pulse. Only 8 mesh points across the gap between the emitting surface and the current valve were used in the simulations to resolve the z-variations of the electric field. The comparison of current, normalized emittance, beam envelope radius, and beam divergence from GYMNOS simulation, experiment[6], and EGUN are given in Table I. The EGUN calculations were done from the immediately downstream side of the current valve mesh to the emittance diagnostics location by Henestroza[7]. To include finite temperature effects in the EGUN calculations, the initial transverse beam velocity distributions at the current valve location must be assumed. Several "reasonable" distribution functions, all with the same transverse temperature as the GYMNOS runs, were chosen to characterize the emittance that is representative of this geometry. The result is the range of emittances that are given for EGUN in Table 1. Figure 3 shows that the beam radius is comparable to the electrodes' aperature size. Hence, the beam experiences a large nonlinear external field and its normalized beam emittance grows from its intrinsic value of 0.05 mm-mr at the source to 0.25 mm-mr at the emittance diagnostics location.

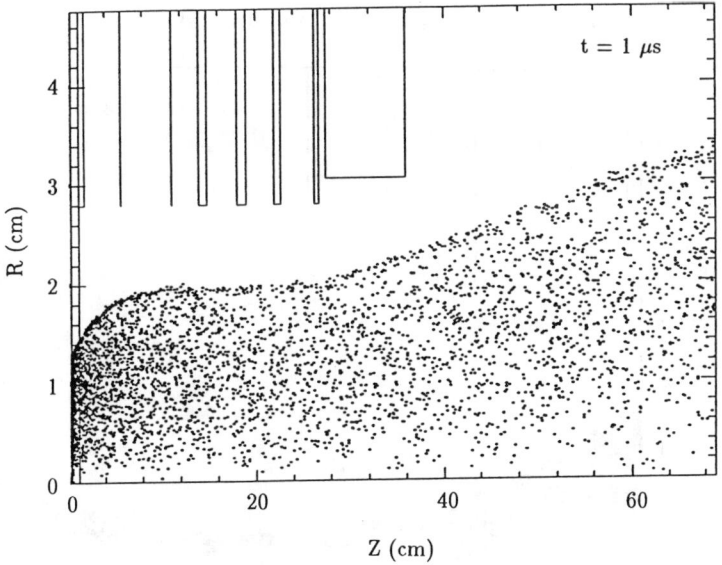

Fig.3 The LBL HIF injector with a current valve

Table I The LBL HIF injector with a current valve

	GYMNOS	EXP[6]	EGUN[7]
Current (mA)	82	80	80
Normalized emittance (mm-mr)	.26	.25	0.07-0.2
Beam radius (mm)	32.5	31.2	31.0
Beam divergence (mr)	34.5	38.4	36.0

The GYMNOS result of the same configuration without the current valve and with the same mesh size in the z direction is shown in Fig. 4. When the current valve mesh was removed, the voltage of the first electrode (at z=1.2 cm in Fig. 4) was the same as that of the ion emitting anode. This voltage arrangement results in curved equipotential surfaces near the anode so that the beam sees a very strong radial focusing force near the ion emitting surface and the first electrode. The beam is pinched and focused roughly to a 1mm radius spot size at the injector exit. The space-charge limited current is then reduced. Since the beam radius is much smaller than the electrodes' aperature size, the external field beam sees is very linear. There is no normalized emittance growth in this case. The radial mesh size used in the simulation was quite coarse ($\Delta r = 0.6$ mm) compared with the 1 mm beam radius. There is not enough resolution to simulate the small beam size and beam divergence properly. Nevertheless, we have obtained very good

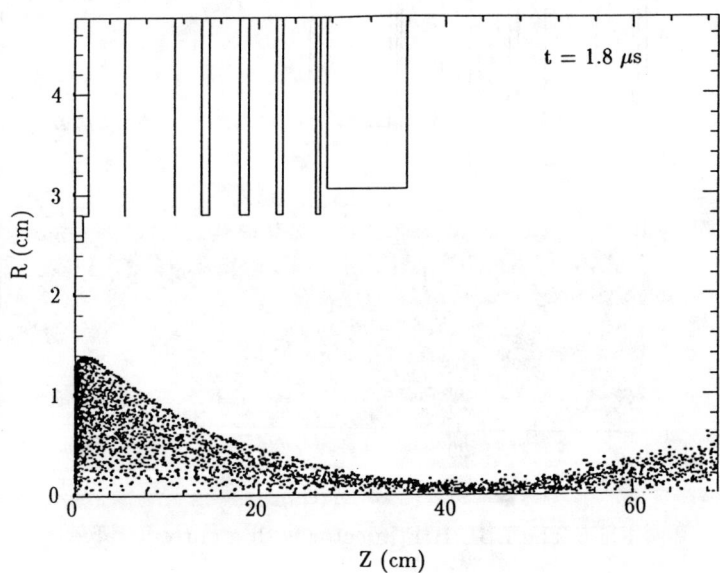

Fig. 4 The LBL HIF injector without a current valve

agreement in the values of current and normalized emittance with experiments and EGUN calculations as given in Table II.

Table II The LBL HIF injector without a current valve

	GYMNOS	EXP[6]	EGUN[7]
Current (mA)	20	> 24	19
Normalized emittance (mm-mr)	.06	.04	0.05
Beam radius (mm)	5.0	1.2	0.9
Beam divergence (mr)	19	6	8

SUMMARY

We have developed and implemented a space charge emission algorithm to the time dependent axisymmetric PIC code, GYMNOS. The algorithm can provide the correct Child-Langmuir current when the number of mesh points in the A-K gap is as little as 8. Comparing with the analytic results, the simulation can also provide the predicted transient behavior. Finally, the GYMNOS simulations of the LBL HIF electrostatic column injector agree with the experimental data and EGUN results quite well.

ACKNOLEGEMENT

The authors would like to thank S.S. Yu and J.J. Barnard for their useful discussions during the course of this study.

REFERENCES

1. W.B. Herrmannsfeldt, "Electron Trajectory Program", SLAC Report No. SLAC-226, UC-28 (A), Nov. 1979.
2. D.W. Hewett and D.J. Larson, "The Best of GYMNOS: A User's Guide", LLNL Report No. UCRL-UR-110499, May 4, 1992.
3. Lample and Tiefenback, Appl. Phys. Lett. Vol. 43, No. 1, p. 1 (1983).
4. H.L. Rutkowski, D.W. Hewett, and S. Humphries Jr., 'Development of Arc Ion Sources for Heavy Ion Fusion", IEEE Transactions on Plasma Science, 19, No. 5, Pg. 782 (1991).
5. D.W. Hewett, M.R. Gibbons, and H.L. Rutkowski, "Extracting Low Emittance Ion Beams through a Current Valve/Switch Mesh", (in preparation, 1993).
6. S. Eylon, private communication, (1992).
7. E. Henestroza, private communication, (1993).

ENVELOPE MODEL OF BEAM TRANSPORT IN ILSE*

W. M. Sharp, J. J. Barnard, D. P. Grote, S. M. Lund, and S. S. Yu[†]

Lawrence Livermore National Laboratory, Livermore, CA 94550

ABSTRACT

CIRCE is an efficient beam dynamics code developed to facilitate the design and analysis of heavy-ion accelerators. The code combines an envelope description of the beam transverse dynamics with a fluid-like treatment of longitudinal dynamics, and terms are included to account for the effects of space charge, emittance, and image forces. CIRCE is currently being adapted to model the Induction Linac Systems Experiments (ILSE) facility, a proposed heavy-ion accelerator designed to test aspects of an inertial-fusion driver. The numerical model in the code is discussed, and changes needed for modeling ILSE are outlined. Preliminary work is presented on beam matching along the ILSE lattice and on transport around the ILSE achromatic bend.

I. INTRODUCTION

The Induction Linac System Experiments (ILSE) is an ion accelerator and series of experiments[1] planned at Lawrence Berkeley Laboratory to study critical physics issues in heavy-ion fusion (HIF) driver. The facility will accelerate four beams simultaneously from an injection energy of 2 MeV to 5 Mev, using electrostatic quadrupoles for transverse focusing. After the four beams are merged in a "combiner," the single beam will be accelerated to 10 MeV. During acceleration, the beam will be compressed from an initial duration of 1 ms to 0.4 ms by imposing a velocity "tilt" along the beam length, with ions at the beam tail being accelerated more than those near the head. Among the planned experiments are a 180° achromatic bend and a final section where the the beam is compressed as it drifts. The planned current and emittance of the ILSE beam are chosen to give the same line-charge density and beam diameter as in a HIF driver, making ILSE a realistic test of the dynamics of space-charge dominated beams.

The ILSE facility requires detailed design of the acceleration schedule. The beam must remain matched to the transverse focusing, and the strong longitudinal space-charge forces near the beam ends must be balanced by external fields as the beam is simultaneously accelerated and compressed. Due to the large parameter space, much of the initial design work will likely be done using the fast-running multiple-slice envelope code CIRCE[2]. The CIRCE model uses an envelope description of the beam transverse dynamics and treats the beam as a Lagrangian fluid in the longitudinal direction. Appropriate terms are included to account for the effects of image forces, beam emittance, and space-charge in the limit of paraxial motion, and the beam is focused and accelerated by a user-specified lattice of electrostatic and magnetic quadrupoles, bending magnets, and acceleration modules. Although the code was described in Ref. 1, the equations are extensively reviewed in Section II because numerous refinements and generalizations have been made since the earlier paper. Further code modifications required for modeling ILSE are then presented in Section II along with preliminary simulations of the ILSE acceleration section and achromatic bend.

II. MODEL

The set of envelope equations used here to model the beam transverse dynamics is a generalization of that formulated first by Kapchinskij and Vladimirskij and later adapted by Lee, Close, and Smith.[2] The beam transverse distribution function is assumed to be uniform and elliptical in each phase-space plane, and the coordinate-space ellipse may be taken to be unskewed

---

* This work was performed under the auspices of the US Department of Energy by Lawrence Livermore National Laboratory under W-7405-ENG-48.

† Present address: Lawrence Berkeley Laboratory, Berkeley, CA 94720

provided that the quadrupoles have the same orientation everywhere and there is no axial magnetic field. Except for image forces, the treatment is first order in the ratio of the beam radii and centroid displacements to the beam-pipe radius. This approximation makes it appropriate to represent bend magnets and quadrupoles by idealized expressions and to neglect higher-order multipole fields. In the present version of the code, only single-function magnets are used, and a simple analytic model may optionally be used to represent the axial fringe fields. Although the ion mass is multiplied where it appears by the Lorentz factor γ, derivatives of γ are small enough to be dropped from the ion-motion equations. Also, we assume a circular beam pipe of infinite conductivity and radius R.

Transverse Dynamics

With these assumptions, coupled envelope equations for the coordinate-space radii are found by differentiating the expressions defining the root mean-squared beam radii. We use a local Cartesian coordinate system centered on the beam-pipe axis, in which the coordinate s is distance around the axis of the lattice, and x and y are respectively the spatial coordinates in and perpendicular to the nominal plane of the lattice. For a beam pipe with a local bend radius ρ and a curvature vector that is rotated in the transverse plane by an angle α from the negative x axis, the envelope equations have the form

$$\frac{d^2 a}{ds^2} + \frac{1}{\beta}\frac{d\beta}{ds}\frac{da}{ds} = \pm\frac{F'}{[B\rho]}a + \frac{\epsilon_x^2}{a^3} + \frac{2K}{a+b} + \frac{f_0 K}{R^2}a - (a\cos\alpha + b\sin\alpha)\frac{\cos\alpha}{\rho^2} \quad (1a)$$

$$\frac{d^2 b}{ds^2} + \frac{1}{\beta}\frac{d\beta}{ds}\frac{db}{ds} = \mp\frac{F'}{[B\rho]}b + \frac{\epsilon_y^2}{b^3} + \frac{2K}{a+b} - \frac{f_0 K}{R^2}b - (a\cos\alpha + b\sin\alpha)\frac{\sin\alpha}{\rho^2}. \quad (1b)$$

Here, a and b are the beam coordinate-space radii respectively in the x and y directions. The quadrupole-field transverse gradient F' is the on-axis magetic-field gradient B'_q for magnetic quadrupoles and $E'_q/\beta c$ for electrostatic ones, with the sign being determined by the quadrupole orientation. The quantity β is the axial velocity scaled by the speed of light c, and the "magnetic rigidity" in the quadrupole-focusing terms is given in SI units by $[B\rho] = \beta\gamma Mc/qe$, where M and q are the ion mass and charge state, and γMc^2 is the total energy of beam ions. The perveance K in the space-charge and image-force terms is defined as

$$K = \frac{1}{4\pi\epsilon_0}\frac{2qeI_b}{(\beta\gamma)^3 Mc^3}, \quad (2)$$

where I_b is the beam current in Amperes, and ϵ_0 is the free-space permittivity. The beam transverse temperature is accounted for in Eq. (1) by the terms proportional to the unnormalized emittances ϵ_x and ϵ_y, which are calculated here by assuming that the normalized-emittance components $\beta\gamma\epsilon_x$ and $\beta\gamma\epsilon_y$ are constant. Following the method in Ref. 3, the image-force terms in Eq. (1) are derived by assuming that the centroid of the elliptical beam is displaced a distance $(X^2+Y^2)^{1/2} \ll R$ from the axis of a straight beam pipe, where X and Y are the beam centroid coordinates. With this assumption, the coefficient f_0 has the form

$$f_0(a,b,X,Y) = \frac{a^2-b^2}{4R^2} + \frac{X^2-Y^2}{R^2} + \frac{3}{2}\frac{X^2+Y^2}{R^2}\left(\frac{a^2-b^2}{R^2}\right) + \frac{3}{8}\frac{X^2-Y^2}{R^2}\left(\frac{a^2-b^2}{R^2}\right)^2. \quad (3)$$

The presumption of a straight beam pipe substantially simplifies the algebra and is appropriate when the bend radius ρ is much larger than R. The $d\beta/ds$ term in Eq. (1) arises from changing variables from t to s, and, as discussed below, an approximate expression for the derivative is obtained directly from one of the equations for longitudinal motion.

Equations for the lattice-plane centroid coordinates X and Y are obtained from distribution averages of the single-particle motion equations. We find

$$\frac{d^2 X}{ds^2} + \frac{1}{\beta}\frac{d\beta}{ds}\frac{dX}{ds} = \pm \frac{F'}{[B\rho]}X + \left(\frac{\cos\alpha}{\rho} - \frac{B_{dy}}{[B\rho]}\right)$$
$$+ \frac{(1+g_0)K}{R^2}X + \frac{g_1 K}{R^2}Y - \left[\frac{X\cos\alpha + Y\sin\alpha}{\rho^2}\right]\cos\alpha, \qquad (4a)$$

$$\frac{d^2 Y}{ds^2} + \frac{1}{\beta}\frac{d\beta}{ds}\frac{dY}{ds} = \mp \frac{F'}{[B\rho]}Y + \left(\frac{\sin\alpha}{\rho} + \frac{B_{dx}}{[B\rho]}\right)$$
$$+ \frac{(1-g_0)K}{R^2}Y + \frac{g_1 K}{R^2}X - \left[\frac{X\cos\alpha + Y\sin\alpha}{\rho^2}\right]\sin\alpha, \qquad (4b)$$

where $\underset{\sim}{B}_d = B_{dx}\hat{x} + B_{dy}\hat{y}$ is the dipole field, and the image-force coefficients g_0 and g_1 are given in the straight-pipe paraxial limit by

$$g_0(a,b,X,Y) = \frac{a^2-b^2}{4R^2} + \frac{(X^2-Y^2)}{R^2} + \frac{3}{4}\frac{(X^2+Y^2)}{R^2}\left(\frac{a^2-b^2}{R^2}\right)$$
$$+ \frac{1}{8}\frac{(X^2-Y^2)}{R^2}\left(\frac{a^2-b^2}{R^2}\right)^2 \qquad (5a)$$

$$g_1(a,b,X,Y) = \frac{2XY}{R^2}\left[1 - \frac{1}{8}\left(\frac{a^2-b^2}{R^2}\right)^2\right]. \qquad (5b)$$

To avoid deflecting the beam from the pipe axis, the dipole-field components should be $B_{dx} = -B_d\sin\alpha$ and $B_{dy} = B_d\cos\alpha$.

Longitudinal Dynamics

To model axial dynamics, we treat slices of the beam as Lagrangian fluid elements characterized by an axial velocity βc and the time τ that the slice arrives at an axial location s. This approach implicitly assumes that the beam has a negligible longitudinal temperature and that the slices remain approximately collinear. If the slice boundaries are presumed to remain perpendicular to the beam-pipe axis, then the equation for τ is found from orbit kinematics to be

$$\frac{d\tau}{ds} = \frac{1}{\beta c}\left(1 + \frac{X\cos\alpha + Y\sin\alpha}{\rho}\right), \qquad (6)$$

where we have again assumed paraxial motion. An approximate β equation is obtained by retaining only the electrostatic force in the single-particle motion equations and averaging the axial component over the beam elliptical cross-section:

$$\frac{d\beta}{ds} = \frac{qe}{\beta Mc^2}\left(1 + \frac{X\cos\alpha + Y\sin\alpha}{\rho}\right)(E_{ext} + E_{sc}). \qquad (7)$$

Here, the average external electric field E_{ext} is approximated only by the voltage across accelerating modules divided by the gap length. The space-charge field is approximated by

$$E_{sc} \approx g\left[\frac{\partial}{\partial \tau}\left(\frac{\lambda}{\beta c}\right) + \frac{\lambda}{\beta}\frac{d\beta}{ds}\right], \qquad (8)$$

where the line-charge density λ for a slice of duration $\delta\tau$ containing charge δQ is estimated by

$$\lambda = \frac{\delta Q}{\beta c \delta\tau}, \qquad (9)$$

and the inductance-like factor g is given by

$$g \approx \frac{1}{4\pi\epsilon_0} \ln\left(\frac{R^2}{ab}\right), \qquad (10)$$

provided the beam has a uniform charge density in the transverse plane. In deriving the space-charge field, the radial electrostatic field is assumed to vary over a much shorter scale length than E_{sc}, and the continuity equation is used to convert derivatives with respect to s into τ derivatives. When Eq. (8) is substituted into the β equation Eq. (7), the resulting equation is trivially rearranged to give an equation for $d\beta/ds$ in terms of E_{ext} and the time derivative of λ/β.

The equations Eqs. (1) - (9) are recast in the code as a set of ten first-order equations and are integrated by a conventional fourth-order Runge-Kutta method. A constant step size in s is used except near the boundaries of lattice elements, where the step is chosen to land on each boundary. The results are found to be insensitive to the choice of step size so long as there are 10 or more integration steps per lattice element. To initialize the equations in equilibrium, we integrate the equations over the first full lattice period and use a vector form of Newton's method to adjust the initial values of the beam radii and their derivatives in each slice until they equal the corresponding final values.

User Interface

An important aspect of the code is the lattice specification. The user may specify an arbitrary number of distinct lattice elements, specifying such properties as length, aperture, strength, bend angle, and the rotation angle in the plane transverse to the beam direction. At present, the element types allowed in the code are drifts, accelerating gaps, sector bend magnets, electrostatic and magnetic quadrupoles, beam-position monitors, and steering stations. Solenoids and higher-order multipoles might also be added, but the assumptions of the model would have to be substantially modified. Each lattice element is given a name by the user, and lattice sections may be defined by listing names of previously defined elements and subsections along with the number of times the listed items are repeated in that section. The final such grouping of subsections is treated as the complete lattice.

To facilitate code use, the code has a lattice "self-design" option. The main assumption used to modify the lattice parameters is that energy is gained linearly in distance s along the lattice. With this assumption, the code can set both the time-averaged voltage across accelerating gaps and the field strengths of any bend magnets. Compensation for longitudinal space charge is possible by imposing an appropriate time variation on the accelerating voltage. In the code, these voltage "ears" are calculated from appropriate numerical derivatives of the beam current and velocity and can optionally be modified to mimic the effects of the pulse-forming lines or field-effect transistor switches that are usually used to generate the waveforms. In addition, CIRCE has a palette of errors in the strength and alignment of magnets, the timing of acceleration fields, and the position reading of monitors that can be introduced to study error sensitivities. These run options, as well as the beam and lattice specifications, can be entered through either a namelist input file or a graphical user interface.

III. APPLICATION TO ILSE

The parameters and acceleration schedule of ILSE require modifications to the CIRCE lattice-design algorithm. The code was originally written to model the recirculating HIF accelerator studied by the Lawrence Livermore National Laboratory.[4] As proposed, this "recirculator" consists of several circular lattices in which pulses would be focused by superconducting quadruples as their energy is increased by about an order of magnitude. The quadrupole occupancy and strength as well as the lattice period would be constant around each ring, and in early designs, most of the acceleration was to be done at constant beam duration. In contrast, the lattice parameters in ILSE will change along the accelerator to maintain the desired phase advance and transverse size, and the beam is continually compressed during acceleration. In addition, the beam combiner and achromatic bend have more complicated design requirements than are presently considered in the lattice-design section. Simple algorithms to handle these ILSE lattice-design functions and beam manipulations have been developed and are presented in the following subsections.

Lattice Transitions

The current ILSE design has several changes in the lattice period and the strength and occupancy of quadrupoles in order to maintain the transverse focusing during acceleration. To avoid mismatch oscillations, which invariably increase the emittance, we need a criterion for keeping the beam near transverse equilibrium at these lattice transitions. A simple criterion is obtained by noting that, for a beam in equilibrium, the beam charge density ρ_c depends only on lattice quantities and, for electrostatic focusing, the axial velocity. This result can be derived from the envelope equation Eq. (1) in the continuum limit, in which a and b are replaced by $\langle a \rangle$, their average value over a lattice period. Assuming $\langle a \rangle$ to be constant in s and neglecting the small image and bend terms, we obtain a quadratic equation for $\langle a \rangle^2$, which in the limit of negligible emittance gives

$$\langle a \rangle^2 \approx \frac{K}{k_\beta^2}, \tag{11}$$

where k_β is the betatron wavenumber in the absence of space charge. This assumption of negligible emittance clearly fails near the beam ends, where the perveance vanishes, but it is typically valid over much of the beam. Eq. (11) is put into a usable form by substituting the perveance expression Eq. (2) and using $k_\beta \approx qe\eta_q F'L/2\gamma\beta Mc$, where η_q is the fraction of the half-lattice period L occupied by the focusing quadrupole. The resulting equation can be rearranged to give

$$\rho_c \approx \frac{\epsilon_0}{2} \frac{qe\gamma}{M} \eta_q^2 F'^2 L^2, \tag{12}$$

where we have used the relation $I_b \approx \pi \langle a \rangle^2 \rho_c \beta c$. This ρ_c expression suggests that a matched condition, with constant or slowly varying charge density, can be maintained by keeping $\eta_q |F'| L$ constant across lattice transitions at which one or more of the factors change abruptly. This criterion is physically reasonable because $\eta_q |F'| L$ is proportional to the integrated focusing force in a half lattice period.

The effectiveness of this matching criterion is seen by comparing the CIRCE results in Fig. 1. The beam radii a and b for the center beam slice are shown in Fig. 1a for a preliminary ILSE lattice, constructed from considerations of phase advance and beam radius only, and significant mismatches are seen to occur at the first two points where the lattice period and the quadrupole strength and occupancy are changed. The corresponding plot is shown in Fig. 1b for a lattice with similar changes in η_q and L, but with the strength chosen to keep $\eta_q |F'| L$ constant. The latter case, while still not perfectly matched, at least shown no major disruption at the transitions.

Beam Compression

A procedure for compressing a beam while maintaining an approximately self-similar axial density profile can be inferred from the longitudinal equations. Neglecting bend terms in Eqs.

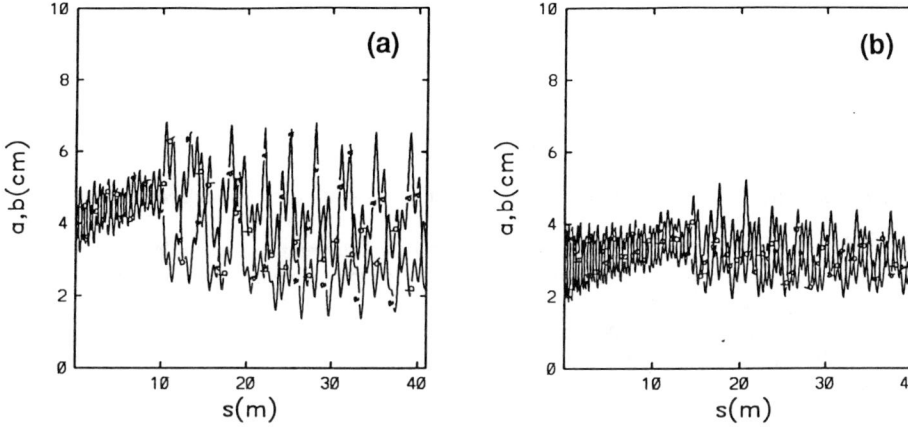

Fig. 1 Radii of the ILSE beam through the acceleration section, calculated by CIRCE, for lattices with (a) discontinuous changes in $\eta_q|F'|L$ at three positions and (b) smoothly varying $\eta_q|F'|L$.

(6) and (7) and the $d\beta/ds$ term in Eq. (8), we may write the longitudinal equations in the form

$$\frac{d\tau}{ds} \approx \frac{1}{\beta c} \qquad (13a)$$

$$\frac{d\beta}{ds} \approx \frac{qe}{\beta Mc^2}\left[E_{ext} + g\frac{\partial}{\partial \tau}\left(\frac{\lambda}{\beta c}\right)\right]. \qquad (13b)$$

The external field E_{ext} is taken to be the sum of a time-independent acceleration field E_{acc}, a compression field E_{comp} that typically has a linear variation in time, and a component with a more complicated time dependence, termed an "ear field" here, that on average balances the longitudinal space-charge force. The ear field may be written

$$E_{ear} \approx -\frac{g}{\eta_g \bar{\beta} c}\frac{\partial \lambda}{\partial \tau}, \qquad (14)$$

where $\bar{\beta}$ is a density-weighted average of β and η_g is the gap occupancy. Since this field approximately cancels the space-charge field arising from variation in λ, the only force between acceleration gaps is due to the remaining $\partial \beta/\partial \tau$ term in Eq. (13b). If we assume that this term is approximately constant along the beam, than Eq. (13) yields equations between gaps for the velocity variation $c\Delta\beta$ from tail to head and the beam duration $\Delta\tau$:

$$\frac{d\Delta\tau}{ds} \approx -\frac{\Delta\beta}{\bar{\beta}^2 c} \qquad (15a)$$

$$\frac{d\Delta\beta}{ds} \approx -\frac{gqeQ}{M(\bar{\beta}c)^4}\frac{\Delta\beta}{(\Delta\tau)^2}. \qquad (15b)$$

Here, λ has been written in terms of the total charge Q using the linear approximation $\lambda \approx Q/(\bar{\beta}c\Delta\tau)$. These equations can be solved exactly, but more useful expressions for $\Delta\beta$ and $\Delta\tau$ are obtained by assuming that the change in $\Delta\tau$ between gaps is small compared with the initial value $\Delta\tau_0$:

$$\frac{\Delta\tau}{\Delta\tau_0} \approx 1 - \frac{\Delta\beta_0}{\beta}\frac{s-s_0}{\bar{\beta}c\Delta\tau_0} \tag{16a}$$

$$\frac{\Delta\beta}{\bar{\beta}} \approx \frac{\Delta\beta_0}{\bar{\beta}}\left[1 - \frac{gqe\lambda_0}{M(\bar{\beta}c)^2}\frac{s-s_0}{\bar{\beta}c\Delta\tau_0}\right]. \tag{16b}$$

The zero subscripts here denote beam quantities at the end of the previous acceleration gap.

With these expressions for the change in $\Delta\beta$ and $\Delta\tau$ between gaps, a simple procedure may be followed to specify suitable compression and ear fields for succeeding gaps. Starting with an initial velocity tilt $\Delta\beta_0$ and beam duration $\Delta\tau_0$, the beam is advanced to the next gap, and expected values of $\Delta\beta$ and $\Delta\tau$ there are calculated from Eq. (16). Also, the new average velocity after the next gap is estimated from $\bar{\beta} \approx (\bar{\beta}_0^2 + 2qeE_{acc}\eta_g L/Mc^2)^{1/2}$. With these expressions, an appropriate ear field at the next gap is given by

$$E_{ear} \approx -\frac{g}{g_0}\frac{\eta_{g0}}{\eta_g}\left(\frac{\bar{\beta}_0\Delta\tau_0}{\bar{\beta}\Delta\tau}\right)^2 E_{ear0}, \tag{17}$$

where E_{ear0} is the prior ear field written in terms of $(\tau - \bar{\tau})/\Delta\tau$. Here, the quantity $\bar{\tau}$ in Eq. (17) is the temporal midpoint of the pulse at the center of the next gap, and we have assumed that the λ profile changes self-similarly during compression. The linearly varying compression field E_{comp} needed to produce a velocity tilt $\Delta\beta_{new}$, which is arbitrary so long as E_{comp} remains physically reasonable, is similarly given by

$$E_{comp} \approx \frac{Mc^2}{qe\eta_g L}\left(\bar{\beta}\Delta\beta_{new} - \bar{\beta}_0\Delta\beta\right)\left(\frac{\tau-\bar{\tau}}{\Delta\tau}\right). \tag{18}$$

This calculation is repeated at each succeeding gaps.

The principal shortcoming of this procedure is the absence of any overall compression schedule. The allowable beam duration at any point in the accelerator is obviously tied to the focusing strength and the beam-pipe radius, but the correlation has not been worked out. Also, a modified procedure for shaping the pulse prior to final compression must be developed.

Beam Combination

As presently conceived, the beam combiner will consist of a series of bends and quadrupole lenses that finally arranges the four ILSE beams into an approximation of a circular cross section. The resulting single beam would then have four times the current of each of the original beams, slightly more than four times the original emittance, and roughly twice the radius. The detailed beam dynamics in this section is difficult to model using CIRCE due to the complicated geometry, the inter-beam forces prior to merging, and the enhanced emittance growth resulting from entrapment of low-density regions of phase space. Instead, the combiner is initially being modeled in the code as a zero-length lattice element in which the current and emittance are simply multiplied by the number of beams N_b being combined, and the radii a and b are multiplied by $N_b^{1/2}$.

Achromatic Bend

The ILSE achromatic bend is intended to change the beam direction by 180° while approximately preserving the phase-space distribution of ions over some range of β values. This achromaticity is necessary because the beam is planned to have a residual head-to-tail velocity

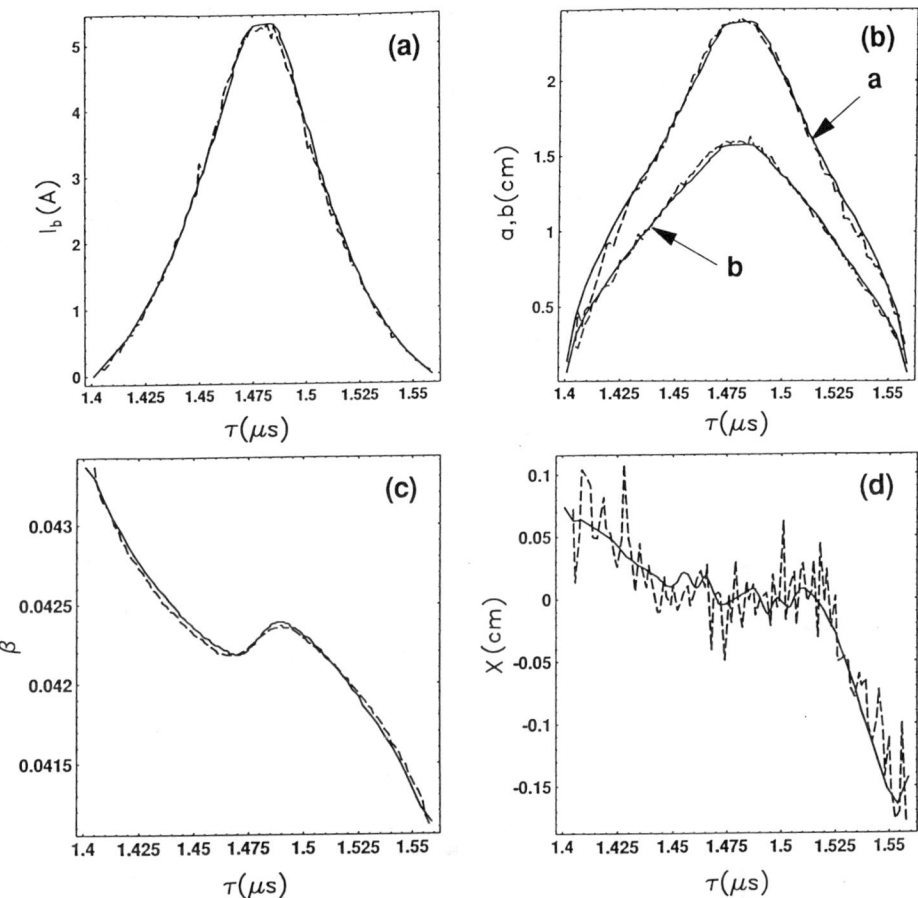

Fig. 2 ILSE beam quantities after traversing a 180° achromatic bend, as calculated by CIRCE (solid line) and WARP3d (dashed line): (a) current; (b) beam radii a and b; (c) axial velocity scaled by c; (d) displacement in x

variation after acceleration of a few percent. Although achromats can readily be designed in the absence of image forces, no systematic procedure including these forces has been devised for beams whose transverse dynamics is dominated by space charge.

To design an achromatic bend for a space-charge dominated beam such as that in ILSE, we plan to adapt an optimization routine developed by Hewett and Bangerter.[5] The technique, which is akin to a multi-dimensional Muller's method, in principle allows an arbitrary number of lattice variables to be adjusted to optimally match user-specified performance criteria. For the ILSE bend, quadrupole strengths and positions could be adjusted to minimize final centroid displacements and transverse velocities of beam slices with initial axial velocities above, at, and below the design velocity. A supplemental criterion would also require that the greatest transverse excursion of the beam slices be simultaneously minimized.

At present, the bend lattice specified in CIRCE is achromatic through first order in position and momentum in the absence of image forces. With this choice, we find that a 10 MeV beam of ^{12}C ions with a linear 3% velocity increase from head to tail develops the transverse displacement shown by the solid curve in Fig. 2d. For this case, the peak beam current increases by 46% due the the velocity tilt to a final value of 5.2 A, as shown in Fig. 2a, while the longitudinal space charge increases the beam duration by 17.5%. The higher line-charge density near the beam midpoint is found to cause an increase in a and b proportional to the square root of the perveance, as expected from Eq. (11), and Fig. 2b shows the corresponding $K^{1/2}$ variation of the final radius profiles. The plot of β for the beam slices in Fig. 2c shows that the beam center is still compressing at the end of the bend, but the ends are expanding due to the longitudinal space charge. The corresponding plots from the three-dimensional particle simulation WARP3d[6] are shown as dashed lines in Figs. 2a-d, with the transverse average of β being plotted in Fig. 2d. The excellent agreement here between CIRCE and WARP3d is nontrivial because the assumptions in CIRCE of zero longitudinal temperature and perpendicular slice boundaries are visibly violated in the WARP3d simulation. Furthermore, the initial line-charge density and velocity profiles must be set up as functions of τ to match the s-dependent profiles used in WARP3d, and the two codes must use similar algorithms for determining the initial emittance variation along the beam.

REFERENCES

[1] T. Fessenden, R. Bangerter, D. Berners, J. Chew, S. Eylon, A. Faltens, W. Fawley, C. Fong, M. Fong. K. Hahn, E. Henestroza, D. Judd, E. Lee, C. Lionberger, S. Mukherjee, C. Peters, C. Pike, G. Raymond, L. Reginato, H. Rutkowski, P. Siedl, L. Smith, D. Vanecek, S. Yu, F. Deadrick, A. Friedman, L. Griffith, D. Hewett, M. Newton, and H. Shay, "ILSE, The Next Step toward a Heavy Ion Induction Accelerator for Inertial Fusion Energy," *Proc. 14th Int. Conf. on Plasma Physics and Controlled Nuclear Fusion Research*, IAEA, Wurzburg, Germany, 30 September - 7 October 1992.

[2] W. M. Sharp, J. J. Barnard, and S. S. Yu, *Part. Accel.* **37-38**, 205 (1992).

[3] E. P. Lee, E. Close, and L. Smith, "Space Charge Effects in a Bending Magnet System," in *Proceeding of the 1987 IEEE Part. Accel. Conf*, E. R. Lindstrom and L. S. Taylor, eds. (IEEE, New Jersey, 1987), p. 1126.

[4] J. J. Barnard, A. L. Brooks, J. P. Clay, F. E. Coffield, F. J. Deadrick, L. V. Griffith, A. R. Harvey, D. L. Judd, H. C. Kirbie, V. K. Neil, M. A. Newton, A. C. Paul, L. L. Reginato, G. E. Russell, W. M. Sharp, H. D. Shay, J. H. Wilson, and S. S. Yu, "Study of Recirculating Induction Accelerators as Drivers for Heavy Ion Fusion," Lawrence Livermore National Laboratory Report UCRL-LR-108095, May 1992.

[5] D. W. Hewett and R. O. Bangerter, private communication.

[6] A. Friedman and D. P. Grote, *Phys. Fluids B* **4**, 2203 (1992).

ENVELOPE CODE FOR ELECTROSTATICALLY ACCELERATED BEAM WITH ESQ FOCUSING

L. Soroka and O. A. Anderson[a]
Lawrence Berkeley Laboratory, Berkeley CA 94720

ABSTRACT

We present a new envelope code, "ESQACL," which we have used for designing ESQ-focused ion or electron accelerators and transport systems. This code uses improved envelope equations and also allows the option of using accurate field maps along the beam axis instead of typical models. We show how the transverse fields are handled in the different cases. ESQACL is designed to interact with a 3-D Laplace solver that provides the fields of the complete system of electrodes including their supporting structure. We show examples that contrast results from this new code with those obtained from traditional envelope equations, from traditional field models, and from a particle code.

INTRODUCTION

The code described here has been used to design a number of accelerators and transport systems, including: a 200-kV CCVV prototype accelerator built and tested by the MFE group at LBL [1]; a 1-MV accelerator for a proposed test facility [2]; a 1.3-MV, 1-A D⁻ accelerator channel for ITER [3]; a high-current electron accelerator for an FEL [4]; and an injector proposed for SSC and CERN [5]. ESQACL would also be suitable for designing ESQ-focused heavy ion accelerators.

In the next section we discuss the special features of our code, including our improved envelope equations [6] and the available field options. The options are: (a) accurate fields from a 3-D Laplace solver, which ESQACL combines in proportion to user-specified electrode voltages; (b) an arbitrary field model supplied by the user; or (c) fields from ESQACL's own associated trapezoidal field module.

We then discuss the treatment of transverse fields for the different field models, the code structure, and examples of applications. The examples contrast results from this new code with those obtained from traditional envelope equations [7,8], from traditional field models [9], and from a particle code [10].

FIELD MODELS AND ENVELOPE EQUATIONS

ESQACL accepts two types of external field maps; alternatively, it can stand alone and generate its own a trapezoidal map. When a certain flag [11] is set to 0, the code takes a set of input field maps (on axis) from a 3-D Laplace solver. For each Laplace calculation used in the set of maps, a single quadrupole pair with support plate is activated with unit voltage while all the other electrodes are grounded. There are as many Laplace field maps as there are independent voltages in the channel. These separate fields are combined by ESQACL in proportion to electrode voltages specified as input.

When the field map flag = 1, the user supplies the fields on axis for the entire system; these could by obtained, for example, from a single Laplace run for fixed electrode voltages.

---

[a] Also affiliated with Particle Beam Consultants, 2910 Benvenue Ave., Berkeley CA 94705.

When the flag = 2, ESQACL calculates the fields internally using the trapezoidal field model [9]. The first and third options allow interactive optimization of the electrode voltages.

The Laplace fields and trapezoidal fields are compared in Fig. 1 for the CCVV prototype accelerator [1], [2] in which the ESQ electrodes are supported by end plates. In the trapezoidal case, the potential along the axis (dotted line in Fig. 1a) is set at the mean electrode potential in the overlapping quadrupole region and ramps linearly to the plate potential in the nonoverlapping part. It ramps linearly again between the support plates for the different cells, which are at different potentials for beam acceleration. The actual Laplace field (the solid line) is, of course, much smoother. Transverse field gradients are shown in Figs. 1b and 1c. The trapezoidal model (1c) shows linear ramps in the nonoverlapping regions and constant values in the overlapping regions. On the other hand, the realistic Laplace maps (1b) show a large shielding effect from the support plates. In a later section we show that the beam envelopes for the fields in Figs. 1b and 1c are not greatly different, providing that our improved envelope equations are used. Presumably, this similarity in the envelopes is due to the fact that if the Laplace gradients G_x and G_y are suitably averaged, they do resemble the trapezoidal models.

The improved envelope equations used by ESQACL were derived in Ref. [6] and are reproduced here in the appendix. These equations describe a relativistic uniformly charged warm beam which experiences a combination of axisymmetric and quadrupole focusing forces as well as *longitudinal acceleration*. The forces are assumed to be electrostatic, although the code could easily be generalized to include magnetic focusing. The derived equations are more general than those previously published. In particular, a focusing term proportional to the second derivative of the potential along the beam axis is included for improved accuracy when field models are used. (This term has been previously omitted in equations involving both ESQ focusing and space charge [7,8,9].) Since the potential on axis is a basic quantity in an electrostatic accelerator, the envelope equations, including relativistic factors, are written directly in terms of the potential (see appendix).

To facilitate comparisons of ESQACL results with those obtained from other envelope equations, field models, and codes [7,8,9,10], we include two switches. One turns on/off the relativistic effects, and the other turns on/off the axial electric field gradient effect. Reference [11] describes other input features of our code.

TRANSVERSE FIELDS: REAL FIELD MAP vs SIMPLIFIED FIELD MODEL

The Laplace and trapezoidal alternatives accepted by ESQACL (discussed above and illustrated in Fig. 1) require two different approaches for calculating the transverse focusing in the paraxial approximation:

(a) <u>Field map</u>: In this mode, the potential and the required gradients along the axis are obtained from an external Laplace solver (or, in principle, from measurements on actual hardware). Near the axis, the transverse fields increase linearly:

$$E_x^{vac} = -G_x(0,0,z)x$$
$$E_y^{vac} = -G_y(0,0,z)y.$$

The Laplace solver calculates G_x and G_y and passes them to ESQACL.

(b) <u>Simplified models</u>: In this mode, the vacuum potential is divided into a part with axisymmetry and a part with quadrupole symmetry; Taylor expansion gives [6]

$$V^{vac}(x,y,z) = V(0,0,z) + G_A(z)\frac{x^2+y^2}{2} + G_Q(z)\frac{x^2-y^2}{2} + \cdots.$$

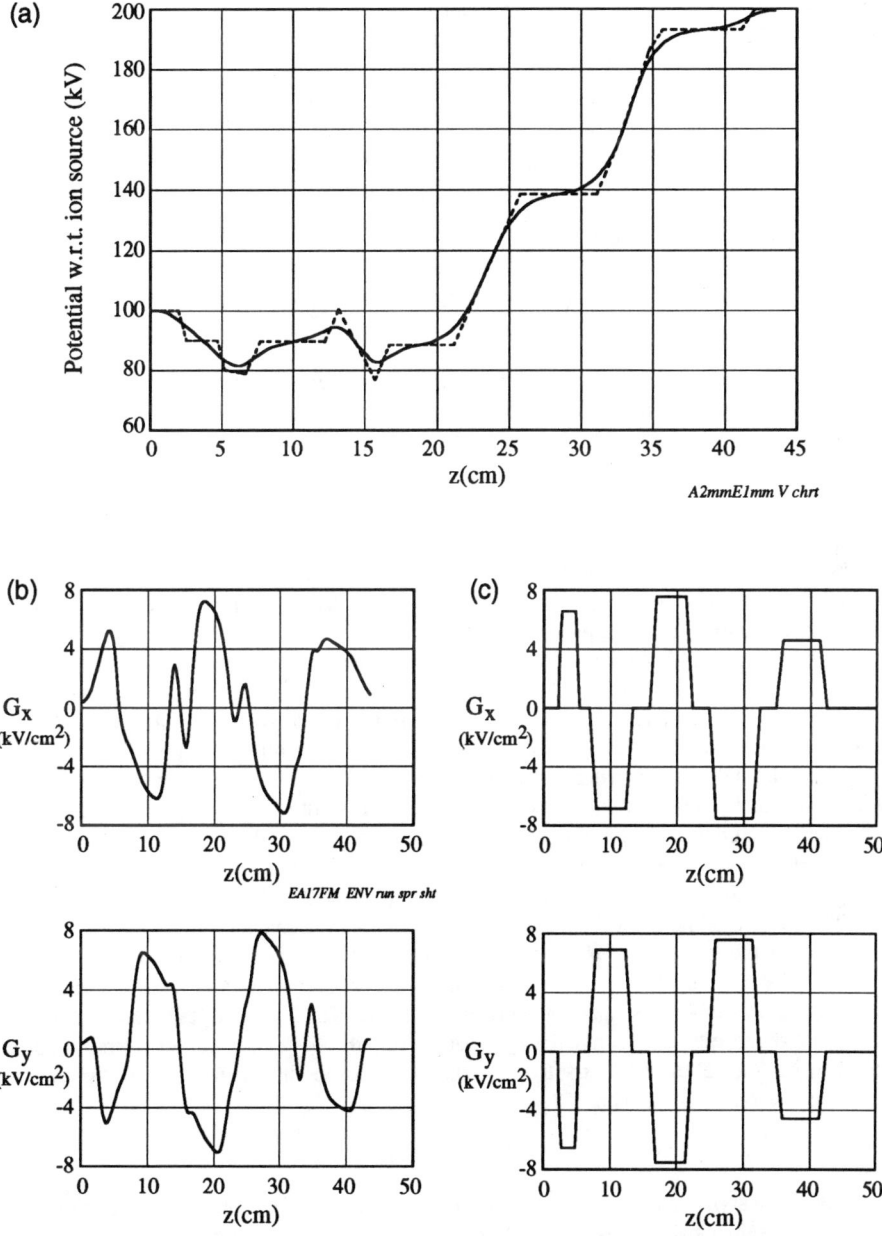

Fig. 1. (a) Potential on axis for CCVV accelerator prototype: actual (solid line); trapezoidal model (dashed). Either is available with ESQACL. (b) Transverse field gradients for actual ESQ-focused accelerator. (c) Transverse field gradients provided by ESQACL's trapezoidal model.

The axisymmetric coefficient G_A arises from the acceleration fields through Laplace's equation. We have [6] $G_A(z) = (1/2)E_z'$, where E_z' stands for $-V'''(0,0,z)$, so that

$$E_x^{vac} = -\frac{1}{2}E_z' x - G_Q(z) x,$$

$$E_y^{vac} = -\frac{1}{2}E_z' y + G_Q(z) y,$$

where the quadrupole gradient is

$$G_Q(z) = \frac{2V_Q}{a_Q^2}$$

in terms of the quadrupole radius a_Q and the quadrupole voltage V_Q. Note that the E_z' term is missing in most treatments that include G_Q, but appears in our envelope equations (see appendix) where the notation $-V''$ is used for E_z'. (This V'' term is not used in the field map mode, because it is already included in the G_x and G_y maps).

THE CODE

ESQACL runs on any Macintosh computer with enough memory to accommodate MacFortran. The code has six principal modules: input, V_Q setup, main module, trapezoidal map generator, output, and graphics.

The basic input file contains lattice information (quadrupole lengths, overlaps, and gaps; support-aperture sizes and gaps); initial integration step; x- and y-plane voltages; beam current and rms emittance; initial radii and entrance angles in the x- and y-planes; mass and charge for the beam particles. If the field map flag = 0 or 1, external maps must be supplied. If the flag = 2, a trapezoidal field map is generated using information from the basic input file. In all cases the lattice geometry input is used for the graphical display of the results of the simulations.

The V_Q setup module computes focusing voltages V_Q from the input voltages specified for each electrode. These are used for the trapezoidal field map and as coefficients for superposition of the multiple maps when the field map flag = 0.

The main module is modular itself. It starts with a pre-cycle stage. Regardless of the model chosen, the result of the pre-cycle stage is always a field map input for the next stage—the cycle. The cycle contains a main z-stepping loop for Runga-Kutta-Gill integration, a type suitable for the variable mesh size typically encountered. The last main stage requests outputs and plots of the results. Interactive dialogue for altering the V_Q's can be used for rapid design optimization. Special modules exist to dump the field map and update the input if it is changed during the interactive design run.

EXAMPLES

Figure 2a shows particle trajectories for the 200 keV CCVV prototype accelerator [1] using the 3-D self-consistent particle code Argus [10]. Argus was then run without beams (a much quicker process) to generate a set of axial vacuum-field maps, one for each pair of electrodes, as described earlier. ESQACL combined these maps in proportion to the electrode voltages used for Fig. 1a, thus producing the envelopes of Fig. 1b. These are nearly identical to the envelopes of the particle trajectories (except that y-plane aberrations at the exit in Fig. 2a create a halo which obscures the bulk of the beam in that region). The comparison between ARGUS and ESQACL is especially stringent because the beam is mismatched in this example.

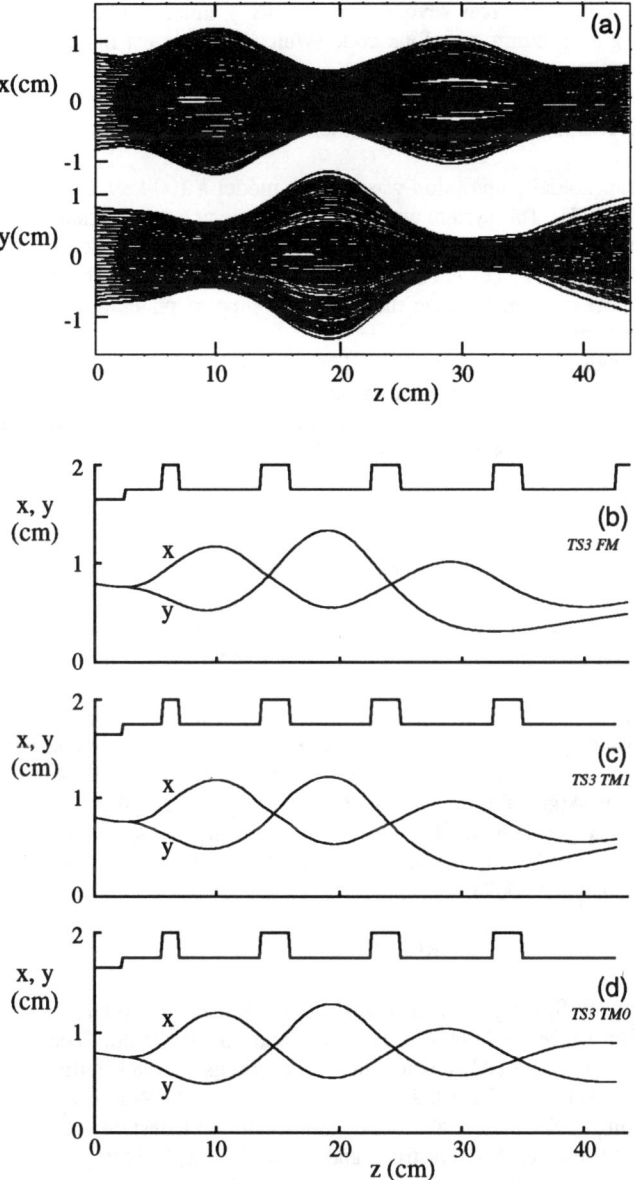

Fig. 2. (a) Particle trajectories in prototype 200 keV CCVV accelerator using Argus code; (b) ESQACL calculation for same electrode voltages using Laplace field maps; (c) same but using the trapezoidal field model; (d) same with V''' term omitted.

Since the particle calculation takes several hours of Cray time, one sees the tremendous advantage of having an accurate envelope code (which runs almost instantaneously on a Macintosh IIci) for optimization runs.

Similar envelopes are obtained (Fig. 2c) from the trapezoidal model when the V" (gap focus) term of ESQACL is included. When that term is dropped (Fig. 2d), the envelopes are quite inaccurate.

ESQACL's trapezoidal map option was used to model a 100-keV LEBT (low-energy beam transport system) [5]. This system was designed to deliver a round beam at the exit. We compare results from ESQACL (Fig. 3a) with results from a previous code [9] where the V" term was omitted (Fig. 3b). The neglect of additional focusing from the V" term is detrimental, even for this transport system, because the electrode support plates and non-overlapping regions produce gap-focus effects.

A 100-A megavolt electron accelerator [4] based on a 1-A D$^-$ accelerator channel [3] is modeled in Fig. 4a. The relativistic corrections involving υ (see appendix) are dominant for this case, as seen by comparison with Fig. 4b where these factors are set equal to unity. Figure 4a shows how easily an ESQ system focuses electrons—the current is 100 times larger than with D$^-$. The V_Q's actually *decrease* at higher beam energies, in a way that can be described analytically [4].

ACKNOWLEDGMENTS

Charles Kim provided his original envelope code and showed us how to use it. (He co-authored several of our papers in 1987 and 1988.) Although we have completely restructured the code and introduced new features such as external field maps, fully relativistic factors, and the gap focusing effect, we have retained most of Dr. Kim's nomenclature. Ken Clubok (an LBL summer student) helped make our code more user friendly, found bugs, produced the multiple field maps for Fig. 2b, and made runs for Fig. 4. John Petillo of SAIC gave valuable help with setting up our Argus run (Fig. 2a). Bill Cooper supported all this work.

This work was supported by the Director, Office of Energy Research, Office of Fusion Energy, Development and Technology Division, of the U.S. Department of Energy under Contract No. DE-AC03-76SF00098.

REFERENCES

[1] O.A. Anderson, W.S. Cooper, W.B. Kunkel, J.W. Kwan, R.P. Wells, C.A. Matuk, P. Purgalis, L. Soroka, M.C. Vella, G.J. De Vries, and L.L. Reginato, "The CCVV High-Current Megavolt Range DC Accelerator," Proceedings of 1989 Particle Accelerator Conf., Chicago, March 20-23, 1989; IEEE Cat. No. 89CH2669-0, p. 1117.

[2] O.A. Anderson, L. Soroka, et al., "Applications of the Constant-Current Variable-Voltage DC Accelerator," Nucl. Instrum. and Meth. B40/41, 877 (1989).

[3] O.A. Anderson, et al., "Negative Ion Source and Accelerator Systems for Neutral Beam Injection in Large Tokamaks," in *Plasma Physics and Controlled Nuclear Fusion Research 1990*, Proc. 13th IAEA Plasma Physics and Controlled Fusion Conf.,Washington DC, International Atomic Energy Agency, Vienna, 1991, p. 503.

[4] O.A. Anderson, L. Soroka, P.W. Van Amersfoort, W.H. Urbanus, "A High-Current MV DC Electron Accelerator," oral presentation at the 1991 IEEE Particle Accelerator Conference, May 6-9, 1991, San Francisco; LBL-29897a.

[5] O.A. Anderson, L. Soroka, J.W. Kwan, and R.P. Wells, "Application of Electrostatic LEBT to High Energy Accelerators," Proc. 2nd European Particle Accelerator

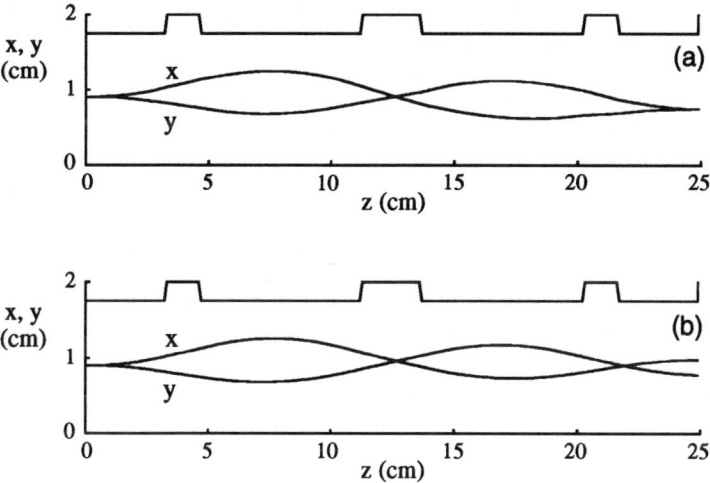

Fig. 3. (a) ESQACL calculation for 100-keV 4-quad LEBT using trapezoidal field model.
(b) Calculation for same case but with V″ term omitted (cf. Appendix A).

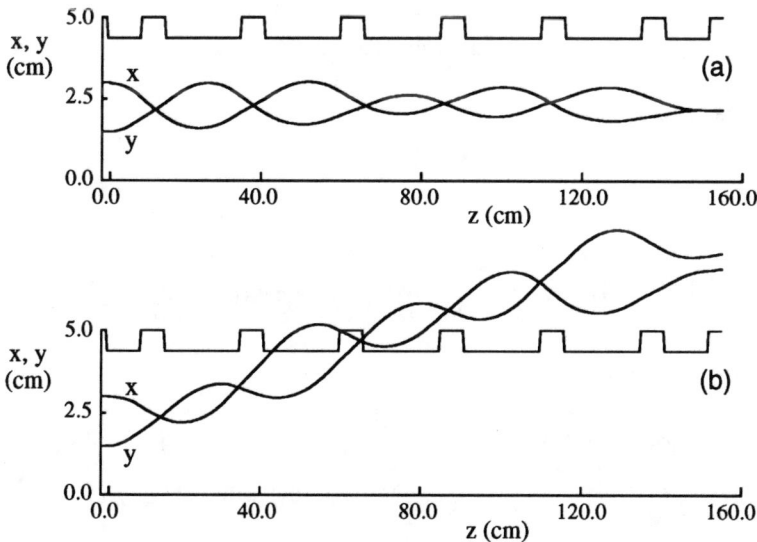

Fig. 4. (a) ESQACL calculation for 1-MeV 100-A electron accelerator.
(b) Calculation for same case but with relativistic factors set equal to unity; see text.

Conference, Nice, France, June 12 - 16, 1990; P. Marin and P. Mandrillon, Editors; Editions Frontières, Gif-sur-Yvette, 1990, p. 1288.

[6] O.A. Anderson, "Envelope Equations for Electrostatically Accelerated Beam with ESQ Focusing," Lawrence Berkeley Laboratory report LBL 30304.

[7] J.D. Lawson, "Space Charge Optics," in *Applied Charged Particle Optics*, A. Septier, Ed., Academic Press, New York, 1983.

[8] J.D. Lawson, *The Physics of Charged-Particle Beams*, Second Edition, Oxford University Press, 1988.

[9] C.H. Kim, Envelope code with acceleration, private communication, Lawrence Berkeley Laboratory, 1987.

[10] A. Mankofsky, "Three-Dimensional Electromagnetic Particle Codes and Applications to Accelerators," Linear Accelerator and Beam Optics Codes, C.R. Eminheizer, Ed., A.I.P. Conf. Proc. No. 177, 1988, p. 137.

[11] L. Soroka and O.A. Anderson, "Users Manual for ESQACL Code," Lawrence Berkeley Laboratory report PUB-3096.

APPENDIX

The envelope equations used by ESQACL are derived in Ref. [6], where the necessary assumptions are discussed. We use the symbols: proton charge = e, particle charge = q = Ze, proton mass = m_p, particle rest mass = m_0 = Am_p, speed of light = c, electrostatic potential *normalized to zero at source* = V, and

$$\upsilon = -\frac{qV}{m_0 c^2} > 0 \quad \text{fractional increase in particle mass (normalized potential);}$$

$$\epsilon_N \equiv \beta\gamma\epsilon \quad \text{normalized emittance of beam.}$$

Our envelope equations are [6]

$$a' = a_p \tag{1a}$$

$$a_p' = -\frac{1+\upsilon}{1+\frac{\upsilon}{2}}\left(\frac{V'a_p}{2V} + \frac{V''a}{4V} - \frac{V_Q a}{V a_Q^2}\right) - \frac{1}{\left(1+\frac{\upsilon}{2}\right)^{\frac{3}{2}}} \frac{1}{\upsilon^{\frac{1}{2}} V} \frac{2^{-\frac{3}{2}}}{\pi\epsilon_0 c} \frac{I}{a+b} + \frac{1}{1+\frac{\upsilon}{2}} \frac{\epsilon_N^2}{2\upsilon a^3} \tag{1b}$$

$$b' = b_p \tag{1c}$$

$$b_p' = -\frac{1+\upsilon}{1+\frac{\upsilon}{2}}\left(\frac{V'b_p}{2V} + \frac{V''b}{4V} + \frac{V_Q b}{V a_Q^2}\right) - \frac{1}{\left(1+\frac{\upsilon}{2}\right)^{\frac{3}{2}}} \frac{1}{\upsilon^{\frac{1}{2}} V} \frac{2^{-\frac{3}{2}}}{\pi\epsilon_0 c} \frac{I}{a+b} + \frac{1}{1+\frac{\upsilon}{2}} \frac{\epsilon_N^2}{2\upsilon b^3}. \tag{1d}$$

The three terms on the right side of (1b) or (1d) represent focusing force, space charge, and emittance pressure. Each term is preceeded by a relativistic correction factor (involving 1+υ/2 or 1+υ) which approaches unity in the nonrelativistic limit. An input switch [11] can set all these relativistic factors equal to unity and allow comparisons with nonrelativistic codes.

INTEGRATION OF OLE INTO THE TACL CONTROL SYSTEM*

B. Bowling, D. Douglas, J. Kewisch, P. Kloeppel, and G. A. Krafft
CEBAF, 12000 Jefferson Ave., Newport News, VA 23606, USA

ABSTRACT

OLE, the On-Line Envelope program, is a first-order optics code which was designed to provide fast lattice transfer functions from actual accelerator magnet and cavity control values. This paper addresses the results of a successful integration of OLE into the CEBAF control system, TACL. This marriage provides the user with the ability for obtaining real-time Twiss parameters and transfer functions which reflect the current operational state of the machine. The resultant OLE calculation provides the analytical core for many control and diagnostic functions used at CEBAF, including focusing corrections, orbit corrections, emittance measurements, and beamline analysis.

INTRODUCTION

CEBAF is a large recirculating linear accelerator with approximately 1600 magnet power supplies in the beam transport system. Control of the CEBAF transport is therefore quite critical, due to the capability of high average beam power. To prevent dangerous beam losses and to prepare optics changes, the control programs must read the magnet power supplies and calculate the optics in a virtually real-time manner.

A program known as OLE (On-Line Envelope) has been developed at CEBAF to give a graphical display of the calculated machine β function or, equivalently, the beam envelope. Emphasis in the design was placed on speed of program execution at the expense of generality of application. As a result, the accelerator operator will be able to alter the magnetic field in any element in the machine, calculate the β functions in both planes at the entrance and exit of each magnet, and display graphs of the functions, all within one second. The time that is required is short enough that the process approximates fairly well real-time operation.

The original version of OLE interfaced to TACL (Thaumaturgic Automated Control Logic) control system using shared memory. A recent upgrade to TACL changed the data handling into a "provide upon demand" system, with a central network data router known to the system simply as the STAR computer. This new networking configuration improved the control system's overall operation and performance, and provided an easy input/output interface for the OLE program.

---

* This work was supported by the U.S. Department of Energy under contract DE-AC05-84ER40150.

OLE DESCRIPTION

The purpose of OLE is to calculate the β function or the beam envelope throughout the accelerator by interfacing with real-time magnet current and accelerating cavity phase and amplitude settings. OLE performs this by representing each element of the machine by an appropriate 2×2 matrix for the horizontal and vertical planes. Because the transfer matrices are of rank 2 rather than 4, we do not allow for the possible existence of horizontal-vertical coupling. The algorithm that is used for matrix multiplication is a special speed-optimized routine that works only for 2×2 matrices.

First-order transfer matrices are developed for quadrupoles and dipoles from magnet operating currents using the measured excitation curve for each device. The accelerating cavity transfer matrix is a special routine developed for the CEBAF cavities which includes focusing effects.[1] This matrix uses the current operating gradient and phase values in the calculation.

OLE propagates initial Twiss parameters (α, β, and ϕ) throughout the defined lattice, storing the results in a local shared-memory segment for both horizontal and vertical planes. The initial parameters are determined experimentally from emittance measurements and back-propagated to the beginning of the lattice. Dispersion parameters are also computed and is available in the shared memory.

TACL CONTROL SYSTEM

CEBAF is currently using TACL as the accelerator control system. In the past, TACL utilized a network configuration known as SuperLan. This connectivity method constantly transported all controllable signals to every supervisor computer using shared memory segments. For small signal loads (i.e. less than 10,000), the SuperLan configuration performed adequately. As the total number of signals increased, however, the network transfer performance degraded, a result which is to be expected. A conceptual change in the control system resulted in the creation of the STAR central computer.

Each front-end computer, known as a local, generates a network connection (via Berkeley sockets) to the STAR computer and delivers a list of available signals. User-interface processes and application codes also connect to the STAR, requesting access to signals residing on the front-end computers. The STAR provides the proper data routing and transfers the signal only when the value of the signal changes. This interface provided OLE with a simple I/O connection, with the OLE process "sleeping" until a magnet or cryomodule signal changes, eliminating the constant polling required for the older shared-memory I/O interface.

The STAR also provides the data routing for the computational results of the OLE calculations. The program OLESPY has been developed to attach to the shared-memory segment created by OLE. It additionally connects to the STAR computer, posting the computed optical parameters on the STAR for availability. Any display screen or application codes which requires optics parameters can request them directly from the STAR, simplifying the I/O interface.

RESULTS AND APPLICATIONS

The first apparent operational improvement of OLE with the new STAR network configuration was that the load factor of the host computer decreased. The earlier polling method noticeably affected the performance and response of the computer that was executing OLE. The STAR allows for the OLE process to sleep until a change in magnet current or cavity control is received, increasing computer performance. The use of OLE-SPY allows anyone easy access to resultant OLE calculations by requesting the values from the STAR.

The optical parameters generated by OLE are used by the orbit correction and autosteering programs being developed at CEBAF. Transfer matrices are easily generated from the Twiss parameters that reflect the current operation of the machine.

The calculational results of the OLE program were tested using two profile measurement devices separated by approximately 25 optical elements, comparing the predicted beam size vs. measured quantities. The agreement in beam size was within 10 percent. This indicates that the program can be effectively used for focusing setup and corrections. A graphical display of the computed beam size can be used for beamline setup and adjustments. Additionally, a program is in development which, using results from OLE, will minimize the computed beta functions for a section of the accelerator.

Another program in development uses OLE optics to predict the source of a beamline disturbance. This method reads the current beam position and, using OLE-based transfer matrices, determines which optical element is the best candidate for creating the measured displacement. One can then apply an orbit correction to that element which should most effectively minimize the orbit.

References
[1] G. A. Krafft, J. Jackson, and D. Douglas, CEBAF TN-90-0287.

EVOLUTION OF SIMPLE PHASE SPACE DISTRIBUTIONS USING THE VLASOV EQUATION

James O'Connell and Scott Butler
Booz, Allen & Hamilton, Inc., Arlington, Virginia 22202
and
Julie O'Connell, American University, Washington, DC 20016

ABSTRACT

VLASOV is a code for Macintosh computers that follows the time evolution of charged- particle phase-space density functions in planar, cylindrically or spherically symmetric geometries. The Vlasov one-dimensional differential equation in radial coordinates is used to propagate different initial space-charge dominated beams in free space or in a linear focusing channel. Color plots of the marginal densities are displayed on the video monitor at selected time intervals. Time dependent moments of the distribution are written to a file for off-line analysis.

1. INTRODUCTION

One dimensional phase space plots of charged particle beams are a traditional analysis tool used to understand and measure the properties of the beam as it propagates through free space, magnetic focusing elements, and accelerating cavities. Most beam dynamics codes are of the particle-in-a-cell (PIC) type that follow the trajectories of a large number (N) of individual particles or macroparticles. When space charge forces are important the particle-particle repulsive force must be computed at each time step for $N(N-1)/2$ pairs. Thus, PIC codes cannot usually be run on small computers if N is large.

An alternative approach is to work with the phase space density function $f(\mathbf{r},\mathbf{v},t)$ treated as a continuous, differentiable function in 6 independent variables. For cw beams with slab or cylindrical symmetry or a spherical bunch the independent variables reduce to 2, radial position, r, and radial velocity. Instead of following particle trajectories, one computes the particle density f in each (r,v) cell as a function of time. A cell continuity equation is used: the change in a cell density is due to the difference of particle flow-in minus flow-out. The continuity equation is driven by the local acceleration function as computed from the net electromagnetic forces (the Vlasov equation). The calculation of acceleration is simplified when Gauss' law can be invoked for beam geometries with a high degree of symmetry.

A grid is constructed (fig. 1) with ranges r = 0 to Rmax, v = Vmin to Vmax chosen large enough to contain the maximum excursions of the evolving $f(r_i, v_j, t_n)$. An initial distribution is chosen (section 2), and the Vlasov equation (section 3) is used to propagate the density using the acceleration calculated (section 4) at each r_i (assuming it is independent of v_j).

Sections 5 & 6 discuss what information can be extracted from the time dependent density plots. Use of the code is illustrated in section 7 with examples in section 8.

The problem of gravitationally self-bound particle swarms can also be addressed by the VLASOV code by changing the sign of the perveance to make the interparticle force attractive.

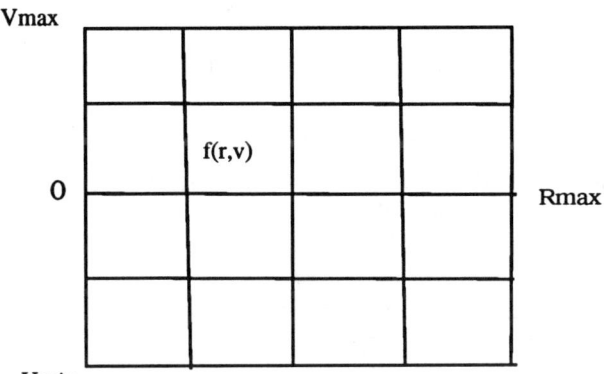

Fig. 1 Example of a phase space grid on which the density function is computed at each time step. Typically, grids are 64 x 64 and 1000 time steps are used.

2. INITIAL DISTRIBUTIONS

Some standard phase space density functions can be chosen as initial distribution functions. Most are products of a radial function and a velocity function. Typical analytic forms are:

Gaussian: $\quad G(x) = \exp(-x^2 / x_o^2)$

Modified Fermi: $\quad F(x) = 1/(1 + \exp[\frac{x^2 - x_o^2}{x_1^2}])$

where x is r or v. For example, an initial density distribution that is Gaussian in r and v would be

$$f_{GG}(r,v) = N\, G(r)\, G(v).$$

A spatial Modified Fermi distribution combined with a Gaussian velocity distribution would be

$$f_{FG} = N\, F(r)\, G(v).$$

An example of a form that is not a product of spatial and velocity distributions is the inverse energy distribution

$$f_E = N \exp(E_o / E)$$

where E(r,v) is the local particle energy

$$E = U(r) + \frac{1}{2}v^2 \qquad (1)$$

with the potential, U(r), chosen to be some reasonable function of radius, e.g.

$$U = 1/(r^2 + r_o^2)^{1/2}.$$

The normalization constant N in each case is chosen to give

$$N \int_o^\infty \int_{-\infty}^\infty f(r,v)\, dv\, dr = 1. \qquad (2)$$

3. VLASOV EQUATION

In one dimension the phase space density continuity equation can be written as

$$-\frac{\partial f(r,v,t)}{\partial t} = v\frac{\partial f}{\partial r} + a\frac{\partial f}{\partial v}. \qquad (3)$$

The right-hand-side (RHS) of eq. (3) can be numerically evaluated at each (r,v) cell in phase space using finite differences, e.g.

$$\frac{\partial f}{\partial r} = \frac{f(r+\Delta r) - f(r-\Delta r)}{2\Delta r}$$

The left-hand-side of eq. (3) can be used to propagate f(r,v,t) to f(r,v,t+Δt). The backward Euler method (ref. 2) is used to keep the propagation stable. Three steps are used to advance f:

1. As a first estimate form

$$f(t+\Delta t) = f(t) + \frac{\partial f(t)}{\partial t}\Delta t$$

using RHS(f(t)) for $-\partial f(t)/\partial t$.

2. Recalculate RHS(f(t+Δt)) for $-\partial f(t+\Delta t)/\partial t$.

3. As a second estimate form

$$f(t+\Delta t) = f(t) + \frac{\partial f(t+\Delta t)}{\partial t}\Delta t.$$

These operations are performed on each phase space (r,v) cell.

4. ACCELERATION

The particle acceleration term in the RHS of eq. 3 contains the physics that drives the density evolution. The general form of the acceleration can be written as a function of radius with internal and external terms

$$\frac{d^2r}{dt^2} = a(r) = \frac{KQ(r)}{r^n} - br^{nb} \qquad (4)$$

where n = 0 for planar symmetry
  = 1 for cylindrical symmetry (cyl.)
  = 2 for spherical symmetry (sph.)

$K = 2Ne^2/(4\pi\varepsilon_o m\gamma^3)$ is a constant related to the beam perveance, $(\beta^2 c^2 K)$

$e, \varepsilon_o, m, c, \beta, \gamma$ have their usual meanings

N = number of particles per unit area for planar beams
  =   "   "   "   "   " length for cyl. beams
  =   "   "   "   "   " volume for sph. beams

b = external focusing strength
nb = external force law power, =1 for linear focusing

The quantity Q(r) is the fraction of particles in the beam at radii equal to or smaller than r,

$$Q(r) = \int_0^r \int_{-\infty}^{\infty} f(r,v)\, dv\, r^n\, dr \;/\; \int_0^{\infty} \int_{-\infty}^{\infty} f(r,v)\, dv\, r^n\, dr \;. \tag{5}$$

$Q(\infty) = 1$. The term KQ/r^n in eq. (4) is an expression of Gauss' law for each symmetry type.

We neglect centrifugal acceleration because we assume angular velocities around the beam axis are small compared to the space charge generated radial velocity. In Cartesian coordinates: x=r, x'=v, y=0, y'=0.

5. MARGINAL PLOTS

Integration of f(r,v) over r or v leads to the marginal functions

$$g(r) = \int_{-\infty}^{\infty} f(r,v)\, dv \tag{6}$$

$$h(v) = \int_0^{\infty} f(r,v)\, dr \;. \tag{7}$$

The radial function g(r) is related to the physical density

$$\rho(r) = g(r) \quad \text{for a plane}$$

$$\rho(r) = \frac{g(r)}{2\pi <r>} \quad \text{for cyl.}$$

$$\rho(r) = \frac{g(r)}{4\pi <r^2>} \quad \text{for sph.}$$

where

$$<r^n> = \iint r^n\, f\, dv\, dr = \int r^n\, g\, dr \;.$$

The velocity function h(v) gives the net distribution in radial velocities from which beam contraction or expansion can be deduced as can the transverse kinetic energy. The g and h marginals normalize to unity since f does. In the code these plots are displayed along with f on the monitor after selected time intervals.

Two additional radial plots are of interest: the acceleration a(r) and the local mean squared velocity

$$v^2(r) = \int_{-\infty}^{\infty} v^2\, f(r,v)\, dv \,/\, g(r) \tag{8}$$

which is a measure of the transverse kinetic energy as a function of radius.

6. MOMENTS

The phase space moment of an operator O is defined as

$$<O> = \int_0^{\infty} \int_{-\infty}^{\infty} O(r,v)\, f(r,v)\, dv\, dr \tag{9}$$

some useful moments are:

mean square radius	$<r^2>$
mean square velocity	$<v^2>$
squared emittance	$\varepsilon^2 = <r^2><v^2> - <rv>^2$
kinetic energy	$KE = \frac{1}{2}<v^2>$
internal potential energy	$PE = K<Q/r>$

564 Evolution of Simple Phase Space Distributions

external potential energy $\quad PEX = \dfrac{b}{2} <r^2>\quad$ (for a linear force)

total energy $\quad E = KE + PE + PEX$

entropy $\quad S = - <\ln f>$.

A study of these moments as a function of time gives insight on the beam dynamics. These moments are written to a file in the code.

7. USING THE CODE

To run VLASOV after it has been compiled and linked on a MAC II or higher the following dialog boxes appear when the program is initiated from the file Vlasov.apl.

Action	Options
Run Vlasov Quit	Run Parameters Initial Distributions Color Output

Clicking on 'Run Parameters' under 'Options' produces the following screen

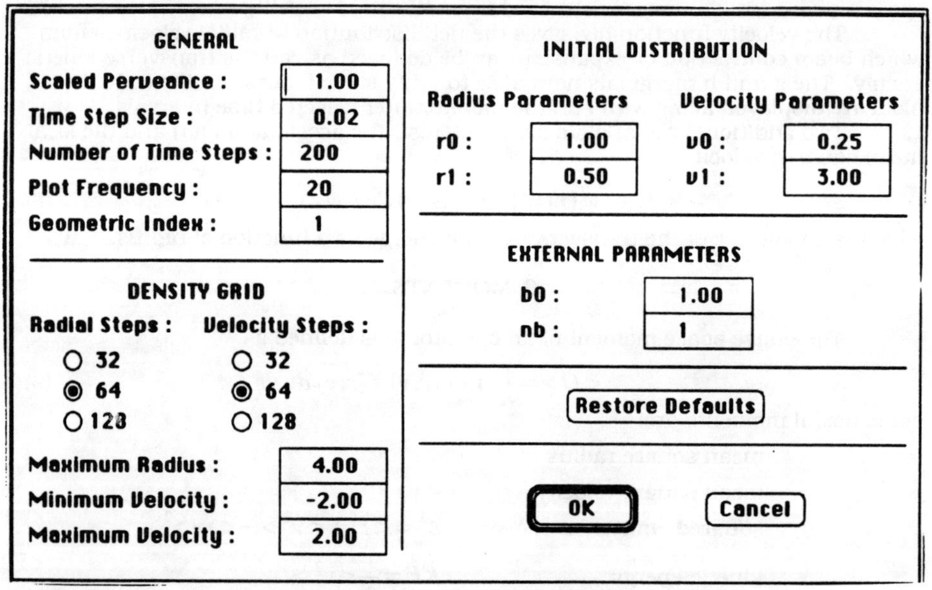

GENERAL

Scaled Perveance :	1.00
Time Step Size :	0.02
Number of Time Steps :	200
Plot Frequency :	20
Geometric Index :	1

DENSITY GRID

Radial Steps : Velocity Steps :
○ 32   ○ 32
● 64   ● 64
○ 128   ○ 128

Maximum Radius :	4.00
Minimum Velocity :	-2.00
Maximum Velocity :	2.00

INITIAL DISTRIBUTION

Radius Parameters   Velocity Parameters

r0 :	1.00	v0 :	0.25
r1 :	0.50	v1 :	3.00

EXTERNAL PARAMETERS

b0 :	1.00
nb :	1

[Restore Defaults]

[OK]   [Cancel]

The numbers shown are the default values; they can be changed by selecting the default value with the mouse and overwriting them.

Clicking on "Initial Distributions' under 'Options' produces the following screen

```
┌─────────────────────────────────────────┐
│  Initial Radial      Initial Velocity   │
│  Distribution        Distribution       │
│                                         │
│  ○ Parabolic         ○ Parabolic        │
│  ○ Gaussian          ○ Gaussian         │
│  ○ Hollow            ○ Hollow           │
│  ○ Fermi             ○ Fermi            │
│  ─────────────────────────────────────  │
│              ● Energy                   │
│  ─────────────────────────────────────  │
│     [  OK  ]          [ Cancel ]        │
└─────────────────────────────────────────┘
```

The Gaussian and modified Fermi distributions were discussed in section 2. The Parabolic $(1-x^2/x_0^2)$ and Hollow $x \exp(-x^2/x_0^2)$ are standard forms. The Energy distribution uses the numerical value of v1 for E_0 in 'Run Parameters'.

8. EXAMPLES

Fig. 2 shows the phase space plots for a cylindrical beam (n=1) with a Gaussian-Gaussian initial distribution (r_0=1, v_0=0.25, K=1., b=1.) and the evolved distribution after 4 time units (200 time steps * 0.02 time step size). Fig. 3 shows the time dependence of the various moments. By watching the evolution of the density and its marginals on the screen one sees a fast initial redistribution of charge to a known (ref. 1,2) stable shape followed by an oscillation of this shape around an equilibrium radius and, in this case, the beginning of filamentation. For certain ranges of the input parameters one can observe filamentation and rms emittance growth. One can also verify which initial shapes and parameters lead to stationary distributions for given values of K and b.

9. REFERENCES

1. J.S. O'Connell, J. Appl. Phys., **70**, 7157 (1991).

2. J.S. O'Connell, 1992 Linear Accelerator Conference Proceedings, p. 522.

566 Evolution of Simple Phase Space Distributions

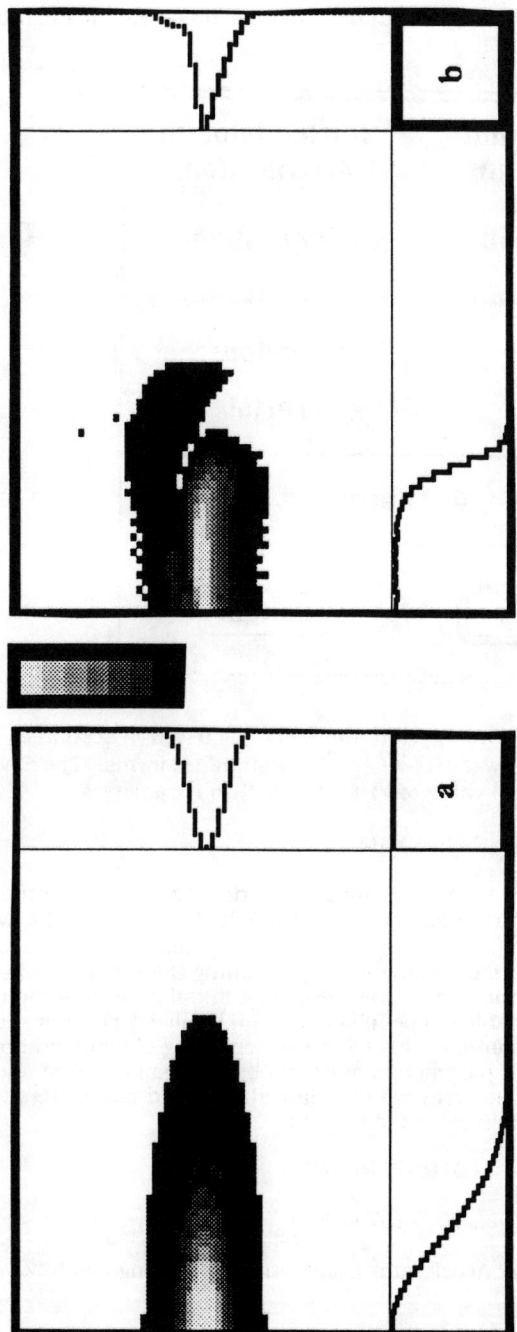

Fig. 2 (a) Initial distribution. The phase space density is in the square plot with v as the vertical axis and r as the horizontal axis. The horizontal line plot is g(r); the vertical line plot is h(v).
(b) The distribution after 4 time units showing that g(r) has evolved from Gaussian to uniform.

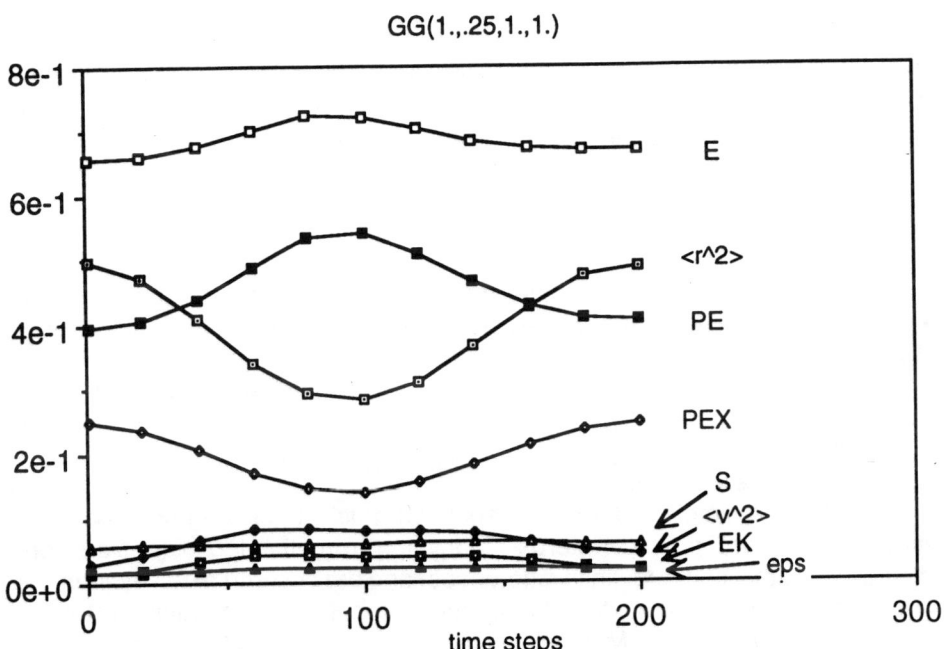

Fig. 3 Moments as a function of time.

SOLUTION OF LAPLACE'S EQUATION BY THE METHOD OF MOMENTS WITH APPLICATIONS TO CHARGED PARTICLE TRANSPORT[*]

C. K. Allen, S. K. Guharay, and M. Reiser
Laboratory for Plasma Research
University of Maryland, College Park, MD 20742

ABSTRACT

A fast approximation method to the 3D electrostatic problem is developed. The method of moments procedure is outlined for the particular case of Laplace's equation. The resulting matrix-vector equation is solved by a conjugate gradient algorithm. These techniques are then implemented with a computer code running on a PC and used to solve example problems.

1. INTRODUCTION

We are interested in the transport of low energy ions beams, specifically, we have been engaged in the design and analysis of Low Energy Beam Transport (LEBT) section of an accelerator column[1,2]. To avoid gas focusing we use electrostatic lenses for the relatively slow moving ions. The action of such lenses is completely characterized by their spatial potential distribution ϕ. Thus, analysis of electrostatic lenses invariably requires the solution of Laplace's equation.

Although simple in form, this equation must be solved numerically for most geometries of practical interest. Many approximation techniques are very successful to this end (e.g. finite differences) and are covered extensively in the literature. However, for a fully 3D treatment computation and machine storage usually become extreme. Manipulation of the solution data also becomes quite cumbersome. Consequently, these situations require special hardware such as a supercomputer. We present an approximation technique which is fully 3D and remarkably computational efficient. Ideally, we wish to implement this technique as CADware for the IBM PC.

The technique relies on a combination of the method of moments and fast iterative techniques for solving linear systems. Specifically, we proceed by reformulating Laplace's equation into an integral equation over the boundary surface, reducing the dimensionality of the original system. The new problem is approximated by the method of moments[3] to yield the matrix-vector equation Ax=y. We then use conjugate gradient algorithms[4] to solve this equation.

---

[*]Supported by ONR/SDIO

2.0 PROBLEM FORMULATION

In this section we clarify the physical system which we wish to model. Figure 1 depicts an abstract geometric representation of the electrostatic problem. We have a closed surface Γ in Euclidean 3-space E^3 which represents the boundary of our problem. Γ separates E^3 into two regions Ω_i and Ω_e representing the (bounded) interior of Γ and the (unbounded) exterior of Γ, respectively.

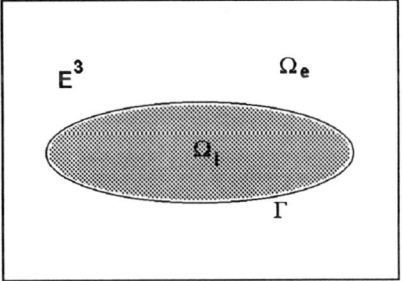

Figure 1: Electrostatic Problem

Typically we associate the union of Γ and Ω_i as a conductor in 3-space while we are interested in the potential ϕ in the region Ω_e external to the conductor. This is an example of the so called *exterior Dirichlet problem* and can be classified mathematically as follows:

$$\nabla^2 \phi(x) = 0 \qquad \forall x \in \Omega_e, \tag{1}$$
$$\phi(x) = f(x) \qquad \forall x \in \Gamma.$$

The function f is the given set of boundary values and it constitutes the data of the problem. For a conductor f is a constant over the boundary Γ. In addition, we usually impose the physical restriction that ϕ vanishes at infinity.

Also of interest is the *interior Dirichlet problem* where we are concerned with the distribution ϕ in the region Ω_i. This situation is formulated mathematically below.

$$\nabla^2 \phi(x) = 0 \qquad \forall x \in \Omega_i, \tag{2}$$
$$\phi(x) = f(x) \qquad \forall x \in \Gamma.$$

Notice that both problems take the same boundary values, namely f. Typically for the interior problem we also require that ϕ remains bounded in Ω_e.

We may reformulate both of these problems into a single integral equation on Γ[5]. The alternate problem is achieved by the introduction of an intermediate function σ defined on Γ.

$$f(x) = \int_\Gamma G(x,x')\sigma(x')dx' \qquad \forall x \in \Gamma \tag{3}$$

where

$$G(x,x') = \frac{1}{4\pi} \frac{1}{|x-x'|}. \tag{4}$$

The potential distribution ϕ is then related to the function σ by the equation

$$\phi(x) = \int_\Gamma G(x,x')\sigma(x')dx' \quad \forall x\in\Omega \tag{5}$$

where Ω is the union of Ω_i and Ω_e. Thus equation (5) is valid in all space except the boundary Γ. We recognize $G(x,x')$ as the free space Green's function for Poisson's equation and equation (5) as the potential due to a surface charge density σ in free space. It is interesting to note that even though problems (1) and (2) are not related in any obvious manner, solution of (3) yields the correct σ in both cases given the same boundary values f.

3.0 THE METHOD OF MOMENTS

We now approximate σ in equation (3) using the method of moments. It will be convenient to rewrite (3) by defining the operator K. We have

$$K\sigma \triangleq \int_\Gamma G(x,x')\sigma(x')dx' = f \quad f\in L_p(\Gamma) \tag{6}$$

where we have chosen f in L_p in order to analyze convergence. We see that K is a compact linear self-adjoint integral operator with kernel $G(x,x')$. Since Γ is compact, we may choose a countable set of functions $\{u_n\}$ which are dense in $L_p(\Gamma)$, that is they form a basis for the vector space $L_p(\Gamma)$. These functions are usually referred to as the set of expansion functions. Accordingly, any element of L_p may be approximated arbitrarily well by a linear combination of these functions. Thus,

$$\sigma(x) = \sum_{n=1}^\infty a_n u_n(x) \tag{7}$$

for some set $\{a_n\}$ of coefficients. By substituting (7) into (6) and using the linearity of K we have

$$\sum_{n=1}^\infty a_n K u_n = f. \tag{8}$$

Now we choose another set of functions $\{v_m\}$ in L_q (where $1/p + 1/q = 1$) to be used as a set of weighting functions. By taking the inner product of (8) with each of the v_m's we get a set of equations of the form

$$\sum_{n=1}^\infty a_n \langle Ku_n, v_m \rangle = \langle f, v_m \rangle \quad m=1,2,3,... \tag{9}$$

where $\langle \, \cdot \, , \, \cdot \, \rangle$ denotes the usual inner product

$$\langle u, v \rangle = \int_\Gamma u(x)\overline{v(x)}dx. \tag{10}$$

Note that since f is given and $\{u_n\}$ and $\{v_m\}$ are selected arbitrarily, the only unknowns are the coefficients $\{a_n\}$. To obtain an approximation for σ we must truncate the indices n and m at a finite value. If we choose the same value for both indices, say N, we are then left with the following matrix-vector equation.

$$\begin{pmatrix} \langle Ku_1, v_1 \rangle & \cdots & \langle Ku_N, v_1 \rangle \\ \vdots & & \vdots \\ \langle Ku_1, v_N \rangle & \cdots & \langle Ku_N, v_N \rangle \end{pmatrix} \begin{pmatrix} a_1 \\ \vdots \\ a_N \end{pmatrix} = \begin{pmatrix} \langle f, v_1 \rangle \\ \vdots \\ \langle f, v_N \rangle \end{pmatrix} \quad (11)$$

which can be written more compactly as

$$Ax = y. \quad (12)$$

The matrix A is known in the literature as the moment matrix for (6), since we take the moments of $K\sigma$ with respect to the weights $\{v_m\}$. Equation (12) may be solved by standard matrix techniques to yield a solution for x, the vector of coefficients $(a_1, \ldots, a_N)^T$. We then have our approximation to σ according to equation (7).

COMMENTS

(1) The method of moments generates solutions that converge in the mean, i.e. in the L_p norm. The exact L_p space where convergence occurs depends on the choice of expansion and weighting functions.

(2) When we chose $\{v_m\} = \{u_m\}$, we have Galerkin's method which is known to be equivalent to the Rayleigh-Ritz variational method[3]. Thus, we see that the method of moments is a generalization of the Rayleigh-Ritz procedure. For Galerkin's method we have convergence in the L_2 norm (mean-squared convergence), where the approximate σ lies in span $\{u_n\}$.

(3) By choosing $\{v_m\} = \{\delta(x-x_m)\}$ where δ is the Dirac delta function and the x_m's are some set of points on Γ, we have a point-matched solution. Such solutions are known to converge in the L_1 norm (pointwise convergence).

(4) For valid solutions we must make sure the set $\{Ku_n\}$ spans the range of K. Otherwise, we converge to solutions to the problem $K\sigma = Pf$ where P is the projection operator onto the space span$\{Ku_n\}$.

(5) K^{-1} is unbounded and there exists f in L_p such that (8) has no solution σ in L_p (such solution may however be interpreted distributionally). However, K is a positive operator, indeed $\langle K\sigma, \sigma \rangle$ is recognized as the electrostatic energy in the system. Consequently, in the discrete approximation, 0 is not an eigenvalue of A (it is in the limit N→∞, though).

4.0 CONJUGATE GRADIENT ALGORITHM

We now turn our attention to the task of solving the matrix-vector equation (12). The traditional approach is to use some direct method such as

LU decomposition or Gaussian elimination. With these techniques the amount of computation necessary to solve an Nth order system is known *a priori*. For example, it is known that inversion of (12) requires $O(N^3)$ operations. Instead, we have chosen an iterative technique where the amount of computation is not known in advance.

The conjugate gradient algorithms are a specialization of the more general technique of conjugate directions methods. These techniques are expressly developed to solve the problem[6]

$$\min_{x \in \mathbb{R}^N} g(x) = \frac{1}{2} x^T A x - y^T x. \qquad (13)$$

It is assumed that the matrix A is positive definite so that a solution does exist. Note that the solution to this problem, obtained by setting the gradient of g to zero, is given by $x = A^{-1}y$. Conjugate direction methods are based on the idea of generating a complete set of linearly independent vectors $\{d_n\}$ which have the property $d_n^T A d_m = 0$ whenever $m \neq n$. This is known as A-orthogonality, or A-conjugacy.

Instead of solving (12) directly, in conjugate gradients we choose to iteratively minimize some functional, for example g in the above equation. The value of x furnishing this minimum is the solution to (12). The method starts with an initial guess for x_0, then it generates a sequence x_i that minimizes the functional. The sequence $\{x_i\}$ will converge to the exact solution x in a finite number of iterations. Even if the matrix A is not invertible in the classical sense (0 is an eigenvalue of A), the conjugate gradient algorithm will converge to a solution in the least squares sense. The algorithm we have chosen was taken from Sarkar *el. al.*[4] The functional which it minimizes is the (l_2) norm squared of the residual $r_i = y - A x_i$ (i.e. minimize $\|y - A x_i\|^2$).

5.0 IMPLEMENTATION

To apply the method of moments it is necessary that we select an appropriate set of expansion functions $\{u_n\}$ and weighting functions $\{v_m\}$. Since our aim is to minimize computation, we have selected a point-matching procedure. Thus, our weighting functions $\{v_n\}$ are given by the set $\{\delta(x-x_n)\}$ where the x_n's are the match points in Γ to be determined. This technique yields the least computing time, and the procedure is straightforward. It also yields good results as long as one is careful in the selection of the match point locations[7,8].

We choose a set of piecewise constant functions for $\{u_n\}$. First, Γ is

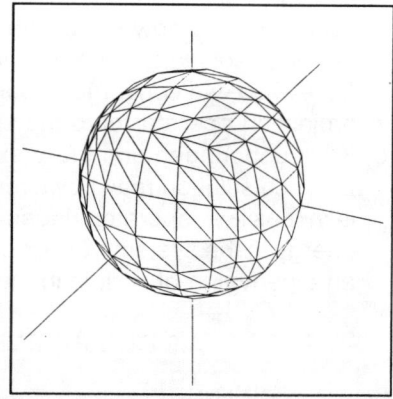

Figure 2: Triangulated Sphere

divided into triangular subdomains, denoted by the set $\{T_n\}$ where n runs from 1 to N. If Γ has curvature then we have an approximation to Γ as well (for example figure 2 depicts the triangulated approximation to a sphere). Now let the u_n's be a set of piecewise constant functions on each T_n.

$$u_n(x) = \begin{cases} 1 & \text{if } x \in T_n \\ 0 & \text{if } x \notin T_n \end{cases} \quad (14)$$

It can be shown that such a set is dense in L_p so σ may be approximated arbitrarily well[9]. Thus, according to comment (4) we must be certain that Pf = $\{f(x_n)\}$ is in the span of $\{Ku_n\}$. We shall see later that this criterion places restrictions on our triangulation of Γ. We have chosen the matching point set $\{x_n\}$ to be the centroids of each triangle T_n. Since this is the center of mass for our constant expansion functions, it seems to be a reasonable approximation to f across the triangle face.

In order to form equation (11) we must determine the inner products $\langle Ku_n, \delta(x-x_n) \rangle$. For our choice of expansion functions, this amounts to the evaluation of the following integral.

$$a_{n,m} = \int_{T_m} G(x_n, x') dx'$$

$$= \frac{1}{4\pi} \int\int_{T_m} \frac{1}{\sqrt{(x_n-x')^2 + (y_n-y')^2 + (z_n-z')^2}} dx' dy' dz' \quad (15)$$

This can be done completely numerically, hybrid analytic and numeric, or completely analytically[10].

6.0 EXAMPLES

We have implemented the above techniques in a computer program written in Borland C++. The platform is an i486 PC operating at 33 MHz and running Windows 3.1 operating system. All examples were run in single precision arithmetic except where noted. We have taken arbitrary units.

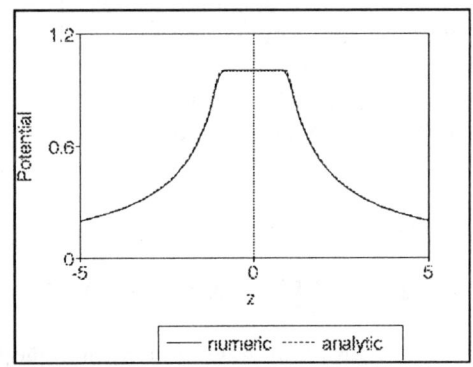

Figure 3: Conducting Sphere Axial Potential

CONDUCTING SPHERE

The analytic potential distribution for a conduction sphere of radius 1 and held at a potential 1 is known to be

$$\phi(x) = \begin{cases} 1 & \text{if } |x| < 1, \\ \dfrac{1}{|x|} & \text{if } |x| \leq 1. \end{cases} \quad (16)$$

Figure 3 shows a comparison of this analytic formula with the numerical results achieved for a triangulated sphere of 432 triangles (shown in figure 2). The solution converges to a norm squared residual error $< 10^{-9}$ in 75 iterations taking about 7 minutes real time (double precision required 42 iterations within 5 minutes).

A single number which is indicative of solution quality is the capacitance. This value may be calculated by recognizing σ as surface charge density. We numerically calculated the capacitance to be 110.2 pF for the above situation in MKS units, the true value is 111.3 pF.

EINZEL LENS

An einzel lens was modeled as two cylindrical pipes of radius 5 and length 10, separated axially by a distance of 3. Both pipes were capped with plates having an inner radius of 1/2. A conducting plate with outer radius 5 and inner radius 1/2 was centered between the two pipes at position z=0. The center plate was driven to a potential of 1 while the two outer pipes were grounded. El-Kareh provides an approximate analytic axial potential distribution given by[11]

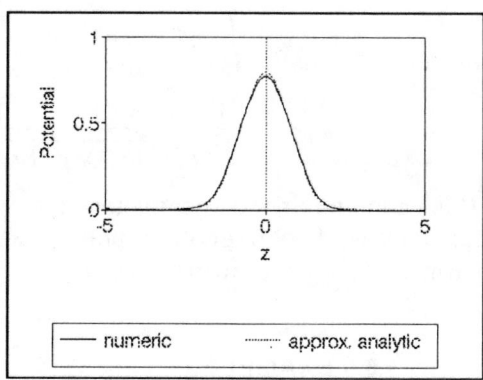

Figure 4: Einzel Lens Axial Potential

$$\phi(z) \approx \frac{2}{3\pi}[\frac{2z+3}{2}\tan^{-1}(2z+3) + \frac{2z-3}{2}\tan^{-1}(2z-3) - 2z\tan^{-1}(2z)]. \quad (17)$$

This approximation assumes three infinite plates of inner radius 1/2 with uniform fields at $z = \pm\infty$. The discretization was made with 464 triangles. This system took 261 iterations to converge to the usual error criterion (about 18 minutes of real time). What is happening here is that our boundary function f of equation (8) is almost outside the range of $\{Ku_n\}$. Fortunately, our iterative technique is indicating this situation by slower convergence! In order to avoid such situations it is necessary to reevaluate the triangulation of the system to insure that $\{Ku_n\}$ spans the space of boundary functions. Even though our patch model was borderline ill-conditioned, the stability of the conjugate gradient method still yielded reasonable results (see figure 4).

7.0 CONCLUSION

The method outlined in this paper exhibits enough computational efficiency to allow practical implementation on a PC. It yields accurate results in a reasonable amount of real computing time. Moreover, the conjugate gradient algorithm furnishes suitable error criterion to judge the quality of the solution. However, the method does require more work to implement than, say, the rather straightforward technique of finite differencing.

REFERENCES

1. S. K. Guharay, C. K. Allen, and M. Reiser, Conf. High-Bright. Beams, College Park, AIP Conf. Proc. 253 (1992), pp. 67-76.

2. S. K. Guharay, C. K. Allen, M. Reiser, K. Saadatmand, and C. R. Chang, AIP Conf. Proc. on Product. and Neut. Negative Ions (1993) (to appear).

3. R. F. Harrington, Field Computation by Moment Methods (Krieger, Malabar, FL, 1968).

4. T. K. Sarkar and E. Arvas, IEEE Trans. Antennas Propagat., vol. AP-33, no. 10, pp. 1058-1066, Oct. 1985.

5. I. Stakgold, Green's Functions and Boundary Value Problems (Wiley, NY), pp. 508-517.

6. D. G. Luenberger, Linear and Nonlinear Programming 2nd Ed. (Addison-Wesley, Reading, MA, 1984), pp. 238-257.

7. T. K. Sarkar, A. R. Djordjević, and E. Arvas, IEEE Trans. Antennas Propagat., vol. AP-33, no. 9, pp. 988-996, Sept. 1985.

8. A. R. Djordjević and T. K. Sarkar, IEEE Trans. Antennas and Propagat., vol. AP-35, no. 3, pp. 353-355, March 1987.

9. H. L. Royden, Real Analysis 3rd Ed. (Macmillan, NY, 1988), p.282.

10. S. M. Rao, A. W. Glisson, and D. R. Wilton, IEEE Trans. Antennas and Propagat. vol. AP-27, no. 5, Sept. 1979, pp. 604-607.

11. A. B. El-Kareh and J. C. J. El-Kareh, Electron Beams, Lenses, and Optics Vol. 1 (Academic Press, NY, 1970), p. 187.

THE SHELL FOR PARTICLE ACCELERATOR RELATED CODES (SPARC)
A Unique Graphical User Interface

George H. Gillespie
G. H. Gillespie Associates, Inc., P.O. Box 2961, Del Mar, CA 92014, U.S.A.

Abstract

A unique graphical user interface (GUI), designed to provide for easy problem set up and data input, has been developed to work with accelerator simulation and analysis codes. Named the Shell for Particle Accelerator Related Codes (SPARC), it includes the basic elements needed to support a GUI: specialized windows, palettes, menus, icons, etc. The approach taken is similar to that suggested by Heighway at the first (1988) conference in this series. The interface is written in C and was developed on the Macintosh personal computer platform. For beam optics programs, such as TRACE 3-D, the configuration of a beam line is set up by selecting icons representing specific transport elements from a palette and dragging them to a document window. Parameter values for the transport elements are entered into data windows for each element in the beam line. The data windows are constructed so that expert system rules may be incorporated. For example, lower and upper limits are displayed for input parameters. The limits can incorporate rules of thumb for the parameter values and then be used to alert the user visually when his or her input data may have impractical consequences. The interface incorporates several other unique features which extend the user's capabilities for a given application beyond that normally found in the mainframe versions. These include: (a) different options for units, including "smart units," (b) multiple beam line document windows open at once, (c) copying and pasting groups of elements between windows, and (d) immediate graphical displays of certain input data, such as phase space ellipses.

Introduction

An interactive GUI has been designed to support standalone simulation and analysis programs used for particle accelerator studies and design efforts. Heighway [1] suggested the usefulness of such a software environment for beam optics design and described a number of desirable features for the environment. A few windows based programs have been developed for specific beam optics codes [2,3], but these do not provide the extensive type of graphical interface proposed by Heighway. This paper describes a new graphical environment developed to be used with a variety of codes.

SPARC Overview

A primary motivating factor in the development of SPARC was to provide a tool which improves the productivity of scientists and engineers involved in the analysis or design of accelerators. The success in achieving this capability has also had a significant additional benefit in reducing the time required to train new researchers in the use of applications which operate within the SPARC environment. Several requirements were imposed on the SPARC development effort from the outset: (1) the use of the C computer language to provide a reasonable basis for platform transportability, (2) the capability to simultaneously interface with different accelerator application codes, (3) the ability to handle multiple problems for each application, (4) the incorporation of expert system features, and (5) the elimination of the need for any command line or text input, i.e. provide for complete graphical problem set up and application execution, to the fullest extent possible.

Figure 1 provides an overview of the SPARC software environment architecture. As the figure suggests, SPARC has been constructed using a modular approach. The figure illustrates a SPARC configuration for two independent codes, simply referred to as applications I and II, although SPARC can incorporate more. It is not possible to describe in this short paper all of the items identified in this figure, but the more important ones will be discussed.

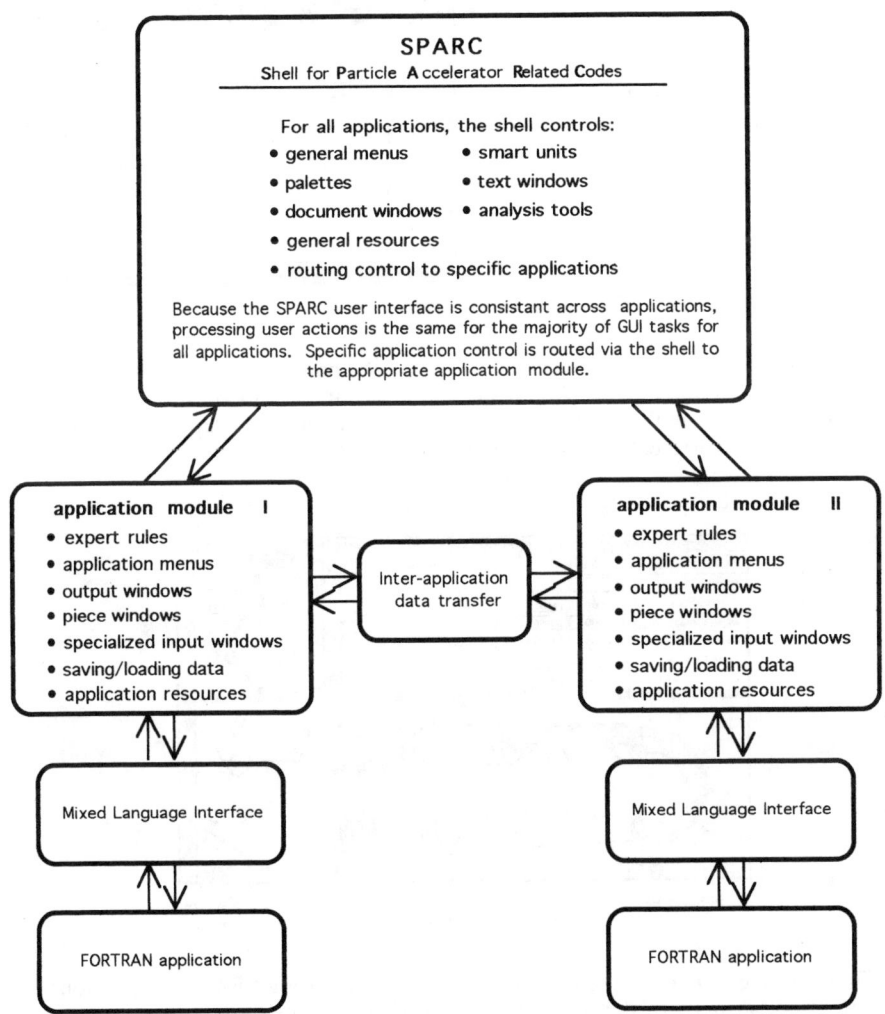

Figure 1. Architecture of the SPARC graphical user interface environment.

THE SPARC INTERFACE

Figure 2 illustrates the SPARC interface screen which appears at start up for the TRACE 3-D application [4,5]. The first three elements indicated in the top block of Figure 1 are shown, as implemented for TRACE 3-D: a Menu Bar, Palette Bar, and Document Window. The Menu Bar contains standard Macintosh options as well as specific menus used to support the application. The Document Window is the primary window used for setting up a beam line, where transport elements are selected from the Palette Bar and dragged to this window.

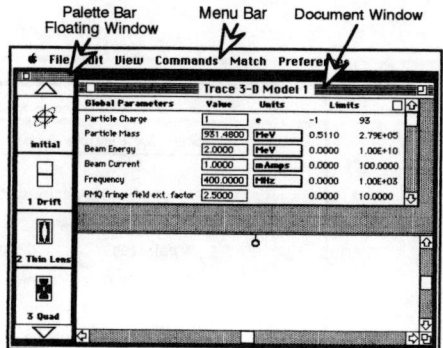

Figure 2. Example of SPARC interface screen for the TRACE 3-D application displaying Menu Bar, Palette Bar, and Document Window.

The Document Window offers considerable flexibility, both in terms of its functionality and in the graphics that may be used with an application. Figure 3 illustrates a generic Document Window with a variety of graphic images used to represent beam line components. The different parts of the window are described next.

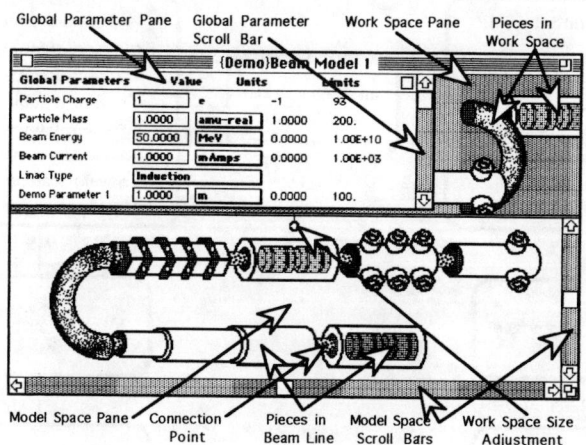

Figure 3. Generic example of a Document Window for a SPARC application. The principal parts of the window are identified and the variety of graphical elements that may used in a Document Window are illustrated.

Three panes of the Document Window are used to support different user functions. A Global Parameter Pane is used to input key parameters needed by the application. A Model Space Pane provides for the graphical definition of the problem which the application will address. A Work Space Pane is a versatile area which uses graphical representations of sub-problems to assist the user in a variety of ways which are discussed later. Each of these panes have unique features which go far beyond the user input and problem formulation environments found in most accelerator codes.

The Global Parameter Pane provides for the user input of "top level" parameters, e.g. parameters which are common to many subroutines of an application, or which may limit, or drive, practical numerical values of other parameters. This pane can be scrolled so that many parameters may be included, but there are typically about a dozen for most applications. For example, the SPARC implementation of TRACE 3-D uses ten Global Parameters [5]. The Global Parameters are ordered according to frequency of use, with default values built in, and often require only minimal attention by the user. As shown in Figures 2 and 3, the Global Parameter Pane displays five fields associated with the input of parameters: parameter name, current value, units of current value, and two guidance limits (lower and upper) for the value. This format is an example of the standard Data Table used for many SPARC input windows and is discussed later.

The beam line configuration for a given problem is set up using the Model Space Pane. Graphical representations of components (e.g. transport elements) or subsystems (e.g. a previously assembled collection of components) are displayed on this pane in a configuration dictated by the application. For example, a beam line for TRACE 3-D is composed of a linear combination of transport elements. When transport elements are selected, dragged and dropped (with the mouse) onto this pane, from either the Palette Bar or from the Work Space Pane, rules are employed to assure that the elements are only placed on a linear beam line. Components are either snapped to one end of the beam line (the closest to the drop point) or inserted into the middle of the beam line if the drop location is a connection point. More sophisticated rules are employed if the problem configuration for an application is more complex, such as the 180 degree bend shown in Figure 3.

The Work Space Pane is a specialized area on the Document Window which allows the user to work on parts of a beam line. This pane can support a variety of user operations, but only a few examples for beam optics applications such as TRACE 3-D are discussed here. Components or a group of connected components may be selected from the Model Space and dragged to this Work Space. The dragging of a selection to the Work Space makes a duplicate of the selection, with the component parameters remaining unchanged. Once on the Work Space, however, parameter values may be changed in the same manner (discussed below) as for components in the beam line on the Model Space. The user can maintain copies of beam line segments, either identical to the original or with modified input parameters. Similarly the user may select a component (or group of connected components) on the Work Space and drag the selection to a new location in the beam line on the Model Space. This provides for an easy, graphical and efficient means of reproducing and incorporating repetitious groups of transport elements into a beam line. Groups of components can be exchanged between the Work Space and Model Space for direct comparisons of alternate designs, providing a simple and intuitive way of comparing options for particular segments of a beam line. Another function of the Work Space is to support the copying and pasting of components or groups of components between different Document Windows.

Numerical input data for a component on the Model Space or Work Space is entered through the use of Piece Windows. Placing the cursor on the icon of the desired component and doubled clicking the mouse button opens the component's Piece Window. Figure 4 displays the Piece Window used for each radiofrequency quadrupole (RFQ) cell of a TRACE 3-D beam line. The same type of Data Table is used as for the Global Parameters input, including the five fields associated with each input parameter. The lower and upper limits for each parameter utilize a set of expert system knowledge base rules developed for the RFQ. (The expert system features of the SPARC Data Tables are discussed briefly near the end of this paper.) In all cases these limits are "soft," that means that any value may be input by the user and that value will be passed to the application, via the data transfer structure shown in Figure 5. However, the user is visually alerted to the fact that his or her input is outside one of the expert system limits (the violated limit is changed to an outline font) indicating that the particular value may have impractical implications.

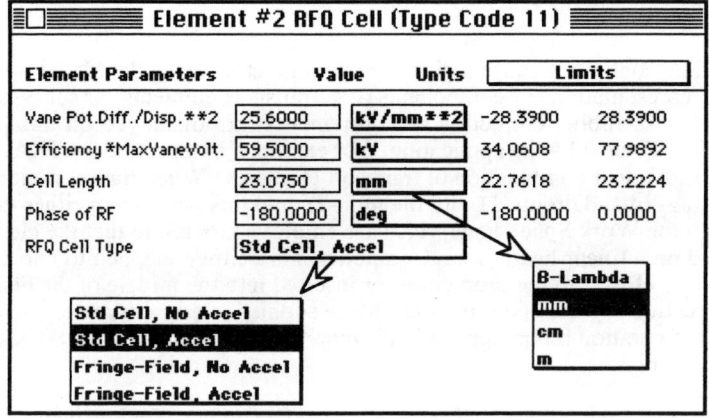

Figure 4. Piece Window for inputting RFQ cell parameters for the TRACE 3-D implementation in SPARC. Shown are the limits developed from expert system knowledge base rules and two examples of pop-up menus.

Also shown in Figure 4 are examples of two pop-up menus that are used in SPARC when a discrete and finite number of input options are appropriate. One pop-up menu is used for inputting the RFQ cell type. SPARC assigns the appropriate integer input needed by TRACE 3-D. The other pop-up menu offers the user a selection of units for the cell length parameter. The units menu shown includes a "smart units" choice, the ability to select the units of $\beta\lambda$ for any input parameter that has the dimensions of length. Whenever a different choice of units is selected the current value of the input parameter and the lower and upper limits are immediately updated to the selected units.

INTERNAL INTERFACES TO APPLICATION PROGRAMS

Once a problem has been set up, the execution of the application commands is accomplished using a pull down menu from the SPARC Menu Bar. The commands are unique to each application and any command may require all or only a part of the input parameters. As shown in Figure 1 there is an internal, mixed language interface between each application and the SPARC GUI which handles the parameter data transfer.

The mixed language interface is used to facilitate the transfer of data between SPARC and each application. A schematic of this internal data interface for TRACE 3-D is shown in Figure 5. All data is stored in a shared (RAM) memory location, with routines written to access the data for both the SPARC and application sides of the interface. A pointer to the location of the shared data, not the data itself, is passed between SPARC and the application. This enhances the speed of execution of application commands and is one of the features which makes the integration of a given application with SPARC into a seamless package.

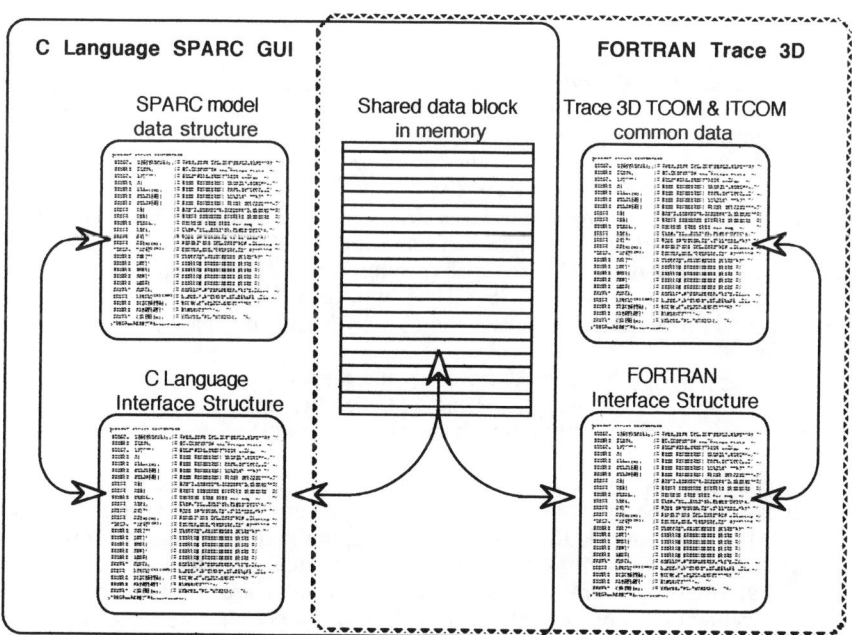

Figure 5. Schematic of the mixed language interface (C and FORTRAN) for the TRACE 3-D implementation in SPARC.

EXPERT SYSTEM FEATURES

The approach used to incorporate expert system features into SPARC is somewhat unconventional. The use of artificial intelligence software environments, such as the Knowledge Engineering Environment (KEE) expert system shell, is one approach that has been used to develop expert systems for both accelerator design codes [2] and control system interfaces [6]. However, the requirement to work within the SPARC GUI precluded the use of such environments. Previous experience had also indicated that an unacceptable overhead, in memory and/or speed, was often associated with such environments. For ease of integration with SPARC it was also desirable to write the expert system rules in C rather than LISP.

The primary objective of incorporating expert system rules is to assist users in the initial set up of problems as well as in running the application. Both beginning and experienced users are intended to benefit from the rules. For novice users, this principally means providing guidance on the input data required, whereas for advanced users the goal is to reduce the number of off-line calculations necessary to define the beam line in detail. Implicit in these considerations is an educational objective as well; the rules should assist in the training of new users.

There are basically two classes of knowledge base rules that are incorporated into SPARC and can be described as (1) problem configuration rules and (2) input parameter rules. Problem configuration rules are concerned with constraints on the arrangement of components. Configuration constraints are implemented in SPARC with rules defining the placement of graphical elements on the Model Space Pane of a Document Window and were briefly discussed previously. The input parameter rules are quantitative and are of two types: those specifying default values of parameters, and those which provide lower and upper limits for user guidance of his or her input. Such rules can be quite sophisticated and are application specific. It is beyond the scope of this paper to discuss these types of rules in any detail so only an overview of the limit rules developed for TRACE 3-D [7] will be given. The limits for the Global Parameters and Piece Parameters are generated by knowledge base rules which are of three origins:

- TRACE 3-D requirements,
- Particle beam optics utility, and
- Practical hardware constraints.

The first type includes some straightforward TRACE 3-D constraints, such as requiring the input parameter for any "identical" element to lie between the first and last element numbers of the beam line. Utility rules for beam optics include things like limits on the strengths, lengths and spacings of quadrupoles in focusing triplets which will assure bidirectional focusing. Practical hardware constraints are derived from specific accelerator technology. For example, practical limits on permanent magnet quadrupole (PMQ) field gradients are directly related to the inner and outer radii of the PMQ magnets. This is used to construct rules for the PMQ limits.

It is clear that input parameter rules of the last type are directly dependent upon particular accelerator technology assumptions. One set of assumptions may not be appropriate for all users. The modular approach used for inputting data via Piece Windows permits multiple technology options to be readily incorporated into SPARC.

Summary

SPARC is a sophisticated graphical user interface that has been developed for use with particle accelerator analysis and simulation programs. Many features are included that significantly enhance the user's productivity. Problem set up and definition are accomplished with a minimal amount of alpha-numeric (keyboard) input. SPARC takes care of setting up arrays and similar bookkeeping. Expert rules can be built into the data input windows to assist users in assigning parameter values, such as options for parameter units, both fixed and scaled smart units, and limit guidelines for each parameter. The interface has been successfully integrated on a Macintosh with the beam optics program TRACE 3-D to form a seamless application which is now available commercially. Other accelerator codes should be integrated with the SPARC software technology in the near future.

Acknowledgements

The author is indebted to Barrey Hill and James Gillespie for their excellent work in developing and programming the SPARC software, and for their assistance in the preparation of several of the figures used in this manuscript.

References

1. E. Heighway, "Magnetic Optics Design (A Designer's View of the Beam Transport Environment)," AIP Conf. Proc. No. 177, 181-203 (1988).

2. R. R. Silbar, "An Interactive Interface to the Beam Optics Code 'TRANSPORT'," AIP Conf. Proc. No. 177, 109-116 (1988).

3. U. Raich, "Beam Modeling with a Window-Oriented User Interface," Nuc. Instr. Meth. Phys. Res. A 293, 450-455 (1990).

4. K. Crandall and D. Rusthoi, "TRACE 3-D Documentation," Los Alamos National Laboratory Report LA-UR-90-4146, 92 pages (1990).

5. G. H. Gillespie and B. W. Hill, "A Graphical User Interface for TRACE 3-D Incorporating Some Expert System Type Features," 1992 Linear Accelerator Conference Proc. (Ottawa) AECL-10728, 787-789 (1992).

6. D. E. Schultz and P. A. Brown, "The Development of an Expert System to Tune a Beam Line," Nuc. Instr. Meth. Phys. Res. A 293 486-490 (1990).

7. G. H. Gillespie and P. K. Van Staagen, "A Knowledge Rule Base for Problem Set Up with the Beam Optics Program TRACE 3-D," presented at the 12th International Conference on the Application of Accelerators in Research and Industry (Denton TX, 1992); and "Knowledge Rule Base for the Beam Optics Program TRACE 3-D," to be presented at the 1993 IEEE Particle Accelerator Conference (Washington DC, 1993)

UPGRADES TO THE LBL LATTICE PROGRAM*

John W. Staples
Building 64, Lawrence Berkeley Laboratory, Berkeley, Cal 94720

ABSTRACT

The LATTICE synchrotron and transport system design code, dating from 1976 and currently running worldwide, is a friendly, interactive first-order design and analysis code with a rich graphics environment. The code is ideal for conceptualizing and designing beam optics systems quickly and intuitively. Recent additions to the code include a tracking module for determination of dynamic aperture and small amplitude tune shift including random element errors to 12-pole, automatic translation of LATTICE data files to input files to several other design codes for higher-order analysis, and additional graphics software. A companion code with the same data and command structure, QUICKTRAN, uses the full 6×6 σ-matrix TRANSPORT formalism to include full phase plane mixing. The two official versions run on Sun and VAX platforms: the code is in Pascal.

INTRODUCTION

LATTICE[1] is a computer program that calculates the first order characteristics of synchrotrons and beam transport systems. It is fully interactive with low and high resolution on-line graphics displays giving the user rapid visualization of the characteristics of the beam line being designed. A powerful editing facility allows beam lines to be quickly reconfigured, and up to nine completely independent problems can be manipulated and stored within the program, even more off-line.

LATTICE has two distinct modes: the *lattice* mode which finds the matched functions of a synchrotron, and the *transport* mode which propagates a beam through a sequence of transport elements. However, each mode can be used for either type of problem: the transport mode may be used to calculate an insertion for a synchrotron lattice, and the lattice mode may be used to calculate the characteristics of a long periodic beam transport system. LATTICE does not support x-y coupling: calculations requiring full 6×6 coupling for solenoids and arbitrary beam line twists are calculated with QUICKTRAN, which uses a first-order TRANSPORT-like σ-matrix formulation[2] within the LATTICE-like user environment.

An optimizer allows up to 10 parameters of a synchrotron or beam line to be set subject to 41 different types of constraints (betatron amplitudes, tunes, magnet positions, matrix elements, e.g.).

The rich graphics environment, both on-line during computation and off-line with graphics post-processors produce printer graphics, Tektronix or Postscript plots of the beam amplitudes, beamline layout, or phase space acceptance diagrams.

LATTICE will carry out two limited types of higher-order calculations. The small-ring chromaticity model (the more difficult one) is calculated correctly by

---

*This work supported by the U. S. Department of Energy contract DE-AC03-76F00098.

LATTICE. A tracking module is included to permit the calculation of the dynamic aperture or amplitude dependent tune shift in the presence of multipole errors up through duodecapole. The higher-order contributions may be random as well as systematic.

LATTICE calculations may be followed up with any of several higher-order codes. LATTICE will produce input files for them, eliminating errors in manual translation. In this way, LATTICE can be used for the original conceptualization of a beam line or synchrotron to first order, and TRANSPORT, SYNCH, MAD, DIMAD or MARYLIE can continue the calculations to higher order.

PROGRAM FEATURES

The minimum input parameters include the beam rigidity, definitions of all the elements, and the ordering of the elements in the beam line. Additionally, emittance, energy spread, comments, periodicity and graph scales can be entered, otherwise the default values are used. The 17 element types presently supported are:

Element	Attributes				
drift	L			a_x	a_y
quadrupole	L	B'		a_x	a_y
skew quadrupole (new)	L	B'		a_x	a_y
sextupole	L	B''		a_x	a_y
octupole	L	B'''		a_x	a_y
decapole	L	B^{iv}		a_x	a_y
duodecapole	L	B^{v}		a_x	a_y
horizontal bend	L	B	n	a_x	a_y
horizontal magnet edge	θ	B	$g/2$	a_x	a_y
vertical bend	L	B	n	a_x	a_y
vertical magnet edge	θ	B	$g/2$	a_x	a_y
einzel-type lens	L	s		a_x	a_y
arbitrary matrix	(mtx)			a_x	a_y
beam line rotation	ϕ			a_x	a_y
solenoid	L	B		a_x	a_y
undulator (new)	L	B	L_c	a_x	a_y
electron accelerator cavity (new)	L	dE/dL		a_x	a_y

The attributes are length, field, angle, half-gap and field index. In addition, each element can have an x- and y-aperture which determines the geometric acceptance of the beam line. The higher order elements and skew quad are significant only when used for amplitude-dependent tune shift or dynamic aperture calculations, otherwise they are replaced by drifts. They may have zero or non-zero lengths. The solenoid and beam rotation elements are significant only in QUICKTRAN, and are replaced by drifts in LATTICE. The anharmonicity, or tune shift normalized by circulating emittance, is calculated to first order. Sextupoles are incorporated in the chromaticity calculation and the strengths to give required chromaticities can be found with the sextupole optimizer module.

586 Upgrades to the LBL Lattice Program

The lattice or beam line is built up by a list of element names, the elements having been defined previously. To manipulate the lattice, editing commands are provided to add, remove or replace sections, invert the beam line, or save segments or even entire beamline descriptions in nine independent storage registers. Any number of external files can be selectively read into or written from the registers during program execution.

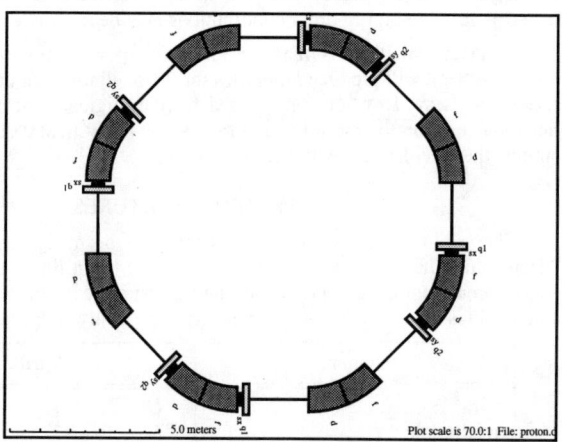

The rich graphics environment for beam line visualization includes on-line printer graphics for minimal terminal environments, or Tektronix graphics for more capable terminals. For publication-quality graphics, post-processors provide plots of beam amplitudes, beamline geometric layout, or phase space acceptance in printer, Tektronix or Postscript format. The geometric layout plot is a 2-D projection scale drawing that can be copied to a transparency and then overlaid onto a building plan. A developmental post-processor, VIEWXYZ provides a 3-D wireframe perspective representation of complex beamlines that occupy 3-space. The upper illustration shows the layout of a small proton synchrotron: the lower one the phase space acceptance of a beam line.

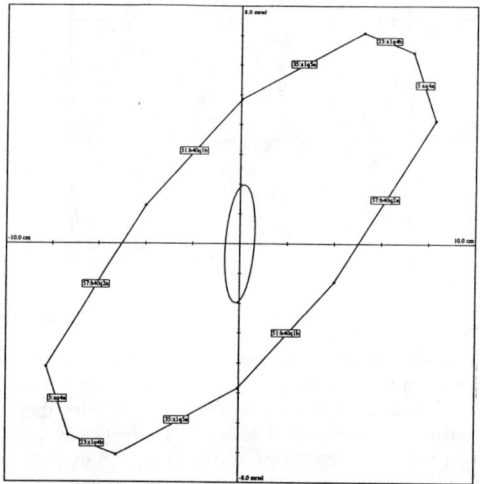

For extraction studies, LATTICE helps prepare an input file for the code EX6, which simulates resonant extraction for half, third, and higher integer orders. The multipole elements are carried over to EX6 from LATTICE so their effects on the extraction orbits may be studied.

The optimizer module allows up to 10 variables to be set to satisfy the same number of constraints (relaxed Newton-Raphson iteration). Any attribute of any first order element (sextupole use a separate optimizer module for chromaticity correction) can be set. The constraints are:

> Matrix elements in x and y subblocks and m_{16}–m_{56} (13)
> $\beta_{x,y}$, $\alpha_{x,y}$, $\eta_{x,y}$ $\eta'_{x,y}$ (8)
> tunes $nux_{x,y}$ (2)
> position and running H and V magnet deflection angle (3)
> γ_{tr} (1)
> Transport variables x, y, x' y', r_{12}, r_{34} (6)
> Composite matrix elements m_{16rev}, m_{26rev}, m_{36rev}, m_{46rev} (4)
> Combinations α_x–α_y, β_x–β_y, trace$_{x,y}$ (4)

where, for example, $m_{16rev} = m_{12}m_{26} - m_{16}m_{22}$ is the m_{16} element of the *reversed* beam line. The constraints can be set at the exit of any element along the beam line.

LATTICE can reverse beam lines, end-for-end, and also chop beam lines up, storing the beam parameters at the beginning and end of each piece for segmentation, and can even store the parameters of reversed beam lines. This is useful when starting a design at each end and working toward the middle.

The phase space acceptance is calculated by transforming all defined apertures to the beginning of the beamline and finding the smallest enclosed polygon. This can also be done for off-momentum beams in which case the effect of dispersion in dipoles may cause an off-axis shift of downstream polygons and chromatic effects in the focusing elements change the aperture transformation to the beginning of the beam line. By reversing the lattice, the acceptance can be projected to the far end of a beam line. A phase space acceptance diagram is shown on a previous page.

The physical layout is calculated by following the beam line axis in 3-space for the on-momentum particle. A direction cosine matrix is calculated (which is stable even if the beam is vertical: TRANSPORT has trouble with this case) which gives the beam direction in terms of three angles. This calculation can produce a file which is used in the 3-D wireframe visualization graphics post-processor as shown on a subsequent page.

NEW CAPABILITY

Since the last report[1], LATTICE has had a number of improvements, some of which were alluded to above. LATTICE has been used to design and verify a FEL line with cavities and an undulator, gantry-based beam transport system for radiotherapy applications, a heavy-ion mass separator, the extraction and beam transport systems of the HIMAC accelerator complex, among others.

A new support code, LAT2X translates a LATTICE file to the input format of the following higher-order codes: SYNCH, MAD, DIMAD, TRANSPORT and MARYLIE. The following example shows the automatic translation for a small synchrotron from LATTICE to SYNCH format.

LATTICE format:
```
* biomed 10
b 8.12600 10.00000 2.00000 0.00000
t off
sc 5.00000 5.00000 5.00000 5.00000
e
b       b 2.1 3.545636  1.200000 0.000000 0.000000 0.000000
l       d 1.1 2.327181  0.000000 0.000000 0.000000 0.000000
o       d 0.0 0.600000  0.000000 0.000000 0.000000 0.000000
f       q 3.2 0.300000  8.237382 0.000000 0.000000 0.000000
d       q 4.2 0.300000 -9.023549 0.000000 0.000000 0.000000
e       e 0.0 0.000000  1.200000 0.000000 0.000000 0.000000
sx      s 0.0 0.600000  3.354135 0.000000 0.000000 0.000000
sy      s 0.0 0.600000 -10.445238 0.000000 0.000000 0.000000
end
l
    l       f       sx      e       b       e       sy
    d       o       e       b       e       o       f
    l       l       d       l
end
p 6
```

SYNCH format:
```
         RUN        biomed 10
C
C Input file biomed10 translated to SYNCH format by lat2x.
C
C23456789012345678901234567890123456789012345678901
C        1         2         3         4         5         6
  BRHO   =           8.12600
  B      MAG         3.54564  -0.00000BRHO         1.20000   0.00000   0.00000
  L      DRF         2.32718
  O      DRF         0.60000
  F      MAG         0.30000   8.23738BRHO
  D      MAG         0.30000  -9.02355BRHO
  SX     SXTP        0.60000   3.35413BRHO
  SY     SXTP        0.60000 -10.44524BRHO
C
  ACC    BML         L    F    SX   B    SY   D    O    B    O    F
                     L         D    L
         CYC    6 ACC
         FIN
         STOP
```

LATTICE is now fully operational on Sun SPARC computers in a C environment. LATTICE is still written in Pascal, but a preliminary version written in C is being developed. The Sun version makes use of the multiple window environment operating under OpenWindows.

The tracking module calculates the dynamic aperture and small amplitude tune shift in the presence of multipole elements up to duodecapole. A small ensemble of particles on the emittance shell at the design momentum are tracked around the ring, with the multipole elements acting as zero-length nonlinear kicks. The resulting tune spectrum is Hanning filtered, FFT analyzed and the shifted tune extracted. The same module calculates dynamic aperture by placing a monitor at each sextupole to detect unstable growth. The anharmonicity, or the tune shift normalized by the circulating emittance, is also calculated.

A thin skew quad element has been added to investigate the effect of x-y coupling on the dynamic aperture and tune shift.

Random perturbations to all elements from quadrupole to duodecapole have recently been added. A peak random field amplitude is assigned to each element which is then multiplied by a random quantity uniformly distributed in the range [-

1..1], unique for each element in the lattice, even if they have the same name. Distributed multipoles in long dipoles can be built up from a number of zero-length multipole kicks spread along a dipole, for example. Systematic errors can be built up in the same way. This way, dynamic aperture can be determined using realistic error models to establish the maximum allowable random and systematic errors allowed.

LATTICE generates an input file for the extraction code EX6 including all of the random and systematic errors defined above. The EX6 code also includes kicker dipoles, and the effect of the random and systematic errors on the extraction trajectories and on the closed orbit can be calculated.

New undulator and electron accelerator ($\beta=1$) elements have been added. The cavity model is very simple: it just models the r.f. defocusing at the gap. The undulator element simulates a periodic dipole sequence, which is a drift in the gap plane, but which acts as a periodic transport system in the perpendicular plane. As this periodic system has a matched betatron amplitude, this matched betatron amplitude can be specified directly, as an alternative to the peak poletip field.

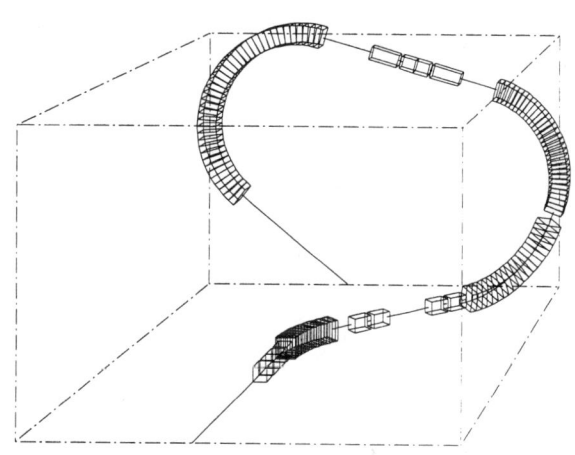

A new off-line graphics processor, VIEWXYZ, provides a wireframe perspective view of beam lines of complex geometry, as shown to the right.

Two new fitting constraints have been added: $\beta_x - \beta_y$ and $\alpha_x - \alpha_y$, used for the purpose of prescribing a round beam without specifying the actual betatron amplitudes. This was needed to fix the beam at a variable twist in a transport system. The flexibility and modularity of LATTICE allows additional complex types of constraints to be easily added to the source code, which must then be recompiled.

REFERENCES

1. LATTICE...A Beam Transport Program, J. Staples, LBL-23939, June 1987
2. A First and Second-Order Matrix Theory..., K. L. Brown, SLAC-75, July 1967

BMAP
DIPOLE MAGNET FIELD ANALYSIS AND ORBIT TRACKING

Stanley Humphries, Jr.
Acceleration Associates, Albuquerque, New Mexico
and
Rose Mary Baltrusaitis, Carl Ekdahl and Carlton Young
Los Alamos National Laboratory, Los Alamos, New Mexico
and
Charles Warn
EG&G Energy Measurements, Las Vegas, Nevada

INTRODUCTION

BMAP is a versatile program for field analysis and orbit tracking in dipole magnets. The program was created to aid the design of charged-particle magnetic spectrometers in Group P-14 of Los Alamos National Laboratory. The program is written in Pascal. The original version handles arrays of up to 40,000 field values and runs on any IBM-PC computer or compatible. A new version, written in machine-transportable Pascal, uses double precision variables and has a capacity of 200,000 mesh points. This program runs on 386/387 or 486 PCs with 8 MB of total memory.

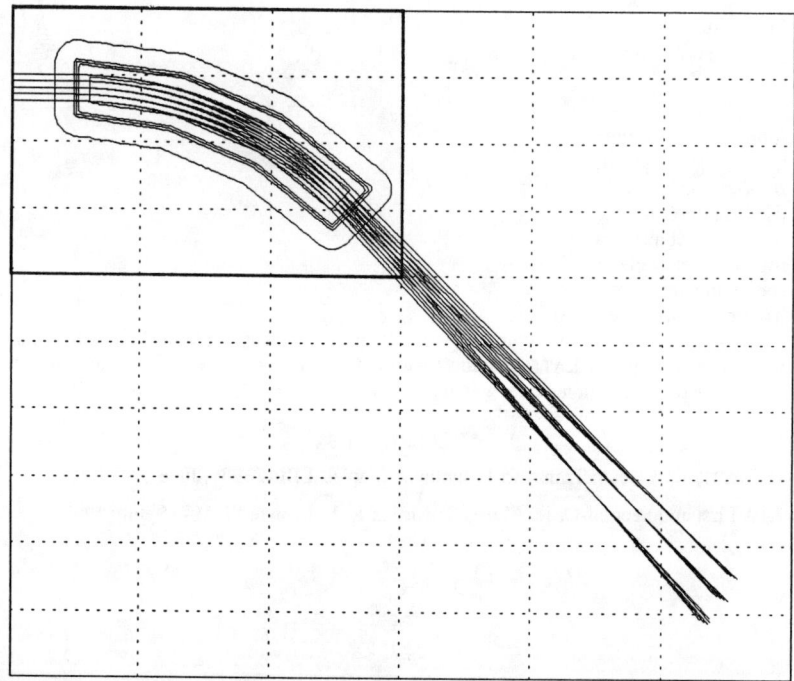

Figure 1. Electron orbits through the 45 degree DARHT spectrometer at 19, 20 and 21 Mev. Parallel input beam. Bold lines show Main Map Outline

BMAP has application to particle orbit calculations in dipole magnets with field symmetric about a midplane. The program can handle complex magnet boundaries, field gradients, and multiple magnets that share a symmetry plane. The basic data structure is the *field map*, an array of vertical field values on a square grid at the magnet midplane. For magnet analysis, the values can be entered from experimental measurements in a simple ASCII format. For design work, the program has a variety of options to simulate realistic maps of magnets with complex boundaries.

As a design tool, BMAP is similar in function to Raytrace[1]. In comparison, Raytrace is more general. It handles several types of optical elements and allows rotations about the main orbit. To achieve this generality, Raytrace may define several local coordinate systems along the main particle trajectory referenced to theoretical magnetic field boundaries. The code user must expend some effort to understand the program and to convert the numerical output to a coordinate system referenced to the physical boundaries of the magnets.

BMAP is not as comprehensive as Raytrace - it is limited to one or more dipole magnets that share a symmetry midplane. Nonetheless, for bending magnets and spectrometers BMAP gives a better picture of particle orbits in physical coordinates. The program uses a fixed reference system and has powerful interactive graphics capabilities. In particular, BMAP can superimpose magnetic contour and orbit plots. Furthermore, BMAP uses modern parsing methods to process the input control file - preparation of run command sequences is straightforward and well-documented. Finally, BMAP is specifically designed for experimentally measured field profiles and makes more accurate orbit calculations.

Section B reviews features and capabilities of BMAP and the utility programs MMAP and BGraph. MMAP creates realistic dipole field maps given an outline for the pole boundary and coefficients for the edge field variation. BGraph is an interactive graphics post-analysis program for constructing screen and hardcopy graphs from BMAP plot files. The user can choose entities and shift plot boundaries for the best display. BGraph makes several types of plots, including field contours, orbits projected in the symmetry plane, vertical motion of charged particles, and particle distributions in an inclined detector plane. Section C shows examples of output and briefly describes spectrometer designs and benchmark tests carried out to date.

BMAP CAPABILITIES

Reference 2 reviews the physical basis of BMAP. The basic assumption is that the dipole magnet produces a predominantly vertical field symmetric about the $z = 0$ plane. BMAP uses a second order field expansion to determine values of $B_x(x,y,z)$, $B_y(x,y,z)$ and $B_z(x,y,z)$ from an array of midplane values $B_z(x,y,0)$ defined on a square mesh. The term *map* designates an array of midplane vertical field values in a rectangular region.

Maps can be created from experimental values or can be constructed for design studies using the MMAP program or a variety of utility commands in BMAP.

The BMAP input control file consists of a sequential array of standard commands with required or optional parameters. The input parser accepts comment lines and a variety of delimiters and formats. As a result, the input files can be structured and self-documenting. The program has comprehensive diagnostics to report syntax errors, missing input files, and incorrect numerical input. The BMAP Reference Manual[2] describes the syntax of each command and also contains a command dictionary grouped according to function. Table 1 shows an input command file to set up the Main Map of Fig. 1.

Table 1
BMAP Command File to Set Up a Field Map

```
DARHT 01. Contour plot to check field map creation
MAINMAP 120 80 0.25
* Read the input map
INPUTFILE C:\BMAP\BUFFER\DMAP.MAP
* Move input map to main map with no changes
TRANSLATE 0.00 0.00 0.00 0
* Reverse direction of field for electrons
MMULT MAIN -1.0
GRAPH ON
* Make a contour map of good field region and edge
CONTOUR MAIN 2 1150.0 2288.5
* Save the map in a binary file
SAVEMAP C:\BMAP\BUFFER\DARHT.BIN
ENDFILE
```

To begin a BMAP run, it is necessary to define the field map. The first step is to enter field values into the Input Map. The Input Map is a dynamic data structure used only during field construction. Values can be obtained from files of experimental values, the MMAP utility, or from internal BMAP commands that define edges and uniform field regions. Once the Input Map is complete, it is transferred to the Main Map with user-defined displacement, rotation and over-ride option. The process can be repeated any number of times. In this way, it is possible to construct complex field patterns or to enter multiple experimental magnets and shift their locations. Once the Main Map is complete, the Input Map is disposed to free memory for orbit calculations. BMAP can make a binary file of the resulting Main Map for quick field definition in following runs.

BMAP can make versatile contour plots and list field values or arrays from both the Input and Main maps to check the field definition process. The program also computes average field values and standard deviations over specified regions. Several contour plot commands with different parameters can be issued. BMAP stores each plot as a numbered entity in a plot file. The BGraph utility can interactively plot these entities, either singly or in combination. Plots of orbits in the z=0 plane, vertical orbits, and particle distributions in a detector plane are handled in the same way.

BMAP can make orbit calculations after setting up the Main Map. There are two modes: ray and distribution tracking. In the ray tracking mode, BMAP follows individual orbits for a beam with a given particle mass, charge and incident kinetic energy. The initial conditions for an individual ray are defined by its positions and inclination in a source plane. The source plane can be either inside the field map or outside at a given distance. For the external option, the program assumes free drift from the source to the map boundary. There are several options to stop rays: 1) maximum total distance, 2) striking a pole face, 3) reaching the map boundary, and 4) crossing a specified inclined detector plane. The last two options use interpolation to a reference plane for high accuracy - the output listing also includes transit time information for isochronous systems. With Option 3, the program can project orbits a given distance beyond the map boundary in a field-free drift region. This is a useful feature to locate horizontal or vertical focal points. With Option 4, the detector plane can be either inside or outside the map, separated by a drift space. Table 2 shows the command file to generate the particle orbits of Fig. 1.

Table 2
BMAP Command File to Track Rays to a Detector Plane

```
DARHT 02. Detector plane identification
* Load the binary file from the previous run
GETMAP C:\BMAP\BUFFER\DARHT.BIN
* Add magnetic field contours to the plot file
CONTOUR MAIN 2 -2288.50 -1150.00
* Set up the main axis and source plane
AXIS INTERNAL 0.00 14.00  0.00
* Stop particles at any boundary and project 35 cm beyond
RAYSTOP BOUNDARY 35.00
* 20 MeV electrons
BEAM 0.0 20.0E6 -1.0
* Track three parallel rays from different horizontal positions
RAYPLOT START
  RAY -1.00  0.00  0.00  0.00  0.00
  RAY  0.00  0.00  0.00  0.00  0.00
  RAY  1.00  0.00  0.00  0.00  0.00
RAYPLOT END
 ...
ENDFILE
```

The distribution mode operates in conjunction with the Detector Plane stopping option. BMAP calculates a main orbit and marks its position in the detector plane. Next, the program generates a large number of rays from a source plane normal to the main axis. There are several options for ray generation, including circular and rectangular sources, individual angular spreads in the horizontal and vertical directions, and random or uniform distributions in position and velocity. There is also an option to generate a uniform spatial grid of orbits to identify optical aberrations. In the distribution mode, BMAP makes plot file entries of ray positions in the source and inclined detector planes. The program also lists the main orbit intersection with the detector plane in absolute coordinates, the relative coordinates and arrival times of other rays, and computes the mean and standard deviation of ray positions about the main orbit in the horizontal and vertical directions.

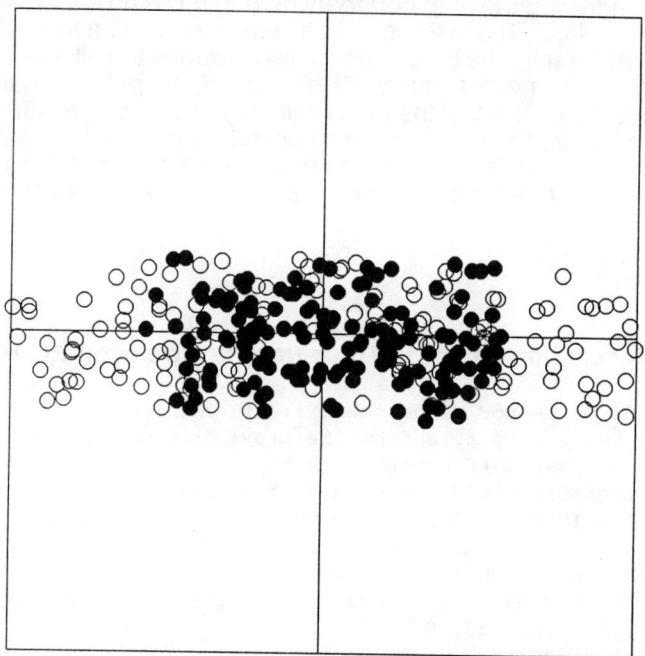

Figure 2. BMAP distribution plot for detector plane geometry of Fig. 1. Horizontal width: 2.0 cm, vertical width: 2.0 cm. Light circles: Incident electron positions. Dark circles: detector plane positions.

APPLICATIONS AND BENCHMARKS

To date, we have used BMAP for three applications. Figure 1 shows calculations for a spectrometer for high-resolution energy measurements on the DARHT accelerator at Los Alamos National Laboratory. The simulated field distribution was created with the MMAP utility using edge field coefficients derived from Poisson calculations. The figure shows superimposed magnetic field contour lines and electron traces. The boundaries of the main map are marked in bold. Using data similar to Fig. 1 to locate a plane at the point of horizontal focus, we could find ray distributions using the detector plane stopping option. These runs gave information on the energy resolution, including effects of optical distortions, beam emittance, and field variation normal to main axis. Figure 2 shows a distribution plot for incident particles randomly distribution over a 2 cm width in the horizontal direction and 0.5 cm width in the vertical directions with a 5 mr horizontal angular divergence. The light circles are particle locations projected in the source plane, while the filled circles are locations in the inclined focal plane of Fig. 1. We made similar design studies for a permanent magnet proton spectrometer for neutron energy distribution measurements.

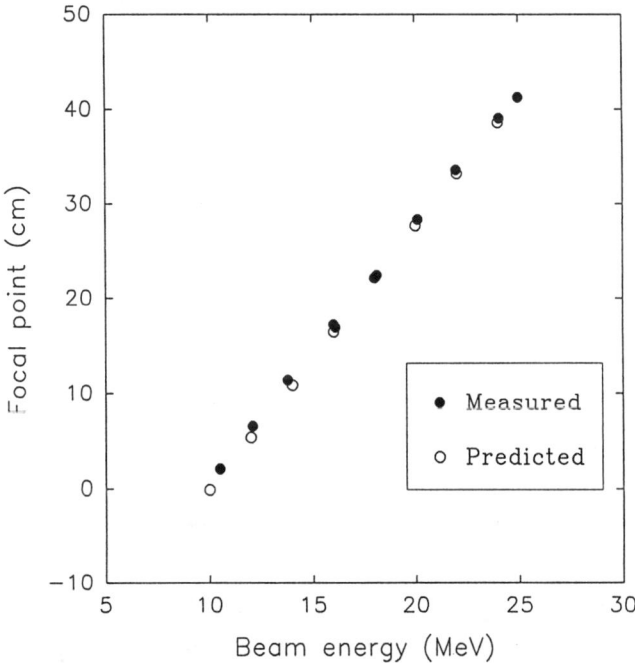

Figure 3. *Measurements and absolute predictions of the horizontal focal position versus electron energy for the WHEBY spectrometer.*

The most extensive use of BMAP was to design a permanent magnet Compton spectrometer for gamma ray energy distributions. Reference 3 gives a detailed description of the WHEBY spectrometer. WHEBY is a large permanent magnet dipole with shaped pole edges. The pole face area was 810 cm^2. Electrons enter and exit at a 45 degree angle relative to the pole faces. The geometry provides good horizontal focusing to an extended detector plane over the broad energy range 10 to 24 MeV. The magnet also gives partial vertical focusing over the full energy range. Fabrication of the

WHEBY magnet was based entirely on BMAP design simulations. After assembly, calibration experiments were carried out at the EG&G electron linac. A scintillator screen, located in the computed detector plane, was observed with a television camera and optical digitizer for precise measurements of beam position and spot size. Figure 3 shows the absolute calibration of horizontal beam position over the full energy range. The measured positions agreed with BMAP predictions to within the accuracy of existing energy diagnostics on the linac. Figure 4 shows predicted and measured values for the vertical beam size. Within the limits of accuracy of beam emittance measurements, the predictions and measurements were in good agreement.

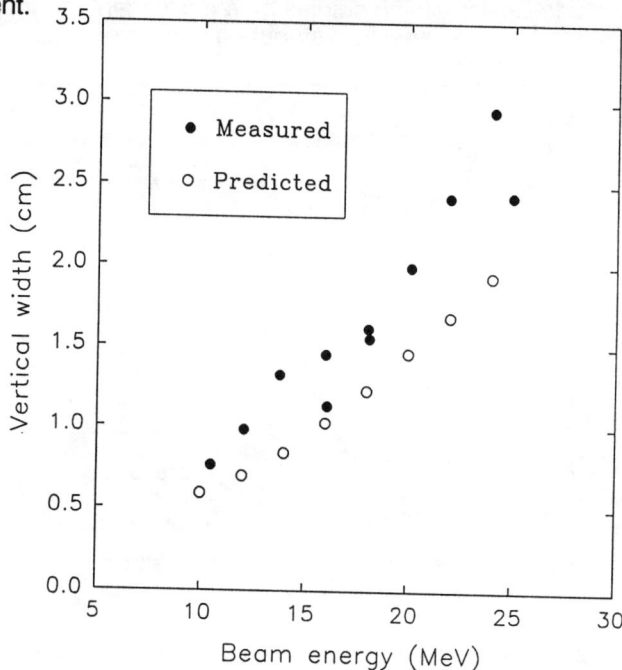

Figure 4. *Measurements and absolute predictions of the vertical beam height in the WHEBY spectrometer as a function of incident electron energy.*

REFERENCES

1. S. Kowalski and H.A. Enge, **Raytrace** (Laboratory for Nuclear Science, Massachusetts Institute of Technology, 1987), unpublished.

2. Stanley Humphries, Jr., **BMAP** (Acceleration Associates, Albuquerque, New Mexico, 1992), unpublished.

3. Stanley Humphries, Jr., Carl Ekdahl, Robert Chrien, Carlton Young, Rose Mary Baltrusaitis, and Charles Warn, **Proc. IEEE Conference on Plasma Science** (Williamsburg, 1991), unpublished.

TRAK
Charged Particle Tracking in Electric and Magnetic Fields

Stanley Humphries, Jr.
Field Precision
PO Box 13595
Albuquerque, New Mexico

INTRODUCTION

TRAK[1] is a versatile program that uses field solutions created by Poisson and Pandira[2] to compute charged particle orbits. The program has application to a wide variety of design problems, including particle accelerators, ion sources, acceleration columns, electrostatic and magnetostatic lenses, vacuum tubes, and electro-optical devices. The present version of **TRAK** is limited to single particle orbits without space charge fields. A version under development will include iterative space-charge correction of electrostatic fields and will have provisions for self-consistent field emission and space-charge limited emission from surfaces. **TRAK** is available in executable form for use with Poisson and Pandira in the 386/486 personal computer packages **EMP 2.0** and **EMP 3.0**. Figure 1 shows graphics output for a test example. Here, a beam of electrons entering from the left is incident on an array of biased wires with alternate polarity. The figure illustrates superposition of field and orbit information and the accuracy of orbit calculations even with strong field variations.

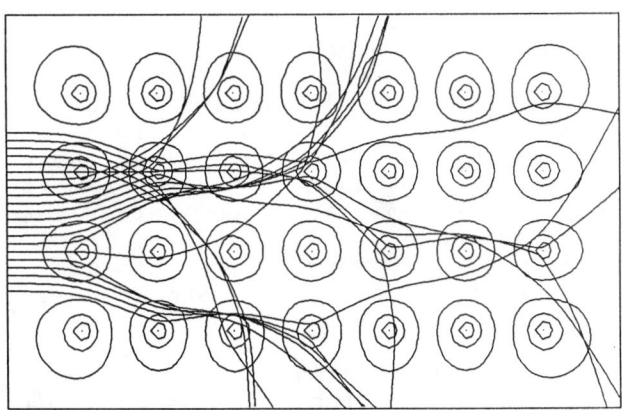

Figure 1. Test example - electron orbits in the electrostatic fields of a biased wire array

Innovations in **TRAK** include the following.

- Full use of conformal triangular meshes of Poisson and Pandira.
- Ability to combine electric and magnetic field solutions even if they use different meshes.
- Orbit computation in three dimensions with an option for perturbation dipole fields.
- Accurate calculations, even near conducting and dielectric boundaries.
- Extensive ASCII output listing options, including elapsed time listing for isochronous systems.
- Interpolation to stopping planes for high-accuracy applications.
- Completely documented in a step-by-step users' manual.
- Options for programmed time variations or low-frequency harmonic modulation of fields.
- Capability to pass orbit information and shift fields for multiple runs on periodic systems.

Because **TRAK** is distributed to individual users for personal computers, it is imperative that the program and its documentation are completely self-explanatory and easy to learn. The next section discusses user interfaces of the program. The final section reviews computational methods in the code and gives examples of output.

USER INTERFACE

Field Precision programs use a standardized input format that has evolved from an active exchange with over 200 users. We believe the most effective method to enter information for the main technical programs is through a free-form input command file in text format. In this way, runs are self-documenting. Most important, when there is a syntax error or change in parameters users can make corrections quickly with a familiar editor rather than struggling through a redundant menu structure. In contrast, all post-processing programs are interactive, with graphics displays, on-screen menus, and mouse-support.

The Input Command Files for Poisson, Pandira, TRAK, and Superfish share a similar structure. The input parser allows comment lines, blank lines, and indentations to create structured, visually-pleasing documents. Active lines consist of a command followed by one or more parameters. The programs give immediate screen messages for unrecognized commands, wrong number of parameters, incorrect numerical format, and other syntax errors. Commands are grouped in function blocks that follow in sequence. Within a function block, commands can be issued in any order. The **TRAK** command file has three blocks.

Block 1. FIELDS
The command file loads an electrostatic field solution and/or a magnetic field solution created by Poisson or Pandira. There are optional commands to adjust the field magnitudes or to add constant magnetic field components.

Block 2. PARTICLES
This section of the command file specifies initial parameters for up to 150 particles and control variables for the orbit integrations. **TRAK** *follows the orbits sequentially. There are several options to end orbits, including accurate interpolation to stopping planes.*

Block 3. DIAGNOSTICS
At the end of a run, **TRAK** *can analyze the initial and final distributions of particles.*

   TRAK gives on-screen error messages if the blocks are out of sequence or if commands appear in the wrong block. The full set of TRAK commands is documented in a TSR (terminate-and-stay-resident) utility. This program can be called by a keyboard interrupt from within other programs such as editors and word processors. In this way, critical information on **TRAK** is immediately available during preparation of the command file. TABLE 1 gives an example of program input for an electrostatic einzel lens. The commented file loads an existing electric field solution from Poisson and tracks several particle orbits from a parallel beam. The particle input parameters can be listed directly in the command file or loaded from another ASCII file created by a spreadsheet or user program. The input quantities, chosen to minimize user calculations, are particle mass in AMU (0.0 signals electron mass), charge, kinetic energy in eV, position in cm, and particle momentum fractions projected along three axis. The program automatically adjusts to cylindrical or rectangular coordinates.

TABLE 1.
Example of TRAK Input Command File

```
EINZEL 01
* ================================================================
* Electron focusing by an einzel lens
* Central electrode is -5.0 kV in EINZLENS.POU
* Adjust to -6.5 kV for solution
* ================================================================
FIELDS
    *     Load electrostatic Poisson solution
    EFILE: EINZLENS
    *     Adjust the magnitude of the fields
    EMULT: 1.30
    *     Set boundaries for plots and particle stopping
    BOUNDARY: 0.00 -2.50  2.50  2.50
    END
* ================================================================
PARTICLES TRACK
    *     Start particles from the following list
    PLIST
*   No  Mass Chrg    Eng      r    theta    z     pr    pthet   pz
*   ==============================================================
     1  0.0  -1.0   10.0E3  0.10   0.0   -2.40   0.0    0.0    1.0
     2  0.0  -1.0   10.0E3  0.15   0.0   -2.40   0.0    0.0    1.0
     3  0.0  -1.0   10.0E3  0.20   0.0   -2.40   0.0    0.0    1.0
     4  0.0  -1.0   10.0E3  0.25   0.0   -2.40   0.0    0.0    1.0
     5  0.0  -1.0   10.0E3  0.30   0.0   -2.40   0.0    0.0    1.0
     6  0.0  -1.0   10.0E3  0.35   0.0   -2.40   0.0    0.0    1.0
     7  0.0  -1.0   10.0E3  0.40   0.0   -2.40   0.0    0.0    1.0
     8  0.0  -1.0   10.0E3  0.45   0.0   -2.40   0.0    0.0    1.0
     9  0.0  -1.0   10.0E3  0.50   0.0   -2.40   0.0    0.0    1.0
    10  0.0  -1.0   10.0E3  0.55   0.0   -2.40   0.0    0.0    1.0
    11  0.0  -1.0   10.0E3  0.60   0.0   -2.40   0.0    0.0    1.0
    12  0.0  -1.0   10.0E3  0.65   0.0   -2.40   0.0    0.0    1.0
    13  0.0  -1.0   10.0E3  0.70   0.0   -2.40   0.0    0.0    1.0
    STOP
    *     Set orbit control parameters
    NSEARCH: 4
    TMAX: 2.0E-9
    *     Interpolate orbit parameters to a plane at z = 2.40
    ZPLANE2: 2.40
END
* ================================================================
DIAGNOSTICS
    *     Make a table of values for a spreadsheet analysis with
    *     comma delimiters
    PTABLE Comma
END
*
ENDFILE
```

Figure 2 shows equipotential lines and orbits computed by TRAK. The graph information was prepared with the VTRAK utility. In this program, the user can load field solution files and access a random access plot file. Features of the graph can be modified interactively: options include zooming plot boundaries, adding equipotential lines over specified intervals, and choosing individual orbits. When the screen graph is satisfactory, the use can make one or more plot files to direct to hardcopy devices.

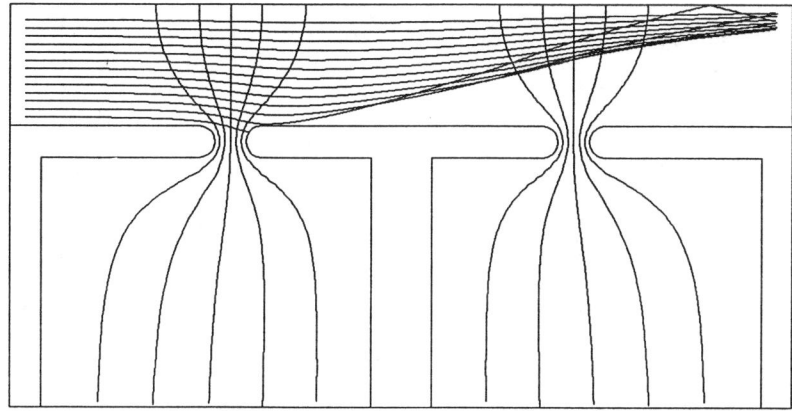

Figure 2. Electron orbits in an einzel lens. Central electrode: -6.5 kV. Incident electron energy: 10 keV.

COMPUTATIONAL METHODS AND EXAMPLES

Field computations in **TRAK** use the same method as VISION and PROBE[3,4]. The method gives accurate field values, even near conducting, dielectric and ferromagnetic boundaries. The first step in calculating the field at the location of a particle is to identify the mesh triangle that contains the target point. This problem can be challenging on a conformal mesh with variable resolution. To initiate a particle orbit, **TRAK** checks all triangles in the mesh until it finds the triangle that contains the starting point. To speed calculations, following searches are performed in the neighborhood of the previous point. The second step is to collect potential values at points on and around the target triangle. The search expands until **TRAK** finds from

6 to 24 valid points. To ensure that all points are on the same side of a material boundary, valid points must be adjacent to at least one triangle that has the same material designation as the target triangle. After collection of potentials at mesh points, **TRAK** makes a least-squares fit to a second order Taylor expansion of the potential about the particle location. The expansion gives an accurate interpolation of the potential and its derivatives to find vector components of electric or magnetic fields.

TRAK solves the equations of particle motion using Cartesian coordinates for both rectangular and cylindrical problems. Although the fields have two-dimensional symmetry, the orbit calculation is fully three-dimensional. When the equations of motion are expressed in cylindrical coordinates, the solution becomes inaccurate if particles cross near the axis. To avoid this problem **TRAK** always internally uses Cartesian coordinates. For cylindrical problems, **TRAK** converts input parameters and fields to Cartesian components, follows orbits, and then changes the results to cylindrical coordinates for output reports and plots.

The input of spatial coordinates to **TRAK** is always in units of centimeters for consistency with the mesh positions recorded in Poisson binary solution files. Internally, **TRAK** solves the relativistic equations of motion in MKS units. These equations are conveniently written in terms of the dimensionless momenta,

$$P_x = p_x/m_0 c, \quad [1]$$

$$P_y = p_y/m_0 c, \quad [2]$$

$$P_z = p_z/m_0 c, \quad [3]$$

and a relativistic energy shift

$$\delta\gamma = \gamma - 1. \quad [4]$$

Here, m_0 is the rest energy of the particle, c is the speed of light and γ is the relativistic energy factor. If the particle has kinetic energy T, then

$$\delta\gamma = T/m_0 c^2. \quad [5]$$

The equations of motion are

$$\delta\gamma = -1 + [1 + P_x^2 + P_y^2 + P_z^2]^{1/2}, \quad [6]$$

$$\frac{dx}{dt} = \frac{cP_x}{1 + \delta\gamma},\qquad [7]$$

$$\frac{dy}{dt} = \frac{cP_y}{1 + \delta\gamma},\qquad [8]$$

$$\frac{dz}{dt} = \frac{cP_z}{1 + \delta\gamma},\qquad [9]$$

$$\frac{dP_x}{dt} = \left[\frac{q}{m_o c}\right] E_x + \left[\frac{q}{m_o(1+\delta\gamma)}\right][P_y B_z - P_z B_y] \qquad [10]$$

$$\frac{dP_y}{dt} = \left[\frac{q}{m_o c}\right] E_y + \left[\frac{q}{m_o(1+\delta\gamma)}\right][P_z B_x - P_x B_z] \qquad [11]$$

$$\frac{dP_z}{dt} = \left[\frac{q}{m_o c}\right] E_z + \left[\frac{q}{m_o(1+\delta\gamma)}\right][P_x B_y - P_y B_x] \qquad [12]$$

Equations 6 through 12 are solved using a standard time-centered two-step difference method.

The quantity $\delta\gamma$ in Eq. 6 gives the total energy of the particle as a function of time. For static field solutions, the kinetic energy factor for a particle is also given by

$$\delta\gamma'(t) = \delta\gamma_o + q[\phi_o - \phi(t)]/m_o c^2. \qquad [13]$$

In Eq. 13, $\delta\gamma_o$ and ϕ_o are the energy shift and electrostatic potential at the initial particle position, while $\phi(t)$ is the electrostatic potential at the current position. During the orbit calculation, **TRAK** uses the relativistic energy shift of Eq. 6 to advance particle orbits. At the end of the calculation, **TRAK** calculates $\delta\gamma'$ from Eq. 13 and compares the two to estimate the accuracy of the calculation.

Figure 3 shows electron orbits in a magnetic mirror field, illustrating **TRAK**'s ability to follow complex orbits. The program stopped orbits after a given elapsed time of if they left the solution region.

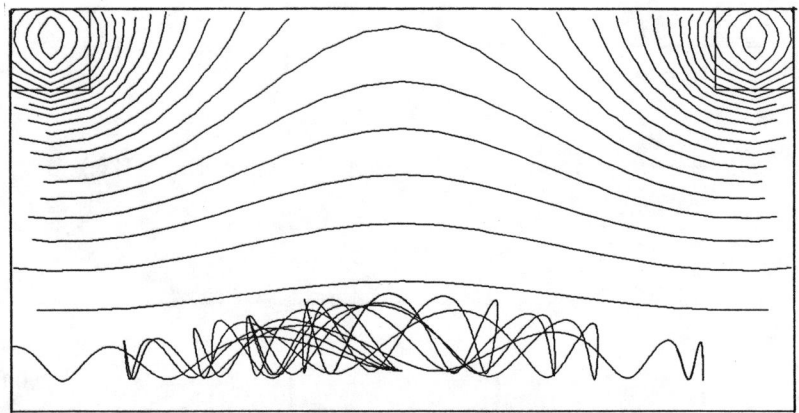

Figure 3. Electron orbits in a magnetic mirror. Peak magnetic field: 500 G. Electron kinetic energy: 5 keV

Figure 4 shows computed magnetic fields and orbits for electrons in a solenoid lens with an iron flux return yoke. In contrast to other orbit tracing codes, **TRAK** can use magnetic field solutions directly and include effects of non-linear materials.

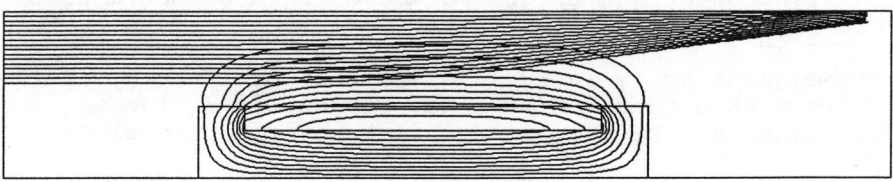

Figure 4. Solenoid lens, field lines and electron traces.

REFERENCES

1. **TRAK Users' Manual** (Field Precision, Albuquerque, New Mexico, 1992), unpublished.
2. **Reference Manual for the Poisson/Superfish Group of Codes** (Los Alamos National Laboratory, LA-UR-87-126, 1987), unpublished.
3. S. Humphries, Jr., "SUPERFISH Post-processor for IBM-PC Compatible Computers," **Proc. Beams 92 Conf.**, 1992.
4. S. Humphries, Jr., "Poisson/Superfish Codes for Personal Computers," **Proc. 16th Int'l. Linac Conf.**, Ottawa, 1992.

Author Index

A

Adams. F. P., 43
Allen, C. K., 568
Anderson, O. A., 549
Arman, M. J., 219

B

Baltrusaitis, R. M., 590
Barnard, J. J., 540
Barts, T., 107, 403
Batygin, Y. K., 196, 419
Becker, U., 477
Berg, J. S., 413
Berz, M., 143, 267, 467
Bondarev, B. I., 131, 377
Bourianoff, G., 19
Bowling, B., 557
Bruhwiler, D. L., 524
Burke, J. J., 27
Butler, S., 560

C

Callahan, D. A., 211, 347
Carey, D. C., 333
Chang, L., 19
Chen, Y.-J., 532
Chin, Y. H., 393
Chou, W., 107
Cole, B., 19
Courant, E. D., 403

D

de Jong, M. S., 43
Dehler, M., 123, 477
Dionne, N., 261
Douglas, D., 557
Drobot, A. T., 303
Durkin, A. P., 131, 377

E

Ekdahl, C., 590
Ellison, J. A., 442
Eppley, K. R., 357
Evans, Jr., K., 450

F

Fiorentini, G. M., 367
Forest, E., 413
Friedman, A., 203, 211, 347

G

Garren, A. A., 403
Gillespie, G. H., 576
Goldmann, E., 143
Goplen, B., 313
Goren, Y., 150, 160
Grimm, T., 182
Grote, H., 408
Grote, D. P., 203, 347, 540
Guharay, S. K., 568

H

Haber, I., 211, 347
Halbach, K., 58
Hall, B. J., 235
Herrmannsfeldt, W. B., 1
Hewett, D. W., 532
Hoffstatter, G. H., 467
Huang, Y., 167
Hulsey, G., 182
Humphries, Jr., S., 590, 597

K

Kenney, A. S., 403
Kewisch, J., 557

Kloeppel, P., 557
Ko, K., 1, 99, 243
Koga, J., 175
Kostas, C., 303, 485
Krafft, G. A., 557
Kress, M. E., 251
Kroll, N., 99
Krueger, W. A., 303

L

Langdon, A. B., 211, 347
Li, M., 188
Liu, H., 385, 508
Ludeking, L., 313
Lund, S. M., 540
Lysenko, W. P., 516

M

Machida, S., 175, 188, 459
MAFIA Collaboration, 291
Mankofsky, A., 303
Mao, N., 135
Merson, J. L., 427
Michelotti, L., 264
Millich, A., 35
Mondelli, A. A., 303, 485
Mottershead, C. T., 435
Murin, B. P., 377

N

Nelson, E. M., 323
Ng., C.-K., 1, 243
Nguyen, K., 313
Nicolaishvili, G. T., 131, 377

O

O'Connell, J., 560
Ohnuma, S., 167
Orthel, J. L., 227

P

Petillo, J. J., 303

R

Reiser, M., 568
Reusch, M. F., 524
Rimmer, R. A., 27
Ruiz, E., 160
Russell, A. D., 403
Rusthoi, D., 115
Ruth, R. D., 413
Rybarcyk, L. J., 427

S

Schiesser, W. E., 442
Schlueter, R. D., 58
Sen, T., 403
Sharp, W. M., 540
Shih, H.-J., 442
Shlygin, O. Yu., 131, 377
Smithe, D., 313
Sohn, E., 251
Soroka, L., 549
Spayd, N., 160
Staples, J. W., 584
Sullivan, D. J., 219
Swatloski, T. L., 500
Syphers, M. J., 403

T

Tajima, T., 175
Tantawi, S., 99
Thoma, P., 66
True, R., 493
Tsang, K. T., 485

V

van Rienen, U., 90
van Zeijts, J., 285

W

Walling, L., 150, 160, 182
Wan, W., 143
Wang, F., 135
Wangler, T. P., 9
Warn, C., 590
Warnock, R. L., 82, 413
Warren, G., 313
Weiland, T., 51, 66, 74, 123, 291, 477
Wolter, H., 74

Y

Yan, Y. T., 279
Young, C., 590
Yu, S., 203, 540

Z

Zhang, M., 51
Zhang, P., 188

AIP Conference Proceedings

		L.C. Number	ISBN
No. 210	Production and Neutralization of Negative Ions and Beams (Brookhaven, NY, 1990)	90-55316	0-88318-786-8
No. 211	High-Energy Astrophysics in the 21st Century (Taos, NM, 1989)	90-55644	0-88318-803-1
No. 212	Accelerator Instrumentation (Brookhaven, NY, 1989)	90-55838	0-88318-645-4
No. 213	Frontiers in Condensed Matter Theory (New York, NY, 1989)	90-6421	0-88318-771-X 0-88318-772-8 (pbk.)
No. 214	Beam Dynamics Issues of High-Luminosity Asymmetric Collider Rings (Berkeley, CA, 1990)	90-55857	0-88318-767-1
No. 215	X-Ray and Inner-Shell Processes (Knoxville, TN, 1990)	90-84700	0-88318-790-6
No. 216	Spectral Line Shapes, Vol. 6 (Austin, TX, 1990)	90-06278	0-88318-791-4
No. 217	Space Nuclear Power Systems (Albuquerque, NM, 1991)	90-56220	0-88318-838-4
No. 218	Positron Beams for Solids and Surfaces (London, Canada, 1990)	90-56407	0-88318-842-2
No. 219	Superconductivity and Its Applications (Buffalo, NY, 1990)	91-55020	0-88318-835-X
No. 220	High Energy Gamma-Ray Astronomy (Ann Arbor, MI, 1990)	91-70876	0-88318-812-0
No. 221	Particle Production Near Threshold (Nashville, IN, 1990)	91-55134	0-88318-829-5
No. 222	After the First Three Minutes (College Park, MD, 1990)	91-55214	0-88318-828-7
No. 223	Polarized Collider Workshop (University Park, PA, 1990)	91-71303	0-88318-826-0
No. 224	LAMPF Workshop on (π, K) Physics (Los Alamos, NM, 1990)	91-71304	0-88318-825-2
No. 225	Half Collision Resonance Phenomena in Molecules (Caracas, Venezuela, 1990)	91-55210	0-88318-840-6
No. 226	The Living Cell in Four Dimensions (Gif sur Yvette, France, 1990)	91-55209	0-88318-794-9

No. 227	Advanced Processing and Characterization Technologies (Clearwater, FL, 1991)	91-55194	0-88318-910-0
No. 228	Anomalous Nuclear Effects in Deuterium/Solid Systems (Provo, UT, 1990)	91-55245	0-88318-833-3
No. 229	Accelerator Instrumentation (Batavia, IL, 1990)	91-55347	0-88318-832-1
No. 230	Nonlinear Dynamics and Particle Acceleration (Tsukuba, Japan, 1990)	91-55348	0-88318-824-4
No. 231	Boron-Rich Solids (Albuquerque, NM, 1990)	91-53024	0-88318-793-4
No. 232	Gamma-Ray Line Astrophysics (Paris-Saclay, France, 1990)	91-55492	0-88318-875-9
No. 233	Atomic Physics 12 (Ann Arbor, MI, 1990)	91-55595	088318-811-2
No. 234	Amorphous Silicon Materials and Solar Cells (Denver, CO, 1991)	91-55575	088318-831-7
No. 235	Physics and Chemistry of MCT and Novel IR Detector Materials (San Francisco, CA, 1990)	91-55493	0-88318-931-3
No. 236	Vacuum Design of Synchrotron Light Sources (Argonne, IL, 1990)	91-55527	0-88318-873-2
No. 237	Kent M. Terwilliger Memorial Symposium (Ann Arbor, MI, 1989)	91-55576	0-88318-788-4
No. 238	Capture Gamma-Ray Spectroscopy (Pacific Grove, CA, 1990)	91-57923	0-88318-830-9
No. 239	Advances in Biomolecular Simulations (Obernai, France, 1991)	91-58106	0-88318-940-2
No. 240	Joint Soviet-American Workshop on the Physics of Semiconductor Lasers (Leningrad, USSR, 1991)	91-58537	0-88318-936-4
No. 241	Scanned Probe Microscopy (Santa Barbara, CA, 1991)	91-76758	0-88318-816-3
No. 242	Strong, Weak, and Electromagnetic Interactions in Nuclei, Atoms, and Astrophysics: A Workshop in Honor of Stewart D. Bloom's Retirement (Livermore, CA, 1991)	91-76876	0-88318-943-7
No. 243	Intersections Between Particle and Nuclear Physics (Tucson, AZ, 1991)	91-77580	0-88318-950-X
No. 244	Radio Frequency Power in Plasmas (Charleston, SC, 1991)	91-77853	0-88318-937-2

No. 245	Basic Space Science (Bangalore, India, 1991)	91-78379	0-88318-951-8
No. 246	Space Nuclear Power Systems (Albuquerque, NM, 1992)	91-58793	1-56396-027-3 1-56396-026-5 (pbk.)
No. 247	Global Warming: Physics and Facts (Washington, DC, 1991)	91-78423	0-88318-932-1
No. 248	Computer-Aided Statistical Physics (Taipei, Taiwan, 1991)	91-78378	0-88318-942-9
No. 249	The Physics of Particle Accelerators (Upton, NY, 1989, 1990)	92-52843	0-88318-789-2
No. 250	Towards a Unified Picture of Nuclear Dynamics (Nikko, Japan, 1991)	92-70143	0-88318-951-8
No. 251	Superconductivity and its Applications (Buffalo, NY, 1991)	92-52726	1-56396-016-8
No. 252	Accelerator Instrumentation (Newport News, VA, 1991)	92-70356	0-88318-934-8
No. 253	High-Brightness Beams for Advanced Accelerator Applications (College Park, MD, 1991)	92-52705	0-88318-947-X
No. 254	Testing the AGN Paradigm (College Park, MD, 1991)	92-52780	1-56396-009-5
No. 255	Advanced Beam Dynamics Workshop on Effects of Errors in Accelerators, Their Diagnosis and Corrections (Corpus Christi, TX, 1991)	92-52842	1-56396-006-0
No. 256	Slow Dynamics in Condensed Matter (Fukuoka, Japan, 1991)	92-53120	0-88318-938-0
No. 257	Atomic Processes in Plasmas (Portland, ME, 1991)	91-08105	0-88318-939-9
No. 258	Synchrotron Radiation and Dynamic Phenomena (Grenoble, France, 1991)	92-53790	1-56396-008-7
No. 259	Future Directions in Nuclear Physics with 4π Gamma Detection Systems of the New Generation (Strasbourg, France, 1991)	92-53222	0-88318-952-6
No. 260	Computational Quantum Physics (Nashville, TN, 1991)	92-71777	0-88318-933-X
No. 261	Rare and Exclusive B&K Decays and Novel Flavor Factories (Santa Monica, CA, 1991)	92-71873	1-56396-055-9

No. 262	Molecular Electronics—Science and Technology (St. Thomas, Virgin Islands, 1991)	92-72210	1-56396-041-9
No. 263	Stress-Induced Phenomena in Metallization: First International Workshop (Ithaca, NY, 1991)	92-72292	1-56396-082-6
No. 264	Particle Acceleration in Cosmic Plasmas (Newark, DE, 1991)	92-73316	0-88318-948-8
No. 265	Gamma-Ray Bursts (Huntsville, AL, 1991)	92-73456	1-56396-018-4
No. 266	Group Theory in Physics (Cocoyoc, Morelos, Mexico, 1991)	92-73457	1-56396-101-6
No. 267	Electromechanical Coupling of the Solar Atmosphere (Capri, Italy, 1991)	92-82717	1-56396-110-5
No. 268	Photovoltaic Advanced Research & Development Project (Denver, CO, 1992)	92-74159	1-56396-056-7
No. 269	CEBAF 1992 Summer Workshop (Newport News, VA, 1992)	92-75403	1-56396-067-2
No. 270	Time Reversal—The Arthur Rich Memorial Symposium (Ann Arbor, MI, 1991)	92-83852	1-56396-105-9
No. 271	Tenth Symposium Space Nuclear Power and Propulsion (Vols. I–III) (Albuquerque, NM, 1993)	92-75162	1-56396-137-7 (set)
No. 272	Proceedings of the XXVI International Conference on High Energy Physics (Vols. I and II) (Dallas, TX, 1992)	93-70412	1-56396-127-X (set)
No. 273	Superconductivity and Its Applications (Buffalo, NY, 1992)	93-70502	1-56396-189-X
No. 274	VIth International Conference on the Physics of Highly Charged Ions (Manhattan, KS, 1992)	93-70577	1-56396-102-4
No. 275	Atomic Physics 13 (Munich, Germany, 1992)	93-70826	1-56396-057-5
No. 276	Very High Energy Cosmic-Ray Interactions: VIIth International Symposium (Ann Arbor, MI, 1992)	93-71342	1-56396-038-9

No. 277	The World at Risk: Natural Hazards and Climate Change (Cambridge, MA, 1992)	93-71333	1-56396-066-4
No. 278	Back to the Galaxy (College Park, MD, 1992)	93-71543	1-56396-227-6
No. 279	Advanced Accelerator Concepts (Port Jefferson, NY, 1992)	93-71773	1-56396-191-1
No. 280	Compton Gamma-Ray Observatory (St. Louis, MO, 1992)	93-71830	1-56396-104-0
No. 281	Accelerator Instrumentation Fourth Annual Workshop (Berkeley, CA, 1992)	93-072110	1-56396-190-3
No. 282	Quantum 1/f Noise & Other Low Frequency Fluctuations in Electronic Devices (St. Louis, MO, 1992)	93-072366	1-56396-252-7
No. 283	Earth and Space Science Information Systems (Pasadena, CA, 1992)	93-072360	1-56396-094-X
No. 284	US-Japan Workshop on Ion Temperature Gradient-Driven Turbulent Transport (Austin, TX, 1993)	93-72460	1-56396-221-7
No. 285	Noise in Physical Systems and 1/f Fluctuations (St. Louis, MO, 1993)	93-72575	1-56396-270-5
No. 286	Ordering Disorder: Prospect and Retrospect in Condensed Matter Physics: Proceedings of the Indo-U.S. Workshop (Hyderabad, India, 1993)	93-072549	1-56396-255-1
No. 287	Production and Neutralization of Negative Ions and Beams: Sixth International Symposium (Upton, NY, 1992)	93-72821	1-56396-103-2
No. 288	Laser Ablation: Mechanisms and Applications-II: Second International Conference (Knoxville, TN, 1993)	93-73040	1-56396-226-8
No. 289	Radio Frequency Power in Plasmas: Tenth Topical Conference (Boston, MA, 1993)	93-72964	1-56396-264-0
No. 290	Laser Spectroscopy: XIth International Conference (Hot Springs, VA, 1993)	93-73050	1-56396-262-4
No. 291	Prairie View Summer Science Academy (Prairie View, TX, 1992)	93-73081	1-56396-133-4